ESSENTIALS OF MODERN
CHEMISTRY

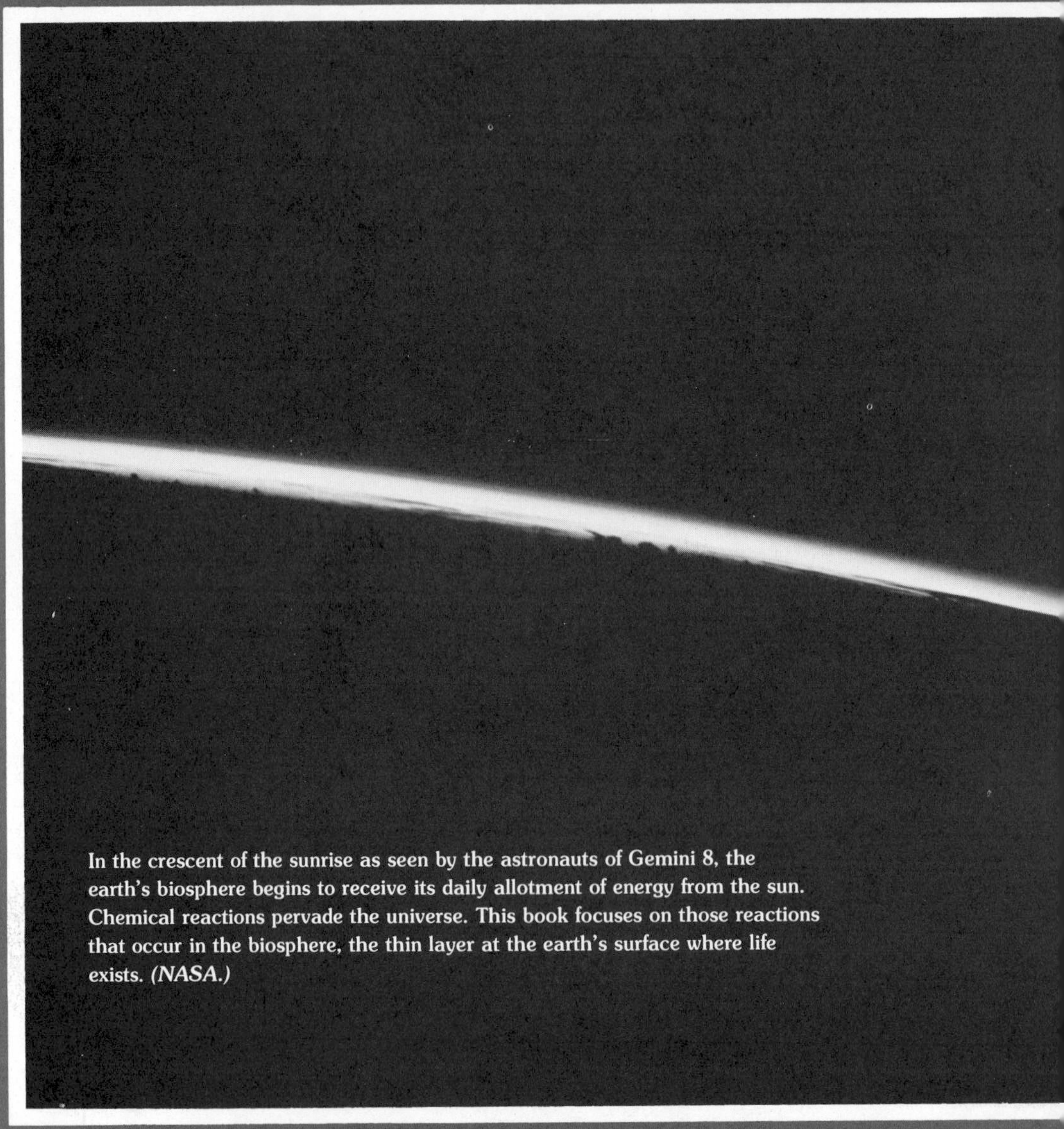

In the crescent of the sunrise as seen by the astronauts of Gemini 8, the earth's biosphere begins to receive its daily allotment of energy from the sun. Chemical reactions pervade the universe. This book focuses on those reactions that occur in the biosphere, the thin layer at the earth's surface where life exists. *(NASA.)*

ESSENTIALS OF MODERN CHEMISTRY

RICHARD H. EASTMAN
Stanford University

Rinehart Press / Holt, Rinehart and Winston
San Francisco

Cover photo: A scanning electron micrograph of human red blood cells (see Photo 24). (Dr. Norman Hodgkin, Micrographics, Newport Beach, California.)

Library of Congress Cataloging in Publication Data

Eastman, Richard H 1918–
Essentials of Modern Chemistry

Includes index.
1. Chemistry. I. Title.
QD31.2.E16 540 74-23604
ISBN 0-03-085694-9

5643 Paradise Drive
Corte Madera, California 94925

A division of Holt, Rinehart and Winston, Inc.

Printed in the United States of America

5 6 7 8 032 9 8 7 6 5 4 3 2

To the Student

I have written this book to help you feel comfortable about learning chemistry and to give you a sufficient grasp of the subject to enable you to share in the excitement that chemists enjoy from their work.

There are exercises both within and at the ends of chapters, and selected answers are given in Appendix 7. You should do all of the exercises because they will increase your understanding of chemical principles. Worked-out examples throughout the chapters provide you with a guide to solving exercises. To help you study, a blue key symbol (⊶) is placed at the beginning and end of every example. You will find explanatory notes in the margins and occasionally at the bottom of the page. Each note is indicated by a hand (☛) pointing right, left, or down, as appropriate.

A summary guide and a list of key words with their meanings appear at the end of each chapter. You should read both of these *before* starting in on the chapter itself, because they will provide you with a road map for your trip through the chapter.

The readings at the end of each chapter provide an opportunity for you to extend and enrich your experience with chemistry. You need not read these materials to learn the essentials of chemistry, but the readings will show you the importance of chemistry in the practical areas of environmental problems and life processes.

In the margins you will occasionally find brief biographical notes of outstanding contributors to chemistry. They are there primarily to show you that chemistry, like all sciences, is a truly international activity that transcends national borders. The biographies will also show you that modern chemistry has developed in the short period of only 200 years.

It has been my experience that the greatest difficulty students have with chemistry is with the meanings of the symbols used in chemical equations and the equations of chemical arithmetic. Form the habit right away of translating symbols into words, groups of words, or pictures in your mind. If you do, what seems at first to be a very complex equation will turn out to be easily understood.

RICHARD H. EASTMAN
Stanford, California

To the Instructor

I have written a book that won't frighten the student with the immensity of chemistry or cheat by oversimplification. I have presented chemistry as a language, describing the essentials of modern chemistry without emphasizing the details of how they were arrived at. By the language approach, currently used with success at Stanford and elsewhere, the student is quickly encouraged to start thinking the way chemists do—about what those tiny particles are doing down there—rather than worrying about the roots of chemical theory. I find that the nonspecialist benefits little from scrutinizing the experiments and rationalizations that led to the concepts and models of today's chemistry. Better he or she accept them and get on to the exciting and rewarding business of applying chemical ideas to interesting and important systems. My aim is to create a level of chemical literacy in the student that will permit him to appreciate the importance of chemistry in the environment, to understand the central role of chemistry in life processes, and to make informed decisions in these areas as a citizen.

Wherever it seems appropriate I have used second person, active voice, rather than the third person passive so common in scientific writings. Second person active voice draws the student into chemistry in a reassuring way that makes both the book and the subject friendlier and more lively. To the same purpose, I have balanced applications of chemistry to environmental problems with applications to human physiology, which, I have found, lend an immediate, personal aspect to the science that students enjoy.

This book is for the first-year, general chemistry course and is particularly directed to students going on into biology, environmental science, medicine, and related fields. It is in no way a reference text, and contains only an amount of material that a conscientious, average student can master in two semesters. With supplementary paperback material judged necessary by the instructor, it can also serve as the core of a majors course. By following the outline on pages 5–7 in the instructor's manual, the book will also fit a one-semester, terminal course. The only prerequisites are arithmetic and high school algebra.

Mathematical operations needed to handle pH and equilibrium problems are developed in text as the need for them arises. Appendixes include a review of algebra, the small amount of physics needed, a qualitative description of the LCAO derivation of molecular orbitals, and definitions of the basic units of the International System (SI) of measurements. The cgs system is used in the text.

The book differs from many of its contemporaries in several ways. Topics that are simply sources of information and not essential to modern chemistry have been omitted. There is no detailed presentation of units of measurement until the text material makes them a logical convenience. Because the text is built on the microparticle-energy view of chemistry, there is no separate chapter on thermodynamics. Instead, thermodynamic state functions are introduced as useful ways of looking at the energetics and efficiencies of such familiar systems as power generating stations, refrigerators, heat pumps, and batteries, and then applied in examples and exercises. Finally, and most importantly, the discussion of bioorganic chemistry starts with the essential ingredient, chirality in molecules (optical isomerism), because of its fundamental influence on the structure of the molecules involved in the chemical reactions of life.

My thanks go to the staff of Rinehart Press, particularly to my editor, Frank Meltzer, for help and suggestions that substantially improved the book. Designer Janet Bollow deserves high marks for the handsome appearance of the book. I also acknowledge the encouragement and help of my colleagues at Stanford. Finally and most important have been the support and efforts of my wife, Patricia, who not only typed the manuscript, but read proof, prepared reading lists and biographical notes, and kept track of the details of the organization of the whole book. For her efforts I am thankful beyond the power of words to express.

RICHARD H. EASTMAN
Stanford, California

Contents

ESSENTIALS OF MODERN CHEMISTRY

The rotor of a steam turbine that will drive a 1,060,000 kilowatt electric generator, enough to fill the needs of 1 million people. The steam to drive the turbine will come from a nuclear heater. *(Photo 1, Pacific Gas and Electric Co.)*

Introduction

Chemists try to understand and explain the properties and behavior of bulk matter by studying the properties and behavior of the atoms that make up that matter. For example, one property of atoms is their vigorous, ceaseless motion. To measure this motion, we measure temperature—the greater the vigor and intensity of the motion, the higher the temperature. We call such motion *thermal energy* or heat.

When combustion occurs in a cylinder of an automobile engine, the moving particles of hot gas push against the piston and do work that moves the car. The energy that does that work is called *kinetic energy*.

Kinetic energy is the capacity a body has for doing work because of its motion.

Atoms constantly attract or repel each other because of the electrical charges they carry. Opposite charges (positive and negative, or + and −) attract each other. Charges having the same sign (both + or both −) repel each other. Because of the attraction and repulsion between charges, atoms possess *potential energy*, energy in storage. The energy we get from food is stored in the form of chemical potential energy.

This introduction will help you understand the various kinds of energy, temperature scales, and attraction and repulsion between charged particles. You will find a few algebraic equations, but it is more important to understand the concepts than the calculations.

Energy and Work Energy is the capacity to do work. We recognize this when we say "I'm full of energy, I'm going to get a lot of work done." But even when we have a large *capacity* to do work, we may not actually do any. We realize this aspect of energy when we say that *work is energy expressed.* If we physically express our capacity for work, by shoveling sand or chopping wood, for instance, our energy drains away and seems to have been lost until it is replaced by food and rest. But *energy is never lost.* The energy we have spent has only been converted through our work into other forms of energy. For example, suppose you spend some of your energy rowing a boat on a lake. You do the mechanical work of pulling on the oars to move the boat through the water. When you return the boat to its dock, everything is returned to its original condition except that you are tired because you have used some of your energy. Where did it go? The answer to questions of this type was found in the 19th century by Rumford, Mayer, and particularly Joule (pronounced "jool"). Joule used a paddlewheel arrangement to show that stirring water increases its temperature. The mechanical energy required to overcome the friction between the paddles and the water is converted into an equivalent amount of heat or *thermal energy* that warms the water. The energy you used to row the boat went into warming the water of the lake a tiny amount through friction of the oars and the boat with the water.

James Prescott Joule (1818–1889) was an English physicist and brewer who established the theory of the equivalence of energy in the forms of work and heat. The *joule*, a unit of work or heat, is named after him.

Count Benjamin Thompson Rumford (1753–1814), a contemporary of Benjamin Franklin, was born in Woburn, Massachusetts. He was made a Count of the Holy Roman Empire for his contributions to Bavaria. He helped to establish the equivalence between heat and work.

Julius von Mayer (1814–1878) was a German doctor and physicist who shared with Joule the discovery of the energy equivalence of mechanical work and heat.

If mechanical energy can be completely converted into heat, can the reverse process be carried out? This is an important question, for modern civilization depends on what are called *heat engines* for much of its mechanical work. The auto engine, the steam engine, and the steam turbine (Photo 1) are heat engines, and it is important to know the efficiency with which they convert the thermal energy released in the burning of fossil fuels (coal, gas, oil) into mechanical work.

Without going into details we can accept the result from the science of thermodynamics that *the conversion of thermal energy into mechanical work can never be complete.* A heat engine operates on the natural or *spontaneous* (meaning from "within"—that is, without any outside help) flow of thermal energy from a heat reservoir at a higher temperature, called the *source*, to a heat reservoir at a lower temperature, called the *sink* (Figure 1). In a steam electric-generating plant, the boiler is the source and the condensers that convert the steam back to water after it passes through the turbines are the sink. Some of the heat energy taken from the source goes into warming up the sink and cannot be used for mechanical work, unless there is another sink at a lower temperature. This inevitable warming of the sink results in what is called *thermal pollution* (Photo 2) by steam-driven electric power generating plants. Some plants use the water of lakes and streams to cool the condensers. A large plant circulates as much as 350 million gallons of cooling water daily for this purpose, discharging it back to the river or stream at a temperature several degrees above normal for the river. That temperature increase may be harmful to the natural fish and plant life of the river. On the other hand, the warmer water can be used in "fish farms" to raise types of edible fish that require warm water. This is being done at Moss Landing, California.

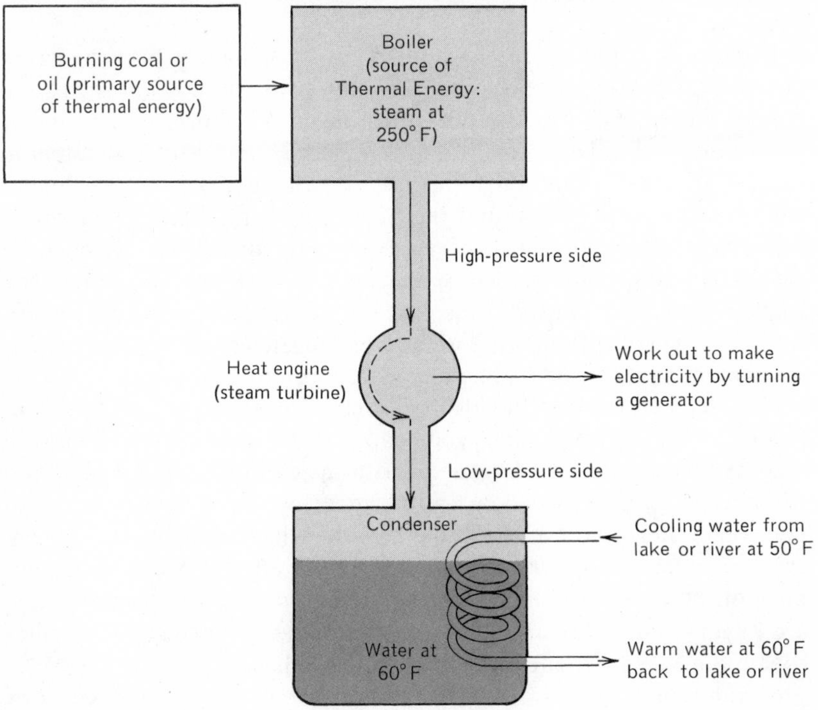

FIG. 1 Principle of heat engine (steam turbine). Arrows show flow of thermal energy.

PHOTO 2 The dark areas in the water just off the beach are thermal and chemical pollution—the waste water from the two factories whose smokestacks you see in the photo.

Because it is impossible to convert heat completely back to work, mechanical energy is said to be *degraded* when it ends up as thermal energy. Thermal energy is often called the "lowest form" of energy for this reason. All other forms of energy can be converted completely to heat. Two other important forms are electric energy from the flow of electric current in wires, and the energy of light. Electric energy is converted into mechanical energy, with only small losses as heat, by electric motors—and completely to heat in electric stoves and heaters. Light energy is converted to heat when the light is absorbed by a surface. The increase in lighting in the Los Angeles area (Photo 3) represents an increase in degrading electric energy to heat.

Now let's go back to the idea of being "full of energy" and think of another way you might use your capacity for doing work. This will help explain *potential energy*. Suppose instead of rowing a boat you work to pump your bicycle up a hill. Ignoring minor heat losses from your body and from friction in the bicycle, what happened to your energy? It is stored as *gravitational potential energy* because of your position on top of the hill. With a rope-and-pulley arrangement (Figure 2) you could take the energy out of storage to pull a slightly lighter friend in a "free" or effortless ride up the hill.

A spider named Arabella wove a perfect web in the zero gravity of Skylab II in 1973. She couldn't have done it if gravity had any serious effect on the chemical reactions of life.

Gravitational potential energy is not important in chemistry. But the potential energy associated with electric forces is very impor-

PHOTO 3 Valley lights of the Los Angeles basin as seen from Mt. Wilson. The upper photo was taken in 1911, the lower in 1965. The increased lighting indicates increased use of electrical energy.

tant. Look at the two *oppositely charged* spheres in Figure 3. Because of the opposite charges, they attract each other. As long as they are restrained from being moved together by the attraction between them, they have *electric potential energy*. You could convert the energy into mechanical work by holding one of the spheres and allowing the motion of the second toward the first to lift a weight through the string-and-pulley arrangement. In the opposite way, energy would have to be provided to do the work of pulling the spheres apart to a greater separation. This type of electric potential energy causes the tiny particles of matter to stick together. It results in chemical "bonds" between atoms. *Energy is needed to break bonds, and energy is released when bonds form.*

With bulk matter, the electric potential energy of the individual atomic particles is *chemical potential energy*. The fossil fuels (coal, natural gas, oil) are storage depots of chemical potential energy that is

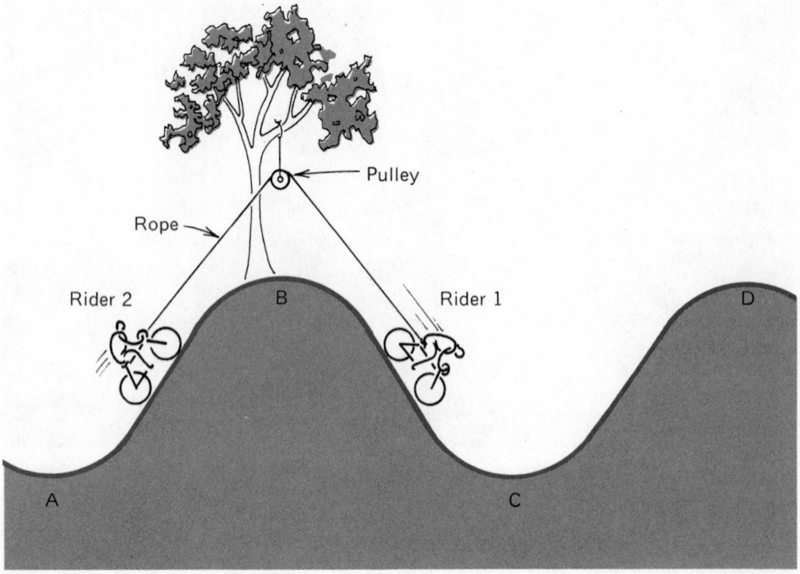

FIG. 2 Rider 1, who has pumped up hill B from valley A, can pull slightly lighter rider 2 up hill B by coasting down to valley C, which is the same depth as valley A.

released in the form of heat when they are burned. The products of the burning (combustion) are "farther down" the chemical potential energy "hill" than the starting materials, and the difference in potential energy appears as the thermal energy of the flame.

One fact about potential energy of any type is particularly important. *Potential energy always exists in relation to a condition that has been assigned a potential energy of zero.* In the case of the cyclist, zero potential energy occurs at the bottom of the hill. For the fossil fuels, zero potential exists in the products of the combustion, carbon

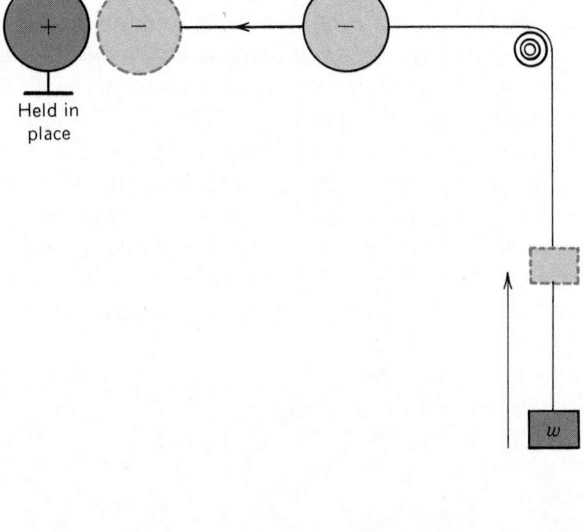

FIG. 3 Attraction between oppositely charged spheres gives system electrostatic potential energy, which can be used to do work of lifting weight W, as shown by dashed lines.

dioxide and water, which can no longer be "burned" to give energy. In the case of the spheres shown in Figure 3, the zero is where the spheres are an infinite distance apart. As the oppositely charged spheres come together, *the electric potential energy becomes less than zero*; that is, it takes on *negative* values. The idea of negative potential energy may seem confusing at first, but it is really commonplace. In Figure 2, the gravitational energy of rider 2 (when he is at the bottom of the hill) is *negative* compared to the gravitational energy of rider 1 (when he is at the top of the hill). The water at the bottom of a well is at a negative level of gravitational potential energy *relative* to the level of the ground because energy is required to pump (lift) it up to ground level. In the same way, *chemical bonds represent conditions of negative potential energy* because energy must be spent to separate the bonded particles:

$$\underset{\text{bond}}{\mathrm{H{-}H}} + \text{energy} \rightarrow \mathrm{H} + \mathrm{H} \text{ (separated atoms)}$$

On the other hand, if the spheres in Figure 3 both carry charges of the same sign (++ or −−), bringing them together from infinite separation would require pushing them together to overcome the repulsive force between them. As the spheres are forced closer together, their electric potential energy rises above zero and takes on positive values equal to the work of pushing. Left to themselves, the spheres will fly apart because of the repulsion between their like charges. It's like a compressed spring (representing positive potential energy) spontaneously expanding to its normal length (zero potential energy) when released.

Let's return to the cyclist who has pedaled his way to the top of a hill. What happens to the gravitational potential energy he has stored if he simply coasts down the hill? At the bottom of the hill his *potential* energy will be zero, but he will be going very fast. Neglecting losses to air resistance and friction, his potential energy has been converted into the *energy of his motion*, or *kinetic* energy. *Kinetic energy is the capacity a body has for doing work as a result of its motion.* The cyclist could use his kinetic energy to coast up the adjacent hill (D in Figure 2). Or, if he brakes to a stop, his kinetic energy is degraded to heat through friction that raises the temperature of his brakes and tires.

Kinetic energy is important in chemistry because *thermal energy* (heat) *is the kinetic energy of the tiny particles of matter in their random chaotic motion.* Temperature is a measure of the average kinetic energy of atom-sized particles of matter, and temperature increases as the average kinetic energy of the particles is increased by input of thermal energy. Our sensations of heat and cold are the response of our nerves to greater or lesser average kinetic energies of microparticles (atoms).

Exercises

1. Two cyclists, one weighing 150 lb and the second weighing 180 lb, pedal up to the top of the same hill. Which would expend more energy? Which would have the greater potential energy at the top? (*Hint*: Gravitational potential energy depends on both the weight of the body and the elevation to which it is raised.)

2. Imagine two hills of exactly the same height separated by a valley (hills D and B in Figure 2). A cyclist pumps to the top of hill B, coasts down the hill, through the valley, and up hill D. If there were no friction, how high up hill D would he get?

3. When would the cyclist in Exercise 2 be at zero potential energy?

4. Where in the cyclist's trip would his potential energy have been converted completely to kinetic energy?

5. A cyclist at the top of a hill coasts very slowly down the hill, applying his brakes all the way, and comes to a stop in the valley. To what form has his original potential energy been converted?

Temperature We are all familiar with the measurement of temperature with a thermometer. But it is important in science to understand clearly just what a "thermometer" measures. It measures, on an arbitrary scale, the *intensity* of thermal energy or molecular motion. A thermometer *does not measure the* **quantity** *of thermal energy in a sample of matter*. The *quantity depends* on the *size* of the sample; temperature does not. A two-pound block of ice at 32°F contains twice as much thermal energy as a one-pound block at the same temperature.

The unit of temperature most used by scientists is the Celsius or *centigrade* degree (°C), defined as $\frac{1}{100}$ of the difference between the temperatures of melting ice and of water boiling at sea level. To make a centigrade thermometer of the type used in the laboratory, mercury is sealed into a glass bulb attached to a glass tube of small and uniform bore (a capillary tube). The thermometer is put in a mixture of ice and water in an insulated container, and a mark is made at the level of the mercury in the tube. The mark is labeled 0°C (Figure 4). The thermometer is then placed in water boiling at sea level. The expansion of the mercury causes it to rise in the tube and a mark labeled 100° is made at the new mercury level. The distance between the two marks is evenly divided into 100°C. A Fahrenheit thermometer is made in the same way except that the temperature of melting ice is taken as 32°F and that of boiling water as 212°F. The Fahrenheit degree is $\frac{1}{180}$ the difference between the melting and boiling points of water (212 − 32 = 180). The two scales are compared (a and b) in Figure 5. Degrees Fahrenheit are converted to centigrade by the equation $°C = \frac{5}{9}(°F - 32)$.

The boiling point of water depends on elevation above sea level. At Denver, Colorado ("the Mile-High City"), water boils at 95°C because of the effect of elevation in lowering atmospheric pressure.

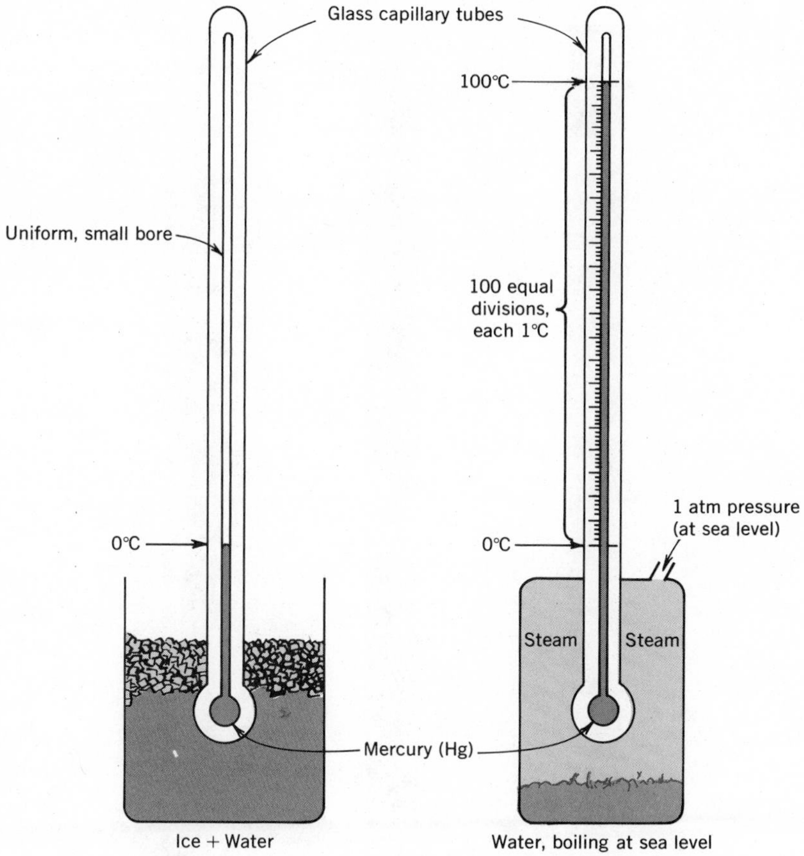

FIG. 4 Celsius scale of temperature in the preparation of a simple thermometer.

Scientists frequently use the *Kelvin* or *absolute* temperature scale (K). This is based on the average kinetic energy of the particles (molecules) of a perfect gas (Chapter 5). The centigrade degree is still the unit of the absolute scale, but the melting point (mp) of ice is taken as 273.16 K and the boiling point (bp) of water is taken one hundred degrees higher at 373.16 K. In this text you can convert from the centigrade to the absolute scale by using the approximate equation $K = {}^\circ C + 273^\circ$. Thus, $0\ K = -273^\circ C$. At 0 K the average kinetic energy of gas molecules is zero, so their random chaotic motion has stopped and the thermal energy content is zero. Temperatures as low as 0.001 K have been reached by special methods. Since the absolute scale is based on the properties of gases, it is used in calculations of the pressure-temperature-volume relationships of gases (Chapter 5).

Exercises

6. Body temperature averages 98.6°F. Calculate body temperature on the centigrade and absolute scales.

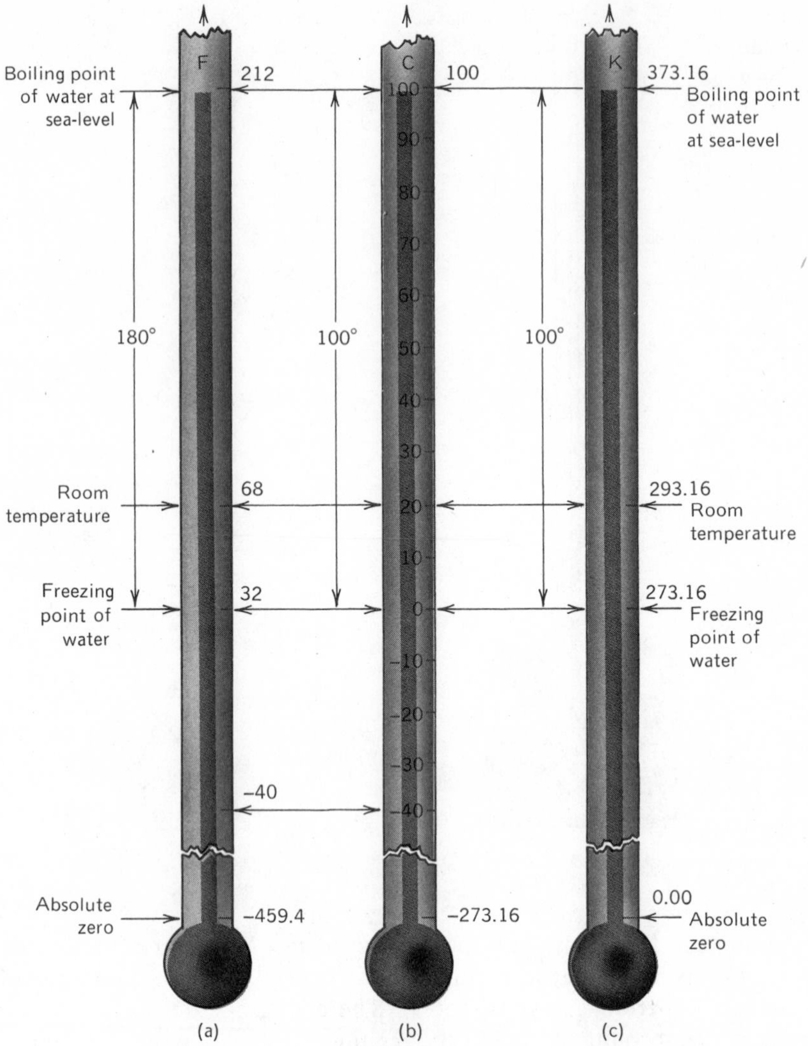

FIG. 5 Comparison of temperature scales: (a) Fahrenheit; (b) Celsius or centigrade; (c) Kelvin or absolute (not to scale).

7. How many Fahrenheit degrees are there in 1°C?

8. Show that the Fahrenheit and centigrade temperatures are the same at −40° on either scale as indicated in Figure 5.

9. Convert 298 K to °C and °F.

Forces and Coulomb's Law We all have a good idea of the meaning of the term "force." When we move an object by lifting or pushing it, we exert a force on it. A force can be exerted without causing any movement, however. Perhaps the most familiar case of this type is the "force of gravity" that holds us and other objects to the earth. Gravitational force causes a ball to fall when we let go of it. The ball falls because there is gravitational attraction *between* it and the earth. The key word here is "between"—the ball attracts the earth and the earth the ball with exactly the same force. The gravitational force between two samples of matter is in proportion to the amount of matter they contain, that is to their *masses*, m_1 and m_2, multiplied together. The force is inversely proportional to the square of the distance (r) between the centers of the masses, so the greater the distance r, the smaller the force between them. The equation for gravitational force is known as Newton's Law of Universal Gravitation:

Sir Isaac Newton (1642–1727) was an English mathematician, astronomer, and natural philosopher. He is best known for discovering the laws of gravitation and mechanics, and for developing calculus.

$$\text{Force} = \frac{G m_1 m_2}{r^2}$$

Here G is the gravitational constant, the value of which depends on the units of measurement in which the masses, the distance, and the force are expressed. Gravitational attraction between bodies of ordinary size is very weak. It is only because of the huge mass (m_2) of the earth that we (m_1) experience the force of gravity at all. Gravitational forces between the tiny particles of matter are really unimportant because their masses are so small.

On the other hand, if particles carry electric charges, the *electric forces* between the particles are large and important. Electric forces regulate the behavior of particles. Ordinarily bulk matter is electrically neutral because it contains equal amounts of positive and negative electricity. Charged matter has more of one type of charge than the other, and the excess charge of either type may be created in a number of ways. One way you know to give matter a charge is simply to remove a Dacron sweater on a dry day. You can even hear tiny sparks, and the sweater clings to your body because of opposite electric charges on sweater and body. Another way to charge matter is to scuff your feet on a wool rug. If you then touch a metal object by chance, you get a shock. The force between two quantities of electric charge q_1 and q_2 concentrated at two points separated by a distance r in a vacuum is given by Coulomb's law:

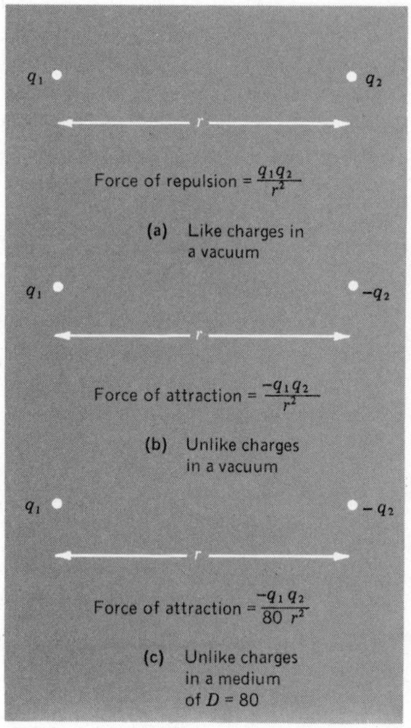

FIG. 6 Coulomb's law: effect of dielectric constant.

Charles A. Coulomb (1736–1806) was a French physicist who discovered the law that gives the force of attraction between electrically charged bodies. The unit of electric charge is named for him.

$$\text{Force} = \frac{q_1 q_2}{r^2}$$

Coulomb's law is illustrated in Figure 6. If the charges q_1 and q_2

If the charges are not in a vacuum, the dielectric constant (D) of the medium between them must be included, and we have this equation:

$$\text{Force} = \frac{q_1 q_2}{D r^2}$$

For air $D = 1.0006$; for water $D = 80.4$; for a vacuum $D = 1$.

have the same sign (++ or −−), the electric force acting between them drives them apart ("like charges repel"). If the charges are of opposite sign (+− or −+), the electric force pushes them together ("unlike charges attract").

Coulomb's law, like the law of gravitation, is an "inverse-square" law because electric force is inversely proportional to the square of the distance between charges; that is, electric force is proportional to $1/r^2$, where r is the distance. This means that doubling the distance between charges reduces the electric force, whether of attraction or repulsion, to one-fourth the original value. To see how this comes about we write equations for the forces f_1 and f_2 with the distance doubled to $2r$ for f_2:

$$f_1 = \frac{q_1q_2}{Dr^2} \qquad f_2 = \frac{q_1q_2}{D(2r)^2} = \frac{q_1q_2}{D4r^2} = \frac{1}{4}\frac{q_1q_2}{Dr^2} = \frac{1}{4}f_1$$

It follows that electric forces *decrease* rapidly as the distance between charges increases. The inverse-square dependence also means that both attraction and repulsion *increase* rapidly as the distance between the charges becomes very small. Decreasing the distance to $\frac{1}{10}$ of its original value causes the force to increase a hundredfold. Using $r/10$ for the decreased distance and Coulomb's law, we obtain:

$$f_1 = \frac{q_1q_2}{Dr^2} \qquad f_2 = \frac{q_1q_2}{D(r/10)^2} = \frac{q_1q_2}{Dr^2/100} = \frac{100q_1q_2}{Dr^2} = 100f_1$$

The distances between the particles of matter are often very small, so when the particles carry electric charge the electric forces between them are very strong. These forces are so strong, in fact, that they are mostly responsible for the behavior of the particles. Electric forces of attraction between unlike charges cause the sticking together or *cohesion* of the particles seen in liquids and solids, and are responsible for chemical bonds.

To understand some chemical applications of Coulomb's law, it is also important to see the way in which the force depends on the quantities of charge q_1 and q_2. If you double *both*, for example, from q_1 and q_2 to $2q_1$ and $2q_2$, the electrostatic force is increased *four* times:

$$f_1 = \frac{q_1q_2}{Dr^2} \qquad f_2 = \frac{2q_12q_2}{Dr^2} = \frac{4q_1q_2}{Dr^2} = 4f_1$$

For our purposes it isn't necessary to know the units of Coulomb forces—we'll be more concerned with the *changes* in the force that go with changes in charge and distance.

Exercises

10. What would be the effect on the attraction between two opposite, equal charges, $+q$ and $-q$, if their values were increased three times?

11. What would be the effect on the repulsion between two like, equal charges if their values were both doubled at the same time the distance (r) between them was doubled?

Systems of Measurement Whenever some property of a sample of matter is described in numbers, some measuring unit is always attached. You are probably most familiar with the "foot/pound/second" system of measurement. Scientists have agreed to use the "meter, kilogram, second" (MKS) units in their measurements. The MKS system that chemists use is in fact based on the International System of Units (SI, from the French *Système Internationale*). This system, which is used in virtually all the world, is described in Appendix 2. The MKS system is a decimal one, like our money system, and the names attached to the subunits have Greek prefixes that describe them. In this system the basic unit of length or distance is the *meter*, defined as the length of the standard meter bar kept in the International Observatory near Sèvres in France. You will find a more fundamental definition of the meter in Appendix 8. Decimal multiples and fractions of the meter and their equivalents in more familiar units are as follows:

1 kilometer (km) = 1000 meters (m) = 0.62137 miles (mi)

1 centimeter (cm) = 1/100 m = 10^{-2} m = 0.3937 inches (in.)

1 millimeter (mm) = 1/1000 m = 10^{-3} m = 10^{-1} cm = 0.03937 in.

1 micron (μ) = 1/1,000,000 m = 10^{-6} m = 10^{-3} mm = 10^{-4} cm

It's convenient to remember that 1 m is about 1 yard (yd) (actually 1 m = 39.37 in.); 1 cm is a shade less than 0.40 in., and 1 mm is about the diameter of an ordinary pin. You should learn the meanings of the Greek prefixes—kilo-, centi-, milli-, and micro-, because they are used throughout the MKS system. Their meanings can be expressed as multiples, as fractions, or in what is called exponential notation. Thus, "kilo" means 1000 or 10^3 times the fundamental unit; "centi" means 1/100, 0.01, or 10^{-2}; "milli" means 1/1000, 0.001, or 10^{-3}; and "micro" means 1/1,000,000, 0.000001, or 10^{-6}. Exponential notation is used often in chemistry and is explained in Appendix 1. If you are not familiar with this way of expressing numbers, you should study that appendix. In exponential notation 1 msec (millisecond) = 10^{-3} sec. If you have had algebra, you will remember that $a^{-3} = 1/a^3$, so 1 msec = $1/10^3$ sec = 1/1000 sec = 0.001 sec.

The fundamental unit of *mass* or quantity of matter is a block of platinum called the *standard kilogram* (kg) kept with the standard meter in France. As its name indicates, the standard kilogram has a mass of 1000 grams (g). One kilogram weighs 2.20 lb on earth and 1 g weighs about as much as a peanut. A kilogram *weighs* only about 0.37 lb on the moon where gravity is only $\frac{1}{6}$ as strong as on earth, but a kilogram of matter has the same *mass* anywhere. Chemists use the terms mass and weight as though they were the same thing. This is because when they "weigh" a sample of matter on a chemical scale, they are really comparing the mass of the sample with the standard masses of gram weights.

The unit of volume used by chemists is the *liter* (l). One liter of volume is the volume occupied by 1 kg of water at a temperature of 4° centigrade (°C). Chemists also use the *milliliter* (ml), which is 0.001 or 10^{-3} liter. Until scientists all agreed to use the MKS system, volume measurements were sometimes given in cubic centimeters (cc or cm^3). One cc is the same as 1 ml of volume.

One second of time is defined fundamentally in terms of the rate of vibration of certain molecules. For our purposes, the second is 1/86,400 of what is called an "ephemeris" day (morning to night).

Exercises

12. Convert your height in feet and inches to centimeters; to meters.

13. Express your weight in kilograms; in grams.

14. Convert the H—H bond length, 0.74×10^{-8} cm, to inches.

15. The speed of light in a vacuum is 186,280 miles per second (miles sec^{-1}). Convert it into cm per sec (cm sec^{-1}) and m per sec (m sec^{-1}).

16. Distances in machine work can be measured to ± 0.00001 in. ($\pm 10^{-5}$ in., 1/100,000 in.). Express this distance in millimeters (mm).

Units in Equations When you solve an algebraic equation like $3x = 57$ and get $x = 19$, you don't ordinarily ask "19 what?" Mathematical equations used in the sciences are not like this—their numbers almost always have units of measurement attached to them, and you must be sure that you have *set up calculations so that the same units are on both sides of the equals sign, before applying algebra to solve the equation.* Not only is proper attention to units necessary; it will also help you to solve problems correctly because the units of the answer tell how to set it up.

EXAMPLE Suppose in a study of environmental pollution by mercury you need to know the volume in milliliters (ml) occupied by 25.00 grams (g) of mercury at 20°C. This volume is not in any table but what *is* tabulated is the *density* of mercury in grams per ml measured at 20°C. It is given in the *Handbook of Chemistry and Physics*, published by the Chemical Rubber Co., as 13.59 g per ml or, using exponential notation, 13.59 g ml^{-1}. The value means that at 20°C each ml of mercury weighs 13.59 g. Now the problem can be restated. How many 13.59-gram masses are there in 25.00 g? You find out by dividing 13.59 into 25.00, just as you would calculate the number of dozens in 132 by dividing 12 into 132. Now, 25.00 divided by 13.59 = (25.00/13.59) = 1.840, and the units are ml, so 25.00 g of mercury occupy a volume of 1.840 ml at 20°C. If you form the habit of attaching the units to the numbers, you will find that the answer has to come out in the right units:

$$\frac{25.0\ \cancel{g}}{13.59\ \cancel{g}\ ml^{-1}} = \frac{25.0}{13.59}\ ml = 1.84\ ml$$

The units of grams divide out as shown and ml^{-1} in the denominator is the same as ml in the numerator ($1/ml^{-1} = ml$). The fact that the units are right means that you set up the calculation correctly. If you mistakenly did the division the other way around, the units of the answer would not be ml as required but ml^{-1} or "per ml," which doesn't make sense as a volume:

$$\frac{13.59\ \cancel{g}\ ml^{-1}}{25.0\ \cancel{g}} = 0.544\ ml^{-1}$$

There are not very many mathematical equations in this book, but you should always check units when you use them, particularly in the gas law problems in Chapter 5.

Exercises

17. The density of water at 25°C is 0.997 g ml^{-1}. Calculate the mass (weight) of 1.00 liter of water at 25°C.

18. Using the fact that 1 in. = 2.54 cm, calculate the number of centimeters in 1 mile (5280 ft).

Limitations in Making Measurements With only one exception, the only numbers having exact values in chemistry are the "counting numbers" like those in chemical formulas. For example, the formula for the water molecule, H_2O, tells us that there are exactly two hydrogen atoms and one oxygen atom in each water molecule. The symbol ^4_2He means, for reasons you will see later, a helium atom made up of exactly two protons, two neutrons, and two electrons. The only number used in chemistry that has an exact value and is not a counting number is the *atomic weight* of $^{12}_6C$ atoms, which is exactly 12.

Exceptions are the values of mathematical numbers such as $\pi = 3.141592654\ldots$ and the natural logarithm base $e = 2.718281828\ldots$, which can be made as accurate as needed.

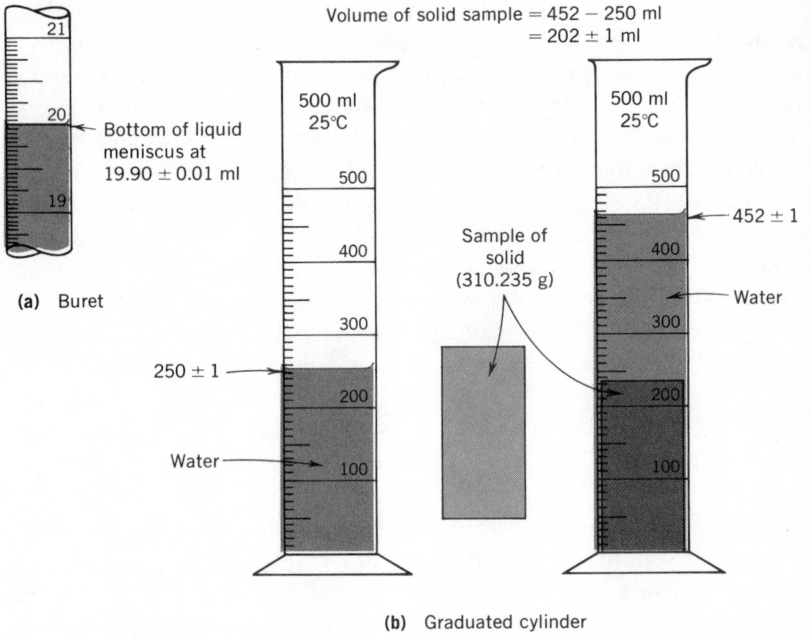

FIG. 7 Accuracy in the measurement of volume using (a) a buret; (b) a graduated cylinder.

Most of the other numbers in chemistry are the result of experimental measurements and therefore contain errors. Even the simple experiment of weighing a block of metal can only be done within certain limits of accuracy. On the very best analytical scale (microbalance) a mass of about 1 gram can be weighed to ± 0.000001 g (10^{-6} g or 1 *micro*gram). Ordinary laboratory balances determine masses within an error of about ± 0.001 g (10^{-3} g, 1 *milli*gram). Knowing that the numbers you are using are of limited accuracy means that you can save time in calculations. There is usually no point in carrying out a calculation (usually a division) to a larger number of decimal places than there are in the numbers you are using. This rule does not apply to counting numbers (integers), which have exact values.

The question of how many decimal places to keep in a calculated value is answered by the *rules of significant figures*. A result of applying these rules is shown by the example of determining the density (weight/volume) of a block of metal by using a graduated cylinder in Figure 7(b). The weight of the solid sample was found to be 310.235 g. This means that the value of the weight is accurate to ± 0.001 g, or the true weight of the sample is between 310.234 g and 310.236 g and therefore accurate to about 1 part in 310,234. The volume, measured by displacement of water in the graduated cylinder, is 202 ± 1 ml, or accurate to only 1 part in 202. Density is defined as weight divided by volume, so the density of the sample is 310.234 g/202 ml. Carrying out the division gives the density as 1.535811881 g ml^{-1}. The calculated value gives the false impression that the density is accurate to 1.535811881 ±0.000000001, or better than 1 part in a billion. However, this accuracy is much greater than the 1 part in 202 ±1 for the volume used in the calculation. To save time in calculation and still have proper regard for accuracy means that you can round off ☛ the weight of the sample so that its accuracy is about that of the volume measurement. Rounding off 310.236 to three decimal places gives 310, a value good to 1 part in 310 and similar to the 1 part in 202 for the volume. Using these figures gives the most accurate value for the density that can be calculated from the experimental data:

Round off to zero when the digit removed is less than 5 and to one when it is 5 or greater. Examples are 310.236 → 310.24 → 310.2 → 310 and 1.5345 → 1.535 → 1.54.

$$\text{Density} = \frac{310 \text{ g}}{202 \text{ ml}} = 1.54 \text{ g ml}^{-1}$$

This result is accurate to 1 part in 154, which is not more accurate than the least accurate number (202) used in the calculations. This is an example of the rule of thumb that *the final result of any calculation should not imply a greater accuracy than the accuracy of the least accurate number used in it.*

Figure 7(a) shows that by using a buret, volumes can be measured to ±0.01 ml in about 20 ml. The value 19.90 has *four significant figures*, because only the last digit (0) is uncertain. Almost all chemical calculations can be carried out by using no more than four significant figures. So, although the accurate value for the atomic weight of oxygen is 15.9994 and has six significant figures, you can use the rounded-off value of 16.00 (four significant figures) in ordinary chemical calculations.

Zeros in numbers are significant except when they give only the magnitude of a number less than 1. For example, in the number 0.000549 there are not seven but only *three* significant figures—5, 4, and 9. The zeros show only that the number is 549 *millionths*. You can see this more easily if the number is written in exponential notation as 549×10^{-6} or 5.49×10^{-4}, both of which clearly have three significant figures (the 10^{-6} and 10^{-4} only place the decimal point properly). Zeros in numbers such as 3750, 2.0003, 1.60×10^2 are significant. In general, it is incorrect to have more significant figures in a calculated answer than there are in the smallest number in the values used. Thus, if 2.3 is multiplied by 3.1416, the answer with the correct number of significant figures is 7.2—not 7.22568. You can save time in calculations by using the idea of significant figures. Counting numbers have an unlimited number of significant figures—2 molecules means 2.000000000 . . . molecules.

Supplementary Exercises

19. Determine the number of significant figures in the following:

a. 22.414	b. 16,000	c. 3×10^{-3}
d. 16	e. 16×10^{4}	f. 0.00297
g. 16.0	h. 16.000×10^{4}	i. 0.002970
j. 16.00	k. 0.003	l. 2.97×10^{-3}

20. The fluorine molecule is made up of two fluorine (F) atoms, so its chemical formula is F_2. One of these molecules has a mass, or "weighs," 6.3092×10^{-23} g. What is the most accurate value for the mass of one fluorine atom that can be calculated from the data? (*Hint*: Remember, 2 is a counting number.)

21. An empty glass bulb weighs 350.00 g. When filled with water at 25°C, where the density of water is 0.99707 g ml^{-1}, the bulb and water weighed 750.00 g. Calculate the most accurate value of the volume of the bulb obtainable from the data, taking significant figures into account and showing units.

22. You are climbing Mt. Whitney and succeed in reaching the top, but only after several tries. Does your potential energy, once you are on the peak, depend on how much work you may have done to get there? What general statement can you make about the relation between potential energy and the way it is created?

23. There are about 1.5×10^{9} cubic kilometers (km^3) of water in the combined oceans of the earth. Convert the volume to cubic miles and to cubic centimeters (cm^3). How does the number of cm^3 compare with the number 6.023×10^{23} (Avogadro number), which is the number of C atoms in 12.01 grams of carbon?

24. One sodium (Na) atom weighs 1.9164 times as much as one $^{12}_{6}C$ atom. If the atomic weight of a $^{12}_{6}C$ atom is 12, what is the atomic weight of an Na atom?

25. Thermodynamics, the science dealing with heat (thermal energy) and work, gives for the maximum possible efficiency of a heat engine converting thermal energy to mechanical work the following equation:

$$\text{maximum percent of thermal energy converted to work} = \frac{T_2 - T_1}{T_2} \times 100$$

Here, T_2 is the *absolute* temperature of the source of thermal energy and T_1 is the absolute temperature of the sink. Calculate the maximum percent of thermal energy converted to work by a steam turbine if the boiler is at 227°C and the condenser at 27°C. (Use 273 for converting °C to K.) What would have to be the value of T_1 to turn 100% of thermal energy into work?

Key Words

Chemical potential energy energy stored in chemical compounds and released in chemical reactions, such as burning oil and gasoline.

Coulomb's law law that gives the force between electrically charged particles.

Electric potential energy ability of electrostatic attraction between charged particles to do work.

Energy capacity of a body to do work because of its motion (kinetic energy) or position (potential energy).

Exponential notation expressing numbers using powers of 10.

SI International System of Units of measurement of which the meter-kilogram-second (MKS) system is a part.

Significant figures meaningful digits in a measured numerical value.

Thermal energy heat energy; kinetic energy of atomic particles.

Suggested Readings

Astin, A. V., "Standard of Measurements," *Sci. Am.*, **218**(6):50 (1968).

Boyle, Robert H., "At the Rate We're Going, It's Good-by Fish," *Reader's Digest*, Jan. 1974, pp. 140–144.

Bright, G. W., and Jones, C., "Teaching Children to Think Metric," *Today's Educ.*, **62**:16–19 (April 1973).

Cairns, John, Jr., "Coping with Heated Waste Water Discharges from Steam-Electric-Power Plants," *BioScience*, **22**:411–419 (July 1972).

Diamond, Henry L., "Beneficial Uses of Thermal Discharges," *Catalyst*, **1**(4):17–21 (Winter 1971).

Duke, James B., "Locomotion: Energy Cost of Swimming, Flying, Running," *Science*, **177**:222–227 (July 21, 1972).

Holden, Constance, "Metrication: Craft Unions Seek to Block Conversion Bill," *Science*, **184**:48–50 (April 5, 1974).

Jefferson, Edward G., "Industry Can Save Energy by Ending Waste," *Catalyst*, **4**(1):16–18 (1974).

Lewicke, Carol K., "Thermal Pollution in Uncharted Waters," *Env. Sci. and Tech.*, **5**:1170–1172 (Dec. 1971).

"Liabilities into Assets," *Env. Sci. and Tech.*, **8**:210–211 (March 1974).

Meyer, C. F., and Todd, D. K., "Conserving Energy with Heat Storage Wells," *Env. Sci. and Tech.*, **7**:512–516 (June 1973).

Murdock, W. W., ed., *Environment: Resources, Pollution, and Society*, Stamford, Ct., Sinauer Associates, 1971.

Paul, Martin A., "International System of Units (SI)," *Chemistry*, **45**:14–18 (Oct. 1972).

Riegel, Kurt W., "Light Pollution," *Science*, **179**:1285–1291 (March 30, 1973).

"The Metrics Are Coming! The Metrics Are Coming!" *Changing Times*, May 1974, pp. 33–34.

chapter 1

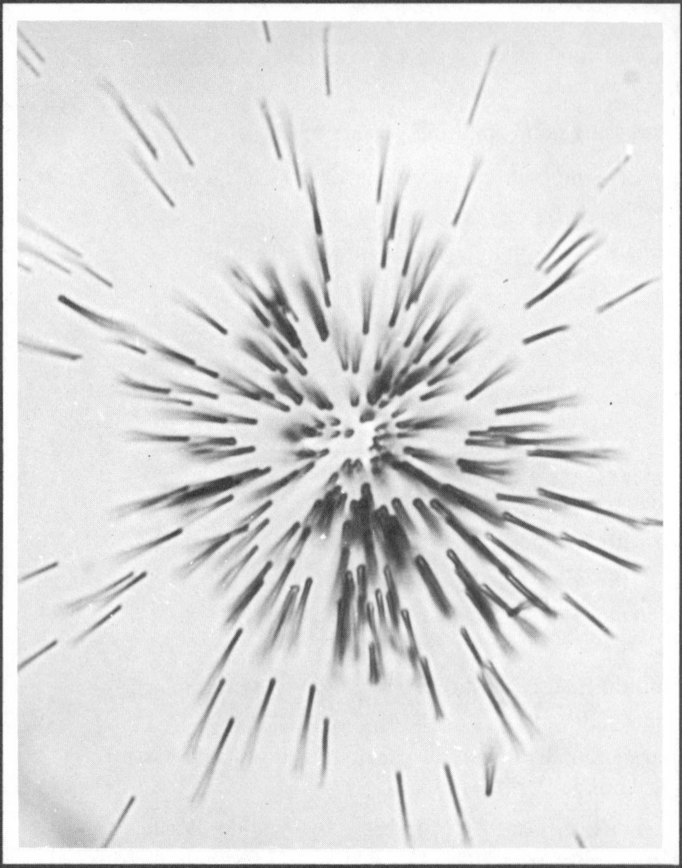

A fission track "sunburst" produced in mica by the paths of microparticles resulting from fission of atoms in a small grain of uranium. These sunbursts allow analysis of dust particles for uranium content. *(Photo 4, Nuclear Tracks in Solids by Robert L. Fleischer, P. Buford Price, and Robert M. Walker, University of California Press, 1975.)*

The Microparticles of Chemistry

1.1 Introduction What is chemistry? How did it start? Where is it going? Why learn something about it? Is it difficult to understand? These are a few of the questions you might ask on beginning your study of this subject. None of these questions has a simple answer. Even a group of chemists would have great difficulty in reaching agreement on the answers. But if you study chemistry, you will find reassuring answers about the nature and importance of the subject. Some of these answers will help you understand the ways we pollute our environment and what we can—and cannot—do to prevent it. Depending on your plans for the future, you may also learn something that will be useful to you in your profession, that will help you to make better decisions.

Chemistry deals mainly with *things*, particularly with matter (solids, liquids, and gases) and the changes that matter undergoes. Those changes can be as simple as the boiling of water or as complicated as the changes in matter that are responsible for our hereditary traits. Chemists invent ideas and concepts that explain in simpler terms the complex changes that matter undergoes. These ideas often give us considerable control over changes in matter, so that we can use changes to our advantage.

The most fundamental and simplifying idea in chemistry is that matter is made up of tiny particles. This idea goes back to the "atoms" of the ancient Greek philosopher Democritus, but was put on a useful basis for chemists in the early 1800s by Dalton and Lavoisier, among others. These *microparticles* are unbelievably tiny—less than one-billionth of an inch in diameter—so there are incredibly large numbers of them in any sample of matter you can see. Furthermore, because the particles are so small that they can't be seen or measured individually, the properties of the particles must be based on the average behavior of the very large numbers of them present in any seeable sample of matter. Sometimes properties *can* be individually measured from the tracks left by single particles even though the particles themselves cannot be seen (Photo 4).

John Dalton (1766–1844), the son of a poor Quaker weaver, was an English chemist, physicist, and teacher who published his atomic theory of matter in 1803. He also published the first list of atomic weights of elements.

Antoine Laurent Lavoisier (1743–1794) was a noted French chemist who has been called the father of modern chemistry. He proved the law of conservation of matter. In spite of his many contributions to French government and to science, he was regarded as an enemy of the French Republic and was put to death by the guillotine.

This process of relating the properties of sizable, seeable, or *macroscopic* samples of matter to the properties and arrangement of its *microparticles* is what chemistry is all about. Most people are satisfied with knowing that sugar tastes sweet and is a food. A chemist wants to know what kinds of atoms there are in sugar, how the atoms are arranged and held together, and even *why* sugar tastes sweet.

1.2 Electron, Proton, and Neutron Knowing that matter is made up of microparticles raises some questions. How many kinds of microparticles are there? How heavy are they? How big are they? Are they all the same size? In answering these questions we will find that there are thousands of different types of microparticles but that some are simpler or more *fundamental* than others and are the building blocks for the more complicated microparticles. The fundamental particles are the electron (symbolized e^-), the proton (p^+), and the neutron (n). By various indirect methods it is possible to determine the *mass* or amount of matter in these particles. The kilogram (see the Introduction) is inconveniently large as a unit in chemistry. Chemists use the gram (g) as their unit of mass ($\frac{1}{1000}$ of a kilogram). But even the small amount of matter in one gram is still incredibly larger than the amount in an electron, proton, or neutron, as shown in the following:

$$\text{Mass of one electron} = 0.0000000000000000000000000009109 \text{ g}$$
$$\text{Mass of one proton} = 0.0000000000000000000000016725 \text{ g}$$
$$\text{Mass of one neutron} = 0.0000000000000000000000016748 \text{ g}$$

Such tiny numbers are hard to work with in ordinary decimal form, so they are usually expressed in exponential notation:

Note that the negative exponent of 10 is the number of zeros preceding the first nonzero digit, *counting the zero in front of the decimal point.*

$$\text{Mass of one electron} = 9.109 \times 10^{-28} \text{ g}$$
$$\text{Mass of one proton} = 1.6725 \times 10^{-24} \text{ g}$$
$$\text{Mass of one neutron} = 1.6748 \times 10^{-24} \text{ g}$$

But these are still very tiny numbers, even if more convenient. We usually express weights or masses in the units that are convenient in size to the body being weighed. We buy butter by the pound, and we express the weights of airliners in tons. Microparticle masses are expressed in *atomic mass units*, described in the next section.

Exercises

1. Convert the following quantities to exponential notation: 0.001, 0.00025, $\frac{1}{4}$, 5280.
2. If a kilogram is equal to 2.2046 lb, how many grams are there in 1 lb?
3. At 16 oz to the pound, how many grams make 1 oz?

The atomic mass unit The unit of convenient size for expressing the masses of microparticles is called the *atomic mass unit*, abbreviated *amu*. One amu equals 1.660×10^{-24} g. To put the masses of the electron, proton, and neutron on the amu scale you simply divide their masses in grams by the mass of one amu in grams:

$$\text{Electron: } \frac{9.109 \times 10^{-28}}{1.660 \times 10^{-24}} = 5.49 \times 10^{-4} \text{ amu} = 0.000549 \text{ amu}$$

$$\text{Proton: } \frac{1.6725 \times 10^{-24}}{1.660 \times 10^{-24}} = 1.008 \text{ amu}$$

$$\text{Neutron: } \frac{1.6748 \times 10^{-24}}{1.660 \times 10^{-24}} = 1.009 \text{ amu}$$

With the values shown in amu you can see that the proton and neutron have about the same mass and that both are much heavier than the electron. If you've already had some chemistry, you may recognize that the atomic weights of the elements are given in amu.

Exercises

4. One sodium atom (Na) weighs 3.817×10^{-23} g. What is the mass of one sodium atom in kg? In amu?

5. Calculate the ratio of the mass of the electron to the mass of the proton, using their masses in grams and in amu. What do you notice about the two ratios?

These macroscopic units are discussed in Appendix 2. We will seldom use them.

The electronic unit of charge The three fundamental particles differ in another important way. The electron and the proton carry electric charges; the neutron does not. Macroscopic quantities of electric charge are measured in electrostatic units (esu) or in coulombs. Just as the masses of the proton and electron are tiny on the macroscopic (gram) scale, so are their charges tiny. The charge on one electron is -4.8030×10^{-10} esu, while the charge on the proton is equal but opposite in sign, or $+4.8030 \times 10^{-10}$ esu. This quantity of charge is termed *one electronic unit* of charge, regardless of its sign. One electronic or atomic unit of negative charge is shown by a minus sign written as a superscript, as in the symbols for the electron, e^-, and the chloride ion, Cl^-. One electronic unit of positive charge is shown by a superscript plus sign, as in p^+, the symbol for the proton. The symbol for the neutron is n, with no charge shown because the neutron is electrically neutral (uncharged). Just as the amu is convenient for measuring the masses of microparticles, so the electronic unit of charge is the right size for expressing the charges on microparticles. Thus, if we wish to indicate that some microparticle (X) carries two electronic units of positive charge, we write X^{2+}, and if another microparticle (Y) carries three units of negative charge we write Y^{3-}, and so on.

1.3 Chemical Symbolism

The microparticles called *atoms* are somewhat more complex than the electron, proton, and neutron. Different kinds of atoms are made up of different numbers of e^-, p^+, and n, and the symbol used for a particular atom gives its composition in terms of the numbers of e^-, p^+, and n it contains. Atoms as a whole are electrically neutral. This means that *the number of protons must be equal to the number of electrons in the atom*, so that their equal but opposite charges cancel. The number of protons (or electrons) in a particular atom is called the *atomic number* (Z) of the atom. The number of neutrons plus the number of protons is given by the *mass number* (M) of the atom. The number of neutrons is not given directly, but you can always find it because it is the difference between the mass number and the atomic number. In other words,

$$\text{Number of } n = M - Z$$

The mass number, M, is placed as a superscript and the atomic number, Z, as a subscript, both before the letter symbol for the atom.

The letter symbol is an abbreviation of the chemical name of the atom—H for hydrogen, O for oxygen, He for helium, Na for sodium (Latin, *natrium*), Ra for radium, and so on. Thus, an oxygen atom that contains $8e^-$, $8p^+$, and $8n$ is represented by the symbol

$$^{16}_{8}\mathrm{O}$$

Other examples are $^{1}_{1}\mathrm{H}$, $^{4}_{2}\mathrm{He}$, $^{23}_{11}\mathrm{Na}$, $^{226}_{88}\mathrm{Ra}$, or in general, $^{M}_{Z}X$. For example, $^{226}_{88}\mathrm{Ra}$ represents a radium atom made up of $88p^+$, $88e^-$, and $(226-88) = 138n$. An atom for which the values of Z and M are given is called a *nuclide*.

Nearly as important as atoms are the building blocks of matter called *ions*. The simplest ions result when atoms lose or gain one or more electrons. Since atoms are electrically neutral, ions carry a net positive or a net negative charge, depending on whether electrons were lost or gained in forming the ion. The following examples show the compositions and chemical symbols for some ions:

$$^{4}_{2}\mathrm{He} \xrightarrow{\text{remove } 1e^-} {}^{4}_{2}\mathrm{He}^{+} + e^-$$

$$\left(\begin{array}{c} 2p^+ \\ 2n \\ 2e^- \end{array}\right) \xrightarrow{\text{remove } 1e^-} \left(\begin{array}{c} 2p^+ \\ 2n \\ 1e^- \end{array}\right)^{+} + e^-$$

$$^{226}_{88}\mathrm{Ra} \xrightarrow{\text{remove } 2e^-} {}^{226}_{88}\mathrm{Ra}^{2+} + 2e^-$$

$$\left(\begin{array}{c} 88p^+ \\ 138n \\ 88e^- \end{array}\right) \xrightarrow{\text{remove } 2e^-} \left(\begin{array}{c} 88p^+ \\ 138n \\ 86e^- \end{array}\right)^{2+} + 2e^-$$

$$^{16}_{8}\mathrm{O} + 2e^- \xrightarrow{\text{add } 2e^-} {}^{16}_{8}\mathrm{O}^{2-}$$

$$\left(\begin{array}{c} 8p^+ \\ 8n \\ 8e^- \end{array}\right) + 2e^- \xrightarrow{\text{add } 2e^-} \left(\begin{array}{c} 8p^+ \\ 8n \\ 10e^- \end{array}\right)^{2-}$$

Notice that *the formation of an ion does not result in any change in the atomic number* (Z). Even though the number of electrons is changed, the number of protons remains the same. Ions that carry a positive charge are called *cations* (pronounced "cat-ion"); those that carry a negative charge are called *anions* ("an-ion"). Thus, $^{226}_{88}\mathrm{Ra}^{2+}$ is a cation, and $^{16}_{8}\mathrm{O}^{2-}$ is an anion. Because of their opposite charges, there is a strong force of attraction between anions and cations. This attraction holds the ions together in a class of solids called *ionic compounds*—"ionic" because they are built up of oppositely charged ions and "compounds" because they are derived from more than one kind of atom. Radium oxide, built up from equal numbers of $^{226}_{88}Ra^{2+}$ and $^{16}_{8}O^{2-}$ ions, is an ionic compound. Because radium is such a scarce element, only a few grams of radium oxide have ever been prepared. In contrast, tons of the ionic compound sodium fluoride, $^{23}_{11}Na^{+19}_{9}F^{-}$, are used to "fluoridate" water supplies to prevent tooth decay.

One cation stands out from the large number of ions possible. This unique ion is produced from the hydrogen atom by removal of its single electron:

$$^{1}_{1}H \xrightarrow{\text{remove } 1e^{-}} {}^{1}_{1}H^{+} + e^{-}$$

$$\left(\begin{array}{c} 1p^{+} \\ 1e^{-} \end{array}\right) \xrightarrow{\text{remove } 1e^{-}} \left(1p^{+}\right) + e^{-}$$

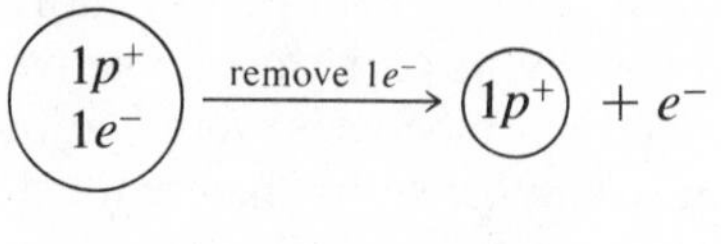

Removal of the electron from a hydrogen atom leaves only the proton! There are three symbols for the proton, p^{+}, $^{1}_{1}H^{+}$, or simply H^{+}. We will use all of these symbols, depending on the subject under discussion. The proton or *hydrogen ion*, as it is often called, is second in importance only to the electron in chemistry.

Exercises **6.** Using the periodic table and the atomic numbers (inside the front cover), find the letter symbols for the following atoms: $^{40}_{20}X$, $^{197}_{79}X$, $^{19}_{9}X$.

7. Give the charges in electronic units and the chemical symbols for the ions made up of $20n$, $20p^{+}$, $18e^{-}$; $16n$, $16p^{+}$, $18e^{-}$; $18n$, $17p^{+}$, $18e^{-}$. Which are anions? Which are cations?

1.4 The Nuclear Atom Up to this point, atoms and ions have been described by giving the numbers of electrons, protons, and neutrons present without any hint about how they are arranged — we have been using a "mashed-potato" model. Actually atoms (and derived ions) have a tiny *nucleus* that contains the protons and neutrons. The nucleus is surrounded by rapidly moving electrons, which create a spherical, smeared-out cloud of negative charge called the *electron cloud* of the atom or ion. The diameter of the electron cloud is about 10,000 times that of the nucleus. On the other hand, since the mass of the electron is only 0.000549 amu, while those of the proton and neutron are slightly more than 1 amu, *most of the mass of an atom is in its nucleus*. The positive charge is also in the nucleus, since the protons are there. The following are some examples of the structures of atoms and ions (not drawn to scale):

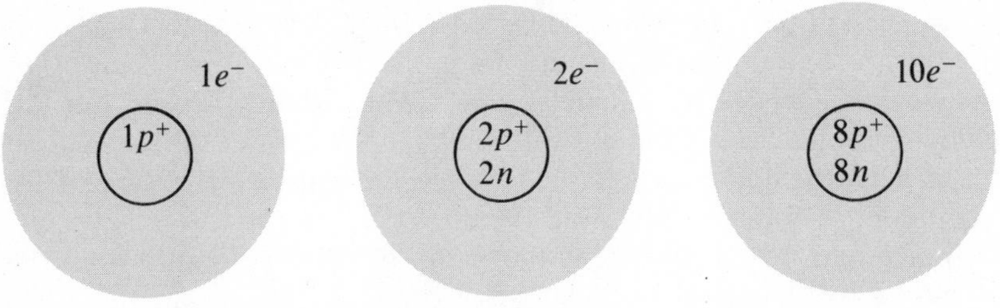

^1_1H ^4_2He $^{16}_8O^{2-}$

Since atoms cannot be seen, how do we know that this arrangement is correct? Credit for the nuclear atom model (1911) goes to Rutherford. He observed (Figure 1.1) that when a stream of very rapidly moving $^4_2He^{2+}$ ions [also called alpha (α) particles] was directed at a very thin sheet of gold, nearly all passed right through and only occasionally was one deflected. Though thin, the gold foil was known to be many thousands of atoms thick. Deflections occurred only when the $^4_2He^{2+}$ ion came close to or collided with something that was massive and positively charged. That something was the gold nucleus—the thin, light, electron cloud of the atoms offered little or no resistance to the passage of the $^4_2He^{2+}$. The experiment showed that most of the volume occupied by an atom is the volume of its electron cloud. In the gold sheet, the atoms are packed together so that their electron clouds

Ernest Rutherford, later Lord Rutherford (1871–1937), was born in New Zealand. He was a physicist whose research into the structure of atoms led to his discovery of the nuclear atom. He received the Nobel prize in chemistry in 1908.

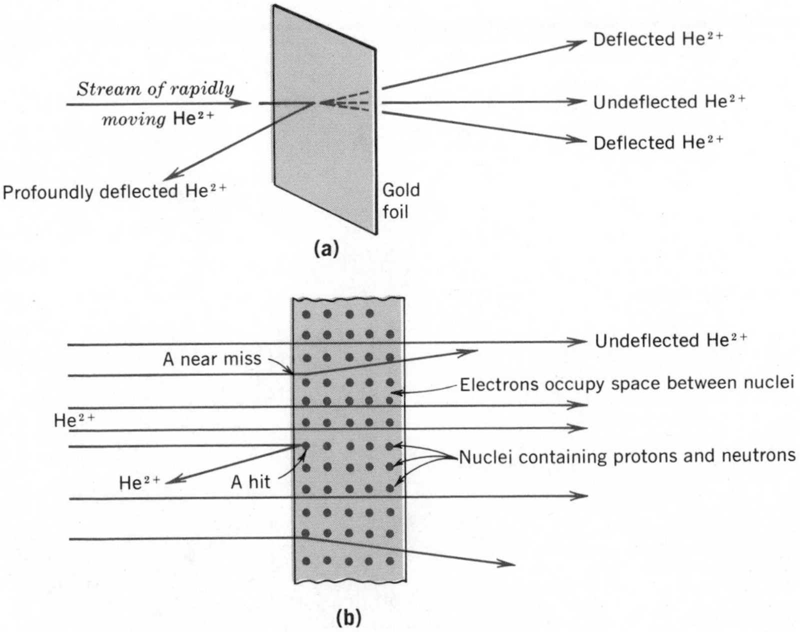

FIG. 1.1 Rutherford's explanation of α-particle scattering: (a) macroscopic observations; (b) microscopic explanation. [The paths of the He^{2+} are detected by the flashes of light produced when He^{2+} particles strike a zinc sulfide (ZnS) screen $^4_2He^{2+}$ abbreviated to He^{2+}.]

touch each other. The repulsion between the like (negative) charges of the electron clouds prevents the atoms from penetrating each other very much and keeps the nuclei apart. The moving electrons of the cloud are held close to the positively charged nucleus by the attractive force between opposite electrical charges. Because the electrons are in motion, they don't "fall" into the nucleus.

Antoine Henri Becquerel (1852–1908) was a French scientist whose discovery of radioactivity led to the era of modern physics. He was awarded the Nobel prize in physics in 1903.

In Rutherford's experiment the paths of the $^{4}_{2}He^{2+}$ ions were determined by the flashes of light they produced on striking a screen coated with zinc sulfide (ZnS). The $^{4}_{2}He^{2+}$ ions came from a sample of radium, the atoms of which are *radioactive*. Radioactivity was discovered in 1896 by Becquerel, and used by Rutherford to unravel secrets of atomic structure about 15 years later. Radioactive atoms give off three kinds of particles—alpha (α), beta (β), and gamma (γ); and the effects of an electric field between charged plates on streams or rays of the particles are shown in Figure 1.2. The stream of α particles ($^{4}_{2}He^{2+}$) is bent toward the negative plate because they carry + charge; β rays (fast-moving electrons) are bent toward the positive plate by their negative charge; and γ rays go straight through because they carry no charge. β and γ rays are discussed in Chapter 16.

The $^{4}_{2}He^{2+}$ ions used by Rutherford came from radioactive radium atoms, according to the following *nuclear* chemical equation:

In writing equations for nuclear chemical reactions it is customary to omit any ionic charges, so $^{4}_{2}He$ is used instead of $^{4}_{2}He^{2+}$ for α particles.

$$^{226}_{88}Ra \rightarrow {}^{4}_{2}He + {}^{222}_{86}Rn$$

Note that the sums of the atomic numbers and mass numbers of the

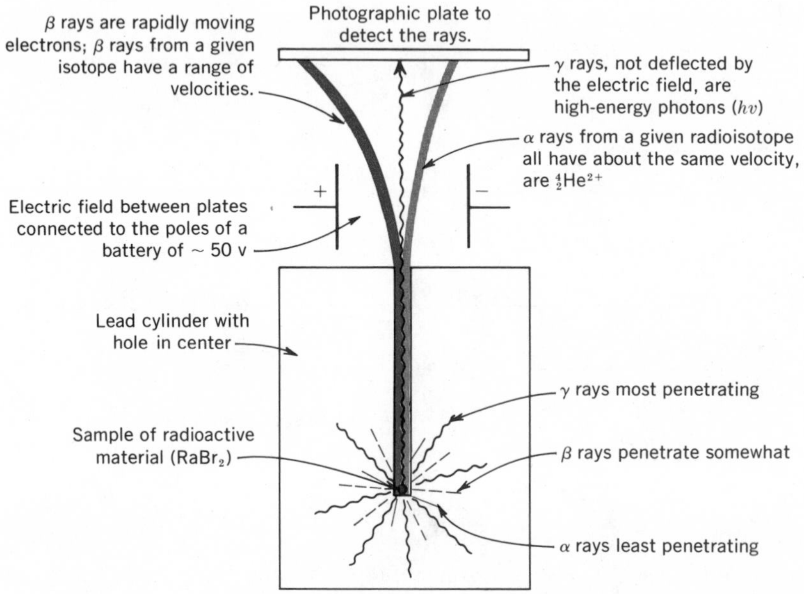
β rays are rapidly moving electrons; β rays from a given isotope have a range of velocities.
Photographic plate to detect the rays.
γ rays, not deflected by the electric field, are high-energy photons (hν)
α rays from a given radioisotope all have about the same velocity, are $^{4}_{2}He^{2+}$
+
−
Electric field between plates connected to the poles of a battery of ~ 50 v
Lead cylinder with hole in center
γ rays most penetrating
Sample of radioactive material (RaBr₂)
β rays penetrate somewhat
α rays least penetrating

FIG. 1.2 Deflection of radioactive rays by an electric field.

products equal the atomic and mass numbers of the reactant ($88 = 2 + 86$, $226 = 4 + 222$).

Exercises

8. Uranium atoms ${}^{238}_{92}U$ are radioactive and give off ${}^{4}_{2}He$ particles. Complete the following nuclear equation by determining M, Z, and X:

$${}^{238}_{92}U \rightarrow {}^{4}_{2}He + {}^{M}_{Z}X$$

9. What nuclide would result from the emission of $2\,{}^{4}_{2}He$ in steps from ${}^{230}_{90}Th$?

EXAMPLE

Advantage was taken of the radioactivity of ${}^{222}_{86}Rn$, which emits ${}^{4}_{2}He$, to actually count the number of atoms in a macroscopic sample of matter. The count was approximate because of experimental difficulties but gave a clear indication of the huge number of atoms in seeable samples of matter.

A sample of radon was sealed in a bulb of very thin glass (Figure 1.3). Like gold foil, the thin glass did not obstruct the passage of the ${}^{4}_{2}He^{2+}$. The number of ${}^{4}_{2}He^{2+}$ emitted per minute was counted, using the flashes of light they produced on striking a zinc sulfide (ZnS) screen. The thin glass bulb was then enclosed in a larger, much thicker glass bulb that stopped all of the ${}^{4}_{2}He^{2+}$, permitting them to capture electrons and so become ${}^{4}_{2}He$ atoms. As the radioactive decay went on, ${}^{4}_{2}He$ atoms, which make up helium gas, collected in the space between the bulbs. After several months, the weight of ${}^{4}_{2}He$ that had accumulated was measured by using atomic spectrometry (Chapter 3) because the amount of helium (${}^{4}_{2}He$) was too small for direct weighing. Knowing the time required to produce a given weight of ${}^{4}_{2}He$ and the number of ${}^{4}_{2}He^{2+}$ per minute coming from the radon in the thin glass bulb, researchers calculated the number of ${}^{4}_{2}He$ atoms in the sample. It was found that during one year 1.0 g of radon emits 1.2×10^{18} ${}^{4}_{2}He^{2+}$ and yields 8.0×10^{-6} g of ${}^{4}_{2}He$. The number of ${}^{4}_{2}He$ atoms per gram of helium is then

$$\frac{1.2 \times 10^{18}}{8.0 \times 10^{-6}} = 1.5 \times 10^{23}\ {}^{4}_{2}He \text{ per gram of helium}$$

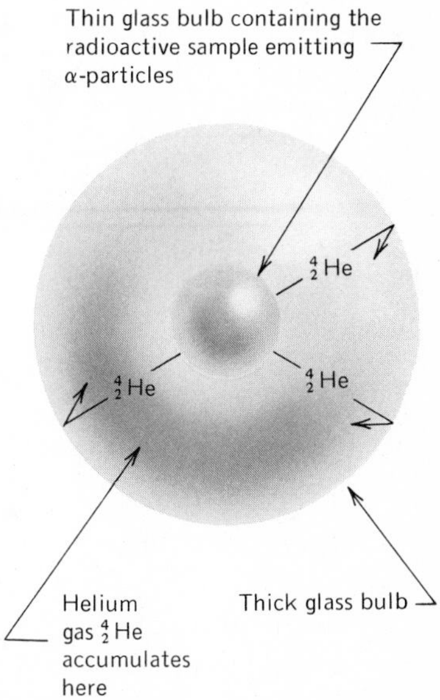

FIG. 1.3 Counting helium atoms from radioactive decay.

Clearly, the number of atoms or other microparticles in a macroscopic sample of matter is so huge as to be beyond practical human comprehension. To put the number 10^{23} in scale, it is approximately the number of grams of water in the combined oceans of the earth.

$1.5 \times 10^{23} = 150{,}000{,}000{,}000{,}000{,}000{,}000{,}000{,}000$. Note that there are 23 places after the decimal point in the exponential notation.

1.5 Molecules and Complex Ions Next to atoms and simple ions in complexity and variety of structure as units of matter are *molecules* and *complex ions*. Molecules are uncharged microparticles containing

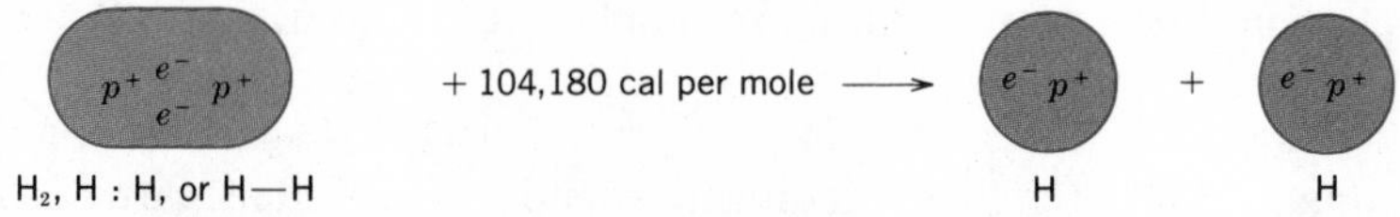

FIG. 1.4 Covalent bond-breaking of the electron-pair bond in H_2.

two or more atomic nuclei held together by a *molecular electron cloud* made from overlapping of atomic electron clouds. The simplest molecule is the hydrogen molecule made up of two protons with two electrons in the binding electron cloud (Figure 1.4). The two protons are held together by their simultaneous attraction to the two rapidly moving electrons. In the absence of the electrons the repulsion between the like charges of the protons would cause them to fly apart (Coulomb's law). The two electrons shared by the protons make up a *two-electron covalent bond*. Several representations of the bond are shown in Figure 1.5, where the mass number and atomic numbers have been omitted for clarity. The electron-dot (c) and covalent-bond (d) representations are the most useful. You can think of the formation of the hydrogen molecule as a result of bringing two hydrogen atoms together until their one-electron atomic clouds overlap and blend to produce the two-electron cloud in which the protons are buried:

$$\mathrm{H}\cdot + .\mathrm{H} \rightarrow \mathrm{H{:}H} + \text{energy}$$

As shown by the equation, the formation of the bond is an *energy-releasing process*. The question is, how much energy is released. The energy unit of the right size for individual bonds is the *electron volt* (eV). This amount of energy is the energy of motion (kinetic energy) of an electron that has been accelerated through a potential of one volt in a vacuum (Figure 1.6). The electron volt is tiny on the macroscopic scale—a person puts out about 6.5×10^{25} electron volts of energy in daily living. The energy released in the formation of one hydrogen molecule from two hydrogen atoms is 4.51 eV, so the electron volt is the right size on the microscopic scale. Applying the law of conservation of energy, which states that *energy may be neither created nor destroyed*, we see that the same amount of energy must be supplied to

An ordinary flashlight battery has a potential (voltage) of 1.5 V.

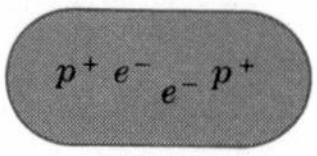

(a) Electron cloud

$H^+ \; {}^{e^-}_{e^-} \; H^+$ **(b)** Two protons—two electrons

H : H **(c)** Electron-dot formula

H—H **(d)** Covalent bond

FIG. 1.5 Different representations of the covalent bond in the hydrogen molecule. The most frequently used are (c) and (d).

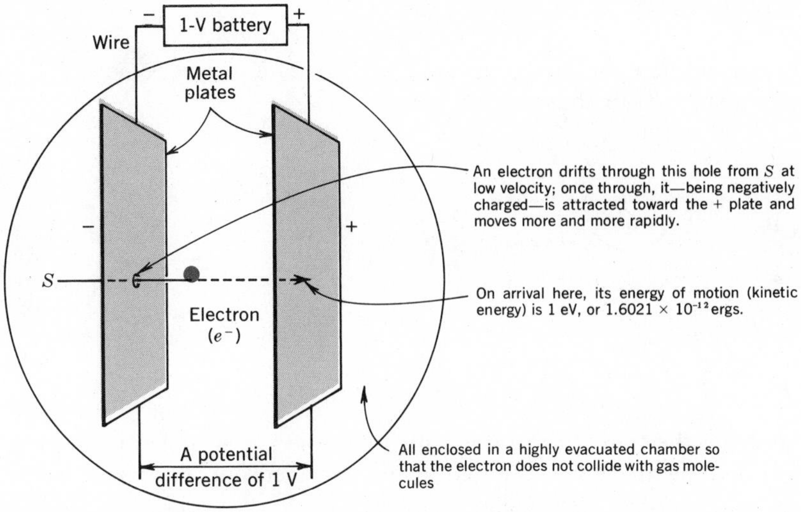
1-V battery
Wire
Metal plates
An electron drifts through this hole from S at low velocity; once through, it—being negatively charged—is attracted toward the + plate and moves more and more rapidly.
S
Electron (e^-)
On arrival here, its energy of motion (kinetic energy) is 1 eV, or 1.6021×10^{-12} ergs.
A potential difference of 1 V
All enclosed in a highly evacuated chamber so that the electron does not collide with gas molecules

FIG. 1.6 Electron volt as an energy unit.

break up the H_2 molecule into its atoms as is released in the formation of the molecule. So, using an arrow (→) to stand for "yields" or "goes to," we can write the two chemical equations:

$$H\cdot + .H \rightarrow H{:}H + 4.51 \text{ eV} \qquad \text{or} \qquad H{:}H + 4.51 \text{ eV} \rightarrow H\cdot + .H$$

The first equation reads, "Two H atoms combine to form one H_2 molecule, releasing 4.51 eV of energy." The second equation says, "Addition of 4.51 eV of energy to one H_2 molecule breaks it apart into two H atoms."

Exercises

10. An average daily diet contains 2500 kilocalories (commonly called "calories") of energy. Assuming that you use 6.5288×10^{25} electron volts of energy per day, calculate the number of kilocalories in one electron volt.

11. Express the H—H bond energy of 4.51 eV in kilocalories per molecule.

12. The radium atom $^{226}_{88}Ra$ is radioactive, giving off α particles with kinetic energy of 4.78 million electron volts (Mev). How many kilocalories is this? (See inside back cover for energy conversions.)

The idea of covalent bonding adds to the chemical symbolism of Section 1.3. Thus, if you want to show that two ^1_1H atoms are covalently bonded to form the hydrogen molecule, place the numeral 2 as a following subscript, $^1_1H_2$. The following are examples of this symbolism for some other molecules:

$^{16}_{8}O_2$ $^{23}_{11}Na_2$ $^{12}_{6}C^{16}_{8}O_2$ $^{19}_{9}F_2$ $^{14}_{7}N_2$ $^{31}_{15}P_4$

The usefulness of this shorthand representation can be seen if you take apart $^{31}_{15}P_4$ to show its full meaning:

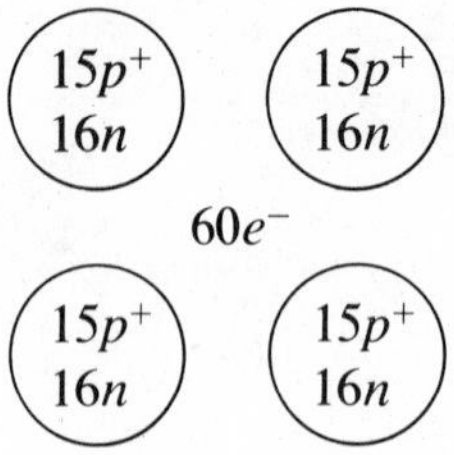

The solid figure that has equilateral triangles for its four faces.

Actually, the four $^{31}_{15}P$ nuclei in this molecule are arranged in space so that they lie at the vertices of a regular tetrahedron, and the nuclei are held together by covalent bonds:

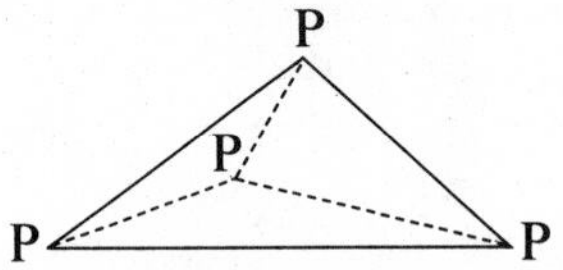

Covalent bonding is very widespread, and there are millions of kinds of molecules known. The large number results from the fact that covalent bonding involving *different* nuclei is also possible. Perhaps the most important example of this type of bonding is provided by the water molecule, usually written H_2O. In the complete symbolism, the water molecule, made up of two protons and one $^{16}_{8}O$ atom, is $^{1}_{1}H_2{}^{16}_{8}O$. There are ten electrons in the cloud of this molecule, eight from the oxygen, one from each hydrogen. Covalent bond theory (Chapter 4) puts four of the electrons in two covalent bonds, leading to the structure (omitting mass and atomic numbers for clarity):

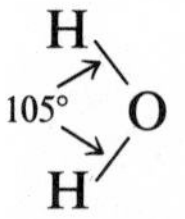

According to theory the remaining six electrons are mainly concentrated about the oxygen nucleus and contribute little to the bonding of the protons. Covalent bonding involving different nuclei is particularly important with atoms of hydrogen (H), nitrogen (N), and oxygen (O), and there are millions of different kinds of molecules made up from these atoms. Many of them form the basis for life.

Complex ions are like molecules in having more than one nucleus and in being held together by covalent bonds. They differ from atoms and molecules by having the number of positive charges on the nuclei not equal to the number of electrons in the cloud. The difference represents the charge on the ion. One of the more important complex ions is the *hydronium ion*. Here three protons and one oxygen nucleus are covalently bonded, but one electron is missing in the molecular electron cloud — instead of eleven electrons (eight from O, one from each H) there are only ten. The formula of the ion is usually written H_3O^+, but if we wish to describe the hydronium ion made up of three 1_1H atoms and one $^{16}_8O$ atom, the symbol is $[^1_1H_3{}^{16}_8O]^+$. The hydronium ion has the bond structure

$$\left[\begin{array}{c} H \\ \\ H \end{array}\!\!\!\!\diagdown\!\!\!\!\diagup O\!-\!H\right]^+$$

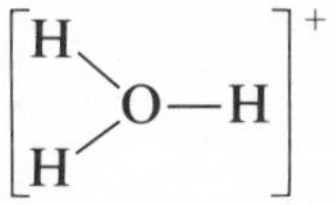

indicating that of the ten electrons present, six are in the three covalent bonds, and the remaining four are mainly concentrated around the oxygen nucleus. The structure of this ion is very similar to that of water, but with an additional covalently bonded proton:

$$\begin{array}{c} H \\ \\ H \end{array}\!\!\!\!\diagdown\!\!\!\!\diagup O + H^+ \rightarrow \left[\begin{array}{c} H \\ \\ H \end{array}\!\!\!\!\diagdown\!\!\!\!\diagup O\!-\!H\right]^+$$

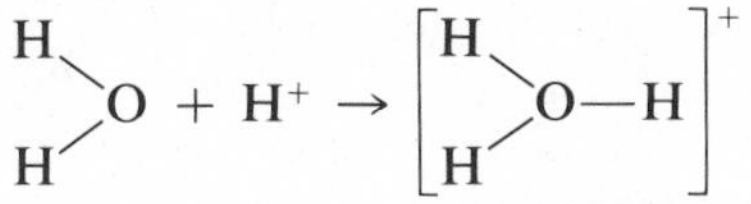

The number of complex ions is very large. Here are some additional examples, with atomic and mass numbers omitted for clarity:

$[OH]^-$	hydroxide ion	$[SO_4]^{2-}$	sulfate ion
$[NH_4]^+$	ammonium ion	$[NO_3]^-$	nitrate ion
$[CO_3]^{2-}$	carbonate ion	$[PO_4^{3-}]$	phosphate ion

The covalent bonding within these ions is discussed in Chapter 4. This completes the list of the types of microparticles making up the macroscopic samples of matter that we will be interested in. In the chapters that follow, we will be concerned mainly with molecules, simple ions, and complex ions as the structural units of matter. The only *atoms* that exist as free (unbonded) microparticles in bulk samples of matter at ordinary temperatures are the atoms of the gases helium (He), neon (Ne), argon (Ar), krypton (Kr), xenon (Xe), and radon (Rn). All other atoms are bonded together in one way or another.

PHOTO 5 H_2O (solid) in a glacier and icebergs in a frozen sea off the coast of northern Greenland. (*C. J. Pings, California Institute of Technology.*)

1.6 Crystal Structures Matter occurs in *three states*, solid, liquid, and gas. Ice, water, and steam provide a familiar example. Most kinds of matter can be changed from one of the states to another by changing the temperature. Thus oxygen, a gas at ordinary temperature, first becomes a liquid (at −183°C) and then a solid (at −218°C). These changes raise an important question—what kinds of microscopic building blocks (atoms, molecules, ions) are found in each of the three states and how are they arranged? Water molecules are the building blocks in ice (Photo 5).

Crystalline solids have the most regular arrangement of units. The nature of the units is determined by chemical analysis, and their arrangement is revealed by *X-ray crystal analysis*. Methods for chemical analysis are explained in Chapter 2. For the time being, let's assume that such analysis has shown the kinds of atoms, molecules, or ions that are present in a crystal and turn our attention to the kinds of arrangements revealed by X-ray studies.

The determination of the arrangement of the units in a crystal—the *crystal structure*—by the use of X rays is a very complicated operation, both experimentally and mathematically. We will only sketch out the broad principles upon which X-ray analysis depends. X rays are electromagnetic radiation ("light") of very short wavelength and considerable penetrating power. This feature plus the fact that X rays can expose a photographic film is what makes skeletal or dental X rays so useful. However, the use of X rays in crystal analysis has little in common with X-ray photography. Atoms, molecules, and ions are far too tiny to register their presence as shadows on a photographic film. Remember that there are about 1.5×10^{23} atoms in one gram of helium. X-ray analysis does not produce an "image" of the particles in a crystal.

When X rays pass through matter they interact with the electron clouds of the atoms, molecules, or ions present. The electron clouds act rather like tiny mirrors that reflect ("scatter" is the correct term) some of the beam of X rays. If *and only if* the electron clouds are arranged in a regularly repeating pattern, the reflections from the individual clouds coincide in direction and add up to produce strongly reflected beams. The angle between the reflected beam and the incoming beam of X rays depends on the distances between the centers of the electron clouds and the wavelength of the X rays. The reflected beams are registered as dots on a photographic plate to give what is called the X-ray diffraction pattern of the crystal (Figure 1.7). ☛

The production of such a regular pattern as that in Figure 1.7 is the surest evidence that a sample of matter is in fact crystalline.

From the spacings and intensities of the dots it is possible to work backward using mathematics and computers to determine the arrangement and size of the electron clouds that produced the pattern. Since the electron clouds are concentrated around the nuclei of the particles in the crystal, X-ray analysis gives the distance between the nuclei ("internuclear distance"), and the arrangement of atoms, molecules, or ions in the crystal is revealed.

Units of length Before we can talk about internuclear distances and structures of typical crystals, we need a unit of length. The *macroscopic* unit in chemistry is the *centimeter*, but like the gram with atomic masses, the centimeter is much too large for conveniently expressing atomic and molecular dimensions, which are of the order of 10^{-10} m or 10^{-8} cm. So the unit of length for atomic and molecular dimensions is the *angstrom unit* (Å), which is 1×10^{-8} cm (1 Å $= 10^{-8}$ cm $= 10^{-10}$ m).

EXAMPLE

To see that the angstrom unit is about the right size for atomic sizes, let's calculate the approximate diameter of one helium atom, assuming closest packing of spheres in the liquid form of the substance. We know that 1 g of helium contains about 1.5×10^{23} atoms (Section 1.4). If we know the volume occupied by 1.0 g of liquid helium, we can find the volume of one helium atom by division. Since $V = \frac{4}{3}\pi r^3$ for the volume of a sphere, we know V, and can thus obtain r and the diameter ($2r$) of the helium atom. Experiment shows that 1 g of liquid helium occupies 6.7 cm³. Following through the calculation, we obtain:

$$\frac{6.7 \text{ cm}^3}{1.5 \times 10^{23}} = 4.5 \times 10^{-23} \text{ cm}^3 \text{ per helium atom}$$

$$V = 4.5 \times 10^{-23} \text{ cm}^3 = \frac{4}{3} \times 3.14 \times r^3$$

Solving for r^3 gives $r^3 = 1.07 \times 10^{-23}$ and

$$r = \sqrt[3]{1.07 \times 10^{-23}}$$
$$= 2.2 \times 10^{-8} \text{ cm}$$

Thus the approximate radius for the helium atom is 2.2×10^{-8} cm or 2.2 Å, and the diameter is about 4.4 Å. More accurate methods give a diameter of about 1 Å.

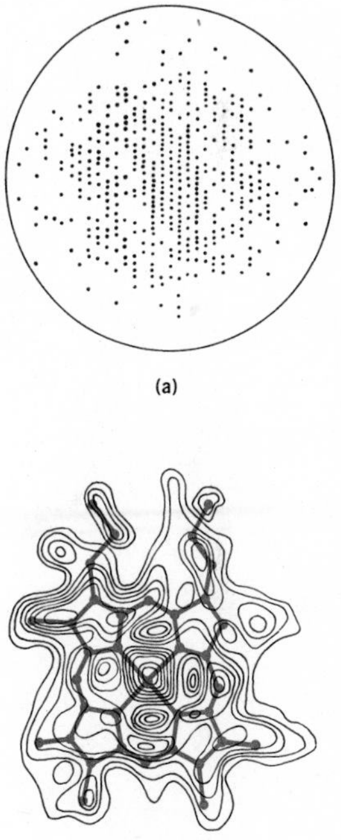

(a) (b)

FIG. 1.7 X-ray crystal analysis: (a) photographic record of the diffraction pattern; (b) derived molecular structure—heavy lines represent covalent bonds between atomic nuclei, light lines are contours of electron density.

Internuclear distances for covalent bonds ("covalent bond lengths") are typically between 1 and 3 Å. The hydrogen molecule has the shortest covalent bond, only 0.74 Å. The H—O bond length in liquid water is 1.01 Å, and the O—O bond length in the oxygen molecule (O_2) is 1.21 Å. You can see that the angstrom unit is the X-ray crystallographic "inch."

Exercises

13. How many angstrom units are there in 1 cm? In 1 m?

14. If 1 in. = 2.54 cm, how many angstrom units are there in 1 in.?

15. How many O—O bond lengths (1.21 Å) make up 1 in.?

16. The distance from the earth to the moon may be taken as 375,000 km (1 km = 1000 m). Compare the distance to the moon in centimeters to the number of angstrom units in 1 cm found in Exercise 13.

In describing the typical crystal structures that follow, the distances between the centers of the atomic nuclei are given in angstrom units. Mass and atomic numbers have been omitted to simplify the representations of structure.

Diamond – a covalent crystal Diamonds are mined in Africa and South America. Rough diamonds bear no resemblance to the "cut" diamonds familiar in jewelry. A rough diamond often appears to be a rather smooth pebble of glass covered with a gray film. For this reason the crystalline nature of a diamond is often difficult to see. But the X-ray diffraction pattern of even a rough diamond shows by the clear pattern of dots that this is a truly crystalline material. On the other hand, even the most highly polished piece of glass fails to give a distinct pattern because there is no regular, repetitive arrangement of structural units. Glass is an example of an *amorphous* ("without structure") type of matter.

Analysis shows that the diamond is made up of carbon atoms only. A diamond is "pure carbon"; that is, it consists of atoms of atomic number 6, having six protons in the nucleus. Most of the carbon atoms in diamond are of the $^{12}_{6}C$ type; that is, they have six protons and six neutrons in the nucleus. About 1 in 100 (1.1% actually) of the carbon

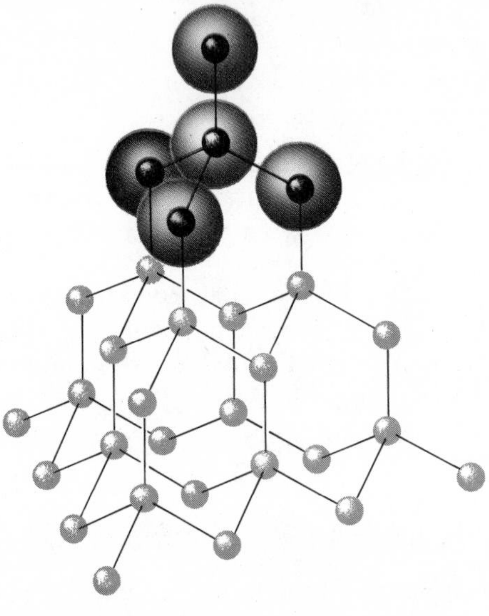

FIG. 1.8 Structure of the carbon (diamond) crystal.

atoms are of the type $^{13}_{6}C$, which have six protons and seven neutrons in the nucleus. Atoms $^{12}_{6}C$ and $^{13}_{6}C$ are isotopes of carbon. (Isotopes are discussed in Chapter 2.) In diamond, each of the carbon nuclei is bonded by four two-electron covalent bonds to four other carbon nuclei, so a diamond is a giant molecule, the molecular "size" being that of the diamond itself (Figure 1.8). The angle between each pair of covalent bonds is 109°28′ (almost 109.5°), the so-called *tetrahedral angle*. This is the angle between the lines joining the center of a regular tetrahedron to its vertices. The bond length is 1.54 Å, and the bonds are very strong. The diamond structure is an example of a *covalent crystal*. To scratch a diamond requires breaking some of the bonds; consequently the diamond is the hardest substance known. This hardness leads to the use of diamonds in special metal- and glass-cutting tools, as well as in jewelry. Carborundum (silicon carbide, SiC) has the same covalent crystal structure as diamond, with silicon atoms alternating with carbon atoms in a regular, repetitive fashion. It is also very hard and is used in grinding wheels and abrasive papers for shaping and polishing steel.

Sodium chloride—a simple ionic crystal Sodium chloride (table salt) is an example of an *ionic crystal*. In it, equal numbers of the simple ions Na^+ and Cl^- are arranged as shown in Figure 1.9. The Na^+ ion results from removal of one electron from a sodium atom, the Cl^- ion from addition of one electron to a chlorine atom. Each sodium ion has six chloride ions as nearest neighbors and each chloride ion has six sodium ions as nearest neighbors. There is only one short sodium-to-chlorine internuclear distance (2.78 Å) because each of the ions "belongs" equally to six ions of opposite charge. For this reason no covalent NaCl *molecule* exists in the crystal.

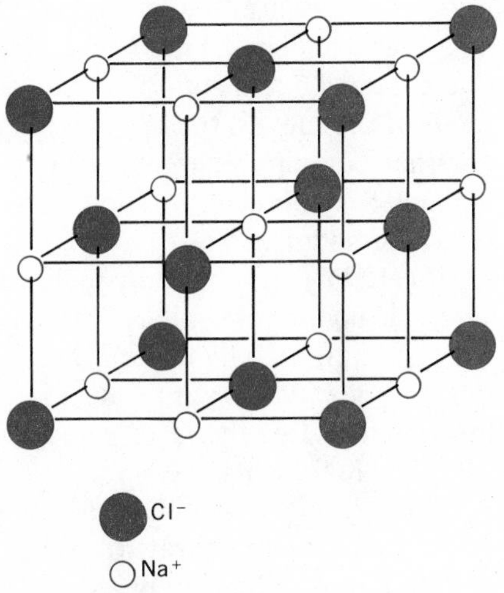

FIG. 1.9 Structure of the NaCl crystal.

Many crystals are made up of simple ions, although not all of them have the same arrangement of ions as sodium chloride. Examples of other simple ionic crystals are potassium iodide, K^+I^- (in "iodized" salt for goiter prevention), sodium fluoride, Na^+F^- (used to "fluoridate" water to prevent tooth decay), potassium bromide, K^+Br^- (used as a sedative in medicine), calcium oxide, $Ca^{2+}O^{2-}$ (lime, used to make plaster and mortar), and magnesium oxide, $Mg^{2+}O^{2-}$ (used to line steel furnaces and to make $Mg(OH)_2$, "milk of magnesia," a laxative). Ionic crystals having ions of different charge types are also common. Examples are calcium chloride, $Ca^{2+}(Cl^-)_2$, which is used as a drying agent because it absorbs water to form $CaCl_2 \cdot 6H_2O$, and magnesium chloride, $Mg^{2+}(Cl^-)_2$, which is prepared from seawater and used in the production of magnesium metal. If the ions have *different* numbers of electronic charges, the relative numbers of ions must be such that the negative and positive charges cancel. You can see this in calcium chloride, where there are two Cl^- ions for each Ca^{2+} ion. The relative numbers are shown by the subscripts in the formulas.

Ionic crystals are reasonably hard in most cases, because of the strong attraction between the oppositely charged ions. Many ionic crystals dissolve readily in water the way salt (Na^+Cl^-) does. Of the examples given earlier, only magnesium oxide and calcium oxide do not readily dissolve in water.

Calcium carbonate—a complex ionic crystal Calcium carbonate $[Ca^{2+}(CO_3)^{2-}]$ is an example of another large class of ionic crystals that contain complex as well as simple ions. Calcium carbonate occurs as limestone, marble, in oyster shells and corals, and in large crystals of the mineral *calcite*. The units in the crystal are Ca^{2+} ions and CO_3^{2-} ions. The carbonate (CO_3^{2-}) ion is a complex ion in which the three oxygen nuclei are bonded to one carbon nucleus by covalent bonds. The four nuclei in the CO_3^{2-} ion all lie in the same plane, and the oxygen nuclei are all equidistant from the central carbon nucleus. The arrangement of the ions in the crystal is shown in Figure 1.10.

Other common examples of complex ions in crystals, written in an expanded symbolism to show the ions clearly, are sodium carbonate (washing soda) $[(Na^+)_2(CO_3)^{2-}]$, sodium bicarbonate (baking soda) $[Na^+(HCO_3)^-]$, sodium hypochlorite (laundry bleach) $[Na^+(ClO)^-]$, and tricalcium phosphate (principal component of bone) $[(Ca^{2+})_3\{(PO_4)^{3-}\}_2]$, in which there are three Ca^{2+} ions for each two PO_4^{3-} ions. Notice that as with simple ions, the positive and negative charges must cancel. For the last example, $3Ca^{2+}$ represents six plus charges, $2(PO_4)^{3-}$ represents six minus charges.

Ionic crystals in which *both* ions are complex are also known. Examples are ammonium nitrate (fertilizer and explosive) $[(NH_4)^+(NO_3)^-]$, ammonium sulfate (prepared from ammonia and sulfuric acid and

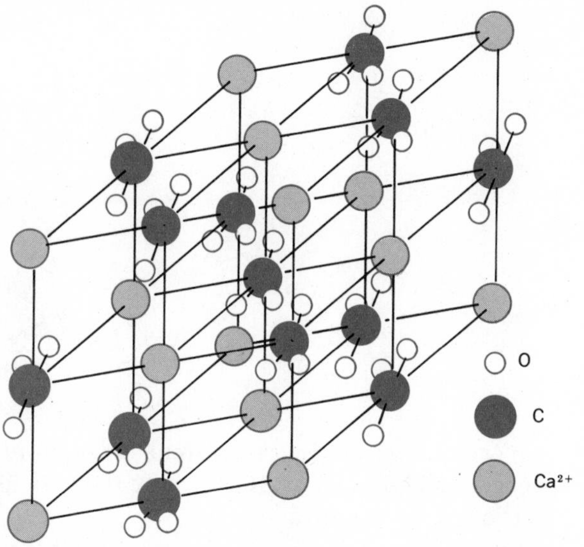

FIG. 1.10 Structure of the $CaCO_3$ (calcite) crystal.

used as a fertilizer) $[(NH_4)_2^+(SO_4)^{2-}]$, and hydronium perchlorate $[(H_3O)^+(ClO_4)^-]$. Or the cation may be complex, the anion simple; ammonium chloride $[(NH_4)^+Cl^-]$ is an example.

Ordinarily the formulas for complex ionic crystals are not written in the expanded symbolism. They are usually written with most of the brackets omitted. Thus $(NH_4)^+(Cl^-)$ becomes $NH_4^+Cl^-$ or even NH_4Cl when attention is not focused on the ionic nature of the crystal. Likewise $(Ca^{2+})_3\{(PO_4)^{3-}\}_2$ becomes $Ca_3^{2+}(PO_4)_2^{3-}$ or even $Ca_3(PO_4)_2$. Complex ionic crystalline substances are often quite soft in contrast to those involving simple ions because the forces of attraction between the oppositely charged ions are not so effective. Many complex ionic substances are quite soluble in water.

Iodine—a molecular crystal Iodine forms blue-black crystals at ordinary temperatures. On being heated the crystals readily vaporize to a purple gas in which the units are iodine molecules, I_2. In the molecule, two iodine nuclei are held together by a covalent bond, making the molecular structure I—I. X-ray analysis of the solid reveals that it consists of a regular packing of I_2 molecules (Figure 1.11). The results show that there are two spacings for the iodine nuclei. The shorter distance of 2.70 Å separates pairs of covalently bonded iodine nuclei. The larger distance (3.84 Å) separates the iodine nuclei in closest-neighbor molecules. The forces between the molecules that are responsible for holding the crystal together are much weaker than both covalent bonds and interionic attractions. They are called *secondary* or *van der Waals* forces, and because they are weak, crystalline iodine is soft and readily converted to a gas (vaporized) of I_2 molecules. Typical

Johannes Diderik van der Waals (1837–1923) was professor of physics at the University of Amsterdam, where he studied the forces between molecules. He was awarded the Nobel prize in physics in 1910.

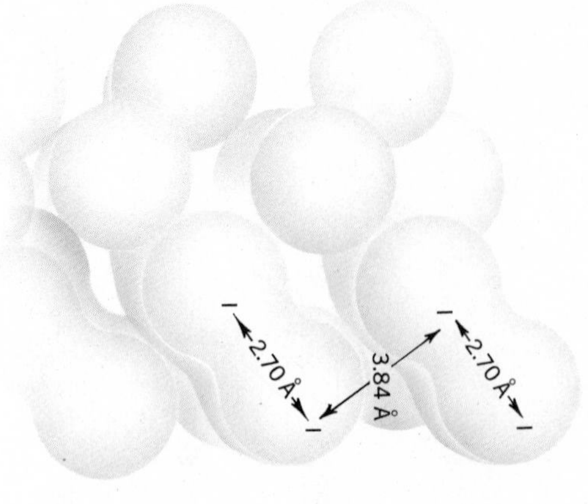

FIG. 1.11 Arrangement of I_2 molecules in crystalline iodine.

ionic crystals are not vaporized until temperatures above a red heat are reached, and covalent crystals require the very high temperature of an electric arc for vaporization.

Molecular crystals are often called van der Waals crystals. They are typically soft and most do not dissolve much in water. Important exceptions to this solubility rule are sugars such as *sucrose* (table sugar) in which the molecular units are $C_{12}H_{22}O_{11}$ covalent molecules, *glucose* (blood sugar) in which the covalent molecular unit is $C_6H_{12}O_6$, and many of the proteins (huge C, H, O, N, S molecules) necessary for life.

Solid carbon dioxide (dry ice) is a molecular crystal in which the units are covalently bonded carbon dioxide molecules (CO_2). Its softness and ready vaporization to a gas of CO_2 molecules at very low temperatures (−78°C) show the weakness of the crystal forces. The covalent bonding within the CO_2 molecule is very strong, however.

1.7 Structure of Liquids and Gases The structural units in most liquids and gases are covalent molecules. In all gases—oxygen (O_2), nitrogen (N_2), methane (natural gas, CH_4), carbon dioxide (CO_2), and hydrogen (H_2) are examples—the molecules are in rapid, chaotic, and random motion so that the gas is structureless (except for the "inner" structure of the moving molecules).

In liquids made up of covalent molecules the molecules are packed more closely together than in gases by the action of van der Waals forces. The molecules are in motion and there is enough free space in the liquid to allow them to tumble and slide past each other so that no one molecule remains at any point for long. The result is lack of regular patterns of spacing for nuclear centers, so the liquid is without structure other than the "inner" structure of its molecular units. Liquids and gases do not yield definite X-ray diffraction patterns because their molecules are always moving around.

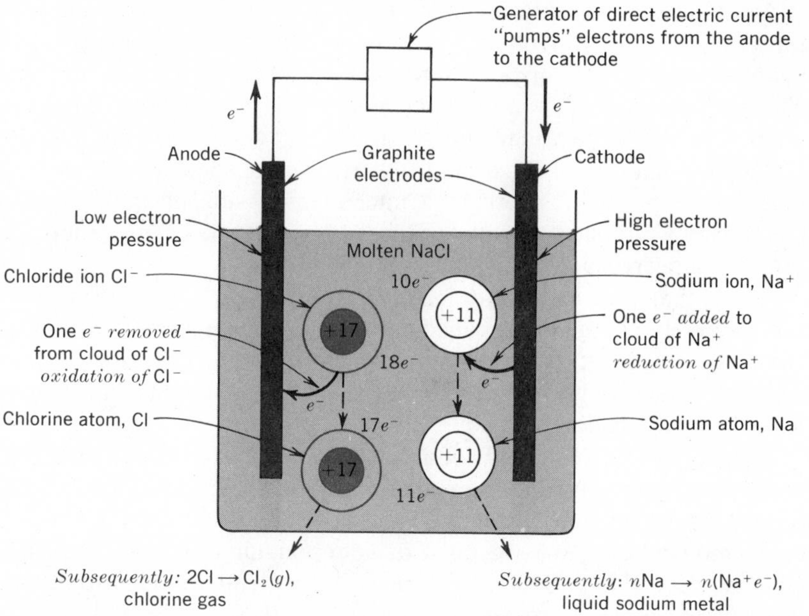

FIG. 1.12 First steps in the electrolysis of molten sodium chloride from an electronic point of view.

At high temperatures most ionic crystals can be melted to liquids, in which the units are ions that are relatively free to move. Evidence for this free motion is the huge increase in electrical conductivity that melting produces. The increase in conductivity happens when the ions become free to move and can transport electric current by migrating through the melt. The positive ions (cations) move toward the negative electrode, the negative ions (anions) move toward the positive electrode, and a movement of electric charge or an electric current results (Figure 1.12).

As you might expect, *solutions* of ionic crystals in water also conduct an electric current strongly. The electrical conductivity results from the freeing of the ions from their fixed positions in the crystal by formation of the solution. Substances that increase the conductivity of water when dissolved in it are called *electrolytes.* Ionic crystals are important members of this class of substances.

1.8 Summary Guide Chemistry deals with changes in the nature and properties of matter. It seeks to explain the macroscopic behavior of matter in terms of the properties of the microparticles of which matter is composed. The most basic microparticles for chemistry are the electron (e^-), the proton (p^+), and the neutron (n). The electron and proton carry equal but opposite electric charges of one electronic unit. The neutron carries no charge. The proton and neutron have similar, but not identical, masses close to 1 atomic mass unit. The electron has a much smaller mass.

Atoms are microparticles made up of a tiny nucleus containing protons and neutrons, surrounded by a much larger cloud of rapidly moving electrons equal in number to the number of protons in the nucleus. The number of protons in the nucleus is the atomic number (Z) of the atom. The sum of the numbers of protons and neutrons in the nucleus is the mass number of the atom (M). An atom of specified atomic and mass numbers ($^{M}_{Z}X$) is termed a nuclide. Emission of an α particle ($^{4}_{2}He$) from a radioactive nuclide causes a decrease of 4 in mass number (M) and a decrease of 2 in atomic number (Z).

Atoms that have lost one or more electrons from their electron cloud are positively charged ions called cations; those that have gained electrons are negatively charged ions termed anions. The electrostatic force of attraction between the opposite charges of anions and cations leads to ionic crystals.

Merging of electron clouds of atoms may result in covalently bonded molecules. When covalently bonded molecules keep their identity but form solids because of attractive van der Waals forces between molecules, molecular crystals result. Covalent bonding may also produce a covalent crystal in which each nucleus is covalently bonded to others throughout the whole crystal.

Complex ions are like molecules in having two or more nuclei covalently bonded, but are unlike molecules in having a net negative or positive charge. The charge results from too many or too few electrons compared to the total positive charge of the nuclei. Complex ions may function as anions or cations in ionic crystals. The hydronium ion, H_3O^+, is one of the more important complex ions.

The fixed arrangements of nuclei in crystalline substances are determined by X-ray crystallography. Internuclear distances in crystals and molecules are expressed in units of 10^{-8} cm or 10^{-10} m, called angstrom units (Å).

Ordinary liquids and gases have covalently bonded molecules as structural units. In liquids the molecules are not fixed as in crystals but, though close-packed, are free to slide past each other, permitting flow of the liquid. In gases the molecules are at relatively large distances from each other and in random, chaotic motion. The melting of an ionic crystal results in a large increase in electrical conductivity.

The energy unit of size appropriate for individual bonds is the electron volt (eV or "volt"). Energy is released when bonds form, and must be supplied to break bonds. Bonded systems possess negative potential energy.

Exercises

17. A fluorine atom (F) weighs almost exactly 19 times as much as a hydrogen (H) atom. There are nine electrons in the electron cloud of the fluorine atom. How many neutrons are there in the nucleus of a fluorine atom?

18. There are two kinds of chlorine atoms, $^{35}_{17}Cl$, atomic mass 34.97, and $^{37}_{17}Cl$, atomic mass 36.96. In chlorine there are 310 $^{35}_{17}Cl$ atoms for every 100 $^{37}_{17}Cl$ atoms. Calculate the mass of an "average" chlorine atom in amu.

19. If 1 amu (1.660×10^{-24} g) is $\frac{1}{12}$ the mass of one $^{12}_{6}C$ atom, what is the mass of one $^{12}_{6}C$ atom in amu? In g?

20. Using data from Exercise 19, calculate the number of $^{12}_{6}C$ atoms in 12.00 g of $^{12}_{6}C$ atoms.

21. The element phosphorus is made up of one kind of atom, $^{31}_{15}P$, with a mass of 30.97 amu. Calculate the mass of one $^{31}_{15}P$ atom in g. How many $^{31}_{15}P$ atoms are there in 30.97 g of $^{31}_{15}P$?

22. Comparing the results of Exercises 20 and 21, what do you conclude about the number of atoms of any element in a sample whose weight in grams is equal to the mass of the atom in amu?

23. The radioactive arsenic isotope $^{74}_{35}As$ is used in locating rapidly growing tumors of the brain. Any rapidly growing tissue concentrates arsenic in the form of arsenate (AsO_4^{3-}) units. By using particle counters to scan the brain for the γ rays given off by $^{74}_{35}As$, the tumor can often be located before brain surgery. By radioactive decay $^{74}_{35}As$ goes to an isotope of the element with $Z = 32$. Give the nuclear composition of $^{74}_{35}As$ and the name of its decay product.

Key Words

α rays $^{4}_{2}He$ nuclei.

β rays fast-moving electrons.

γ rays electromagnetic radiation ("light") of very high energy.

Å unit 10^{-8} cm, 10^{-10} meter, unit of length for atoms, bonds, and crystal structures.

Atomic mass unit 1.660×10^{-24} g, $\frac{1}{12}$ the mass of one $^{12}_{6}C$ atom.

Atomic number (Z) number of protons in the nucleus of an atom.

Complex ion an ion made up of two or more atoms bonded together.

Complex ionic crystals crystals made up of ions in which either the cation or anion (or both) are complex ions made up of more than one kind of atom.

Compounds matter made up of more than one kind of atom, ionically or covalently bonded.

Covalent bonding sharing of electrons between two atomic nuclei that holds the nuclei together.

Covalent crystals a solid with a regular pattern of atoms held together by covalent bonds, as in diamond.

Crystal structure regular pattern of arrangement of atoms, molecules, or ions in a solid.

Electrolyte any substance that increases the electrical conductivity of water when dissolved in water, by forming free ions.

Electron (e^-) fundamental negatively charged microparticle, in electron cloud of an atom.

Electron volt (eV) an energy quantity equal to the kinetic energy of an electron that has been accelerated by a potential difference of 1 V.

Ion an atom or molecule that carries a positive (cation) or negative (anion) charge because of a deficiency or excess of electrons.

Ionic crystals solid materials made up of anions and cations.

Macroscopic large enough to see or weigh.

Mass number (M) sum of numbers of protons and neutrons in the nucleus of an atom.

Matter anything in the form of a liquid, solid, or gas that has mass and occupies space.

Microparticles the unseeable, tiny units of matter—electrons, protons, neutrons, atoms, molecules, ions.

Molecular crystals solids in which molecules are held together by secondary forces; same as van der Waals crystals.

Molecules uncharged microparticles made up of two or more atoms bonded together.

Neutron (n) fundamental uncharged microparticle, in the nucleus of an atom.

Nuclide an atom for which the values of Z and M are given.

Proton (p^+, ${}^1_1H^+$) fundamental positively charged microparticle, in the nucleus of an atom.

Radioactivity emission of α, β, or γ rays from the nucleus of an atom.

Suggested Readings

Ashford, Theodore A., "Rutherford's Theory of the Nuclear Atom," in *The Mystery of Matter*, Louise B. Young, ed., Oxford Univ. Press, New York, 1965, Part 2, pp. 84–95.

Branscomb, Lewis M., "Science and the American Society," *Am. Sci.*, **61**:38–41 (Jan.–Feb. 1973).

Cook, C. Sharp, "Energy: Planning for the Future," *Am. Sci.*, **61**:61–65 (Jan.–Feb. 1973).

Feinberg, Gerald, "Ordinary Matter," *Sci. Am.*, **216**:126 (May 1967).

Holden, Alan, and Singer, Phyllis, *Crystals and Crystal Growing,* New York, Anchor Books, Doubleday, 1960.

Hunt, C. A., and Garrels, R. M., *Water: The Web of Life*, New York, W. W. Norton and Co., Inc., 1972.

Keller, J., ed., "Man and the Universe," Part II, World of the Atom. *Chemistry,* **45**:9–12 (July–Aug. 1972).

Lapp, R. E., *Matter,* New York, Time-Life Books, 1969.

Metz, William D., "Helium Conservation Program: Casting It to the Winds," *Science,* **183**:59–63 (Jan. 11, 1974).

Moore, Walter J., *Seven Solid States,* New York, W. A. Benjamin, Inc., 1967.

Rose, D., "Controlled Nuclear Fusion: Status and Outlook," *Science,* **172**:797–808 (May 21, 1971).

Science, Vol. 8, No. 4134 (April 19, 1974) is devoted to the subject of energy.

Shaheen, Esbar I., "The Energy Crisis: Is It Fabrication or Miscalculation?" *Env. Sci. and Tech.,* **8**(4):316–320 (April 1974).

"The Nature of the Atom," *Life,* May 16, 1949, pp. 68–88.

chapter 2

The surface of the moon as seen by the astronauts of Apollo 17. Matter at the surface of the moon is not geonormal in its makeup (Section 2.4). *(Photo 6, NASA AS17-140-21496.)*

Elements and Compounds

2.1 Introduction Chemists experiment with macroscopic amounts of matter; they explain the results by using microparticle theory. Usually the samples of matter worked with are those that occur naturally on earth, so chemistry deals mainly with *geonormal matter*—matter of the type normally found on earth (*geos* is the Greek word meaning "earth"). This matter is made up of atoms, molecules, and ions. In this chapter we take up the relationships between the masses of individual microparticles (in amu) and the masses of the macroscopic samples (in grams) that chemists work with.

Macroscopic samples of matter are classified as either *homogeneous* or *heterogeneous*. A sample of matter is homogeneous if small samples taken from it are alike in all ways except for their size or mass. The samples are compared in terms of their *macroscopic intensive properties*. These are properties that are independent of the size or mass of the sample. Most important among the intensive properties is the *chemical composition* of the sample in terms of the *relative numbers* of the different kinds of microparticles (atoms, molecules, ions) present. Other intensive properties are color, density (weight per unit volume), taste, odor, crystal structure, melting point if solid, boiling point if liquid, and chemical reactivity.

Suppose, for example, you compare a cup of distilled water with a drop or two taken from any place in the cup. All of the macroscopic intensive properties of the two samples of matter are identical. Both taste the same, are colorless, boil at 100°C (212°F), and are composed of water molecules in which hydrogen and oxygen atoms are combined in the 2:1 numerical ratio given by its formula, H_2O. Furthermore, both have the same density, provided that they are at the same temperature. The only differences between the samples are their volume and weight. Volume and weight (or mass) are called *extensive properties*, since they are determined by the extent or size of the sample.

Now suppose you are given a sample of matter prepared by mixing some table sugar and salt. With a magnifying glass you could see that the sample was *heterogeneous*, that is, made up of two kinds of white crystals that differ in shape. Using fine tweezers and a magnifying glass you could separate the two kinds of crystals. By taste alone you could show that you had separated a mixture of sugar and salt. By chemical analysis and crystal structure determination, the separated sugar and salt crystals could be shown to be homogeneous, the sugar composed of $C_{12}H_{22}O_{11}$ molecules, the salt of Na^+ and Cl^- ions in a 1:1 ratio.

Solutions present a special case of homogeneous matter. Take a sample of matter prepared by dissolving some sugar in water. As in the case of distilled water, the macroscopic intensive properties of the whole sample are identical with those of any macroscopic portion so the solution is homogeneous. The key word here is *macroscopic*, which means you can see or weigh the sample. Homogeneity does not apply at the microparticle level of individual atoms, molecules, and ions. The number of sugar and water molecues in even the tiniest drop of a sugar solution is so huge that the composition of the drop in terms of relative numbers of sugar and water molecules is the same as that in the whole solution. It is impossible to take a *macroscopic* sample that contains just a few sugar and water molecules, and so by chance obtain a sample that has a different composition in terms of the molecules present. The tiny size and mass of microparticles make it certain that they will appear in the same proportions in any macroscopic sample. You could show that you had a solution by boiling off the water until the sugar crystallized.

Gases are also examples of homogeneous matter. A sample of the atmosphere, a mixture of (mainly) oxygen (O_2) and nitrogen (N_2), is homogeneous provided that it is man-sized. A cubic mile of the atmosphere would *not* be homogeneous. For one thing, it would be more dense at the bottom than at the top because of the effect of gravity. Samples like this, in which the properties change in a regular or smooth way from one point to another, are called *continuous*. However, a laboratory-size sample of dust-free air is homogeneous by every experimental test.

2.2 Systems of Matter—Phases Chemists usually examine matter and its changes in *closed systems*. A closed system of matter is a sample separated from other matter by the wall of a flask or by other means so that no other matter may enter or leave the sample. *Any and all homogeneous parts of a system having the same intensive properties constitute a* **phase** *of the* **system**. The system in Figure 2.1(a) has three phases: gas, liquid, and solid. A flask like that in Figure 2.1(a) but containing *separated* pieces of ice and *separated* drops of water would still have only three phases—liquid, solid, and gas. This is the meaning of "any and all parts" in the definition of phase. Only one gas phase is possible in a system, but there may be several liquid and solid phases, as shown in Figure 2.1(b). A solid may also have more than one phase, as shown in the magnified view of Figure 2.1(c).

Chemists use *phase-transfer* operations to separate solutions and heterogeneous samples of different kinds of matter into their homogeneous components. Prominent among these operations is the process of *distillation*, shown in Figure 2.2, in which a liquid phase is converted to a gas by heating and back to a liquid by cooling the gas.

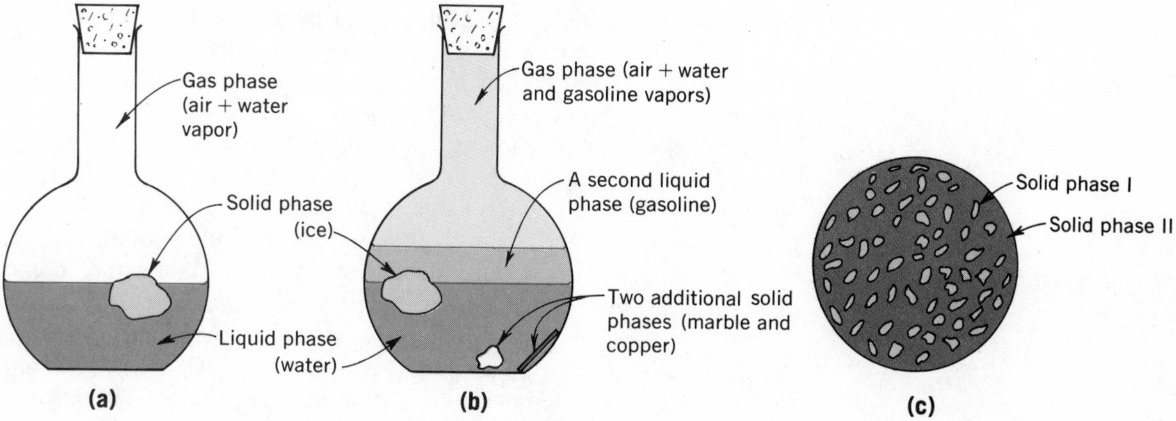

FIG. 2.1 States of matter and phases: (a) the three states of water; (b) a system of several phases; (c) appearance of a solid of two phases seen through a microscope.

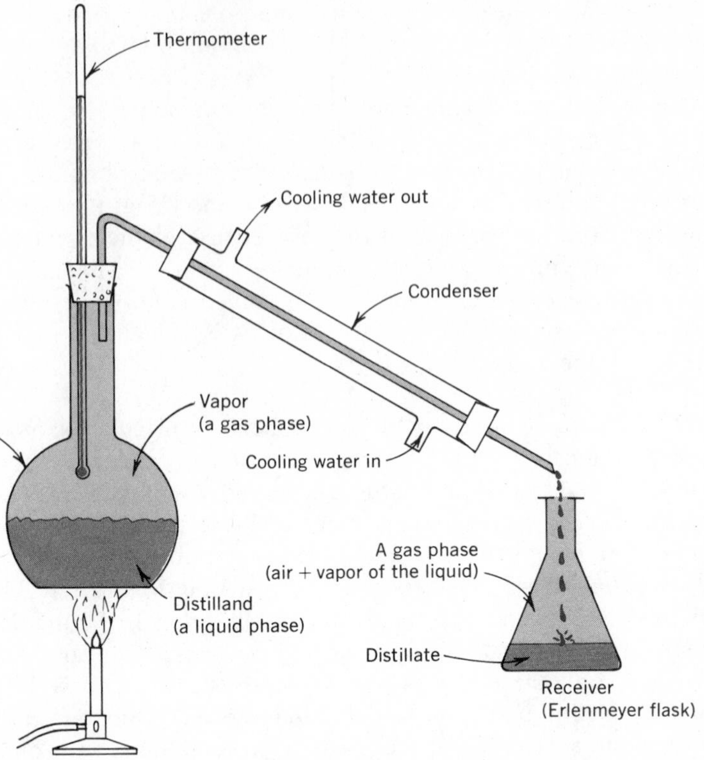

FIG. 2.2 Simple distillation apparatus.

Distillation of a solution of salt in water separates it into water (distillate) and leaves the nonvolatile salt as a crystalline residue in the distilling flask. You can shake a mixture of sugar and sand with water, then filter it to obtain the water-insoluble sand as residue and a water solution of sugar. The sand and sugar have been separated by the phase-transfer process of *extraction*. Then if you distill off most of the water from the sugar solution, you get a separation of sugar and water and obtain a residue of crystalline sugar in the process of *crystallization*. The sugar can be further purified by again dissolving it in water, followed by evaporation of most of the water until crystals appear. This process is called *recrystallization*.

When repetition of phase-transfer operations produces no further change in the intensive properties of a substance, the substance is said to be chemically pure. Table sugar (sucrose) is obtained from the juice of sugar cane and sugar beets by the process of crystallization and recrystallization. It is a pure substance. Pure substances are homogeneous and are further classified either as *elementary substances* ("elements") or as *compound substances* ("compounds"). Water (H_2O), sugar ($C_{12}H_{22}O_{11}$), sand (SiO_2), and salt (NaCl) are examples of compounds. Compounds can be separated into their elements by chemical means, but not by phase-transfer operations. A diamond is a sample of the element carbon. A few other elements occur as pure substances in nature; copper (Cu), silver (Ag), gold (Au), and platinum (Pt) are important examples. Most other elements occur as compounds or in mixtures. For example, uranium, the fuel for atomic reactors, occurs as the oxide U_3O_8 in the uranium ore called pitchblende.

Other phase-transfer operations for separating complex samples of matter into their simpler, homogeneous components (elements or compounds) are *sublimation* and *chromatography*. In sublimation, a volatile (readily vaporized) solid is separated from a nonvolatile one by heating the mixture to convert the volatile component directly to its gas form, then cooling the gas to get the volatile component back in solid form (Figure 2.3c). The heating is done under conditions such that the solid does not melt; otherwise the process is similar to distillation. The principles of chromatography will be described later.

2.3 Elements and Compounds We saw in Section 2.2 that pure substances are classified as either elements or compounds. We can now give microparticle definitions of these two basic kinds of matter. A pure substance is *an element if all its atomic nuclei have the same atomic number (Z)*. A pure substance is *a compound if the atomic nuclei have different atomic numbers*. In a compound the different nuclei are present in the fixed numerical ratios given by its formula. It was pointed

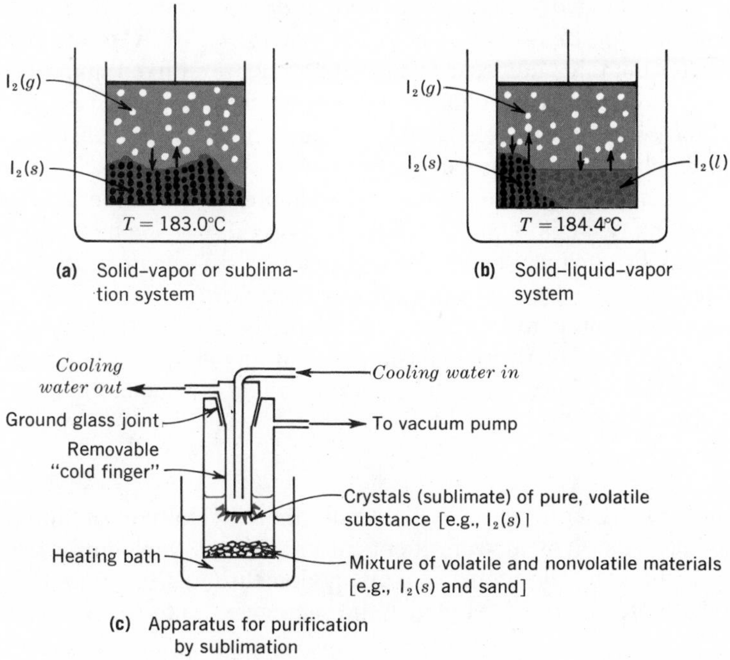

FIG. 2.3 Systems for iodine: (a) solid–vapor; (b) solid–liquid–vapor; (c) apparatus for purification by sublimation. (I_2 molecules shown as spheres for convenience.)

TABLE 2.1 Estimated percentages of the commoner elements in chemical combination in the rocks of the earth's crust.

Name of Element	*Symbol*	*Weight %*
Oxygen	O	46.6
Silicon	Si	27.7
Aluminum	Al	8.1
Iron	Fe	5.0
Calcium	Ca	3.6
Sodium	Na	2.8
Potassium	K	2.6
Magnesium	Mg	2.1
Hydrogen	H	0.1

out earlier that of the 92 elements that occur naturally on earth, only 6, the gases helium (He), neon (Ne), argon (Ar), krypton (Kr), xenon (Xe), and radon (Rn), are composed of separate atoms. In all other elements the nuclei are bonded together by merging of their electron clouds to form covalent molecules (H_2, N_2, O_2, F_2, Cl_2, Br_2, I_2, P_4, S_8), covalent crystals (diamond, C_n), or metallic crystals (Chapter 4). None of the elements is ionically bonded. Appendix 3 gives the names, atomic numbers, and symbols of the elements. Table 2.1 lists the commoner elements with their estimated amounts in the earth's crust.

In all compounds, the electron clouds interact to bond the nuclei, either covalently (H_2O, HCl, $C_{12}H_{22}O_{11}$, . . . ,) to produce molecules or ionically (Na^+Cl^-, $Mg^{2+}O^{2-}$, . . . ,) to produce ionic crystals. It is important to note that *a sample of matter must be a pure substance for the definition of a compound to apply.* A solution of salt in water contains different nuclei, and they are in fixed numerical ratios for the particular solution. However, solutions are not pure substances, as we saw in Section 2.2. They are not pure substances because, in contrast to compounds, their composition can be changed.

The number of compounds possible is unlimited. The number known is approaching two million and is added to every day by chemists using the methods of chemical synthesis and analysis. In recent years thirteen elements (transuranium elements) that are not found naturally on earth have been prepared by using the methods of nuclear chemistry (Chapter 16). Most of them were obtained in tiny amounts, but some are prepared in sufficient quantity to serve as fuel for nuclear reactors and as atomic explosives.

2.4 Isotopes According to early atomic theory (Dalton, 1808), the atoms of a given element were identical in all ways, including mass. For this to be true, the atoms of a given element must have not only the same atomic number (Z) but also the same mass number (M). That is, they must have identical electron clouds *and nuclei.* We now know from positive ray analysis (to be discussed shortly) that only a few elements are made up of only one kind of atom. Important among these elements are fluorine ($^{19}_{9}F$), sodium ($^{23}_{11}Na$), aluminum ($^{27}_{13}Al$), phosphorus ($^{31}_{15}P$), manganese ($^{55}_{22}Mn$), cobalt ($^{59}_{27}Co$), and gold ($^{197}_{79}Au$). The majority of elements are collections of atoms having the same atomic number but *different* mass numbers. The different mass numbers result from different numbers of neutrons in the nucleus. *Nuclides having the same atomic number but different mass numbers are called the* ***isotopes*** *of the element designated by the atomic number.* Examples are the element helium ($Z = 2$), which has isotopes $^{3}_{2}He$ and $^{4}_{2}He$; oxygen with isotopes $^{16}_{8}O$, $^{17}_{8}O$, and $^{18}_{8}O$; and the hydrogen isotopes $^{1}_{1}H$ and $^{2}_{1}H$. The stable (nonradioactive) isotopes of the first ten ($Z = 1$–10) elements with their nuclear compositions are listed in Table 2.2.

TABLE 2.2 Fundamental particles and stable isotopes of the first ten elements

	Atomic Number Z	*Isotopic Mass (amu)*	*Mass Number* M	*Abundance (atom %)*	*Number of* *Protons*	*Neutrons*	*Symbol*
Electron	–	0.0005486					e^-
Neutron	0	1.00867					n
Proton	1	1.00728					p^+ or p
Protium	1	1.00783	1	99.98	1	0	^{1}H
Deuterium	1	2.01410	2	0.02	1	1	^{2}H or D
Helium	2	3.01603	3	10^{-4}	2	1	^{3}He
	2	4.00260	4	100	2	2	^{4}He
Lithium	3	6.01513	6	7.4	3	3	^{6}Li
	3	7.01601	7	92.6	3	4	^{7}Li
Beryllium	4	9.01219	9	100	4	5	^{9}Be
Boron	5	10.01294	10	19.6	5	5	^{10}B
	5	11.00931	11	80.4	5	6	^{11}B
Carbon	6	12[a]	12	98.9	6	6	^{12}C
	6	13.00335	13	1.1	6	7	^{13}C
Nitrogen	7	14.00307	14	99.6	7	7	^{14}N
	7	15.00011	15	0.4	7	8	^{15}N
Oxygen	8	15.99491	16	99.76	8	8	^{16}O
	8	16.99914	17	0.04	8	9	^{17}O
	8	17.99916	18	0.20	8	10	^{18}O
Fluorine	9	18.99840	19	100	9	10	^{19}F
Neon	10	19.99244	20	90.9	10	10	^{20}Ne
	10	20.99395	21	0.3	10	11	^{21}Ne
	10	21.99138	22	8.8	10	12	^{22}Ne

[a] By definition, 12.00000 . . . .

Carbon from dust and rocks brought back from the moon (Photo 6) contains a larger proportion of the isotope ${}^{13}_{6}\mathrm{C}$ than does carbon on the earth (Table 2.2). This phenomenon is not yet completely understood and is under investigation.

The isotopes of an element form the same kinds of compounds because *the chemistry of an element is determined by the number of electrons in the cloud of its atom.* The chemistry is not influenced much by the mass of the nucleus, but the existence of isotopes makes chemical symbolism using microparticles very complicated. For example, if you could somehow "see" the molecules in a sample of pure water (prepared by repeated distillations) you would find all of the following, different, molecules:

$${}^{1}_{1}\mathrm{H}\text{—}{}^{16}_{8}\mathrm{O}\text{—}{}^{1}_{1}\mathrm{H} \qquad {}^{1}_{1}\mathrm{H}\text{—}{}^{17}_{8}\mathrm{O}\text{—}{}^{1}_{1}\mathrm{H} \qquad {}^{1}_{1}\mathrm{H}\text{—}{}^{18}_{8}\mathrm{O}\text{—}{}^{1}_{1}\mathrm{H}$$

$${}^{2}_{1}\mathrm{H}\text{—}{}^{16}_{8}\mathrm{O}\text{—}{}^{1}_{1}\mathrm{H} \qquad {}^{2}_{1}\mathrm{H}\text{—}{}^{17}_{8}\mathrm{O}\text{—}{}^{1}_{1}\mathrm{H} \qquad {}^{2}_{1}\mathrm{H}\text{—}{}^{18}_{8}\mathrm{O}\text{—}{}^{1}_{1}\mathrm{H}$$

$${}^{2}_{1}\mathrm{H}\text{—}{}^{16}_{8}\mathrm{O}\text{—}{}^{2}_{1}\mathrm{H} \qquad {}^{2}_{1}\mathrm{H}\text{—}{}^{17}_{8}\mathrm{O}\text{—}{}^{2}_{1}\mathrm{H} \qquad {}^{2}_{1}\mathrm{H}\text{—}{}^{18}_{8}\mathrm{O}\text{—}{}^{2}_{1}\mathrm{H}$$

Thus "chemically pure H_2O" is a *mixture of nine different kinds of molecules*! Because of the relative proportions of the isotopes of hydrogen and oxygen (Table 2.2), molecules of the type ${}^{1}_{1}\mathrm{H}\text{—}{}^{16}_{8}\mathrm{O}\text{—}{}^{1}_{1}\mathrm{H}$ make up more than 99% of water, but the other isotopes are not to be ignored, especially ${}^{2}_{1}\mathrm{H}$ (deuterium). If the process of *nuclear fusion* (Chapter 16) can be put on a practical basis, deuterium from the oceans will provide man with a source of energy for heat and power for millions of years.

Exercises

1. The element chlorine has isotopes ${}^{35}_{17}\mathrm{Cl}$ and ${}^{37}_{17}\mathrm{Cl}$. It exists as diatomic molecules Cl_2. Give the formulas of the different kinds of Cl_2 molecules.

2. Hydrogen chloride, molecular formula HCl, is a gas at ordinary temperatures and pressures. Give the formulas of all of the molecules possible, taking the isotopes of chlorine and hydrogen into account.

Chemists avoid such complex sets of formulas for elements and compounds by *letting the letter symbol for an element without mass and atomic numbers represent the naturally occurring mixture of the isotopes of the element.* The letter symbol for an element stands for a *class* of microparticles: H stands for ${}^{1}_{1}\mathrm{H}$ and ${}^{2}_{1}\mathrm{H}$ in the proportion 99.98 to 0.02; O stands for ${}^{16}_{8}\mathrm{O}$, ${}^{17}_{8}\mathrm{O}$, ${}^{18}_{8}\mathrm{O}$ in the proportions in which they occur on earth; and so on. Only for the elements without isotopes,

In Greek, "isotope" means *same place*. The term was chosen because the isotopes of an element, having the same chemistry, occupy the same place in the Periodic Table, in which elements are grouped according to properties (Chapter 3).

such as sodium, fluorine, and the others given earlier, does the unadorned symbol for the element stand for a distinct microparticle: Na means $^{23}_{11}Na$ because only one kind of sodium atom occurs naturally on earth.

When the symbols without mass and atomic numbers attached are used in formulas, the formulas also define *classes* of microparticles, not individual particles. The symbol H_2O stands for the class of microparticles made up of the subclasses H and O in a 2:1 ratio. It stands for the mixture of all of the nine kinds of water molecules given earlier. Another way of putting it is that H_2O means the class of molecules made up of atoms of atomic numbers 1 and 8 combined in a 2:1 numerical ratio. Similarly, the formula H_2 means $^1_1H—^1_1H$, $^2_1H—^1_1H$, $^2_1H—^2_1H$ in their naturally occurring (geonormal) proportions.

You must always keep in mind that the bare letter symbols for the majority of elements stand for a kind of "average" atom, which has no more real existence as a microparticle than the "average man" as a person. ☛ This simplification does not lead to any difficulties in ordinary chemical work because the number of atoms in any macroscopic sample of matter you might work with is so huge that the "averaging" is perfect. The idea of the "average" atom *cannot* be used, however, in discussing radioactivity, because radioactivity is a *property of the nucleus*. The different isotopes of a radioactive element have different radioactive properties and must therefore be discussed individually (Chapter 16).

The only exceptions are the symbols for the elements that have no isotopes (such as Na).

Positive ray analysis *Positive ray analysis* or *mass spectrometry* shows that isotopes exist and gives their proportions and masses. In a mass spectrometer (Figure 2.4), atoms of an element are first converted into their positive ions by knocking out an electron through bombardment with high-velocity electrons. The ions are accelerated into a beam by an electric field and then enter a magnetic field where

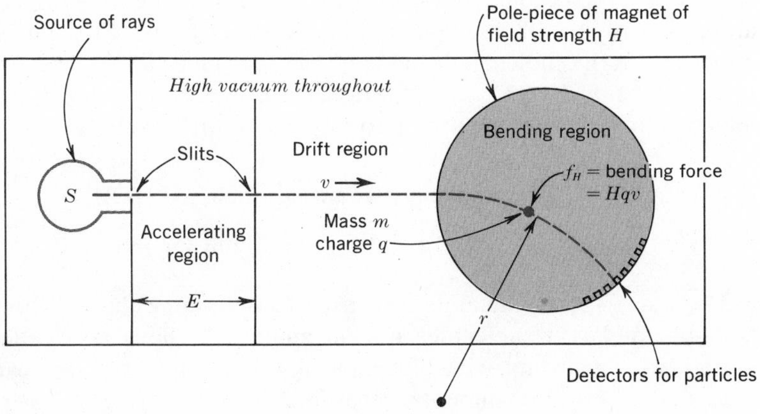

FIG. 2.4 Schematic representation of a simple mass spectrometer.

TABLE 2.3 Spectrometric data for neon

Isotope	*Isotopic Mass (amu)*	*Proportion (atom %)*	*Nuclear Composition Protons*	*Neutrons*
$^{20}_{10}Ne$	19.99244	90.92	10	10
$^{21}_{10}Ne$	20.99395	0.26	10	11
$^{22}_{10}Ne$	21.99138	8.82	10	12

Atoms per hundred atoms. The data for neon mean that, of 10,000 atoms in a sample of geonormal neon, 9092 are $^{20}_{10}Ne$, 26 are $^{21}_{10}Ne$, and 882 are $^{22}_{10}Ne$ atoms.

their paths are altered according to their *mass-to-charge* ratio (M/q). The arrangements are such that all ions having the same mass-to-charge ratio follow the same curved path in the bending region and arrive at the same detector. Taking the element neon (Ne) as an example, beams of singly charged ions of three different mass-to-charge ratios are observed: $^{20}_{10}Ne^+$, $^{21}_{10}Ne^+$, $^{22}_{10}Ne^+$. From the location of the detectors where the three kinds of ions arrive, we can calculate very accurately the masses in amu of their isotopic atoms. From the relative intensities of the three beams we can also obtain the proportions of the isotopes. These are given in Table 2.3.

These data for neon are typical of the sensitivity and accuracy of mass spectrometric analysis. They are also examples of the rule that *the mass number of an atom is the whole number closest to the mass of the atom expressed in amu.* This relationship between mass number and isotopic mass results because the masses of both proton and neutron are very close to 1 amu (Table 2.2) and the mass of the electron is very much smaller, so it makes little contribution to the mass of the atom.

The positive rays of nearly all the elements have been analyzed to obtain their isotopic compositions. Mass spectrometry also provides convincing evidence that molecules such as H_2, O_2, N_2, F_2, and Cl_2 exist in gaseous elements. In fluorine, for example, the molecular ion $^{19}_{9}F_2^+$ is observed along with the simple ion $^{19}_{9}F^+$. If fluorine molecules are bombarded with low-energy electrons, only one electron is removed from the molecule, producing the molecular ion F_2^+, according to the equation $F_2 \rightarrow F_2^+ + e^-$. Higher-energy electrons break up the molecule completely, according to the equation $F_2 \rightarrow 2F^+ + 2e^-$.

Exercises

3. What types of positive rays would be predicted for $^1_1H_2$?

4. What types of positive rays would be found for ^4_2He?

The mass spectrometer is also used to determine the weights and amounts of larger molecules on the amu scale. This application is particularly important with biochemically active material that can be obtained only in tiny amounts, because only microgram (10^{-6} g) quantities are needed in mass spectrometry. Once the molecular weight of a complex molecule on the amu scale is known, the chemist can start to determine its composition and structure.

EXAMPLE

Some biologically active compounds are almost unbelievably potent. The sex hormones (Section 14.1) are good examples. The normal requirement for the female hormone estradiol is about 50 micrograms (0.00050 g) per day. So potent is the related female hormone ethynylestradiol that even the tiny amounts absorbed through the skin of male laboratory workers has caused their "feminization" (enlargement of breasts, lack of beard growth). The first pure sample of estradiol amounted to 12 milligrams (mg). It was isolated in 1930 by extraction of 4 *tons* of the sex glands (ovaries) of female pigs. Androsterone, a male hormone, was obtained in 1931 in the amount of 15 mg by extraction of 10,000 liters (2640 gallons) of male urine. The concentration of the hormone in the urine was only 0.015×10^{-4} g/l or 1.5×10^{-6} g/l. Today mass spectrometry can be used to analyze the urine of *one* person for sex hormones and many other biologically active materials to assist in the diagnosis of disease.

2.5 Atomic Weights *The atomic weight of an element is the average weight of atoms of the element in atomic mass units (amu), taking the geonormal isotopic composition of the element into account.* In the mass spectrometric method for obtaining atomic weights, the average weights of atoms are calculated in a direct way by weighting the masses of the isotopes according to their proportions. Using the data for the neon (Ne) isotopes given before, we calculate the atomic weight of neon by weighting the masses of each of its three isotopes in amu by their percentage in geonormal neon:

$$\text{At. wt of Ne} = \left(19.99244 \times \frac{90.92}{100.00}\right) + \left(20.99395 \times \frac{0.26}{100.00}\right) + \left(21.99138 \times \frac{8.82}{100.00}\right) = 20.18 \text{ amu}$$

For most purposes, atomic weights with two digits following the decimal point are satisfactory. Isotopic weights should be used with the accuracy shown, however.

You can make this type of calculation for all elements for which mass spectral data are available. (The *Handbook of Chemistry and Physics*, published by the Chemical Rubber Company, has complete lists of both nonradioactive and radioactive isotopes with their mass and proportions.) But notice that it is the *accurate values of the isotopic masses that are averaged, not their mass numbers* (*M*). This is because the mass numbers are the *whole* numbers closest to the actual isotopic masses.

In Section 1.2 the atomic mass unit was given as 1.660×10^{-24} g. Although this gram value is correct, it depends on being able to count the number of atoms in a weighed, macroscopic sample of an element. To avoid the experimental errors involved in counting microparticles, the atomic mass unit is properly defined in the following, more fundamental way:

$$1 \text{ amu} = \frac{\text{mass of one } ^{12}_{6}\text{C atom}}{12}$$

The mass of one $^{12}_{6}$C atom is taken as the standard of atomic masses, just as the kilogram is the standard for ordinary masses. Using this new definition of the atomic mass unit (amu) as $\frac{1}{12}$ the mass of one $^{12}_{6}$C atom, we give the atomic weight of an element as:

$$\text{At. wt of an element} = 12 \times \frac{\text{average atomic mass for the element}}{\text{mass of one } ^{12}_{6}\text{C atom}}$$

The choice of $^{12}_{6}$C as the standard for atomic masses avoids counting atoms and establishes the atomic weight scale as a truly relative one, independent of any particular mass unit. The assignment of exactly 12 amu to the carbon isotope $^{12}_{6}$C is arbitrary but helpful, since it makes all atomic weights greater than 1 amu and many of them close to whole numbers. It also permits simple calculation of the number of neutrons and protons in the nucleus of an atom for which the atomic (Z) and mass (M) numbers are known.

Exercises

5. How many neutrons are there in an atom for which $Z = 20$ and $M = 40$? What is the element?

6. Using data from Table 2.2, calculate the atomic weight of O.

7. Using data from the same table, account for the atomic weight of C being 12.01 amu. (*Hint*: C is not $^{12}_{6}$C.)

The *formula-composition method* for determining atomic weights averages the isotopic masses for the atoms of an element in an indirect way by using macroscopic samples of the element and its compounds. The huge numbers of atoms in any macroscopic sample guarantee that the relative masses of the atoms found will be the relative masses of the "average" atoms of the elements. This method requires that you know the *formula* of a compound of the element, the *weight composition* of the compound in terms of the elements combined in it, and, equally important, *the arbitrary assignment of a relative weight in amu to one of the atoms in the compound.* Before the modern $^{12}_{6}$C scale was set up in 1961, the average mass of oxygen atoms was chosen as standard and assigned the atomic weight of exactly

16.0000 . . . amu. ☛ Oxygen was chosen originally as the standard because it forms compounds with nearly all elements, and the value of 16 amu for its atomic weight was chosen because it resulted in nearly whole number values for the atomic weights of many elements. To illustrate the formula-composition method, we can apply it to the determination of the atomic weight of nitrogen (N).

The atomic weight of oxygen on the $^{12}_{6}C$ scale is 15.9994.

Nitric oxide is known to have the formula NO, and decomposition of a 10.000-g sample of the oxide into its elements yields 4.668 g of nitrogen and 5.332 g of oxygen. The 10.000-g sample of the oxide contains a definite number of NO molecules; let's call this number *n*. Even though we do not know the value of *n*, we can let m_{NO} represent the mass of *one* NO molecule and write:

$$10.000\ g = nm_{NO}$$

Since the formula NO tells us that nitrogen atoms and oxygen atoms are present in a 1:1 ratio in the compound, there must be *n* nitrogen atoms in the 4.668 g of nitrogen obtained from 10.000 g of NO and the *same* number *n* of oxygen atoms in the 5.332 g of oxygen obtained. Letting m_N and m_O represent the masses of the respective individual atoms, we can now write two equations:

$$nm_N = 4.668\ g$$
$$nm_O = 5.332\ g$$

Dividing one of these equations by the other gives:

$$\frac{nm_N}{nm_O} = \frac{4.668}{5.332} = 0.8755$$

Since the *n*'s have the same value, they divide out, leaving

$$\frac{m_N}{m_O} = 0.8755$$

This ratio tells us that the mass of one (average) ☛ nitrogen atom is 0.8755 times the mass of one (average) oxygen atom. The method is shown graphically in Figure 2.5. If we now decide on a mass of 16.0000 . . . amu for the oxygen atom, the atomic weight of N on the O = 16 scale is simply

"Average" because we are dealing with macroscopic samples of the elements.

$$\text{At. wt of N} = 0.8755 \times 16.0000\ .\ .\ . = 14.01$$

On the $^{12}_{6}C = 12.0000$ . . . scale now used, the atomic weight of nitrogen is 14.0067, or 14.01 to four significant figures.

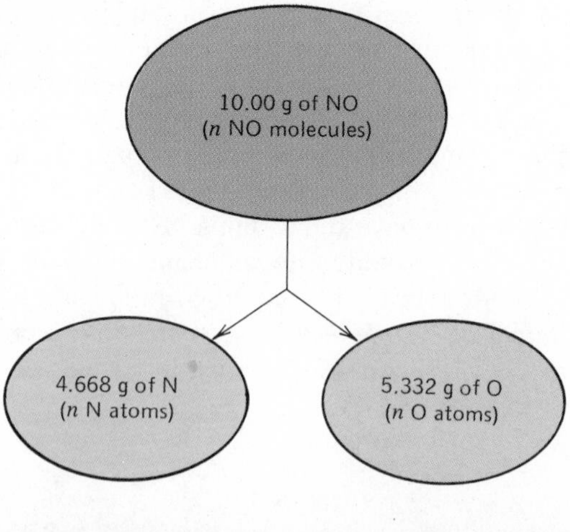

$$\frac{\text{wt of } n \text{ N atoms}}{\text{wt of } n \text{ O atoms}} = \frac{4.668 \text{ g}}{5.332 \text{ g}} = \frac{\text{wt of one N atom}}{\text{wt of one O atom}} = 0.8755$$

FIG. 2.5 Relation between the masses of N and O atoms.

EXAMPLE *The atomic weight of uranium.* An important application of the formula-composition method is finding the atomic weight of uranium. Uranium, a metal, is the source of nuclear fuel for atomic reactors and atomic bombs. Uranium trioxide, UO_3, is 83.22% U and 16.78% O by weight. When the composition is given in percentages, it's convenient to think of a 100.00-g sample of the compound. A 100.00-g sample of UO_3 contains some number, call it n, of UO_3 formula units. Letting m_{UO_3}, m_U, and m_O stand for the respective weights of one UO_3 unit, one U atom, and one O atom, you can write the following equation:

$$nm_{UO_3} = 100.00 \text{ g} = nm_U + 3nm_O = 83.22 \text{ g} + 16.78 \text{ g} \tag{1}$$

The weight of oxygen is $3nm_O$ because the formula UO_3 shows that for every U atom in the compound there are three O atoms. Continuing, we have:

$$nm_U = 83.22 \text{ g} \tag{2}$$

$$3nm_O = 16.78 \text{ g} \tag{3}$$

Dividing Equation (2) by Equation (3) gives:

$$\frac{nm_U}{3nm_O} = \frac{83.22}{16.78} \quad \text{or} \quad \frac{\not{n}m_U}{\not{n}m_O} = \frac{3 \times 83.22}{16.78} = \frac{249.66}{16.78} \tag{4}$$

The n's divide out, giving the ratio of the weight of one U atom to the weight of one O atom:

$$\frac{m_U}{m_O} = \frac{249.66}{16.78} = 14.88$$

If the atomic weight of O is taken as 16.00 amu, we have:

$$\text{At. wt of U} = 14.88 \times 16.00 \text{ amu} = 238.08 \text{ amu}$$

Like most elements, uranium is a mixture of isotopes, and mass spectrometry provides the following data on its geonormal isotopic composition:

Isotope	*Mass in amu*	*Proportion of atoms* (%)
$^{234}_{92}U$	234.0409	0.0057
$^{235}_{92}U$	235.0439	0.72
$^{238}_{92}U$	238.0508	99.27

The atomic weight of uranium is calculated from the mass spectral data in the way shown earlier for neon:

$$\begin{aligned}\text{At. wt of U} &= \left(\frac{0.0057}{100}\right)(234.0409) + \left(\frac{0.72}{100}\right)(235.0439) \\ &\quad + \left(\frac{99.27}{100}\right)(238.0508) \\ &= 238.02 \text{ amu}\end{aligned}$$

The accepted value for the atomic weight of uranium is a compromise between the values from the two methods and is 238.03 amu. The atomic weight of uranium is important because it must be known to make calculations of the weights of geonormal uranium required in the reactions used to prepare uranium compounds for isotope separation. The isotope $^{235}_{92}U$ undergoes nuclear fission and was the explosive in the first atomic bombs. Isotope $^{238}_{92}U$ is used to provide fuel in "breeder reactors" (Chapter 16).

The formula-composition method was used to determine atomic weights long before the invention of the mass spectrometer and the discovery of isotopes. The formula-composition method is still used today together with mass spectral data to obtain the accurate values for the atomic weights of the elements listed in the Periodic Table on the inside front cover.

Exercises

8. Why could the formula-composition method for atomic weights be used before the discovery of isotopes?

9. Carbon dioxide has the molecular formula CO_2 and is 27.29% C and 72.71% O. Calculate the atomic weight of oxygen (O) on the modern scale, $^{12}_{6}C = 12$ exactly. (*Hint*: CO_2 is not just $^{12}_{6}CO_2$.)

Nuclear binding energy It is an experimental fact that the *mass of an atom determined by mass spectrometry is always less than the sum of the masses of the protons, neutrons, and electrons of which it is made*. This means that the mass of an atom cannot be calculated accurately from knowing the numbers of protons, neutrons, and electrons present. As an example, consider $^{4}_{2}He$, made up of two protons, two neutrons, and two electrons. Using data from Table 2.2, we obtain

$$\begin{aligned} \text{Mass of 2 protons} &= 2 \times 1.00728 = 2.01456 \text{ amu} \\ \text{Mass of 2 neutrons} &= 2 \times 1.00867 = 2.01734 \text{ amu} \\ \text{Mass of 2 electrons} &= 2 \times 0.00055 = 0.00110 \text{ amu} \\ \hline \text{Sum of particle masses} &= 4.03300 \text{ amu} \end{aligned}$$

But the mass of one $^{4}_{2}He$ atom determined by mass spectrometry is only 4.00260 amu (Table 2.2), which is $(4.03300 - 4.00260) = 0.030400$ amu *less* than the sum of the masses of its particles. All atoms except $^{1}_{1}H$ have actual masses less than the calculated value. The difference between the sum of the particle masses and the actual measured mass for a given atom is called the *mass defect* for the atom. The mass defect is accounted for by the fact, discovered by Einstein, that *energy has mass*. Large amounts of energy are *released* in the formation of an atom from neutrons, protons, and electrons, and this energy accounts for the mass defect. The energy released is termed the *nuclear binding energy* of the atom. According to Einstein's theory, mass is conserved (neither created nor destroyed) if we write for the formation of one $^{4}_{2}He$ atom:

Albert Einstein (1879–1955) was the American physicist whose theory of relativity is one of the greatest intellectual achievements in history. He was born in Germany but fled from the Nazis and became an American citizen in 1940. He was awarded the Nobel prize in physics in 1921 for his theory of relativity and his discovery that energy has mass.

$$\begin{aligned} \underset{(2p^+ + 2n + 2e^-)}{4.03300 \text{ amu of matter}} &\rightarrow 4.00260 \text{ amu of } ^{4}_{2}\text{He (matter)} \\ &\quad + 0.03040 \text{ amu of energy} \\ \hline 4.03300 \text{ amu} &= 4.03300 \text{ amu} \end{aligned}$$

One amu (1.660×10^{-24} g) of energy is 9.31×10^{8} electron volts of energy (Chapter 16), so the nuclear binding energy of $^{4}_{2}He$ is $(0.03040 \times 9.31 \times 10^{8}) = 2.83 \times 10^{7}$ eV or 28,300,000 eV. The same amount of energy would be required to blast one $^{4}_{2}He$ atom apart into separated protons, neutrons, and electrons. The nuclear binding energy of $^{4}_{2}He$ can be related to the energies of ordinary chemical reactions by comparing it with the covalent bond energy of the hydrogen molecule (Section 1.5):

$$\mathrm{H}\cdot + .\mathrm{H} \rightarrow \mathrm{H{:}H} + 4.52 \text{ eV}$$
$$2p^{+} + 2n + 2e^{-} \rightarrow {}^{4}_{2}\mathrm{He} + 28{,}300{,}000 \text{ eV} \quad (2.83 \times 10^{7} \text{ eV})$$

We see that nuclear binding energies are tens of millions of times larger than ordinary chemical bond energies, and that atomic nuclei are very stable indeed. The huge size of nuclear binding energies accounts for the ability of the sun and other stars to radiate astronomical quantities of energy for billions of years. In the interior, atoms are constantly being formed with the release of their nuclear binding energies in the form of heat and light. In the sun the process is believed to occur by a series of complex steps, but the overall process is the conversion of four hydrogen atoms to one ${}^{4}_{2}He$ atom. Hydrogen is therefore the fuel for the nuclear furnace in the sun. Hydrogen is also a fuel for rockets used in space exploration, but here the energy is provided by its *chemical* reaction with oxygen to form water. The energy output is tiny compared to that of the *nuclear* process in the sun and stars (Photo 7):

PHOTO 7 The "Whirlpool Galaxy" lies thousands and thousands of light years distant. All of the light from this huge system, as with all galaxies, comes from the conversion of matter mass to an equal amount of energy mass in the form of light and other types of radiation.

$$4\mathrm{H} \rightarrow {}^{4}_{2}\mathrm{He} + 27{,}000{,}000 \text{ eV (sun)}$$
$$2\mathrm{H_2}(g) + \mathrm{O_2}(g) \rightarrow 2\mathrm{H_2O}(g) + 5.02 \text{ eV (rocket)}$$

Atomic and hydrogen bombs, and the generation of power by nuclear reactors, also depend on the conversion of matter into energy by nuclear processes. The details are discussed in Chapter 16.

Exercises

10. Calculate, using data from Table 2.2, the nuclear binding energies of ${}^{7}_{3}Li$, ${}^{12}_{6}C$, and ${}^{13}_{6}C$.

11. Why is it meaningless to speak of the nuclear binding energy of one oxygen atom but correct to speak of the nuclear binding energy of one sodium atom or one fluorine atom? (*Hint*: Think of isotopes.)

The importance of the mass defect in our study of atomic weights is that the atomic weight of an element such as sodium (${}^{23}_{11}Na$), which has no natural isotopes, ☛ cannot be calculated from its mass number (M) and atomic number (Z). The total mass of 11 protons, 12 neutrons, and 11 electrons is 23.19 amu, but the experimentally determined (true) atomic weight of sodium is 22.99 amu, giving a mass defect of 0.20 amu.

Isotopes of Na, such as ${}^{22}_{11}Na$, can be prepared in nuclear reactors (Chapter 16), but they are radioactive and of short life, so they are not present in geonormal Na.

In fact, to find the nuclear composition of an atom we work in the other direction. The atomic number and mass in amu of the atom are determined by experiment. The mass *number* is then taken as the whole number closest to the mass of the atom on the amu scale, and finally the number of neutrons in the nucleus is found by using the equation $N = M - Z$.

2.6 Determination of Formulas To use the formula-composition method for determining atomic weights, you need to know the formulas of compounds. How are formulas obtained?

The formulas of the molecules of gaseous compounds produced by the reaction of gaseous elements are obtained by using the method invented by Avogadro. Avogadro proposed that *equal volumes of different gases measured at the same temperature and pressure contain equal numbers of molecules*. He came to this totally new idea about the nature of gases after some hard thinking about such experimentally observed volume relationships in gaseous reactions as the following:

Amedeo Avogadro (1776–1856) was an Italian physicist and chemist. He published his theory in 1811 but other scientists ignored his work for about 50 years.

hydrogen gas 1 volume	+	chlorine gas 1 volume	→	hydrogen chloride gas 2 volumes
nitrogen gas 1 volume	+	hydrogen gas 3 volumes	→	ammonia gas 2 volumes
nitrogen gas 1 volume	+	oxygen gas 1 volume	→	nitric oxide gas 2 volumes
hydrogen gas 2 volumes	+	oxygen gas 1 volume	→	steam 2 volumes

You can think of "1 volume" as being 1 quart or 1 liter—it doesn't matter. The important fact is that the volume of gas produced is not always the sum of the volumes of reacting gases.

Results like these were summarized in Gay-Lussac's law of combining volumes: *when gases react, the proportions by volume* (measured at the same temperature and pressure) *stand in the ratios of small whole numbers*. The whole-number volume ratios can be explained only if it is assumed (1) that gases have molecules made up of two or more atoms as their structural units and (2) that equal volumes of different gases contain the same number of molecules. From these two ideas Avogadro was able to provide a beautiful and correct explanation for the small numerical ratios.

Joseph Louis Gay-Lussac (1778–1850) was a French chemist and physicist who investigated the upper atmosphere. He rose 21,000 feet above sea level in a hydrogen-filled balloon to collect samples of air to find out what substances it was composed of.

Taking the formation of nitric oxide as an example, his argument goes something like this: One volume, say 1 liter, of nitrogen gas contains some definite number, call it n, of nitrogen molecules. The same number of oxygen molecules must be present in 1 liter of oxygen gas, according to his principle that equal volumes of different gases, measured at the same T and P, contain equal numbers of molecules. And, since the volume of the compound, nitric oxide gas, is *twice* the volumes of the nitrogen and oxygen used, the one volume of nitric oxide must contain twice as many or $2n$ molecules of nitric oxide. To summarize to this point:

	nitrogen molecules	+	oxygen molecules	→	nitric oxide molecules
Experiment:	1 volume		1 volume		2 volumes
Avogadro:	*n* molecules		*n* molecules		2*n* molecules

Because nitric oxide is a *compound* of nitrogen and oxygen atoms, its molecules must contain *at least* one nitrogen and one oxygen atom – it could contain more of each, but let's *assume* that the simplest possible formula, NO, is correct. Letting the formulas of the nitrogen molecule and the oxygen molecule be N_a and O_b, where *a* and *b* are the numbers of atoms in their molecules, we can write:

n nitrogen molecules + *n* oxygen molecules → 2*n* nitric oxide molecules

n N_a molecules + n O_b molecules → $2n$ NO molecules

Now, 2*n* NO molecules contain 2*n* atoms of nitrogen in combination. If these 2*n* nitrogen atoms come from only *n* N_a molecules, *a* must be 2 and the formula of the nitrogen molecule is N_2. By the same argument, *b* must be 2 if the 2*n* oxygen atoms in the NO molecules came from only *n* O_b molecules. To summarize, we write the *molecular* equation for the reaction:

$$N_2(g) + O_2(g) \rightarrow 2NO(g)$$

Avogadro's argument can also be shown with boxes to represent the volumes of the gases:

Of course, there are many more than three molecules in any macroscopic gas sample, but this fact does not change the logic of Avogadro's argument.

N_2		O_2		NO	NO
N_2	+	O_2	→	NO	NO
N_2		O_2		NO	NO
1 volume (3 molecules)		1 volume (3 molecules)		2 volumes (6 molecules)	

But, you say, we haven't proved anything, because we *assumed* that the molecular formula of nitric oxide is NO. It's true that the formula NO has not been proved – in fact you can write a balanced molecular equation that fits the 1:1:2 volume relationships by assuming *any* molecular formula that has an even number of N atoms. The same applies to O_b. Here are some examples:

$$N_4(g) + O_2(g) \rightarrow 2N_2O(g)$$

$$N_6(g) + O_2(g) \rightarrow 2N_3O(g)$$

$$N_6(g) + O_{10}(g) \rightarrow 2N_3O_5(g)$$

However, we *have* proved one thing. *No matter how many equations you write, a and b in the formulas* N_a *and* O_b *must be even numbers to give the 1:1:2 volume ratio found by experiment.* No balanced molecular equation that fits the volume ratio can be written in which either the nitrogen molecule or the oxygen molecule contains an *odd* number of atoms.

That the formula of nitric oxide is in fact NO (and nitrogen N_2, oxygen O_2) is shown by the following additional experimental results: *in no reaction involving oxygen does* ***one*** *volume of oxygen give more than* ***two*** *volumes of a gaseous product containing oxygen atoms in combination; and in no reaction involving nitrogen gas does* ***one*** *volume of nitrogen give more than* ***two*** *volumes of a gaseous compound of nitrogen.* Thus *no* reactions of the following types are known:

$$\underset{\text{1 volume}}{N_2} + \text{any gas } X_n \rightarrow \text{more than 2 volumes of gaseous product}$$

$$\underset{\text{1 volume}}{O_2} + \text{any gas } X_n \rightarrow \text{more than 2 volumes of gaseous product}$$

It is therefore never necessary in explaining reacting volumes to assume that either oxygen or nitrogen molecules contain more than two atoms. Avogadro's argument, applied to the reactions of hydrogen with chlorine, nitrogen, and oxygen, leads to the following molecular equations for the other reactions given earlier:

$$\underset{\text{1 volume}}{H_2(g)} + \underset{\text{1 volume}}{Cl_2(g)} \rightarrow \underset{\text{2 volumes}}{2HCl(g)} \text{ (hydrogen chloride)}$$

$$\underset{\text{1 volume}}{N_2(g)} + \underset{\text{3 volumes}}{3H_2(g)} \rightarrow \underset{\text{2 volumes}}{2NH_3(g)} \text{ (ammonia)}$$

$$\underset{\text{2 volumes}}{2H_2(g)} + \underset{\text{1 volume}}{O_2(g)} \rightarrow \underset{\text{2 volumes}}{2H_2O(g)} \text{ (steam)}$$

Once the formulas of water (H_2O), ammonia (NH_3), nitric oxide (NO), and hydrogen chloride (HCl) are known, and their compositions in terms of the weights of combined elements have been obtained, the atomic weights of H, N, and Cl can be calculated by the formula-composition method described in Section 2.5.

Exercises

12. Passing O_2 gas through an electric spark converts it to ozone, which is made up of only O atoms. The volume relationships in the reaction are

$$\underset{\text{3 volumes}}{\text{oxygen}} \xrightarrow[\text{spark}]{\text{electric}} \underset{\text{2 volumes}}{\text{ozone}}$$

What is the simplest molecular equation for the reaction?

13. Natural gas is largely methane, and the skeleton equation for its combustion (burning) is $CH_4 + O_2 \nrightarrow CO_2 + H_2O$. Balance the equation. What volume relationships would be observed if the reaction were carried out under conditions of T and P in which all reactants and products were gases?

The Avogadro method for determining formulas works only for elements and compounds that are gases. Nevertheless, Avogadro's method was very important in establishing the atomic weight scale. For elements that do not form gaseous compounds, other methods for determining formulas are used. These will be described in a more appropriate place. Now let's look at the uses for the atomic weights of the elements. To do so we will need the concept of the *mole*, a very useful chemical idea.

2.7 The Mole The idea of a *mole*, or molar weight of an element or compound, provides a simple way to go from the masses of atoms in amu to the masses in grams of the macroscopic samples used in laboratory experiments. A mole is a macroscopic sample of matter having a definite mass or weight that depends on the atomic composition of the matter. For elements, the definition of the mole is very simple:

*One mole of an element is one atomic weight of the element but taken in **grams** rather than in amu.*

According to the definition, one mole of calcium (Ca, at. wt 40.08 amu) is simply 40.08 g of calcium, one mole of oxygen (O, at. wt 16.00 amu) is 16.00 g of oxygen, one mole of sodium (Na, at. wt 22.99 amu) is 22.99 g of sodium, and so on. A mole of an element is sometimes called a gram-atomic weight or "gram-atom" of the element. Mass units other than grams may be used for moles. In industry, the pound-mole and ton-mole are more convenient, since large samples of matter are dealt with. In chemistry, gram-mole is shortened to mole—just remember that the mass unit is the gram, and moles *in grams* are meant.

You can see the convenience of the mole idea by applying it to a laboratory problem. The elements zinc (Zn) and sulfur (S) combine, when heated, to form the compound zinc sulfide, for which the formula is ZnS. What weights of the two elements should you heat together if the zinc sulfide obtained is to be pure, that is, free of any leftover zinc or sulfur? Since the formula ZnS tells us that zinc and sulfur atoms are combined in a 1:1 ratio, the weights of zinc and sulfur used must be the weights that contain *equal* numbers of atoms. From the definition of atomic weights as the *relative* masses of the atoms, and the definition of the mole, *the number of zinc atoms in one mole of zinc is equal to the number of sulfur atoms in one mole of sulfur.* To see this, let's calculate the actual numbers of atoms. Since (to five significant figures) 1 amu $= 1.6604 \times 10^{-24}$ g, we have, for the masses of the individual atoms:

$$\text{Mass of one Zn atom} = 65.37 \times 1.6604 \times 10^{-24} = 1.0854 \times 10^{-22}\ \text{g}$$
$$\text{Mass of one S atom} = 32.06 \times 1.6604 \times 10^{-24} = 5.3232 \times 10^{-23}\ \text{g}$$

For the molar amounts of the elements, the number of atoms present is the mass of one mole divided by the mass of one atom:

$$\text{Number of Zn atoms in 65.37 g of Zn} = \frac{65.37}{1.0854 \times 10^{-22}} = 6.023 \times 10^{23}\ \text{atoms}$$

$$\text{Number of S atoms in 32.06 g of S} = \frac{32.06}{5.3232 \times 10^{-23}} = 6.023 \times 10^{23}\ \text{atoms}$$

Pure zinc sulfide would result from heating together one mole each of zinc and sulfur. Applying the law of conservation of mass gives us this equation:

$Zn(s)$	+	$S(s)$	→	$ZnS(s)$
1 mole		1 mole		
65.37 g		32.06 g		(65.37 + 32.06) = 97.43 g
6.023×10^{23} atoms		6.023×10^{23} atoms		6.023×10^{23} ZnS units

Of course, you don't have to run the reaction on a full-mole scale. If less zinc sulfide is needed, you can take 0.10 mole each of zinc (6.537 g) and sulfur (3.206 g), to get $0.10 \times 97.43 = 9.743$ g of zinc sulfide. In fact, *any* weights of zinc and sulfur taken in *any* weight unit can be used, as long as they stand in the *ratio* of their molar (or atomic) weights, 65.37/32.06. This means that heating together 65.37 lb of zinc and 32.06 lb of sulfur would give 97.43 lb of zinc sulfide. One great advantage of the mole idea is easily seen—because atomic weights are relative weights, any mass unit can be used. It is customary and convenient to use the gram only in the laboratory.

You can use the idea of the mole for compounds as well. Just as one mole of an element is one atomic weight in grams, *one mole of a compound is one formula weight in grams.* The formula weight of ZnS is $(65.37 + 32.06) = 97.43$ amu, and one mole of ZnS is 97.43 g. For tin oxide, SnO_2 (stannic oxide), the formula weight is the sum of the atomic weight of tin and *twice* the atomic weight of oxygen because there are two oxygen atoms per formula unit. One mole of SnO_2 is therefore $[114.82 + (2 \times 16.00)] = 146.82$ g.

Exercises

14. Zinc reacts with oxygen when heated to give zinc oxide, ZnO. What weight of zinc would be required for complete reaction with 4.00 g of oxygen?

15. Tin reacts with oxygen to form stannic oxide, SnO_2. Using 118.7 for the atomic weight of Sn, what weights of Sn and O_2 should you combine to obtain 10.0 g of SnO_2?

16. Calculate the number of H atoms in 2.016 g of H_2 and the number of oxygen atoms in 32.00 g of O_2. (At. wt H = 1.008, at. wt O = 16.00, 1 amu = 1.6604×10^{-24} g.)

If the formula for a compound is the formula of a covalent molecule, then the formula weight is called the *molecular weight* of the compound. The formula for the glucose molecule is $C_6H_{12}O_6$. Its molecular weight is calculated as follows:

$$\begin{array}{rr} 6 \times 12.01 = & 72.06 \text{ amu} \\ 12 \times 1.008 = & 12.10 \text{ amu} \\ 6 \times 16.00 = & 96.00 \text{ amu} \\ \hline \text{Molecular wt} = & 180.16 \text{ amu} \end{array}$$

One mole of glucose is 180.16 g, one molecular weight in amu but taken in grams. The word mole suggests molecule, but it is perfectly correct to speak of a mole or molar weight of a compound, even if the compound is ionic like ZnS and SnO_2 of which, therefore, no molecule exists. It is also correct to speak of a mole of electrons (0.0005486 g), a mole of protons (1.00728 g), a mole of neutrons (1.00867 g) (Table 2.2), or a mole of any other particle.

Returning to the zinc sulfide case, we saw that one mole of zinc contains the same number of atoms as one mole of sulfur, namely, 6.023×10^{23}. There is nothing unusual about these elements—one mole of any element is 6.023×10^{23} atoms of the element. In fact, *one mole of any specified particle contains 6.023×10^{23} or one Avogadro number* ☛ *of the particles.*

The value of the Avogadro number depends on the unit of weight used for one mole. Since 1 lb = 453.59 g, there are ($453.59 \times 6.023 \times 10^{23}$) = 2.732×10^{26} particles in a pound-mole, in contrast to the 6.032×10^{24} particles in a gram-mole.

In using the mole idea it is absolutely necessary to give the formula of the particle. Some examples will make this clear:

$$\begin{aligned} \text{One mole of oxygen atoms (O)} &= 16.00 \text{ g} \\ \text{One mole of oxygen molecules } (O_2) &= 32.00 \text{ g} \\ \text{One mole of ozone molecules } (O_3) &= 48.00 \text{ g} \end{aligned}$$

All three of these forms of oxygen are known, but oxygen gas has O_2 molecules as the structural unit. On the other hand, one mole of neon (Ne, at. wt 20.18) is 20.18 g and contains 6.023×10^{23} neon *atoms* as structural units. One mole of sulfur *atoms* is contained in 32.06 g of S, but X-ray crystallography shows that the sulfur atoms in the yellow, crystalline form of the element are in fact covalently bonded together to form S_8, ring-shaped molecules. One mole of the specified particle S is 32.06 g and contains $\frac{1}{8} \times 6.023 \times 10^{23}$ S_8 molecules.

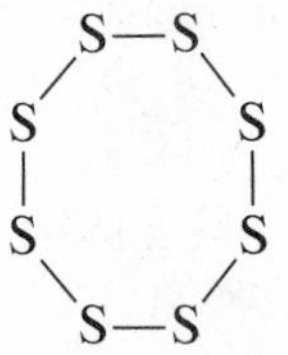

S_8 molecule

One mole of S_8 is $(8 \times 32.06) = 256.48$ g and contains 6.023×10^{23} S_8 molecules. With this additional information about the molecular nature of sulfur, the equation for the formation of zinc sulfide could better be written:

$8Zn(s)$	+	$S_8(s)$	→	$8ZnS(s)$
8 moles of Zn		1 mole of S_8		8 moles of ZnS
8×65.37 g		8×32.06 g		$8 \times (65.37 + 32.06)$ g

The weight ratio of the elements is the same as for the equation $Zn(s) + S(s) \rightarrow ZnS(s)$:

$$\frac{\text{wt of Zn}}{\text{wt of S}} = \frac{\cancel{8} \times 65.37}{\cancel{8} \times 32.06} = \frac{65.37}{32.06}$$

Whenever you know the molecular formulas, they and the corresponding molecular weights should be used. Thus, although the weight relations for the formation of water from hydrogen and oxygen are correctly given by

$2H$	+	O	→	H_2O
2 moles of H		1 mole of O		1 mole of H_2O
2×1.008 g		16.00 g		$(2.016 + 16.00)$ g

the preferred equation takes account of the molecular nature of the microparticles found by the Avogadro method:

$2H_2$	+	O_2	→	$2H_2O$
2 moles of H_2		1 mole of O_2		2 moles of H_2O
$2\,(2 \times 1.008)$ g		2×16.00 g		$2\,(16.00 + 2.016)$ g
4.032 g		32.00 g		36.032 g

The mole idea is also useful in calculations with complex ions. For example, you might wish to know how many moles of SO_4^{2-} ions there are in 30.00 g of the ionic compound $Ca^{2+}SO_4^{2-}$. The formula tells us that there is one mole of SO_4^{2-} in each mole of $Ca^{2+}SO_4^{2-}$. One mole of $Ca^{2+}SO_4^{2-}$ is (at. wt of Ca + at. wt of S + 4 × at. wt of O) in grams or 40.08 + 32.06 + 4(16.00) = 136.14 g. The 30.00-g sample represents (30.00/136.14) = 0.2204 mole of the compound and contains 0.2204 mole of SO_4^{2-} and 0.2204 mole of Ca^{2+}.

EXAMPLE

A somewhat more difficult example is the following: How many moles of Ca^{2+} and PO_4^{3-} are there in 90.00 g of the ionic compound tricalcium phosphate, $Ca_3(PO_4)_2$, made up of Ca^{2+} ions and PO_4^{3-} ions in a 3:2 ratio? The formula weight of $Ca_3(PO_4)_2$ is calculated as follows:

$$\begin{aligned} 3 \times \text{at. wt of Ca} &= 3 \times 40.08 = 120.24 \text{ amu} \\ 2 \times \text{at. wt of P} &= 2 \times 30.97 = 61.94 \text{ amu} \\ 8 \times \text{at. wt of O} &= 8 \times 16.00 = 128.00 \text{ amu} \\ \hline \text{Formula wt of } Ca_3(PO_4)_2 &= 310.18 \text{ amu} \end{aligned}$$

One mole of $Ca_3(PO_4)_2$ is therefore 310.18 g, and the 90.00-g sample is (90.00/310.18) = 0.2902 mole of the compound. Since the formula shows that there are three moles of Ca^{2+} and two moles of PO_4^{3-} in each mole of $Ca_3(PO_4)_2$, the 90.00-g sample contains (3 × 0.2902) = 0.8706 mole of Ca^{2+} and (2 × 0.2902) = 0.5804 mole of PO_4^{3-}.

Exercises

17. Copper sulfate pentahydrate forms beautiful blue crystals having the formula $CuSO_4(H_2O)_5$, often written $CuSO_4 \cdot 5H_2O$. How many moles of H_2O are there in one mole of the pentahydrate? In 100.00 g of the pentahydrate?

18. Hydrogen peroxide, a waterlike liquid having molecular formula H_2O_2, slowly decomposes into water and oxygen gas. Write the molecular equation for the decomposition. What weight of oxygen gas (O_2) would be obtained from the decomposition of 18.0 g of H_2O_2? How many moles of O_2 does this weight represent? How many moles of O? (Use at. wt O = 16, at. wt H = 1.01.)

Calculating formulas The mole concept is also useful in determining the formula of a compound from its elementary composition and the atomic weights of its constituent elements.

A 2.000-g sample of a compound of magnesium (Mg, at. wt 24.31) and chlorine (Cl, at. wt 35.45), decomposed into its elements, gave 0.5106 g of magnesium and 1.4894 g of chlorine. To obtain the formula of the compound, you first calculate the numbers of moles of Mg and Cl produced. Since one mole of Mg is 24.31 g and one mole of Cl is 35.45 g, we have:

$$\text{Number of moles of Mg} = \frac{0.5106}{24.31} = 2.10 \times 10^{-2} \text{ mole}$$

$$\text{Number of moles of Cl} = \frac{1.4894}{35.45} = 4.20 \times 10^{-2} \text{ mole}$$

One mole of any specified particle contains Avogadro's number, 6.023×10^{23}, of particles, so the numbers of atoms represented by the weights of magnesium and chlorine obtained are as follows:

$$\begin{aligned}\text{Number of Mg atoms} &= 2.10 \times 10^{-2} \times 6.023 \times 10^{23} \\ &= 1.265 \times 10^{22} \text{ Mg atoms} \\ \text{Number of Cl atoms} &= 4.20 \times 10^{-2} \times 6.023 \times 10^{23} \\ &= 2.530 \times 10^{22} \text{ Cl atoms}\end{aligned}$$

You can see that there are two chlorine atoms for each magnesium atom (2.530/1.265 = 2), so the formula of the compound is $MgCl_2$.

Formulas found in this way are called *empirical* formulas—they give only the *numerical ratios* in which the atoms are combined. They do not necessarily give the formula of an actual microparticle, that is, of a molecule. Magnesium chloride is an ionic compound made up of Mg^{2+} ions and Cl^- ions in a 1:2 ratio.

To see the difference between the empirical formula and the molecular formula for a compound, take the compounds formaldehyde, molecular formula CH_2O, and glucose, molecular formula $C_6H_{12}O_6$. They both have carbon, hydrogen, and oxygen atoms in a 1:2:1 ratio, as you can see by dividing through the glucose formula by 6:

$$C_{6/6}H_{12/6}O_{6/6} = CH_2O$$

Both compounds have the same composition in terms of their elements, and therefore they both have the same *empirical* formula, CH_2O. To determine a molecular formula from an empirical formula, you need to know the "size" or number of empirical formula units in the molecule. This is given by the molecular weight. In many cases the molecular weight of a covalent compound can be found by mass spectrometry of the positive rays from the compound. In the positive rays from formaldehyde there are those of the formaldehyde molecular ion CH_2O^+, molecular weight 30, and in those from glucose there are rays of the ion $C_6H_{12}O_6^+$, molecular weight 180. If you know the molecular weight from mass spectrometry or other data source, you can find the number of empirical formula units in the molecule by dividing the molecular weight by the empirical formula weight:

$$\text{Number of empirical formula units in the molecule} = \frac{\text{molecular wt}}{\text{empirical formula wt}}$$

For example, if mass spectrometry of a compound known from analysis to have the empirical formula CH (formula wt 13) showed that its molecular weight was 78, then the number of empirical formula units in the molecule = 78/13 = 6, and the molecular formula is $(CH)_6$ or C_6H_6.

Exercises

19. Acetic acid, which gives vinegar its sour taste, has the composition 40.0% C, 6.67% H, and 53.33% O. What is the empirical formula of acetic acid? Do you know of any other compounds that have the same empirical formula? How many are possible? Mass spectrometry gives 60.0 for the molecular weight of acetic acid. What is the molecular formula of acetic acid?

20. A compound of hydrogen and oxygen, sometimes used to bleach hair, has the composition by weight 5.88% H, 94.12% O. What is the empirical formula of the compound? Do you know the name and molecular formula of the compound?

When you calculate the atomic ratios for a compound from composition and atomic weights, they don't always come out in whole numbers, as the following example will show: A compound of potassium (K, at. wt 39.10), chromium (Cr, at. wt 52.00), and oxygen (O, at. wt 16.00) has the elementary composition 26.58% K, 35.35% Cr, 38.07% O. (When the composition of a compound is given in percent, you can think of a 100-g sample of the compound, because then the percentages of the elements become their weights in grams.) In 100.00 g of the compound there are 26.58 g of K, 35.35 g of Cr, and 38.07 g of O in combination. You get the number of moles of each element in 100.00 g of the compound by dividing the weight of the element by its molar weight:

Moles in 100.00 g of compound	*Atom ratios*
$\frac{26.58}{39.10} = 0.6798$ mole of K	$\frac{0.6798}{0.6798} = 1.000$
$\frac{35.35}{52.00} = 0.6798$ mole of Cr	$\frac{0.6798}{0.6798} = 1.000$
$\frac{38.07}{16.00} = 2.379$ moles of O	$\frac{2.379}{0.6798} = 3.500$

The atom ratios are 1:1:3.5, since for each mole of K and Cr there are 3.500 moles of O. The formula could be written $K_{1.000}Cr_{1.000}O_{3.500}$ but atoms are not divisible, and you get the correct empirical formula by multiplying through by 2, giving the formula $K_2Cr_2O_7$, which has whole numbers of atoms.

Calculation of percentage composition Often we want the percentage of an element in a compound. We might, for example, in studying air pollution, want to know the percentage of the poisonous metal lead (Pb, at. wt 207.19) in tetraethyl lead [$Pb(C_2H_5)_4$], or PbC_8H_{20}, a compound that has been added to gasoline for many years to improve the performance of auto engines. The molecular weight of tetraethyl lead is calculated in the usual way:

$$\begin{aligned} 1 \times \text{at. wt of Pb} &= 1 \times 207.19 = 207.19 \text{ amu} \\ 8 \times \text{at. wt of C} &= 8 \times 12.01 = 96.08 \text{ amu} \\ 20 \times \text{at. wt of H} &= 20 \times 1.008 = 20.16 \text{ amu} \\ \hline \text{Molecular wt of } Pb(C_2H_5)_4 &= 323.43 \text{ amu} \end{aligned}$$

Each mole of tetraethyl lead contains one mole of Pb or 207.19 g of Pb, so the percentage of lead in $Pb(C_2H_5)_4 = 207.19/323.43 \times 100 = 64.06\%$. At one time about four grams of the lead compound were added to each gallon of gasoline. In an urban area where 10 million gallons of leaded gasoline were consumed per day the weight of lead discharged with the automobile exhausts was

$$10 \times 10^6 \times 4 \times \frac{64.06}{100} = 2.56 \times 10^7 \text{ g per day}$$

$$= 28.2 \text{ tons per day}$$

PHOTO 8 Smoke containing lead results from burning old automobile batteries. (*EPA-DOCUMERICA—Marc St. Gil.*)

Because of the hazards to health from discharging such large amounts of lead into the environment, the use of tetraethyl lead in gasoline was being phased out in 1974. Photo 8 shows another way pollution by lead can occur.

2.8 Molar Energy Quantities You saw earlier that the covalent bond energy of the hydrogen molecule is 4.51 electron volts, that is, 4.51 eV of energy are released when two hydrogen atoms combine to form the molecule. Suppose you want to calculate the energy released when two *moles* of H atoms are combined to form one mole of H_2 molecules (2.016 g). Because Avogadro's number of hydrogen molecules is formed, the energy released per mole is the product of Avogadro's number and the bond energy:

$$6.023 \times 10^{23} \times 4.51 \text{ eV} = 2.72 \times 10^{24} \text{ eV}$$

This is about the size of the energies involved in chemical reactions as carried out on a laboratory scale. The electron volt is too small an energy unit for convenience on the gram scale, so chemists use the kilocalorie ☛ instead. One *kilocalorie* (kcal) = 2.6116×10^{22} eV. Thus the covalent bond energy of the hydrogen molecule is $(2.72 \times 10^{24}/2.6116 \times 10^{22}) = 104$ kcal *per mole* of bonds. We therefore have the following two equations:

1 kcal is the amount of heat energy required to increase the temperature of 1 kilogram (1000 g, 2.2 lb) of water from 14.5 to 15.5°C.

Microparticle scale

H	+	H	→	H_2 + 4.51 eV
one atom		one atom		one molecule

Laboratory scale

H	+	H	→	H_2 + 104 kcal
one mole		one mole		one mole of molecules (6.023×10^{23} of them)

A reaction like this that releases energy to the environment is called an *exothermic* ("heat-out") reaction. Familiar examples are the burning of wood, coal, and oil. Here are some other examples:

$$2H_2(g) + O_2(g) \rightarrow 2H_2O(l) + 136.6 \text{ kcal}$$
$$H_2(g) + Cl_2(g) \rightarrow 2HCl(g) + 44.1 \text{ kcal}$$
$$N_2(g) + 3H_2(g) \rightarrow 2NH_3(g) + 22.1 \text{ kcal}$$

In all these examples the quantity of energy released corresponds to the formation of *two* moles of product, according to the equations. Photo 9 compares energy release from burning of coal and fission of uranium, both exothermic reactions—one chemical, one nuclear.

PHOTO 9 Two forms of energy. The block of granite weighs the same as the block of coal. Both can yield the same amount of energy—the coal on burning, the granite on nuclear fission of the 0.00004 percent of uranium it contains. (*Weinberg-Hammond, "Limits to the Use of Energy,"* American Scientist, *58,* August *1970, ORNL Photo No. 89622.)*

A reaction that absorbs energy from the environment as it proceeds is called an *endothermic* ("heat-in") reaction. An example is the formation of nitric oxide (NO):

$$N_2(g) + O_2(g) + 43.2 \text{ kcal} \rightarrow 2NO(g)$$

This reaction occurs to some extent in the explosion of the gasoline and air mixture in the cylinder of the internal combustion engine. The high temperature of the explosion provides the necessary thermal energy. The nitric oxide, a poisonous gas, is emitted in the exhaust and is a major participant in the reactions that produce photochemical smog (Chapters 7 and 15).

EXAMPLE Molecular equations and the molar energy quantities associated with them have many practical applications. One such application is to the economics of replacing natural gas (methane, CH_4) as fuel for home heating and cooking by hydrogen (H_2), as a part of a "hydrogen energy economy." The reserves of natural gas underground in gas fields are large but not unlimited and will run out in the next hundred years or so, if consumption continues to increase. The relevant molar quantities in the replacement of methane by hydrogen as a fuel are their *molar heats of combustion*. The molar heat of combustion of a compound is the quantity of heat energy released in the burning of *one mole* of the compound in oxygen or air. The symbol ΔH_{comb} is used for heat of combustion. Here are the equation and heat of combustion for methane:

$$CH_4(g) + 2O_2(g) \rightarrow CO_2(g) + 2H_2O(l) \qquad \Delta H_{comb} = 213 \text{ kcal/mole}$$

The equation given earlier for the combustion of hydrogen, $2H_2(g) + O_2(g) \rightarrow 2H_2O(l) + 136.6$ kcal is for the burning of *two* moles of H_2, not *one* as required for ΔH_{comb}, so the equation *and* the energy must be divided by 2:

$$\frac{2H_2}{2}(g) + \frac{1}{2}O_2(g) \rightarrow \frac{2}{2}H_2O(l) + \frac{136.6}{2}\text{ kcal}$$

or

$$H_2(g) + \frac{1}{2}O_2(g) \rightarrow H_2O(l) \qquad \Delta H_{comb} = 68.3\text{ kcal/mole}$$

Hydrogen comes out a bad second on a *molar* basis with one mole of CH_4 yielding more than three times as much thermal energy as one mole of H_2 ($213/68.3 = 3.1$). On a *weight* basis, however, H_2 is the winner. One mole of H_2 is $2 \times 1.008\text{ g} = 2.016\text{ g}$ and one mole of CH_4 is $12.01 + 4(1.008\text{ g}) = 16.04\text{ g}$, so the weight of one mole of CH_4 corresponds to $16.04/2.016 = 7.96$ moles of H_2. At 68.3 kcal/mole, 7.96 moles of H_2 will yield $7.96 \times 68.3 = 543$ kcal, more than $2\frac{1}{2}$ times the thermal energy obtained from an equal weight of methane. At present the main factor in the cost of natural gas is the cost of the pipe system that delivers it. Hydrogen could be sent through the same distribution system and most likely will be when natural gas supplies are exhausted. At present hydrogen is produced by the electrolysis of water [$2H_2O(l) \rightarrow 2H_2(g) + O_2(g)$] or from methane along with carbon monoxide, in the water gas reaction that uses carbon in the form of coal as starting material:

$$C(s) + H_2O(g) \rightarrow CO(g) + H_2(g)$$

At the present price level for natural gas, H_2 from either of these sources is far too expensive as a replacement. In fact, the *cheapest* H_2 is presently *made* from natural gas by a reaction with steam at high temperatures:

$$CH_4(g) + H_2O(g) \rightarrow CO(g) + 3H_2(g)$$

Clearly, as long as supplies of natural gas remain, hydrogen cannot have any economic advantage. But hydrogen will have its day.

2.9 Summary Guide A sample of matter is *homogeneous* if macroscopic subsamples of it have the same intensive properties. A homogeneous sample of matter is chemically *pure* when repeated phase-transfer operations such as distillation, sublimation, and crystallization produce no further change in its intensive properties. Pure samples of matter are *elements* if they are made up of atoms having the same atomic number (Z). *Compounds* are made up of atoms of different atomic numbers in fixed, numerical proportions. *Solutions* are homogeneous but differ from compounds in having proportions of atoms or molecules that may be changed.

Most elements are a mixture of atoms called the *isotopes* of the element. Isotopes differ in mass number (M) because they have different numbers of neutrons in their nuclei. The electron clouds of isotopic atoms are identical, so the isotopes of an element all have the same chemical reactions. The symbols for elements without atomic (Z) and mass (M) numbers attached stand for the geonormal isotopic mixtures of atoms of the elements, and are ordinarily used in writing chemical formulas.

The *atomic weight* of an element is the average mass of its atom in atomic mass units, taking its isotopic composition into account. The unit of the atomic weight scale, the *atomic mass unit* (amu), is $\frac{1}{12}$ the mass of one ${}^{12}_{6}C$ atom. Atomic weights are determined from mass spectral data or by the formula-composition method. The *mass number* (M) of an isotope (nuclide) is the whole number closest to its mass in amu.

The sum of the masses of the protons, neutrons, and electrons in an atom is always greater than the actual mass of the atom. The difference is called the *mass defect* and is the mass of the nuclear binding energy released in the formation of the atom from its separated electrons, protons, and neutrons. Nuclear binding energies are millions of times greater than covalent and ionic bond energies.

The molecular formulas for elementary gases and gaseous compounds formed from them are determined from reacting volumes of gases by using Avogadro's hypothesis that equal volumes of different gases measured at the same temperature and pressure contain equal numbers of molecules.

One *mole* of an element or compound is a macroscopic quantity numerically equal to the mass of its formula in amu, but taken in grams. One mole contains Avogadro's number (6.023×10^{23}) of the specified formula units. For a molecular substance one mole is its molecular weight. The mole idea is used in calculating the weights of reactants needed for complete chemical reaction, to obtain the percentage composition of compounds of known formula, and in determining empirical formulas of compounds, using their composition and the atomic weights of the elements combined in them.

Empirical formulas give only the simplest whole-number ratios of atoms making up a compound. *Molecular* formulas are some whole-number multiple of the empirical formula and give the actual numbers of atoms of each kind in one molecule of the compound. To get molecular formulas we need to know the value of the molecular weight of the compound, obtained by mass spectroscopy or other means.

Energies for exothermic and endothermic reactions between individual atoms, molecules, and ions are expressed in units of electron volts. Where moles of particles are involved, energy quantities are expressed in kilocalories.

Exercises

21. In radiocarbon dating (Chapter 16) of samples of wood by analysis for $^{14}_{6}C$, the wood is burned to carbon dioxide which is absorbed in barium hydroxide solution to produce solid barium carbonate:

$$(C_6H_{10}O_5)_n + 6nO_2 \rightarrow 6nCO_2 + 5nH_2O + \text{heat}$$
$$CO_2 + Ba(OH)_2 \rightarrow BaCO_3 + H_2O$$

Calculate the percentage of C in $BaCO_3$ and in $Ba^{14}CO_3$. (For $^{14}_{6}C$ use 14.00 amu.)

22. Electrolysis of a silver nitrate ($AgNO_3$) solution causes silver metal to separate at the cathode (compare Figure 1.12). The weight of silver metal that separates is a measure of the amount of electricity that passed through the cell, and 1 coulomb of electricity is the amount that plates out 0.0011180 g of Ag. Using the following mass spectrometric data for the isotopic composition of silver, calculate the atomic weight of silver and the number of coulombs required to plate out one mole of Ag:

$^{107}_{47}Ag$ (106.9041 amu, 51.82 atom %)
$^{109}_{47}Ag$ (108.9047 amu, 48.18 atom %)

23. Radioactive decay of one $^{226}_{88}Ra$ atom yields 4.78 MeV (million electron volts) of energy. What quantity of energy expressed in kcal is released in the decay of one mole (226.0 g) of $^{226}_{88}Ra$ atoms?

24. Hydrazine (N_2H_4) and liquid NO_2 have been tested as possible rocket fuels. When brought together they spontaneously burst into flame. Balance the equation for their spontaneous reaction:

$$NO_2 + N_2H_4 \nrightarrow N_2 + H_2O + \text{energy}$$

Reaction of one mole of each yields 136 kcal of energy in the form of heat. Calculate the energy (in kcal) released by reaction of 226 g of an equimolar mixture of NO_2 and N_2H_4, and compare the result with that in Exercise 23.

25. The strontium radioisotope $^{90}_{38}Sr$ (at. wt 89.9 amu) is produced in nuclear bomb explosions. After the explosion, the isotope formed reacts with oxygen in air to form strontium oxide (SrO), which further reacts with water vapor in the atmosphere, giving strontium hydroxide $[Sr(OH)_2]$. Calculate the weight of $^{90}Sr(OH)_2$ in the "fallout" of a bomb explosion that produced 300 g of $^{90}_{38}Sr$.

Key Words

Atomic mass unit (amu) $\frac{1}{12}$ the mass of one $^{12}_{6}C$ atom or 1.660×10^{-24} g.

Atomic weight average weight of atoms of an element in atomic mass units (amu), taking account of the isotopic composition of the element.

Compound substance or compound pure substance in which there are atoms of different atomic numbers.

Elementary substance or element pure substance in which the atoms have the same atomic number.

Empirical formula simplest chemical formula for a compound—gives only the simplest atomic proportions.

Endothermic reaction a reaction that absorbs heat from the surroundings as it progresses.

Exothermic reaction a reaction that releases thermal energy (heat) as it progresses.

Extensive properties properties of matter determined by the size of the sample, such as mass and volume.

Heat of combustion heat released on burning one mole of an element or compound.

Heterogeneous matter mixtures for which small samples differ in properties.

Homogeneous matter elements, compounds, or solutions in which samples have identical properties.

Intensive properties properties of matter that are independent of the size of the sample, such as chemical composition, color, and density.

Isotopes atoms (nuclides) having the same atomic number but different mass numbers.

Kilocalorie heat required to raise the temperature of 1 kg of water from 14.5 to 15.5°C—equal to 2.6116×10^{22} eV.

Mass defect difference between the total mass of the particles making up an atom and the measured mass of the atom.

Mole a macroscopic sample of matter having a definite weight (or mass) determined by the atomic composition of the matter—6.023×10^{23} particles of specified chemical formula, or the mass of that number of particles.

Molecular formula atomic composition of an actual molecule—some whole-number multiple of the empirical formula of a compound.

Nuclear binding energy energy released in the formation of an atom from protons, neutrons, and electrons. It accounts for the mass defect.

Phase-transfer operations procedures, such as distillation, extraction, crystallization, sublimation, and chromatography, used to separate mixtures into their elements or compounds.

Positive ray analysis or mass spectrometry method for determining the isotopic composition of elements and the molecular weights of compounds.

Pure substance homogeneous matter whose intensive properties are not changed by repetition of phase-transfer operations.

Suggested Readings

Cahn, R. W., "The Wind of Hydrogen and of Change Blew Gentle, Clean, and Persistent at Miami," *Nature*, **248**:628–629 (April 19, 1974).

Gillette, Robert, "Synthetic Fuels: Will Government Lend the Oil Industry a Hand?" *Science,* **183**:641–643 (Feb. 15, 1974).

Graham, F., "The Infernal Smog Machine," in Love, G. H., and Love, R. M., eds., *Ecological Crisis: Readings for Survival,* New York, Harcourt Brace Jovanovich Inc., 1970, pp. 205–218.

Hall, Stephen K., "Sulfur Compounds in the Atmosphere," *Chemistry,* **45**:16–18 (March 1972).

Kaplan, Martin, "Environmental Hazards for Human Health," *World Health,* May 1972, pp. 4–11.

Keller, Eugenia, "Man and the Universe," *Chemistry,* **45**:4–22 (July–Aug. 1972).

Kieffer, William F., *The Mole Concept in Chemistry,* New York, Van Nostrand-Reinhold, 1962.

"Lead Pollution: Records in Southern Calif. Coastal Sediments," *Science,* **181**:551–552 (Aug. 10, 1973).

Nash, L. K., *Stoichiometry*, Reading, Mass., Addison-Wesley Publishing Co., 1966.

Schramm, David N., "The Age of the Elements," *Sci. Am.,* **230**:69–77 (Jan. 1974).

Sisler, Harry H., *Electronic Structure, Properties, and Periodic Law,* New York, Reinhold Publishing Corp., 1963.

Weeks, Mary Elvira, and Leicester, Henry M., *Discovery of the Elements,* 7th Ed., Easton, Pa., *Journal of Chem. Ed.,* 1968.

Weinberg, Alvin M., and Hammond, R. Philip, "Limits to the Use of Energy," *Am. Sci.,* **58**:412–418 (July–Aug. 1970).

Winchester, James H., "Here Comes the Hydrogen Era," *Reader's Digest,* Dec. 1973, pp. 144–147.

Winsche, W. E., Hoffman, K. C., and Salzano, F. J., "Hydrogen: Its Future Role in the Nation's Energy Economy," *Science,* **180**:1325–1332 (June 29, 1973).

chapter 3

Developing salamanders *(Ambystroma maculatum)* in their eggs under water. Only the jellylike "shell" in which they are embedded protects them from pollutants in the water. *(Photo 10, Henry M. Wilbur, Department of Zoology, Duke University.)*

Electronic Structures of Atoms and Simple Ions

3.1 Introduction Chapters 1 and 2 were mainly about the nuclear part of atoms, where their mass and positive electric charge are located. This chapter is about the electron cloud, where the negative charge is located, but whose mass is less than 0.06% of the total mass of the atom. In spite of its small contribution to the mass of the atom, it is mainly responsible for the chemical reactions of the atom. As we have seen, X-ray analysis gives accurate determination of internuclear distances in crystals and molecules. In diamond, for example, the distance between nearest-neighbor carbon nuclei is 1.545 Å (1.545×10^{-8} cm).

Can equally accurate measurements be made of the distances between the nucleus of an atom and the electrons moving in its cloud? In other words, what can we learn about the arrangements of the electrons in the cloud of an atom? These are important questions because the ways atoms interact to form molecules and crystals depend on the structure of their electron clouds. For example, helium atoms (He, $Z = 2$), with two electrons in their cloud, show no tendency to form He_2 molecules, yet, as we have already seen, hydrogen atoms (H, $Z = 1$), with one electron in their cloud, spontaneously form covalently bonded H_2 molecules with energy release. So, using formulas that show electrons as dots, we can write:

$$\text{H}\cdot + .\text{H} \rightarrow \text{H:H} + \text{energy}$$

but

$$.\text{He}\cdot + .\text{He}\cdot \not\rightarrow .\text{He:He}.$$

This and the following chapter try to explain differences like these in behavior of seemingly very similar atoms.

It's impossible to locate electrons in the cloud with anywhere near the accuracy with which nuclei can be located in a crystal. In the first place, the electrons in the cloud are in rapid motion and are not in fixed positions like the nuclei in crystals. The positive charge due to the protons in the nucleus keeps the electrons close to the nucleus, but their motion prevents their being captured by the nucleus. The second factor that makes it difficult to determine cloud structure is the very small mass of the electron. To observe the path of a moving particle it is necessary to make two or more successive observations of its position. We determine the path of a baseball or other moving macroscopic objects by observing the light waves (electromagnetic radiation) reflected from it at various points along the path. The "push" given to the baseball as the light wave bounces off of it is so tiny that the light wave has no measurable effect on the path. The situation is quite different for the electron; the light wave knocks the electron completely out of the atom and the additional observations needed to measure its natural path in the atom cannot be made. Let's analyze what happens in the two cases: trying to observe the position of a macroscopic object, and trying to observe the position of an electron particle.

It is a rule of optics that a particle may be "seen" by reflection of electromagnetic radiation (light) only if the wavelength of the radiation is *smaller* than the size of the object. You can imagine electromagnetic waves to be like waves on water—one wavelength is the distance between neighboring crests (or troughs). Visible light has wavelengths in the range 4000–7000 Å ($4.0–7.0 \times 10^{-5}$ cm), so it is suitable for viewing macroscopic particles, even down to tiny bacteria cells ($\sim 10^{-3}$ cm, 10^5 Å), using a microscope. However, to observe a particle as tiny as the electron ($\sim 10^{-13}$ cm) would require radiation of very much shorter wavelength ($\sim 10^{-6}$ Å or 10^{-14} cm) than visible light. Unfortunately for the attempt to observe an electron, *the energy of a light wave increases as its wavelength decreases*, and the energy of radiation having a suitably short wavelength to reflect from an electron knocks the electron completely out of the atom, leaving no hint of its position or path in the electron cloud.

With these limitations, it might seem that very little could be known about the structures of the electron clouds of atoms. Nevertheless, though electron paths and positions cannot be determined, the science of quantum mechanics yields highly accurate data on the *probable* or average behavior of electrons in the cloud. And analysis of the wavelengths of light emitted from atoms that have been brought to a high-energy state by strong heating gives information on the *energy-level structure* of the electron cloud. The results in both cases are exact for the hydrogen atom, and the quantum mechanical analysis of this, the simplest of atoms, provides the framework for understanding the relation between electron-cloud structures and chemical properties of all the atoms.

3.2 Energy Levels for the Hydrogen Atom We all have seen that as the temperature of a sample of matter is increased it begins to give off light. A piece of iron (Fe) heated in a flame begins to glow red, then white as its temperature is raised. The "white" light from an incandescent light bulb comes from its tungsten (W) filament, heated to a high temperature by the passage of electric current. What our eyes see as "white" light is a mixture of all wavelengths of radiation in the region 4000–7000 Å. Passing visible light through the prism of a spectrometer separates it into bands of its component colors (Figure 3.1) that merge into one another to give the *continuous spectrum* of the rainbow. But when you pass the light from hydrogen gas (H_2) heated to a high temperature by an electric spark through the spectrometer, the results are quite different. When light from this source is passed through, it gives—instead of a continuous spectrum—a series of *spectral lines* of separated wavelengths (Figure 3.2), called the *line spectrum* of the hydrogen atom.

According to quantum theory, the high temperature of the electric discharge first breaks the hydrogen molecules (H_2) up into hydrogen atoms ($H^{\cdot}$). The hydrogen atoms then absorb additional energy from the spark and are promoted by it into various energy-rich states. In these "excited" states the average distance between electron and proton is greater than in the lowest or *ground energy state*. As the electron

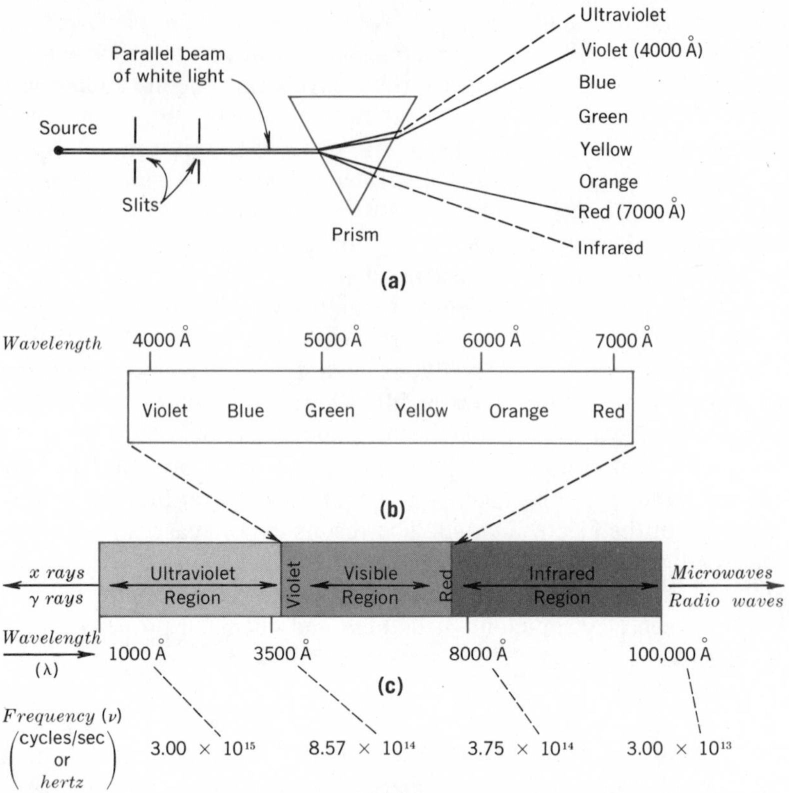

$$\nu = \frac{c}{\lambda} = \frac{2.998 \times 10^{10}\ \text{cm sec}^{-1}}{\lambda\ \text{cm}}$$

FIG. 3.1 (a) Production of the continuous spectrum of visible light from an incandescent source; (b) visible portion of (c) the electromagnetic spectrum.

returns to the ground energy state, the energy released appears as light having sharply separated wavelengths. These make up the *line spectrum* of the hydrogen atom.

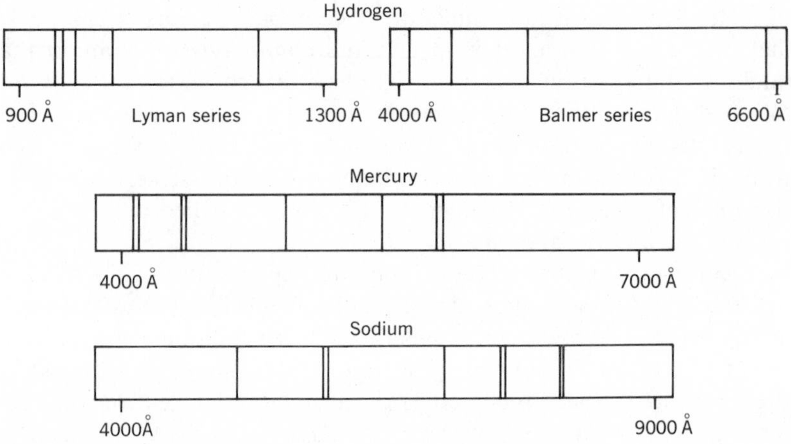

FIG. 3.2 Portions of the line spectra of some elements (not all lines are shown).

According to the quantum theory of radiation, light is not only waveform radiation but also may be regarded as a stream of energy capsules or packets called *photons*. Each atom emits one photon as it goes from a higher to a lower energy state. The streams of photons make up the wavelengths in the line spectrum. The energy, E, in kcal per mole of photons in light of wavelength λ in angstrom units is given by $E = (2.8597 \times 10^5)/\lambda$. Notice that the *shorter* the wavelength of light the *greater* the energy of its photons. When we look at the wavelengths in the spectrum of atomic hydrogen, we see the *energy differences* between the excited states of the hydrogen atom and the ground state or other excited states. The most remarkable result of this analysis is that only *certain definite values of energy are permitted for the hydrogen atom*. These values are termed the *energy levels or energy states for the atom* and are shown in Figure 3.3, along

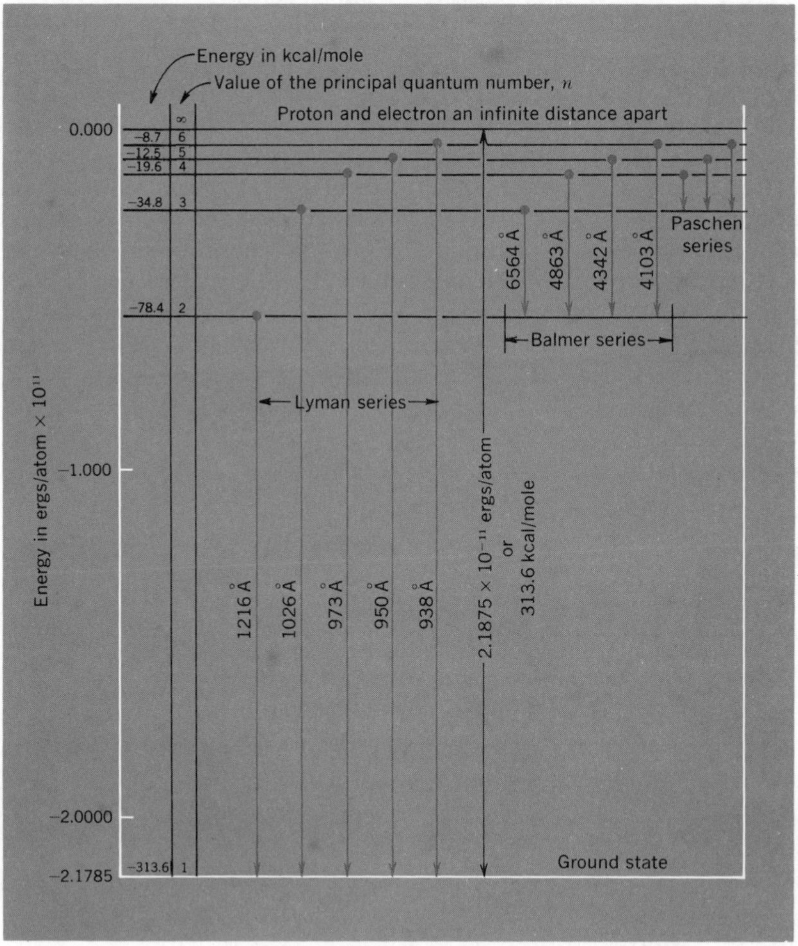

FIG. 3.3 The first six energy levels and spectral lines in portions of three spectral series for the hydrogen atom.

with the transitions between energy levels that give rise to some of the wavelengths in the line spectrum of hydrogen.

Exercises

1. Calculate the energy in kcal/mole of photons in light of wavelength 1216 Å and compare it with the energy difference between the $n = 1$ and $n = 2$ levels for the hydrogen atom in Figure 3.3.

2. There is an energy-level difference of 113 kcal/mole for an electron in the mercury atom (Hg). To what wavelength of light does it correspond? In what region of the electromagnetic spectrum would the corresponding spectral line appear? Light of this wavelength is sometimes called "black light" or "invisible light." Why?

The erg is a quantity of energy equivalent to 6.242×10^{11} electron volts or 2.390×10^{-11} kcal.

The energy levels for the hydrogen atom are described by the value assigned to the *principal quantum number* (n) *of the level.* Values for n are 1, 2, 3, . . . , but not zero, and the energy E_n of the nth level is given simply by $E_n = E_1/n^2$. Here E_1 is the *ground state* or lowest energy level possible for the atom. It has the value -2.1785×10^{-11} ergs per atom, or -313.61 kcal per mole of atoms (Figure 3.3). The higher levels have energies $\frac{1}{4}, \frac{1}{9}, \frac{1}{16}, \ldots$ of this value. The idea of an atom having a negative energy may seem confusing but comes from taking the zero of energy to be for infinitely separated electron and proton (Figure 3.4). (You saw a similar situation with oppositely charged spheres in the Introduction.) As the atom is formed, the attractive force between the oppositely charged proton and electron pulls them together and energy is *released* as radiation, just as energy is released when the atom goes from a level of higher n value to one of lower n value, where the electron is closer to the nucleus. Let's write it as an equation, with $H^{\cdot}$ representing a ground-state hydrogen atom:

$$p^+ + e^- \rightarrow H^{\cdot} + 313.61 \text{ kcal/mole (released)}$$

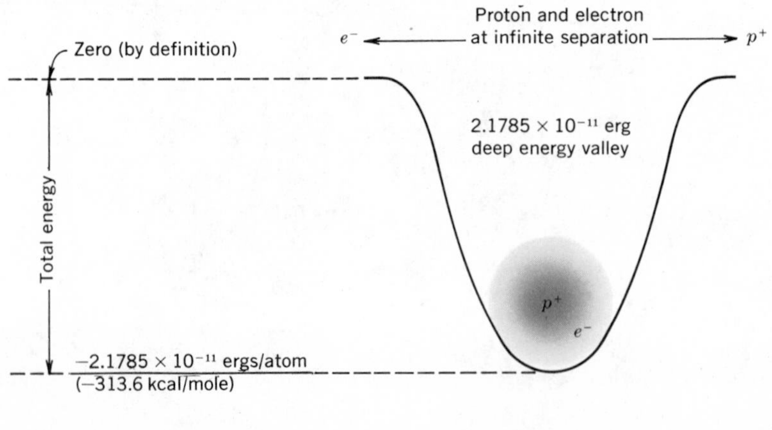

FIG. 3.4 Formation of a hydrogen atom in its lowest, most stable energy state.

Thus a hydrogen atom, left to itself, is a stable or *bound* system, and 313.61 kcal of energy must be *supplied* to take one mole of hydrogen atoms apart into separated electrons and protons. This process of separation is called *ionization* of the hydrogen atom, since it produces a hydrogen ion (H^+ or p^+), and the 313.61 kcal/mole needed for separation is called the *ionization energy* or *ionization potential* of the hydrogen atom. Ionization energies are usually expressed in electron volts; for the hydrogen atom the value is 13.60 eV. The meaning of this value is that a free electron that has been accelerated through a voltage difference of 13.60 volts (compare Figure 1.6) is moving just rapidly enough to knock the bound electron completely out of one ground-state hydrogen atom on collision.

The hydrogen atom is not alone in undergoing ionization when the proper amount of energy is supplied. Ionization energies for nearly all of the elements have been measured (Appendix 4), and we will see that they are powerful clues to understanding the relationship between the chemistry of an element and the electronic, energy-level structure of its atoms.

Exercises

3. In what wavelength of light do the photons have just enough energy to ionize the hydrogen atom from its ground state?

4. What wavelength of light would just ionize a hydrogen atom if the electron were in the level $n = 4$? In the level $n = 6$? Account for the difference.

3.3 Electron Orbitals for the Hydrogen Atom The first model for the hydrogen atom (H), proposed by Bohr, was like a miniature solar system with the electron circulating in orbit around the proton. The electrostatic attraction between the opposite charges of proton and electron played the part of gravitation, which keeps the planets in the solar system from flying off into space. According to Bohr's model only certain values were permitted for the radii of the orbits—this explained the energy levels discussed in Section 3.2. According to his theory, a photon is emitted each time an electron falls from an outer orbit into an inner one. In the ground-state atom, the electron was in the orbit of smallest possible radius, 0.53 Å. Although a giant step forward, the Bohr model turned out to be incorrect in some of its predictions and failed to predict the line spectra of atoms with more than one electron, so use of the Bohr model was discontinued.

Niels Bohr (1885–1962) was a Danish physicist whose theory of the H atom was published when he was only 27 years old. He was 37 when he received the Nobel prize in physics in 1922, the youngest winner of the prize up to that time.

We have already seen that determining the path of an electron is impossible, so the idea of fixed orbits is not realistic. In wave mechanics the statistical idea of the *orbital* replaces the fixed orbits of the Bohr theory. Electron positions about the nuclei are treated from a *probability* point of view, and the electron orbitals give this information.

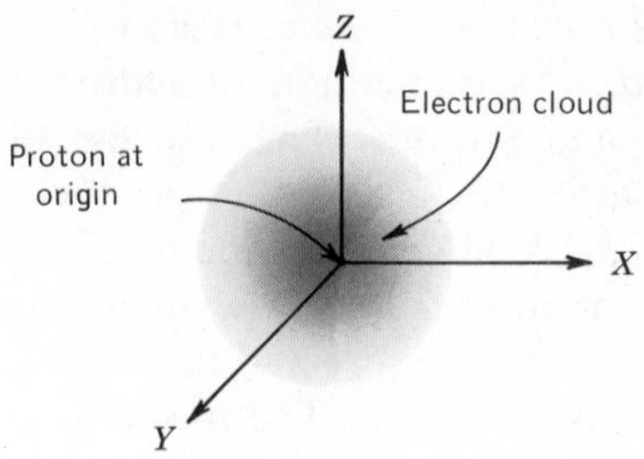

FIG. 3.5 The 1s orbital.

The 1*s* orbital

We will use a picture treatment of orbitals instead of a mathematical one because the equations that describe orbitals are very complicated.

For the hydrogen atom in its ground or lowest energy level ($n = 1$), quantum mechanics gives only one probability distribution or orbital for the electron. It is called the 1*s* orbital. This orbital is spherical about the proton as shown by the "charge cloud" diagram in Figure 3.5.

For the next energy state of hydrogen ($n = 2$), wave mechanics gives four orbitals of equal energy. One of these is like the 1*s* orbital in being spherical but differs from it in having a *spherical nodal* surface where the probability of finding the electron is zero (Figure 3.6). The other three orbitals are identical in having a sort of dumbbell shape, but are directed at right angles to each other. They are called the $2p_x$, $2p_y$, and $2p_z$ orbitals, and like the 2*s* orbital each has a nodal surface, but it is flat and perpendicular to the axis of the dumbbell that passes through the proton (Figure 3.7).

For energy level 3 ($n = 3$), there are one 3*s* orbital, three 3*p* orbitals ($3p_x$, $3p_y$, $3p_z$), and five 3*d* orbitals having more complicated patterns (Figure 3.8). In fact, the total number of orbitals in any one energy level is simply n^2, where n is the quantum number of the level. These energy levels calculated by quantum mechanics for the hydrogen atom are summarized in Figure 3.9(a). For hydrogen, as the diagram shows,

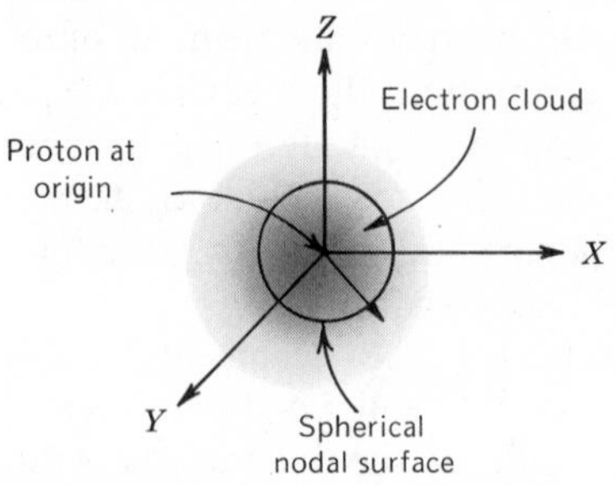

FIG. 3.6 The 2s orbital.

The 2*s* orbital

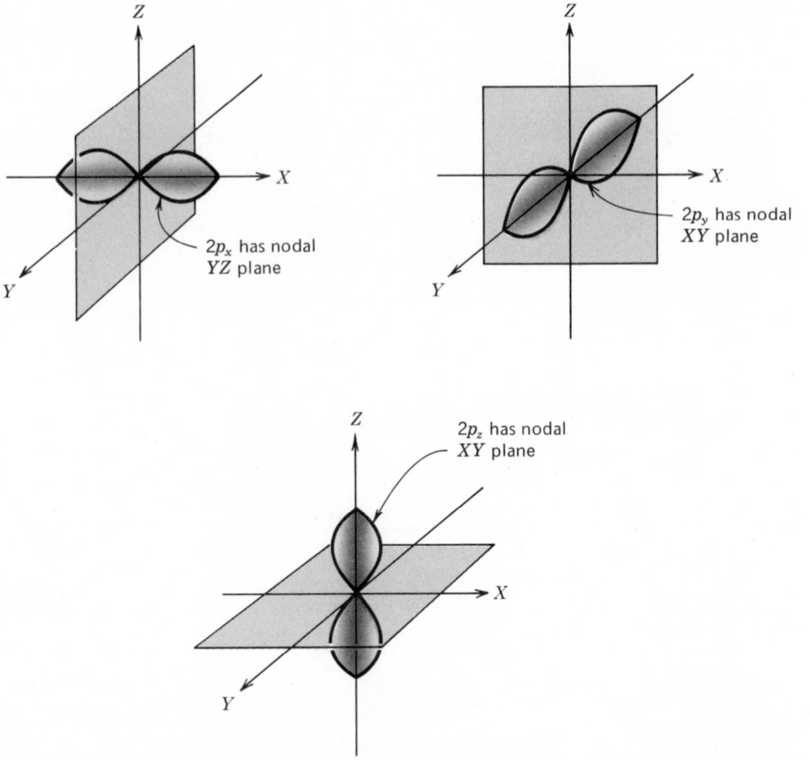

FIG. 3.7 The three 2*p* orbitals.

all of the orbitals of any one level (same value of *n*) have the same energy. For example, if a hydrogen atom is raised by input of energy from its ground state ($n = 1$) to the level for which $n = 3$, there are nine different orbitals for the electron to go into that have the same energy.

3.4 Building Up Atoms The ground electric energy states for the atoms of the remaining elements are built up in the thought experiment of adding to the hydrogen nucleus one proton at a time, then the neutrons necessary to make up the mass number of the atom being built, and finally one electron at a time to the cloud to keep the atom electrically neutral as a whole. The *Pauli exclusion principle* says each orbital of the cloud can contain only two electrons and they must be of opposite spin (usually indicated by arrows ↑ and ↓). ☛ Thus the $^4_2$He (helium) atom is imagined to result from the addition of one proton and two neutrons to the hydrogen atom nucleus—giving $Z = 2$ and $M = 4$ as required—and then addition of a second electron, of spin opposite to that already present, to the 1*s* orbital of the hydrogen atom. We write, in symbols, using dots to keep track of electrons:

Wolfgang Pauli (1900–1958), an Austrian physicist, won the Nobel prize in physics in 1945. Among other discoveries, he saw that no more than two electrons can occupy an orbital, so the exclusion principle bears his name.

For elementary purposes electron spin may be likened to the spinning of the electron on its axis as the earth does. Quantum mechanics requires that spin must be in either of two directions, indicated by the arrows ↑ and ↓.

$$\mathrm{H}^{\cdot} + 1p^+ + 1e^- + 2n \rightarrow {}^4_2\mathrm{He}\!:$$

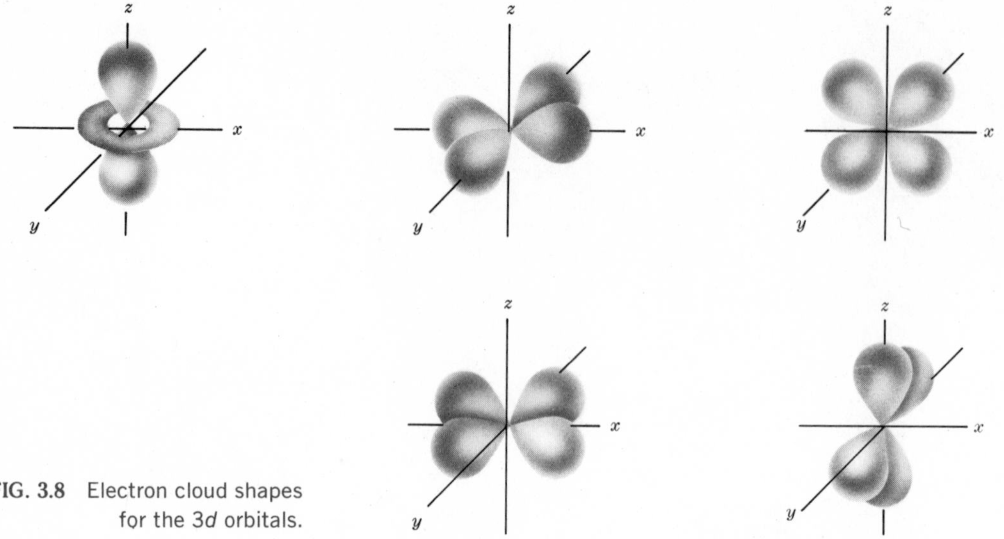

FIG. 3.8 Electron cloud shapes for the 3*d* orbitals.

or, to show the pairing of electron spins in the l*s* orbital required by the Pauli principle:

$$\underset{\text{H}}{\underset{1s}{\underline{\uparrow}}} \xrightarrow{\text{add } e^-} \underset{\text{He}}{\underset{1s}{\underline{\uparrow\downarrow}}}$$

In the latter (He) picture, addition of the protons and neutrons necessary to form the new nucleus is assumed and is not shown specifically. This shouldn't cause any difficulties and even has the advantage of emphasizing that *the electron-cloud structure is the same for all isotopes of a given element*, because isotopes differ only in the number of neutrons in the nucleus. In shorthand notation, using superscripts to indicate orbital occupancy, the electron *configuration* (arrangement) of the ground state of the hydrogen atom is $1s^1$, that of the helium atom is $1s^2$, regardless of whether we are dealing with isotope $^3_2$He or $^4_2$He (Table 2.2). But why does the electron added in going from H to He enter the 1*s* orbital instead of the 2*s* or one of the 2*p* or other orbitals that are unoccupied? The electron enters the 1*s* orbital because in building up ground states for atoms *the most energetically stable or "close-in"* (smaller *n* value) *orbitals are filled first.*

EXAMPLE ⊸ What would happen to a helium atom with electron configuration $1s^12s^1$ or $1s^12p^1$? These configurations put the second electron on the average farther from the nucleus than it would be if it paired up with the one already in the "closer-in" 1*s* orbital. Because of the attractive electrostatic force between the electron and the positively charged

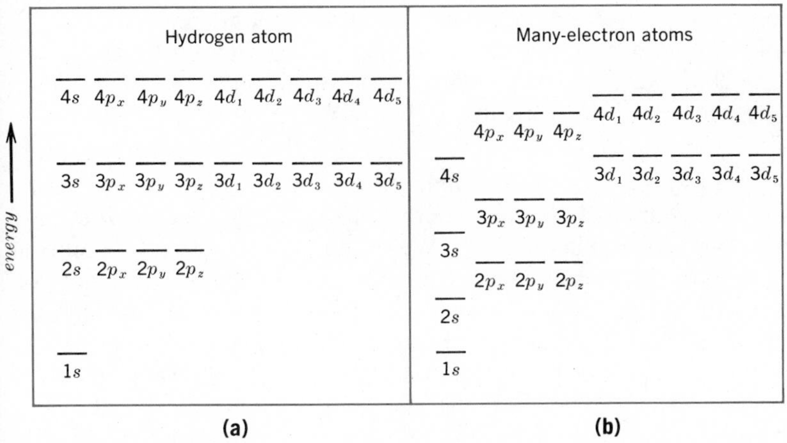

FIG. 3.9 Energy sequence of orbitals for hydrogen (a) and many-electron atoms (b). (Schematic and not to scale; the 4f orbitals have been omitted.)

nucleus, the $1s^1 2s^1$ and $1s^1 2p^1$ arrangements have extra (less negative) electrostatic potential energy. The situation of the electron in the $2s$ or $2p$ orbital is like that of the bicyclist at the top of the hill in Figure 2 of the Introduction. But there are no "brakes" on the electron, so it falls down into the "energy valley" of the $1s$ orbital to give the most stable He atom possible. The electric potential energy released appears as light (electromagnetic radiation) that produces one of the lines in the line spectrum of the helium atom. Atoms that are not in their lowest energy state are sometimes described as "photoexcited." They always go very rapidly ($\sim 10^{-9}$ sec) to their ground state with the release of electromagnetic radiation. ☞○

The next element is the metal lithium (Li, $Z = 3$), obtained (in the thought experiment) from He by adding a proton, the necessary neutrons, and an electron to the cloud. Now the added electron must enter one of the orbitals of level $n = 2$ because the $1s$ orbital is already filled at He (Pauli principle). But here a difficulty arises. According to the orbital diagram for the hydrogen atom [Figure 3.9(a)] the $2s$, $2p_x$, $2p_y$, and $2p_z$ orbitals all have *the same energy*, and we are left with no choice on stability grounds among the following possibilities for the ground-state electron configuration ☞ of Li

It makes no difference at this point whether the additional electron has spin ↑ or spin ↓. The energies would be the same.

$$\frac{\uparrow}{2s}\ \frac{}{2p_x}\ \frac{}{2p_y}\ \frac{}{2p_z} \quad \text{or} \quad \frac{}{2s}\ \frac{\uparrow}{2p_x}\ \frac{}{2p_y}\ \frac{}{2p_z} \quad \text{or} \quad \frac{}{2s}\ \frac{}{2p_x}\ \frac{\uparrow}{2p_y}\ \frac{}{2p_z}$$

$$\frac{\uparrow\downarrow}{1s} \qquad\qquad\qquad\quad \frac{\uparrow\downarrow}{1s} \qquad\qquad\qquad\quad \frac{\uparrow\downarrow}{1s}$$

or $\quad \frac{}{2s}\ \frac{}{2p_x}\ \frac{}{2p_y}\ \frac{\uparrow}{2p_z}$

$\quad\ \frac{\uparrow\downarrow}{1s}$

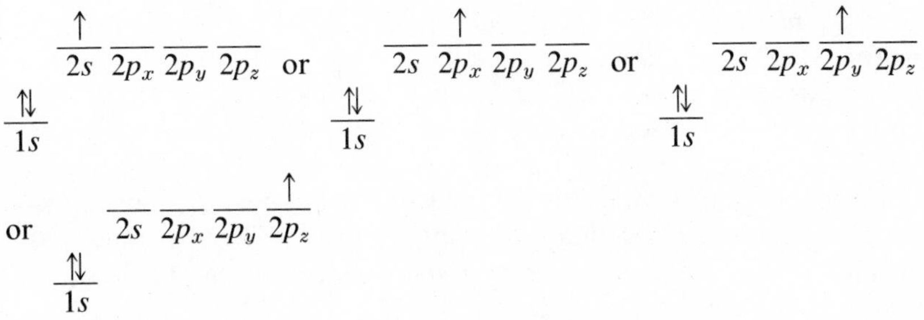

The ground-state electron configuration of the lithium atom is known from the energy levels revealed by its line spectrum to be $1s^2 2s^1$ and not $1s^2 2p_x^1$, $1s^2 2p_y^1$, or $1s^2 2p_z^1$. This fact is explained by the penetration to the nucleus of electrons in *s* orbitals, or the "penetration effect."

The penetration effect The orbitals in any one quantum level for the hydrogen atom have identical energies [Figure 3.9(a)] because there is only one electron in the cloud. When additional electrons are present they tend to shield each other from the nucleus—that is, each electron "feels" the attracting positive charge of the nucleus only through the cloud of all the other electrons present. As pointed out earlier, the 2*p* orbitals have a nodal plane of zero probability through the nucleus. On the other hand, the value of the probability for the 2*s* orbital is at a maximum at the nucleus (Figure 3.6). For this reason the third electron in lithium has a better chance of being close to the nucleus and at lower energy if it is in the 2*s* orbital than if it is in one of the 2*p* orbitals. An electron in a 2*p* orbital not only suffers screening from the nucleus by the inner $1s^2$ pair, but is also kept away from the nucleus by the nodal plane of the orbital. Because the most stable arrangement has all of the electrons as close to the attracting nucleus as possible, the third electron in lithium is in the 2*s* orbital, not in one of the 2*p* orbitals. The situation is summarized by saying that "electrons in *s* orbitals penetrate to the nucleus."

Penetration splits the simple orbital energy-level pattern for hydrogen [Figure 3.9(a)] into the more complex one in Figure 3.9(b) for the heavier atoms. The effect becomes so pronounced, as more electrons are added, that the *s* orbital of the fourth level falls slightly below the *d* orbitals of the third energy level as shown. It follows that in building up the atoms, the 4*s* and the five 3*d* orbitals will be filled before any electrons enter the 4*p* orbitals. This has important chemical consequences, as we shall see shortly.

But let's return to building in level 2, using the new orbital energy sequence of Figure 3.9(b). (The sequence for the hydrogen atom has served its purpose by providing the broad outline for the arrangement of electrons in orbitals.) After lithium, with configuration $1s^2 2s^1$, comes the metal beryllium (Be, $Z = 4$), with configuration $1s^2 2s^2$, the added electron pairing spin with the one already present in the 2*s* orbital. Boron (B, $Z = 5$), a nonmetal, is next and for it we can write the noncommittal configuration $1s^2 2s^2 2p^1$ without worrying about which of the 2*p* orbitals the electron enters because they are all at the same energy level.

Which of the 2*p* orbitals are used makes no difference, since they all are at the same energy level. For the same reason both unpaired spins in arrangement (c) can point up or down.

The next atom is carbon (C, $Z = 6$), and here new problems arise in assigning electrons to orbitals. Starting with the $1s^2 2s^2 2p^1$ arrangement in boron, one additional electron is to be placed in one of the 2*p* orbitals. There are three choices: ☛

$$
\begin{array}{cccc}
\underset{2s}{\uparrow\downarrow} & \underset{2p_x}{\uparrow\downarrow}\ \underset{2p_y}{__}\ \underset{2p_z}{__} \\[6pt]
\underset{1s}{\uparrow\downarrow} \\[6pt]
\text{(a)}
\end{array}
\qquad
\begin{array}{cccc}
\underset{2s}{\uparrow\downarrow} & \underset{2p_x}{\uparrow}\ \underset{2p_y}{\downarrow}\ \underset{2p_z}{__} \\[6pt]
\underset{1s}{\uparrow\downarrow} \\[6pt]
\text{(b)}
\end{array}
\qquad
\begin{array}{cccc}
\underset{2s}{\uparrow\downarrow} & \underset{2p_x}{\uparrow}\ \underset{2p_y}{\uparrow}\ \underset{2p_z}{__} \\[6pt]
\underset{1s}{\uparrow\downarrow} \\[6pt]
\text{(c)}
\end{array}
$$

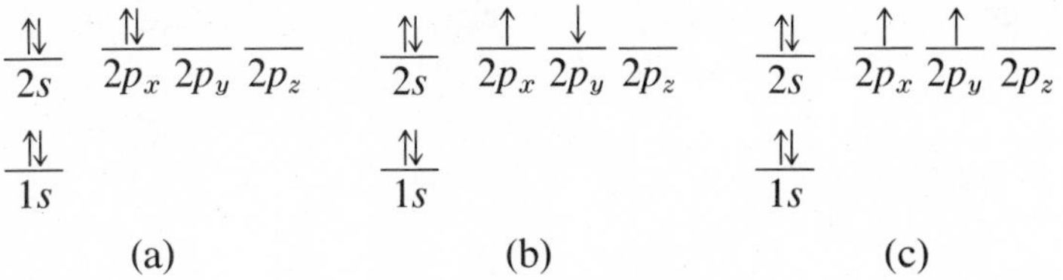

In (a) the additional electron has paired spin with the one already present and enters the same ($2p_x$) orbital; in (b) spins are paired but the extra electron has entered an unoccupied $2p$ orbital; and in (c) each electron is in its own $2p$ orbital but the spins are not paired.

The atomic spectrum of carbon shows that (c) is the ground-state arrangement of electrons in the orbitals of the atom. Alternative (a) is higher in energy (less stable) because the two mutually repelling electrons are forced into the same region of space (same orbital), and (c) is more stable than (b) for the quantum mechanical reason that two electrons in separate orbitals of the same energy move in a way that keeps them as far apart as possible *if their spins are not paired*. The larger distance between them lowers the repulsion between their like charges, giving a more stable (lower energy) atom. In (b), although the two electrons are in separate orbitals, their spins are paired, and for quantum mechanical reasons they move in a way that keeps them close together and hence at higher energy because of repulsion between their like charges.

Carbon serves as an example of a general rule called *Hund's rule*. The rule is that *electrons avoid each other by going into separate orbitals when empty orbitals of equal energy are available, and they avoid pairing spins as much as possible*. The rule gives the following configuration for the nitrogen atom (N, $Z = 7$):

$$
\begin{array}{cccc}
\underset{2s}{\uparrow\downarrow} & \underset{2p_x}{\uparrow\downarrow}\ \underset{2p_y}{\uparrow}\ \underset{2p_z}{\uparrow} \\[6pt]
\underset{1s}{\uparrow\downarrow}
\end{array}
$$

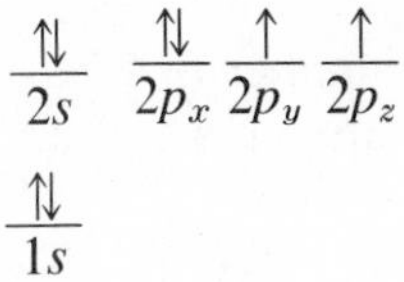

Hund's rule may be stated more accurately by using the idea of a *subshell* of orbitals. *A subshell contains all orbitals of the same value of n and the same letter designation.* The $2p$ subshell contains the $2p_x, 2p_y,$ and $2p_z$ orbitals, the $3d$ subshell contains the five $3d$ orbitals, the $4f$ subshell contains seven $4f$ orbitals, and so on. In terms of subshells, the Hund rule for ground states is: *two electrons do not enter the same orbital of one subshell until all of the orbitals of the subshell contain one electron*, and *single electrons in separate orbitals of a given subshell have unpaired spins*.

With oxygen (O, $Z = 8$) the added electron *must* pair with one already in the $2p$ orbitals to give this arrangement:

↑↓	↑↓	↑	↑
$2s$	$2p_x$	$2p_y$	$2p_z$
↑↓			
$1s$			

Fluorine (F, $Z = 9$) and neon (Ne, $Z = 10$) complete the filling of the orbitals of the second energy level [Figure 3.9(b)]:

↑↓	↑↓	↑↓	↑	↑↓	↑↓	↑↓	↑↓
$2s$	$2p_x$	$2p_y$	$2p_z$	$2s$	$2p_x$	$2p_y$	$2p_z$
↑↓				↑↓			
$1s$				$1s$			
	F				Ne		

Exercises

5. Give the electron configurations *by orbitals*, showing electron spin with arrows, for Al, Si, and P.

6. Spectra show that the electron most easily removed from the oxygen atom to give the O^{+1} ion is one from the pair in the $2p_x$ orbital. Explain why this electron has the lowest ionization energy.

7. The first ionization energies for atoms generally increase from left to right across a row of the Periodic Table (Table 3.1, page 102), but the value for oxygen (13.61 eV) is *smaller* than the value for nitrogen (14.54 eV). Why?

8. What is the numerical relationship between the number of *types* of orbitals (s, p, d, and so on) and the principal quantum number n?

3.5 Electronic Structures and Chemical Reactivity Two of the atoms discussed in Section 3.4 stand out from the other eight; these are helium ($1s^2$) and neon ($1s^2 2s^2 2p^6$). They are unusual because they do not form diatomic molecules with themselves or compounds with any other element. The other eight (H, Li, Be, B, C, N, O, F) do one or both. The reason helium and neon do not form any kind of bonds results from what are called *filled-shell* electron configurations. *A shell is made up of all of the orbitals having the same value for the principal quantum number n.* Shells are labeled K, L, M, N, O corresponding to values 1, 2, 3, 4, 5, . . . , for n. Thus the K shell contains only the $1s$ orbital. The K shell is filled for helium ($1s^2$). The L shell ($n = 2$) is made up of four orbitals: one $2s$ and three $2p$; it is filled for neon ($1s^2 2s^2 2p^6$). For $n = 3$ (M shell) things become more complicated. The M shell has nine orbitals: one $3s$, three $3p$, and five $3d$, and space for $9 \times 2 = 18$ electrons. The element with filled K, L, and M shells would have 28 electrons – 2 from the K shell, 8 from the L shell, and 18 from the M shell. The element with $Z = 28$ is nickel (Ni), and its electron configuration is *not* the filled-shell one, $1s^2 2s^2$-$2p^2 3s^2 3p^6 3d^{10}$, but is instead $1s^2 2s^2 2p^6 3s^2 3p^6 4s^2 3d^8$. This configuration is of lower energy because the penetration effect discussed earlier drops the $4s$ orbital of the N shell ($n = 4$) slightly below the $3d$ orbitals in energy [Figure 3.9(b)], so the $4s$ fills first. There is in fact no *atom* that has only filled K, L, and M shells in the ground state. A zinc (Zn, $Z = 30$) *ion*, Zn^{2+}, obtained from the zinc atom ($1s^2 2s^2 2p^6 3s^2 3p^6 4s^2 3d^{10}$) by removal of the two $4s$ electrons, has a filled M shell.

Exercises

9. What is the relationship between the value of the principal quantum number n for an atomic shell and the total number of orbitals of all letter types in the shell?

10. What is the relationship between the value of n for a shell and the total number of electrons required to fill the shell?

A low level of chemical reactivity goes along with the outer configuration $ns^2 np^6$ of eight electrons, called an *octet*. An octet is first seen in neon ($n = 2$) but it also occurs in argon ($n = 3$), krypton ($n = 4$), xenon ($n = 5$), and radon ($n = 6$) (Table 3.1, page 102). These elements are all monatomic (one-atom-molecule) gases and, although krypton, xenon, and radon form a few compounds, the elements are generally quite unreactive chemically. They are called the noble gases for this reason. ☛ Recognition of the particular stability of the $ns^2 np^6$ octet by Lewis and Langmuir led to the octet theory of valence (combining power) for the elements.

Similarly, gold, silver, and platinum, which are also quite unreactive, are called the noble metals.

Irving Langmuir (1881–1957) was the second American to win the Nobel prize in chemistry, in 1932. He received the award for his studies in the electron theory of matter and surface chemistry.

Gilbert N. Lewis (1875–1946) was an American chemist at the University of California at Berkeley.

Octet theory for ionic compounds Octet theory separates the electrons of an atom into *core* or inner electrons and *valence* or outer electrons. Core electrons are in orbitals too close to the nucleus and therefore too strongly attracted to it to participate in chemical reaction by sharing with or transfer to other atoms. *Chemical reactions involve rearrangements in the distribution of the outer or* ***valence*** *electrons*. Generally the valence electrons are in the orbitals of highest principal quantum number n. The core electrons usually represent the configuration of a noble gas, but may include completely filled d orbitals. In writing electron configurations for atoms so as to show their core and valence electrons, the symbols for the noble gases (He, Ne, Ar, Kr, Xe, Ra) in brackets [] stand for their $1s^2$ or $ns^2 np^6$ electron configurations. Thus the fluorine atom (F, $Z = 9$), with electron configuration

$1s^2 2s^2 2p^5$, is written $[\text{He}]2s^2 2p^5$ to emphasize that it has seven *valence* electrons and the helium $1s^2$ *core*. Similarly, sodium (Na, $Z = 11$), with configuration $1s^2 2s^2 2p^6 3s^1$, becomes $[\text{Ne}]3s^1$, because the configuration of Ne is $1s^2 2s^2 2p^6$. In using symbols such as [He] and [Ne] to represent core electrons remember that *they do not stand for their respective atoms but only for the electron clouds of their atoms*. The symbol $[\text{Ne}]3s^1$ for the sodium atom *does not mean a neon atom with an added 3s electron*—that would be the unknown Ne^- ion. The symbol $[\text{Ne}]3s^1$ means that *the sodium atom has the electron configuration* of the neon atom ($1s^2 2s^2 2p^6$) as core plus one electron in the $3s$ orbital.

Octet theory states that atoms react to form compounds by producing octets of electrons in the ns^2np^6 arrangement. Lithium fluoride (Li^+F^-) is an ionic crystal, and the theory pictures the formation of one Li^+F^- unit by the transfer of one electron from a lithium atom to a fluorine atom. The lithium atom has ground-state electron configuration $1s^2 2s^1$ or $[\text{He}]2s^1$; that of the fluorine atom is $1s^2 2s^2 2p^5$ or $[\text{He}]2s^2 2p^5$. Transferring the single $2s$ electron of lithium to the half-filled $2p$ orbital of fluorine "completes the octet" of the fluorine atom, giving it the stable $2s^2 2p^6$ outer configuration of neon.

$$\underset{\text{Li}}{[\text{He}]2s^{①}} \quad \underset{\text{F}}{[\text{He}]2s^2p^5} \quad \rightarrow \quad \underset{\text{Li}^+}{[\text{He}]^+} \quad \underset{\text{F}^-}{[\text{He}]2s^2p^{6-}}$$

The lithium *ion* carries a net positive charge of one electronic unit because it has three protons ($Z = 3$) in its nucleus but, like helium, Li^+ has only two electrons in its cloud. The fluoride ion has a net negative unit of charge because there are ten electrons in its cloud but only nine protons in its nucleus ($Z = 9$ for F).

It is important to see that the compound between lithium and fluoride has a 1:1 ratio of ions and formula Li^+F^- *because the lithium atom has one valence electron to donate and the fluorine atom needs one electron to complete its octet*. Octet theory, coupled with electron configurations based on the energy sequence of orbitals, thus predicts the correct formula for lithium fluoride. And octet theory makes correct predictions in thousands of other cases.

EXAMPLE It helps when using octet theory to predict chemical formulas if you show valence electrons as crosses and dots. What does octet theory predict for the formula of the compound between beryllium (Be, $Z = 4$) and fluorine (F, $Z = 9$)? Putting in the electrons gives:

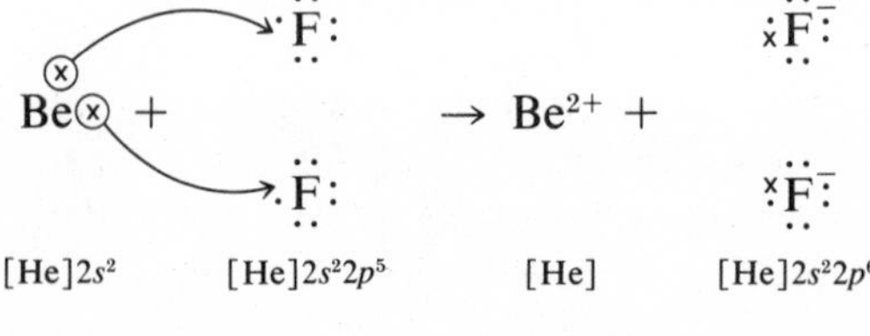

[He]$2s^2$ [He]$2s^2 2p^5$ [He] [He]$2s^2 2p^6$

The formula predicted is $Be^{2+}(:\ddot{\underset{..}{F}}:^{-})_2$ because one Be atom with two valence electrons can complete the octets of two fluorine atoms, which have only seven valence electrons. The predicted formula is correct, but is ordinarily written BeF_2 unless you want to show that the bonding is of the ionic type.

What would the theory predict for the formula of the compound between aluminum (Al, $Z = 13$) and oxygen (O, $Z = 8$)? The electron configuration of the aluminum atom is $1s^22s^22p^63s^23p^1$ or $[Ne]3s^23p^1$ with three valence electrons. Oxygen with configuration $[He]2s^22p^4$ has six valence electrons and needs two to complete its octet. Two aluminum atoms can provide $2 \times 3 = 6$ electrons, enough to fill the octets of three oxygen atoms, so the formula is $(Al^{3+})_2(O^{2-})_3$, usually written Al_2O_3:

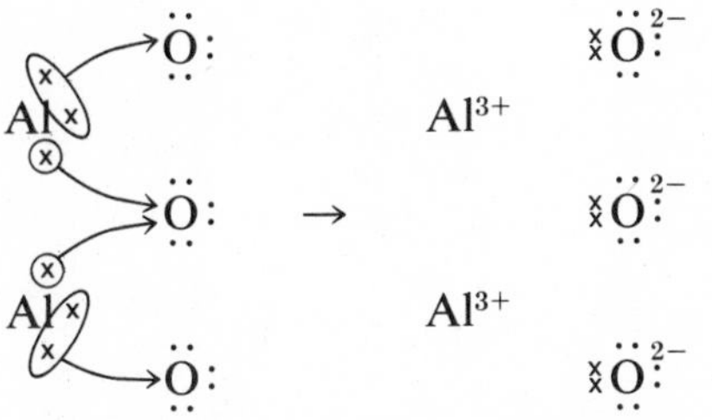

In these examples, as with all simple ionic compounds, the atoms that are closest to already having octets ($\cdot\ddot{\underset{..}{F}}:$ and $\ddot{\underset{..}{O}}:$) get the electrons to complete their octets. The positive ions (Be^{2+} and Al^{3+}), left after the electron transfers, have the configuration of a noble gas, $1s^2$ or [He] for Be^{2+} and $1s^22s^22p^6$ or [Ne] for Al^{3+}.

The formulas of all the simple ionic compounds between the elements of Groups 1a and 2a in the Periodic Table (inside front cover), on the one hand, and the elements in Groups 6a and 7a on the other can be predicted correctly by octet theory.

Octet theory predicts the formulas of many covalent compounds as well as those of ionic type. *In covalence, octets are produced by a sharing of electrons rather than complete transfer.* The water molecule is a good example of electron sharing to make an octet:

$$2H^{\times} + \ddot{\underset{..}{O}}: \rightarrow H\overset{\times}{\ddot{O}}: \quad \text{or} \quad H{-}\ddot{O}:$$
$$\qquad\qquad\qquad\quad\ \underset{H}{\overset{\times\cdot}{}} \qquad\qquad\quad \underset{H}{|}$$

By sharing, the oxygen atom gets eight valence electrons around it and each hydrogen atom gets two. Covalence is discussed in detail in Chapter 4.

TABLE 3.1 Electron configurations and first ionization potentials[a] of gaseous atoms of elements in the Octet Periodic Table[b]

Groups ⟶	*1a*	*2a*		*3a*	*4a*	*5a*	*6a*	*7a*	*0*
First period[c]	$_{1}$H $1s^1$ 13.60								$_{2}$He $1s^2$ 24.58 [$1s^2$ = He]
Second period	$_{3}$Li $2s^1$ 5.39 [He]	$_{4}$Be $2s^2$ 9.32 [He]		$_{5}$B $2s^2 2p^1$ 8.30 [He]	$_{6}$C $2s^2 2p^2$ 11.26 [He]	$_{7}$N $2s^2 2p^3$ 14.54 [He]	$_{8}$O $2s^2 2p^4$ 13.61 [He]	$_{9}$F $2s^2 2p^5$ 17.42 [He]	$_{10}$Ne $2s^2 2p^6$ 21.56 [He $2s^2 2p^6$ = Ne]
Third period	$_{11}$Na $3s^1$ 5.14 [Ne]	$_{12}$Mg $3s^2$ 7.64 [Ne]		$_{13}$Al $3s^2 3p^1$ 5.98 [Ne]	$_{14}$Si $3s^2 3p^2$ 8.15 [Ne]	$_{15}$P $3s^2 3p^3$ 11.00 [Ne]	$_{16}$S $3s^2 3p^4$ 10.36 [Ne]	$_{17}$Cl $3s^2 3p^5$ 13.01 [Ne]	$_{18}$Ar $3s^2 3p^6$ 15.76 [Ne $3s^2 3p^6$ = Ar]
Fourth period	$_{19}$K $4s^1$ 4.34 [Ar]	$_{20}$Ca $4s^2$ 6.11 [Ar]	↑	$_{31}$Ga $4s^2 4p^1$ 6.00 [Ar $3d^{10}$]	$_{32}$Ge $4s^2 4p^2$ 7.88 [Ar $3d^{10}$]	$_{33}$As $4s^2 4p^3$ 9.81 [Ar $3d^{10}$]	$_{34}$Se $4s^2 4p^4$ 9.75 [Ar $3d^{10}$]	$_{35}$Br $4s^2 4p^5$ 11.84 [Ar $3d^{10}$]	$_{36}$Kr $4s^2 4p^6$ 14.00 [Ar $3d^{10} 4s^2 4p^6$ = Kr]
Fifth period	$_{37}$Rb $5s^1$ 4.18 [Kr]	$_{38}$Sr $5s^2$ 5.69 [Kr]	*transition metals*	$_{49}$In $5s^2 5p^1$ 5.79 [Kr $4d^{10}$]	$_{50}$Sn $5s^2 5p^2$ 7.34 [Kr $4d^{10}$]	$_{51}$Sb $5s^2 5p^3$ 8.64 [Kr $4d^{10}$]	$_{52}$Te $5s^2 5p^4$ 9.01 [Kr $4d^{10}$]	$_{53}$I $5s^2 5p^5$ 10.45 [Kr $4d^{10}$]	$_{54}$Xe $5s^2 5p^6$ 12.13 [Kr $4d^{10} 5s^2 5p^6$ = Xe]
Sixth period	$_{55}$Cs $6s^1$ 3.89 [Xe]	$_{56}$Ba $6s^2$ 5.21 [Xe]		$_{81}$Tl $6s^2 6p^1$ 6.11 [Xe $4f^{14} 5d^{10}$]	$_{82}$Pb $6s^2 6p^2$ 7.42 [Xe $4f^{14} 5d^{10}$]	$_{83}$Bi $6s^2 6p^3$ 7.29 [Xe $4f^{14} 5d^{10}$]	$_{84}$Po $6s^2 6p^4$ 8.43 [Xe $4f^{14} 5d^{10}$]	$_{85}$At $6s^2 6p^5$ — [Xe $4f^{14} 5d^{10}$]	$_{86}$Rn $6s^2 6p^6$ 10.75 [Xe $4f^{14} 5d^{10} 6s^2 6p^6$ = Rn]
Seventh period	$_{87}$Fr $7s^1$ — [Rn]	$_{88}$Ra $7s^2$ 5.28 [Rn]	↓						

[a]Ionization potentials in electron volts.
[b]Configurations and potentials for all of the elements are in Appendix 4.
[c]H and He are not octet elements, but are included for comparison.

TABLE 3.2 Oxides, hydroxides, oxyacids, and halides of Octet Periodic Table elements

Groups ⟶	*1a*	*2a*	*3a*	*4a*	*5a*	*6a*	*7a*	*0*
Oxide ⟶	M_2O	MO	M_2O_3	MO_2	M_2O_5	MO_3	M_2O_7	MO_4
	Lithium	**Beryllium**	**Boron**	**Carbon**	**Nitrogen**	**Oxygen**	**Fluorine**	**Neon**
Second period	Li_2O	BeO	B_2O_3	**CO_2** (g)	**N_2O_5** (l)		**(F_2O)**[b] (g)	
	$LiOH$	$Be(OH)_2$	**H_3BO_3**	$[H_2CO_3]$[a]	**HNO_3**	—	—	
	$LiCl$	$BeCl_2$	**BCl_3** (l)	**CCl_4** (l)	**(NF_3)** (g)	**(OF_2)**[b] (g)	—	
	Sodium	**Magnesium**	**Aluminum**	**Silicon**	**Phosphorus**	**Sulfur**	**Chlorine**	**Argon**
Third period	Na_2O	MgO	Al_2O_3	SiO_2	**P_4O_{10}**	**SO_3**	**Cl_2O_7**	
	$NaOH$	$Mg(OH)_2$	$Al(OH)_3$	$[H_2SiO_3]$[a]	**$H_3P_3O_9$**	**H_2SO_4**	**$HClO_4$**	
	$NaCl$	$MgCl_2$	**Al_2Cl_6**	**$SiCl_4$** (l)	**PCl_5**	**SF_6** (g)	**(ClF_3)** (g)	
	Potassium	**Calcium**	**Gallium**	**Germanium**	**Arsenic**	**Selenium**	**Bromine**	**Krypton**
Fourth period	K_2O	CaO	Ga_2O_3	GeO_2	As_2O_5	**SeO_3**	$(Br_3O_8)_n$[b]	
	KOH	$Ca(OH)_2$	$Ga(OH)_3$	$[H_2GeO_3]$[a]	$HAsO_3$	H_2SeO_4	$[HBrO_4]$	
	KCl	$CaCl_2$	**Ga_2Cl_6**	**$GeCl_4$** (l)	**AsF_5** (g)	**SeF_6** (g)	**(BrF_5)**[b] (l)	**(KrF_2)**
	Rubidium	**Strontium**	**Indium**	**Tin**	**Antimony**	**Tellurium**	**Iodine**	**Xenon**
Fifth period	Rb_2O	SrO	In_2O_3	SnO_2	Sb_2O_5	TeO_3	(I_2O_5)[b]	**(XeO_3)**
	$RbOH$	$Sr(OH)_2$	$In(OH)_3$	$[H_2Sn(OH)_6]$[a]	$[HSb(OH)_6]$[a]	**H_6TeO_6**	HIO_4	**$[H_4XeO_6]$**
	$RbCl$	$SrCl_2$	$InCl_3$	**$SnCl_4$** (l)	**$SbCl_5$** (l)	**TeF_6** (g)	**IF_7** (l)	**(XeF_6)**
	Cesium	**Barium**	**Thallium**	**Lead**	**Bismuth**	**(Polonium)**[c]	**(Astatine)**[c]	**(Radon)**[c]
Sixth period	Cs_2O	BaO	Tl_2O_3	PbO_2	Bi_2O_5	PoO_3		
	$CsOH$	$Ba(OH)_2$	$Tl(OH)_3$	$[H_2Pb(OH)_6]$[a]	$[HBiO_3]$[a]			
	$CsCl$	$BaCl_2$	$TlCl_3$	**$PbCl_4$** (l)	**BiF_5**			
	(Francium)[c]	**(Radium)**[c]						
Seventh period		$RaBr_2$						

Notes

[a]Brackets signify that the free acid is unstable—however, the salts [e.g., Na_2CO_3, $Na_2Sn(OH)_6$] are known.
[b]Formulas in parentheses where simple periodic law does not hold.
[c]Elements in parentheses are rare and radioactive and, except for radium, they have been little studied.
Unless indicated by (l) or (g), compounds are solids at or near room temperature (below 35°C).
Molecular formulas in boldface type; all other formulas are empirical.

Exercises

11. Using octet theory, predict the empirical formula for the ionic compound formed between silicon ([Ne]$3s^2 3p^2$) and oxygen ([He]$2s^2 2p^6$).

12. Magnesium (Mg) reacts with phosphorus (P) to give the ionic compound magnesium phosphide, Mg_3P_2. Using octet theory, explain the formation of magnesium phosphide.

13. Explain why the formula of calcium oxide ("quicklime") is CaO and not Ca_2O.

Dmitri Mendelyeef (1834–1907) was a great Russian chemist who was born in Siberia and was the youngest of 17 children. He published the first widely used periodic table and predicted from it the discovery of several chemical elements, including gallium and germanium.

Lothar Meyer (1830–1895) was a German chemist who developed a periodic table independently of Mendelyeef but didn't get as much publicity or credit for his discovery.

3.6 Periodicity in the Properties of the Elements About 1870 (nearly 30 years before the discovery of the electron in 1897), Mendelyeef and Meyer each observed that if the 60-odd elements known at the time were arranged in order of increasing atomic weights there was a repetition or periodicity in their chemical and physical properties. Figure 3.10 shows periodicity in a physical property, the density of the elements. It is now known that periodicity is a function of atomic number (Z), not of atomic weight, but the two quantities parallel each other with few exceptions (Te and I, Ar and K among them), so the idea of a relationship between atomic weight and properties of the elements was close to the truth. If the symbols of the elements are printed on a long strip of paper in order of increasing atomic number and the strip is cut and parts are arranged so that elements having similar physical and chemical properties fall in vertical columns, the complete Periodic Table (see inside front cover) results. Our objective now is to explain the regular repetition or periodicity of properties with increasing atomic number.

Periodicity in the properties of the elements is most easily explained in terms of the electronic structures of their atoms if we limit ourselves to what are termed the *representative elements*. The periodic arrangement of these elements with their electron configurations by core and valence electrons is shown in Table 3.1, and is called the *Octet Periodic Table*. If you look at any vertical column (called a *group*) you will see that all elements in it have *the same outer or valence electron configurations in s and p orbitals except for the value of the principal quantum number n*. Thus all Group 1a elements have one valence electron in an s orbital, all Group 2a elements have the valence electron configuration ns^2, Groups 3a has the configuration ns^2np^1, and so on. In fact, for *Octet Table elements the number of valence electrons is the same as the group number of the element in the table*. The elements in each of the horizontal rows in the table belong to a *period* of elements that begins with an element having the valence configuration ns^1 and, except for the first period, ends with a noble gas having octet configuration ns^2np^6, with n being the number of the period. The first period

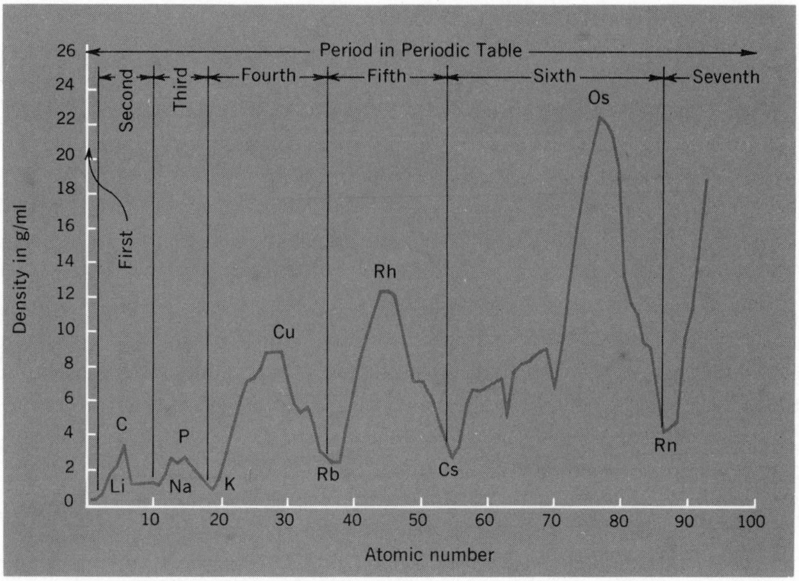

FIG. 3.10 Periodicity in the density of the elements. (Densities for elements that are gases under ordinary conditions are those of the liquid elements near their normal boiling points; all other densities at 20°C.)

is exceptional in containing only the two elements hydrogen and helium and ending with the $1s^2$ configuration. You can see the relationship between valence electron configuration and chemical properties if you compare the electron configurations in Table 3.1 with the formulas of typical compounds of the elements in Table 3.2. The Group 1a elements (Li, Na, K, Rb, Cs, Fr), called the *alkali metals*, all form *oxides* of formula type M_2O, compounds called *hydroxides* of formula MOH, and *chlorides* of formula MCl. Group 2a, the *alkaline earth metals*, form oxides MO, hydroxides $M(OH)_2$, and chlorides MCl_2. The electronic explanation for the formulas of the oxides follows the pattern for Al_2O_3 given earlier. The formulas for the chlorides follow the pattern established in Section 3.5 for the fluorides—the chlorine atom, like the fluorine atom, lacks one electron for its octet, so magnesium chloride is $MgCl_2$. Regularities in formula type continue through Group 5a and into the lower members of Group 6a.

Why do valence electron configurations and chemical properties repeat periodically with the octet elements? According to quantum mechanics, periodicity is a result of the *repetition of orbital types* as the principal quantum number n increases (Table 3.3, p. 106). Excluding H and He, each value of $n(2, 3, 4, 5, \ldots,)$ yields orbitals of the s and p types. This repetition results again and again in elements that, in spite of increasingly large core electron arrangements, nevertheless have the *same number of* valence electrons in the same types (s and p) of *orbitals*. They therefore have similar chemical and physical properties. For example, all of the alkali metals (Group 1a) are soft enough to be cut with a knife, and are excellent conductors. Further,

TABLE 3.3 Repetition of orbital types

Shell	*n*	*Orbital Types in Energy Level*
K	1	*s*
L	2	*sp*
M	3	*spd*
N	4	*spdf*
O	5	*spdfg*
P	6	*spdfgh*
Q	7	*spdfghi*

all react vigorously with water to form their hydroxides and hydrogen gas, and with chlorine to form chlorides, according to the general equations:

TABLE 3.4

First ionization potentials, electronegativities, and electron affinities of representative elements

					2.2 H 13.6 (0.75)					
— He 24.6	1.0[a] Li 5.4[b] (0.54)[c]	1.6 Be 9.3 (−0.13)		2.0 B 8.3 (0.33)	2.6 C 11.3 (1.12)	3.0 N 14.5 (0.05)	3.4 C 13.6 (1.47)	4.0 F 17.4 (3.63)		
— Ne 21.6 (−0.57)	0.9 Na 5.1 (0.47)	1.3 Mg 7.6 (−0.32)		1.6 Al 6.0 (0.52)	1.9 Si 8.2 (1.46)	2.2 P 11.0 (0.77)	2.6 S 10.4 (2.07)	3.2 Cl 13.0 (3.78)		
— Ar 15.8	0.8 K 4.3	1.0 Ca 6.1		1.8 Ga 6.0	2.0 Ge 7.9	2.2 As 9.8	2.6 Se 9.8	3.0 Br 11.8 (3.54)	1.4 Sc 6.6	1.5 Ti 6.8
— Kr 14.0	0.8 Rb 4.2	1.0 Sr 5.7	transition and inner transition metals	1.8 In 5.8	2.0 Sn 7.3	2.1 Sb 8.6	2.1 Te 9.0	2.7 I 10.5 (3.24)	1.3 Y 6.5	1.6 Zr 7.0
— Xe 12.1	0.8 Cs 3.9	0.9 Ba 5.2		2.0 Tl 6.1	2.3 Pb 7.4	2.0 Bi 7.3				

[a]Electronegativity (Pauling scale).
[b]First ionization potential in volts.
[c]Electron affinity.

Noble gases
Octet metals
Metalloids
Nonmetals
Typical transition metals

$$2M(s) + 2H_2O(l) \rightarrow 2M^+OH^-(s) + H_2(g) + \text{energy}$$
$$2M(s) + Cl_2(g) \rightarrow 2M^+Cl^-(s) + \text{energy}$$

The chlorides are simple ionic compounds as indicated; the hydroxides are complex ionic compounds with equal numbers of M^+ and covalently bonded hydroxide O—H^- ions for which the octet structure is $:\ddot{\underset{..}{O}}:H^-$. The alkali metal hydroxides are the most important sources of OH^- ions, which they release along with the M^+ ion when dissolved in water. The presence of OH^- ions makes a solution *alkaline* or *basic* (Chapter 8).

The metallic character (shown by high electrical conductivity) of the Group 1a metals extends to the alkaline earths (Group 2a), to aluminum (Al), gallium (Ga), indium (In), and thallium (Tl) in Group 3a, and to tin (Sn) and lead (Pb) in Group 4a. Metallic properties are largely absent from the lighter elements of Group 4a and essentially disappear in the elements of Groups 5a, 6a, and 7a. These latter elements are appropriately called *nonmetals,* and the halogens, Group 7a (F, Cl, Br, I, At), are the most nonmetallic of the elements.

In going across any period there is a stepwise transition from metals to nonmetals. Thus sodium, magnesium, and aluminum are metals; silicon has weak metallic properties; and phosphorus, sulfur, and chlorine are typical nonmetals. Metallic character also increases going down through any group. Boron, carbon, silicon, and germanium have weak metallic properties (poor electrical conductors), but the elements below them in their groups (Al and Sn) are metals. The trend from metallic to nonmetallic character is shown by shading in Table 3.4. Elements that have properties between those of metals and nonmetals are called metalloids (Chapter 11).

Ions of octet elements in biochemistry One usually thinks of biochemistry as being concerned mainly with the reactions of the covalent compounds of carbon with hydrogen, oxygen, nitrogen, and a few other elements. Increasingly, in recent years it has become clear that many metal ions and elements not previously thought to participate in the reactions of life do in fact play key roles there.

Each of the simple ions of the representative (octet) elements in any chemical group of the Periodic Table has the electron configuration of the noble gas with atomic number closest to its own. Thus Li^+ and Be^{2+} have the helium configuration [He]; O^{2-} and F^- have the neon configuration [Ne]. Furthermore, the ions of any group have the same charge. The similarities in electron configuration and charge are emphasized by the following arrangement for the ions of Groups 1a, 2a, 3a, 6a, and 7a:

Electron Configuration of Ions	*Group 1a*	*Group 2a*	*Group 3a*	*Electron Configuration of Ions*	*Group 6a*	*Group 7a*
[He]	Li^+	Be^{2+}	B^{3+}	[Ne]	O^{2-}	$\mathbf{F}^-$
[Ne]	$\mathbf{Na}^+$	$\mathbf{Mg}^{2+}$	Al^{3+}	[Ar]	$\mathbf{S}^{2-}$	$\mathbf{Cl}^-$
[Ar]	$\mathbf{K}^+$	$\mathbf{Ca}^{2+}$	Ga^{3+}	[Kr]	$\mathbf{Se}^{2-}$	Br^-
[Kr]	Rb^+	Sr^{2+}	In^{3+}	[Xe]	Te^{2-}	$\mathbf{I}^-$
[Xe]	Cs^+	Ba^{2+}	Tl^{3+}	[Rn]		At^-
[Rn]	Fr^+	Ra^{2+}				

The ions of any group all face their environment with the same charge and a noble gas octet of electrons (duet for Li^+, Be^{2+}, B^{3+}), and you might therefore think that the ions of any group would behave similarly in the chemical reactions of life processes. Nothing could be further from the truth. The biochemical reactions in living things—plant and animal alike—are very selective in the ions they use. The ions that appear in boldface type in the table are known to play essential roles in the reactions of life. Those in italics are found in living tissue and are believed to be vital, but their role is not yet understood.

EXAMPLES Definite amounts of sodium, potassium, and calcium ions in blood and tissue fluids are absolutely essential to our lives. Those amounts result in what is called "electrolyte balance," and this balance is carefully maintained in the sick and injured by additions to their diets or by intravenous injection. A balanced diet provides plenty of these vital ions, and any excess is excreted in the urine. The transmission of nerve impulses depends on Na^+, K^+, and Ca^{2+} ions with Cl^- ions acting as "bystanders" to neutralize their positive charges. Chloride ions pass freely through cell membranes; the positive ions (cations) do not. The cells in the body contain more K^+ ions; the blood and other tissue fluids contain more Na^+ (Table 11.2). A deficiency of Na^+ produces muscle cramps and mental confusion, and lack of K^+ leads to muscular cramps and paralysis. On the other hand, an excess of K^+ produces irregularities in heart action. A large decrease in Ca^{2+} results in convulsions because nerve transmission is overstimulated. None of these conditions will arise unless a person is seriously ill from such diseases as kidney failure or loss of body fluid from large areas of burned tissue.

The chloride ion has been regarded for a long time as one of the anions that accompany the cations (Na^+, K^+, Ca^{2+}) to maintain electrical charge neutrality. Chloride ions are now believed to play an essential part in the process by which images on the retina at the back of the eye are converted to nerve signals to the brain, resulting in vision.

In contrast to the role of the chloride ion, the bromide ion acts as a sedative and like fluoride and iodide ions (F^-, I^-) is poisonous in amounts of a gram or so. In spite of their toxicity in large doses, traces of fluoride ion are necessary for the proper development of tooth enamel, and our bodies need daily about a microgram (10^{-6} g) of iodide ion (or iodine) for incorporation into the hormone thyroxine, which controls the rate of resting body activity (basal metabolism). About one gram of fluoride (as Na^+F^-) per 10^6 g of water is added to the domestic water supplies of most cities, and traces of iodide (as K^+I^-) are added to salt to produce "iodized salt."

The calcium ion plays an essential part in bone, which is mainly $Ca_3(PO_4)_2$ in various crystalline and hydrated forms. Strontium 90 ($^{90}_{38}Sr$) in the "fallout" debris from atomic bomb explosions can replace Ca^{2+} in bone, but $^{90}_{38}Sr$ is radioactive and the radioactive rays can cause leukemia and bone cancer. The beryllium ion can also take the place of Ca^{2+} but not without toxic reactions and even the production of cancer.

As research continues, more and more elements once regarded either as poisonous or unnecessary for living systems are found to be essential. In this regard it is a good idea to eat a diet that is not only balanced in terms of fat, carbohydrate, protein, and vitamins but that also has a variety of meats, breads, fruits, and vegetables. Each plant and animal concentrates some of the trace elements, and eating a variety of foods gives you an adequate supply of essential "minerals" or elements.

Occurrence of ionic and covalent bonding Ionic bonding with the production of ionic crystals *is found in compounds between metals and nonmetals*. Among the examples encountered before are Li^+F^-, Na^+Cl^-, and $Mg^{2+}O^{2-}$. If we let M stand for a Group 1a element and X_2 for the diatomic, covalent molecule of a Group 7a element, the reactions to form the ionic compounds called *alkali metal halides* can be summarized in a single equation:

$$2M + X_2 \rightarrow 2M^+X^- + \text{energy}$$

For the Group 2a elements, the equation for the formation of the ionic *alkaline earth halides* becomes

$$M + X_2 \rightarrow M^{2+}(X^-)_2$$

where M now stands for the element with the ns^2 valence electron configuration.

Covalent bonding with the formation of molecules or covalent crystals *is found mainly between the atoms of nonmetals.* Examples are F_2, Cl_2, Br_2, I_2, O_2, N_2, S_8, P_4, SCl_2, PCl_3, PCl_5, SO_2, P_4O_{10}, N_2O_5, $SiCl_4$, and many others shown in boldface type in Table 3.2. Looking at the ionization potentials in Table 3.1, you can see that ionic bonding is generally found between elements of low ionization potential (metals) and those of high ionization potential (nonmetals). Covalent bonding, with the formation of molecules, occurs between atoms of similar or identical ionization potential. The explanation for the different kinds of bonding is not difficult. Metals, because of their low ionization potentials, readily give up their valence electrons to complete the octets of the nonmetals, and ionic bonding results.

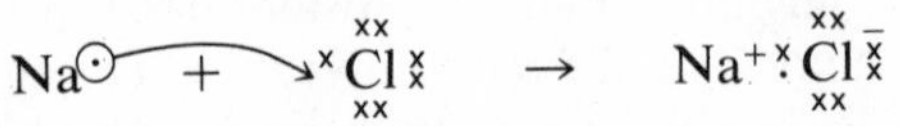

$IP = 5.14$ eV $IP = 13.01$ eV   ionic

When the electrons are almost equally held (similar ionization potentials), the octet is achieved by *sharing* electrons. Thus the compound iodine chloride is made up of covalent :I—Cl: molecules whose formation from atoms may be seen by using dot structures as follows:

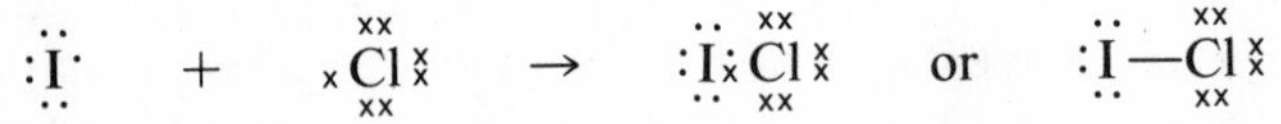

$IP = 10.45$ eV $IP = 13.01$ eV   covalent

In ICl, as in most covalent compounds, each nucleus has, in addition to its core electrons, an outer octet of electrons *when shared pairs are counted for both.*

3.7 The Periodic Arrangement of the Elements The periodic arrangement of the elements or Periodic Table (see inside front cover) contains all the known elements, not just the representative ones of the Octet Periodic Table. In the Octet Table electrons are added only to *s* and *p* orbitals in going across any period. The other orbitals, *d* and *f*, that may be available are either completely empty or completely filled. The complete Periodic Table (Table 3.5) includes elements whose electrons are going into *d* and *f* orbitals. The filling of the *d* orbitals produces what are called the *transition* metals, and begins first with scandium (Sc, $Z = 21$), with the production of the *first transition metal series* of ten elements (Sc, Ti, V, Cr, Mn, Fe, Co, Ni, Cu, Zn). There are ten elements in the series because there are five 3*d* orbitals to be filled with one pair of electrons each. The way in which the orbitals are filled is shown in Table 3.6. The Hund rule applies, and in addition, the configurations of Cu and Cr show that states having

TABLE 3.5 Periodic table with transition and inner transition elements

Periodic shell

Shell	Elements						Subshell
Q	Fr 87	Ra 88					← 7*s*
	Ti 81	Pb 82	Bi° 83	Po* 84	At* 85	Rn 86	← 6*p*
P	Cs 55	Ba 56					← 6*s*
	In 49	Sn 50	Sb° 51	Tc° 52	I* 53	Xe 54	← 5*p*
O	Rb 37	Sr 38					← 5*s*
	Ga 31	Ge° 32	As° 33	Se° 34	Br* 35	Kr 36	← 4*p*
N	K 19	Ca 20					← 4*s*
	Al 13	Si° 14	P* 15	S* 16	Cl* 17	Ar 18	← 3*p*
M	Na 11	Mg 12					← 3*s*
	B° 5	C° 6	N* 7	O* 8	F* 9	Ne 10	← 2*p*
L	Li 3	Be 4					← 2*s*
K	H 1	He 2					← 1*s*

← *octet periodic table* →

Ac 89 ← 6*d*

Inner transition metals, II. Uranium metals or actinides

Th	Pa	U	Np	Pu	Am	Cm	Bk	Cf	E	Fm	Mv	No	Lw	
90	91	92	93	94	95	96	97	98	99	100	101	102	103	← 5*f*

Transition metals, III

La	Hf	Ta	V	Re	Os	Ir	Pt	Au	Hg	
57	72	73	74	75	76	77	78	79	80	← 5*d*

Inner transition metals, I. Rare earths or lanthanides

Ce	Pr	Nd	Pm	Sm	Eu	Gd	Tb	Dy	Ho	Er	Tm	Yb	Lu	
58	59	60	61	62	63	64	65	66	67	68	69	70	71	← 4*f*

Transition metals, II

Y	Zr	Nb	Mo	Tc	Ru	Rh	Pd	Aq	Cd	
39	40	41	42	43	44	45	46	47	48	← 4*d*

Transition metals, I

Sc	Ti	V	Cr	Mn	Fe	Co	Ni	Cu	Zn	
21	22	23	24	25	26	27	28	29	30	← 3*d*

Subshells:

Type	*Number of Orbitals*	*Number of Electrons to Fill*
s	1	2
p	3	6
d	5	10
f	7	14

← *transition metals and inner transition metals* →

Note: The atomic numbers are written below the symbol for the element. *Transition metals* having but one electron in the *s* orbital of quantum number one greater than the *d* level being filled are marked with a single bar beneath the atomic number, those with no electrons in the outer *s* orbital with a double bar. *Noble gases* are boxed, as He. All *other nonmetals* are marked with an asterisk, as P*. Elements with both *metallic* and *nonmetallic* properties are marked with °, as B°.

filled or half-filled subshells are especially stable. The metallic character of the transition metals is due to their having one or two electrons in the 4*s* orbital, like the metallic Group 1a and 2a elements of the Octet Table. Once the configuration $[Ar]4s^2 3d^{10}$ is reached with Zn, electrons begin to enter the 4*p* orbitals, producing the progressively less metallic elements gallium (Ga), germanium (Ge), arsenic (As), and bromine (Br). The period terminates at krypton (Kr) with configuration $[Ar]3d^{10}4s^2 4p^6$ (Table 3.1). The chemical inertness of krypton shows that electrons in *filled* 3*d* orbitals behave like core electrons.

The presence of electrons in *partly filled d* orbitals gives special properties to many of the transition metals and their compounds that are not shown by octet metals. In many transition metals, the presence of *d* electrons with unpaired spins confers magnetic properties on both the metal (ferromagnetism) and its compounds (paramagnetism). Furthermore, the energy-level spacings of the orbitals are so small that the low-energy photons of visible light are absorbed by compounds of many transition metals. This absorption results in color. These properties of transition metals and their compounds are discussed in Chapter 11.

Transition metals in life processes The ions of several transition metal elements, like those of the octet metals, play vital roles in the chemical reactions in plants and animals. Among the vital transition metal ions are the ions of vanadium (V), chromium (Cr), manganese (Mn), iron (Fe), cobalt (Co), nickel (Ni), copper (Cu), and zinc (Zn) from the first transition series; and molybdenum (Mo) from the second transition series. The transition metal ions are built into large molecules of C, H, O, N, S atoms to make complex ions called enzymes. En-

TABLE 3.6 Electron configurations in the 4*s* and 3*d* orbitals and first ionization potentials for the first transition series[a]

Atom	*Electron Configuration* by orbitals						*by subshells*	*First Ionization Potential* (eV)
	$4s$	$3d_1$	$3d_2$	$3d_3$	$3d_4$	$3d_5$		
$_{21}$Sc	⇅	↑					$4s^2 3d^1$	6.6
$_{22}$Ti	⇅	↑	↑				$4s^2 3d^2$	6.8
$_{23}$V	⇅	↑	↑	↑			$4s^2 3d^3$	6.7
$_{24}$Cr	↑	↑	↑	↑	↑	↑	$4s^1 3d^5$	6.8
$_{25}$Mn	⇅	↑	↑	↑	↑	↑	$4s^2 3d^5$	7.4
$_{26}$Fe	⇅	⇅	↑	↑	↑	↑	$4s^2 3d^6$	7.9
$_{27}$Co	⇅	⇅	⇅	↑	↑	↑	$4s^2 3d^7$	7.9
$_{28}$Ni	⇅	⇅	⇅	⇅	↑	↑	$4s^2 3d^8$	7.6
$_{29}$Cu	↑	⇅	⇅	⇅	⇅	⇅	$4s^1 3d^{10}$	7.7
$_{30}$Zn	⇅	⇅	⇅	⇅	⇅	⇅	$4s^2 3d^{10}$	9.4

[a]All atoms have the Ar core.

zymes are substances our bodies need for carrying out the complex reactions of life at ordinary temperatures. Enzymes are "catalysts" because they promote vital reactions, such as the oxidation of glucose, without themselves being rapidly used up or generating high temperatures because of the energy released. The energy keeps our bodies warm and can be used for muscular activity and thinking.

$$\underset{\text{glucose}}{C_6H_{12}O_6} + 6O_2 \xrightarrow[\text{and steps}]{\text{many enzymes}} 6CO_2 + 6H_2O + 686 \text{ kcal}$$

Enzymes are examined more closely in Section 14.6.

There is an iron atom at the "working center" of the red pigment *hemoglobin* found in red blood cells. Hemoglobin carries oxygen from the lungs to the tissue cells where the reaction of glucose with oxygen is going on. There is about enough iron in your hemoglobin to make a small nail. Lack of iron in the diet results in too little hemoglobin and the condition called *anemia*, with its symptoms of fatigue and loss of mental alertness, because the cells aren't getting enough O_2. In the plant world, a magnesium ion occupies a place like that of iron in hemoglobin but in a complex structure called *chlorophyll.* Chlorophyll is green in color and absorbs energy from sunlight to drive the reaction of photosynthesis, by which glucose and starch are made from water and atmospheric carbon dioxide.

$$686 \text{ kcal} + 6CO_2 + 6H_2O \xrightarrow[\substack{\text{many enzymes} \\ \text{many steps}}]{\text{photosynthesis}} C_6H_{12}O_6$$

With the exception of iron, the amounts of transition metal ions needed by plants and animals are very small—in the microgram range per day.

Albert Eschenmoser (b. 1926) received the International R. A. Welch Award in 1974 for his highly creative contributions to synthetic chemistry.

Robert Burns Woodward (b. 1917) won the Nobel prize in chemistry in 1965 for his work as the outstanding architect in the design of syntheses of biologically vital, complex molecules, including cholesterol, chlorophyll and vitamin B_{12}.

EXAMPLES

⊸ Vitamin B_{12} contains one cobalt atom, as shown by its molecular formula, $C_{63}H_{88}N_{14}O_{14}PCo$. Its complete bond structure was determined by chemical and X-ray crystal analysis, and the molecule has been synthesized from smaller molecules by two teams of chemists. One team under the direction of Eschenmoser in Switzerland made one half of the molecule, and the other half was made by an American research team under the direction of Woodward. The two halves fitted together to produce the vitamin. Vitamin B_{12} occurs in meat, especially liver, and the daily requirement is about 1 microgram (μg). The vitamin is effective in the treatment of *pernicious anemia* in which a lack of red blood cells occurs because the bone marrow, where they are made, ceases to function. There is sufficient B_{12} in a balanced diet to prevent any deficiency. In cases of pernicious anemia, the basic cause of the disease appears to be a failure of the B_{12} in the diet to be absorbed through the wall of the small intestines. In treating the disease, the B_{12} is usually injected hypodermically.

Related by John C. Bailar, Jr.

You can get some idea of what is meant by a "trace" amount of an element from the experience of sheepherders in certain areas of Australia. In these areas sheep were sickly and did not grow well. The problem was traced to a lack of cobalt in the soil, and therefore in the grass the sheep were eating. The cobalt shortage was eliminated by forcing the sheep to swallow a small pellet of cobalt metal and a steel screw. Both pellet and screw remain in the stomach of the sheep (actually the "second stomach" or rumen where the "cud" is stored for rechewing). The purpose of the screw is to scrape off any coating that develops on the cobalt pellet. A coating would prevent slow dissolving of the cobalt in digestive juices to produce cobalt ions (Co^{2+}) needed for vitamin B_{12}. When the sheep is slaughtered, the pellet is recovered and can be used again many times.

The group of transition metal ions Zn^{2+}, Cd^{2+}, and Hg^{2+} (from the elements in Group 2b) provides an example of vital and toxic trace elements from transition metals. All three of these ions have the same outer electron configuration $ns^2(n-1)d^{10}$ in which only the value of n differs. Like the octet ions of any group, they face their environment with the same charge and highly similar outer electron arrangement. The zinc ion (Zn^{2+}) is needed in tiny amounts because it is essential in the use of vitamin A in the body. The cadmium ion (Cd^{2+}) and mercuric ion (Hg^{2+}) are strong poisons. The cadmium ion replaces Ca^{2+} ions in bones, causing them to soften and break easily, and in higher proportions inactivates enzymes using Zn^{2+}. The mercuric ion, Hg^{2+}, is poisonous to all forms of life (Photo 10). It is used as a germicide in hospitals for sterilizing instruments.

A more insidious form of mercury poisoning results from bacterial action in lakes that converts metallic mercury and mercury compounds to methyl mercury, CH_3—Hg—CH_3. Substantial amounts (tons per year) of mercury and its compounds are released into the aquatic environment by industry each year. The methyl mercury produced is taken up by lower forms of life (plankton and algae) to which it is not so toxic, and ends up in the flesh of fish that eat the aquatic plants and algae. The fish, like us, have no way of getting rid of the mercury compounds in their food, so the amount builds up to toxic levels (Photo 11). At the time of this writing large lots of canned salmon and tuna were being removed from the stores because of this high mercury content.

Inner transition metals The order in which the subshells of orbitals are filled up to and beyond krypton is summarized in Table 3.5. A second series of ten transition metals begins at yttrium (Y) in period 5 and is completed with cadmium (Cd). Then building in the 5p orbitals

PHOTO 11 Industrially polluted area 100 miles south of Houston, Texas, on the Gulf Coast. (*EPA-DOCUMERICA—Marc St. Gil.*)

commences and generates the elements indium (In) through xenon (Xe) to complete period 5. The third series of transition metals begins in period 6 with lanthanum (La), but is immediately interrupted as building in the seven 4*f* orbitals takes over to produce the set of 14 *inner transition metals* (Ce through Lu) called the *lanthanides* or *rare earths*. After lutecium (Lu), building resumes in the 5*d* orbitals to complete the third series of transition metals with mercury (Hg). Building then returns to the 6*p* orbitals and generates the octet elements thallium (Tl) through radon (Rn) to complete period 6.

Period 7 begins like period 6, but is incomplete because there are not enough elements known to complete the transition series that starts at element 89, actinium (Ac). The series of 14 inner transition elements following actinium is called the *actinide series*. It includes many elements not found on earth that have been prepared by the methods of nuclear chemistry (Chapter 16).

Whole books are devoted to the chemistry of the transition and inner transition metals and we must be satisfied here with a few generalizations. The inner transition metals (lanthanides and actinides) tend to have similar properties because electrons in *f* orbitals are seldom valence electrons and do not influence the chemistry of the atom very much. On the other hand, there are pronounced differences in properties for the transition metals going across periods. This is because electrons in *d* orbitals *do* act as valence electrons in many cases, especially when the *d* orbitals are only partially filled. Thus scandium forms only the ion Sc^{3+} but iron forms Fe^{2+} and Fe^{3+}, copper Cu^{+} and Cu^{2+}. Reference to Table 3.6 will show that *d* electrons must be removed to form some of these ions. The formation of Cu^{2+} even involves loss of one electron from the usually stable $3d^{10}$ configuration.

Compared with the properties of octet elements going down through a given group, the properties of transition elements going down the columns of "*b* groups" change in a much more pronounced manner (and in many cases in the opposite direction). A representative transition metal set is nickel (Ni), palladium (Pd), and platinum (Pt). Nickel is an "active metal," so called because it dissolves in acidic solutions with hydrogen gas evolution, according to the equation

Note that (*aq*) stands for "aqueous," meaning "solution in water."

$$Ni(s) + 2HCl(aq) \rightarrow NiCl_2(aq) + H_2(g)$$

But palladium and platinum, in the same group, do not react under the same conditions. The inertness of platinum accounts for its use in jewelry. Copper, silver, and gold show a similar marked downward trend in reactivity.

It is important to realize that the transition and inner transition elements are all *metals*. Since they make up 69 of the 103 elements, they represent more than half. When the octet metals and metalloids are added to the list, the nonmetallic elements are seen to be few in number. There are only 16 important nonmetals among all of the elements. They are hydrogen, carbon, nitrogen, oxygen, fluorine, phosphorus, sulfur, chlorine, bromine, iodine, and the six noble gases. If the nonmetals were capable of forming only their *simple* ions (F^-, Cl^-, O^{2-}, S^{2-}, N^{3-}, and so on) the number of compounds between metals and nonmetals would be rather limited. The fact that the nonmetals (and most transition metals) form *complex ions,* such as OH^-, ClO^-, ClO_2^-, ClO_3^-, ClO_4^-, SO_4^{2-}, NO_3^-, BrO_3^-, IO_3^-, IO_4^-, MnO_4^-, and countless others, tremendously increases the variety and number of compounds possible. The electronic structure of complex ions is discussed in Chapter 4.

Selenium is placed with the nonmetals in some listings that also include the very rare elements polonium and astatine for completeness.

At this point the utility of the Periodic Table and octet theory in organizing the chemistry of the elements should be clear to you. It is one of the great achievements of science that much of the arrangement can be explained by the electronic structures of atoms given in quantum theory.

3.8 Summary Guide

Because it is impossible to observe the paths of individual electrons in atoms, the most probable shapes of electron clouds around the nucleus are given by probability distributions called orbitals. The orbital is the basic unit for building up the electron configurations of the gaseous atoms of elements. Quantum mechanical calculations and the line spectra of the elements show that the orbitals are arranged in an energy pattern. The energy level or principal shell for an orbital is given by the value of the quantum number n for the level. The values permitted for n are the integers 1, 2, 3, 4, 5, . . . , and the number of orbitals in energy level n is equal to n^2. Thus the first (most stable) energy level has $n = 1$ and contains one orbital, the 1*s*. For $n = 2$ there are four orbitals, the 2*s* and three 2*p* orbitals. The third level contains nine orbitals – the 3*s*, three 3*p*, and five 3*d* orbitals – and so on. Each level repeats the orbital types of the preceding level but adds one new type. This repetition of orbital types (*s*, *p*, *d*, *f*) results periodically in atoms that have similar valence electron configurations and provides the theoretical explanation for the periodic arrangement of the elements.

Because of the penetration effect, there is a small separation between the energies of the orbitals of a given level. This results in the approximate energy sequence of orbitals 1*s*2*s*2*p*3*s*3*p*4*s*3*d*4*p*5*s*4*d*5*p*6*s*4*f*5*d*6*p*-7*s*5*f*6*d*. In building up atoms, the orbitals are filled with electrons starting with the 1*s* but limited by the Pauli exclusion principle that an orbital is filled by a pair of electrons of opposite spin. The Hund rule also applies: electrons avoid pairing of spins and avoid occupancy of the same orbital if empty orbitals of the same energy are available.

The outer octet (ns^2np^6) of electrons found for the noble gases corresponds to a condition of particular stability and low reactivity, recognized in the octet theory of valence. Completion of the octet of a nonmetal atom by transfer of electrons from an atom of a metal leads to ionic bonding between simple ions. The formulas of simple ionic compounds can be predicted from the number of valence electrons the metal atom has to donate and the number of electrons the nonmetal atom needs to complete its octet.

In the Octet Periodic Table, elements in vertical columns (groups) have the same valence electron configuration except for the value of the principal quantum number. Hence they show similar chemical reactivity and form the same types of compounds. The table divides the representative (octet) elements, in which building is going on in the *ns* and *np* orbitals, into metals, metalloids, and nonmetals. Metallic character increases in going down a column or group and decreases in going left to right across a horizontal row or period. Metals have low ionization potentials, nonmetals higher ionization potentials. Ionic bonding is found between metals and nonmetals. Covalent bonding is common between nonmetals.

The complete periodic arrangement of the elements contains three sets of ten transition metals in which electrons are entering *d* orbitals, and two sets of fourteen inner transition metals (lanthanides and actinides) in which they are entering *f* orbitals. Paramagnetic properties are common among the compounds of the transition and inner transition metals because of the presence of unpaired electron spins. Color is also common because energy-level spacings for electrons frequently correspond to the wavelengths of visible light.

Most of the elements are metals because they have one or two valence electrons in an outer *s* orbital. There are only a few important nonmetals; consequently the number of ionically bonded compounds would be limited were it not for the ability of nonmetals (and many transition metals) to form complex ions using covalently bonded oxygen atoms.

Exercises

14. The ionization energy for the removal of a second electron from an atom is always much larger than that for removal of the first electron. Why?

15. Using octet theory, explain the formation of the ionic compound lithium deuteride, $Li^{2}_{1}H$, a possible source of nuclear fusion energy (Chapter 16).

16. There is a cobalt (II) ion Co^{2+} in vitamin B_{12}, the anti-anemia vitamin. Give the electron configuration in the 4*s* and 3*d* orbitals of Co^{2+}.

17. Which electrons are removed from Cd to produce Cd^{2+}? (*Hint*: compare $Zn \rightarrow Zn^{2+}$.)

18. Selenium in trace amounts is a vital element; in larger amounts it is poisonous. For Se, $Z = 34$. Give the electron configuration by orbitals for the Se atom and the Se^{2-} ion.

19. Predict the type of bonding (ionic or covalent) expected between pairs of elements of the following atomic numbers without consulting the Periodic Table: (a) 3 and 17; (b) 6 and 9; (c) 11 and 16; (d) 14 and 12; (e) 8 and 16. (*Hint*: figure out the electron configurations by subshells.)

20. Without consulting the Periodic Table, tell whether the elements with the following atomic numbers are metals or nonmetals: (a) 31; (b) 35; (c) 38; (d) 21; (e) 29.

21. The element of atomic number 16 is used to link protein chains together (Chapter 14). Give the electron configuration by subshells for the atom of the element and predict the type of bonds it forms with carbon atoms and hydrogen atoms.

22. The Mg^{2+} ion is at the center of the chlorophyll molecule that is needed for photosynthesis. Give the electron configuration of Mg^{2+} ($Z = 12$).

Key Words

Core electrons electrons in orbitals close to the nucleus that aren't involved in chemical reactions.

Covalence bonding due to sharing of electrons by atoms to produce octets.

Diatomic molecule molecule made up of two atoms such as N_2, H_2, HCl.

Electromagnetic radiation the form of energy that is transmitted through space as wave motion or as streams of energy packets or photons. It includes light, radio waves, and X rays, as well as other forms of radiation.

Electron configuration arrangement of electrons in the orbitals of an atom.

Electron spin a property of electrons in addition to their charge, mass, and energy.

Energy levels definite quantities of energy allowed in atoms depending on location of electrons in orbitals.

Erg energy quantity equal to 6.242×10^{11} electron volts or 2.390×10^{-11} kilocalories.

Excited state any energy state of an atom except the ground state.

Ground energy state state of an atom that has the largest negative potential energy—most stable state; one in which electrons are as close to the nucleus as possible.

Ionization production of charged particles by removal of one or more electrons from an atom or molecule.

Orbital region in space around the nucleus of an atom that may be occupied by one or a pair of electrons.

Penetration effect tendency of electrons in *s* orbitals to be close to the nucleus and therefore of lower energy than electrons in other kinds of orbitals.

Periodicity repetition of chemical properties of atoms with increasing atomic number.

Photons energy capsules or packets in electromagnetic radiation.

Principal shell made up of all orbitals with the same principal quantum number (n) (K, L, M, N, and so on).

Quantum number whole numbers (1, 2, 3, . . . ,) used to describe the principal energy levels for electrons in atoms.

Simple ionic bonding bonds between positively and negatively charged ions resulting from transfer of electrons from one atom to another, with production of noble gas electron configurations (duet, octet).

Spectrometer instrument used to show the wavelengths of electromagnetic radiation coming from a source of radiation.

Spectrum pattern of wavelengths in electromagnetic radiation.

Subshell made up of all orbitals with the same principal quantum number (n) and the same letter designation—*s*, *p*, *d*, or *f*.

Valence electrons electrons in outer orbitals that may be donated to or shared with other atomic nuclei – the "chemical" electrons that are responsible for chemical reactions.

Wavelength the distance between neighboring crests (or troughs) of waves.

Suggested Readings

Frieden, Earl, "The Chemical Elements of Life," *Sci. Am.*, **227**:52–60 (July 1972).

"How Much Metal Is There in Our Waters?" *Env. Sci. and Tech.*, **8**:112–113 (Feb. 1974).

Maugh, Thomas H., II, "Trace Elements: A Growing Appreciation of Their Effects on Man," *Science*, **181**:253–254 (July 20, 1973).

Pettijohn, W. A., "Trace Elements and Health," *Sci. Teach.*, **39**:37–40 (May 1972).

Ratcliff, J. D., "I Am Joe's Thighbone," *Reader's Digest*, June 1973, pp. 85–88.

Seaborg, Glenn T., *Man-Made Transuranium Elements*, Englewood Cliffs, N.J., Prentice-Hall, 1963.

Technology Report, "Effects of Pollutants on Marine Life Probed," *Chem. and Eng. News*, Dec. 17, 1973, p. 17.

Univ. of Baghdad and Univ. of Rochester (an interuniversity report), "Methylmercury Poisoning in Iraq," *Science*, **181**:230–241 (July 20, 1973).

Wilson, J. D., and Baker, S. D., *Physical Science: Readings on the Environment*, Lexington, Mass., D. C. Heath and Co., 1974.

Wolff, P., *Breakthroughs in Chemistry*, New York, Signet Science Library, 1967.

"Zinc: A Trace Element Essential in Vitamin A Metabolism," *Science*, **181**:954–955 (Sept. 7, 1973).

chapter 4

Stone sculpture, exterior of Herten Castle, near the city of Westphalia, West Germany. The left-hand photo was taken in 1908, 200 years after the statue was completed. The right-hand photo, taken in 1969, shows almost complete destruction from acidic smog. (*Photo 12, E. M. Winkler,* Stone: Properties, Durability in Man's Environment, *Springer-Verlag, 1973.*)

Electronic Structures of Molecules and Complex Ions

4.1 Introduction Important as it is, ionic bonding is only one of three ways attractions between microparticles bind atoms into larger structural units. The other two are *covalent bonding* and *metallic bonding*. Covalent bonding between carbon, hydrogen, oxygen, nitrogen, and sulfur is particularly important because it provides the complex molecules necessary for life.

At the heart of covalent bonding is the idea of the *electron-pair bond*, in which a pair of valence electrons is *shared* between two atoms to make a bond. The differences between an ionic bond, a covalent bond,

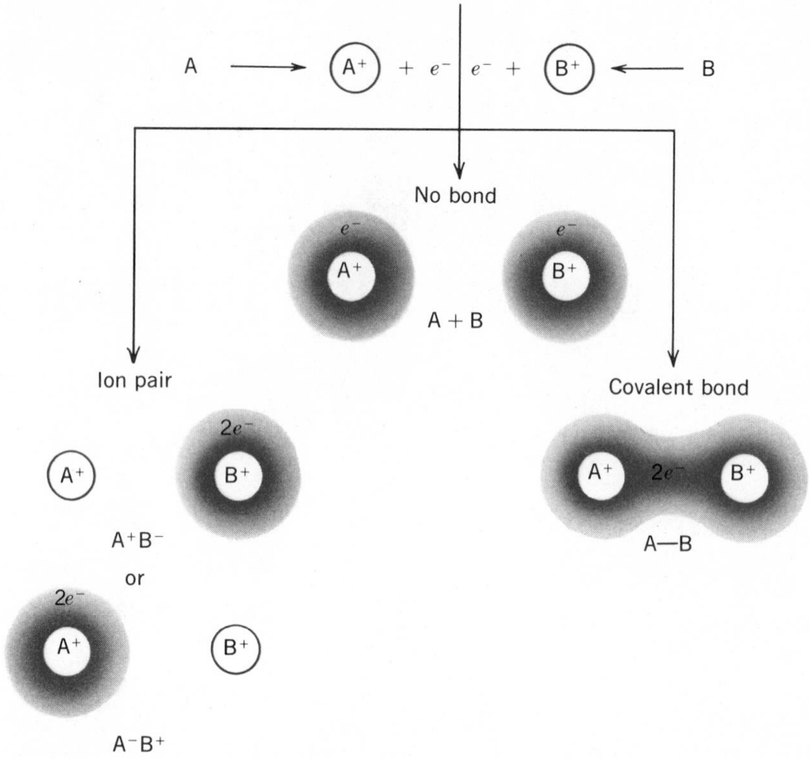

FIG. 4.1 Ion-pair and covalent bonds formation in the process $A^+ + B^+ + 2e^-$.

and no bond are shown in Figure 4.1. We imagine two separated atoms A and B to be ionized by the removal of one electron from each. The resulting ions A^+ and B^+ are then imagined to be brought within a typical bond-length distance (1–2 Å) of each other. Both of these steps, ionization and forcing the like-charged ions together, require an *input* of energy. Then the two electrons removed by ionization are returned to the A^+B^+ unit. The three basic fates of the pair of electrons are shown in the figure. If they return to the atomic orbitals of A^+ and B^+, atoms A and B are re-formed and no bond results. If both of them enter an atomic orbital of either A^+ or B^+, an ion pair (A^+B^- or A^-B^+) results, as shown. If ion pairs are formed, they join together to produce an ionic crystal. Finally, if the two electrons enter a *molecular orbital* that extends over both A^+ and B^+, a covalent bond (A—B) is formed with the production of a covalent molecule. If the temperature is low enough, the molecules will condense to form a liquid or a molecular crystal; otherwise they remain in the gaseous state. For example, at room temperature, fluorine atoms join to form covalent F_2 molecules that remain in the gas state, but iodine atoms form I_2 molecules that condense to solid iodine.

It is essential to bond formation of either type (ion pair or covalent) *that the energy released when the two electrons are returned to the* A^+B^+ *unit be greater than that put in to ionize atoms* A *and* B *and bring the resulting* A^+ *and* B^+ *ions within bonding distance*. If the energy released is just equal to that put in, no bond results. When a bond of either type *is* formed, the difference between the energy released and that put in is called the *bond dissociation energy* (D_{AB}).

EXAMPLE

⊸ The bond dissociation energy $D_{H—H}$ for the hydrogen molecule is 104 kcal per mole of bonds or 4.5 eV per bond. To see how this value arises we can put the steps in Figure 4.1 on a quantitative basis for H_2.

The first step is to ionize two hydrogen atoms: $2H^{\cdot} \rightarrow 2H^+ + 2e^-$. This step requires an input of energy equal to twice the ionization potential of the hydrogen atom or 2×13.6 eV = 27.2 eV. The second step, bringing the $2H^+$ to the H—H bond distance of 0.74 Å against the electrostatic force of repulsion between their charges, requires a further input of 19.7 eV. The last step, adding the $2e^-$ to the H^+H^+ unit, releases 51.4 eV because of the strong force of attraction between the opposite charges of protons and electrons. To summarize the energies:

Ionization of 2H atoms	+27.2 eV
To bring the H^+ together	+19.7 eV
Total energy put in	+46.9 eV
Energy released when $2e^-$ are put back	−51.4 eV
Total energy change	− 4.5 eV

The negative sign for the total energy change shows that 4.5 eV was *released* in forming the molecule. This is the bond dissociation energy $D_{H—H}$ for H_2. It is the amount of energy required to take a hydrogen atom apart into separated hydrogen atoms:

$$H—H + 4.5\ eV \rightarrow H^{\cdot} + H^{\cdot}$$

This calculation reveals an interesting feature of bond energies—they are small differences between very large energy quantities (compare 4.5 with 46.9 and 51.4). It has been said that measuring bond dissociation energies in this way is like weighing the captain of a ship by measuring the difference in the amount of water displaced by the ship when the captain is on board and on shore. Only a very small part of the total electronic potential energy is in the bond. ⊸

Bond dissociation energies in kcal per mole and bond lengths (distance between the centers of the nuclei of A^+ and B^+) for a number of covalent molecules are given in Table 4.1. Most of the molecules listed are of the type in which A and B are identical atoms. Not all of them are held together by *single* electron-pair bonds. Several (O_2, N_2, S_2, Te_2, C_2, CO, P_2, and As_2) involve *multiple covalent bonding* (Section 4.3) in which two or more electron-pair bonds are present. In any case, the bond dissociation energy is the energy *input* required to separate the molecule into its gaseous atoms. The values for bond dissociation energies are obtained from *molecular spectra* in a manner somewhat similar to the way energy levels for electrons in atoms are obtained from atomic spectra (Section 3.2). Two generalizations from the data in Table 4.1 are that short bonds tend to be strong bonds, and that multiple bonding is stronger than single bonding. The H—H single bond is exceptionally strong because it is exceptionally short. It is the shortest covalent bond.

Any theory of covalent bonding must answer the questions why do covalent bonds form at all and why is there a limit to the number of covalent bonds an atom may form? For example, why do hydrogen atoms form H_2 molecules and not H_3, H_4, . . . , and why do helium, neon, and the other noble gas atoms fail to form even the molecules He_2, Ne_2, . . . ? The theory must also provide an explanation for the three-dimensional geometry or *shapes* of molecules. Some of these questions are answered by *Lewis octet theory* for covalence, some

TABLE 4.1 Values of bond dissociation energies and bond lengths for diatomic molecules

Molecule	*States in Which Molecule Exists*	D_{AB} (kcal/mole)	*Bond Length* (Å)
H_2	*s, l, g*	104	0.74
O_2	*s, l, g*	118	1.21
N_2	*s, l, g*	225	1.10
F_2	*s, l, g*	36	1.42
Cl_2	*s, l, g*	57	1.99
Br_2	*s, l, g*	46	2.28
I_2	*s, l, g*	36	2.67
Li_2	*g*, $T > 1317$°C	25	2.67
Na_2	*g*, $T > 883$°C	17	3.08
S_2	*g*, $T > 445$°C	83	1.89
Te_2	*g*, $T > 990$°C	53	2.59
C_2	*g*, $T > 4200$°C	150	1.31
CO	*s, l, g*	256	1.13
HF	*s, l, g*	134	0.92
P_2	*g*, $T > 900$°C	116	1.89
As_2	*g*, $T > 615$°C	91	2.49
ICl	*s, l, g*	50	2.32
IF	*s, l, g*	46	1.99

require *molecular orbital theory*. These aspects of covalent bond theory are taken up in Sections 4.2 and 4.3.

4.2 Lewis Octet Theory for Covalent Bonding In Section 3.5 we saw how ionic bonding results from forming octets of electrons by the complete *transfer* of valence electrons from atoms of metals to atoms of nonmetals. According to Lewis theory, *covalent* bonding results when valence electrons are *shared* to obtain octets. For example, two fluorine atoms, with electron configuration $[He]2s^22p^5$, have seven valence electrons each. Each can get an octet if they share a pair of electrons and the shared pair is counted for both of them, as shown:

$$:\ddot{F}\cdot \ \ + \ \ ^{xx}_{xx}\!\!\overset{xx}{F}x \ \ \rightarrow \ \ \left(:\ddot{F}\overset{xx}{\underset{xx}{x}}Fx\right)$$

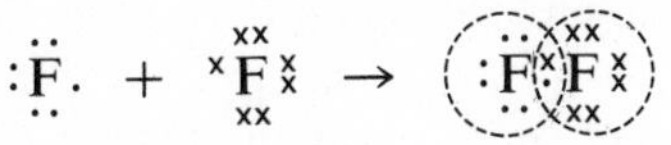

You need three pieces of information before even starting to construct the Lewis structure of a polyatomic molecule. First, you must know the *molecular* formula. Second, you need the number of valence electrons for each of the atoms in the molecule, because *all valence electrons must be shown in the structure.* (For Octet Table elements the number of valence electrons is simply their group number in the table. ☞ And finally, you must know *the order of bonding of the atoms in the molecule.*

As an example, take the covalent molecule ethanol. Analysis and molecular weight determination give C_2H_6O as the molecular formula. The total number of valence electrons to be shown is 20 — each carbon atom contributes 4, each hydrogen 1, and the oxygen 6. But you can't make any progress toward the dot structure of the compound unless you know that three of the hydrogens are bonded to one carbon, two of the hydrogens are bonded to the second carbon, the two carbons are bonded to one another, and the oxygen is bonded to one of the carbons and to one of the hydrogens. In other words, you have to know that the molecular skeleton is (a), not (b) or any other arrangement:

Octet theory is not very useful for the transition and inner transition metals.

```
 H  H                    H       H

H  C  C  O  H          H  C  O  C  H

 H  H                    H       H

   (a)                       (b)
```

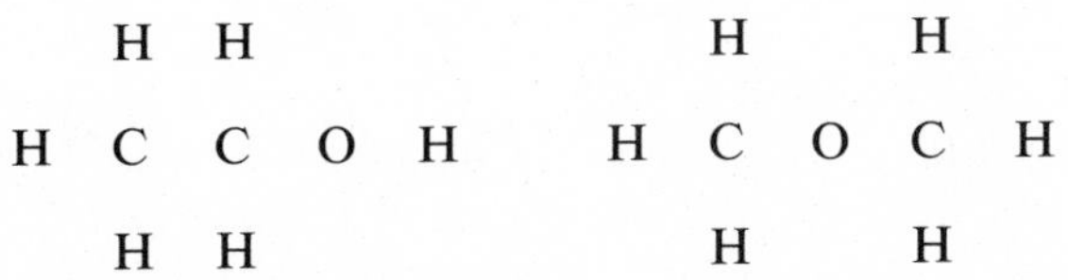

Information about molecular skeletons comes from X-ray crystal analysis, molecular spectrometry, and chemical properties. In the discussion that follows, it will be assumed that the skeleton structures have been determined.

Remembering that two electrons constitute a covalent bond, we can put twenty electrons into skeleton (a) in only one way that gives the carbon and oxygen atoms the octets required by Lewis theory.

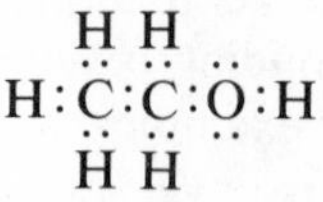

Dot structure

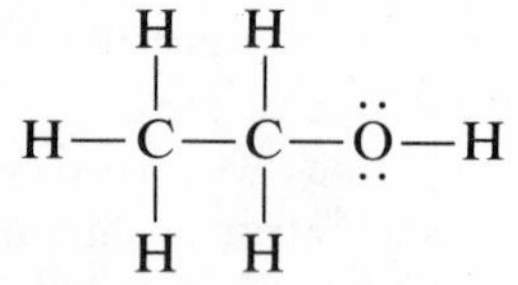

Corresponding bond structure

In counting the octet, you count the pairs of electrons in bonds, as well as those unshared ("unshared pairs"), for each atom. According to the theory, the combining power of the hydrogen atom is satisfied when it has a duet of electrons counted in the usual way. The bond structure results from replacing each shared pair by a dash and writing any unshared or "nonbonding" electrons as dots.

The structure of ethanol can be derived in another way. We do the thought experiment of putting the molecule together from its separated atoms written to show their valence electrons. For clarity the electrons may be distinguished by using different symbols to represent them:

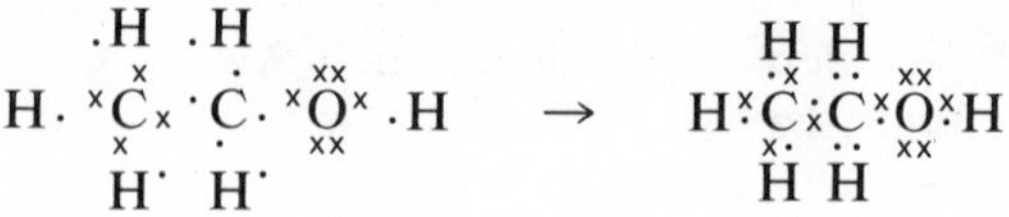

Ethanol

Here are other examples:

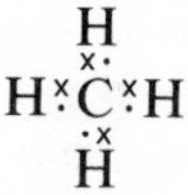

Methane

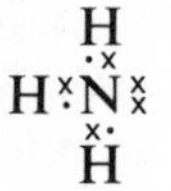

Ammonia

Hydrogen chloride

Hydrogen sulfide

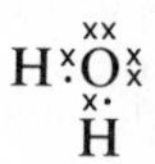

Water

A close look at these structures shows that the letter symbols for the atoms undergo subtle but important changes in meaning when used in Lewis structures. Taken by itself, the letter symbol C, for example, stands for a carbon atom *complete with its electron cloud* ($1s^2 2s^2 2p^2$). When put into a formula, such as that for ethanol, the letter symbol alone now stands for a carbon atom that has been stripped of its *valence* electrons ($2s^2 2p^2$) because these are now shown separately as dots. In formulas, the symbol C stands for a carbon nucleus surrounded by its *core* electrons only. An atom stripped of its valence electrons this way is called the *kernel* of the atom, and *the symbols for the elements stand for kernels in Lewis structures.* Kernels of atoms carry a positive charge equal to the number of valence electrons removed – the kernel of the carbon atom is C^{4+}, that of hydrogen is H^{+}, that of oxygen is O^{6+}, and so on. These charges are neutralized when the electrons are put back to form the bonds. To see this, note that in ethanol each hydrogen "owns" half of its shared pair or one electron to neutralize its +1 charge, each carbon owns half (four) of the eight electrons around it, just neutralizing the +4 charge on its kernel, and the oxygen atom owns four electrons outright in the unshared pairs and two more from the two shared pairs, making a total of six to neutralize the +6 charge of its kernel.

Structural isomerism Let's look now at the other skeleton (b) for C_2H_6O, which shows another way of arranging the atoms. Can the 20 electrons be put into this skeleton to give a Lewis structure? Indeed they can:

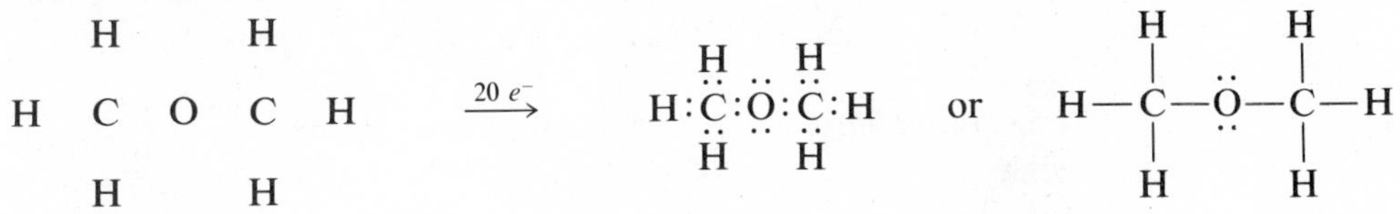

(b) Methyl ether

The theory predicts that there are *two* structures corresponding to the molecular formula C_2H_6O. You may ask are there *two* different compounds, distinguishable by their properties, that have the molecular formula C_2H_6O? The answer is yes. Ethanol is a liquid at ordinary temperature; the compound (b) in which the carbon atoms are linked through oxygen is called *methyl ether* and is a gas at ordinary temperatures. Ethanol and methyl ether are called *structural isomers. Structural isomers are compounds having the same molecular formula but different orders of attachment of their atoms.* Structural isomers are particularly important in organic compounds – the compounds of carbon (Chapter 13).

You may wonder whether structures like the following represent additional isomers of the formula C_2H_6O:

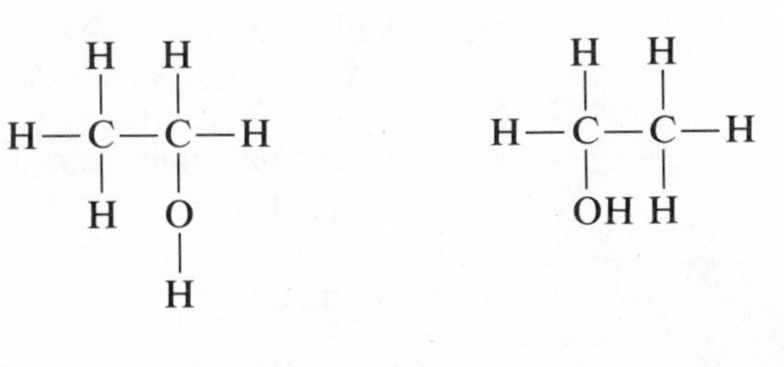

At first sight they certainly seem to; in fact they do not, and that they don't represent additional isomers of the formula C_2H_6O reveals one of the weaknesses of octet theory—*it gives no hint of the geometry of molecules*, that is, of the way in which bonds are directed in space. As we shall see in Section 4.4, bonds from carbon are directed in space in such a way that only one ethanol molecule and only one methyl ether molecule are possible, and these two molecules are the *only* isomers of formula C_2H_6O. If you build the two molecules, using ball-and-stick atomic models, you will quickly see that this is true, assuming free rotation about single bonds, which goes on at any except the lowest temperatures (−270°C).

Multiple bonding It was pointed out earlier that some of the molecules in Table 4.1 involve multiple bonds rather than single electron-pair bonds. Multiple bonding is present in the nitrogen molecule (N_2). Each nitrogen atom has five valence electrons, and you get an octet about each nitrogen if you put six of the ten valence electrons in three electron-pair bonds between the nitrogens and leave one unshared or nonbonding pair on each. The result is $^{x}_{x}N\overset{xxx}{\underset{\cdot\cdot\cdot}{}}N{:}$ or $^{x}_{x}N{\equiv}N{:}$.

EXAMPLES ⊸ Some other important molecules having multiple bonds are the following:

Name	*Structure*		*Importance*
Carbon monoxide CO	$:C:::O:$,	$:C{\equiv}O:$	Poisonous gas in auto exhaust in particular and smog in general
Carbon dioxide CO_2	$\underset{\cdot\cdot}{\ddot{O}}::C::\underset{\cdot\cdot}{\ddot{O}}$,	$\underset{\cdot\cdot}{\ddot{O}}{=}C{=}\underset{\cdot\cdot}{\ddot{O}}$	Product of combustion and respiration. Source of carbon for photosynthesis by plants. Reacts with H_2O to form carbonic acid: $CO_2 + H_2O \rightarrow H_2CO_3$
Ethene (ethylene) C_2H_4	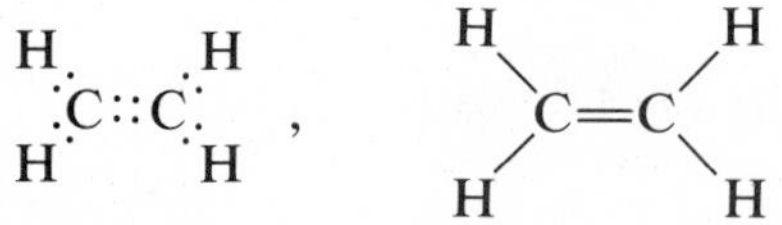		Product of petroleum refining, raw material for polyethylene plastic and other chemicals
Ethyne (acetylene) C_2H_2	$H:C:::C:H$,	$H{-}C{\equiv}C{-}H$	Used in oxyacetylene welding, and as raw material for plastics
Sulfur dioxide SO_2	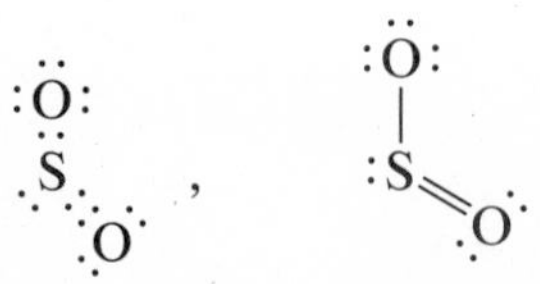		Atmospheric pollutant from burning high-sulfur coal and fuel oil. In stack gases from copper refineries

Sulfur trioxide
SO_3

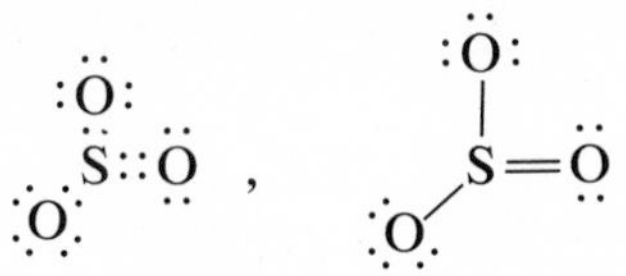

Produced by reaction of SO_2 with oxygen: $2SO_2 + O_2 \rightarrow 2SO_3$. Atmospheric pollutant

Hydrogen cyanide
HCN

H:C:::N: , H—C≡N:

Very poisonous gas, used for executions in "gas chambers." Used to make sodium cyanide for electroplating baths: $HCN + Na^+OH^- \rightarrow Na^+CN^- + H_2O$

Nitric acid
HNO_3

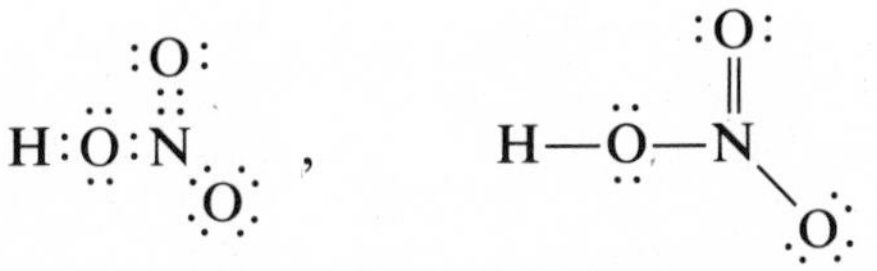

Strong acid. Used to make explosives (dynamite, TNT, RDX), fertilizer ($NH_4^+NO_3^-$, ammonium nitrate), and many dyes for cloth

Multiple bonding is particularly important in carbon compounds. Of all the atoms, the carbon atom shows the greatest tendency to form double and triple covalent bonds with itself and other atoms. This property, combined with isomerism, permits the formation of the complex molecules necessary for life. Simple examples of multiple bonding are found in acetic acid, the sour-tasting component of vinegar, and in the *peptide linkage* of proteins.

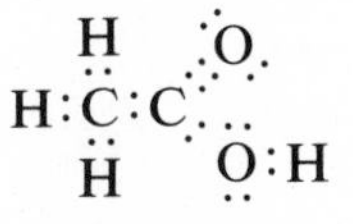

or

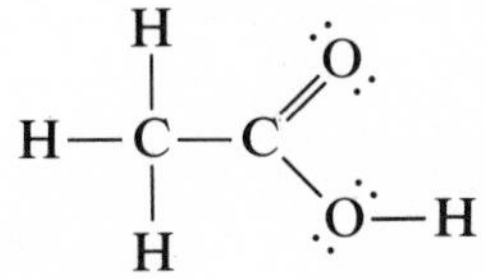

Acetic acid

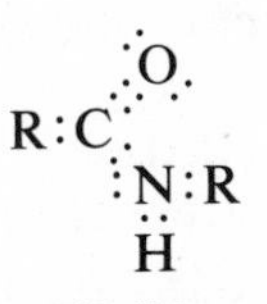

or

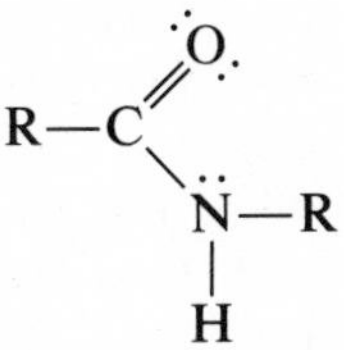

Peptide linkage

The R's in the structure of the peptide linkage stand for complex, covalently bonded groups of atoms.

Exercises

1. The organic compound allene has its atomic kernels arranged as follows:

H H

 C C C

H H

Give the dot and covalent bond structures for allene.

2. Correct any of the following Lewis structures that are incorrect, remembering that all valence electrons must be shown.

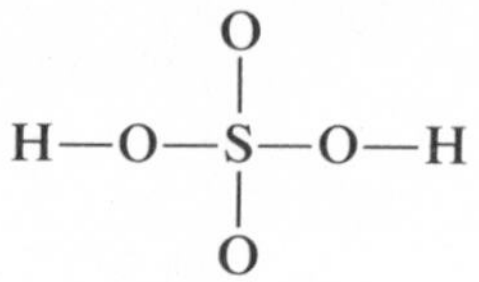

O
|
H—O—S—O—H
|
O

Sulfuric acid

$$\begin{array}{c} H \\ \diagdown \\ C{=}\ddot{\underset{\cdot\cdot}{O}} \\ \diagup \\ H \end{array}$$

Formaldehyde

$$C{=}N^{1-}$$

Cyanide ion

$$\begin{array}{c} H \\ \diagdown \\ \ddot{O}{-}H \\ \diagup \\ H \end{array}$$

Hydronium ion

:O:
‖
H—O—N

Nitrous acid

$$\begin{array}{c} H \quad\quad H \\ \diagdown \quad\quad \diagup \\ H{-}C{-}C{-}H \\ \diagup \quad\quad \diagdown \\ H \quad\quad H \end{array}$$

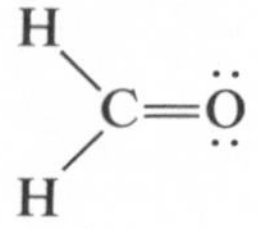

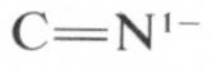

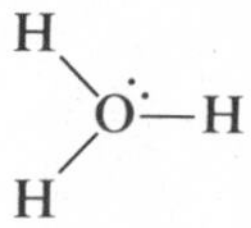

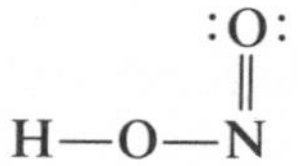

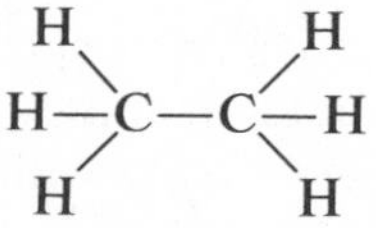

Ethane

Resonance theory When writing Lewis structures for molecules or complex ions, you may sometimes find structures that *differ only in the position of the electrons.* Thus for the sulfur dioxide (SO_2) molecule you can write two dot structures:

$$:\!S\!\overset{\cdot\cdot}{\underset{\cdot\cdot}{O}}\!\!\!\diagup\quad\text{or}\quad :\!S\overset{O}{\diagup}\!\!\!\diagdown_{O}\quad\text{and}\quad :\!S\overset{O}{\diagup}\!\!\!\diagdown_{O}\quad\text{or}\quad :\!S\overset{O}{\diagdown}\!\!\!\diagdown_{O}$$

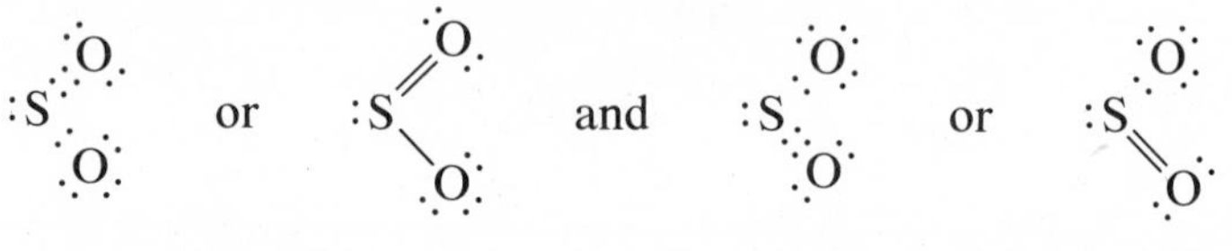

Which is correct? Neither, yet both! Experiment shows that there is only *one* sulfur-to-oxygen distance in this molecule, 1.432 Å. Since double bonds are known to be shorter than single bonds, neither of the structures with single and double bonds can be correct. In cases of this kind, the true structure of the molecule is a *blend* or *resonance hybrid* of the written structures. This doesn't mean that sulfur dioxide is a *mixture* of two kinds of molecules, but just that simple octet theory does not account for there being only one S–to–O bond distance. Electrons are very mobile, and in the SO_2 molecule they distribute themselves evenly with the result that there are on the average *three* electrons bonding each oxygen to the sulfur. This kind of bonding can be shown by using a dashed bond to represent covalent bonding by *one* electron while the familiar full dash stands for a bonding pair. Using this symbolism, we write the structure of SO_2 as

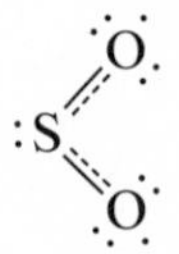

Here each kernel is surrounded by an octet. Another way to show the hybrid nature of the bonding is to use a double-headed arrow between the Lewis structures:

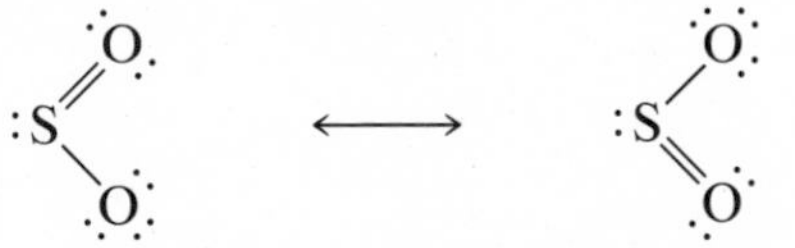

It is basic to understanding this use of formulas to realize that *the arrow does not imply any motion of nuclei or electrons*, but rather that the two structures taken together show the actual arrangement of electrons in bonds in the molecule. The dashed-line structure expresses this idea more clearly, but the resonance hybridization using the double-headed arrow is often used. The structures making up the hybrid are often called *canonical* structures, from "canon," meaning law or rule—in this case the rules for writing Lewis structures.

3. The arrangement of atoms in phosphoric acid, H_3PO_4, is Exercise

O H

O P O H

O H

How many electrons must be shown in the Lewis dot structure? Give the dot structure for H_3PO_4.

Complex ions Many complex ions involve both multiple bonding and resonance hybridization. A good example is the carbonate ion CO_3^{2-}, found in marble, limestone, pearls, and oyster shells in the form of calcium carbonate ($Ca^{2+}CO_3^{2-}$). Putting the carbonate ion together from its atoms and the excess of two electrons responsible for the 2− charge on the ion will give you a review of octet theory and another example of resonance. X-ray crystallography shows that the CO_3^{2-} ion is flat, the carbon atom is central, the oxygen atoms are all at a distance of 1.31 Å from the carbon atom, and the O—C—O angles are all exactly 120°. Thus we start with the skeleton

$$
\begin{array}{cc}
\text{O} & ^{2-} \\
\text{C} & \\
\text{O} \quad \text{O} &
\end{array}
$$

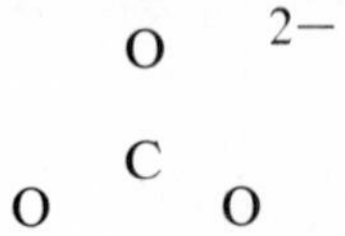

The three oxygen atoms contribute 18 valence electrons, the carbon atom contributes 4, *and the double negative charge on the ion contributes an additional 2*, making a total of 24 electrons to be shown in the dot structure. They may be placed to give octets for all the kernels involved as follows:

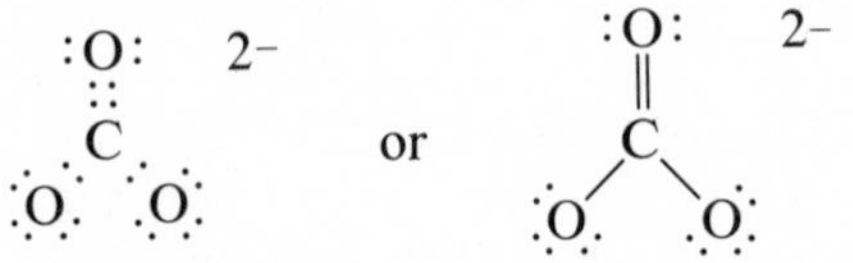

The fact that there is only one C—O distance in the ion means that it must be a hybrid of three equivalent structures:

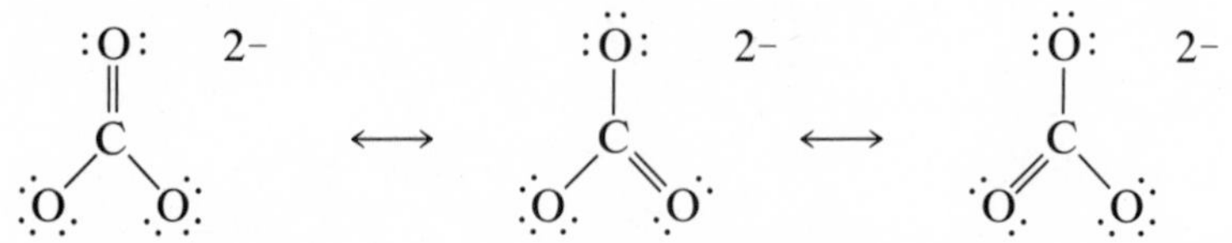

EXAMPLE ☛ The carbonate ion CO_3^{2-} has come down from ancient times in the form of statues and buildings made of marble, a white, crystalline form of $Ca^{2+}CO_3^{2-}$. Many of these objects of art have been damaged or destroyed by only a few years of exposure to the polluted air in heavily industrialized regions. The pollutants responsible for the damage are sulfurous acid and sulfuric acid, coming by a series of reactions from burning coal and fuel oil that contain sulfur compounds. As the coal or oil burns, the sulfur compounds are converted to gaseous sulfur dioxide and sulfur trioxide. Using *thiophene* as the sulfur-containing compound in fuel oil, we obtain the equations:

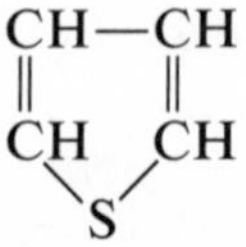

$$CH\!-\!CH$$
$$\|\qquad\|$$
$$CH\quad CH + 6O_2 \rightarrow 4CO_2 + 2H_2O + SO_2, \quad 2SO_2 + O_2 \rightarrow 2SO_3$$
$$\diagdown_{S}\diagup$$

Thiophene

The SO_3 produced reacts with water vapor in the air to produce tiny droplets of sulfuric acid in the smoggy atmosphere:

$$SO_3 + H_2O \rightarrow H_2SO_4$$

The sulfuric acid is brought back to earth in the rain, and attacks the marble by destroying the CO_3^{2-} ions, replacing them with sulfate, SO_4^{2-}, ions. The carbonate ions are converted to carbonic acid, which decomposes to CO_2 and water:

$$Ca^{2+}CO_3^{2-} + H_2SO_4 \rightarrow Ca^{2+}SO_4^{2-} + H_2CO_3$$
$$H_2CO_3 \rightarrow CO_2 + H_2O$$

By these reactions the surface of the marble is slowly eaten away. (Photo 12 demonstrates what an acidic atmosphere does in 60 years.) There are two solutions to this problem—paint or coat the marble with a plastic to keep the H_2SO_4 out, or design methods for "scrubbing" SO_2 out of chimney gases and auto and truck exhausts where sulfur-containing fossil fuels are used.

The structures and geometries of some other molecules and complex ions that are resonance hybrids are in Figure 4.2, page 136. Note that for negative ions the total number of electrons shown is the sum of the number of valence electrons of the component atoms *and* the charge on the ion.

The classic case of resonance hybridization is presented by the benzene molecule, which is part of hundreds of molecules important in biochemistry, drugs, and plastics. Using bond structures, we represent benzene as a hybrid of two structures:

Frederick A. Kekulé (1829–1896), professor at the Universities of Ghent and Bonn, assigned the correct canonical structure for benzene more than fifty years before the theory of resonance was invented. He first studied architecture, later turned to chemistry, and laid the foundations of structural theory of organic chemistry.

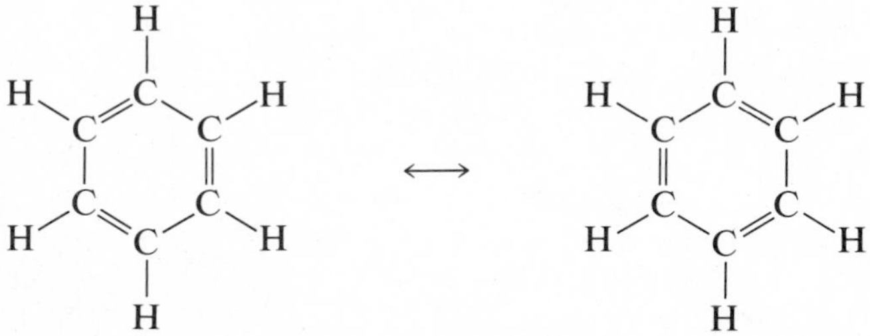

In accord with theory, the molecule does not have alternating single and double bonds, but instead has three electrons bonding each of the carbon atoms to its neighbors to give the C—C distance 1.39 Å. This value lies between that in two-electron-bonded ethane and four-electron-bonded ethene.

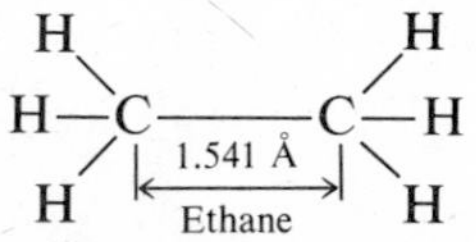

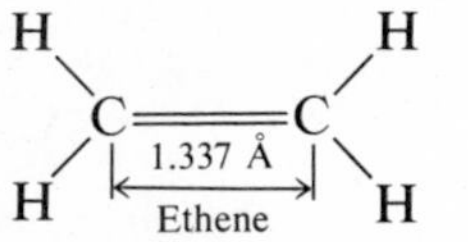

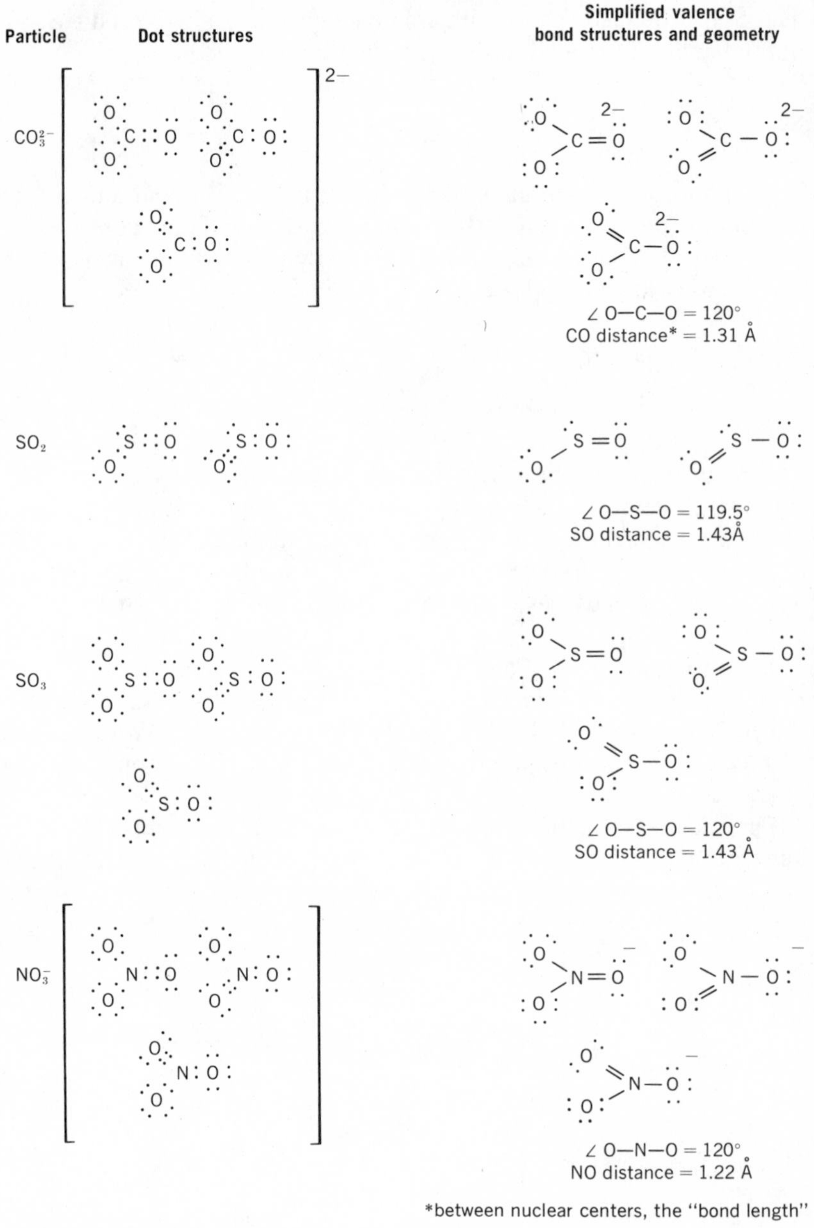

FIG. 4.2 Canonical structures for some molecules and ions: resonance hybrids.

To show that the bonding between the carbon atoms is identical, the structure of benzene is often written as follows:

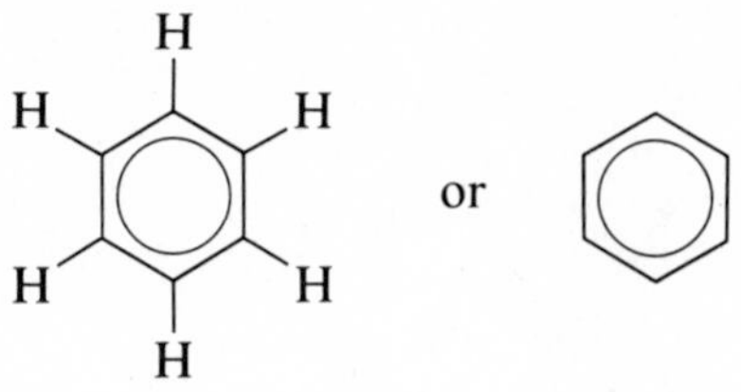

One important result of the theory of resonance is that *the real molecule is always more stable than any of the canonical structures used in the hybridization.* How this applies to benzene is shown in Figure 4.3. The benzene ring or benzene "nucleus," as it is termed, is an important structural unit in many compounds such as vitamins and hormones. Bond structures for some examples are in Figure 4.4.

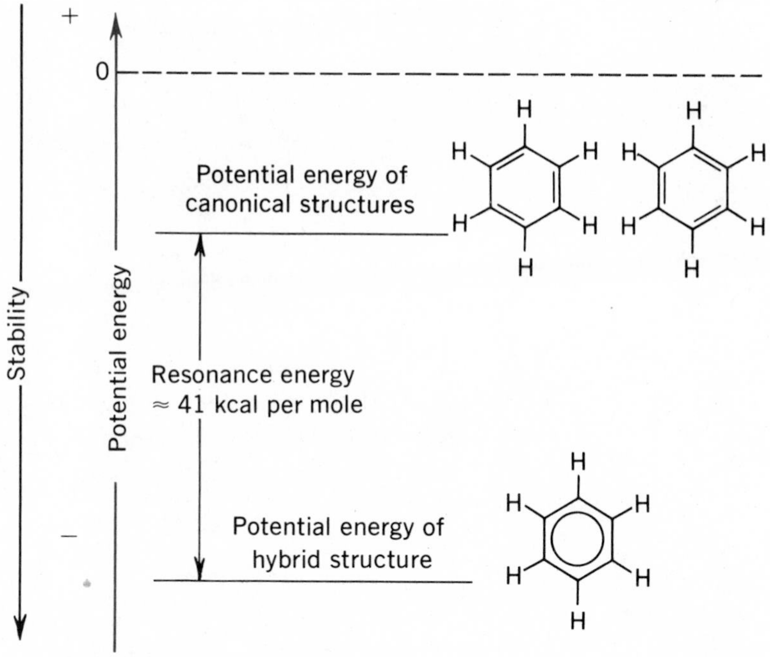

FIG. 4.3 Meaning of resonance energy for the benzene molecule (note abbreviated symbols for the benzene ring).

Lewis theory combined with resonance still leaves important questions about bonding unanswered. If H_2, why not He_2; if F_2, why not Ne_2? Also, how may the geometry of molecules be explained or predicted? These are taken up in Sections 4.3 and 4.4.

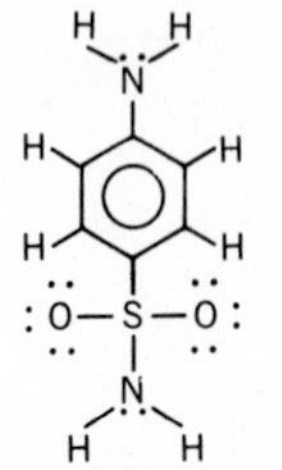

Sulfanilamide
(antibacterial agent)

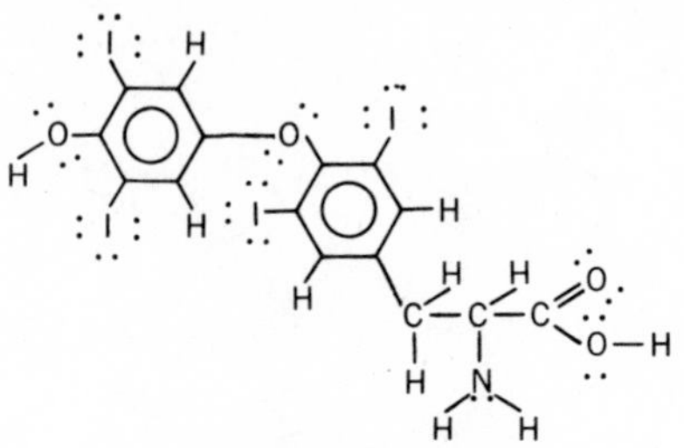

Thyroxine
(hormone of the thyroid gland)

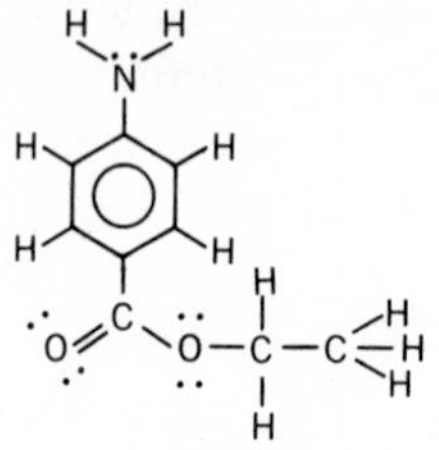

Benzocaine
(local anesthetic)

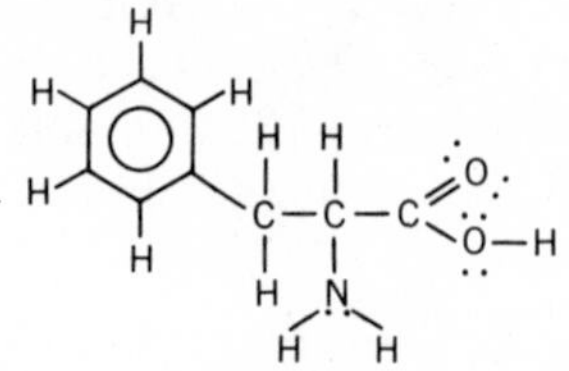

Phenylalanine
(aminoacid)

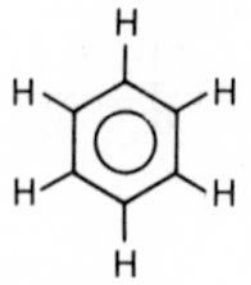

= Benzene

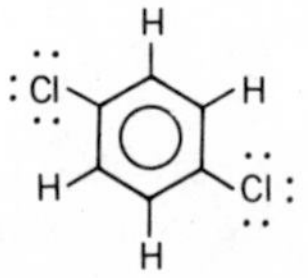

Paradichlorobenzene
(moth crystals)

FIG. 4.4 Some examples of molecules from the multitude containing the benzene-ring system (abbreviated notation, carbon atoms in ring not shown explicitly).

Exercises

4. One Lewis structure for the bicarbonate ion (hydrogen carbonate ion), the anion in sodium bicarbonate (baking soda, $Na^+HCO_3^-$), is

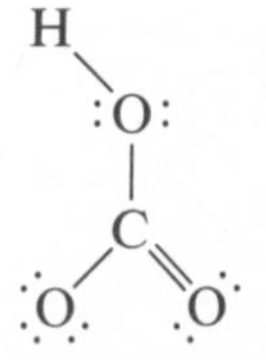

Give another structure. Is there resonance hybridization?

5. Naphthalene, $C_{10}H_8$, has the structure of two benzene rings fused together:

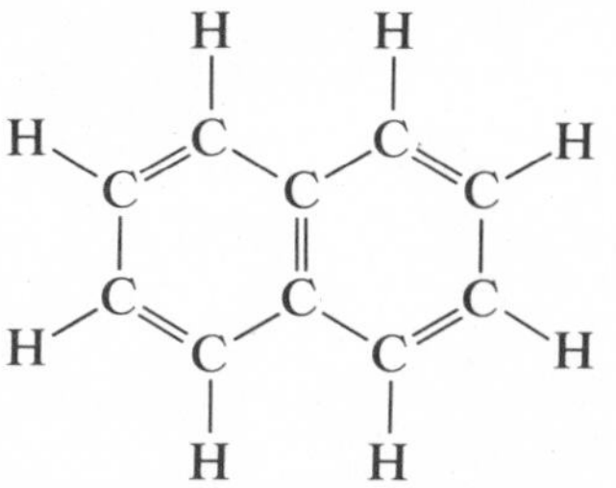

Is this the only canonical structure for naphthalene or should it be considered a resonance hybrid?

6. Should the hydroxide ion: $\ddot{\underset{\cdot\cdot}{O}}$—$H^{1-}$ be represented as a resonance hybrid?

4.3 Molecular Orbital Theory Lewis structures are static representations because the electrons are represented as stationary dots. Actually the electrons are in rapid motion, and molecular orbital theory attempts to describe the regions in space (molecular orbitals) in which they are most likely to be found. As with atomic orbitals, there are several types of molecular orbitals. The most important types are sigma (σ) and pi (π).

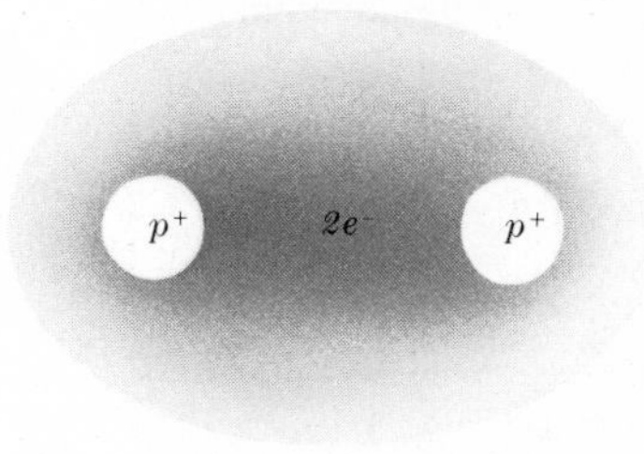

FIG. 4.5 Sigma molecular orbital for H—H.

The differences between sigma and pi molecular orbitals are seen by comparing Figures 4.5 and 4.6. The differences are similar to the differences between *s* and *p* atomic orbitals. Atomic orbitals of the *p* type have a nodal plane through the nucleus of the atom. Molecular orbitals of the π type have a nodal plane that passes through the *two* nuclei that are bonded. Because of the nodal plane, the π electrons have a high probability of being on the two sides of the plane through the atomic nuclei. Electrons in sigma molecular orbitals have a high probability of being found in the space right between the bonded nuclei. Sigma orbitals have an ellipsoidal or egg shape, just as their atomic counterparts are spherical or ball-shaped.

Sigma and pi molecular orbitals are further classified as *bonding* or *antibonding*. The orbitals in Figures 4.5, 4.6, and 4.7(a) are of the bonding type. Antibonding orbitals have a nodal plane (plane of zero probability of finding the electrons) that is *perpendicular* to the line joining the nuclei at its midpoint, as shown in Figure 4.7(b). This nodal plane keeps electrons out of the bonding region between the nuclei, and so the term *antibonding* is used. Bonding orbitals are labeled σ (sigma) and π (pi); the antibonding orbitals are shown by a star, as in $\sigma^\star$ and $\pi^\star$. Electrons in antibonding orbitals cancel the bonding resulting from electrons in bonding orbitals.

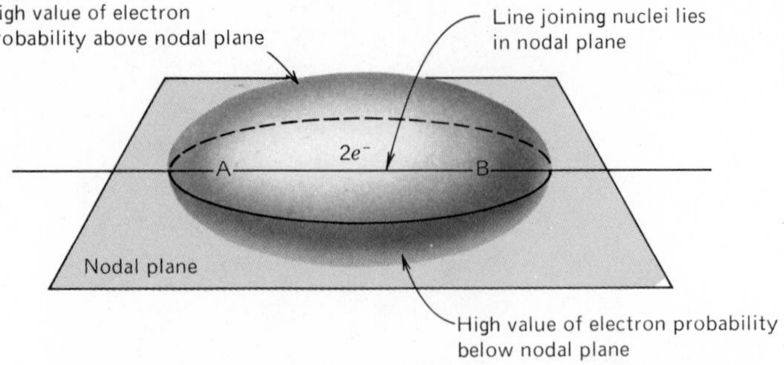

FIG. 4.6 Nodal plane through nuclei A and B producing a π orbital.

The way the energy sequence of the orbitals and their shapes are derived from atomic orbitals is described in Appendix 5.

Molecular orbitals for two-atom (diatomic) molecules are arranged in an energy or stability sequence, just as atomic orbitals are. This energy sequence is shown schematically in Figure 4.8. The rules for filling molecular orbitals are the same as the rules for filling atomic orbitals. Molecular orbitals are filled with two electrons of opposite

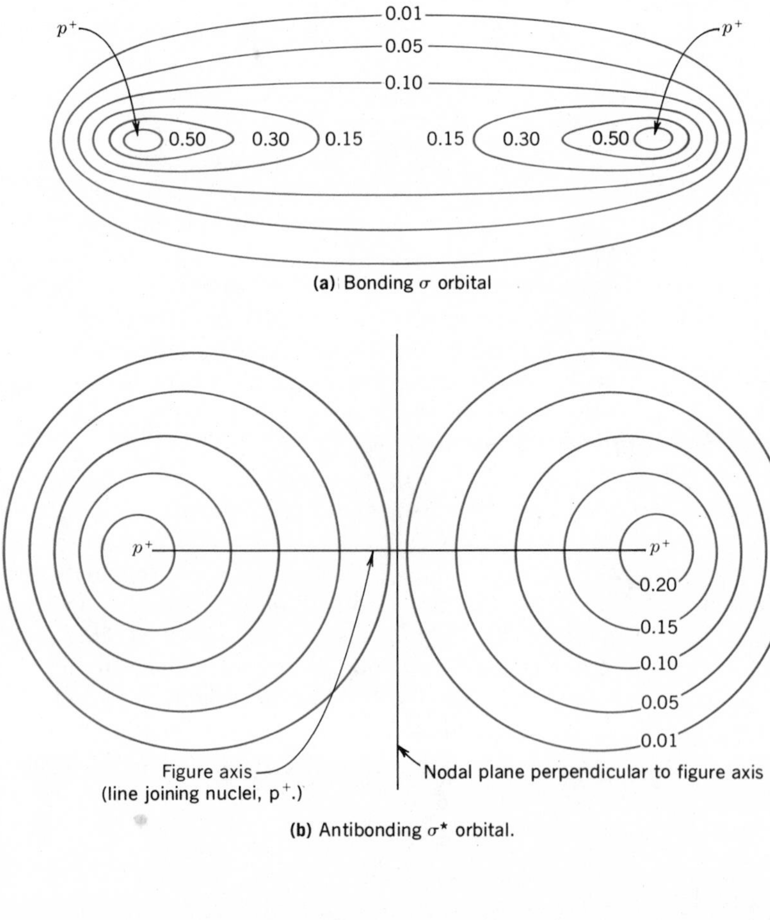

FIG. 4.7 Contour lines for electron probability in σ and σ^* orbitals as seen in a cross section through the protons in H_2.

spin (Pauli exclusion principle) in order of decreasing stability—that is, starting from the bottom in Figure 4.8. Hund's rule also applies—if empty orbitals of the same energy are available, electrons enter them separately and without pairing spins as long as possible.

Building up diatomic molecules Molecular orbital theory for diatomic molecules explains why H_2 is a stable molecule yet He_2 is not in the following way: each hydrogen atom has one 1*s* electron, so when two H atoms are brought together, the two electrons pair their spins and enter the σ_1 bonding orbital to produce the H_2 molecule. Two helium atoms, He $(Z = 2)$, provide four electrons to be located in molecular

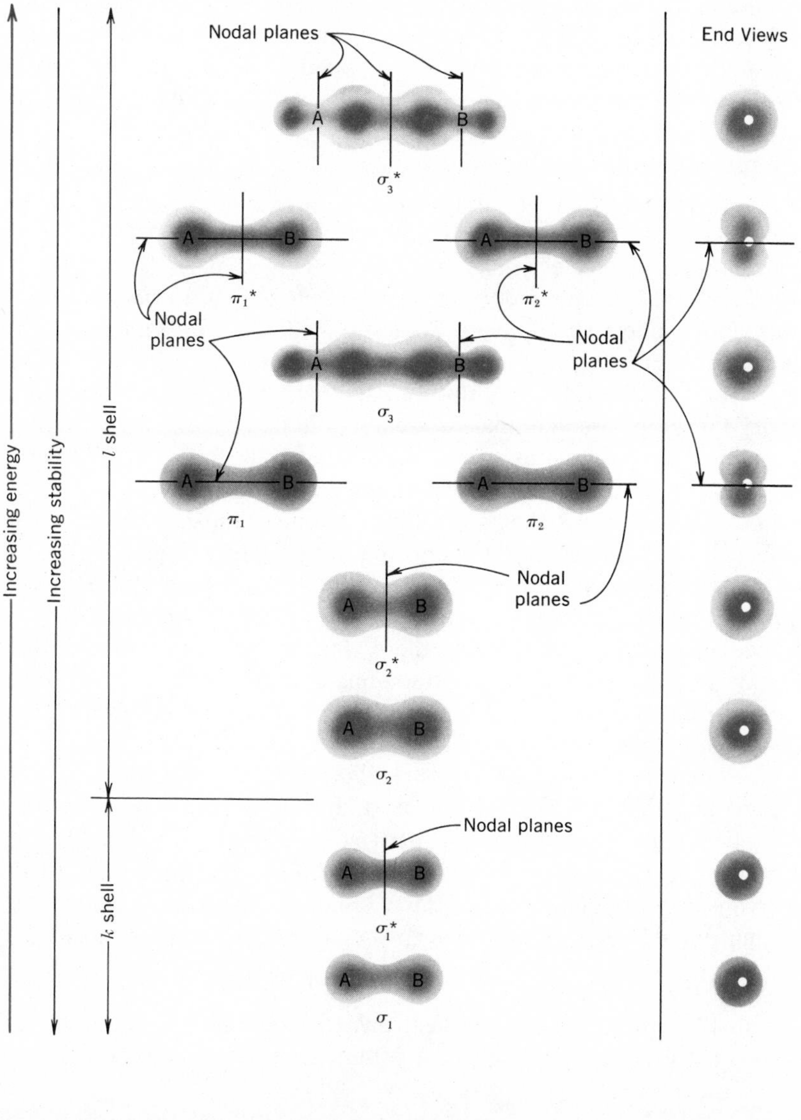

FIG. 4.8 Cross section of molecular orbitals for homopolar diatomic molecules (heavy lines stand for nodal planes; electron-cloud representation).

orbitals. One pair goes into the σ_1 bonding orbital, as with hydrogen, but the second pair has to go into the $\sigma_1^\star$ *antibonding* orbital, since the $\sigma_1^\star$ orbital is next in the energy sequence. The presence of the two electrons there cancels the bonding of the two electrons in the σ_1 orbital, so no He_2 molecule is formed from two He atoms.

Note that according to molecular orbital theory *all* electrons—valence *and* core—are put in molecular orbitals.

The nitrogen molecule, for which the Lewis structure is :N:::N: or :N ≡ N:, is neatly described by molecular orbital theory. Since each nitrogen atom ($Z = 7$) contributes seven electrons there are fourteen to be placed in molecular orbitals. Starting with σ_1, the arrangement of the fourteen electrons that results is $\sigma_1^2[\sigma_1^\star]^2\sigma_2^2[\sigma_2^\star]^2\pi_1^2\pi_2^2\sigma_3^2$. The bonding provided by the filled σ_1 and σ_2 bonding orbitals is neutralized by the filled $\sigma_1^\star$ and $\sigma_2^\star$ antibonding orbitals. After this cancellation a net of three covalent bonds is left, given by π_1^2, π_2^2, and σ_3^2. Molecular orbital theory and octet theory thus come to the same result—the nitrogen atoms in N_2 are strongly held together by a trio of electron-pair covalent bonds. Molecules of N_2 make up about $\frac{4}{5}$ of the molecules in air. The other $\frac{1}{5}$ is mainly O_2 molecules.

Surprisingly, the two theories do *not* come to identical results for the oxygen molecule, O_2. The dot structure $\ddot{\underset{..}{\mathrm{O}}}::\ddot{\underset{..}{\mathrm{O}}}$ satisfies octet theory. However, molecular orbital theory gives the following arrangement for the sixteen electrons of the two oxygen atoms: $\sigma_1^2[\sigma_1^\star]^2\sigma_2^2[\sigma_2^\star]^2\pi_1^2\pi_2^2\sigma_3^2[\pi_1^\star]^1[\pi_2^\star]^1$. The important thing to note is that the last two electrons added obey Hund's rule by going into *separate* molecular orbitals of the same energy, the $\pi_1^\star$ and $\pi_2^\star$ antibonding orbitals. Further, in accord with the rule, they do not pair their spins. Molecular orbital theory therefore predicts that oxygen should be attracted by a magnet because of the "atomic magnets" produced by the two unpaired electron spins in its molecules. Oxygen molecules are attracted by a magnet, and the prediction that they should be was one of the great triumphs of molecular orbital theory. Although the dot structure does not predict the proper magnetic properties, it is correct in showing that the oxygens are bonded by four electrons in two bonds. According to the molecular orbital occupancies, there are ten electrons in bonding orbitals and six electrons in antibonding orbitals. There is therefore a net of four electrons, equivalent to two electron-pair bonds, holding the oxygen molecule together.

Molecular orbital theory also predicts that F_2 should be singly bonded and that Ne_2 should not exist. For F_2 with 18 electrons the orbital occupancies are $\sigma_1^2[\sigma_1^\star]^2\sigma_2^2[\sigma_2^\star]^2\pi_1^2\pi_2^2\sigma_3^2[\pi_1^\star]^2[\pi_2^\star]^2$. In this arrangement there are ten electrons in bonding orbitals and eight electrons in antibonding ones. Two bonding electrons, σ_3^2, are left, making one covalent bond between the fluorines. The octet structure $:\ddot{\underset{..}{\mathrm{F}}}:\ddot{\underset{..}{\mathrm{F}}}:$ is therefore correct in showing one bonding pair.

With Ne_2, there are twenty electrons to accommodate. They fill all of the ten orbitals available in the orbital scheme, giving the arrangement $\sigma_1^2[\sigma_1^*]^2\sigma_2^2[\sigma_2^*]^2\pi_1^2\pi_2^2\sigma_3^2[\pi_1^*]^2[\pi_2^*]^2[\sigma_3^*]^2$. Ten of the electrons are in bonding orbitals, ten in antibonding orbitals, so there is no net bonding to hold an Ne_2 molecule together. There is no evidence that Ne_2 exists.

In these ways, molecular orbital theory answers the question posed earlier: if H_2 and F_2 form, why don't He_2 and Ne_2? They don't because filled orbital "shells" have equal numbers of electrons in bonding and antibonding orbitals. But molecular orbital theory does much more, as you can see from the contents of Table 4.2. There the electron arrangements for all the homonuclear (same nucleus) diatomic (two-atom) molecules and some molecular ions for the elements of periods 1 and 2 are shown. The observed bond dissociation energies and magnetic properties are also given.

⊸ Although the O_2^+ ion is found only fleetingly in electric discharges EXAMPLE
through oxygen gas, it is no stranger to the upper atmosphere, where the intense radiation from the sun ionizes the O_2 molecule:

$$O_2(g) + 288 \text{ kcal/mole} \rightarrow O_2^+ + e^-$$

Spectroscopic studies of O_2 and O_2^+ reveal the remarkable fact that O_2^+ is *more* stable than O_2.

$$O_2(g) + 118 \text{ kcal/mole} \rightarrow 2 : \ddot{O} : (g)$$

$$O_2^+(g) + 149 \text{ kcal/mole} \rightarrow : \ddot{O} : (g) + : \ddot{O} .^+(g)$$

Octet theory predicts the opposite order of stabilities, O_2 more stable than O_2^+. The electron removed from O_2 has to break the octet of one oxygen or both, depending on where it is taken from.

$$\ddot{O} :: \ddot{O} \text{---} \left\langle \begin{array}{l} \rightarrow \ddot{O} :: \dot{O}^+ + e^- \\ \rightarrow \ddot{O} :. \ddot{O}^+ + e^- \end{array} \right.$$

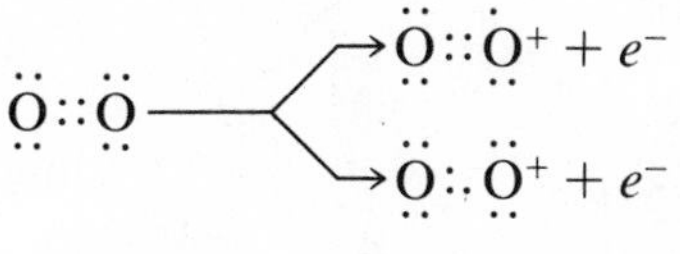

Destruction of the octet(s) should make O_2^+ *less* stable. You can see immediately why O_2^+ is more stable (with fewer electrons) than O_2 if you compare their molecular orbital electron configurations:

O_2 $\sigma_1^2[\sigma_1^*]^2\sigma_2^2[\sigma_2^*]^2\pi_1^2\pi_2^2\sigma_3^2[\pi_1^*]^1[\pi_2^*]^1$

O_2^+ $\sigma_1^2[\sigma_1^*]^2\sigma_2^2[\sigma_2^*]^2\pi_1^2\pi_2^2\sigma_3^2[\pi_1^*]^1$

The electron removed going from O_2 to O_2^+ comes from one of the π^* orbitals and they are *anti*bonding. Decreasing the number of electrons

TABLE 4.2 Ground-state electron configurations of homonuclear diatomic molecules of elements of periods 1 and 2

	Molecular Orbitals	H_2^+	H_2	He_2^+	"He_2"	Li_2	"Be_2"	B_2	C_2	N_2^+	N_2	O_2^+	O_2	F_2	"Ne_2"
L shell	σ_3^*														⇅
	$\pi_1^*\pi_2^*$											↑	↑ ↑	⇅ ⇅	⇅ ⇅
	σ_3									↑	⇅	⇅	⇅	⇅	⇅
	$\pi_1\pi_2$							↑ ↑	⇅ ⇅	⇅ ⇅	⇅ ⇅	⇅ ⇅	⇅ ⇅	⇅ ⇅	⇅ ⇅
	σ_2^*						⇅	⇅	⇅	⇅	⇅	⇅	⇅	⇅	⇅
	σ_2					⇅	⇅	⇅	⇅	⇅	⇅	⇅	⇅	⇅	⇅
K shell	σ_1^*			↑	⇅	⇅	⇅	⇅	⇅	⇅	⇅	⇅	⇅	⇅	⇅
	σ_1	↑	⇅	⇅	⇅	⇅	⇅	⇅	⇅	⇅	⇅	⇅	⇅	⇅	⇅
Bonding **valence** e^-		1	2	2	2	2	2	4	6	7	8	8	8	8	8
Antibonding **valence** e^-		0	0	1	2	0	2	2	2	2	2	3	4	6	8
Bond order		$\frac{1}{2}$	1	$\frac{1}{2}$	0	1	0	1	2	$2\frac{1}{2}$	3	$2\frac{1}{2}$	2	1	0
Lewis structure		H·H⁺	H:H	·He·He⁺·	no bond	Li:Li	no bond	:B:B:	:C::C:	:N:·:N:⁺	:N:::N:	:O::·O:⁺	:O::O:	:F:F:	no bond
Dissociation energy[a]		61	103	58	0	25	0	69	150	146	225	149	118	36	0
Paramagnetic		yes	no	yes	no	no	no	yes	no	yes	no	yes	yes	no	no

[a]Spectrometric value in kcal per mole.

in antibonding orbitals leads to greater bonding strength in O_2^+, as observed.

Exercises

7. Without consulting Table 4.2, give the molecular orbital description of the bonding in Be_2. Would the molecule be stable?

8. Assuming that the energy sequence of molecular orbitals for the $m(n=3)$ molecular shell is like that for the L shell, give the molecular orbital description for S_2. Would it be paramagnetic?

Polar molecules—dipole moments Molecular orbitals for diatomic molecules having identical nuclei (H_2, Cl_2, F_2, O_2, and so on) are symmetrical in such a way that the center of negative charge of their electrons is at the same point as the center of positive charge located midway between the atomic nuclei [Figure 4.9(a)]. Molecules in which the centers of + and − charge coincide this way are called nonpolar. On the other hand, if the nuclei are not identical, one of them usually has a greater attracting power for electrons that distorts the orbital in such a way that the centers of negative and positive charge do not coincide [Figure 4.9(b)]. In this case a *polar* molecule results. Polar molecules have *permanent electric dipole moments*, and the size of the dipole moment (μ), measured in *debye* units, is a measure of the distortion of the electron cloud and the polarity of the molecule. Molecules with permanent dipole moments act like atomic sticks with positive and negative charged ends, and they *tend* to orient themselves slightly when placed between plates bearing opposite electric charges. Their positive ends tend to point toward the negative plate and negative ends toward the positive plate (Figure 4.10). This tendency to line up weakens the electric field between the plates, and the amount of weakening is used to determine the value of the dipole moment of the molecule. There is only a tendency to line up; random thermal motion prevents permanent alignment.

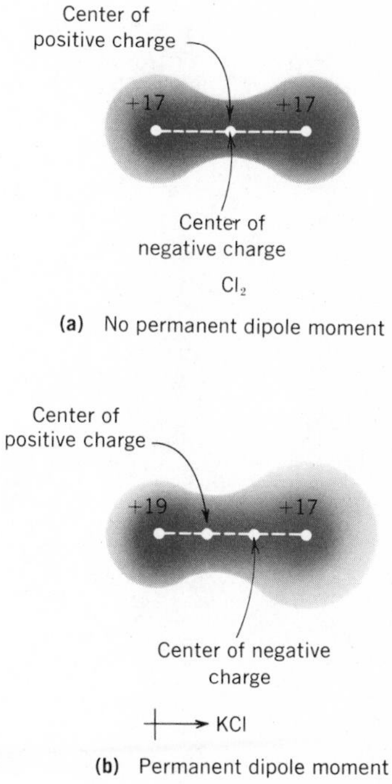

FIG. 4.9 Charge distributions and dipole moment: (a) homonuclear case; (b) heteronuclear case—the head of the arrow is the negative end of the dipole.

Dipole moments for some diatomic covalent molecules and ion-pair molecules (K^+Cl^-, Cs^+Cl^-, Cs^+I^-) are listed in Table 4.3. The values for ion-pair molecules are very large, as would be expected, since their electron clouds are the most distorted because of actual electron transfer. Of the covalent molecules listed, hydrogen fluoride has the largest dipole moment and is therefore the most polar. Covalent molecules with dipole moments *tend to cluster* with the negative end of one close to the positive end of a second and vice versa. This tendency to clump holds the molecules together in the liquid form of a substance, and it has the effect of raising the boiling point. You can see this effect by comparing the dipole moments in Table 4.3 with the

The ion-pair molecules result from strongly heating (1000°C) the ionic solids in a vacuum so that a gas of them forms.

One debye unit is 10^{-18} esu cm and a dipole moment $\mu = 1.00$ debye corresponds to 1×10^{-10} esu of charge separated by a distance of 10^{-8} cm (1 Å); $1 \times 10^{-10} \times 1 \times 10^{-8} = 1 \times 10^{-18}$ esu cm = 1 debye.

TABLE 4.3 Dipole moments (μ) and internuclear distances for diatomic molecules in the gas phase

Molecule[a]	*Temperature* (°C)	μ (debye units[b])	*Internuclear Distance* (Å)
HF	35°	1.91	0.92
HCl	35°	1.08	1.27
HBr	25°	0.80	1.41
HI	25°	0.42	1.61
KCl	750°	8.00	2.67
CsCl	676°	10.50	2.91
CsI	600°	10.20	3.32
BrCl	25°	0.57	2.14
BrF	25°	1.29	1.76
ClF	25°	0.83	1.62

[a]The negative end of the dipole points to the right in the formulas as written.
[b]One debye unit is 10^{-18} esu cm.

boiling points of HF, HCl, HBr, and HI, which are respectively 19.5°C, −114.8°C, −88.5°C, and −50.8°C.

Whether a molecule has a dipole moment or not depends on its shape or geometry and the *electronegativity* of its component atoms. *Electronegativity* (χ) *measures the attraction for bonding electrons by bonded atoms.* Values obtained from bond energy data (Pauling scale) are listed above the symbols for the elements in Table 3.4. Diatomic molecules of atoms with different electronegativities *always* have dipole moments. Polyatomic molecules (more than two atoms) of atoms having different electronegativities have dipole moments if their geometries are correct. Two cases for comparison are the mercuric chloride molecule ($HgCl_2$) and the water molecule. The electronegativity of Hg is 1.9; that of Cl is 3.2. These values show that chlorine attracts bonding electrons much more strongly than mercury. It follows that electrons in mercury-to-chlorine bonds will be displaced toward the chlorine; that is, such bonds should have a *bond dipole moment* in the direction $\overset{+\longrightarrow}{\mathrm{Hg{-}Cl}}$. It also follows that the $HgCl_2$ molecule should have a *resultant dipole moment* if the molecule has the bent shape in Figure 4.11. Only if the molecule is *linear* will the bond dipoles cancel to give a zero dipole moment for $HgCl_2$. The dipole moment is zero, proving that the linear structure is correct.

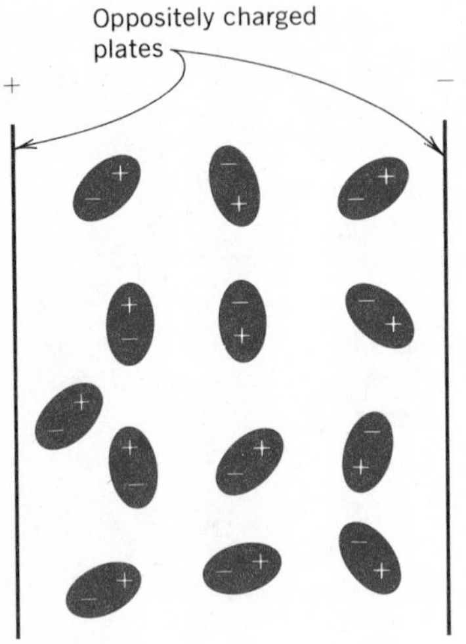

FIG. 4.10 Tendency of molecules having a permanent dipole moment to become oriented in an electric field.

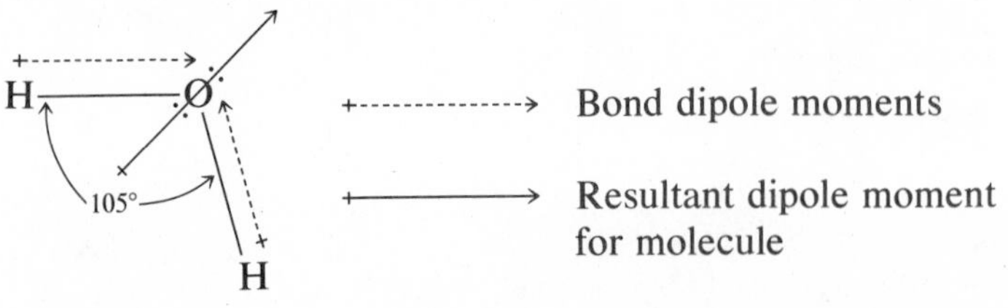

The electronegativities of H and O are 2.2 and 3.4, respectively, so if the water molecule were linear, the H—O bond dipoles would cancel to give a resultant dipole of zero, as in $HgCl_2$. The fact that the water molecule has a dipole moment of 1.85 debye units means that it must have a "bent" geometry. It does, and molecular spectra give the angle as 105°:
The unusual solvent properties of water depend largely on its polar character.

EXAMPLE

Water is sometimes called the "universal solvent." Of course, water doesn't dissolve everything—this would be a strange world if it did—but water is the best solvent for dissolving ionic compounds. Its secret of success lies in the large value of its dielectric constant, D, which is 80 at 20°C. Coulomb's law (Introduction) for the attractive force between oppositely charged particles accounts for the structure of ionic crystals, as we have seen. The value of the dielectric constant (D) of the matter separating charged particles appears in the denominator of the equation for Coulomb's law:

$$\text{Force} = \frac{q_1 q_2}{D r^2}$$

For the ions Na^+ and Cl^- at a distance $r = 10$ Å (10^{-7} cm) in a vacuum, where $D = 1$, the force of attraction between them is

$$\text{Force} = \frac{(4.803 \times 10^{-10})^2}{1 \times (10^{-7})^2} = 2.31 \times 10^{-5} \text{ dyne}$$

One dyne is the force that, acting on a 1-gram mass for 1 second, gives the mass a velocity of 1 cm sec^{-1}. The value 4.803×10^{-10} is the electronic charge in electrostatic units (esu) carried by one Na^+ ion or one Cl^- ion, with sign disregarded. If, now, the vacuum between the Na^+ and Cl^- ions is replaced by H_2O molecules, attractive force between the ions falls to 1/80 the value in a vacuum.

$$\text{Force} = \frac{1}{80} \times \frac{q_1 q_2}{r^2} = \frac{1}{80} \times \frac{(4.803 \times 10^{-10})^2}{(10^{-7})^2} = \frac{1}{80} \times 2.31 \times 10^{-5} \text{ dyne}$$

$$= 2.89 \times 10^{-7} \text{ dyne}$$

This weakening of attraction between ions is only one reason why water is such a good solvent for ionic compounds. Its remarkable solvent properties for certain kinds of covalent compounds depend on its ability to form hydrogen bonds (Section 4.6). Water is the solvent for the solutions in which the reactions of life go on. These reactions involve both ionic and covalent compounds, as we have seen.

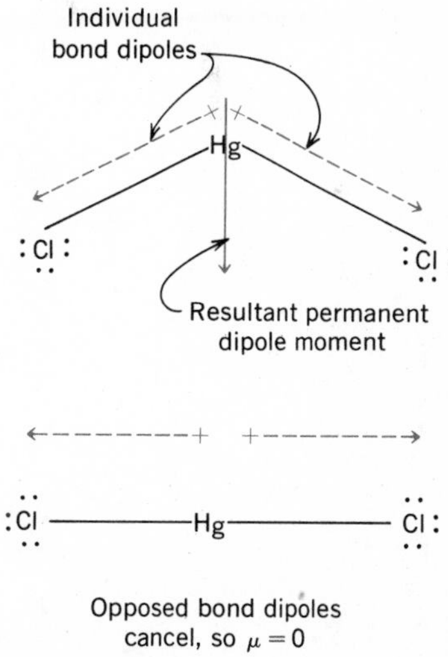

FIG. 4.11 Dipole moments predicted for $HgCl_2$ on bent and linear models.

TABLE 4.4 Dipole moments and electronegativity differences, $\chi_B - \chi_A$.

	μ (debye units)	$\chi_B - \chi_A$
A—B molecules		
HF	1.91	1.8
HCl	1.08	1.0
HBr	0.80	0.8
HI	0.42	0.5
KCl	8.00	2.4
CsCl	10.50	2.4
BrF	1.29	1.0
ClF	0.83	0.8
IF	0.57	1.0
A_nB_m molecules		
H_2O	1.85	1.2
H_2S	0.92	0.4
NH_3	1.47	0.8
PH_3	0.55	0.0
SO_2	1.63	0.8
SO_3	0.00[a]	0.8
BF_3	0.00[a]	2.0
CCl_4	0.00[a]	1.4
$HgCl_2$	0.00[a]	1.2

[a]Zero because of molecular symmetry.

It follows from the two examples, $HgCl_2$ and H_2O, that *dipole moment measurements give clues to the geometry of molecules*. Further examples are given in Table 4.4, where electronegativity differences and dipole moments are related. The moments of the last four entries are zero because of molecular geometry that cancels the bond dipoles, as in $HgCl_2$. The sulfur trioxide and boron trifluoride molecules have a planar arrangement of nuclei with bond angles of 120°. Carbon tetrachloride has tetrahedral geometry (see Section 4.4), which results in cancellation of the C—Cl bond dipoles.

Differences in electronegativities ($\chi_A - \chi_B$) are more reliable than ionization potentials (Section 3.6) in predicting whether a substance will be covalently or ionically bonded. The data in Table 4.5 show that covalent bonding goes over to ionic when the electronegativity difference between the bonded atoms is about 2.0, or greater, as it is between metals and nonmetals.

Molecular orbital theory explains dipole moments as a result of distortion of the electron cloud of a molecular orbital. Resonance theory explains dipole moments by adding ionic structures to the resonance hybrid representing the molecule. Thus, to explain the 1.08 debye moment of hydrogen chloride, the actual bonding is considered to be a hybrid of purely covalent and ionic structures:

TABLE 4.5 Electronegativity differences and bond type

Substance A_nB_m	*Electronegativity Difference* $\chi_B - \chi_A$	*Predominant Bond Type*
Li_2O	2.4	ionic
BeO	1.8	ionic
B_2O_3	1.4	covalent
CO_2	0.8	covalent
NO	0.4	covalent
OF_2	0.6	covalent
Na_2O	2.5	ionic
Al_2O_3	1.8	ionic
Al_2Cl_6	1.6	covalent
SO_2	0.8	covalent
ScO	2.0	ionic
BaO	2.5	ionic
H_2O	1.2	covalent
HCl	1.0	covalent
H_2S	0.4	covalent
PH_3	0.0	covalent
BF_3	2.0	covalent
CH_4	−0.4	covalent
CCl_4	0.6	covalent
ICl	0.5	covalent
CsF	3.2	ionic
HF	1.8	covalent

$H—\ddot{\underset{\cdot\cdot}{Cl}}:$ ↔ $H^+:\ddot{\underset{\cdot\cdot}{Cl}}:^-$

Purely covalent (no μ) — Purely ionic (large μ)

Remember that these symbols *do not mean that the compound hydrogen chloride is a physical mixture of covalent and ionic molecules.* The two formulas mean that the electron distribution in the bond is intermediate between equal sharing of the bonding pair and complete possession by the chlorine.

Exercises

9. Which of the following molecules would have the largest permanent dipole moment?

$:\ddot{\underset{\cdot\cdot}{Cl}}:Hg:Hg:\ddot{\underset{\cdot\cdot}{Cl}}:$ (linear molecule)

$H:\ddot{\underset{\cdot\cdot}{F}}:$

$:\ddot{\underset{\cdot\cdot}{I}}:\ddot{\underset{\cdot\cdot}{Cl}}:$

10. Would carbon tetrafluoride, CF_4, have a permanent dipole moment?

4.4 Molecular Geometry When chemists talk about molecular geometry, they mean the bond lengths and bond angles of a molecule or complex ion. Bond lengths are measured in angstrom units (1 Å = 10^{-8} cm) from the centers of the nuclei of the bonded atoms, and bond angles are measured in degrees between the lines joining the bonded nuclei. As pointed out in Chapter 1, bond lengths and angles in crystalline solids are determined by X-ray crystallography. For gases, information on bond lengths and angles comes from molecular spectrometry, particularly in the infrared spectral region, and from electron or neutron diffraction. All of these techniques are highly specialized and involve complex mathematical analyses, so we must be content with their results. Although the best ways to determine the geometry of a molecule are the direct experimental methods, bond theory provides a way of predicting the geometries observed in many cases. If you learn to do this, you will be able to "lift" the flat structures from the printed page and imagine them in their three-dimensional shapes. Being able to "see" molecules in "3-D" makes all chemistry, but particularly organic and biochemistry, more understandable and interesting.

One way of predicting bond angles begins with Lewis octet structures and the ideas of *ligands* and *unshared pairs of electrons*. Unshared pairs are pairs of valence electrons that are not in bonds in the Lewis structure. Ligands are the atoms bonded to a *central atom* of interest. In counting ligands, it doesn't matter how many covalent bonds hold the ligand to the atom of interest. Thus in the formaldehyde molecule the central carbon atom has three ligands and no unshared pairs; the oxygen has one ligand and two unshared pairs:

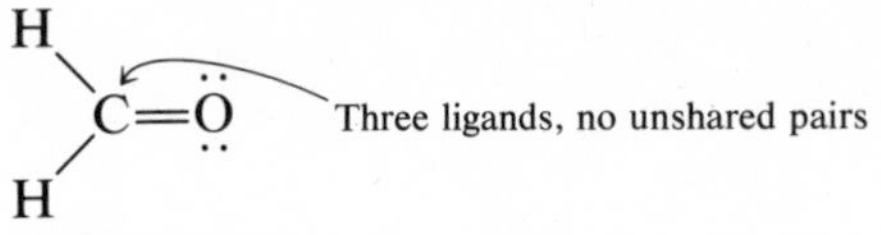

The acetic acid molecule (CH_3COOH) illustrates some other combinations:

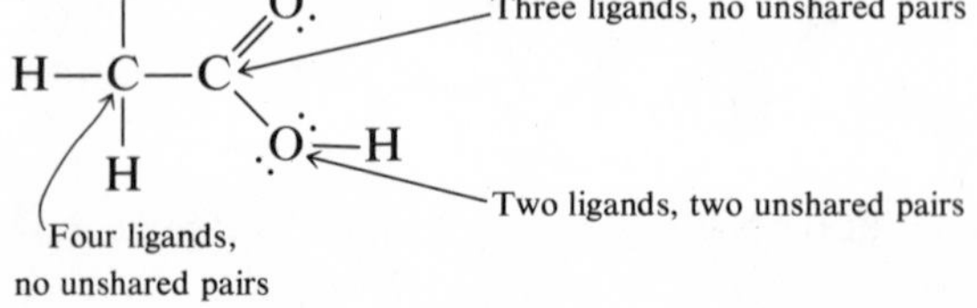

Once the numbers of ligands and unshared pairs have been counted, you can predict the bond angles by applying the following rules:

Ligands	*Unshared pairs*	*Predicted bond angle*
4	0	109.5° (tetrahedral)
3	1	109.5° (tetrahedral)
2	2	109.5° (tetrahedral)
3	0	120.0° (trigonal or "flat")
2	1	120.0° (trigonal or "flat")
2	0	180.0° (linear)

Applying these rules to acetic acid (CH_3COOH) gives the indicated bond angles:

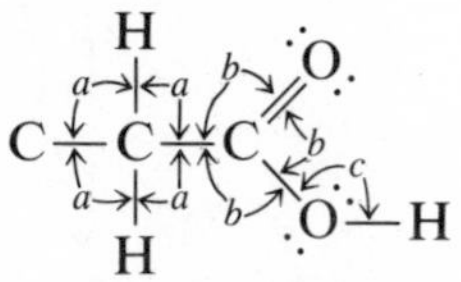

Angles a = 109.5° (four ligands, no unshared pairs)

Angles b = 120.0° (three ligands, no unshared pairs)

Angle c = 109.5° (two ligands, two unshared pairs)

The angles found for acetic acid by spectrometry and X-ray analysis are within a few degrees of those predicted. The combinations of ligands and unshared pairs in the rules that are not shown by acetic acid are seen in ammonia (NH_3), ozone (O_3), and carbon dioxide (CO_2):

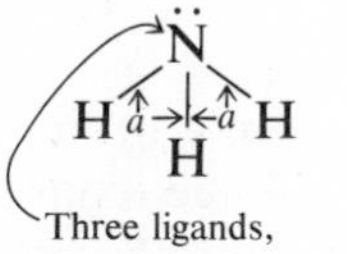

Three ligands,
one unshared pair
angle a = 109.5°

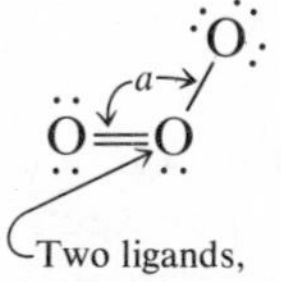

Two ligands,
one unshared pair
angle a = 120.0°

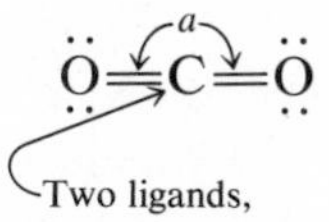

Two ligands,
no unshared pairs
angle a = 180.0°

The angles found experimentally are those predicted or close to them: CO_2 is linear (angle a = 180°), in O_3 angle a is 127°, and in NH_3 angle a is 107°. The water molecule, 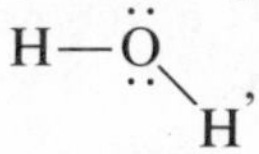, with two ligands and two unshared pairs, has a bond angle of 105°, only slightly less than the predicted 109.5°.

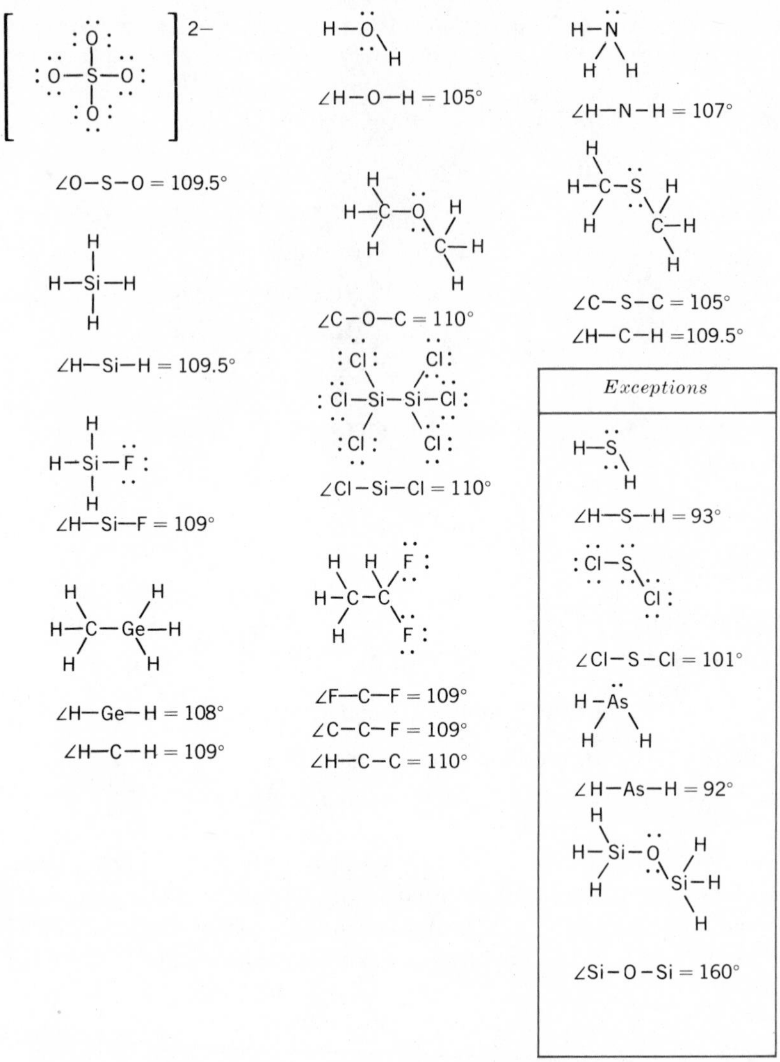

FIG. 4.12 Experimentally determined bond angles in molecules and ions predicted to have tetrahedral (109.5°) bond angles.

The rules work quite well for compounds of the elements in the second period (Li → F), as can be seen in Figure 4.12. Exceptions appear when they are applied to elements further down in the Periodic Table.

The rules can be explained rather simply by considering the repulsive forces that exist between pairs of electrons in orbitals. The CO_2 molecule is straight or linear rather than "bent" because in the linear arrangement the electron clouds of the double bonds are as far away from each other as they can get.

Consequently the repulsion between the electron clouds of the bonds is least in the linear form, and the atoms take up that arrangement. If there are no unshared pairs and three ligands the electron clouds of the bonds are most separated when the bond angle is 120°. The boron trifluoride (BF_3) molecule is an example. It is a resonance hybrid of the following canonical structures, all of which should have 120° angles, according to the rules:

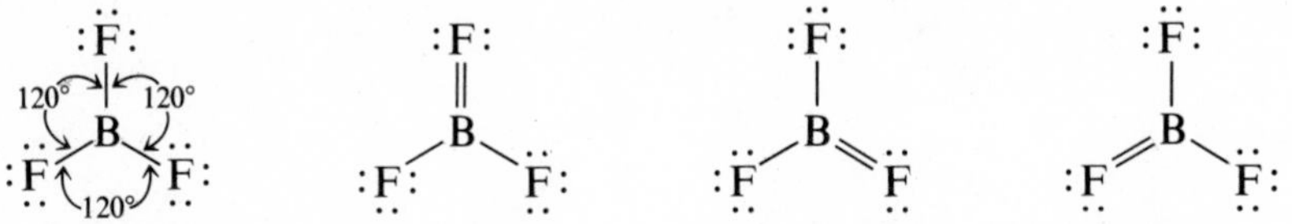

The prediction of a 120° angle for the case of two ligands and one unshared pair follows the electronic repulsion argument if you assume that unshared pairs of electrons in nonbonding orbitals repel electrons in bonding orbitals. If there weren't an unshared pair on the central oxygen of ozone (O_3), the molecule would be linear like CO_2. To make room for the unshared pair the bonding pairs are forced out of the linear arrangement, giving an angle of 127°, close to the 120° predicted.

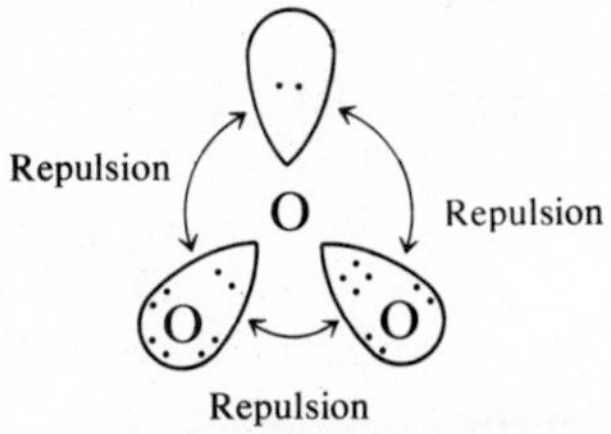

The tetrahedral angle of 109.5° is found for any combination of ligands and unshared pairs (including none) that totals four. Only if the axes of the four bonding electron clouds are at 109.5° to each other is the distance between them at a maximum and the repulsion between them smallest. Methane (CH_4), ammonia (NH_3), and water (H_2O) are examples:

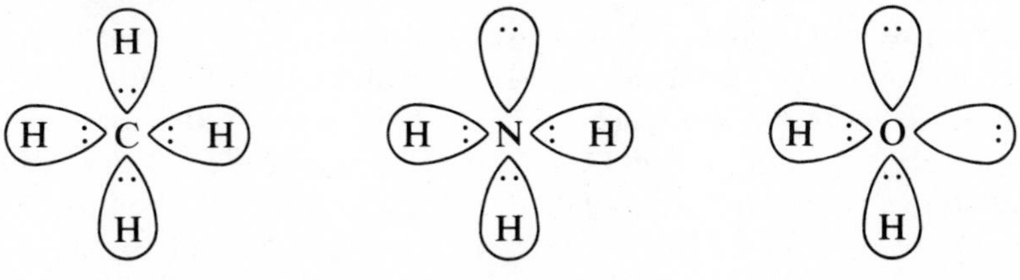

Methane Ammonia Water

11. Predict the geometries of acetylene, H—C≡C—H; hydrogen cyanide, H—C≡N:; and formic acid, H—C(=Ö:)—Ö—H. Exercises

12. The peptide link is all-important in determining the overall geometry of protein molecules. Predict the angles around carbon and around nitrogen for a peptide link:

$$R—C\overset{\overset{\displaystyle ..\!O\!..}{\diagup\diagdown}}{\underset{\underset{\displaystyle H}{|}}{N—R}}$$

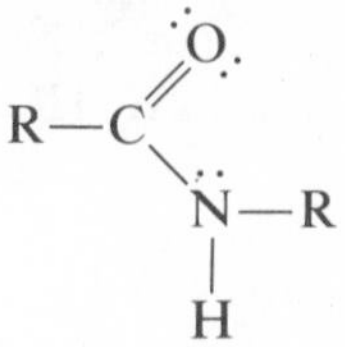

Peptide link

13. The dot structure for NO_2 has one unshared electron on nitrogen:

$$\begin{array}{c} :O: \\ :: \\ \cdot N \\ :O: \end{array}$$

Would you expect the ONO angle to be greater or less than 120°? Would you expect NO_2 to be paramagnetic?

Molecular orbital theory for polyatomic molecules Molecular orbital theory for covalent bonds also explains the geometry of polyatomic (more than two-atom) molecules, but in a different way than by considering the repulsions between electrons in bonds to ligands and in unshared pairs, as we have just done. The principle of overlapping or blending of atomic orbitals to make molecular orbitals is used in making up the molecular orbitals for polyatomic molecules. Atomic orbitals belonging to the separate atoms are first combined or mixed to produce *hybrid atomic orbitals*, and it is the blending or overlap of these hybrid atomic orbitals for the different atoms that yields the molecular orbitals.

To get the orbitals for methane, CH_4, the equations for the $2s$ and $2p$ carbon atomic orbitals are mixed mathematically to produce four identical sp^3 hybrid atomic orbitals extending from the carbon atom. These four hybrid orbitals turn out from the mixing to have their axes (center lines) directed at angles of 109.5° to each other, as shown for two of the new orbitals in Figure 4.13. The molecular orbitals for each of the C—H bonds in methane are formed by the overlap of one sp^3 hybrid atomic orbital from the carbon with the $1s$ orbital of the hydrogen atom as shown in Figure 4.14. The result for the four bonds in methane is shown in Figure 4.15. Hybridizing atomic orbitals according to the sp^3 "recipe" leads in this way to 109.5° angles between the C—H bonds in methane. This is the same value obtained by considering electron repulsions as we did earlier. Whenever a carbon atom has four ligands attached to it, the sp^3, tetrahedral, hybrid atomic orbitals

109°28'

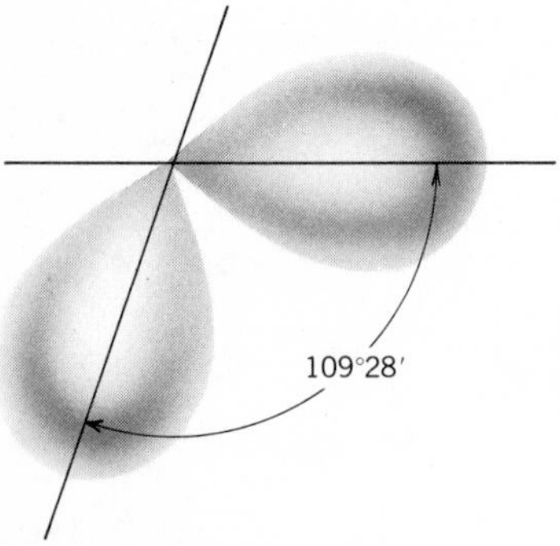

Fig. 4.13 Two of the four tetrahedral hybrid atomic orbitals, showing angle between axes.

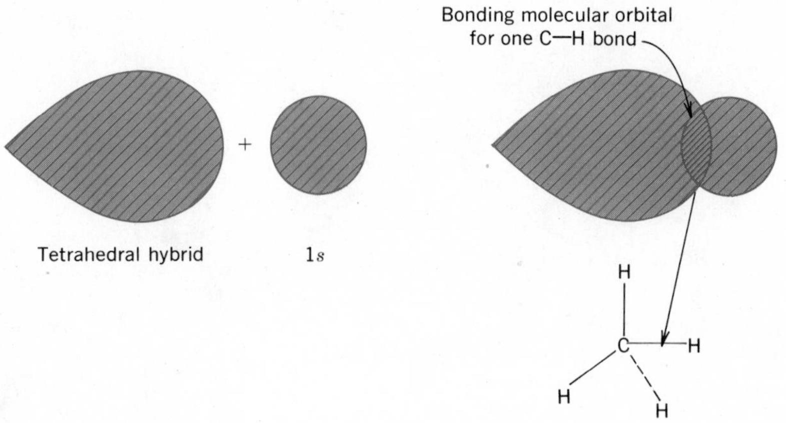

FIG. 4.14 Production of the molecular orbital for a C—H bond in methane.

are used to overlap with orbitals from the ligands. This means that in the examples ethane (CH_3—CH_3), propane ($CH_3CH_2CH_3$), and acetic acid ($CH_3C(=O)$—OH)—where the carbon atoms have four ligands—the bond angles indicated are 109.5°.

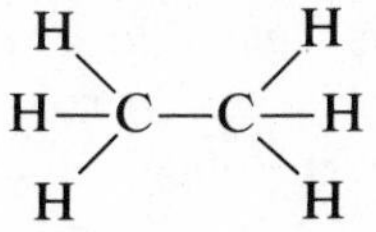

∠HCH = ∠CCH = 109.5°
Ethane

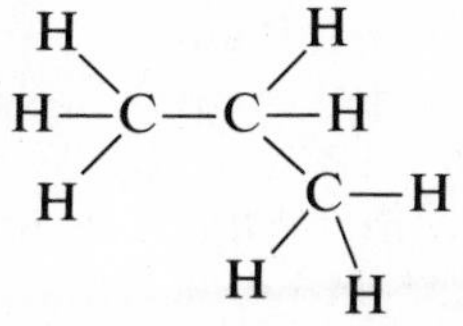

∠HCH = ∠CCC = 109.5°
Propane

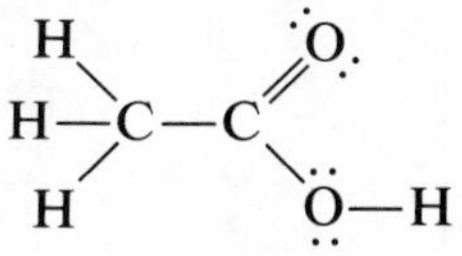

∠HCH = 109.5°
Acetic acid

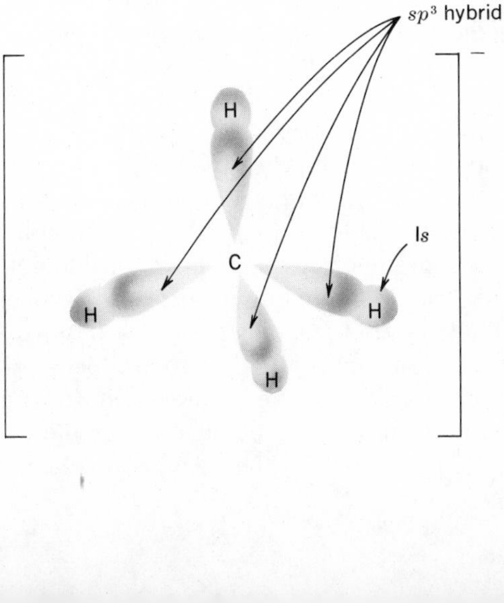

FIG. 4.15 Tetrahedral structure of CH_4.

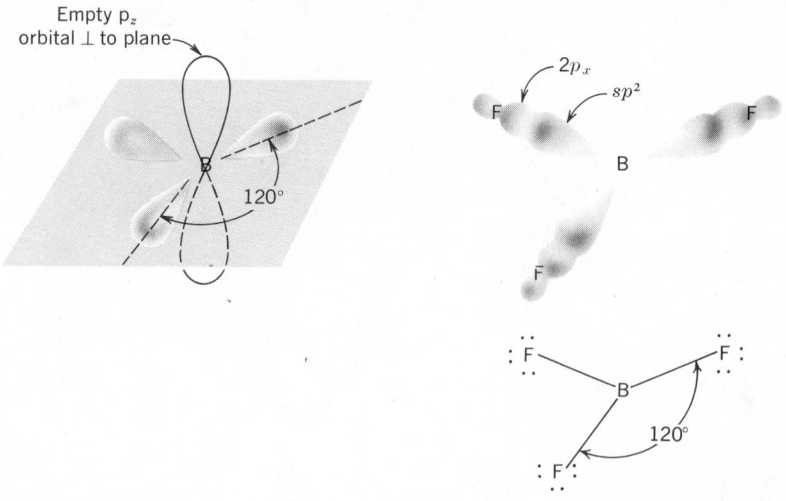

FIG. 4.16 Trigonal hybrid sp^2 orbitals and the BF_3 molecule.

The molecular orbitals for the C—C bonds in ethane and propane are produced by overlapping of sp^3 hybrid atomic orbitals on carbon; the C—H orbitals are sp^3–$1s$ overlaps.

Angles of 120° between the axes of hybrid orbitals result from combining the $2s$ atomic orbital with two (instead of three) $2p$ atomic orbitals. This sp^2 hybridization of atomic orbitals is called *trigonal* hybridization and is exemplified by the boron trifluoride (BF_3) molecule shown in Figure 4.16. In BF_3 the boron and fluorine nuclei all lie in the same plane. The $2p_z$ orbital that was not used up in the sp^2 hybridization remains unoccupied by electrons and is perpendicular to the plane containing the B and 3F nuclei. Again, molecular orbital theory comes to the same geometry the electrostatic rules predict—120° bond angles for three ligands and no unshared pairs.

In developing the molecular orbitals for diatomic molecules two kinds of orbitals resulted—sigma (σ) orbitals where electron probability is highest between the bonded atoms, and pi (π) orbitals where the probability is high on both sides of a plane containing the bonded nuclei (Figures 4.5 and 4.6). Bonding orbitals of the π type are also found in certain polyatomic molecules—those that contain multiple (double or triple) bonds. In ethene, $CH_2{=}CH_2$, for example, one bond of the double bond is of the σ type and one is of the π type. Figure 4.17 shows a carbon atom with three sp^2 hybrid atomic orbitals containing one electron each and one electron in the $2p_z$ orbital not used in the hybridization. If two such carbon atoms are bonded together by overlap of one sp^2 orbital from each, making a σ bond, the two $2p_z$ atomic orbitals can overlap as shown in Figure 4.18 to produce a π-bonding orbital, using the two electrons from the atomic orbitals. The four C—H bonds in ethene are of the sigma type, made from the overlap of an sp^2 hybrid orbital from carbon with the $1s$ atomic orbital of the hydrogen atom.

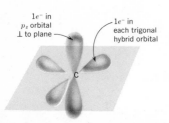

FIG. 4.17 Trigonal hybridized carbon atom.

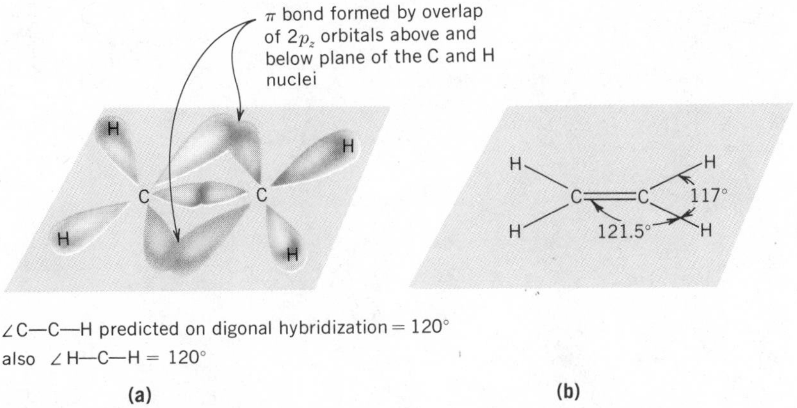

FIG. 4.18 (a) Molecular orbital structure for the ethylene molecule; (b) geometry of ethylene from spectroscopy.

The bond angle predicted between sp^2 hybrids is 120° (as in BF_3), and the angles in ethene shown in Figure 4.18(b) are satisfactorily close to the predicted values. The C—C bond in acetic acid is produced by an sp^3–sp^2 overlap, making the angles around the —C(=O)— also 120°.

The molecular orbital model for double bonds explains why there are two isomeric molecular structures for 1,2-dichloroethene ClCH=CHCl (Figure 4.19). To convert the *cis*-isomer (a) in the figure to the *trans* (c), the π bond must be broken by twisting one CHCl group against the other as shown at (b). This twisting requires high temperatures to provide the energy to break the π bond, so *cis*- and *trans*-1,2-dichloroethene are different compounds having the same molecular formula, and like ethanol and methyl ether (Section 4.2), the two dichloroethenes are a pair of isomers. Isomers resulting in this way from lack of "free" rotation of doubly bonded groups are called *geometric isomers* or "*cis–trans*" isomers. This kind of isomerism is very important in organic and bioorganic chemistry (Chapters 13 and 14). Your ability to see in dim light depends on a *cis–trans* isomerization using energy from light (Section 14.3).

Other mixtures of atomic orbitals yield hybrid orbitals with other characteristic angles, and explain the geometry of most molecules. In acetylene, H—C≡C—H, one 2*s* and one 2*p* orbital of each carbon atom are mixed to produce two *sp* hybrid orbitals. The axes of *sp* hy-

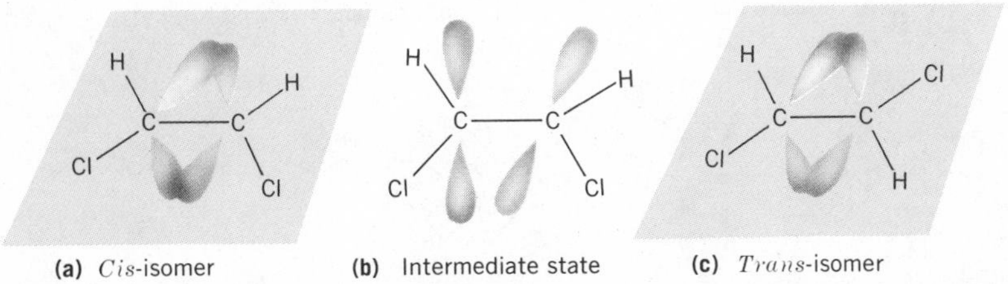

FIG. 4.19 Isomeric forms of the 1,2-dichloroethylene molecule: (a) *cis*; (c) *trans*; and (b) an intermediate state in their conversion.

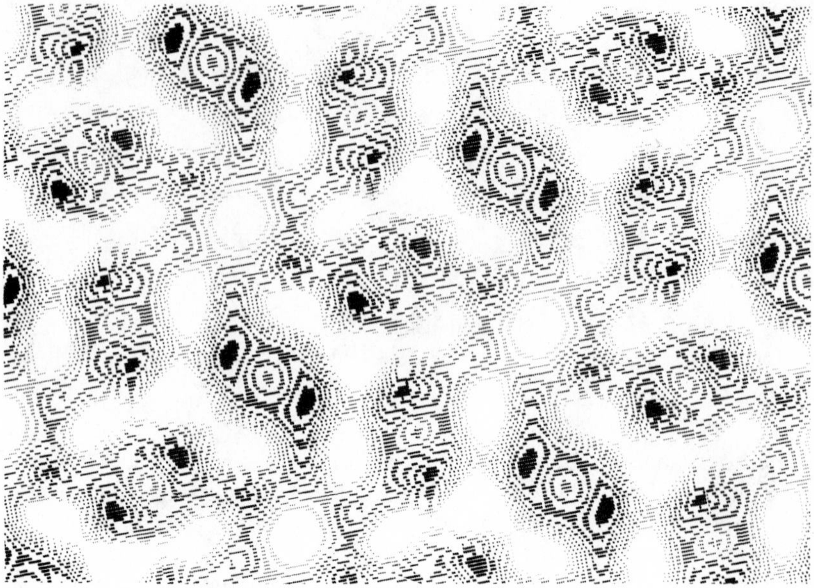

PHOTO 13 Computer-calculated electron isodensity map of the caspid structure of *Caulobacter crescentus* bacteriophage. Dark areas represent high electron density in bonding regions. (*James A. Lake and Kevin R. Leonard,* Science, **183**: *744;* 22 February 1974. *Copyright 1974 by the American Association for the Advancement of Science.*)

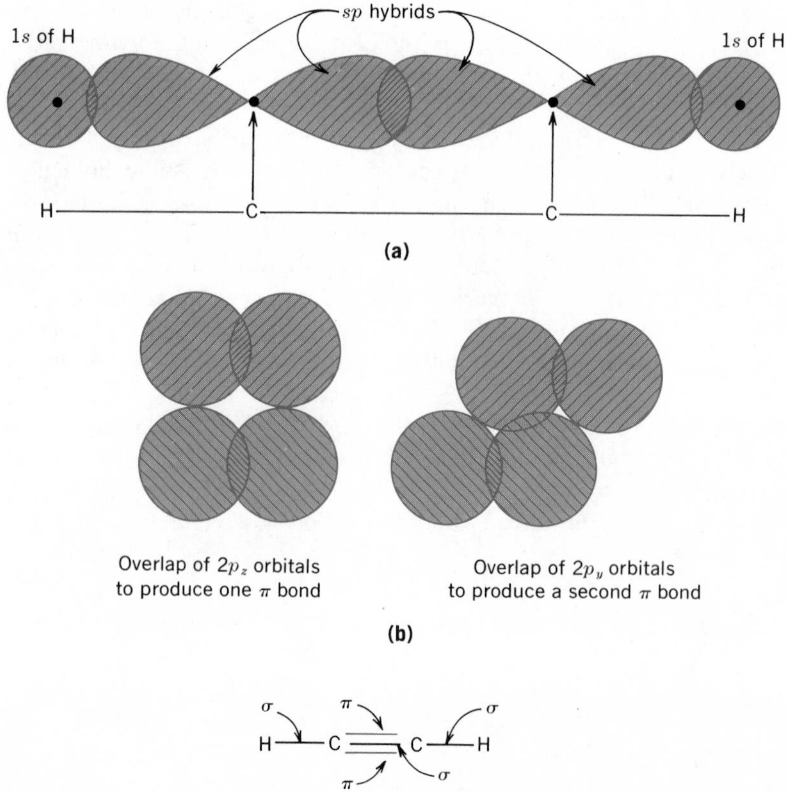

FIG. 4.20 Molecular orbital structure for the acetylene molecule.

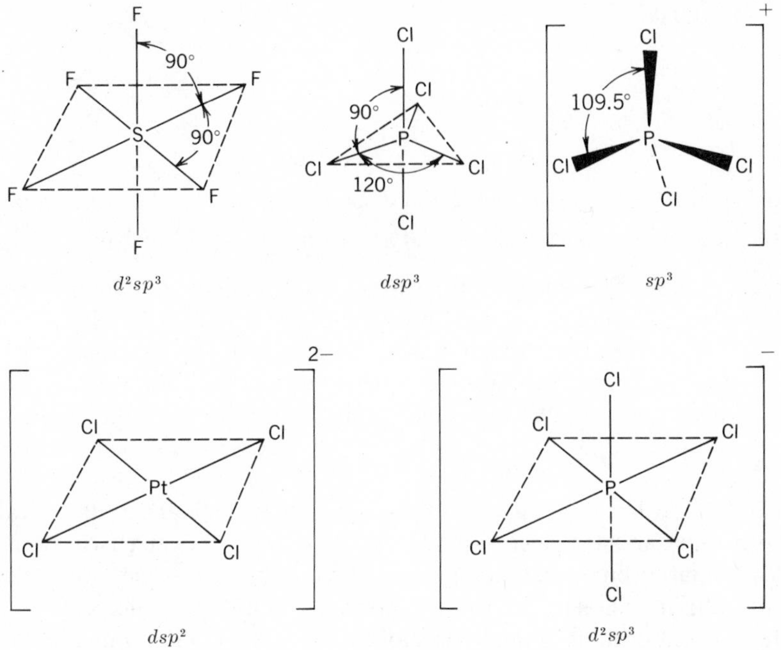

FIG. 4.21 Octahedral (d^2sp^3), square planar (dsp^2), and tetrahedral (sp^3) hybridization of atomic orbitals in some molecules and ions.

brid orbitals are at 180°, so the acetylene molecule has all of its atoms in a straight line. Figure 4.20 shows how the molecule is put together according to molecular orbital theory. Note that there are two $2p$ orbitals left on each carbon after the sp hybridization and that the two unused $2p$ orbitals overlap to form two π bonds. The carbon–carbon σ bond results from the overlap of two sp hybrids as shown, the C—H bonds from sp–$1s$ overlaps.

The electron density map in Photo 13 shows how the geometry of bonding at the molecular level can result in crystal-like regularity of structure at the size level of hundreds of Å.

With elements of the third (and higher) periods the d orbitals come into play and can be mixed in various ways with s and p orbitals to give a variety of geometries. Examples of molecules and complex ions with various geometries are given in Figure 4.21, along with the descriptions of the hybrid orbitals used in their formation.

Exercises

14. Derive the molecular orbital structure for the cyanide ion :C:::N:$^-$ using hybrid atomic orbitals. Do the same for N_2 and compare the result with the bonding for N_2 shown in Table 4.2.

15. Which of the following molecules would show isomerism of the *cis–trans* type? (Remember, C=C is flat.)

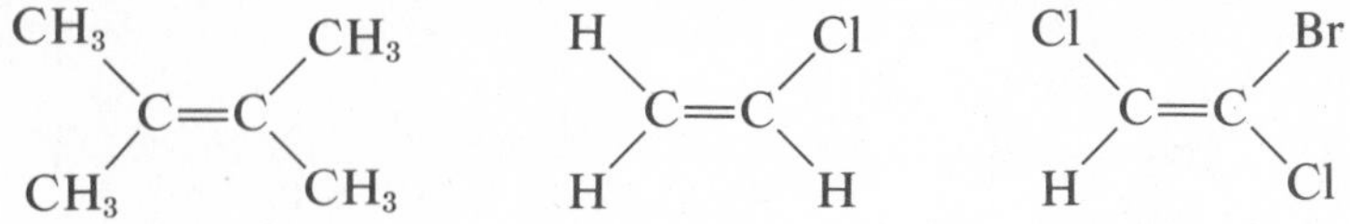

$$CH_3 \atop CH_3 \Big\rangle C=C \Big\langle {CH_3 \atop CH_3}$$

(a)

$$H \atop H \Big\rangle C=C \Big\langle {Cl \atop H}$$

(b)

$$Cl \atop H \Big\rangle C=C \Big\langle {Br \atop Cl}$$

(c)

16. What bond angle is predicted for H—Ö—H in the water molecule if the unshared pairs of electrons are in sp^3 hybrid orbitals?

17. Molten cryolite, Na_3AlF_6, is used as a solvent for aluminum oxide, Al_2O_3, in the electrolytic production of aluminum (Section 11.4). Cryolite is ionic, being made of three Na^+ ions and one AlF_6^{3-} complex ion. Predict the geometry for the AlF_6^{3-} (hexafluoaluminate) ion and give molecular orbital descriptions of the overlaps producing the Al—F covalent bonds (see SF_6 in Figure 4.21).

4.5 Delocalized Molecular Orbitals The molecular orbitals dealt with thus far have been located or "localized" between two atoms. Molecular orbitals extending over three or more atoms are called *delocalized* orbitals. The most important example of a molecule with delocalized orbitals is the hydrocarbon *benzene* (C_6H_6), found in coal tar and petroleum, that serves as the starting material for making thousands of compounds, from vitamins to synthetic rubber.

According to molecular orbital theory, the framework of the six-membered carbon ring in the benzene molecule consists of sp^2 hybrid orbitals from the carbon atoms [Figure 4.22(a)]. The result is a "sigma framework," the regular, plane hexagon at (b) in the figure. The C—C bonds are formed by overlapping sp^2 hybrid atomic orbitals from adjacent carbon atoms, and the C—H bonds are sp^2–$1s$ overlaps. There is one $2p_z$ orbital left on each of the six carbons of the hexagonal framework. The total number of valence electrons in benzene (C_6H_6) is $(6 \times 4) + (6 \times 1) = 30$. There are 24 electrons in the 12 sigma bonds of the framework, so $(30 - 24) = 6$ electrons are unaccounted for. In the Lewis structure for benzene the six electrons are in the second (π) bonds of the three double bonds:

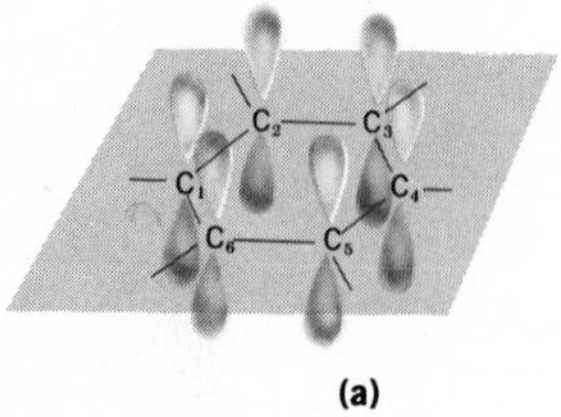

(a)

sp_2–sp_2 overlap

sp_2–$1s$ overlap

120°

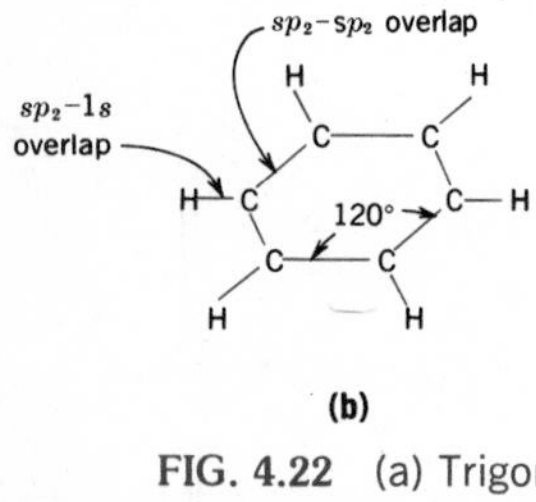

(b)

FIG. 4.22 (a) Trigonally hybridized carbon atoms; (b) sigma framework for the benzene molecule.

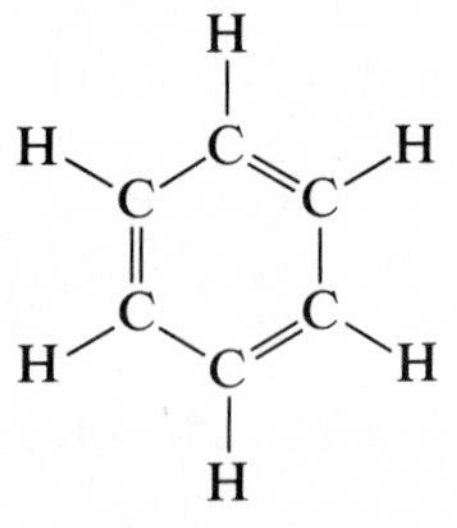

Lewis structure for benzene

According to molecular orbital theory the six electrons are in three delocalized molecular orbitals that extend over all six atoms of the ring. The delocalized orbitals are produced by overlapping the $2p_z$ orbitals in various combinations. The three delocalized orbitals that take up the six electrons are shown in Figure 4.23, where the bonds to hydrogen and symbols for the carbons have been omitted for clarity. All three of the orbitals are of the pi (π) type with their nodal planes lying in the plane of the ring. The delocalized orbital of lowest energy (π_a) bonds all carbon atoms equally with *one* pair of electrons. Two orbitals (π_b, π_c) of higher energy take up the remaining four electrons. These two orbitals are at the *same* energy and each has an additional nodal

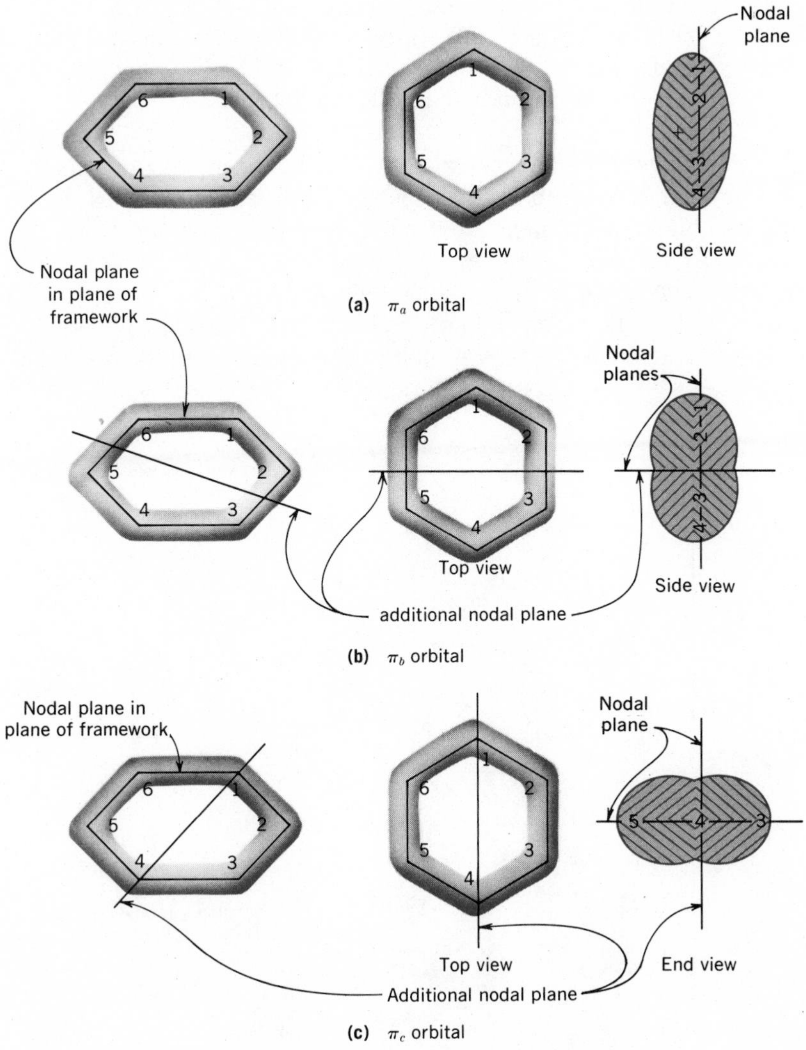

FIG. 4.23 Delocalized pi molecular orbitals for the benzene molecule.

plane that is perpendicular to the plane of the ring and cuts the ring skeleton as shown. The π_b orbital bonds mainly carbon atoms 1–2, 6–1, 3–4, and 4–5, while bonding by the π_c orbital is concentrated between carbons 2–3 and 6–5. The overall result is $1\frac{1}{2}$ bonds (three electrons) between each of the carbons, just as found by resonance theory (Section 4.2). One of the bonds is the two-electron sigma (σ) bond between carbons that holds the ring together. The one additional electron bonding the carbons to one another makes up the additional half bond and is the total effect of the six electrons in the three delocalized π orbitals. For each of the bonding π orbitals there is a corresponding antibonding π orbital of higher energy, as shown in Figure 4.24. Benzene is unusually stable for a molecule with π bonds, much more stable than ethene, for instance. The unusual stability of benzene results from three filled π *bonding* orbitals with no electrons in antibonding π orbitals. The nitrogen molecule, also stable, has filled π orbitals with unoccupied antibonding ones (Table 4.2).

EXAMPLE Nature uses the unusual stability of the benzene ring or "nucleus" in forming many durable molecules that enter into life processes. Hormones, vitamins, proteins, penicillin, and thousands of other "biochemicals" have a benzene ring as part of their covalent bond structure. People have taken advantage of the durability of the benzene ring to produce dyes that don't fade; durable plastics such as polystyrene; many drugs, including aspirin and sulfanilamide; and even synthetic rubber, which outwears the natural form.

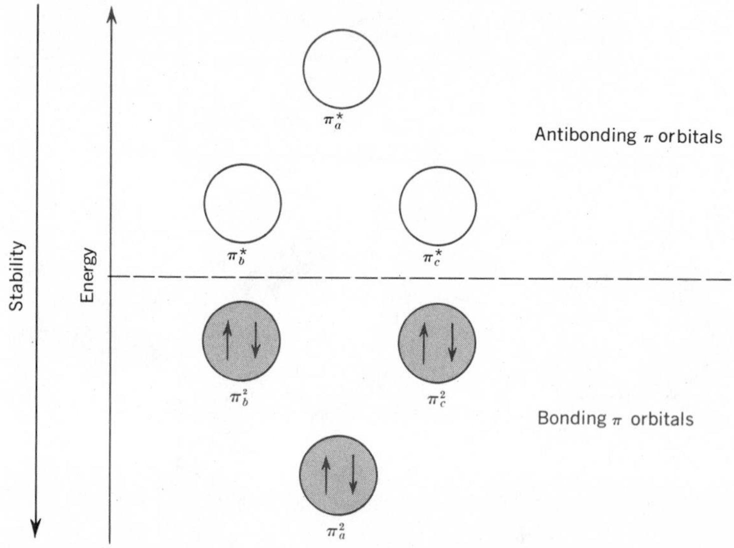

FIG. 4.24 Energy-level sequence of delocalized pi orbitals for the benzene molecule and their occupancy by electrons.

But the stability that makes the benzene ring resistant to attack and destruction by chemicals and microorganisms in our environment has, in the case of DDT, created a worldwide health problem. DDT is an abbreviation of the name of one of the most widely used insecticides, dichlorodiphenyltrichloroethane.

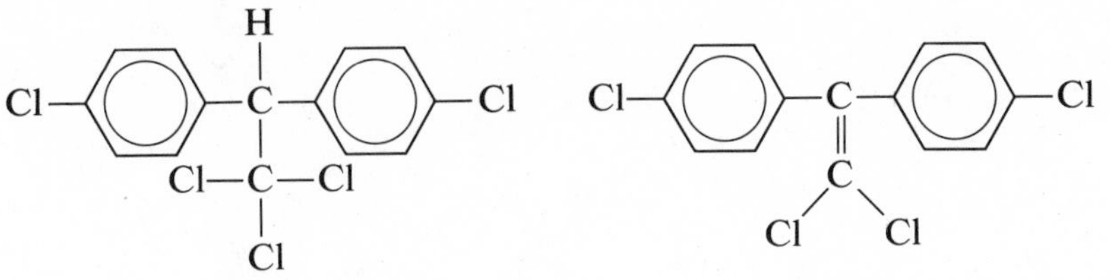

DDT DDE

The benzene rings in DDT make the molecule so stable that, once introduced into the environment, DDT persists for decades. Widespread use of DDT to kill malaria mosquitoes and insects that prey on agricultural crops has put more than a million tons of DDT into our environment. All of us (all mammals, in fact) carry DDT in our livers and body fat (as much as 1g/10^6 g) because it comes in with our food, and our bodies have no way of breaking it down so that it can be excreted. DDT is metabolized to DDE (dichlorodiphenyldichloroethene) in our bodies, but DDE is even *more* durable than DDT. Toxic reactions have not been observed in humans, but several species of birds, including the California brown pelican, are in serious danger of becoming extinct because the DDT in their diets has resulted in egg shells so fragile that they crack when the mother bird nests on them. Efforts have been made to stop or at least minimize the use of DDT. A simple worldwide ban on the use of DDT would have serious effects in decreasing agricultural production, and unless DDT were replaced by an equally effective agent for killing mosquitoes, a resurgence of malaria in tropical regions would certainly result. DDT pollution and some alternatives to the use of DDT are discussed in detail in Chapter 15. Photo 14 shows the devastation that can result from an insect "population explosion" uncontrolled by insecticides.

PHOTO 14 Forest on Cape Cod, Massachusetts, photographed July 1970. Gypsy moths have stripped the leaves from all the trees, and birds and other wildlife have left the area. Ways of attacking these insects are discussed in Chapter 15. (*U.S. Department of Agriculture.*)

Delocalized orbitals of the π type explain bond distances for many molecules and ions (Figure 4.25). Thus there is only one C—O bond distance in the carbonate ion (CO_3^{2-}) because there is a pair of electrons in a four-center, delocalized π orbital, as shown. This orbital adds equally to the bonding by the sigma C—O bonds, so that there is only one C—O distance. The result is the same as that provided by resonance theory, which treats the carbonate ion as a resonance hybrid. Molecular orbital theory gives a more direct view of electron distribution.

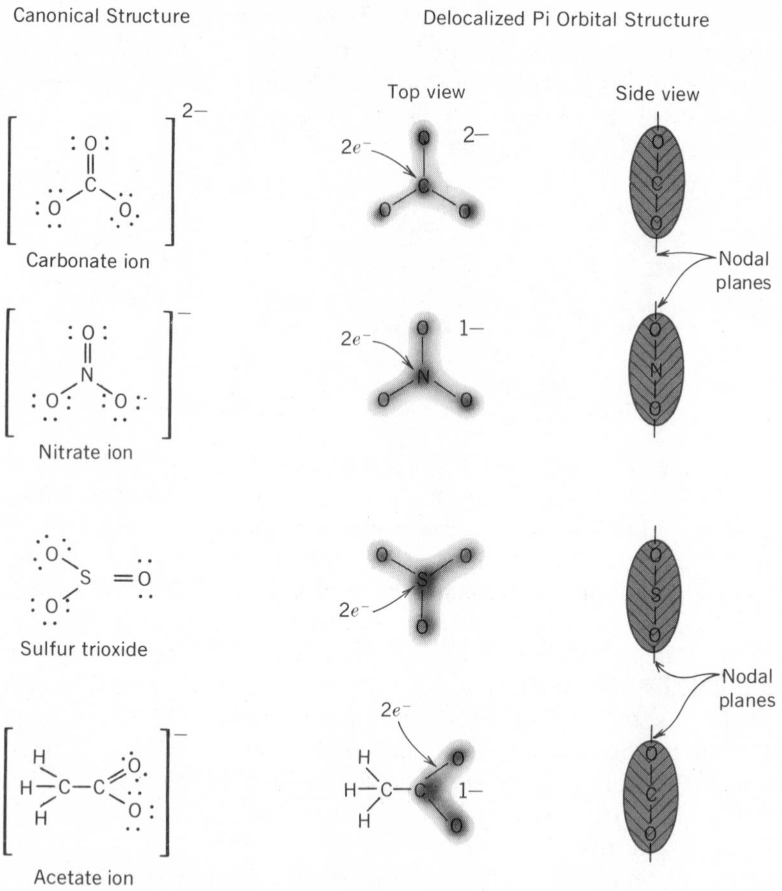

FIG. 4.25 Examples of tetranuclear, or four-center, delocalized pi orbitals and one three-center one.

Delocalized orbitals also explain the bonding that holds metals together and the *high electrical conductivity* of metals. Here the delocalized orbitals of the bonds may be thought of as resulting from the overlapping of the atomic orbitals occupied by the valence electrons of the metal (Figure 4.26). The bonding orbitals are delocalized over all of the atomic kernels in the sample, so electrons in the orbitals are free to move under an applied voltage. For this reason metals conduct electricity very well.

In the *metalloids* or *semiconductors*, such as silicon and germanium, most of the valence electrons are in orbitals localized between atomic kernels and only a few are in orbitals of the delocalized, metallic type. Correspondingly these elements are poor conductors of electricity at ordinary temperatures. An increase in temperature raises the energy of the localized electrons, enabling them to move into the higher-energy, delocalized orbitals or *conduction bands* (Figure 4.27). Thus the electrical conductivity of a semiconductor increases as its temperature

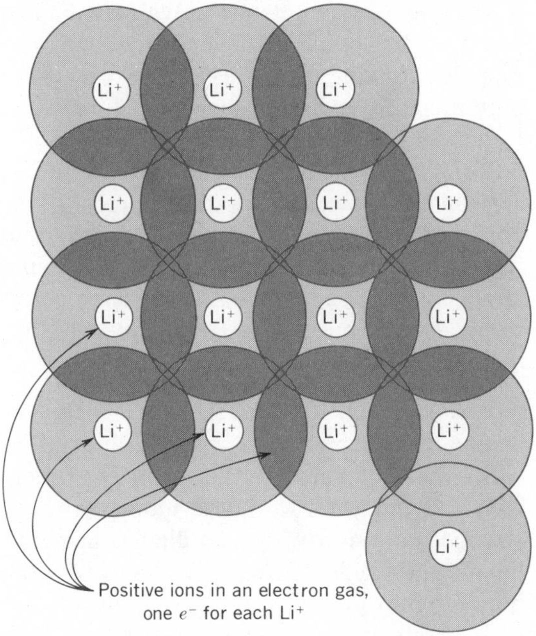

FIG. 4.26 Region in a crystal of solid lithium metal showing creation of delocalized orbitals for valence electrons by multiple overlapping of *s* atomic orbitals.

is raised. Carbon in the form of diamond is an insulator at room temperature but becomes a semiconductor when heated to about 700°C.

4.6 Secondary Forces Between Microparticles Secondary forces are weak forces of attraction between atoms and molecules. They are responsible for existence of the liquid and solid states (molecular crys-

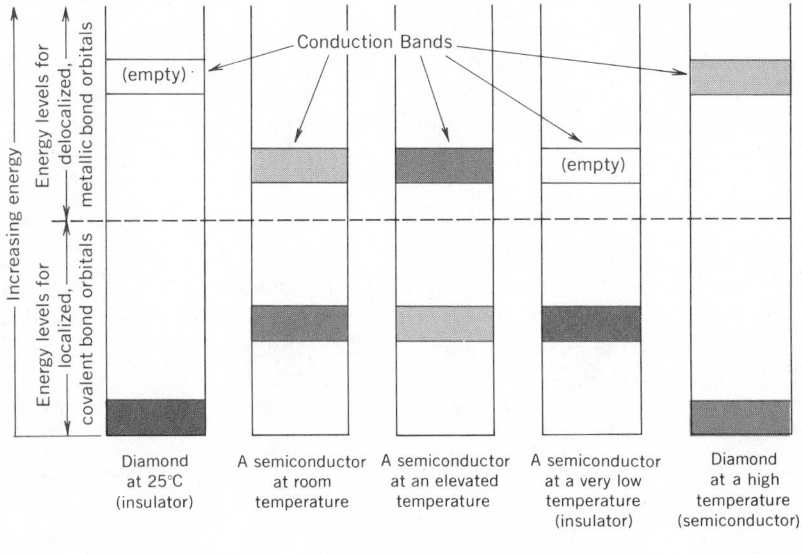

FIG. 4.27 Schematic representation of population of localized and delocalized orbitals by electrons in insulators and semiconductors. Depth of shading indicates electron population.

tals) of many elements and covalent compounds. They are weak compared to the much stronger forces that result in covalent, ionic, or metallic bonding. Some examples will illustrate the difference in energy required to disrupt secondary forces on the one hand and valence forces on the other.

You know that thermal energy must be supplied to boil a liquid such as water. The energy input required (9.7 kcal/mole) goes into overcoming the secondary forces holding the molecules close together in the liquid with the result that the molecules can take up the much larger average separation found in the gas form (steam). By contrast, to break even one of the H—O covalent bonds in the water molecule and produce gaseous hydrogen atoms and $\cdot\underset{\cdot\cdot}{\ddot{\mathrm{O}}}\!:\!\mathrm{H}$ particles (hydroxyl radicals) requires the much larger energy input of 110 kcal/mole resulting from temperatures of thousands of degrees.

Secondary forces are the result of attractive forces between molecules having electric dipole moments (Section 4.3). They result in stronger cohesion when the dipoles are close together, as they are in liquids and solids. Figure 4.28 shows why this is so. At large distances between the dipoles (c), the attraction between their oppositely charged ends is offset by the repulsion between their like-charged ends. As a gas of polar molecules is cooled and compressed, however, its molecules move closer together, the attractive forces between its molecules become greater, and they "clump" or *condense* to a liquid or solid. However, even molecules such as O_2, CH_4, N_2, F_2, . . . , which have no *permanent* electric dipole moment can also be obtained as liquids and solids if cooled sufficiently. The molecules condense because of the existence of *transitory dipole moments* (as opposed to permanent ones) that result from fluctuations in their electron clouds (Figure 4.29). For the briefest instant the positive nuclear charge is not screened by its own electrons and it "sees" the negatively charged electrons in an adjacent molecule. The momentary attraction results in weak cohesion. Cohesive forces due to both kinds of dipoles (transitory and permanent) are often called van der Waals forces.

Hydrogen bonding One kind of cohesion due to the attraction between permanent electric dipoles in liquids and solids is particularly strong. This is when the dipole results from *X*—H bonds, where *X* is a strongly electronegative element (F, O, N). This type is termed *hydrogen bonding*. Water is the prime example, and the hydrogen bonding is often symbolized by a dashed line:

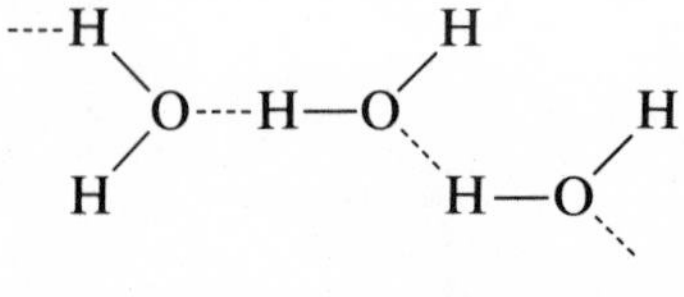

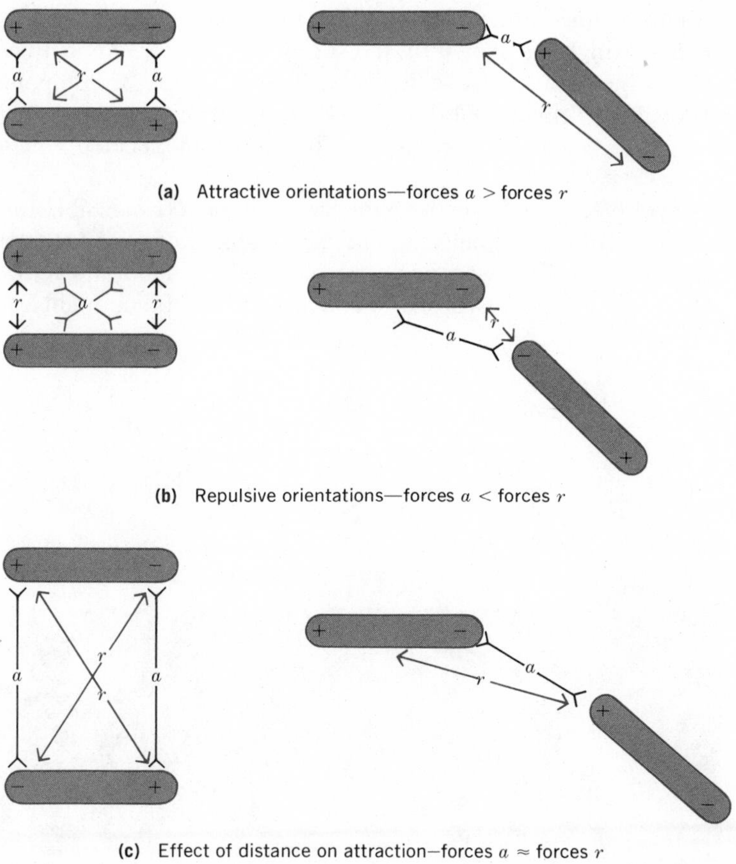

FIG. 4.28 Electrostatic interactions between dipoles.

In ice the covalent H—O bond distances are 1.00 Å, and the hydrogen bonds $\rangle$O----H— are 1.76 Å in length. Hydrogen bonds, being longer, are weak (~10 kcal/mole) and, unlike the covalent O—H bonds (110 kcal/mole), are constantly being made and broken in liquid water at ordinary temperatures as the molecules tumble around. In spite of its weakness, hydrogen bonding plays a large part in determining the properties of many liquids and the structures of molecular crystals.

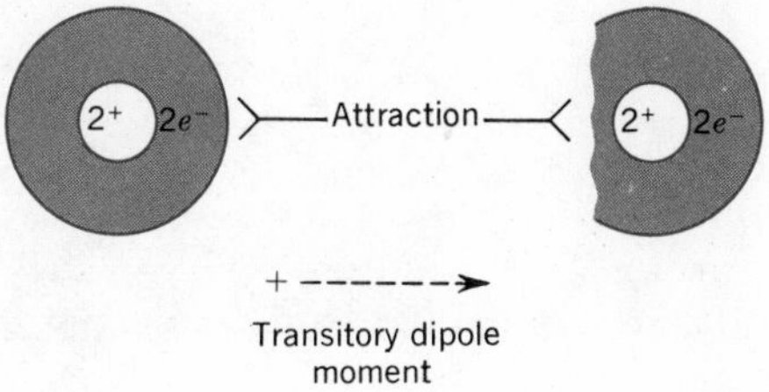

FIG. 4.29 Attraction between two helium atoms from a transitory dipole.

The boiling points of liquids generally increase with increasing molecular weight as shown by the series Ne, Ar, Kr, Xe in Figure 4.30. The same upward trend is shown by the series of compounds H_2S through H_2Te and HCl through HI. But the first members of their two series, H_2O and HF, are unusual in having unexpectedly high boiling points.

Hydrogen bonding is responsible. It is also responsible for the difference in boiling points for the two isomers of C_2H_6O, ethanol and methyl ether discussed in Section 4.2. The ethanol molecule, having an O—H group, is hydrogen bonded and has boiling point 78.5°C.

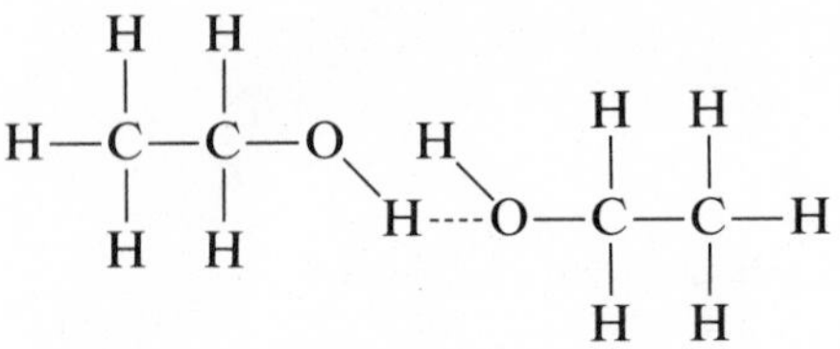

Ethanol

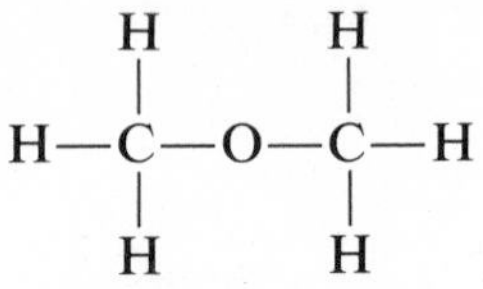

Methyl ether

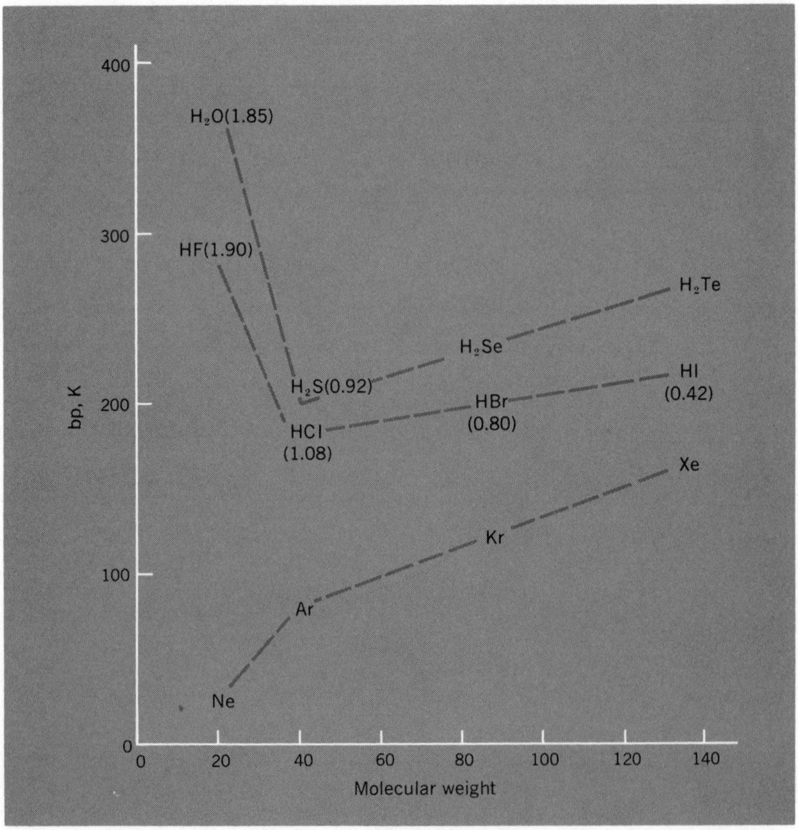

FIG. 4.30 Influence of permanent dipole moment on boiling points of isoelectronic molecules (dipole moments in debye units in parentheses, where known).

Methyl ether has no O—H group for hydrogen bonding, and the C—H bonds do not participate in hydrogen bonding because carbon is not sufficiently electronegative. Consequently methyl ether is a gas at ordinary temperatures and condenses to a liquid only at −25°C.

Hydrogen bonding is mainly responsible for what is termed *secondary structure* in the proteins and nucleic acids vital to living systems. The way in which the covalently bonded chains or *primary structures* of these huge molecules are held together by hydrogen bonds in the secondary structure is indicated in Figure 4.31. The importance of hydrogen bonds in biological systems depends on the fact that hydrogen bonds can be readily broken and re-formed at ordinary temperatures. Hydrogen bonding provides a cohesive force of just the right strength to hold together covalently bonded molecules in the larger units needed by living systems. It is vital to life, particularly to sexual reproduction (Section 14.8).

Ordinary van der Waals forces can also operate to bind chain molecules together. The properties of many plastics depend on cohesion of this type. An example is the attraction between chains in polyethylene, the plastic familiar in the form of the "squeeze bottle." The secondary bonding forces accumulate between the chains, which are millions of carbon atoms in length, so that if an attempt is made to vaporize the plastic, the carbon–carbon bonds break first, resulting in

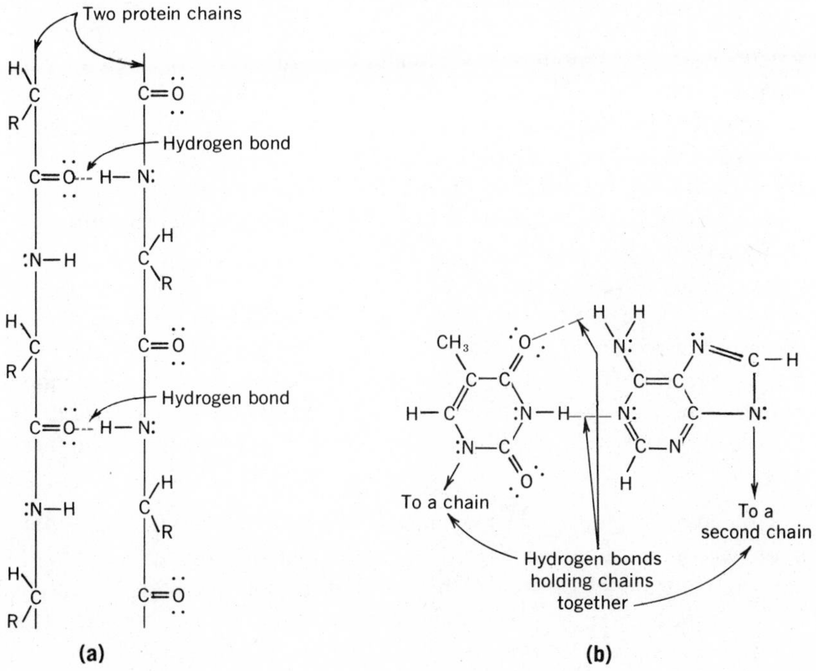

FIG. 4.31 Schematic representation of hydrogen bonding between adjacent chains in (a) a protein, (b) a nucleic acid.

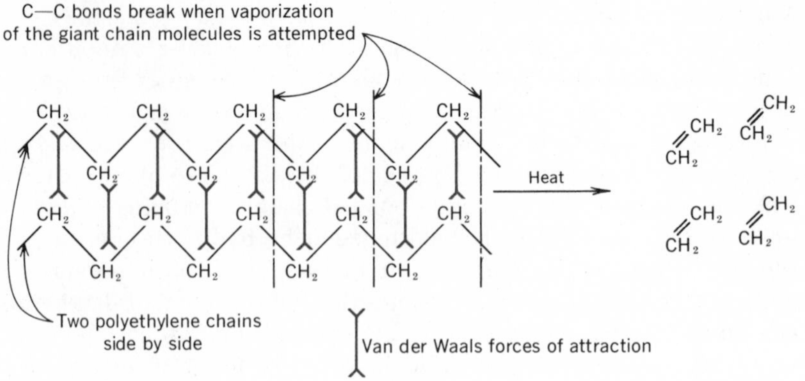

FIG. 4.32 Accumulation of van der Waals forces and depolymerization on being heated strongly.

depolymerization or a splitting up of the chain into smaller molecules (Figure 4.32).

4.7 Summary Guide Ionic bonding results from attraction between oppositely charged ions produced by complete transfer of electrons from one atom or group of atoms to another, most often with the production of octets of electrons. Ionic bonding is found between metals and nonmetals and between metal ions and complex ions.

According to Lewis theory, covalent bonding results when two or more atoms share electrons to achieve an octet (duet for H). Electronegativity is a measure of the attraction of atoms for electrons in bonds. Identical electronegativities for bonded atoms result in nonpolar covalent molecules; modest electronegativity differences result in polar covalent molecules; and large differences produce ionic bonding. Polar covalent molecules have permanent electric dipole moments unless individual bond dipoles cancel because of molecular symmetry. By counting the numbers of ligands and unshared pairs of electrons in Lewis structures, approximate bond angles for many molecules can be predicted.

According to resonance theory, if more than one Lewis electron-dot structure can be written for a molecule, the actual distribution of electrons in the molecule corresponds to a blending or hybridization of all dot structures. The actual electron distribution results in stronger bonding than corresponds to any of the individual Lewis structures. Dipole moments are accounted for by ionic contributing structures, as in $H—\ddot{\underset{..}{F}}: \leftrightarrow H^+:\ddot{\underset{..}{F}}:^-$.

Molecular orbitals give the shapes of electron clouds that extend over two or more atoms. Sigma (σ) molecular orbitals are egg-shaped and concentrate electrons in the bonding region between the atoms. Pi (π) orbitals have a nodal plane that passes through the nuclei, concentrating electrons above and below them. Molecular orbitals may be bonding or antibonding. Antibonding orbitals are indicated by a star as in σ^* or π^*. They have a nodal plane between the atoms that keeps electrons away from the bonding region. The energy sequence of molecular orbitals for homonuclear diatomic molecules (H_2 to F_2) is $\sigma_1 < \sigma_1^* < \sigma_2 < \sigma_2^* < \pi_1 = \pi_2 < \sigma_3 < \pi_1^* = \pi_2^* < \sigma_3^*$ with σ_1 most stable (lowest in energy). The orbitals are filled with electrons starting with σ_1, following the Pauli exclusion principle and Hund's rule. Each antibonding orbital filled cancels the bonding effect of a filled bonding orbital. Since there are equal numbers of the two types of orbitals in He_2 and Ne_2, these molecules do not exist. Molecular orbital theory and Lewis theory both predict that there should be a triple bond in N_2, but only molecular orbital theory correctly predicts that molecular oxygen gas should be attracted by a magnet.

Molecular orbital theory accounts for the bonding and geometry of many polyatomic molecules and complex ions. The theory uses the ideas of hybridizing (mixing) atomic orbitals and overlapping of the hybrid atomic orbitals produced to form molecular orbitals of both σ and π types.

Molecular orbital theory for the structure of the benzene molecule replaces the two contributing Lewis structures of resonance theory with three delocalized π orbitals on a sigma bond framework. Both theories predict $1\frac{1}{2}$ bonds between adjacent carbon atoms, plane hexagonal geometry, and the unusual stability of benzene, used to advantage by nature and humans.

The high electrical conductivity of metals is a result of the valence electrons of the metal atoms occupying completely delocalized orbitals (conduction bands). Semiconductors have most of their electrons in localized bond orbitals. As the temperature of a semiconductor is increased, its electrical conductivity increases because some of the electrons localized in bonds move into higher-energy metallic (conduction band) orbitals.

Secondary forces between atoms and molecules are much weaker than ionic, covalent, and metallic bonding, yet they are responsible for the existence of many liquids and solids at room temperature. Secondary forces result from attractions between molecules with permanent electric dipole moments, and between molecules with transitory dipole moments that result from short-lived fluctuations in electron clouds.

Hydrogen bonding is an unusually effective dipole–dipole attraction found principally with covalent compounds containing NH and OH groups. It is responsible for the secondary structure in proteins and nucleic acids.

Exercises

18. Methyl alcohol, CH_3OH, has bp 65°C and mol. wt 32.04, but Ar, mol. wt = 39.95, has bp −186°C. Explain the difference in boiling points.

19. Acetic acid ($CH_3—\overset{\ddot{O}:}{\overset{/\!/}{C}}—\ddot{O}H$) forms hydrogen-bonded dimers $(CH_3COOH)_2$. Give the structure of the dimer.

20. The carbon disulfide molecule CS_2 has its atoms arranged S C S. Give a covalent bond structure for CS_2.

21. The dipole moments of the hydrogen halide molecules (HF, HCl, HBr, HI) decrease with increasing atomic number of the halogens. Explain.

22. If someone suggested to you that the F_2^- ion might exist as a stable species, could you use molecular orbital theory to support his idea?

23. In urea, NH_2CONH_2, excreted in the urine of mammals and made synthetically for use as a fertilizer, the arrangement of atoms is

```
H     O     H
   N  C  N
H           H
```

Give a bond structure for urea. Should urea be regarded as a resonance hybrid?

24. In diphosphoric acid ("pyrophosphoric acid"), a participant in biochemical reactions (Chapter 14), the atoms are arranged as follows:

```
      O     O

H  O  P  O  P  O  H

      O     O

      H     H
```

Give electron-dot and bond structures for diphosphoric acid.

25. Nitrous oxide, N_2O, is widely used for inhalation anesthesia, particularly for minor operations and in dental surgery. It was one of the first anesthetics discovered and earned the name "laughing gas" because of the exhilaration it produces when first inhaled. Its use is not without danger, however, because N_2O (like O_2) supports combustion. The arrangement of atoms in the

N_2O molecule is N N O. Give an electron-dot and a bond structure for N_2O, and use them to explain the electron shifts in the reaction $N_2O + Ca \rightarrow N_2 + Ca^{2+}O^{2-} + 171$ kcal.

Key Words

Benzene nucleus a benzene ring with attached covalent groups replacing hydrogens.

Bond dissociation energy (D_{AB}) energy input needed to break a covalent (A—B) bond.

Canonical structures Lewis or bond structures that contribute to a resonance hybrid molecule.

Covalent bonding bonding due to simultaneous attraction by nuclei for electrons shared between atoms.

Debye unit unit measuring the strength of dipole moments – 10^{-18} esu cm.

Delocalized orbitals molecular orbitals that extend over three or more atomic nuclei.

Dipole moment (μ) separation of electric charge in a molecule multiplied by the distance of separation.

Electronegativity a numerical measure of the attraction of bonded atomic nuclei for electrons.

Geometric isomerism (*cis–trans* isomerism) isomerism due to lack of free rotation about double bonds when the lack results in different arrangements of the groups attached to the doubly bonded atoms.

Hydrogen bonding strong secondary forces between molecules having an H atom bonded to a strongly electronegative atom.

Ionic bonding bonding of atoms in compounds due to attraction between cations and anions – formation involves electron transfer.

Ligands atoms or atomic groups bonded to a central atom upon which attention is being focused for bond geometrical considerations.

Metallic bonding bonding by valence electrons in completely delocalized orbitals of metals.

Molecular orbitals descriptions of probability distributions for electrons in covalent molecules.

Multiple bonding joining of atoms by two or more pairs of electrons in covalent bonds between the atoms.

Pi (π) orbitals molecular orbitals in which electron probability is concentrated above and below a plane of no probability passing through the nuclei of covalently bonded atoms.

Polar molecules molecules in which the centers of gravity of electronic (negative) charge and nuclear (positive) charge do not coincide.

Resonance energy negative potential energy difference between actual structure of a resonance hybrid molecule and the energy of the most stable Lewis or dot structure.

Resonance theory superposition or blending of electron-dot structures to show actual distribution of bonding electrons.

Secondary forces of attraction (van der Waals forces) weak attractive forces between molecules due to dipole moments or transitory dipole moments – responsible for liquid and solid forms of covalent molecules.

Sigma (σ) orbitals molecular orbitals in which electron probability is concentrated in the space between covalently bonded atoms.

Structural isomers molecules that have the same molecular formula but are different because of the order in which their atoms are bonded to one another.

Suggested Readings

Bolmar, Y., "Photochemical Ozone Formation in the Atmosphere over Southern England," *Nature,* **241**:341–343 (Feb. 2, 1973).

Devlin, Robert M., "DDT: A Renaissance?" *Env. Sci. and Tech.,* **8**(4): 322–325 (April 1974).

Gillett, J. D., "The Mosquito: Still Man's Worst Enemy," *Am. Sci.,* **61**: 430–436 (July–Aug. 1973).

Maugh, Thomas H., II, "DDT: An Unrecognized Source of Polychlorinated Biphenyls," *Science*, **180**:578–579 (May 11, 1973).

Peakall, David B., "Pesticides and the Reproduction of Birds," "Chemistry in the Environment," *Readings from Scientific American*, San Francisco, W. H. Freeman Co., 1973, pp. 113–119.

Raw, I., and Holleman, G. W., "Water–Energy for Life," *Chemistry*, **46**:6–11 (May 1973).

Rudd, R., "Pesticides," in Murdoch, W. W., ed., *Environment: Resources, Pollution, and Society*, Stamford, Ct., Sinauer Associates, 1971, pp. 279–301.

Ryschkewitsch, George E., *Chemical Bonding and the Geometry of Molecules*, New York, Reinhold Publishing Corp., 1963.

"The Plight of the Tussock Moth," *Env. Sci. and Tech.*, **8**(6):506–507 (June 1974).

Van der Bosch, Robert, "Insecticide Crisis Accents Need of Biological Controls," *Catalyst* **1**(4):14–16 (1971).

Winkler, E. M., *Stone, Properties, Durability in Man's Environment,* New York, Springer-Verlag, 1973, Ch. 14.

chapter 5

The reaction $2H_2 + O_2 \rightarrow 2H_2O$ + energy out of control and on a huge scale. The burning of the hydrogen-filled dirigible *Hindenburg* at Lakehurst, New Jersey, May 6, 1937. What ignited the hydrogen remains a mystery. (*Photo 15, World Wide Photos.*)

Gases and Liquids

5.1 Introduction Gases have the property of expanding uniformly to fill any container in which they are placed. The particles flying around in a random, chaotic manner in a gas are *molecules*, and the energy of their motion is the thermal energy of the sample. Molecules are made up of two or more covalently bound atoms, examples being H_2, N_2, O_2, NH_3, H_2S, and CH_4. Unless the temperature of the sample is high enough to result in molecular collisions that are vigorous enough to break the bonds, the molecules remain as the structural units in the gas.

Molecules are also the structural units of most substances that are liquids at ordinary temperatures. When a liquid is placed in a container of volume larger than the sample, some of the molecules of the liquid *evaporate* to form a gas phase above the liquid. The molecules of a liquid are also in a random, chaotic motion, but their motion is restrained by the cohesive action of secondary forces. Only the molecules in the liquid having large kinetic energies escape to form the gas phase. As evaporation of a liquid goes on in a closed container, molecules in the gas phase begin to return to the liquid phase in the process of *condensation*. When the number of molecules escaping from the liquid is equal to the number returning to it from the gas phase, a state of *dynamic equilibrium* exists. The gas phase is called the *vapor* of the liquid, and the equilibrium is called the *liquid–vapor equilibrium* and shown by double arrows:

$$H_2O(l) \rightleftarrows H_2O(g)$$

You can't tell from looking at a system in dynamic equilibrium whether anything is going on in the system or not. Its dynamic character is at the molecular level.

Psi stands for pounds per square inch.

When the molecules of a gas rebound from the wall of the containing vessel, they exert a force on the wall in the same way that a moving tennis ball exerts a force in rebounding from a stationary racquet. The force exerted by molecular rebounds on unit area (cm^2 or $in.^2$) of the containing vessel is called the *pressure* of the gas. The rebounds of the molecules in the atmosphere (a mixture of gases) exert a force of about 14.7 psi on our bodies and all objects at sea level, but we are unaware of any force because it is exerted uniformly in all directions. The force pressing on one side of an $8\frac{1}{2}$-in. by 11-in. sheet of paper at sea level is $8.5 \times 11 \times 14.7 = 1375$ pounds or slightly over half a ton. It is pressing equally on the other side, so the paper doesn't move and we are unaware of the force. The pressure of the atmosphere will push a column of mercury up about 76 cm or 760 mm at sea level, depending on the weather. The mercury barometer (Figure 5.1) is used to measure atmospheric pressure, and the *standard atmospheric pressure* (1 atm) is defined as the pressure that supports a column of mercury 76 cm or 760 mm long. Gas pressures are ordinarily expressed in millimeters of mercury (mm Hg) usually called *torr*, honoring Torricelli, the inventor of the mercury barometer. The pressure of the atmosphere decreases with increasing altitude as shown by the shading in Figure 5.2. The reason for the decrease is that the number of molecules (mainly N_2 and O_2) per unit volume decreases as the atmosphere thins out with increasing altitude. As a result, there are fewer and fewer molecules to make the collisions responsible for the pressure, and it decreases accordingly.

Evangelista Torricelli (1608–1647) was an Italian physicist and mathematician who discovered the principle of the barometer.

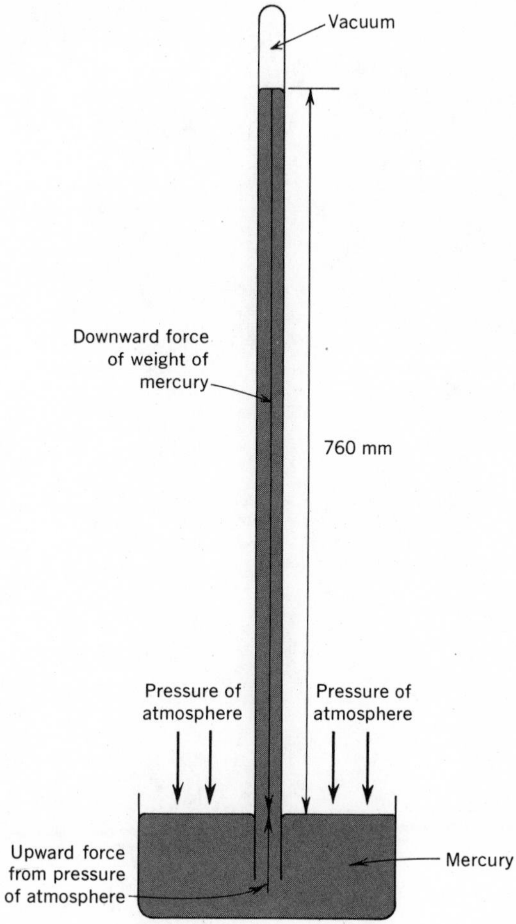

FIG. 5.1 Mercury barometer showing standard (sea-level) atmospheric pressure.

The gas pressure over a liquid–vapor equilibrium is called the *vapor pressure* of the liquid (Figure 5.3). Its value increases as the temperature is raised by heating. The temperature at which the vapor pressure over a liquid reaches 1 atm (760 torr) is called the *normal boiling point* or bp of the liquid. The dependence of vapor pressure on temperature and the boiling points of some liquids are shown in Figure 5.4.

It is essential to understand that the gas pressures we have been discussing are *absolute gas pressures*, not the excess of pressure over the prevailing barometric pressure, which is measured by open-tube manometers. The distinction is made in Figure 5.5 and Figure 5.6. Closed-tube manometers and barometers measure absolute pressures. Whenever the pressure of a gas appears in chemical calculations, it is the absolute pressure that must be used.

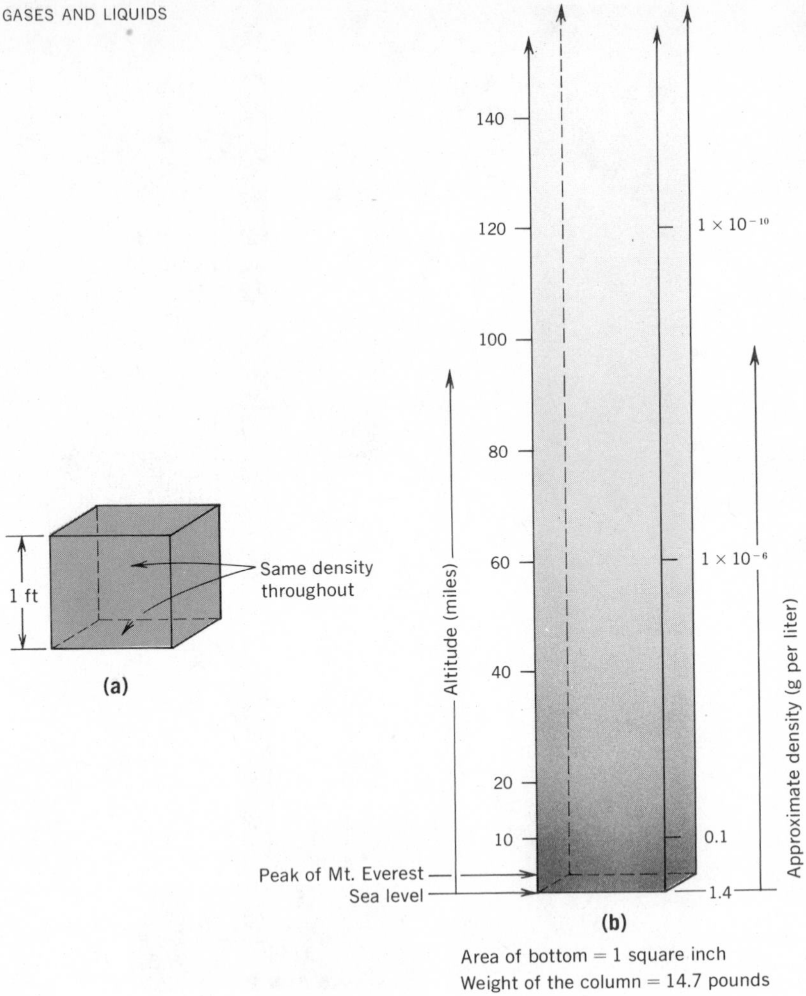

FIG. 5.2 Uniformity of gas samples and the standard pressure of the atmosphere (14.70 psi): (a) laboratory sample; (b) 1-inch-square column of the atmosphere.

5.2 Equation for a Perfect Gas The equation for a perfect gas relates the pressure, volume, temperature, and molar amount of an idealized or perfect gas that fits the following description:

1. There are no attractive forces between the gas molecules.
2. The electron clouds of the gas molecules are of negligible volume compared to the volume of the gas sample.
3. Collisions between molecules and with the walls of the container are perfectly elastic.
4. The thermal energy of the gas is the kinetic energy of the random motion of the gas molecules.

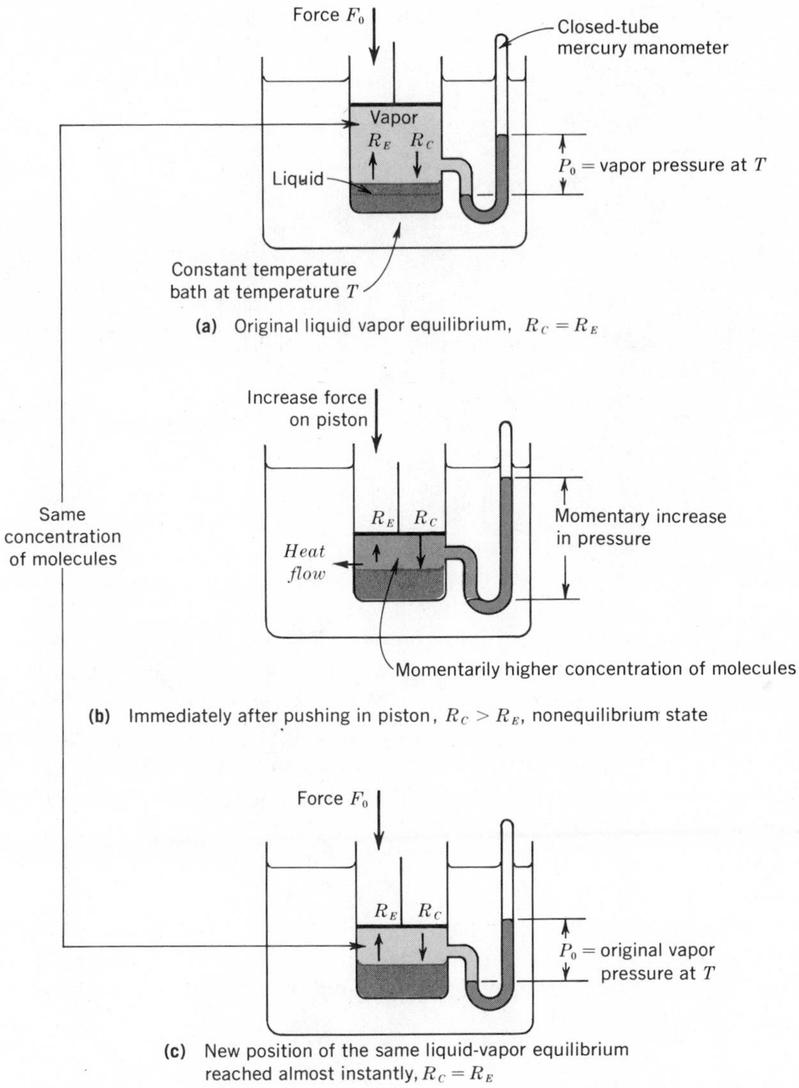

FIG. 5.3 Liquid–vapor equilibrium in an isothermal system.

The equation for a perfect gas, from kinetic molecular theory, is

$$PV = nRT$$

where P is the absolute pressure of the gas sample, V is the volume of the sample, n is the number of molar weights ("moles") of gas in the sample, R is a constant of proportionality having the same value for all gases, and T is the absolute or Kelvin temperature (K) of the sample (see Introduction).

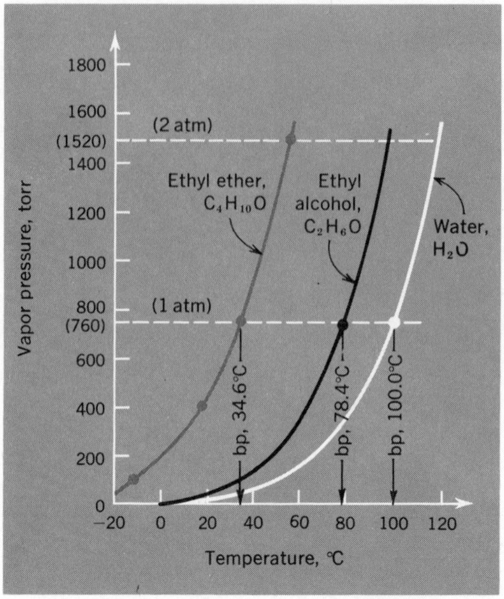

Vapor pressure data for ethyl ether, $C_4H_{10}O$

Temperature (°C)	*Vapor* *P* (torr)
−11.5	100
17.9	400
34.6	760
56.0	1520

FIG. 5.4 Vapor pressure as a function of temperature for some liquids (experimental data for ethyl ether tabulated).

The units must always be the same on both sides of an equation. One liter is the volume occupied by 1000 g (1 kilogram, 2.2 lb) of water at 3.98°C. One liter = 1.06 quarts.

The value used for the *gas constant R* depends on the measuring units for the pressure and the volume. When the gas pressure is measured in atmospheres (1 atm = 760 torr) and the volume of the gas sample is measured in liters, the value of R that you use is 0.08205 and the units for R are liter atm mole^{-1} deg^{-1}, with the understading that it is gram-moles and K that are meant. The equation has broad application because if you know the values of any *three* of P, V, n, and T for a gas sample, you can calculate the value of the fourth.

EXAMPLE

Suppose you want to know the volume occupied by one mole of hydrogen gas (H_2) at 0°C if the pressure of the gas is 1 atm. Keeping track of the units of measurement, and remembering that K = °C + 273, you write

$$1 \text{ (atm)} \times V \text{ (liter)} = 1 \text{ (mole)} \times 0.08205 \text{ (liter atm mole}^{-1}\text{ deg}^{-1}) \times (0 + 273) \text{ deg}$$

First look at the units of measurement. The expressions mole^{-1} and deg^{-1} mean 1/mole and 1/deg respectively. Moles and degrees divide out on the right side of the equation:

$$(\text{mole} \times \text{mole}^{-1}) = (\text{mole/mole}) = 1$$
$$(\text{deg} \times \text{deg}^{-1}) = (\text{deg/deg}) = 1$$

leaving liter atmospheres. These are also the units on the left side of the equation, so the units are correct and you can solve for the volume:

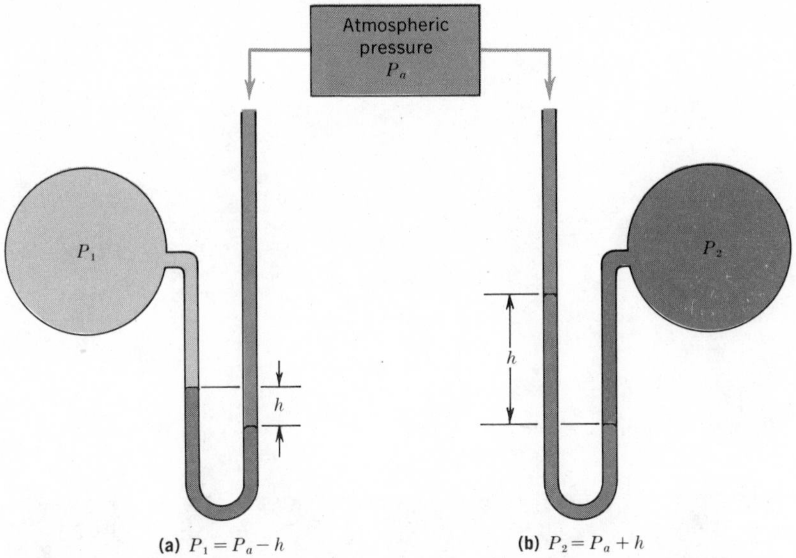

FIG. 5.5 Use of open-tube manometers to measure gas pressure: (a) pressure less than atmospheric; (b) pressure greater than atmospheric.

$$V = \frac{1 \times 0.08205 \times 273}{1} = 22.4 \text{ liters}$$

Thus, if hydrogen is admitted to an evacuated ("pumped-out") 22.4-liter flask held at 0°C (273 K), the pressure will rise to reach 1 atm when one mole (2.016 g) of H_2 has entered the flask. Alternatively, if you have one mole (2.016 g) of hydrogen in a piston–cylinder arrangement held at 0°C (273 K) [Figure 5.7(a)], the pressure, P, exerted by the hydrogen gas will be 1 atm when the piston has been moved so as to make the gas volume 22.4 liters. ⊸

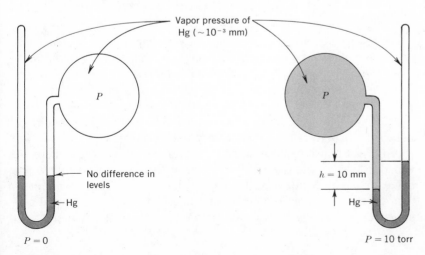

Vapor pressure of
Hg (~10⁻³ mm)
P
P
No difference in
levels
Hg
h = 10 mm
Hg
$P = 0$
P = 10 torr

FIG. 5.6 Closed-tube manometers. (The very small vapor pressure of mercury can be ignored for all practical purposes.)

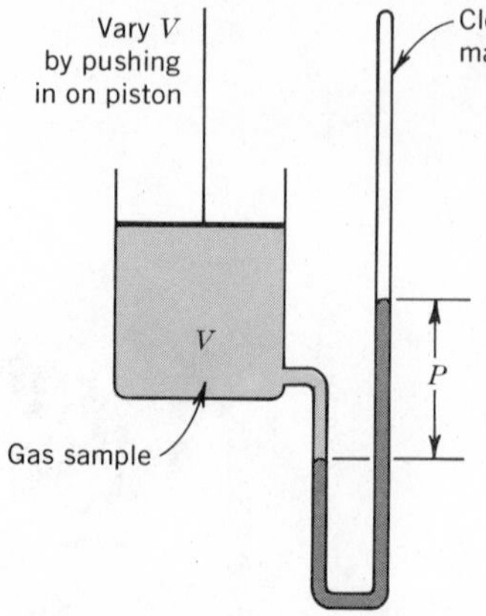

(a) Apparatus at constant and uniform temperature

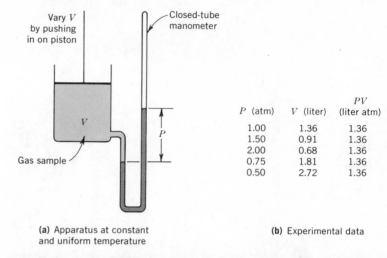

P (atm)	V (liter)	PV (liter atm)
1.00	1.36	1.36
1.50	0.91	1.36
2.00	0.68	1.36
0.75	1.81	1.36
0.50	2.72	1.36

(b) Experimental data

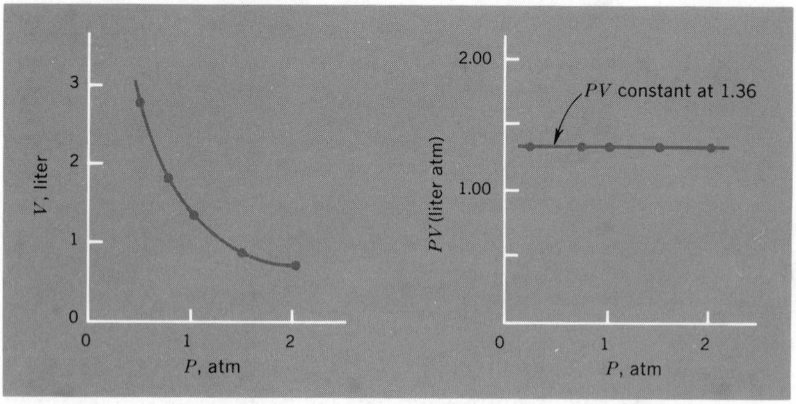

(c) Graph of V vs. P (W, M, t constant)

(d) Graph of PV vs. P (W, M, t constant)

FIG. 5.7 Boyle's law: apparatus and results.

$$V \propto \frac{1}{P} \quad \text{or} \quad PV = \text{constant value}$$

Hidden in the foregoing example is a very important fact. *The value of R is independent of the nature of the gas*—that is, *you use the same value* (with proper respect for units) *for all gases.* This means that the volumes occupied by one mole of all gases at 0°C and 1 atm pressure are the same, namely, 22.4 liters. This volume of 22.4 liters for one mole of any gas at 0°C and 1 atm pressure is called the *gram molecular volume* of a perfect gas. It is a very useful quantity, as we will see shortly. But first the question, how well do real gases (where there are van der Waals attractive forces between the molecules) follow the behavior predicted for a perfect gas? Table 5.1 shows the experimentally determined molar volumes for some real gases. They are determined by measuring the density of the gas in grams per liter at

If W grams of gas occupy a volume of V liters at 0°C and 1 atm pressure, then the density (d) = W/V grams per liter.

0°C and 1 atm pressure, and then dividing the density into the molecular weight of the gas to get the volume that would be occupied by one mole of the gas:

$$\text{Molar volume} = \frac{\text{molecular weight}}{\text{density in g/liter at 0°C, 1 atm}}$$

For ammonia (NH_3) the calculation is

$$\text{Molar volume} = \frac{[14.01 + 3(1.008)]\,g}{0.7712\ g/liter^{-1}} = 22.088\ \text{liters}$$

From the data in Table 5.1, you can see that most gases do not behave "perfectly." Nevertheless, the value of 22.4 liters for the gram molecular volume of a gas at the *standard conditions* (standard temperature and pressure, or STP) of 0°C (273 K) and 1 atm is accurate enough for most chemical calculations.

EXAMPLE

We saw in the example in Section 2.8 that on a *weight* basis, $H_2(g)$ as a fuel is $2\frac{1}{2}$ times as rich in energy as $CH_4(g)$ (methane, natural gas). For a pipeline to deliver the same *weight* of hydrogen as it does of methane without increasing the volume of gas flow, hydrogen must be carried at a higher pressure. Using the gas law, you can calculate the extra pressure needed for hydrogen to meet the requirement for delivery of equal amounts of fuel energy from hydrogen and from methane through the same pipeline.

One mole of methane produces 213 kcal of heat on combustion, one mole of hydrogen only 68.3 kcal. It follows that (213/68.3) = 3.12 moles of $H_2(g)$ can replace one mole of $CH_4(g)$ as fuel. The pressure of a gas at constant temperature and volume is directly proportional to the number of moles of gas according to the gas law:

TABLE 5.1 Molar volumes of real gases

Gas	*Experimentally Determined Density at* STP[a] (g/liter, 0°C, 1 atm)	*Actual Molar Volume at* STP[a] (liters)
He	0.1785	22.423
H_2	0.08988	22.430
N_2	1.2506	22.405
O_2	1.429	22.393
CO_2	1.9767	22.264
NH_3	0.7712	22.083
CH_3Cl	2.3076	21.880

[a]STP stands for standard temperature (0°C) and pressure (1 atm).

$$PV = nRT \qquad P = n\left(\frac{RT}{V}\right)$$

and (RT/V) has a constant value as long as T and V are constant, so $P = \text{const} \times n$. To crowd 3.12 moles of H_2 into the volume occupied by 1.0 mole of CH_4, the direct proportionality says the pressure of H_2 must be increased 3.12 times. If $CH_4(g)$ is being carried at 100 psi in a pipeline, $H_2(g)$ must be carried at $100 \times 3.12 = 312$ psi for the pipeline to deliver equal amounts of fuel energy without increasing the rate of flow of $H_2(g)$. Because of the larger heat of combustion of methane, the flaming destruction of the dirigible Hindenburg (Photo 15) would have been even more spectacular had it involved the burning of an equal volume of methane.

5.3 Boyle's and Charles's Laws The equation for a perfect gas contains two laws describing the behavior of gases that were discovered by experiments before the development of kinetic molecular theory. These are Boyle's law, giving the effect on volume of changes in pressure, and Charles's law, which gives the effect of temperature on volume. Boyle found that if a given gas sample were held at a constant temperature while its volume was varied, as by moving the piston in the apparatus in Figure 5.7(a), the product of the pressure and the volume was a constant number [Figure 5.7(b) and (d)]. That is, if P_1 and V_1 stand for one set of values for the pressure and volume and P_2 and V_2 for a second, then $P_1V_1 = P_2V_2$. Another way of stating his result is that "the volume of a given gas sample held at constant temperature is inversely proportional to the pressure exerted by the gas." Figure 5.7(c) shows the inverse relationship—halving the volume doubles the pressure, and so on.

Robert Boyle (1627–1691) was an Irish chemist and physicist whose experimental approach did much to bring about modern scientific methods.

Jacques Alexander César Charles (1746–1823) was a French mathematician and physicist. He was the first (in 1773) to use hydrogen for the inflation of balloons. He shared honors with Gay-Lussac for his work.

To discover Boyle's law, in the form PV = a constant, in the equation for a perfect gas, we first note that, for a given weight of a gas, the number of moles (n in the equation) has a constant value equal to the weight of the gas sample divided by the molecular weight of the gas, or $n = W/M$. If the temperature of the gas is held constant, as required by Boyle's law, the right-hand side of the equation $PV = nRT$ is a number whose value is determined by the weight of the gas sample, its molecular weight, the constant value of R, and the temperature of the gas. That is, $PV = (W/M)RT$ = a constant number at fixed values of W, M, R, and T, or PV = a constant, as Boyle found experimentally. You can use Boyle's law in the form $P_1V_1 = P_2V_2$ for calculating the pressure–volume relationships for gas samples held at constant temperature. Suppose the volume of a gas sample at 740 torr is 2.50 liters and you want to know what the volume will be at 1 atm (760 torr) pressure if the temperature is not changed. Letting $P_1 = 740$ torr, $V_1 = 2.50$ liters, and $P_2 = 760$ torr, you write ☛

Both pressures must be expressed in the same units, 1 atm = 760 torr.

$$740 \times 2.50 = V_2 \times 760$$

or

$$V_2 = \frac{740 \times 2.50}{760} = 2.43 \text{ liters}$$

You can check the result by noting that the *increase* in pressure (from 740 to 760 torr) caused the *decrease* in volume expected from the *inverse* relationship between volume and pressure. In terms of kinetic theory, the decrease in volume increased the number of gas molecules per unit volume and therefore the number of collisions with the wall. The increase in pressure results.

Exercises

1. With temperature constant, an initial volume of 2.5 liters of nitrogen gas is changed to 3.0 liters by pulling out the piston in an apparatus like that in Figure 5.7(a). If the initial pressure was 740 torr, what is the pressure in torr after the expansion? What is the pressure in atmospheres after the expansion?

2. The pressure of a gas sample is 450 torr. What volume change will be required to change the pressure to 900 torr if temperature is held constant?

Charles's law in its original form was mathematically complicated because it used the centigrade temperature scale. It becomes very simple if the absolute temperature scale (K) is used: *The volume of a given gas sample held at constant pressure is directly proportional to the absolute temperature of the sample.* In an equation the law takes the form $V = \text{constant} \times T$. To discover Charles's law in the equation for a perfect gas, we rearrange the equation to

$$V = \frac{nRT}{P}$$

or, since $n = W/M$,

$$V = \frac{WRT}{MP}$$

For a given sample of a gas W and M are fixed in value, R has its constant value, and if the pressure is held constant, P also has a constant value, so we obtain Charles's law:

$$V = \frac{WR}{MP} T = \text{constant} \times T$$

Charles's law may be applied to the variations in volume with temperature for given gas samples *held at constant pressure*. For this purpose it has the form

$$\frac{V_1}{T_1} = \frac{V_2}{T_2} \qquad \text{(at constant } P, M, \text{ and } W\text{)}$$

where V_1 and T_1 are one set of conditions, V_2 and T_2 a second. For example, suppose the volume of a sample of nitrogen gas is 25.00 ml at 27°C at some (undetermined) pressure and you want to know what its volume would be at 0°C if the pressure remained unchanged. Remembering that *absolute temperatures must be used*, you let $V_1 = 25.00$ ml, $T_1 = (27 + 273)$, and $T_2 = (0 + 273)$, and write

$$\frac{25.00}{27 + 273} = \frac{V_2}{0 + 273}$$

or

$$V_2 = \frac{273}{300} \times 25.00 = 22.75 \text{ ml}$$

Checking, you can see that a *decrease* in temperature resulted in a *decrease* in volume as required by the direct proportionality between volume and temperature. According to kinetic theory, the volume decrease was necessary to keep the pressure from falling, because the temperature decrease lowered the vigor of molecular collisions with the wall of the vessel. Only by decreasing the volume, so that there are more molecules per unit volume and therefore more collisions with the wall, can the lowered vigor of the collisions be offset to keep the pressure constant.

Boyle's and Charles's laws are put together in what is called the *combined gas law*. This equation is very useful for studying a given gas sample (W and M constant) under different sets of pressure and temperature conditions. The combined gas law is

$$\frac{P_1V_1}{T_1} = \frac{P_2V_2}{T_2}$$

The most common use of this expression is to calculate the volume at STP that a gas sample collected under laboratory conditions would occupy. For example, suppose you want to know the volume at STP for a 2.00-liter sample of oxygen you collected at 25°C under atmospheric pressure when the barometer read 765 torr. Letting $P_1 = 765$ torr, $V_1 = 2.00$ liters, and $T_1 = (25 + 273)$, and assigning P_2 and T_2 their standard values of 760 torr and 273 K, respectively, you substitute in the combined gas law to obtain:

$$\frac{765 \times 2.00}{25 + 273} = \frac{760 V_2}{273}$$

or

$$V_2 = \frac{765}{760} \times \frac{273}{298} \times 2.00 = 1.84 \text{ liters}$$

Checking the separate influences of the pressure and temperature changes on the volume, you can see that the pressure factor (765/760) is greater than 1 and acts to increase the volume, as it must, since lowering the pressure from 765 to 760 torr results in expansion. The effect of the temperature factor is opposite, since it (273/298) has a value less than 1, as it must, because lowering the temperature of a gas (from 298 to 273 K in this case) causes it to contract. To prevent mistakes in substituting numerical values in the combined gas law, it is always a good idea to make these checks on the separate effects of the pressure and temperature factors.

Exercises

3. A 10.0-liter sample of hydrogen gas is originally at 0°C in a piston–cylinder arrangement like that in Figure 5.7(a). The gas is now cooled, while pushing in the piston to maintain pressure constant, until a final volume of 1.0 liter is reached. Calculate the final temperature in K and in °C.

4. To what centigrade temperature must a gas sample originally at 273 K be raised to double its volume if pressure is held constant?

Clearly the combined gas law can be used only when the weight of the gas sample and its molecular weight do not change. If either of these values changes, then the entire equation for a perfect gas must be used. Most commonly this happens when a chemical reaction among gases is involved. Examples will follow shortly.

Avogadro's discovery that "equal volumes of different gases measured at the same temperature and pressure contain equal numbers of molecules" is also contained in the equation for a perfect gas. For two gases A and B we may write, using subscripts to identify the quantities for the two gases:

$$P_A V_A = n_A R T_A \qquad P_B V_B = n_B R T_B$$

Rearranging these equations to give the numbers of moles of A and B, we obtain

$$\text{Number of moles of } A = n_A = \frac{P_A V_A}{RT_A}$$

$$\text{Number of moles of } B = n_B = \frac{P_B V_B}{RT_B}$$

Now, since $P_A = P_B$ (same pressure for the two gases), $V_A = V_B$ (equal volumes), $T_A = T_B$ (same temperature), and R has the same value for all gases,

$$n_A = \frac{P_A V_A}{RT_A} = \frac{P_B V_B}{RT_B} = n_B$$

The number of molecules in n moles of a gas is $n \times L$, where L is the Avogadro number (the number of molecules in one mole); thus $n_A L = n_B L$. In words, the numbers of molecules in the equal volumes are the same.

For a given sample of any gas, experiment shows that the pressure exerted by the gas is directly proportional to the absolute temperature as long as the volume is held constant. That is, $P =$ a constant $\times T$. This law is also contained in the equation for a perfect gas, as can be seen by writing it in the rearranged form

$$P = \left(\frac{nR}{V}\right) \times T = \text{constant} \times T \qquad (n \text{ and } V \text{ constant})$$

This means that if the volume of a gas sample is held constant while its temperature is lowered, its pressure falls accordingly. In particular, when T reaches absolute zero (0 K) the pressure will be zero:

$$P = \text{constant} \times T = \text{constant} \times 0 = 0$$

This result agrees with the ideas of temperature being a measure of the intensity of molecular motion and pressure being due to collisions of gas molecules with the walls of the container. At absolute zero, molecular motion ceases and so do the collisions responsible for the pressure. Of course, real gases become solids by the time a temperature close to absolute zero is reached because of the cohesive forces between their molecules.

Helium is an exception, remaining a liquid at -272.2°C or 0.96 K.

Exercises

5. A 0.25-mole sample of nitrous oxide gas (N_2O) is admitted to an evacuated 10.0-liter flask at 25°C. What is the pressure exerted by the N_2O in torr? In cm Hg?

6. A 40.36-g sample of a gaseous element exerts a pressure of 2.00 atm in a 22.4-liter flask at 273 K. What is the molecular weight of the gas? Which element is it? (See Periodic Table.)

7. A gas sample has a volume of 24.00 ml at 22.5°C. When heated to 72.0°C, it occupies 28.00 ml. Pressure is constant. Calculate the centigrade temperature corresponding to absolute zero from these data.

5.4 Dalton's Law of Partial Pressures Dalton's law of partial pressures states that the total pressure (P_{total}) of a mixture of gases is the sum of the pressures that each gas in the mixture would exert if it alone occupied the volume of the sample. The pressure exerted by each individual gas is called the *partial pressure* of the gas (p_i, where i is the formula of the gas). Thus, for a mixture of the gases nitrogen (N_2), oxygen (O_2), and argon (Ar), we have $P_{\text{total}} = p_{N_2} + p_{O_2} + p_{Ar}$. Applying kinetic theory, we see that the partial pressure of a gas is the contribution to the total pressure made by the collisions of its molecules with the walls of the container. That is, each gas acts independently of the others in making up the total pressure. The contribution of each can be calculated by applying the equation for a perfect gas separately to each kind of gas present. Thus, suppose you have a mixture of gases composed of 3 moles of nitrogen, 2 moles of oxygen, and 1 mole of argon in a 20.00-liter container at 27°C and want the total pressure of the mixture. If the nitrogen alone occupied the container, it would exert a pressure p_{N_2}, given by the equation for a perfect gas in the rearranged form $P = (n/V)RT$ or, putting in the numbers for N_2,

$$p_{N_2} = \frac{n_{N_2}}{V}RT = \frac{3}{20} \times 0.08205 \times (27 + 273) = 3.69 \text{ atm}$$

The partial pressure (p_{N_2}) of nitrogen is 3.69 atm. Similarly, the partial pressures of oxygen and argon are given by

$$p_{O_2} = \frac{n_{O_2}}{V}RT = \frac{2}{20} \times 0.08205 \times (27 + 273) = 2.46 \text{ atm}$$

$$p_{Ar} = \frac{n_{Ar}}{V}RT = \frac{1}{20} \times 0.08205 \times (27 + 273) = 1.23 \text{ atm}$$

The total pressure is the sum of the partial pressures:

$$P_{\text{total}} = p_{N_2} + p_{O_2} + p_{Ar} = 3.69 + 2.46 + 1.23 = 7.38 \text{ atm}$$

Another form of Dalton's law is useful when the total pressure of the gas mixture remains constant at the prevailing atmospheric pressure (~ 1 atm at sea level) in spite of changes in both the temperature and composition of the air. This form of Dalton's law is given as

$$p_i = X_i P_{\text{total}}$$

In this equation p_i is the partial pressure of the ith gas in the mixture, P_{total} is the total pressure of the gas mixture, and X_i is the fraction of all the molecules in the gas mixture represented by those of the ith kind. This fraction is called the *mole fraction* of the given kind of molecule, and is equal to the number of moles of the given gas divided by the total number of moles of all gases in the mixture. For example, suppose you want the mole fractions of the principal components, O_2 and N_2, in dry air, given that its composition is 76.8% N_2 and 23.2% O_2 by weight. ☛ Take 100.0 g of air; it contains 76.8 g of N_2 and 23.2 g of O_2. The number of moles of each in the 100.0-g sample is obtained by dividing the weight of the gas by its molecular weight:

Air actually contains water vapor, CO_2, Ar, and other trace gases amounting to 1.3% by weight. They have been included with the figure for N_2 in this approximate calculation.

$$n_{O_2} = \frac{23.2}{32.0} = 0.725 \text{ mole} \qquad n_{N_2} = \frac{76.8}{28.0} = 2.743 \text{ moles}$$

The total number of moles in the sample is $n_{O_2} + n_{N_2}$ or $0.725 + 2.743 = 3.468$, and the mole fractions are

$$X_{O_2} = \frac{n_{O_2}}{n_{O_2} + n_{N_2}} = \frac{0.725}{3.468} = 0.209$$

$$X_{N_2} = \frac{n_{N_2}}{n_{O_2} + n_{N_2}} = \frac{2.743}{3.468} = 0.791$$

If the barometer reading is 760 torr, the partial pressures in dry air are given by Dalton's law as

$$p_{O_2} = X_{O_2} P_{\text{total}} = 0.209 \times 760 = 159 \text{ torr}$$
$$p_{N_2} = X_{N_2} P_{\text{total}} = 0.791 \times 760 = 601 \text{ torr}$$
$$P_{\text{total}} = 760 \text{ torr}$$

After making this kind of calculation, check it this way to make sure that the sum of the partial pressures is equal to the total pressure as required by Dalton's law.

Exercises

8. A 25.0-liter flask contains 0.25 mole of He and 0.75 mole of nitrogen (N_2). What are the total pressure and partial pressures of He and N_2 if the flask is at 100°C?

9. A 20.0-liter flask at 27°C contains 0.25 mole of oxygen gas. What

weight of nitrogen gas (N_2, mol. wt 28.02) must be pumped into the flask to make the total gas pressure 2.00 atm?

10. A mixture of N_2, H_2, and Ar in a 5.00-liter flask at 20°C exerts a total pressure of 1.0 atm. What is the total number of moles of gases in the flask? What would be the effect on the partial pressures of the gases in the mixture if the temperature of the flask were raised to 313°C?

Dalton's law in the form $p_i = X_i P_{\text{total}}$ is useful for understanding what happens to the composition of air as we breathe it.

The air we breathe is not dry. It contains a small amount of water vapor, depending on its humidity, and a trace of carbon dioxide. Representative partial pressures of the gases in the air you inhale and exhale are given in Table 5.2. In your lungs, inhaled air is exposed to a very large area (about 1000 sq ft) of gas-permeable membrane that has circulating blood on the inner side. The membrane is kept moist by secretion of water from the blood so that the air in the lungs becomes *saturated* with water vapor. By *saturated* we mean that the partial pressure of water vapor in the gas mixture is equal to the vapor pressure of water at the prevailing temperature. At body temperature (37°C) the vapor pressure of water is 47.0 torr, so p_{H_2O} in exhaled air has this value, as shown in Table 5.2. During ordinary quiet breathing about 7.5 liters of air enter and leave the lungs per minute. You can calculate the amount of water that leaves your body by way of exhaled air in one minute by using the perfect gas law in the form $PV = (W/M)RT$, with $p_{H_2O} = 47$ torr, $V = 7.5$ liters, $R = 0.08205$ liter atm mole^{-1} deg^{-1}, and $T = (273 + 37)$. Converting the partial pressure of water to atmospheres by dividing by 760, as required by the units of R, and using 18 for the molecular weight of H_2O, we obtain

$$\frac{47}{760} \times 7.5 = \frac{W}{18} \times 0.08205 \times (273 + 37)$$

from which $W = 0.33$ g per minute. Thus every 3 minutes about 1 gram of water leaves the body in exhaled air. In 24 hours (1440 min), $0.33 \times 1440 = 475$ g or slightly more than 1 pint of water is exhaled as water vapor. Of course, some of the water in the exhaled air was in the inhaled air (Table 5.2), and your body doesn't have to provide the total, unless you are in the desert, where the humidity is close to zero. Under conditions of very low humidity, water loss by this route becomes very important, particularly if your respiration rate is increased by exercise. The unusual thirst experienced by skiers and mountain climbers in cold, dry alpine regions is due in part to water loss through exhaled breath.

EXAMPLE

Although a very minor component of air, CO_2 is the source of carbon for all growing plants, where it is converted to sugar, starches, and cellulose by photosynthesis. CO_2 is thus the ultimate source of everything we eat.

TABLE 5.2 Representative partial pressures of gases (in torr) in inhaled and exhaled air

Gas	*Inhaled Air*	*Exhaled Air*
O_2	158.0	116.0
N_2	596.4	568.0
H_2O	5.3	47.0
CO_2	0.3	29.0
Total	760.0	760.0

p_{H_2O} over ice at −10°C (14°F) is only 2.00 torr, so air in equilibrium with it is quite dry.

Oxygen is necessary for life, yet comparison of the oxygen content of inhaled and exhaled air in Table 5.2 shows that only 26.6% of the amount in inhaled air enters the bloodstream to support life processes:

$$\frac{158.0 - 116.0}{158.0} \times 100 = 26.6\%$$

As we go to higher altitudes our need for oxygen does not lessen but the supply available in one breath decreases as the barometric (total) pressure decreases. So we pant, even with mild exercise, at high elevations. At the elevation of Mount Everest (29,028 ft) in the Himalayas, the pressure of the atmosphere has fallen to 250 torr. The mole fractions of O_2 and N_2 remain the same as at sea level, but their partial pressures have fallen correspondingly:

$$p_{O_2} = X_{O_2}P_{\text{total}} = 0.209 \times 250 = 52.3 \text{ torr}$$
$$p_{N_2} = X_{N_2}P_{\text{total}} = 0.791 \times 250 = 197.7 \text{ torr}$$
$$P_{\text{total}} = 250.0 \text{ torr}$$

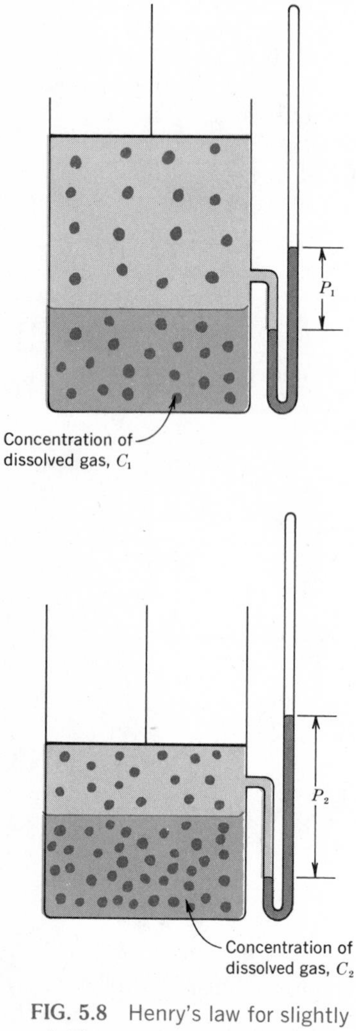

FIG. 5.8 Henry's law for slightly soluble gases: $P_2 = 2P_1$, $C_2 = 2C_1$.

The partial pressure of oxygen in the air is so low that even the most rapid rates of breathing can't provide enough oxygen, so oxygen masks fed by pure oxygen from steel tanks, where it is stored under high pressure, are needed by mountain climbers tackling Mount Everest.

At the extreme altitudes used by jet aircraft the atmospheric pressure is so low that even if pure oxygen ($X_{O_2} = 1$) were supplied to a face mask at the prevailing atmospheric pressure the lungs could not absorb enough to sustain life for long. To solve this problem, the sealed interior of the aircraft is pressurized by pumping in the thin air to give an interior pressure equal to the atmospheric pressure at about 8000 ft. In the rare event of loss of cabin pressure, individual masks supplying pure oxygen are automatically provided to tide the passengers over while the pilot reduces the altitude of the plane until p_{O_2} is sufficient to sustain life.

Although the body cannot adjust quickly to lowered partial pressures of oxygen, considerable adjustment occurs with the passage of time. The adjustment results from changes in the depth of breathing, in the rate of circulation of the blood, and most importantly, by an increase in the number of red cells (erythrocytes) in the blood. The red cells carry oxygen from the lungs to the tissue cells, where it is needed for the oxidation of glucose to maintain body temperature and provide muscular energy. An increase in red-cell count produces a corresponding increase in the percentage of the oxygen absorbed from the inhaled air. The process takes time—weeks to months, depending on the altitude—and is called *acclimatization*. Native Peruvians can work in mines at elevations as high as 14,000 ft above sea level because they have grown up there and are acclimatized. A person reared at sea level and abruptly transported to such elevations finds himself literally "gasping for air" at the slightest exertion, and suffers from weakness, headache, nausea, vomiting, and other symptoms of "altitude sickness." With the passage of time these symptoms diminish and pass away as the body produces additional red cells. Depending on the altitude, the number of red cells may rise from the sea level normal of about 5,000,000 per cubic millimeter to as many as 8,500,000 as acclimatization occurs. But even acclimatization fails to provide sufficient oxygen-carrying capacity in the blood at elevations above about 18,000 feet, the maximum elevation of even semipermanent human habitation. ☛

It's true that Annapurna (26,504 ft) was first climbed without the use of oxygen equipment, but although the climbers survived, their ability to function normally seems to have been somewhat impaired while they were at extremely high altitudes.

Even at sea level, pure oxygen is often supplied to heart and pneumonia patients to ease the strain of breathing. Oxygen used in resuscitators contains about 5% of CO_2, which stimulates breathing. It is also used to help athletes "catch their breath" when on the bench.

Henry's law All gases dissolve to some extent in ordinary liquids. Fish can live without breathing air because the oxygen dissolved in the water is absorbed through their gills. Henry's law states that *the concentration* (moles/liter) *of a gas dissolved in a liquid is proportional to the partial pressure of the gas above the liquid.* ☛ This means that water in contact with air at a pressure of two atmospheres will contain twice the concentrations of dissolved oxygen and nitrogen that are found in water exposed to air at one atmosphere pressure (Figure 5.8).

Henry's law does not apply if the liquid is the liquid form of the gas above it, or if the gas reacts chemically with the liquid.

EXAMPLE

Henry's law has profound consequences for divers doing underwater work. With ordinary diving gear (helmet or face mask) the air supplied must be at the pressure of the water surrounding the diver. Otherwise the hydrostatic pressure of the water exerted on the chest prevents breathing. You can quickly convince yourself of this fact by trying to breathe through a rubber hose open to the surface air while at a depth of 5 ft or so in a swimming pool. A 33-ft column of water exerts a pressure of one atmosphere. This means that the total pressure at a water depth of 33 ft is two atmospheres, one atmosphere due to barometric pressure and one atmosphere from the hydrostatic pressure of the water. At 66 ft, the pressure is three atmospheres, and so on. Suppose you're a diver working 66 ft below the surface. The gas pressure of the air in your mask or helmet has to be three atmospheres for you to be able to breathe. By Henry's law, your blood will contain three times as much dissolved oxygen and nitrogen as it would if you were at the surface. If you return to the surface too rapidly, there is not enough time for the excess dissolved nitrogen to be excreted through your lungs, and it forms tiny bubbles in your bloodstream. The situation is like a bottle of a carbonated drink being uncapped. The liquid

Joseph Henry (1797–1878), born in Albany, New York, was a physicist whose experiments played a large part in the development of the telegraph, telephone, radio, and dynamo.

fizzes because the liquid was "carbonated" (saturated with CO_2) at a partial pressure of CO_2 thousands of times that in the atmosphere to which the bottle is opened.

The bubbles of nitrogen that form in the blood obstruct its flow through the capillaries and produce the painful condition known as "the bends." The term arises because the pain, particularly in the bone joints, causes the diver to bend over and curl up. Brain damage and death may follow unless prompt action is taken. One way out is for the diver to return immediately to the depth at which he was working. There the increased pressure forces the bubbles of nitrogen back into solution in his blood. Then, if he ascends slowly in about 10-ft stages, allowing a scheduled wait at each stage, ☛ time is provided for the excess nitrogen to be excreted in the breath without bubble formation. Another solution is to enter a "decompression" chamber. This is a heavy steel tank with an airtight door. Air is pumped rapidly into the chamber until it reaches the pressure at which the diver was working. After this has been done, the pressure is lowered very slowly by releasing air from the tank through a valve until the pressure has reached atmospheric. As the pressure slowly falls, the excess nitrogen in the blood is excreted through the lungs without the formation of bubbles. The "decompression" of a diver who has been working at extreme depths may take many hours.

Scuba divers (Photo 16) are familiar with the timing of ascents from various depths.

At increased pressures the amount of oxygen dissolved in the blood is also increased according to Henry's law. The excess presents no serious "bubble" problem because it can be used up in the metabolism

PHOTO 16 Scuba divers descending an anchor line near Grand Cayman Island in the British West Indies. (*Kent Schellenger, Sierra Club.*)

of body cells or passed quickly by the red blood cells to the lungs for excretion.

Exercises

11. The solubility of oxygen gas in water at 25°C is 3.16 ml per 100 g of water when $p_{O_2} = 1.00$ atm. Taking p_{O_2} in air as 0.20 atm, what volume of O_2 would be dissolved in 100 g of water at 25°C exposed to the atmosphere?

12. Helium is without physiological effects, and its solubility in water at $p_{He} = 1.0$ atm is 0.94 ml per 100 g of H_2O. Nitrogen dissolves in water when $p_{N_2} = 1.00$ atm to the extent of 1.80 ml per 100 g of H_2O. Would there be an advantage in having deep-sea divers or "Sea Lab" occupants breathe a compressed mixture of 80 volume percent helium, 20 volume percent oxygen instead of compressed air?

5.5 Gas Laws and the Internal Combustion Engine Basically, an internal combustion engine of the type found in automobiles consists of a set of cylinders containing pistons that are connected to a crank for converting their up-and-down motion into the rotary motion of the crankshaft. ☛ In the most common type (four-stroke or four-cycle) of engine, ☛ the intake valve opens at the top of a piston stroke (Figure 5.9). As the piston moves down, the pressure inside the cylinder falls (Boyle's law) and a mixture of gasoline vapor produced in the carburetor is pushed into the cylinder by the external atmospheric pressure. At the bottom of the stroke the intake valve closes and the piston starts upward, decreasing the volume of the cylinder and increasing the pressure of the gasoline–air mixture again according to Boyle's law. The pressure when the piston reaches the top of its stroke is typically about 9 atm or 132 psi. The spark plug then fires to ignite the gasoline–air mixture. It burns rapidly, releasing thermal energy and with an increase in the number of moles of gas in the cylinder. Here is a typical equation for the combustion reaction:

Rotary engines are mechanically different but the same gas laws apply to them.

Technically, a "four-stroke-cycle" engine. A "two-stroke" engine has two strokes per cycle.

$$\underset{\text{Octane}}{2C_8H_{18}(g)} + 25O_2(g) \rightarrow 16CO_2(g) + 18H_2O(g) + 2446 \text{ kcal}$$

As the mixture burns, the gas pressure in the cylinder rises rapidly to several hundred pounds per square inch, for two reasons. First, the total number of moles of gases present is increased somewhat. According to the equation for the combustion reaction, there is an increase from $(2 + 25) = 27$ moles to $(16 + 18) = 44$ moles.

However, the explosive mixture is made by mixing gasoline vapor (C_8H_{18}), not with *pure* O_2, but rather with air. In air there are about 4

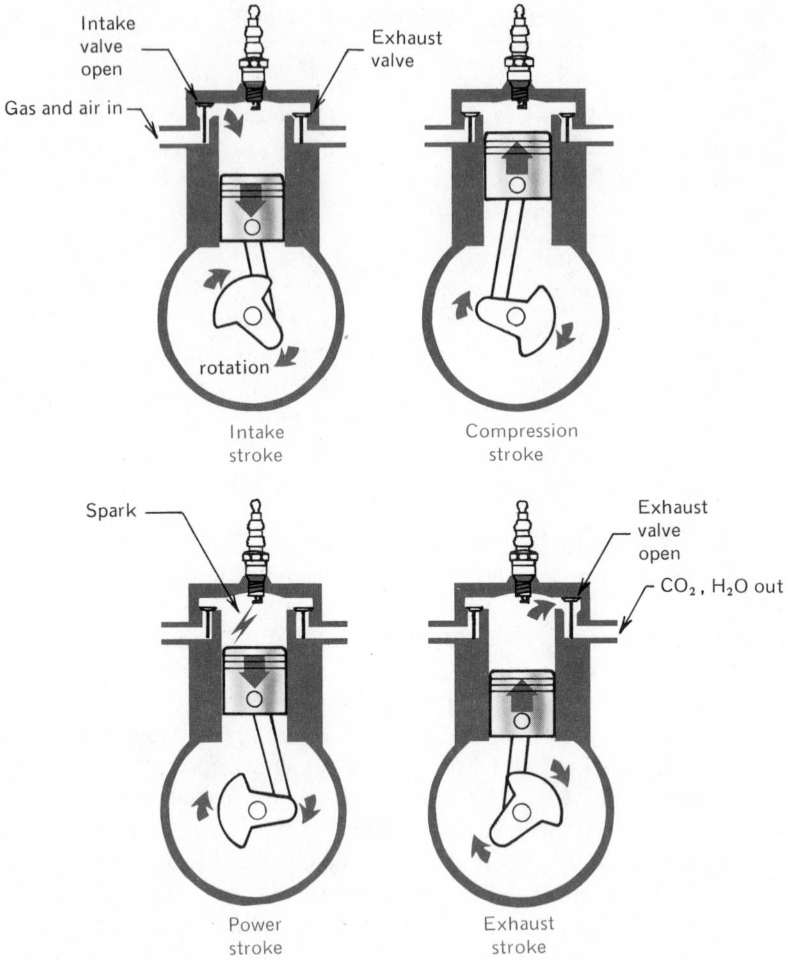

FIG. 5.9 Four-stroke gasoline engine.

moles of N_2 for each mole of O_2, so the change in the number of moles caused by combustion is from 127 to 134, estimated as follows:

$$2C_8H_{18}(g) + 25O_2(g) + 100N_2(g) \rightarrow 100N_2(g) + 16CO_2(g) + 18H_2O(g)$$
$$(2 + 25 + 100) = 127 \text{ moles} \rightarrow (100 + 16 + 18) = 134 \text{ moles}$$

Since (according to the gas law $PV = nRT$) the pressure is directly proportional to the number of moles present, the original pressure of 132 psi is increased to $132 \times (134/127) = 140$ psi simply by the increase in the number of moles due to combustion. But far more important in increasing the pressure is the effect of the thermal energy released to heat up the gas mixture. Typically, temperatures of about 1200 K are reached. According to Charles's law an increase in the absolute temperature of a gas sample from 373 K to 1200 K increases the pressure by a factor of 1200/373 = 3.2. This means that the effect of the temperature change increases the pressure on the piston from 140 to (3.2 × 140) = 448 psi. In a cylinder with a 4-in. bore the piston face has an area ($A = \pi r^2$) of 12.6 sq. in., so the force (area × pressure) pushing against the piston at the instant of combustion would be 448 × 12.6 = 5645 lb if the piston were not free to move. The piston *is* free to move, however, and the large force due to the increased gas pressure in the cylinder is converted into rotation of the crankshaft and, through the transmission, to the rotation of the wheels to propel the automobile. After the power stroke the exhaust valve opens and the piston is driven back into the cylinder by the flywheel to force the spent combustion mixture out to the exhaust pipe. The exhaust valve then closes, and the four cycles are repeated.

During the power stroke, the combustion mixture cools as it expands with the motion (downward in Figure 5.9) of the piston. It cools because the expanding gas is doing work on the piston, and the energy corresponding to that work is provided by the kinetic energy of the gas molecules. At the molecular level, molecules colliding with the surface of the piston rebound with slightly lowered speed (v) because the piston is moving away from them during the collision. The decrease in speed lowers the kinetic energy ($\frac{1}{2} mv^2$) of the molecules, and since temperature is a measure of the average kinetic energy of the molecules, the temperature of the gas goes down. Ideally, for maximum efficiency, the length of the cylinder should be enough so that at the end of the power stroke the gaseous combustion mixture would be at atmospheric pressure and ordinary temperature. However, for practical reasons the pressure is considerably above atmospheric (hence the exhaust noise as the exhaust valve opens), and the temperature of the gas is about 500°C. Furthermore, less than a tenth of the chemical potential energy stored in the fuel mixture is converted into useful work in driving the automobile. This is because the cylinders must be cooled by circulating either water or air around them to prevent the high temperature of the combustion mixture from melting the valves and pistons. Much of the energy available from the combustion is therefore lost as heat into the atmosphere by the cooling required to keep the engine parts from melting. Depending on conditions, only about 10% of the potential energy in the fuel is converted into the useful work of moving the auto. In this light it can be seen that transportation by conventional autos is very wasteful of our limited natural supply of petroleum.

For the most part, the nitrogen in the combustion mixture is just a bystander. However, at the high temperature of the combustion, a small portion reacts with oxygen to form nitric oxide (NO):

$$N_2(g) + O_2(g) + \text{energy} \rightarrow 2NO(g)$$

The nitric oxide enters the atmosphere through the exhaust and is one participant in the formation of the photochemical smog that plagues large metropolitan areas. It, along with unburned hydrocarbons from auto exhausts, ozone, and sunlight, forms the intensely eye-irritating compound *peracetylnitrate*:

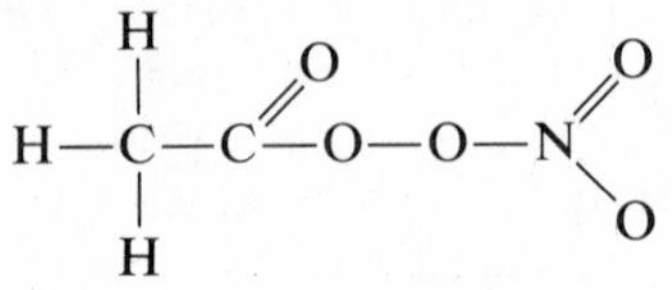

Peracetylnitrate

The formation of NO in the auto engine is discussed in detail in Chapter 9. Photo 17 shows what NO, hydrocarbons, and sunlight can do to the air you may have to breathe.

Exercises

13. Balance the skeleton equation $C_8H_{18}(g) + O_2(g) \leftrightarrow CO_2(g) + H_2O(g)$ for the combustion of gasoline. Does an automobile use a greater weight of O_2 or of gasoline in its operation? What are the volume relationships in the reaction?

14. The NO discharged into the atmosphere in auto exhausts is ultimately converted into nitric acid (HNO_3) by reactions that have the following skeleton equations:

$$NO(g) + O_2(g) \nrightarrow NO_2(g)$$

$$NO_2(g) + H_2O(l) \nrightarrow HNO_3(l) + NO(g)$$

Balance these equations. What weight of nitric acid would result from 3.00 g of NO?

5.6 Liquid Air The production of the liquid form of air by cooling and compressing is carried out on a huge scale to obtain the elementary gases of the atmosphere for industrial and medical use. The composition of the atmosphere is given, along with the boiling points of the gases, in Table 5.3. When dealing with gas mixtures, such as the atmosphere, that contain trace components, more convenient numbers for the composition result if concentrations are expressed in parts per million (ppm) rather than in percent (parts per hundred). One ppm = 0.0001 percent. The composition is given by *volume* rather than by *weight* because volume concentrations are directly convertible to mole fractions. This is so because equal *volumes* of different gases, measured at the same temperature and pressure, contain equal numbers of mole-

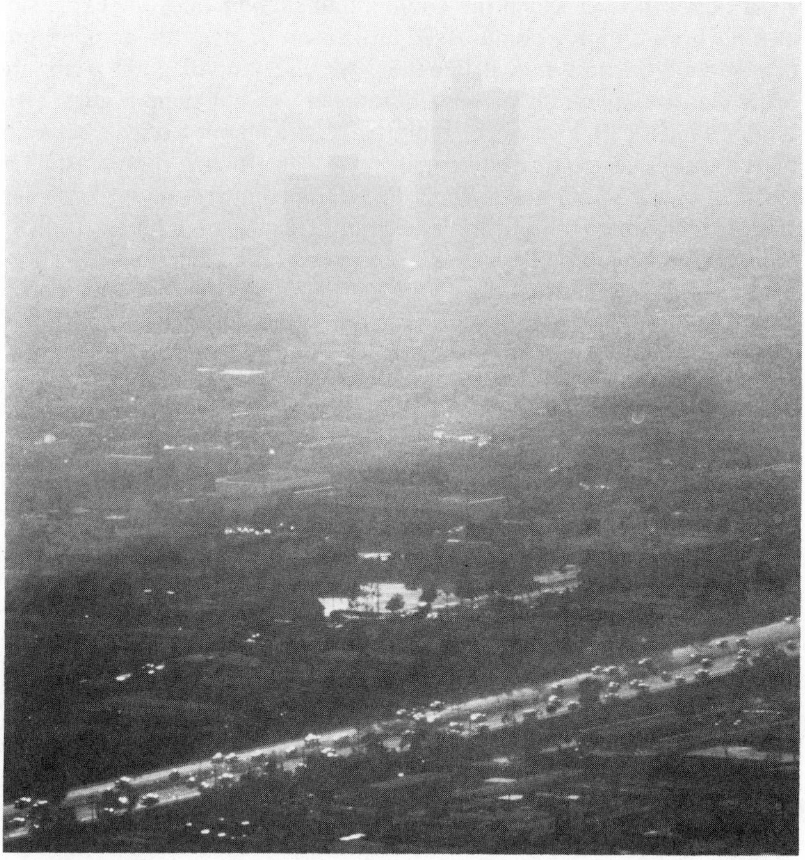

PHOTO 17 Exhaust "garbage" from the internal-combustion engine. (*EPA-DOCUMERICA—Gene Daniels.*)

cules (Avogadro's law). Thus the statement that air contains 780,840 ppm of N_2 by volume means that it contains (780,840 × 0.0001) = 78.0840% N_2 by volume, and that the mole fraction of N_2 in air is 0.780840.

To liquefy air, it must be cooled to the very low temperature of −196°C (77 K). The cooling required is obtained by a sort of "boot-

TABLE 5.3 Composition of dry air

	Gas	*bp* (°C)	*Concentration by Volume* (ppm)
	N_2	−196	780,840
	O_2	−183	209,460
	Ar	−186	9,340
	CO_2	−57	330
	Ne	−246	18
	He	−269	5
	Kr	−152	1
Other gases	(H_2, CH_4, N_2O)		6

strap" operation in which the cooling that accompanies the expansion of a portion of highly compressed air is used to lower the temperature of a second portion until it liquefies. A schematic diagram giving the principles of operation of an air liquefier is shown in Figure 5.10. Understanding how it works will give you examples of the behavior of real gases as compared to perfect gases. In the compressor, air freed of water vapor and carbon dioxide is compressed until its pressure reaches about 200 atm. The high-pressure air coming from the compressor is quite hot because work (force × distance) has been done on it by the piston during the compression, and the work done appears as the thermal energy of the gas, heating it up.

After leaving the compressor, the hot compressed air is cooled to about room temperature by passage through pipes surrounded by running water. It is then piped to a well-insulated expansion chamber, where it passes through a tiny hole (needle valve) and is allowed to expand into the chamber until its pressure falls from 200 atm to about 20 atm. During the expansion the air cools down, and the arrangements are such that the cooling effect is transmitted to the incoming high-pressure air. With the passage of time the incoming air becomes colder and colder until a temperature of about −196°C is reached and a portion of it liquefies. The liquid air collects at the bottom of the expansion

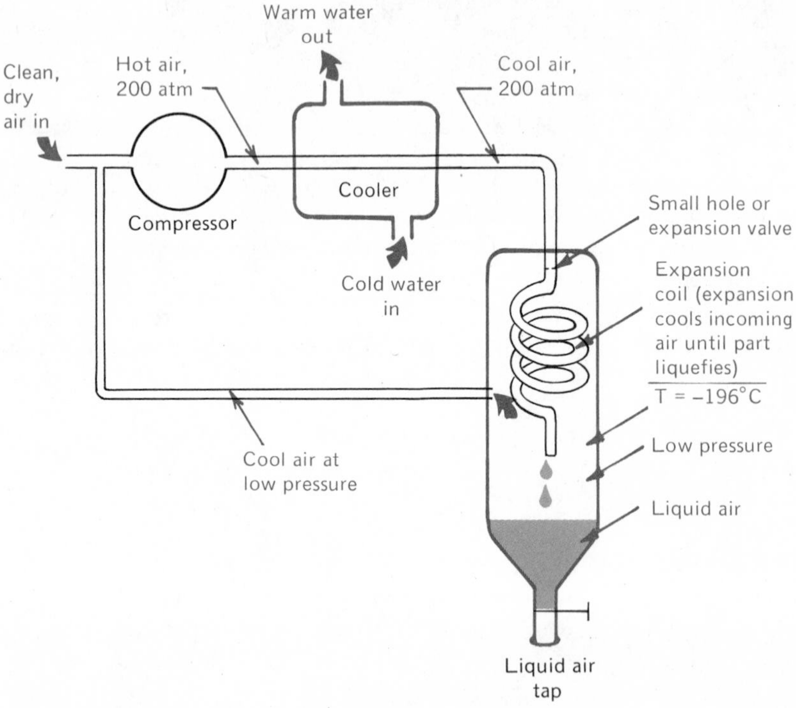

FIG. 5.10 Principle of air liquefier.

chamber and is drawn off into extremely well-insulated containers (essentially thermos bottles).

Why does the air cool down as it expands? The situation is *not* like that encountered earlier in the power stroke of the internal combustion engine, where the gas cools *because it does work* in pushing the piston as it expands. In the *free* expansion in the liquefier, the gas does no work in any *external* sense. However, because there are forces of attraction between real gas molecules, these forces must be overcome as the average distance between molecules increases during the free expansion. The only source of energy available to do the work of overcoming the attractive forces is the kinetic energy of the gas itself. As it is used up in overcoming the attractions, the average kinetic energy of the gas molecules goes down and, correspondingly, the temperature decreases.

EXAMPLE

Liquid air is a pale-blue liquid of high fluidity and low density. Once obtained, it is distilled in high-efficiency equipment to separate it into its component elements. The elements may be stored as liquids in insulated containers or as gases under high pressure in thick-walled, steel "gas bottles." Liquid nitrogen and oxygen are shipped in insulated railroad tank cars. The nitrogen is used mainly in the synthesis of ammonia (Haber process) for use as fertilizer:

$$N_2(g) + 3H_2(g) \rightarrow 2NH_3(g)$$

Increasingly, liquid nitrogen is being used as a refrigerant in the shipping of perishable foods. In this application, a thermostat in the insulated chamber of a truck or freight car controls a valve that admits liquid nitrogen as a spray as needed to keep the food cold. As it evaporates it removes thermal energy from the environment to provide for its *heat of vaporization*. Cooling by evaporation is familiar to anyone who has come out from swimming into a strong wind or used the "wet finger test" to determine which way the wind is blowing. Liquid nitrogen is also used for the quick-freezing of tissue samples removed from tumors before slicing them for microscopic examination to see if cancer is present.

Oxygen from liquid air has many uses. In gas bottles it is used in oxyacetylene welding and to provide oxygen when respiration is weak or atmospheric pressures are lowered, as at high altitudes. The liquid form (lox) is contained in huge tanks in rockets for space exploration. In the rocket motor it supports the combustion of the kerosene or other fuel used to drive the rocket. It is also stored in liquid form in space capsules to support the respiration of the astronauts and, through the hydrogen–oxygen fuel cell (Chapter 10), to provide the electrical energy to run the equipment of the space vehicle.

Oxygen, when liquid or under high pressure, is a hazardous chemical. If it comes in contact with combustible materials such as grease or oil, the rate of the (often spontaneous) burning may be so rapid that explosions or disastrous fires result. In producing steel advantage is taken of the effect of heightened oxygen concentrations in raising the temperature of burning fuel. By adding oxygen to the air stream of a steel furnace before it meets the fuel, much higher temperatures and consequently shortened time for the melt are obtained. Huge quantities of oxygen from liquid air are used for this purpose. (Production of liquid oxygen in the U.S. was 26 billion pounds in 1971.)

Even the noble gases, though present in small amounts in air (Table 5.3), are separated from liquid air by distillation. Argon is used to fill tungsten-filament incandescent light bulbs. It is superior to nitrogen, which was formerly used, because its chemical inertness prolongs the life of the tungsten filament. Neon, xenon, and krypton are separated for chemical uses and for lighted signs of the neon type. Here the gas is contained at low pressure in a glass tube fitted with sealed-in electrodes at the ends. Upon the passage of electric current at high voltage (25,000 V) through the gas, its atoms are energetically excited and emit light of the wavelengths of the line spectrum (Chapter 3) of the element.

At present helium is separated from the natural gas [largely methane (CH_4)] from wells in one locality in Texas. But as this resource is expended, the atmosphere will one day become the only source for helium, as it now is for the other noble gases. Helium is used to provide an inert atmosphere for the welding of aluminum and, mixed with oxygen, to provide "air" for breathing in underwater laboratories and deep-diving operations. The latter use depends on the low solubility of helium in the blood compared to that of nitrogen. Underwater workers can ascend to the surface much more rapidly, without getting the bends (due to nitrogen bubbles), if they breathe a compressed helium–oxygen mixture instead of compressed air (Exercise 12).

Exercises

15. The heat liberated when one mole of kerosene ($C_{12}H_{26}$) is burned in oxygen is 1810 kilocalories. Give the balanced equation for the reaction. What quantity of thermal energy would be provided by the burning of the amount of kerosene that appears in the balanced equation?

16. A typical oxygen gas cylinder when filled contains 220 cu ft of the gas measured at STP. If the volume of the cylinder is 45 liters, what is the pressure of oxygen in psi and in atm in a filled cylinder at 25°C? How many moles of O_2 does the cylinder contain? (1 ft^3 = 28.32 liters)

17. Suppose the filled oxygen cylinder in Exercise 16 weighs 100 lb and gets knocked over so that the outlet valve is broken off, leaving a hole of 1 in. square area open to the interior. What would happen to the cylinder?

5.7 Heat Pumps, Enthalpy, and Entropy Heat pumps are devices for transferring thermal energy from one region to another. The most familiar machine of this type is the electric refrigerator, in which, to provide a cold region, thermal energy is "pumped" from the interior out to the surrounding air. An understanding of how this works will give you some understanding of the important thermodynamic ideas of *enthalpy* (heat content) and *entropy*.

You know that to keep a liquid boiling (for example, water in a pan on the stove) a supply of thermal energy (heat) must constantly be provided. This is true no matter what the boiling temperature of the liquid, and the thermal energy required to "boil off" (evaporate) one mole of a liquid against atmospheric pressure is called the *molar heat of vaporization* of the liquid, abbreviated ΔH_v ("delta H vee"). ☛ The input of thermal energy is needed to overcome the attractive forces between the close-packed molecules of the liquid and separate the molecules to the large average distances of the gaseous state. Molar heats of vaporization for some liquids are in Table 5.4. For many liquids, dividing the molar heat of vaporization by the boiling point in K leads to a value of about 22 cal deg^{-1} mole^{-1}; that is, $(\Delta H_v/T_{bp}) \approx 22$ cal deg^{-1} mole^{-1}. This is known as Trouton's rule and is useful for estimating the heats of vaporization of ordinary liquids when their boiling points are known. The rule fails for hydrogen-bonded liquids (Section 4.6) such as water and ethyl alcohol because they have unusually large heats of vaporization. When they are vaporized, the hydrogen bonds must be broken and ordinary van der Waals forces overcome.

The Greek letter Δ is used to indicate change in the value of a variable.

TABLE 5.4 Molar heats of vaporization and boiling points

Liquid Substance	ΔH_v (cal/mole)	*Boiling Point* (°C)	$\Delta H_v/T_{bp}$
Helium (He)	22	−269	6
Hydrogen (H_2)	216	−253	11
Oxygen (O_2)	1,610	−183	18
Carbon disulfide (CS_2)	6,490	46	20
Ethyl ether ($C_4H_{10}O$)	6,566	35	21
Benzene (C_6H_6)	7,497	80	21
Ethyl alcohol (C_2H_6O)	9,488	79	26
Water (H_2O)	9,700	100	26
Mercury (Hg)	14,200	357	23
Sodium (Na)	23,300	883	20
Potassium chloride (KCl)	40,500	1500	23

From the point of view of thermodynamics, the mole of gas resulting from the evaporation of a mole of liquid contains the thermal energy of the heat of vaporization. The mole of gas has a *heat content* or *enthalpy* (H_g) greater than that of the liquid (H_l) from which it arose by just the amount of the molar heat of vaporization:

$$\Delta H_v = H_g - H_l$$

Since energy may be neither created nor destroyed, it follows that to condense the mole of gas back to liquid, thermal energy equal to ΔH_v must be *removed*. The reason steam produces such scalding burns is that it condenses to water on striking the relatively cool surface of the skin, giving up its heat of vaporization to the flesh in the process. The blast of thermal energy breaks hydrogen bonds, destroying the vital secondary structures of proteins in the tissue cells, killing the cells, and producing the burn.

Contained in Trouton's rule is a second important thermodynamic quantity called *entropy* (S). Entropy is defined as Q/T, where Q is the amount of thermal energy put into a system and T is the absolute temperature at which the thermal energy is transferred. For the process of evaporation, Q is equal to ΔH_v and $T = T_{\text{bp}}$, so Trouton's rule, in the context of entropy, states that the molar entropies of evaporation ($\Delta S_v = \Delta H_v/T_{\text{bp}}$) for many liquids are close to 22 cal deg^{-1} mole^{-1}. As we have seen, thermal energy is energy of molecular motion. What, then, is entropy at the molecular level? For introductory purposes, let's say that the entropy of a sample of matter is a measure of the randomness in the arrangement of its molecules. Crystals, having no randomness in the arrangement of their microparticles, have low entropies. Thus the molar entropy of diamond, the crystalline form of carbon, is only 0.58 cal deg^{-1} mole^{-1} at 25°C. The processes of melting and evaporation both result in increases in entropy as the regular structure of the crystal goes first to the loose structure of the liquid, and finally to the chaotic, structureless gas. For ice at 0°C, the entropy is 10 cal deg^{-1} mole^{-1}; for liquid water at 25°C and 1 atm it is 16 cal deg^{-1} mole^{-1}. But water vapor under the same conditions has the much larger entropy of 45 cal deg^{-1} mole^{-1}.

You saw earlier that the amount of energy in a closed system remains the same during chemical changes, although the form of the energy may change (nuclear processes excluded). The important difference with entropy is that it *increases whenever a process occurs spontaneously* ("by itself" or naturally). We are familiar with the spontaneous flow of thermal energy from a hot to a cold body until both are in thermal equilibrium (at the same temperature). To see how entropy increases in this natural flow of heat, imagine two blocks of iron, one at 100°C (373 K) and one at 0°C (273 K), both contained in a perfectly insulated, evacuated box (Figure 5.11). Now imagine that the blocks are brought into contact. Immediately thermal energy will start to flow from the hotter to the colder one. Suppose that the arrangements are such that after a short time one calorie of thermal energy has passed from the hot block to the cold one. And suppose further that the blocks are large enough that the flow of thermal energy produces no measurable change in the temperature of either block. Then, remembering that $\Delta S = Q/T$, and $Q = 1$ calorie, you can write:

$$\text{Entropy lost by hot block} = \frac{Q}{T_{\text{hot}}} = \frac{1}{373} = 0.00268 \text{ cal deg}^{-1}$$

$$\text{Entropy gained by cold block} = \frac{Q}{T_{\text{cold}}} = \frac{1}{273} = 0.00366 \text{ cal deg}^{-1}$$

For the system of two blocks the entropy change is an increase of $(0.00366 - 0.00268) = 0.00098$ cal deg^{-1}. Now, even though the temperatures of the two blocks will come closer and closer together as they approach the same temperature at thermal equilibrium, the entropy lost by the hot block in any time interval will be *less* than that received by the cold block in the same interval. This is because $T_{\text{hot block}}$ is always greater than $T_{\text{cold block}}$, Q is the same for both, and T appears in the denominator of the definition (Q/T) of the entropy changes they undergo. Thus, although the total quantity of thermal

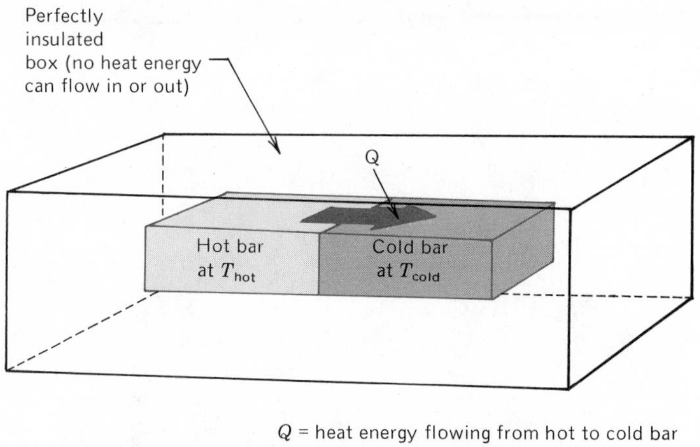

Q = heat energy flowing from hot to cold bar

$$\text{Entropy lost by hot bar} = \frac{Q}{T_{\text{hot}}} = -S_1$$

$$\text{Entropy received by cold bar} = \frac{Q}{T_{\text{cold}}} = S_2$$

$$T_{\text{hot}} > T_{\text{cold}}$$

$$S_1 < S_2$$

$$-S_1 + S_2 > 0$$

$$\Delta S = -S_1 + S_2 = \text{a positive quantity}$$

FIG. 5.11 Positive entropy change accompanies spontaneous flow of heat energy from high-temperature bar to low-temperature bar (Q, the heat flow, is too small to change either T_{hot} or T_{cold} appreciably).

energy in the system of blocks remains unchanged, there is an increase in entropy as the flow of thermal energy goes on. The "one-way" street for entropy changes—always positive in any spontaneous process—becomes clearer if you remember from experience that when the two blocks have finally come to thermal equilibrium at the same temperature, the heat flow will *never automatically* (spontaneously) reverse itself so that one block becomes hotter at the expense of the other becoming cooler. The results of this thought experiment have been generalized in the statement that the *energy* of the universe, though it may take different forms, is constant, but the *entropy* of the universe is constantly increasing as a result of the occurrence of spontaneous changes.

EXAMPLE

But what about a refrigerator? In it, thermal energy is moved ("pumped") from the lower temperature of the interior to the higher temperature of the surrounding air. There seems to be an entropy decrease: the natural flow of thermal energy from hot to cold is reversed. To see that there is actually an increase in overall entropy with the operation of the refrigerator requires a closer look at how the machine works.

The Freon of molecular formula CCl_2F_2 and bp −38°C is commonly used.

A refrigeration system (Figure 5.12) or heat pump consists of a *compressor*, a *condenser*, and an *evaporator*, all operating in a closed system containing a substance of low boiling point (refrigerant), usually a compound of carbon, chlorine, and fluorine called a Freon. In the

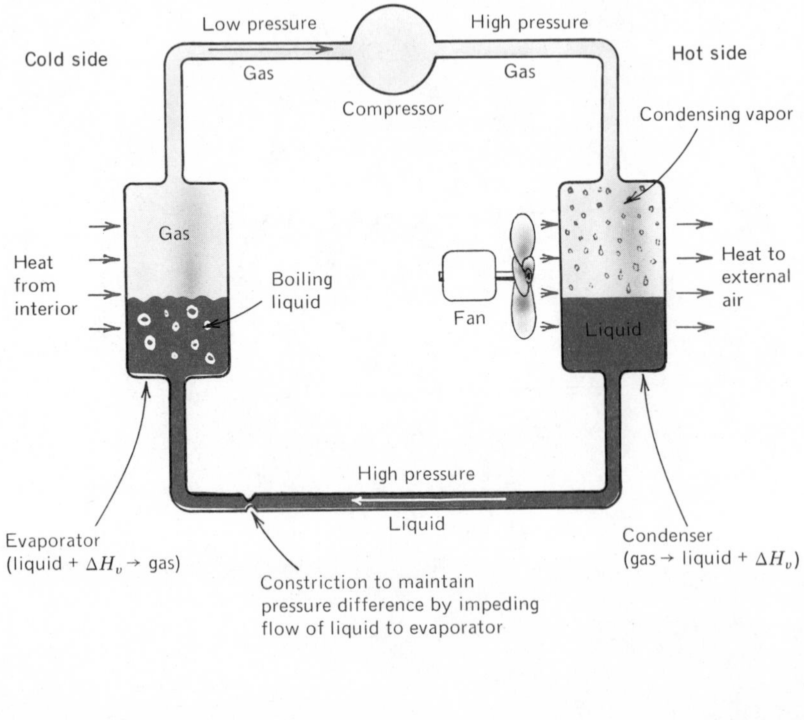

FIG. 5.12 Principles of operation of a heat pump.

compressor, the gas form of the refrigerant is compressed until it condenses to liquid in the condenser, giving up its heat of vaporization (ΔH_v) to the surrounding air. Usually a fan blows room air across the outside of the condenser to make the removal of the heat liberated on condensation more efficient. The liquid refrigerant goes through a limiting valve to the evaporator, which is *inside* the refrigerator. Because of the limiting valve, the pressure in the evaporator is lower than the vapor pressure of the refrigerant at the temperature of the interior. Correspondingly, the liquid evaporates ("boils"), extracting its heat of vaporization (ΔH_v) from the interior and cooling the interior in the process. The gaseous refrigerant is then returned to the compressor, where the cycle is repeated. The overall effect is to take ΔH_v from the interior and deliver it to the exterior. Since T_{interior} is *less than* T_{exterior}, the entropy ($\Delta H_v/T$) taken from the interior is greater than that delivered to the exterior, and the spontaneous flow of thermal energy with its positive entropy change has been reversed. But we have forgotten the compressor. It doesn't run by itself, but requires electrical energy. The electrical energy must be produced by some spontaneous process, whether the fall of water in a hydroelectric plant or the combustion of fuel in a steam-generating plant. Without going into the complicated details of these agencies, we can say that the entropy released in the production of the power to run the motor exceeds the entropy decrease in the refrigeration system. So, overall there is an entropy increase. The lesson here is that entropy changes can be reversed *locally* by pumping thermal energy "uphill" from a cold to a hot region, but the pump must be driven by a spontaneous process somewhere that results in an overall increase in entropy. ⊸

Exercises

18. Freon −22, molecular formula CCl_2F_2 and bp −38°C, is the heat-transfer liquid in some refrigerators and air conditioners. Using Trouton's rule ($\Delta H_v/T_{\text{bp}} = 22$ cal deg^{-1} mole^{-1}), calculate the molar heat of vaporization of Freon −22.

19. Using data from Exercise 18 and taking the heat of fusion of ice as 1438 cal per mole, calculate the number of moles of Freon −22 that must be evaporated to freeze one mole of ice from water at 0°C. How many grams to freeze one kilogram of water?

All human activities result in transferring to our environment thermal energy stored in the form of chemical potential energy in food or fuels. All of this potential energy that ends up as heat can be traced back through photosynthesis to the sun. (Food, oil, gas, and coal all have their origin in living plants that got their energy to grow from the sun.) The thermal effect of human activities is to release the energy of sunlight accumulated over years of photosynthesis in the case of foods, and millions of years in the cases of the fossil fuels, coal, oil, and natural gas. We are releasing "stored sunlight" and in effect making the sun shine brighter, and so warming the earth.

EXAMPLE ⊸ A large increase in the amount of thermal energy in the environment would have disastrous consequences for life on earth. Even a small increase in the average temperature of the earth would result in melting the polar ice caps, producing great changes in climate. Furthermore, the water released from melting would raise the levels of the oceans as much as 200 ft so that large, low-lying coastal areas would be flooded by the sea. It may very well turn out that, as underdeveloped countries become technologically advanced, restrictions will have to be imposed on the quantities of thermal energy poured into the environment if the earth is to remain in its present condition. What the limitations may be are as yet unknown, but the question is an important one to many scientists and other informed people concerned about the impact of human activities on the environment.

To make matters worse there is the effect (the "greenhouse effect") of carbon dioxide released by the combustion of fuels in decreasing the rate of heat loss from the earth to space. Most of this heat loss is through the radiation into space of energy in the infrared region of the electromagnetic spectrum (Figure 3.1). Carbon dioxide molecules *absorb* infrared radiation and convert it into thermal energy that increases the temperature of the atmosphere. Since the effect depends on the amount of carbon dioxide in the atmosphere, any large increase in carbon dioxide content will result in a warming of the atmosphere. Offsetting any increase would be an increase in the amount of photosynthesis on earth, since carbon dioxide is absorbed in the process, but the effectiveness of this is questionable. Some scientists see evidence that the greenhouse effect is already operating, but data for scientific proof are hard to obtain and not yet conclusive. ⊸

5.8 Summary Guide All gases can be liquefied if compressed and cooled enough. A liquid in contact with its vapor at a fixed temperature in a closed system is an example of the dynamic equilibrium, liquid $\rightleftarrows$ vapor, where evaporation and condensation are occurring at the same rate. The pressure of the vapor over its liquid at a specified temperature is called the vapor pressure of the liquid at that temperature. The temperature at which the vapor pressure becomes one atmosphere (760 torr) is the normal boiling point of the liquid.

According to kinetic theory, the pressure exerted by a gas is due to the collisions of its molecules with the walls of the container, and increases with temperature if volume is held constant. The equation describing the behavior of a perfect gas is $PV = nRT$, and some real gases behave almost perfectly. The value of the gas constant (R) is the same for all gases, but when using the equation remember that the units of R must agree with the units of P and V. If they are atmospheres and liters, then $R = 0.08205$ liter atm mole^{-1} deg^{-1}. Boyle's, Charles's, and Avogadro's laws are all contained in the equation for a perfect gas. Boyle's and Charles's laws are combined in the equation $(P_1V_1/T_1) = (P_2V_2/T_2)$, which is used for calculating the effects of pressure and temperature on the volume of gas samples where there are no changes in the number of moles of the gas or its molecular weight. If there are changes in these numbers, then $PV = nRT$ must be used.

Dalton's law of partial pressure states, first, that the partial pressure of a gas in a mixture of gases is the pressure the gas would exert if it alone occupied the volume of the mixture, and second, that the total gas pressure of the mixture (P_{total}) is the sum of the partial pressures of the gases in the mixture. The partial pressure p_A of a gas A in a mixture of gases is given by $p_AV = n_ART$, where n_A is the number of moles of A present. It is also given by $p_A = X_AP_{\text{total}}$, where X_A is the mole fraction of gas A in the mixture. Henry's law states that the concentration of dissolved gas in a liquid is proportional to the partial pressure of the gas above the liquid. Dalton's and Henry's laws are needed for understanding the effects of pressure changes on respiration.

The enthalpy (heat content) of a system in liquid–vapor equilibrium at the boiling point is increased by the molar heat or enthalpy of vaporization, ΔH_v, when one mole of liquid is converted to vapor by input of heat energy, pressure remaining constant. To convert a mole of vapor to liquid, thermal energy equal to ΔH_v must be removed. Refrigerators and air conditioners take advantage of the thermal energy changes in evaporation and condensation to "pump" heat from one location to another.

Entropy is a measure of randomness in the arrangement of the atoms, molecules, or ions of a system. The entropy change (ΔS) for a process occurring at fixed temperature and pressure is $\Delta S = \Delta H/T$, where ΔH is the number of calories of thermal energy entering the system at constant absolute temperature T. For liquids, the entropy change (ΔS_v) accompanying evaporation of one mole of liquid at its boiling point is given by $\Delta S_v = \Delta H_v/T_{\text{bp}}$. For most liquids ΔS_v is in the range 21–23 cal mole^{-1} deg^{-1} (Trouton's rule). For hydrogen-bonded liquids such as water, ΔS_v is larger, 24–27 cal mole^{-1} deg^{-1}.

Entropy increases in all spontaneous (natural) processes. Although it is possible "locally" to reverse spontaneous processes and so decrease entropy, the reverse process must be driven by some spontaneous process somewhere so that an overall entropy increase results. Most human activities return to the environment in the form of heat—energy from the sun stored through photosynthesis in foods, coal, natural gas, and petroleum.

Air may be liquefied by cooling and compressing it. Huge amounts of liquid nitrogen (for refrigeration), oxygen (for steel furnaces), and smaller amounts of the noble gases are separated from liquid air by fractional distillation.

Exercises

20. Hyperoxia is the condition in which the bloodstream is carrying more oxygen (dissolved and bound to hemoglobin) than normal. Hyperoxia can be produced by too rapid breathing of ordinary air (p_{O_2} = 159 torr at sea level) or breathing, normally, air enriched with oxygen or under pressure so that p_{O_2} is greater than 159 torr. Hyperoxia causes dizziness at first, then unconsciousness. Would hyperoxia be a problem for "oceanauts" in a Sea Lab operating 370 ft below sea level, where the pressure ☛ is 12 atm, if ordinary air were pumped down to them at 12 atm? (There is a hatch open to the sea in the floor, so at any depth the gas pressure inside must balance the water pressure outside to keep the sea out.)

Absolute pressure, 11 atm hydrostatic pressure from the water, 1 atm pressing on the surface.

21. To prevent hyperoxia in Sea Labs the occupants breathe a "synthetic" atmosphere made up of helium (He) and O_2. Calculate the mole fractions of O_2 and He in a synthetic atmosphere made up so as to prevent hyperoxia in a Sea Lab operating at a depth at which the total gas pressure is 12 atm.

22. The cook has great difficulties because the bp of water in a Sea Lab at a depth of 370 ft is 370°F (188°C). Why is the boiling point so high?

23. A mountain climber starts out with the pressure gage on his portable oxygen tank reading 1000 psi. Assuming no temperature change, how will the pressure gage reading change (if at all) as he goes up the mountain, assuming he does not use any oxygen from the tank?

24. The burning of one mole of H_2 yields 68.3 kcal, and the burning of one mole of octane (C_8H_{18}, "gasoline") yields 1226 kcal. Which is the more energy-rich fuel for a gas engine on an *equal weight basis*?

25. The boiling point of phenol (hydroxybenzene, "carbolic acid," C_6H_5OH) is 182°C, and its heat of vaporization is 12.5×10^3 cal/mole. Is phenol an associated (hydrogen-bonded) liquid?

Key Words

Boyle's law volume of a gas sample held at constant temperature is inversely proportional to the pressure exerted by the gas: $P_1V_1 = P_2V_2$ at T = constant.

Charles's law volume of a gas sample held at constant pressure is directly proportional to the absolute temperature of the gas: $V_1/T_1 = V_2/T_2$.

Combined gas law for a given sample of a gas under two sets of conditions, P_1, V_1, T_1 and P_2, V_2, T_2, we have $P_1V_1/T_1 = P_2V_2/T_2$.

Condensation movement of molecules from the gas phase to a liquid phase—reverse of evaporation.

Dalton's law of partial pressures the total pressure (P_{total}) of a mixture of gases is the sum of the partial pressures ($p_i = n_iRT/V$ or $p_i = X_iP_{total}$) that each gas would exert if it alone occupied the volume V.

Δ Greek letter delta, indicating a change in quantity.

Density weight of a sample of matter divided by the volume of the sample at a specified temperature and pressure; units are g ml^{-1}, g l^{-1}.

Dynamic equilibrium state or condition of a closed system in which opposed reactions or processes are proceeding at the same rate.

Enthalpy of vaporization, ΔH_v amount of heat energy required to convert one mole of a liquid compound to vapor at the boiling point of the liquid.

Entropy Q/T, where Q is the amount of heat energy put into a system and T is the absolute temperature at which the heat energy is transferred.

Entropy of vaporization of a liquid, ΔS_v molar heat of vaporization of a liquid divided by the absolute temperature of the boiling point of the liquid, or $\Delta H_v/T_{bp}$, has the value of about 22 cal deg^{-1} $mole^{-1}$ for most liquids.

Equation for a perfect gas $PV = nRT = (W/\text{mol. wt})RT$.

Evaporation movement of molecules from a liquid phase to a gas phase.

Gas constant, R 0.08205 liter atm deg^{-1} $mole^{-1}$, 1.987 cal deg^{-1} $mole^{-1}$, 8.314 joule deg^{-1} $mole^{-1}$.

Gas pressure force per unit area (kg m^{-2}, psi) exerted by molecules as they rebound from the walls containing the gas.

Henry's law the concentration (moles per liter) at equilibrium of a gas dissolved in a liquid is proportional to the partial pressure of the gas above the liquid.

Standard atm pressure gas pressure that will support a column of mercury 760 mm (76.0 cm) long at 0°C: 14.70 psi, 1.003 kg cm^{-2}.

Vapor the gas phase in equilibrium with a liquid phase.

Suggested Readings

Bent, Henry A., "Haste Makes Waste—Pollution and Entropy," *Chemistry*, October 1971, pp. 6–15.

Bond, George F., "The Seas—Beginning or End," *Catalyst*, **1**(1):9–13 (Spring 1970).

Eaton, John W., Skelton, T. D., and Berger, Elaine, "Survival at Extreme Altitude: Protective Effect of Increased Hemoglobin–Oxygen Affinity," *Science,* **183**:743–746 (Feb. 22, 1974).

Gregory, Derek P., "The Hydrogen Economy," "Chemistry in the Environment," *Readings from Scientific American,* San Francisco, W. H. Freeman Co., 1973, pp. 219–227.

Halstead, Bruce W., "The Sea: Danger Signals for Man," *World Health,* May 1972, pp. 16–21.

Hobbs, P. V., and Harrison, Robinson E., "Atmospheric Effects of Pollutants," *Science,* **183**:909–914 (March 8, 1974).

Hussain, Farooq, *Living under Water,* New York, Praeger, 1971.

Metz, William D., "Helium Conservation Program: Casting It to the Winds," *Science,* **183**:59–63 (Jan. 11, 1974).

Miller, Keith W., "Inert Gas Narcosis, the High Pressure Neurological Syndrome, and the Critical Volume Hypothesis," *Science,* **185**:867–869 (Sept. 6, 1974).

Perry, H., and Berkson, H., "Must Fossil Fuels Pollute?" in Wilson, J. D., and Baker, S. D., *Physical Science: Readings on the Environment,* Lexington, Mass., D. C. Heath and Co., 1974, pp. 156–170.

Stewart-Gordon, James, "The Wet World of Jacques-Yves Cousteau," *Reader's Digest,* Nov. 1973, pp. 161–165.

Whitfield, M., "Accumulation of Fossil CO_2 in the Atmosphere and in the Sea," *Nature,* Feb. 22, 1974, pp. 523–525.

chapter 6

Gaseous and liquid solutions in contact—detergent froth, Lake Tahoe, on the California-Nevada border. *(Photo 18, EPA—DOCUMERICA-Belinda Rain.)*

Solutions and Their Properties

6.1 Introduction Thus far we have dealt mainly with the structures and properties of pure substances—homogeneous samples of matter of fixed composition. Solutions are also homogeneous but differ from pure substances by having the possibility of *variable* composition. For example, the composition of pure ethanol is given by its formula C_2H_5OH, but we may prepare solutions of ethanol in water of any desired composition by varying the relative amounts of ethanol and water. Solutions are distinguished from *suspensions* by their consisting of a single phase—suspensions are made up of two or more phases. Clean air is a gaseous solution. Smoky air is a suspension of solid particles in clean air and has solid and gas phases. Clean seawater is a liquid solution, muddy water a suspension.

In solutions the component present in greater quantity is called the *solvent* and the component present in lesser quantity is called the *solute*. With solutions made by mixing liquids it is also convenient to separate liquids that may be combined in any proportions with the formation of only one phase from liquids that yield two phases when combined. Liquids that yield only one phase when combined are termed *miscible* liquids. Liquids that form two phases on being mixed are termed *immiscible*. Methanol (CH_3OH), used as antifreeze in automobile radiators, and water are miscible. Octane (C_8H_{18}, a component of gasoline) and water are immiscible. Methanol and octane are "partially miscible." As methanol is added to octane there is only one phase at the start, a solution of methanol in octane, but with further additions two liquid phases appear. One is a solution of methanol in octane as solvent, the other a solution of octane in methanol as solvent.

Solutions in which water is the solvent are called *aqueous* (*aq*) solutions and are the most important type of liquid solutions. Life is believed to have originated in the sea, an aqueous solution of mainly sodium and magnesium chlorides (NaCl and $MgCl_2$), and all life processes go on in aqueous solutions of one sort or another.

Why are solutions so important? Because they permit the intimate contact between microparticles that is needed if chemical reaction between the particles is to occur. For example, no matter how finely the two solids sodium chloride (NaCl) and silver nitrate ($AgNO_3$) are ground together, there is very little occurrence of the reaction

$$NaCl(s) + AgNO_3(s) \rightarrow NaNO_3(s) + AgCl(s)$$

No amount of mechanical grinding will convert the solids to particles of molecular size. The tiniest particles from grinding are still *macroscopic*, and although some reaction takes place where their surfaces come in contact, the interiors of the particles remain inaccessible for reaction. On the other hand, if you mix a *solution* of sodium chloride in water, NaCl(*aq*), with a *solution* of silver nitrate in water, $AgNO_3(aq)$, reaction is virtually instantaneous according to the equation

$$NaCl(aq) + AgNO_3(aq) \rightarrow NaNO_3(aq) + AgCl(s)$$

Viewed at the microparticle level, aqueous solutions of sodium chloride and silver nitrate have their respective ions Na^+, Cl^- and Ag^+, NO_3^- swimming around as structural units, separated by water molecules. When the solutions are mixed, the Ag^+ and Cl^- ions can easily get together to form AgCl(*s*), which does not dissolve much in water, so it separates as a solid, or *precipitates*. The Na^+ and NO_3^- ions do not participate in the reaction and remain in the solution described as $NaNO_3(aq)$ in the equation. Reactions of this type are called precipitation reactions and are one kind of *ion-combination* reaction. They are very important in solution chemistry.

The properties of solutions may or may not be a simple blend of the properties of their components. If they are a simple blend, the solution is called "simple" or ideal. As you saw with gases, the ideal case is the exception rather than the rule. But also, as with gases, there is much to be learned about the behavior of real solutions from the study of ideal ones. In gaseous solutions in which no chemical reactions are going on, the properties of the solution—such as density and composition—are the weighted average of the corresponding properties of the component gases. On the other hand, when reactions do occur on mixing gases, the properties of the products cannot be predicted from the properties of the reactants. An extreme example is the formation of *solid* ammonium chloride (NH_4Cl) when ammonia (NH_3) gas and hydrogen chloride gas (HCl) are brought together:

$$NH_3(g) + HCl(g) \rightarrow NH_4Cl(s)$$

If equal molar amounts of ammonia gas and hydrogen chloride gas are combined, the result is no gas phase at all, but a solid phase of ammonium chloride.

Exercises

1. Taking air as 78% N_2, 21% O_2, and 1% Ar by volume, calculate the density of air at 0°C, 1 atm.
2. Calculate the weight of the NH_4Cl produced when 1.00 liter each of HCl and NH_3 gases measured at 0°C, 1 atm, are brought together. (*Hint*: remember that one mole of any gas has a volume of 22.4 liters at STP.)

6.2 Ways of Expressing Concentration Before you can begin to examine the properties of liquid solutions, you must learn ways to express their concentration, because the properties of solutions depend on the relative amounts (concentrations) of solute and solvent. The most common ways of expressing the composition or concentration of a solution are *weight percent* (% *w/w*), *mole fraction* (X_i), and *molarity* (*M* or *C*). Usually solutions are made up of two kinds of chemicals, solute and solvent. Such solutions are called *binary solutions*, and though made up of only two kinds of chemicals, they may contain a larger number of types of microparticles. You've already seen an example in aqueous sodium chloride solution. This is made up by dissolving the chemical NaCl in the chemical H_2O, but the solution contains sodium ions (Na^+), chloride ions (Cl^-), and H_2O molecules. A solution of sulfuric acid (H_2SO_4) in water contains no H_2SO_4 molecules but rather the ions H_3O^+ (hydronium), HSO_4^- (hydrogen sulfate), and SO_4^{2-} (sulfate), as well as H_2O molecules. The ions result from the reactions

$$H_2SO_4(l) + H_2O(l) \rightarrow H_3O^+(aq) + HSO_4^-(aq)$$
$$HSO_4^-(aq) + H_2O(l) \rightleftarrows H_3O^+(aq) + SO_4^{2-}(aq)$$

The first of these is complete, accounting for the absence of H_2SO_4 molecules in the solution. The second reaction is not complete but is in equilibrium, as indicated by the double arrows ($\rightleftarrows$). Details like these must be considered if you want to understand what is really going on at the microparticle level in a solution. They will be discussed shortly in this chapter and in greater detail in the chapters that follow.

Photo 18 provides a dramatic example of the effect of traces of impurities from sewage on the properties of water in Lake Tahoe, before modern sewage treatment was instituted.

Chemists usually write "%" with "parts by weight" being understood.

Weight percent Weight percent (% *w/w*) is the simplest way of giving the concentration of a solution. It is really a "nonchemical" way of describing a solution because it ignores the formulas of solvent and solute and any reactions that may occur between them. It can be expressed in an equation as follows:

$$\%\ \text{solute} = \frac{\text{wt of solute}}{\text{wt of solute} + \text{wt of solvent}} \times 100$$

EXAMPLE

Suppose you add 10.00 g of pure H_2SO_4 to 100.00 g of water:

$$\%\ H_2SO_4 = \frac{10.00}{10.00 + 100.00} \times 100 = 9.09\%$$

The container would be labeled 9.09% H_2SO_4, even though you just saw that there are no H_2SO_4 *molecules* in the solution. To take another example, suppose you wanted to prepare one kilogram of 5.00% glucose solution for intravenous injection. To make up the solution you would dissolve 50.00 g of glucose in 950.00 g of water:

$$\%\ \text{glucose} = \frac{50.00}{50.00 + 950.00} \times 100 = 5.00\%$$

Notice that combining 1000.00 g of water with 50.00 g of glucose would *not* yield the 5.00% solution desired, because for this solution,

$$\% \text{ glucose} = \frac{50.00}{50.00 + 1000.00} \times 100 = 4.76\%$$

Another way of looking at weight percent is to say that it is the number of grams of solute in 100.00 g of solution. ⊸

It is very important to realize that the *concentration of a solution, no matter how expressed, does not depend in any way on the size of the sample of the solution.* Thus, if you have a 150.00-g sample of a 5.00% glucose solution in water prepared by combining 7.50 g of glucose with 142.50 g of water, it will have the same properties, excepting size, as the kilogram sample in the example just described. Here are the calculations for the 150.00-g sample:

$$\% \text{ glucose} = \frac{7.50}{7.50 + 142.50} \times 100 = 5.00\%$$

Concentration, no matter how expressed, is an intensive rather than an extensive property. ☛

Solutions used in medicine are often described by weight/volume percent: "23 milligrams %" means 0.023 g of solute in 100 ml of water or other liquid.

Exercises

3. What weight of water must be added to 10.00 g of sugar to prepare a 2% solution?

4. "Concentrated sulfuric acid" is 95% H_2SO_4 by weight, the remaining 5% being water. How many pounds of water are there in 10 lb of the concentrated acid?

EXAMPLE

⊸ Drugs for hypodermic injection are often packaged in dry form in sterile bottles closed by rubber caps. The label on the bottle tells how much sterile water to add (using a hypodermic syringe) to prepare the solution for injection. Such instructions must be followed carefully. A nurse was once asked to prepare 10 ml of a 1% solution of Novocain for injection as a local anesthetic. The label read, "Add 99 ml (99 g) of sterile water to make a 1% solution." Since only 10 ml of solution was needed, the nurse added only 9.9 ml (9.9 g) of water. The result was a 10% Novocain solution—not a 1% solution. After the injection, artificial respiration had to be used to save the patient's life. This is not an isolated example. A similar mistake in the concentration of a solution led to irreparable damage to the brain of a girl in San Jose, California. Many preventable accidents of this sort go unreported. ⊸

Mole fraction Mole fraction of a component in a gaseous solution was defined in Chapter 4 as the number of moles of the component divided by the total number of moles of all components in the solution. The definition is the same for liquid solutions. Samples of 100.00 g each of benzene and toluene, two liquid hydrocarbons, are combined. What are the mole fractions X_{benzene} and X_{toluene} in the solution? Right away you can see that some chemical information is needed, namely, the molecular weights of the hydrocarbons. Without the molecular weights you can't calculate the number of moles of the components, and you need these numbers for calculating the mole fractions. Benzene is C_6H_6, mol. wt = (72.00 + 6.00) = 78.00, toluene is C_7H_8, mol. wt = (84.00 + 8.00) = 92.00. First calculate the number of moles of benzene and toluene in the 100.00-g samples of the substances:

$$n_{C_6H_6} = \frac{100.00}{78.11} = 1.280$$

$$n_{C_7H_8} = \frac{100.00}{92.13} = 1.085$$

The mole fractions are, by definition,

$$X_{C_6H_6} = \frac{n_{C_6H_6}}{n_{C_6H_6} + n_{C_7H_8}} = \frac{1.280}{1.280 + 1.085} = 0.5412$$

$$X_{C_7H_8} = \frac{n_{C_7H_8}}{n_{C_6H_6} + n_{C_7H_8}} = \frac{1.085}{1.280 + 1.085} = 0.4588$$

You should check to see that the sum of the mole fractions is 1, as it must be (0.5412 + 0.4588 = 1.0000). Although equal *weights* of the hydrocarbons were used to prepare the solution, the fraction of benzene molecules is greater than the fraction of toluene molecules because benzene molecules are somewhat lighter and you get more of them per gram.

Exercises

5. Vinegar is essentially a 5% solution of acetic acid (CH_3COOH) in water. What is the mole fraction of acetic acid in vinegar?

6. "Isotonic saline," used medically to replace lost body fluid, contains 0.90 g of NaCl per 100 g of H_2O. Calculate the weight percent and mole fraction of NaCl in the solution.

Molarity Molarity (M or C), defined as *the number of moles of solute per liter of solution,* is a way of expressing concentration that is very useful in the laboratory. This is because if you know the molarity of a solution of a reagent you can easily calculate the number of moles of reagent in any measured volume of the solution. Volumes of liquid solutions can be quickly and accurately measured with burets or pipets (Figure 6.1), so desired amounts of reagent can be obtained without the tedious operations of many individual weighings.

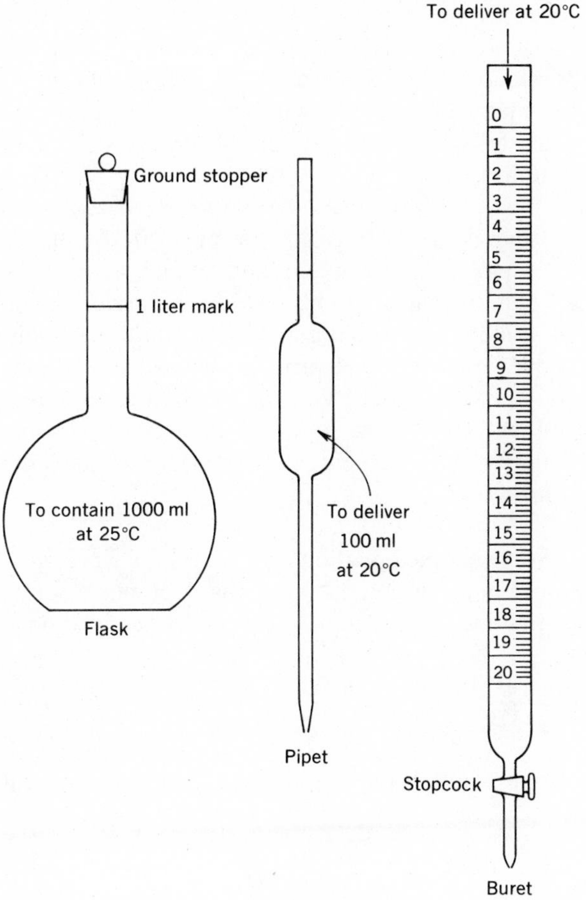

FIG. 6.1 Volumetric glassware. (The temperature is specified because the volumes change as a result of expansion or contraction of the glass.)

EXAMPLE

Suppose, for testing Fehling's solution (Section 11.7), used to analyze the urine of diabetics for glucose, you need aqueous solutions of glucose containing 0.001, 0.002, 0.010, and 0.015 mole of glucose. One way to prepare them would be to weigh out 0.180-, 0.360-, 1.800-, and 2.700-g samples of glucose ($C_6H_{12}O_6$, M = 180.16) and dissolve them separately in water. Four time-consuming weighings would be necessary. Suppose, instead, you made one careful weighing of one mole (180.16 g) of glucose, dissolved it in a volume of water less than 1 liter, transferred the resulting solution to a 1.00-liter volumetric flask (Figure 6.1), and then added water to the 1.00-liter mark. The 1 liter of solution would contain one mole of glucose, and since 1 ml = 0.001 liter, each milliliter of it would contain 0.001 mole of glucose. With a buret (Figure 6.1) you could quickly measure out 1.00-ml, 2.00-ml, 10.00-ml, and 15.00-ml samples of the solution. These would contain, respectively, 0.001, 0.002, 0.010, and 0.015 mole of glucose, the amounts needed for testing the Fehling's solution. Furthermore, most of the glucose solution, which would be labeled "1.00 *M* glucose" or "glucose, $C = 1.00$," would remain and could be used for many future tests without any further weighings. Thus molarity provides a handy way of "weighing out" reagents by measuring volumes.

It is important to realize that a 1.00-*M* solution cannot be prepared by combining one mole of solute with 1 liter of water. The volume of a solution made this way would be greater than 1 liter (Figure 6.2), so 1 liter of it would contain less than one mole of glucose, and the convenience of "weighing" by measuring volumes would be lost. Of course, if the molarity of the solution were determined by some independent means, it could then be used. Solutions for which the molarity of a reagent is known are called *standard* or *volumetric solutions*.

A measured volume of the solution could be evaporated to dryness and the solid residue of solute weighed to obtain its concentration.

Some simple equations are useful for working with molar concentration. For a solution that contains *W* grams of solute of molecular weight M in *V* liters of *solution*, the molar concentration *C* of solute is given by

$$C = \frac{W}{\mathrm{M}V}$$

or, since $(W/\mathrm{M}) = n$, the number of moles of solute,

$$C = \frac{n}{V}$$

If volumes are expressed in milliliters (V_{ml}) instead of liters, the above equations become

$$C = \frac{1000W}{V_{\mathrm{ml}}\,\mathrm{M}} \qquad \text{and} \qquad C = \frac{1000}{V_{\mathrm{ml}}} \times n$$

Another useful relationship gives you the number of moles of reagent contained in V_{ml} of *C* molar solution:

$$\text{Moles of reagent in } V_{\mathrm{ml}} \text{ of } C \text{ molar solution} = \frac{V_{\mathrm{ml}}}{1000} \times C$$

Thus 25.00 ml of 1.15-*M* sucrose ($C_{12}H_{22}O_{11}$) solution contains

$$\frac{25.00}{1000} \times 1.15 = 0.0288 \text{ mole of sucrose}$$

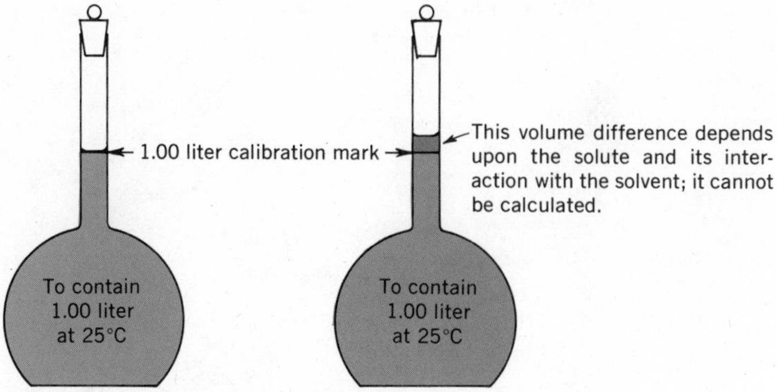
1.00 liter calibration mark
This volume difference depends upon the solute and its interaction with the solvent; it cannot be calculated.
To contain 1.00 liter at 25°C
To contain 1.00 liter at 25°C

(a) Volumetric solution prepared by *diluting to* 1.00 liter; one liter of solution contains all of the sample of solute

(b) Nonvolumetric solution prepared by dissolving weighed sample of solute in 1.00 liter of water; one liter of solution contains less than the weighed sample of solute

FIG. 6.2 Volumetric and nonvolumetric solutions.

Furthermore, the number of *grams* of reagent in V_{ml} of C molar solution of solute having molar weight M is given by

$$\text{Weight in g of reagent in } V_{\mathrm{ml}} \text{ of } C \text{ molar reagent} = \frac{V_{\mathrm{ml}}}{1000} \times C \times \mathrm{M}$$

Thus 35.00 ml of 0.9050-*M* ethanol (C_2H_6O, M = 46.07) solution contains

$$\frac{35.00}{1000} \times 0.9050 \times 46.07 = 1.459 \text{ g of ethanol}$$

However, molarity is more than a convenience in the laboratory. Much of the experimental work in chemistry aims at determining the molar concentrations of the microparticles or species *actually present* in aqueous solutions, and relating their amounts to the macroscopic properties of the solution. To understand this application, it is helpful to distinguish between two kinds of molarity—*component* molarity and *species* molarity. This is most readily done by first distinguishing between *chemicals* and *microparticle species*. Chemicals are substances that can be obtained in bulk at the chemistry stockroom or from suppliers. Examples of solid chemicals are glucose ($C_6H_{12}O_6$), salt (NaCl), aluminum chloride (Al_2Cl_6); examples of liquids are water, ethylene glycol ($C_2H_6O_2$), methanol (CH_4O); and examples of gases (compressed in cylinders) are hydrogen, oxygen, hydrogen sulfide (H_2S), and hydrogen chloride (HCl). In expressing the *component* molarity of a solution the number of moles of the *chemical* per liter of solution is the only concern. Thus you will see bottles of reagents in

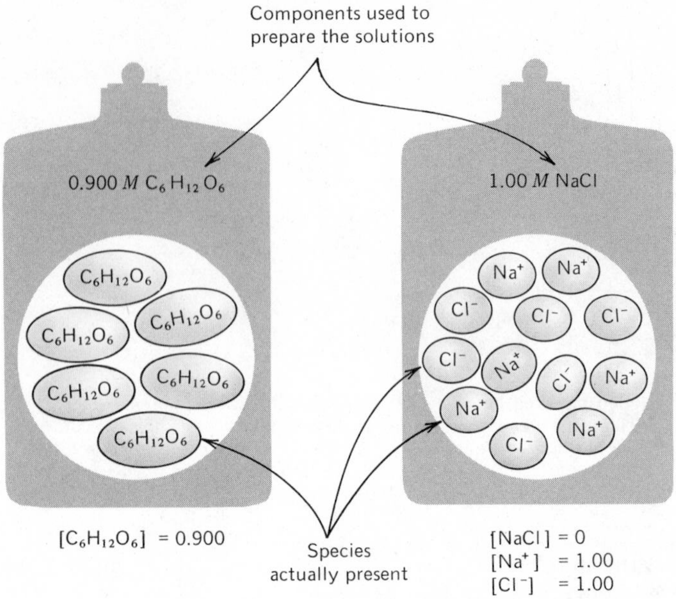

FIG. 6.3 Comparison of component and species molarities (solvent molecules omitted).

the laboratory labeled 1 *M* NaCl, 0.5 *M* H_2SO_4, 0.900 *M* glucose, and so on. These are component molarities—there is no concern for what microparticle species are actually present in the solution. Contrast in Figure 6.3 the solutions labeled 0.900 *M* glucose and 1 *M* NaCl. The properties (low electrical conductivity among them) of the glucose solution show that the solute species are $C_6H_{12}O_6$ molecules. But the sodium chloride solution conducts an electric current very well, indicating the presence of Na^+ and Cl^- ions as species. Quantitative measurements show that there are no NaCl *molecules* in 1 *M* NaCl. Instead there is one mole each of Na^+ and Cl^- per liter of the solution. To distinguish between component and species molarity, brackets [ ] enclosing the species are used.

When the brackets are used, it is understood that concentrations are in moles per liter of solution.

EXAMPLE

In 1 *M* glucose, we have $[C_6H_{12}O_6] = 1$; but in 1 *M* NaCl, $[NaCl] = 0$, $[Na^+] = 1$, $[Cl^-] = 1$. When covalent hydrogen chloride (HCl) molecules dissolve in water they react according to the equation

$$HCl(g) + H_2O(l) \rightarrow H_3O^+(aq) + Cl^-(aq)$$

Since one mole of HCl gives one mole each of H_3O^+ and Cl^- ions, their species molarities are equal to the component molarity of the solution. Thus, in 0.50 *M* HCl, $[HCl] = 0$, $[H_3O^+] = 0.50$, and $[Cl^-] = 0.50$. The two kinds of molarity for the ionic compound $CaCl_2$ are compared in Figure 6.4.

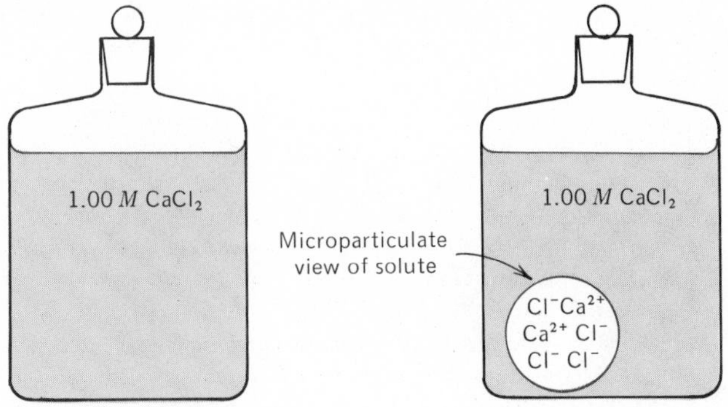

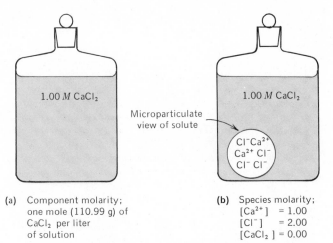

(a) Component molarity; one mole (110.99 g) of $CaCl_2$ per liter of solution

(b) Species molarity;
$[Ca^{2+}] = 1.00$
$[Cl^-] = 2.00$
$[CaCl_2] = 0.00$

FIG. 6.4 Component and species molarities.

Exercises

7. Aluminum chloride, used as a catalyst in organic chemistry (Section 13.9) (Al_2Cl_6, mol. wt = 266.7), reacts violently and completely with water according to the skeleton equation

$$Al_2Cl_6(s) + H_2O(l) \not\longrightarrow Al(H_2O)_6^{3+}(aq) + Cl^-(aq) + \text{heat}$$

Balance the skeleton equation. To make 1.00 liter of 0.200-*M* Al_2Cl_6 solution, what weight of Al_2Cl_6 would be needed? What are $[Al_2Cl_6]$, $[Al(H_2O)_6^{3+}]$, and $[Cl^-]$ in the 0.200-*M* solution?

8. What volume of 1.00-*M* sodium hydroxide solution (NaOH) would provide the amount of NaOH needed to react completely with 50.00 ml of 1.00-*M* H_2SO_4 solution according to the following equation?

$$2NaOH + H_2SO_4 \rightarrow Na_2SO_4 + 2H_2O$$

6.3 Raoult's Law When two miscible liquids that have appreciable vapor pressures are the components of a liquid solution, there will be partial pressures of both components in the gas phase above the liquid (Figure 6.5). Raoult's law states that the equilibrium partial pressures are proportional to the mole fractions of the components in the solution, that is, $p_A \propto X_A$, ☛ or, in an equation, $p_A = kX_A$, where k is a proportionality constant. It also applies to a *single* liquid substance (element or compound) in equilibrium at a given temperature with its vapor—the liquid ⇄ vapor or vapor-pressure equilibrium described in Chapter 5 (Figure 5.3). Here the mole fraction of the substance (X_A)

Francois Marie Raoult (1830–1901) was a French chemist who was professor of chemistry at Grenoble.

α means "is proportional to."

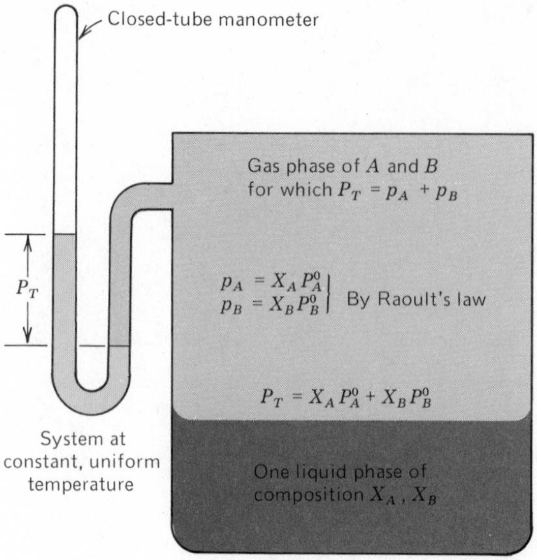

FIG. 6.5 Equilibrium in a system of two miscible liquids.

in the liquid is 1.00 because no other substances are present, so $p_A = kX_A = k \times 1.00$, and you can see that the proportionality constant k is the vapor pressure, written P_A^0, of the pure substance at the temperature of the system. In this case, $p_A = P_A^0$ because $X_A = 1$. On the other hand, if there are two kinds of molecules (A and B) in a *single* liquid phase (A and B miscible), then the partial pressures of A and B in the gas above the liquid are given by

$$p_A = X_A P_A^0 \qquad \text{and} \qquad p_B = X_B P_B^0$$

where X_A and X_B are the mole fractions of the components A and B in the liquid phase, and P_A^0 and P_B^0 are the vapor pressures of the pure components at the temperature of the system. Figure 6.5 describes the situation, and Figure 6.6 is the graphic representation for solutions of the miscible liquids hexane (C_6H_{14}) and heptane (C_7H_{16}). At point C on the graph, $X_{\text{hexane}} = 0.250$, $X_{\text{heptane}} = 0.750$, and the partial pressures of hexane and heptane over the liquid are

$$p_{\text{hexane}} = 0.250 \times 400 = 100 \text{ torr}$$
$$p_{\text{heptane}} = 0.750 \times 150 = 113 \text{ torr}$$

The total pressure is the sum of the partial pressures, $p_{\text{hexane}} + p_{\text{heptane}}$, or $(100 + 113) = 213$ torr. The *dashed* lines for the partial pressures show that the addition of the second component (say, heptane) to the first (hexane) reduces the partial pressure of the first, and vice versa.

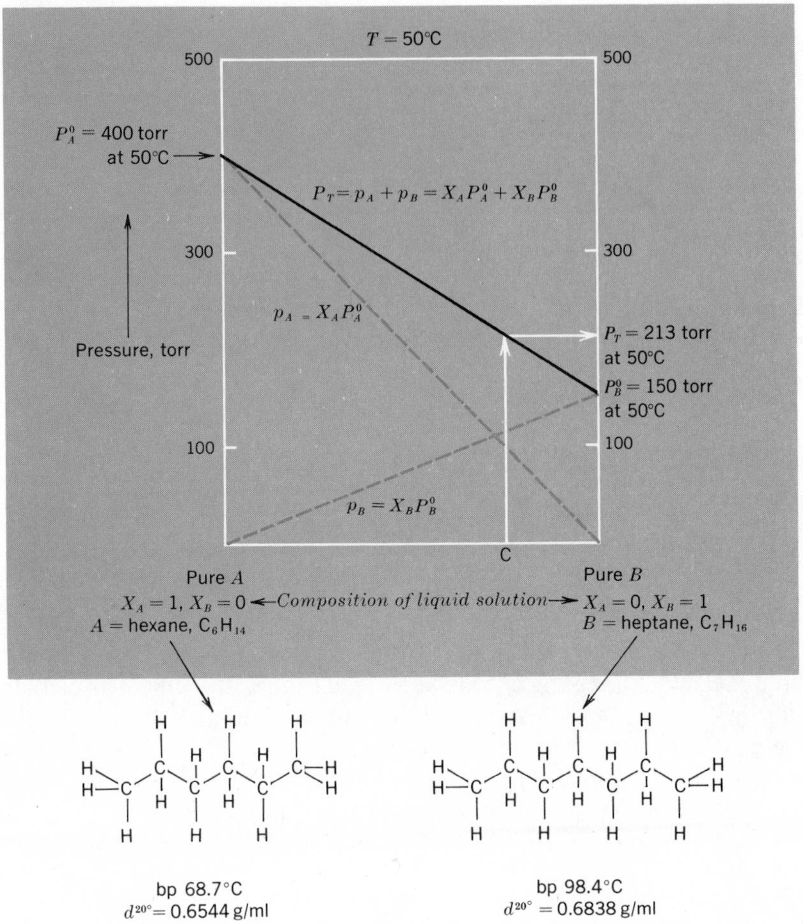

FIG. 6.6 Pressure relationships in a system of two miscible liquids (ideal solution case).

Exercises

9. Make a graph like that in Figure 6.6 showing total and partial pressures for the case of two miscible liquids (A and B) that have the same vapor pressure ($P_A^0 = P_B^0$) at the temperature of the system.

10. Now make a graph like that in Figure 6.6 but for the case $P_A^0 = 100$ torr, $P_B^0 = 0$ at the temperature of the system.

But suppose the two liquids are not miscible (Figure 6.7); then *two* liquid phases will result. If the liquids are immiscible, then each of the liquid phases will be a pure substance, independently exerting its own vapor pressure, and the total pressure will be the sum of the vapor pressures of the pure liquids at the temperature of the system, as shown.

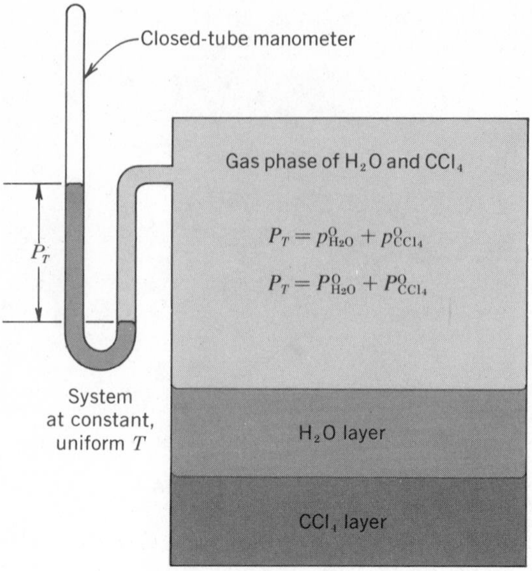

FIG. 6.7 Equilibrium in a system of two immiscible liquids.

EXAMPLE Raoult's law is important for understanding the process of *fractional distillation* by which paint thinner ("mineral spirits"), gasoline, and fuel and lubricating oils are separated from petroleum. The elementary gases N_2, O_2, and Ar are obtained from liquid air, and alcohol is separated from fermentation broths (about 14% alcohol in water), also by fractional distillation. Thinking only about a two-component (A and B) system of miscible liquids, where $p_A = X_A P^0_A$ and $p_B = X_B P^0_B$, we write, by dividing p_A by p_B,

$$\frac{p_A}{p_B} = \frac{X_A P^0_A}{X_B P^0_B}$$

But according to Dalton's law of partial pressures (Section 5.4), in the gas phase $p_A = X^g_A P_{\text{total}}$ and $p_B = X^g_B P_{\text{total}}$, where X^g_A and X^g_B are the mole fractions of components A and B in the *gas* phase (not the liquid) and P_{total} is the total gas pressure. Substituting these expressions for p_A and p_B in the equation for p_A/p_B gives

$$\frac{X^g_A P_{\text{total}}}{X^g_B P_{\text{total}}} = \frac{X_A}{X_B} \times \frac{P^0_A}{P^0_B} \quad \text{or} \quad \frac{X^g_A}{X^g_B} = \frac{X_A}{X_B} \times \frac{P^0_A}{P^0_B}$$

since P_{total} divides out. In words, the ratio of the mole fractions in the *gas* phase is equal to the ratio of mole fractions in the *liquid* phase multiplied by the ratio of the vapor pressures of the pure components. Now suppose that *component A is more volatile than B*. This means that at any temperature, P^0_A is greater than P^0_B, so P^0_A/P^0_B must be greater than 1. With this the case, X^g_A/X^g_B must always be greater than X_A/X_B, because, according to the equation, X_A^g/X_B^g is equal to X_A/X_B *times* P_A^0/P_B^0, and P_A^0/P_B^0 is greater than 1. This means that *the mole fraction of the more volatile component in the* ***gas*** *phase must always be greater than its mole fraction in the* ***liquid*** *phase*, no matter what the mole fractions are in the liquid phase. In other words, the gas phase is always richer in the more volatile component than is the liquid with which it is in equilibrium. If the gas in equilibrium with the liquid is drawn off and condensed to a liquid by cooling, the liquid obtained will contain a greater proportion of the more volatile component than the liquid from which it was distilled. Fractional distillation columns ("fractionating columns") are used to carry out multiple separations of this sort on a continuous basis (Figure 6.8).

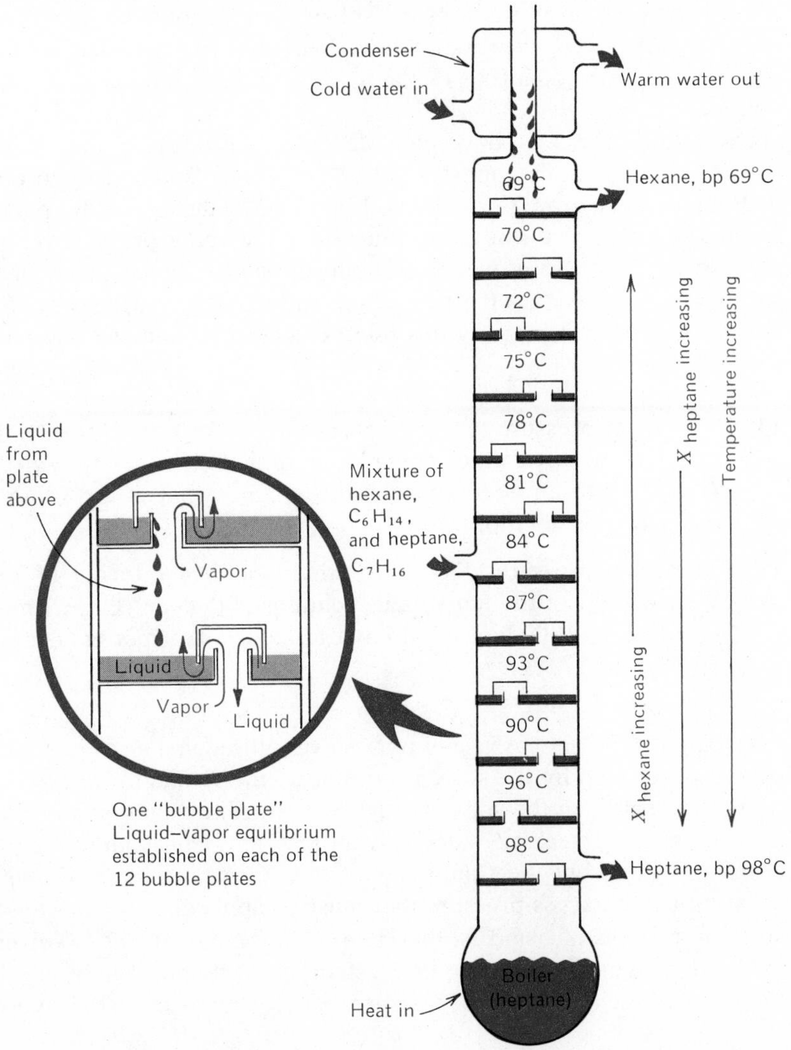

FIG. 6.8 A fractionating column for separating liquids by fractional distillation. A much longer column with many more plates would be needed to separate hexane and heptane completely.

Figure 6.8 shows a fractionating column separating a two-component liquid solution, such as hexane plus heptane, into its pure components. Of all industrial processes, fractional distillation is the most important method of separation and is used with the greatest tonnage of chemicals. ⊸

Vapor-pressure lowering—osmotic pressure Raoult's law also explains *osmosis* and *osmotic pressure*. Osmosis is the process, essential to life, by which water passes through the membranes of living cells. Osmotic pressure is equally important in living systems because it regulates the transport of water through cell membranes, maintaining their vital turgidity (filled-out condition). To understand these processes we start by taking a look at the effect of a nonvolatile solute in reducing the vapor pressure of the solvent over a solution. The effect is often called "vapor-pressure lowering." Returning to Figure 6.6, suppose that component A is volatile ($P_A^0 \neq 0$) but component B is not volatile, meaning that $P_B^0 = 0$. Under these circumstances Figure 6.6 changes to Figure 6.9. As nonvolatile solute B is added to the volatile solvent A, it makes the mole fraction X_A of solvent smaller. According to Raoult's law, $p_A = X_A P_A^0$, the addition of B reduces p_A, the partial pressure of A, over the solution. But, since the vapor pressure (P_B^0) of pure solute B is zero, it makes no contribution to the gas-phase pressure ($p_B = X_B P_B^0 = X_B \times 0 = 0$) and solvent vapor-pressure lowering is the result. You will see how this relates to osmosis and osmotic pressure shortly.

Exercises

11. The vapor pressure ($P_{H_2O}^0$) of pure water at 37°C is 47 torr. If the mole fraction of the nonvolatile antifreeze ethylene glycol ($C_2H_6O_2$) in an aqueous solution is 0.50, what is the value of p_{H_2O} over the solution at 37°C?

12. The vapor pressure ($P_{C_4H_{10}O}^0$) of ethyl ether is 400 torr at 17.9°C. What weight of the nonvolatile solute naphthalene ($C_{10}H_8$) must be added to 1000 g of ethyl ether to reduce its vapor pressure to 300 torr at 17.9°C?

Osmosis is the net passage of pure solvent through a membrane that can be penetrated by the solvent but not by the solute, *toward* a solution of the solute in the solvent (Figure 6.10). The membrane is often termed "semipermeable," since solvent may move through it freely in either direction, but the solute cannot pass through it at all. Osmotic pressure is the excess pressure that must be applied to the *solution* to stop the process of osmosis, that is, to stop the movement of solvent toward the solution and bring the system into osmotic equilibrium. In Figure 6.10(c), the osmotic pressure (π) is shown to be the hydro-

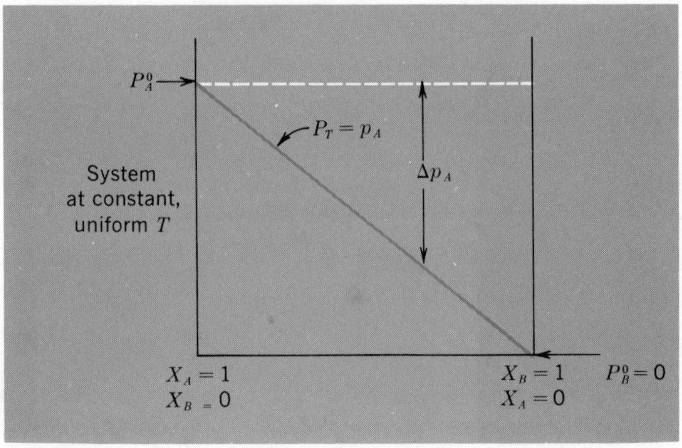

FIG. 6.9 Lowering of vapor pressure of a volatile solvent by a nonvolatile solute according to Raoult's law.

static pressure of the column of liquid exerted on the solution contained by the semipermeable membrane. The whole apparatus is open to the atmosphere, so the osmotic pressure is the excess pressure above the atmospheric.

As a model for osmosis, think about the apparatus in Figure 6.11 and what happens as time passes. Because of vapor-pressure lowering by the nonvolatile solute, the vapor pressure of solvent (p_A) over the solution in the right-hand leg will always be less than that over the pure solvent (P_A^0) in the left, as shown at (a). Solvent molecules can

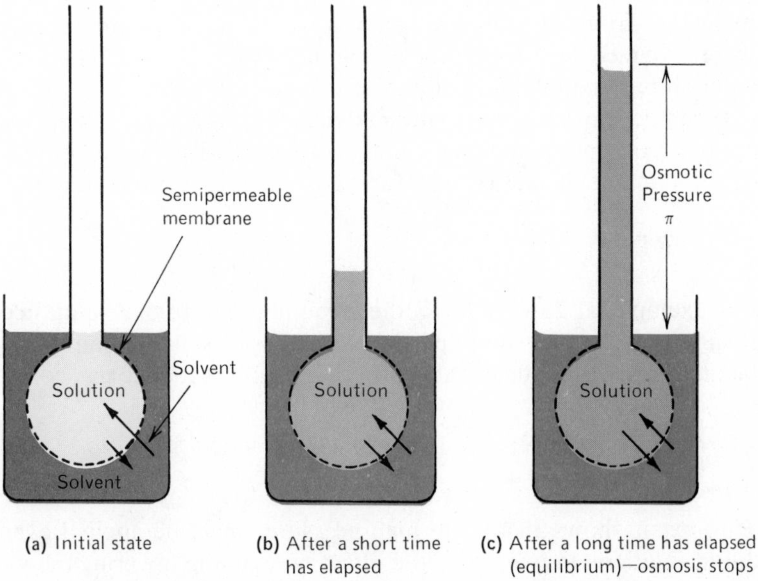

FIG. 6.10 Osmosis and osmotic pressure. Lengths of arrows show relative rates of movement of solvent through the membrane.

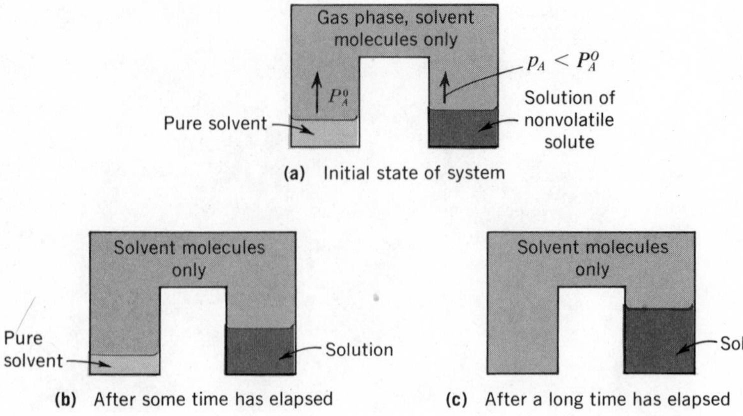

FIG. 6.11 Apparatus for demonstrating movement of solvent from a pure solvent toward a solution of nonvolatile solute as a consequence of vapor-pressure lowering.

move freely through the gas phase in either direction because the solvent is volatile, but solute molecules cannot escape from the solution at all because the solute has no vapor pressure. The gas phase in the model replaces the semipermeable membrane in osmosis. Because P_A^0 is greater than p_A, there will be a *net* movement of solvent *toward the solution,* as shown at (b), and it will go on until all of the solvent has moved into the solution (c). Vapor-pressure lowering provides the driving force for the movement of solvent. In osmosis, the solute to which the membrane is impermeable lowers the vapor pressure or "escaping tendency" of solvent for the solution, and movement of solvent into the solution results. Were it not for the osmotic pressure [Figure 6.10(c)], which *increases* the escaping tendency of solvent molecules from the solution, osmosis would go on as long as solvent were provided to the exterior of the membrane (in principle until the solution became infinitely dilute).

It can be shown by using thermodynamics that the osmotic pressure (π) of a solution containing n moles of solute particles in V liters of solution at absolute temperature T is given by the equation

The solute particles may be molecules or ions. The only requirement is that they be unable to pass through the semipermeable membrane.

$$\pi V = nRT$$

For example, at 25°C (298 K) the osmotic pressure of a solution containing one mole of solute particles per liter of solution can be calculated. Using $R = 0.08205$ liter atm deg^{-1} mole^{-1} for the gas constant,

$$\pi = \frac{n}{V}RT = \frac{1}{1} \times 0.08205 \times 298 = 24.45 \text{ atm} = 359 \text{ psi}$$

This result shows that quite high pressures must be applied even to dilute solutions to overcome the effect of solute in lowering the vapor pressure of solvent over the solution, and so establish osmotic equilibrium. In the example of Figure 6.10(c), if the concentration of the solution is one mole per liter when equilibrium is reached, the column of water supplying the osmotic pressure to the solution is 807 ft high. (A 33-ft column of water exerts a pressure of 1 atm at its base, and 24.45 atm $\times$ 33 ft atm^{-1} = 807 ft.) Think about what will happen to the semipermeable membrane as osmosis goes on. Unless very strong, the membrane will burst long before the osmotic pressure needed for equilibrium is reached. This is exactly what happens to a red blood cell if you put it in water. The concentration of solutes inside a red blood cell is large enough to give an osmotic pressure that is more than enough to rupture the fragile, semipermeable membrane of the cell wall. When a red blood cell is put in water it simply swells until the cell membrane bursts. An immediate, practical consequence of this is that water alone cannot be used in an intravenous injection to make up for blood volume lost by heavy bleeding. Instead a solution that has the same osmotic pressure as the cell contents must be used. Such solutions are termed *isotonic*. "Isotonic saline," often used to make up for blood loss, contains 9.00 g of NaCl per liter of solution. Cell membranes are essentially impermeable to Na^+ ions.

The sodium chloride in normal blood plasma ☛ contributes to the isotonicity of blood, but high-molecular-weight proteins (albumin and globulin) are also present and are the main contributors to the osmotic pressure. Excess sodium chloride in the blood is rather rapidly excreted into the urine by the kidneys, so when blood loss is very large, blood plasma (or whole blood) instead of isotonic saline is injected intravenously to restore blood volume.

The clear fluid that remains after removal of blood cells by centrifugation. It contains about 8 g of NaCl and 2 g of KCl per liter.

So you can see that a very delicate balance of osmotic pressures must be maintained between the plasma, the contents of blood and tissue cells, and the extracellular tissue fluids. For the long term, it is the plasma proteins, mainly albumin, that maintain the balance because protein molecules are too large to pass through the membrane of the kidney into the urine. In certain types of kidney disease, however, the kidney membrane "leaks" and albumin appears in the urine and is excreted (albuminuria). Unless the loss can be made up by synthesis of more protein in the body, the osmotic pressure of the blood plasma decreases. When this happens, osmosis of water *into* the tissue cells and intercellular spaces occurs. The result is a swelling or puffiness, usually beginning at the ankles or below the eyes, termed *edema*. Edema can be treated by administering a class of compounds called *diuretics* that increase the excretion of water in the urine and so raise the concentration of protein solutes in the blood, restoring its osmotic pressure. But the treatment is temporary in most cases, and the swelling disappears only when the diseased kidneys recover and can again retain plasma proteins.

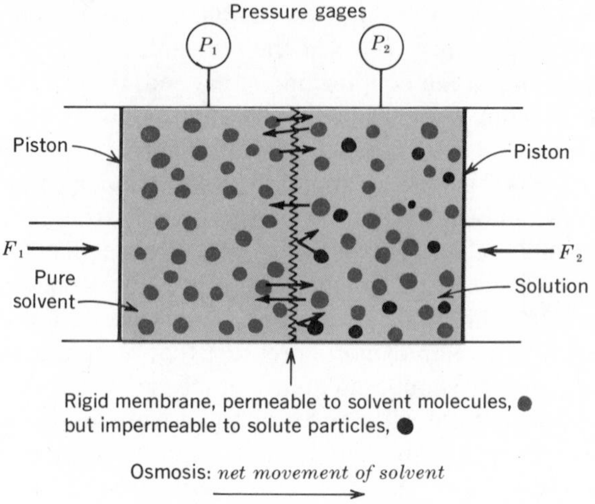

FIG. 6.12 Apparatus for the measurement of osmotic pressure.

Reverse osmosis The spontaneous direction of osmosis *from* solvent *to* solution can be reversed by applying a pressure to the solution that is larger than its osmotic pressure. If this is done, solvent is forced out of the solution toward pure solvent, and *reverse osmosis* occurs (Figure 6.12). The process is used to remove salt and other solutes from seawater to produce fresh water. In this application seawater pumped to very high pressure (400 psi) circulates on one side of a strong membrane permeable to water molecules but not to the solute particles (Na^+, Cl^-, Mg^{2+}, . . . ,) in the seawater. Because the high pressure of the seawater greatly exceeds its osmotic pressure, pure water appears on the other side of the membrane. The membrane filters out the unwanted solutes that make the seawater "salty." Reverse osmosis for the *desalination* of seawater is not a practical process because it still is too slow and expensive in electric power for the pumps. The process has been used to desalinate brackish waters and could be extended to seawater if the proper membranes could be made. Photo 19 is a microscopic view of a reverse osmosis membrane.

Reverse osmosis explains the physiological condition known as *pulmonary edema,* which results from certain types of heart disease. Here the pressure of the blood in the capillaries of the lungs rises so far above normal values that the escaping tendency of water from the blood plasma exceeds that from the tissue fluid. Osmotic equilibrium between tissue fluid and plasma is destroyed, and reverse osmosis of water into the tissue spaces occurs, causing edema or swelling of the lungs.

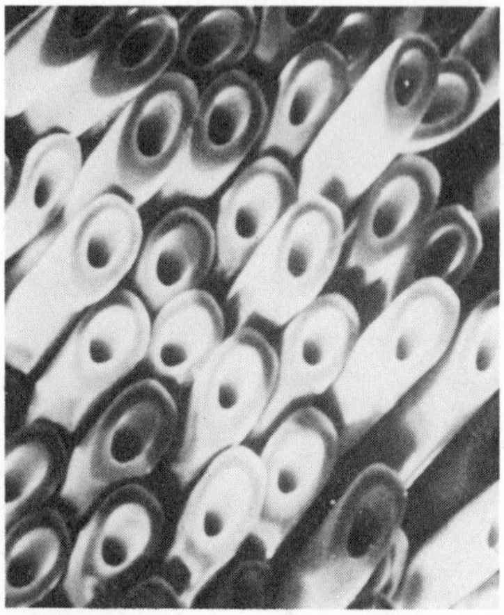

PHOTO 19 Stereo scan of hollow polyamide plastic fibers in a membrane used in reverse osmosis for desalination of brackish waters. Each fiber is approximately 80 microns (80×10^{-6} m, or 0.0032 in.) in diameter. (*Dupont Company.*)

Drinking salt water increases thirst, rather than decreasing it, because the excess salt introduced into the bloodstream must be excreted by urination. The urine volume exceeds the volume of seawater drunk, so net dehydration of the body fluid occurs.

EXAMPLE

⊸ The "osmotic pump" is a fascinating theoretical application of reverse osmosis for desalting seawater. One end of a long pipe is covered with a membrane that is permeable to water but not to the ions (Na^+, Mg^{2+}, Cl^-) in seawater. The pipe is then lowered, membrane end down, into a deep spot in the ocean (Figure 6.13). As the pipe goes deeper, the pressure of the seawater on the outside of the membrane will increase until the pressure is sufficient to start reverse osmosis of salt-free water through the membrane. Reverse osmosis will stop only when the column of fresh water inside the pipe becomes long enough to balance the pressure outside the membrane. Taking account of the salt content, density, and temperature of the ocean, it can be shown that reverse osmosis will just begin when the membrane end of the pipe is about 230 m (525 ft) below sea level. The pressure on the outside of the membrane is 24 atm (353 psi) at this depth, so a sturdy membrane would be required.

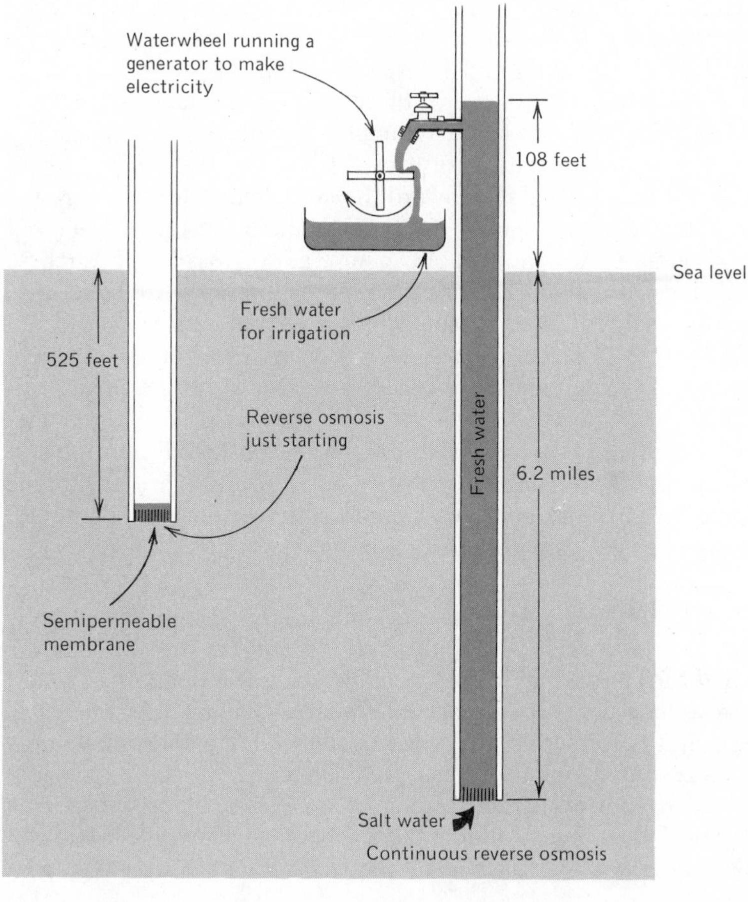

FIG. 6.13 Principle of the "osmotic pump," a theoretical device for desalination of seawater (not to scale).

Seawater, because of its salt content, is slightly more dense than fresh water. As a result, when the pipe is pushed deeper into the sea the length of the column of fresh water inside the pipe increases more rapidly than the length of pipe from sea level down to the membrane end. If the pipe goes deep enough, fresh water will start to flow continuously out of the top of the pipe. In fact, if the membrane end is 10,000 m (6.2 mi) below sea level, fresh water will reach a level in the pipe of 33 m (108 ft) above the level of the sea. Cutting the pipe, say, 30 m (98 ft) above sea level would cause a continuous flow of fresh water to move out of the pipe. The flow of fresh water could even be used to generate power by letting the water move over a paddle wheel or through a hydroelectric turbine, before the water is diverted for irrigation. The depth of 10,000 m required is close to the deepest spots in the ocean, and the pressure exerted on the semipermeable membrane would be very large. But, given a strong enough membrane, the osmotic pump would work in the way described.

The osmotic pump looks like a kind of perpetual motion machine, producing energy from nothing, as well as fresh water. The second law of thermodynamics can be stated "it is not possible to get work (energy) out of a system that is at equilibrium." But the oceans are not at equilibrium. If they were, the osmotic pump could be shown by further analysis not to work. Without going into details, we can say that the circulation of the ocean waters from near the equator where they are warmed by the sun toward the cold polar regions where the water cools and sinks (Section 15.4) keeps the oceans mixed and not at equilibrium. So for the osmotic pump, as for fossil fuels such as coal, oil, and natural gas, the prime source of energy is sunlight.

There is no theoretical reason why an osmotic pump would not work in the "real" ocean. But as with ordinary desalination, efficient semipermeable membranes capable of standing up to the huge pressures in the osmotic pump have yet to be developed. When they are, it might actually be less expensive to pump water taken from the surface of the sea to the high pressures required for ordinary membrane desalination. When this book was being written, distillation (Photo 20) was the only large-scale desalination process. ⊸

Freezing-point lowering Almost everyone in this automotive age is familiar with the use of an "antifreeze" to prevent the water in the car radiator from freezing in cold weather. Ice is less dense than water, so when water freezes the ice occupies a larger volume than the water from which it is formed. If the water is confined to a constant volume as it freezes, tremendous forces are developed by the expansion. The forces are powerful enough to crack the engine block, so freezing of water in the cooling system of the engine must be avoided. How an antifreeze operates is explained by the same Raoult's law for vapor-

PHOTO 20 Multistage distillation plant on the island of Malta. This plant can produce 1,200,000 gallons of electrolyte-free (fresh) water daily from the salt water of the Mediterranean Sea. (*Water Technologies Division, Aqua-Chem, Inc., Milwaukee, Wisconsin.*)

pressure lowering by a nonvolatile solvent that we have just used to explain osmotic phenomena.

Although you might not realize it, there is a vapor pressure due to water molecules over ice. This means that ice can "evaporate" without ever melting, by a process called "subliming." If this didn't happen, the polar ice caps and the snow on permanently snow-covered mountains would grow in quantity indefinitely as snow continued to fall on them. Now just as the vapor pressure of H_2O over liquid water decreases as the temperature is lowered, so does the vapor pressure of H_2O over ice, though at a different rate. Suppose you put a small piece of ice at 0°C into water also held at 0°C. Since ice and liquid water are in equilibrium at this temperature, nothing will happen to the ice. Another way of putting it is that the vapor pressures of water molecules over ice and liquid water at 0°C are the same. Or, the *escaping tendencies* (as measured by the vapor pressures) of water from the solid and liquid phases at 0°C are equal, so nothing is seen to happen.

Now imagine instead that the piece of ice at 0°C is placed in a solution of a nonvolatile solute in water, also at 0°C. Because of vapor-pressure lowering, the escaping tendency of water molecules from the solution will be *less* than their escaping tendency from the ice. Therefore the ice will "evaporate" into the solution, or melt. Now it is a fact that the vapor pressure of water molecules over ice decreases more rapidly than the vapor pressure over a water solution as temperature is lowered (Figure 6.14). It follows that there must be some temperature below 0°C at which the vapor pressure of water over ice becomes equal to that over the solution. That is, the vapor-pressure curves for ice and solution must intersect at some temperature below 0°C, as shown in Figure 6.14. At this temperature, ice and solution can be together in equilibrium because of balance between escaping tendencies. It follows that if the solution is simply cooled down without adding any ice, the first crystals of ice will appear at the temperature at which the vapor-pressure curves for water and solution intersect. The solute, through lowering the vapor pressure or "escaping tendency" of solvent molecules from the solution, has *lowered the freezing point* of the solution. The solute most commonly used to lower the freezing point of the fluid in the car radiator is ethylene glycol (Prestone, $C_2H_6O_2$). A nonvolatile solute also increases the boiling point of a solution, as shown in Figure 6.14.

Numerically, one mole of solute particles per kilogram of water lowers the freezing point to −1.86°C and raises the boiling point to 100.52°C. Such a solution is said to be 1 molal ☛ *in solute particles.*

To be distinguished from a 1-mol*ar* solution, in which there is one mole of solute per *liter* of solution.

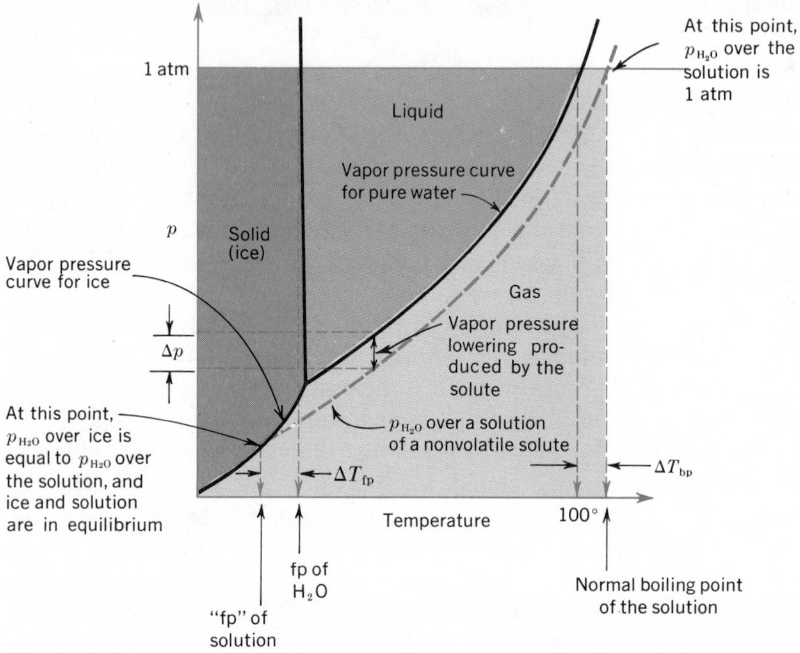

FIG. 6.14 Phase diagram showing how vapor-pressure lowering by a solute produces freezing-point lowering (ΔT_{fp}) and boiling-point elevation (ΔT_{bp}) for the solution (simplified and not to scale).

The lowering is proportional to the concentration of solute, so a 2-molal solution begins to freeze at $2 \times (-1.86°C) = -3.72°C$, and so on. The size of the effect depends on the total concentration of solute *particles*. For ethylene glycol, the solute particles are $C_2H_6O_2$ molecules, so the 1-molal solution begins to freeze at −1.86°C. But in a 1-molal solution of sodium chloride, there are *two* moles of particles, one mole of Na^+ ions and one of Cl^- ions, per kilogram of water, so you would predict a lowering to −3.72° (twice −1.86°C). The actual lowering is slightly less than this (to −3.35°C) because the solution does not behave ideally. Freezing-point lowering, boiling-point elevation, osmotic pressure, and vapor-pressure lowering are called *colligative* effects because their magnitudes are determined by the number of solute particles, and not by the nature of the particles. Gas pressure, is also a colligative property.

Freezing-point lowering explains the use of calcium chloride ($CaCl_2$) to melt the ice on roads during winter, and the use of a mixture of salt and ice ("freezing mixture") to produce the low temperature necessary for the operation of the home ice cream freezer.

6.4 Solutions of Electrolytes A solute is called an *electrolyte* in a solvent if its solution in the solvent conducts an electric current more strongly than does the solvent alone. The high conductivity of electrolyte solutions is due to the presence of ions. Water is the most common solvent, and electrolytes are conveniently classified as *acids*, *bases* (or alkalis), and *salts*. Acids produce hydronium ions (H_3O^+) ☛ and their corresponding anion when dissolved in water, according to the general equation

The hydronium ion in water has three water molecules solvating it, as indicated by the formulas $H_3O^+ \cdot 3H_2O$ or $H_9O_4^+$; we use $H_3O^+(aq)$ as a convenient abbreviation.

$$HA(aq) + H_2O(l) \rightarrow H_3O^+(aq) + A^-(aq)$$

Solutions of acids have in common a sour taste, the ability to dissolve certain metals ("active metals") such as zinc (Zn), iron (Fe), and tin (Sn) with hydrogen evolution, and the property of turning litmus paper red. All of these properties are due to the H_3O^+ ion. Hydrochloric acid, a solution made by dissolving hydrogen chloride gas in water, is a typical acid:

$$HCl(g) + H_2O(l) \rightarrow H_3O^+(aq) + Cl^-(aq)$$

The stoichiometric ☛ equation for the dissolving of zinc metal by hydrochloric acid is

Stoichiometric equations give the formulas and through them the weights of reactant and product chemicals. They are often written with an equal sign (=) rather than an arrow.

$$Zn + 2HCl = ZnCl_2 + H_2$$

but at the microparticle level the reaction is described by the equation

$$Zn(s) + 2H_3O^+(aq) + 2Cl^-(aq) \rightarrow Zn^{2+}(aq) + 2Cl^-(aq) + 2H_2O(l) + H_2(g)$$

This equation indicates that Cl^- ions are present in the solution before and after the reaction and so undergo no real change. They are only "bystanders" needed to make the solution electrically neutral. We therefore write the *net ionic* equation for the reaction by leaving them out:

$$Zn(s) + 2H_3O^+(aq) \rightarrow Zn^{2+}(aq) + 2H_2O(l) + H_2(g)$$

This equation describes the reaction of zinc with any acidic solution that contains a significant (say, 1 molar) concentration of H_3O^+ ions. Here are some other examples of stoichiometric reactions having the same net ionic equation:

$$\underset{\text{Perchloric acid}}{Zn + 2HClO_4} = \underset{\text{Zinc perchlorate}}{Zn(ClO_4)_2} + H_2$$

$$Zn(s) + 2H_3O^+(aq) + \cancel{2ClO_4^-}(aq) \rightarrow Zn^{2+}(aq) + \cancel{2ClO_4^-}(aq) + 2H_2O(l) + H_2(g)$$

$$\underset{\text{Sulfuric acid}}{Zn + H_2SO_4} = \underset{\text{Zinc sulfate}}{ZnSO_4} + H_2$$

$$Zn(s) + 2H_3O^+(aq) + \cancel{SO_4^{2-}}(aq) \rightarrow Zn^{2+}(aq) + \cancel{SO_4^{2-}}(aq) + 2H_2O(l) + H_2(g)$$

$$\underset{\text{Hydrobromic acid}}{Zn + 2HBr} = \underset{\text{Zinc bromide}}{ZnBr_2} + H_2$$

$$Zn(s) + 2H_3O^+(aq) + \cancel{2Br^-}(aq) \rightarrow Zn^{2+}(aq) + \cancel{2Br^-}(aq) + 2H_2O(l) + H_2(g)$$

When the same anions appear on both sides of the equations, they are canceled, and the same net ionic equation is obtained for all of the reactions.

Solutions of acids turn litmus red because the H_3O^+ ion transfers a proton to the litmus molecule, a complex organic molecule that we will represent simply as Lit:

$$H_3O^+(aq) + \underset{\text{Purple}}{Lit} \rightarrow \underset{\text{Red}}{HLit^+(aq)} + H_2O$$

The proton transfer changes the electronic structure of the litmus molecule in $HLit^+$ so that it absorbs light in a different region of the visible spectrum than the Lit molecule—hence the color change. Substances that change color in this way on protonation are called *indicators*. You may already have discovered that many of the dyes used in clothing are indicators when a color change resulted from acid spillage.

The reactions of acids with water are frequently not complete; that is, an *equilibrium* or balanced condition results with the effect that the solution contains a significant concentration of the molecular form of the acid HA, as well as H_3O^+ and A^-. This situation is indicated by the use of double arrows ($\rightleftarrows$) in the equation

$$HA(aq) + H_2O(l) \rightleftarrows H_3O^+(aq) + A^-(aq)$$

The strength of an acid is measured by the position of equilibrium (or point of balance) for the reaction. If, in a 1.00-*M* solution of HA, the molar concentration of HA molecules is very small or zero at equilibrium, the acid is said to be *strong*. Perchloric acid, $HClO_4$, is a strong acid—there are no $HClO_4$ molecules in its solution:

$$HClO_4(aq) + H_2O(l) \rightarrow H_3O^+(aq) + ClO_4^-(aq)$$

Hydrochloric (HCl), hydrobromic (HBr), and hydriodic (HI) acids are also strong, because their reactions with water are complete, so $[H_3O^+] = 1$ in their 1-*M* solutions.

On the other hand, if in a 1.00-*M* solution of an acid there is a significant concentration of the molecular form (HA) of the acid, it is called a *weak* acid. Acetic acid, CH_3COOH, is a typical weak acid. In a 1.00-*M* solution at 25°C, the concentration of CH_3COOH is 0.996 mole per liter and those of H_3O^+ and the anion CH_3COO^- are both 0.004 mole per liter. Notice that the proton transferred (the acidic proton) is the one attached to oxygen in the acetic acid molecule:

$$\underset{\text{Acetic acid}}{CH_3COOH(aq)} + H_2O(l) \rightleftarrows H_3O^+(aq) + \underset{\text{Acetate ion}}{CH_3COO^-(aq)}$$

Exercises

13. Carbolic acid ($C_6H_5O\mathbf{H}$, proper name "phenol") is a weak acid that will lose the proton represented in boldface type to a water molecule. Write an equation for the reaction.

14. What would you predict to occur if equal volumes of 1.00-*M* HCl and 1.00-*M* sodium acetate ($CH_3COO^-Na^+$) solutions are combined? (*Hint*: remember that acetic acid is weak.)

Bases and salts Just as an acid is a proton donor, so *a base is a proton acceptor*. Among the common bases, the hydroxide ion OH^- is the most important. This ion is responsible for the soapy feeling and bitter taste of basic or alkaline solutions, as well as their turning litmus paper blue. Solutions of hydroxides such as Na^+OH^- (lye, drain cleaner) are very dangerous to skin and particularly to the eyes, where only a drop can cause scarring of the cornea and irreparable damage to your sight. High concentrations of hydroxide ion are present in solutions of the Group 1a hydroxides, LiOH, NaOH, KOH, RbOH, CsOH. These are all ionic compounds of the general formula M^+OH^-, and when they dissolve in water, the ions are freed from their ionic bonding, becoming solvated, separate species in the solution:

$$NaOH(s) + (x + y)H_2O(l) \rightarrow Na^+(H_2O)_x + OH^-(H_2O)_y$$

The extent of solvation by water molecules (values of x and y) is not fixed but has a range of about 4–6. The reaction between the proton donor (acid) H_3O^+ and hydroxide ion in water is almost complete:

The long arrow means that the position of equilibrium is far to the $2H_2O$ side.

$$H_3O^+(aq) + OH^-(aq) \rightleftarrows 2H_2O(l)$$

In pure water the concentrations of H_3O^+ and OH^- resulting from the equilibrium nature of the reaction are both only 1×10^{-7} mole per liter at 25°C. This means that when you combine equal volumes of solutions of equal molarity in hydrochloric acid and sodium hydroxide, the reaction is almost completely described by the equation

$$\underbrace{H_3O^+(aq) + Cl^-(aq)}_{\text{Hydrochloric acid solution}} + \underbrace{Na^+(aq) + OH^-(aq)}_{\text{Sodium hydroxide solution}} \rightarrow 2H_2O(l) + Na^+(aq) + Cl^-(aq)$$

The net ionic equation is simply $H_3O^+(aq) + OH^-(aq) \rightarrow 2H_2O(aq)$ given before. But the whole equation, which describes the *neutralization* of hydrochloric acid by sodium hydroxide, is also revealing. Suppose after the neutralization you boil the water off. You would get as a nonvolatile residue the ionic compound Na^+Cl^- or "salt." This is an example of a general type of reaction:

Acid + base → salt + water

The term "salt" is used to describe not only sodium chloride but also in a general sense the ionic compound that results from the neutralization of any acid with any base. The stoichiometric (as opposed to the net ionic) equation for the reaction of hydrochloric acid with sodium hydroxide is

$$HCl + NaOH = NaCl + H_2O$$

Other examples of salt formation in neutralization reactions are

$$\underset{\text{Nitric acid}}{HNO_3} + NaOH = \underset{\text{Sodium nitrate}}{NaNO_3} + H_2O$$

$$\underset{\text{Sulfuric acid}}{H_2SO_4} + 2NaOH = \underset{\text{Sodium sulfate}}{Na_2SO_4} + 2H_2O$$

$$\underset{\text{Hydriodic acid}}{HI} + KOH = \underset{\text{Potassium iodide}}{KI} + H_2O$$

$$\underset{\text{Acetic acid}}{CH_3COOH} + LiOH = \underset{\text{Lithium acetate}}{CH_3COOLi} + H_2O$$

Like acids, bases are classified as strong or weak, depending on their *affinity for protons*. The strength of a weak base (B) in water is measured by the extent of its reaction with water itself acting as the proton donor (acid) according to the equation

$$B(aq) + H_2O(l) \rightleftarrows HB^+(aq) + OH^-(aq)$$

Ammonia (NH_3) is a typical weak base in water. In a 1.00-*M* NH_3 solution the concentrations of NH_4^+ and OH^- are both 0.0042 mole per liter and that of NH_3 is 0.9958, so the reaction $NH_3(aq) + H_2O(l) \rightleftarrows NH_4^+(aq) + OH^-(aq)$ does not proceed to the right to any great extent.

Although weak bases do not react to any great extent with the very weak proton donor H_2O, they may react quite completely with the much stronger proton donor H_3O^+. Thus ammonia and hydrochloric acid react to give the salt called ammonium chloride:

$$NH_3(aq) + \underbrace{H_3O^+(aq) + Cl^-(aq)}_{\text{Hydrochloric acid}} \rightarrow \underbrace{NH_4^+(aq) + Cl^-(aq)}_{\text{Ammonium chloride}} + H_2O(l)$$

The stoichiometric equation for this additional example of a neutralization reaction is $HCl + NH_3 = NH_4Cl$.

Strong acids, strong bases, and salts are termed *strong electrolytes* because they produce high concentrations of ions when dissolved in water, so their solutions conduct electric current very well. Solutions of weak acids and weak bases conduct poorly, because they contain smaller concentrations of ions. They are *weak electrolytes*. Both types of electrolytes are very important industrially, and as substances in blood and other body fluids they are essential to life. In the latter regard, the common salts sodium chloride (NaCl) and sodium bicarbonate ($NaHCO_3$) and the weak acid carbonic acid (H_2CO_3) play a vital part in the lives of blood and tissue cells (Section 8.5).

Salt (NaCl) is more than flavoring; a regular supply of it is absolutely necessary for our lives. Much of it comes from the food we eat, but since it is excreted in the urine and in perspiration, additional supplies may be needed if the natural sources are insufficient to make up for losses. It is for this reason that athletes and workers who perspire heavily are often supplied with salt tablets. A small amount (about 10^{-6} g daily) of the salt sodium iodide (NaI) or other iodides is also necessary for the prevention of the thyroid gland enlargement called simple goiter. Iodine is a component atom in the thyroid hormone *thyroxine* (Figure 4.4), which is vital to our lives and controls the rate of our metabolism in an important way. Sodium iodide is often added in trace amounts to table salt for use where natural sources are low in iodides.

Finally, a source of fluoride ion, such as the salt sodium fluoride (NaF), is necessary for the proper development of hard tooth enamel so that decay does not develop. Fluorides in quantity are quite poisonous, but about 1 ppm in drinking water is enough to promote the development of sound teeth without toxic effects. Stannous fluoride (SnF_2) is added in small amounts to several toothpastes to prevent decay.

6.5 Summary Guide Like pure substances, solutions are homogeneous, but they differ because their composition can be varied, while those of pure substances are given by their formulas and can't be varied. Aqueous (water) solutions are the most important type. Three methods are used to describe the composition of solutions:

1. Weight percent (% *w*/*w*) = grams of solute per 100 g of solution.
2. Mole fraction (X) = moles of solute divided by moles of solute + moles of solvent.
3. Molarity = moles of specified solute per liter of solution.

Molarity is of two types, component and species. The components of a solution are the chemicals used in its preparation. Component molarity (C, or a number followed by M) means moles of specified chemical solute per liter of solution. Component molarity ignores any reaction between solvent and solute. The species in a solution are the microparticles (atoms, molecules, ions) actually present in the solution. Species molarity ([ ]) = moles of specified microparticles per liter of solution, taking into account any reactions between solvent and solute. In V_{ml} milliliters of a C molar solution of solute having molecular weight M, the number of moles of solute = $V_{\text{ml}}C/1000$, and the number of grams of solute = $V_{\text{ml}}C\text{M}/1000$.

Raoult's law in an equation is $p_A = X_A P_A^0$, where p_A is the partial pressure of component A over a solution in which the mole fraction of A is X_A and P_A^0 is the vapor pressure of pure A at the temperature of the solution. Raoult's law explains why fractional distillation works, and why there is a lowering of the vapor pressure of a volatile solvent upon adding a nonvolatile solute. Lowering of vapor pressure by this means accounts for freezing-point lowering and osmosis. The excess pressure that must be applied to a solution to overcome the lowering by solute is the osmotic pressure of the solution.

A substance is an electrolyte if its addition to water gives a solution that conducts electricity better than pure water. Solutions of strong electrolytes have high concentrations of ions; those of weak electrolytes have low concentrations of ions. Soluble ionic compounds ("salts") and covalent substances, such as $HCl(g)$, $H_2SO_4(l)$, $Al_2Cl_6(s)$, that react completely with water to give ions are strong electrolytes. Weak bases such as ammonia (NH_3) and weak acids such as acetic (CH_3COOH) are weak electrolytes.

Net ionic equations focus attention on the reactions that occur in liquid solutions. They are obtained from overall or stoichiometric equations by eliminating species that appear in equal amounts on both sides of the equation.

Exercises

15. Gastric juice formed in the lining of the stomach to promote digestion contains enzymes (Chapter 14) and hydrochloric acid. To measure the HCl content of gastric juice a 10.00-ml sample of the juice is titrated, using a buret (Figure 6.1) to measure the amount of 0.10 *M* NaOH required to neutralize the HCl present. Phenolphthalein is used as an indicator to show when enough NaOH solution has been added to complete the neutralization reaction $HCl + NaOH \rightarrow NaCl + H_2O$. If 5.00 ml of the 0.10-*M* NaOH solution was required in a test, what is the concentration of HCl in moles per liter of gastric juice? (The result is within the normal range.)

16. The normal range of concentration of glucose in the blood is 80–120 mg per 100 ml. What is the range expressed in concentration of glucose in moles per liter of blood? Taking average blood volume as 5000 ml, how many moles of glucose are there in a person's blood, assuming 100 mg per 100 ml? How many grams?

17. Hemoglobin $(C_{738}H_{1166}FeN_{203}O_{208}S_2)_4$ has mol. wt 65,000 and is normally present to the extent of 15 g per 100 ml of blood. How many moles of hemoglobin are present in 5.00 liters of blood (average human blood volume)? How many hemoglobin molecules?

18. Assuming that each hemoglobin molecule carries on the average four O_2 molecules, what volume of O_2 measured at 0°C, 1 atm, is carried by the hemoglobin in the blood of an average person? (Refer to Exercise 17.)

19. In making up a batch of auto radiator antifreeze, equal volumes of ethylene glycol ($HO—CH_2—CH_2—OH$, mol. wt 62.1, $d^{20°}$ = 1.11 g/ml) and water ($d^{20°}$ = 0.998 g/ml) are combined. Calculate the mole fractions of water and ethylene glycol in the solution. What would be p_{H_2O} over the solution at 100°C?

20. The 1:1 by volume solution of ethylene glycol and water described in Exercise 19 begins to boil (p_{H_2O} = 1 atm) at 110°C. What advantage, if any, is there to using the solution instead of water alone in the radiator of an automobile? (See Introduction, Exercise 25.)

21. The "5%" glucose solution used for intravenous injection contains 51 g of glucose ($C_6H_{12}O_6$) per liter. Calculate the osmotic pressure of this solution in atmospheres and torr at 37°C (body temperature). The osmotic pressure of blood serum (the cell-free liquid of blood) is 25 torr at 37°C. Explain why 5% glucose can be "dripped" into a vein without danger. (*Hint*: consider cell-membrane permeability; glucose is "food" for all cells.)

22. A typical "brackish" water contains 1% NaCl by weight. Calculate the minimum pressure in excess of atmospheric pressure at sea level (1 atm) necessary to just start reverse osmosis of this brackish water at 20°C (d^{20} of 1% NaCl = 1.007 g/ml).

23. Taking the contents of the cells in the root tip of a plant as a solution of 1 g of glucose (to which the plant-root outer membrane is impermeable) in 100 ml of water, calculate the osmotic pressure of the cell contents at 20°C. What is the maximum force that 1 sq in. of root tip could exert as a result of osmosis of groundwater into the root tip?

24. Calcium chloride ($CaCl_2$) behaves in a solution in water as if it were 85% ionized into Ca^{2+} and $2Cl^-$ ions. Calculate the freezing point of a 1-molal solution of $CaCl_2$ in water. (The fp depression constant for water is −1.86°C per mole of solute particles in 1 kg of water.)

25. The liquid in a fully charged lead-acid storage battery of the type used in automobiles is 3.73 *M* H_2SO_4. The "concentrated" sulfuric acid available from chemical supply houses is 17.8 *M* H_2SO_4. What volume of concentrated sulfuric acid should be *added to* water (*never* do the reverse) to make up 5.00 liters of 3.73-*M* H_2SO_4 battery acid?

Key Words

Acid proton donor.

Aqueous solutions solutions in which water is the solvent.

Base proton acceptor.

Component molarity (*C*) number of moles of a chemical per liter of solution.

Electrolyte a solute that yields solutions containing ions—acids, bases, and salts.

Fractional distillation separation of liquids by repeated distillations.

Freezing-point lowering use of a solute to lower the vapor pressure of a solvent and thereby lower the freezing point of a solution.

Immiscible liquids liquids that form two liquid phases when combined.

Isotonic solution a solution that has the same osmotic pressure as blood plasma.

Miscible liquids liquids that yield only one phase when combined.

Molarity (*C* or a number followed by *M*) the number of moles of specified solute particles per liter of solution.

Net ionic equation equation for a reaction involving ions obtained by canceling ions that appear on both sides of the equation ("bystander" ions).

Osmosis movement of solvent through a semipermeable membrane toward a solution of a solute to which the membrane is not permeable.

Precipitation formation of a solid phase when liquid solutions are combined.

Raoult's law $p_A = X_A P_A^0$, where p_A = partial pressure of component A over a solution, X_A = mole fraction of A in the solution, and P_A^0 = vapor pressure of pure component A at the temperature of the solution.

Reverse osmosis process of forcing solvent out of a solution through a semipermeable membrane by applying to the solution a pressure greater than its osmotic pressure.

Salt compounds, usually ionic, resulting from reaction of an acid with a base.

Solute the component of a solution present in lesser quantity.

Solutions homogeneous (single-phase) samples of matter having the possibility of variable composition.

Solvent the component of a solution present in greater quantity.

Species molarity ([ ]) number of moles of atoms, molecules, or ions per liter of solution.

Standard or volumetric solutions solutions for which the molarity of a reagent is accurately known.

Stoichiometric equations equations that give the formulas and weights of the reactants and products of a reaction.

Suspensions gas or liquid solutions containing solid particles.

Weight percent (% *w*/*w*) number of grams of solute in 100 grams of solution.

Suggested Readings

Eastman, Peter F., M.D., "Breathing Again Isn't Enough," *Sea*, July 1974, pp. 42–43.

Higdon, Ralph E., "Understanding Reverse Osmosis," *Am. City,* **87**:55–56 (Dec. 1972).

Hollahan, J. R., and Wyderen, Theodore, "Synthesis of Reverse Osmosis Membranes by Plasma Polymerization of Allylamine," *Science,* **179**:500–501 (Feb. 2, 1973).

Hunt, C. A., and Garrels, R. M., *Water: The Web of Life*, New York, W. W. Norton and Co., Inc., 1972.

Levenspiel, Octave, and deNevers, Noel, "The Osmotic Pump," *Science,* **183**:157–160 (Jan. 18, 1974).

Probstein, Ronald F., "Desalination," *Am. Sci.,* **61**:280–293 (May–June 1973).

Ratcliff, J. D., "I Am Joe's Kidney," *Reader's Digest,* May 1970, pp. 98–102.

chapter 7

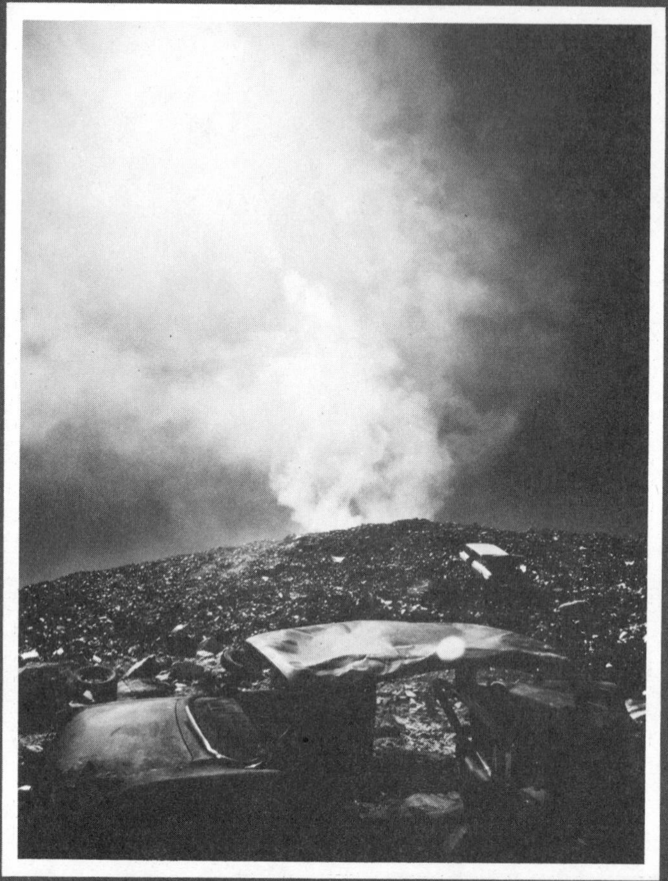

An open system trying to reach equilibrium—fire in the town dump in Moab, Utah. *(Photo 21, EPA—DOCUMERICA—David Hiser.)*

Chemical Equilibrium

7.1 Introduction Ordinarily the term "equilibrium" is used to describe a state of balance in which "nothing is going on." For example, when you step on the bathroom scales, the pointer bounces around for a few seconds and then comes to rest at the equilibrium reading that shows your weight. This mechanical equilibrium results from a balance of forces—your downward weight opposing an equal force exerted upward on the platform by the compressed spring in the body of the scales. When reactive chemicals are brought together, evidence of something going on might be gas bubbles, a color change, or heat or light given off. And, like the mechanical scales, the chemical system settles down after a time and nothing further appears to be happening. We say the system has reached *chemical equilibrium.* Unlike the mechanical equilibrium of the scales, which is motionless, *the chemical equilibrium is dynamic—viewed at the molecular level, chemical equilibrium is an equality or balance of reaction rates.* Reactants are going to products at the same rate that products are going back to reactants. It is only at the macroscopic level that you can't see anything happening. One of the simplest equilibria, equal rates of evaporation and condensation in the liquid $\rightleftarrows$ vapor equilibrium, was discussed at the beginning of Chapter 5. An example of equilibrium in which the rates of forward and reverse *chemical* reactions are equal is the well-studied case of the hydrogen–iodine–hydrogen iodide equilibrium: $H_2(g) + I_2(g) \rightleftarrows 2HI(g)$, which exists at temperatures of 400–500°C.

Both hydrogen (H_2) and hydrogen iodide (HI) are colorless gases at 400°C. Iodine (I_2) is also a gas at this temperature, but has a beautiful violet color. If one mole each of H_2 and I_2 are placed in an empty flask, which is then sealed, and its temperature is raised to about 400°C, the purple iodine color begins to fade as iodine is converted to hydrogen iodide by the reaction $H_2(g) + I_2(g) \rightarrow 2HI(g)$ (Figure 7.1). But, and this is the important point, *the iodine color never disappears completely,* no matter how long you heat the flask. The reaction goes on for a while and then mysteriously stops before all the iodine has been consumed. Equally strange is the behavior of two moles of HI when heated to 400°C in a separate, identical flask. Though colorless at first, the contents gradually take on the purple color of gaseous iodine. But the iodine color never reaches the depth corresponding to the one mole of I_2 required by *complete* reaction according to the equation

$$\underset{\text{2 moles}}{2HI(g)} \rightarrow \underset{\text{1 mole}}{I_2(g)} + \underset{\text{1 mole}}{H_2(g)}$$

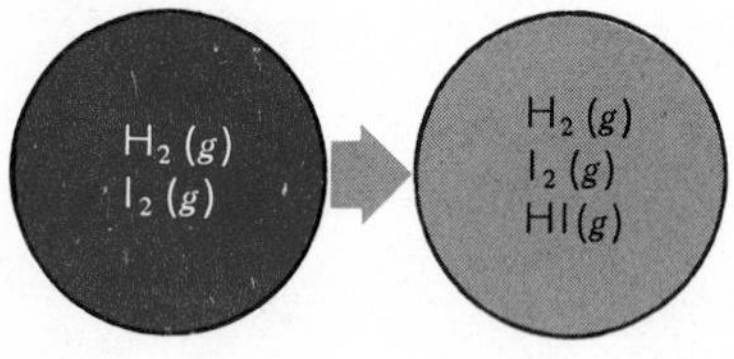

Formation of HI

$H_2(g) + I_2(g) \rightarrow 2HI(g)$

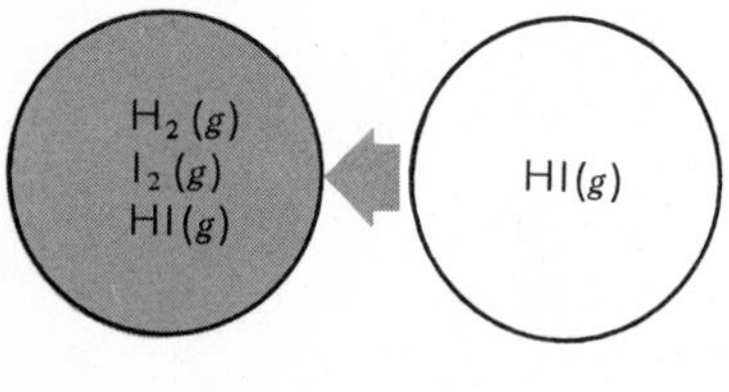

Decomposition of HI

$I_2(g) + H_2(g) \leftarrow 2HI(g)$

FIG. 7.1 Two approaches to the $H_2(g) + I_2(g) \rightleftharpoons 2HI(g)$ equilibrium.

Like the formation reaction, $H_2(g) + I_2(g) \rightarrow 2HI(g)$, the decomposition reaction, $2HI(g) \rightarrow I_2(g) + H_2(g)$, appears to go part way, then stop. A vital clue to the mystery is the *identical* colors of the flasks shown in Figure 7.1, once the iodine color has stopped fading (formation reaction) or deepening (decomposition reaction). At this point the same chemical equilibrium has been reached in both flasks. Both must contain the same amount, or number of moles, of I_2 because they are the same color. Using the equations for formation and decomposition, we can show that the amounts of H_2 and HI are also identical in the two flasks. Let x represent the number of moles of I_2 in either flask at equilibrium. (It's the same amount in both the flasks.) Now you can write

Formation reaction flask

	$I_2(g)$ +	$H_2(g)$ →	$2HI(g)$
Initially	1 mole	1 mole	0 mole
At equilibrium	x mole	x mole	$2(1-x)$ mole

At equilibrium the number of moles of H_2 remaining must be equal to the number of moles of I_2 remaining, because every time one I_2 molecule reacts to form 2HI, it takes one H_2 molecule with it, and we started with equal numbers of H_2 and I_2 molecules. That is what the balanced equation tells us. The amount of I_2 (or H_2) that has *reacted* is

Moles of I_2 initially − moles of I_2 remaining $= 1 - x$ moles

Now, since the equation tells us that the number of moles of HI formed is *twice* the number of moles of I_2 that react, the amount of HI at equilibrium is $2(1 - x)$. Now look at the decomposition reaction in the other flask, remembering that x, the number of moles of I_2, is the same as in the formation flask because both flasks are the same color.

Decomposition reaction flask

	$2HI(g)$ →	$I_2(g)$ +	$H_2(g)$
Initially	2 moles	0 moles	0 moles
At equilibrium	$2-2x$ moles	x moles	x moles

The number of moles of HI remaining is $2 - 2x$ because the formation of *each* mole of I_2 and H_2 requires *two* moles of HI. Thus, if x moles of I_2 and H_2 are to appear, $2x$ moles of HI must disappear, according to the balanced equation. Since two moles of HI were initially present, $(2 - 2x)$ or, factoring, $2(1-x)$ moles of HI will remain at equilibrium.

Comparing the results for the two flasks, we find that if x moles of I_2 are present in each flask at equilibrium, there will be x moles of H_2 and $2(1-x)$ moles of HI in each. In short, *the systems are identical.* In fact, if the flasks weren't labeled, you couldn't tell from any examination of their contents which flask contained the formation reaction and which the decomposition reaction. This represents a very important feature of equilibrium systems—*the same state of equilibrium can be reached from either side of the equation for the equilibrium.* To indicate this, double arrows $\rightleftarrows$ are used to show an equilibrium:

$$I_2(g) + H_2(g) \rightleftarrows 2HI(g)$$

The way the equation is written is a matter of convenience. It could just as easily be written

$$2HI(g) \rightleftarrows I_2(g) + H_2(g)$$

The way the equation is written does, however, *decide which compounds are called "reactants" (on left) and which are called "products" (on right)* in discussing the equilibrium.

Equilibrium can be approached from either side. Does this mean that the opposed reactions stop when equilibrium is reached? No. Chemical equilibrium is *dynamic*. Equilibrium is reached when the *rates* of the forward (→) reaction and the reverse reaction (←) become the same. In the hydrogen–iodine–hydrogen iodide case, equilibrium is reached when the rate of formation of HI from H_2 and I_2, in molecules per second, equals the rate of decomposition of HI into H_2 and I_2 in molecules per second. Because the rates are equal, the amounts of the participants, H_2, I_2, and HI, do not change as time passes—each is being used up at the same rate it is being formed, only the *macroscopic* reaction has stopped. The *microscopic,* opposed reactions continue at equal rates.

Exercises

Write balanced equations for the following industrially important equilibria:

1. Sulfur dioxide (SO_2) + oxygen gas in equilibrium with sulfur trioxide (SO_3). (Manufacture of SO_3 and sulfuric acid.)

2. Nitrogen gas + hydrogen gas in equilibrium with ammonia (NH_3). (Synthesis of ammonia for fertilizer and explosives.)

3. Limestone (calcium carbonate, $CaCO_3$) in equilibrium with lime (CaO) and carbon dioxide. (High temperature required. Used in the production of lime for mortar.)

7.2 Tests for the Existence of Equilibrium How can you tell whether a system is at equilibrium? It is not enough that "nothing appears to be going on" at the macroscopic level. A mixture of air and gasoline vapor (octane, C_8H_{18}) seems to be at equilibrium because it may be kept indefinitely without any evidence of reaction. But strike a spark and the mixture goes explosively to equilibrium according to the equation

$$2C_8H_{18}(g) + 25O_2(g) \rightleftarrows 16CO_2(g) + 18H_2O(g) + \text{energy}$$

$$\Delta H = -2446 \text{ kcal}$$

Remember, ΔH is negative in sign when heat leaves the system (Section 5.7).

The lesson here is that *equilibrium must always be with relation to a specific chemical equation.*

Energy in one form or another is always *released* to the environment when chemical systems go spontaneously (by themselves) to equilibrium from a nonequilibrium condition (Photo 21). *The larger the energy release, the more complete the reaction.* The combustion of two moles of octane according to the equation just given releases 2446 kcal of thermal energy (2,446,000 calories) to the environment, and the reaction to form CO_2 and H_2O is complete, as judged by our failure to detect any leftover reactants.

The discharge of a car battery, as when the car is parked with the lights left on, is an example of an approach to equilibrium by release of electrical rather than thermal energy. The charged battery represents a nonequilibrium condition; the completely discharged battery is at equilibrium in accordance with the equation

$$Pb(s) + PbO_2(s) + 2H_2SO_4(aq) \rightleftarrows 2PbSO_4(s) + 2H_2O(l) \qquad \Delta H = -96.4 \text{ kcal}$$

The electrical energy released to run the lights as the battery is discharged is equivalent to 96.4 kcal of thermal energy. Again, the large energy release means that the products of the reaction ($PbSO_4$ and H_2O) are strongly favored at equilibrium.

Once it is realized that equilibrium must always be with respect to a particular reaction, the *rules for the existence of a dynamic equilibrium are surprisingly simple*:

1. The system must be *closed* to transfers of matter. This means that no matter may enter or leave the system.
2. The temperature *must be the same everywhere in the system* (isothermal ☛ system).
3. The pressure must be the same everywhere in the system (isopiestic system).
4. All phases possible under the prevailing condition of temperature and pressure must be present.
5. There is no change in the chemical composition or amount of any of the phases as time passes.
6. A small change in the conditions under which the equilibrium exists produces a small change in the composition or extent of one or more phases. This small change corresponds to a small occurrence of the reaction for the equilibrium, to either right or left in the chemical equation.

"Iso" means "same"; an *isosceles* triangle, for example, has two *equal* sides.

Chemists use these rules to determine whether a system is at equilibrium. *It is important and necessary for you to understand them.* Some examples will show how powerful they are.

Figure 7.2 shows water boiling at 100°C (sea level pressure, 1 atm) in an open flask. Is the system at equilibrium? It is isothermal at 100°C, it is isopiestic at 1 atm, but it is *open* because matter is leaving the flask as steam. Also, the amount of the liquid phase (water) will decrease as time passes. So it is not an equilibrium system. Now imagine the flask to be stoppered, removed from the heater, and placed in a separate container of water boiling at 100°C. Immediately, the dynamic equilibrium $H_2O(l) \rightleftarrows H_2O(g)$ is established in the flask, since all the rules for an equilibrium are satisfied once the system is closed.

Now think about liquid water at 0°C in a closed flask with a vapor space above the liquid. This system is *not* at equilibrium either, because ice, as well as liquid water and water vapor (three phases), can exist together at 0°C. A piece of ice must be added before three-way equilibrium is established:

$$\begin{array}{ccc} H_2O(l) & \rightleftarrows & H_2O(g) \\ & \nwarrow\!\searrow \quad \nearrow\!\swarrow & \\ & H_2O(s) & \end{array}$$

In the formation reaction for hydrogen iodide the system is initially isothermal, closed, and isopiestic, and the only phase (gas) possible under the conditions is present. But the *chemical composition* of the gas phase changes as time passes because the amounts of H_2 and I_2 decrease and the amount of HI increases. So the system is *not* at equilibrium. It comes to the equilibrium $H_2(g) + I_2(g) \rightleftarrows 2HI$ when the composition of the gas phase no longer changes with time—that

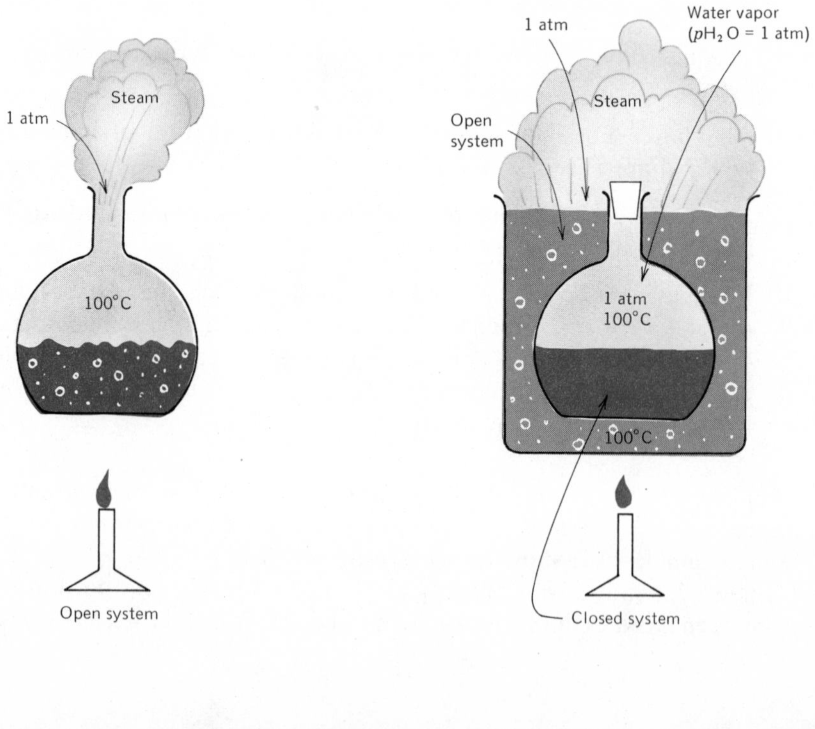

FIG. 7.2 Water and vapor at 100°C, 1 atm, in open (nonequilibrium) and closed (equilibrium) systems.

is, when the rate of formation of hydrogen iodide is equal to the rate of its decomposition.

Exercises

4. Is the earth's atmosphere an example of equilibrium? Explain your answer.

5. You put some benzene (C_6H_6) and water, which are immiscible liquids, in a beaker. Is the system at equilibrium? You pour the contents of the beaker into a flask, which is then stoppered. Is the system at equilibrium now? What processes are involved?

6. Are our bodies at equilibrium when we are asleep? Explain your answer.

7.3 The Equilibrium Constant Equation Up to now, you have seen equilibrium from a nonnumerical or qualitative point of view. The equilibrium constant equation gives a quantitative (numerical) description of equilibrium, and you can use it to calculate the amounts of chemical species (atoms, molecules, or ions) present in an equilibrium reaction. For the generalized equilibrium reaction described by the balanced equation

$$aA + bB + \cdot\cdot\cdot + \rightleftarrows cC + dD + \cdot\cdot\cdot +$$

the equilibrium constant equation is

$$K = \frac{[C]^c[D]^d \cdot\cdot\cdot}{[A]^a[B]^b \cdot\cdot\cdot}$$

Here the brackets [ ] are used to mean the concentrations in moles per liter of the participating species they enclose *when the system is at equilibrium*. The equilibrium constant K has a numerical value for a reaction, and its value for any one reaction *depends only on the temperature of the reaction.*

In writing equilibrium constant equations you must always put the concentrations of products (right-hand side of the chemical equation) in the numerator, and those of the reactants (left-hand side) in the denominator. Then you raise each concentration to the power of the coefficient ($a, b, c, d, \ldots,$) of its species in the balanced chemical equation for the equilibrium reaction. Thus for the gas-phase equilibrium

$$H_2(g) + I_2(g) \rightleftarrows 2HI(g)$$

at 425°C,

$$K = \frac{[HI]^2}{[H_2][I_2]} = 55$$

The value 55 is the *equilibrium constant* for the reaction at 425°C. Since products appear in the numerator, *large values of K indicate reactions in which the products are favored at equilibrium; small values of K indicate that the reactant side is favored.* Thus large values of K show reactions that "go"; small values indicate reactions that do not proceed very far before equilibrium is reached. Changing the temperature changes the value of K. For the H_2–I_2–HI case, $K = 67$ at 357°C. Here lowering the temperature from 425°C to 357°C increases the value of K from 55 to 67 and favors the product side of the chemical equation more strongly. So if the objective is to prepare HI from H_2 and I_2, the reaction should be run at the lowest temperature practicable. Already, you can see one practical aspect of the equilibrium constant. A second is that if the value of K is large, large amounts of useful energy (thermal or electrical) may be obtained by allowing reactants to proceed to equilibrium. Most of the energy people use for their bodily activities and to run machines results from chemical reactions that are moving toward equilibrium. The burning of fossil fuels (coal, oil, and gas) and the oxidation of glucose in body cells are examples.

Exercise **7.** Balance the following skeleton equations and write the equilibrium constant equations for them:

a. $H_2(g) + N_2(g) \rightleftarrows NH_3(g)$
b. $N_2(g) + O_2(g) \rightleftarrows NO(g)$
c. $SO_2(g) + O_2(g) \rightleftarrows SO_3(g)$
d. $H_2SO_4(l) + H_2O(l) \rightleftarrows H_3O^+(aq) + SO_4^{2-}(aq)$
e. $NO(g) + O_2(g) \rightleftarrows NO_2(g)$

The value of K at a given temperature is found by accurate measurement of equilibrium concentrations and substitution of their values into the equilibrium constant equation for the reaction. Figure 7.3 summarizes the determination of K at 425°C for the H_2–I_2–HI equilibrium. *Once the value of K has been found at a given temperature, it can be used for the quantitative (numerical) treatment of all examples of the equilibrium at that temperature.*

To learn more about the equilibrium equation we will apply it to the reaction called *esterification*. In this reaction, very important in organic and biochemistry, a carboxylic acid RCOOH ☛ reacts with an alcohol R′OH to produce an *ester* RCOOR′ and water:

The symbols *R* and *R′* are used to represent the rest of the molecule. For acetic acid $R = CH_3$; for ethanol $R = CH_3—CH_2$ or C_2H_5.

$$\text{RCOOH} + \text{R'OH} \rightleftarrows \text{RCOOR'} + H_2O$$

Carboxylic acid + Alcohol ⇄ Ester + Water

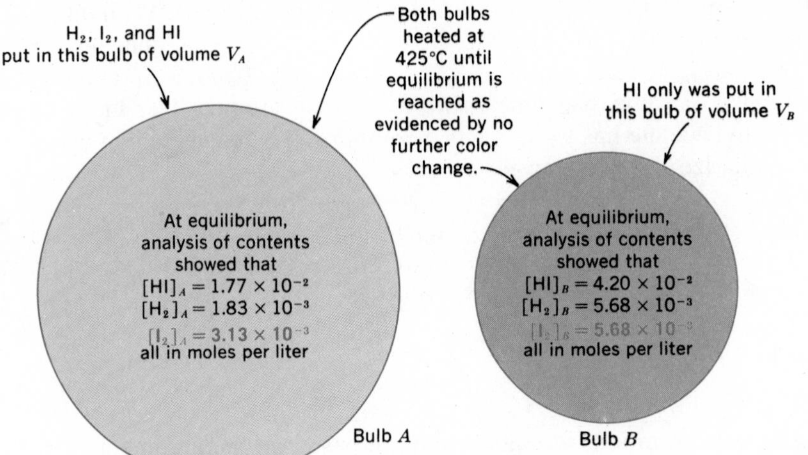

$$K = \frac{[\mathrm{HI}]_A^2}{[\mathrm{H_2}]_A[\mathrm{I_2}]_A} = \frac{(1.77 \times 10^{-2})^2}{(1.83 \times 10^{-3})(3.13 \times 10^{-3})} = 55$$

$$= \frac{(4.20 \times 10^{-2})^2}{(5.68 \times 10^{-3})(5.68 \times 10^{-3})} = \frac{[\mathrm{HI}]_B^2}{[\mathrm{H_2}]_B[\mathrm{I_2}]_B}$$

FIG. 7.3 Two examples of the equilibrium $H_2(g) + I_2(g) \rightleftharpoons 2HI(g)$ at 425°C.

For the simple case of acetic acid (active principle of vinegar) and ethanol (ethyl alcohol, beverage alcohol) the structural details are as follows:

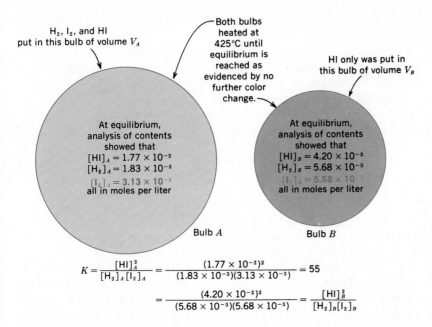

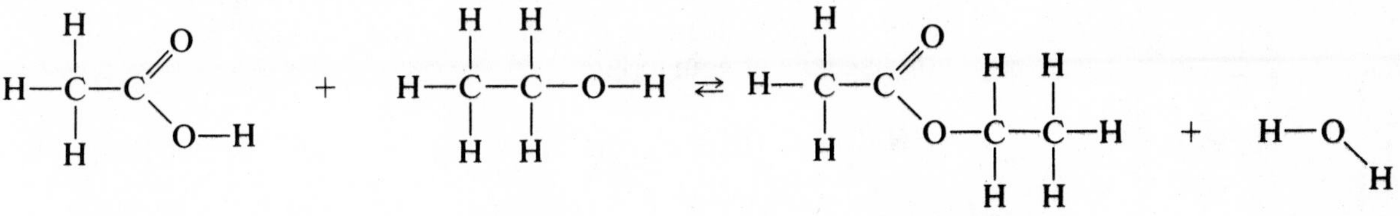

Acetic acid or CH_3COOH | Ethanol C_2H_5OH | Ethyl acetate $CH_3COOC_2H_5$ | Water H_2O

For this reaction,

$$K = \frac{[\mathrm{CH_3COOC_2H_5}][\mathrm{H_2O}]}{[\mathrm{CH_3COOH}][\mathrm{C_2H_5OH}]} = 4 \qquad \text{at } 20°\mathrm{C}$$

To see how this equation for K can be used, suppose you pour 1.00 liter of CH_3COOH into 0.50 liter of C_2H_5OH and want to know how much ethyl acetate will be present at equilibrium. This is a very practical question for anyone making ethyl acetate, of which millions of pounds are made and sold each year for use as a solvent. The first step in a problem of this sort is to calculate the molar concentrations (moles/liter) of reactants (CH_3COOH and C_2H_5OH) *before* any products ($CH_3COOC_2H_5$ and H_2O) have been formed. Without going into the detailed arithmetic, we can give the original concentrations as $C_{CH_3COOH} = 11.70$ moles/liter and $C_{C_2H_5OH} = 5.70$ moles/liter. The next step is to assume that some *unknown* concentration, call it y moles/liter, of ethyl acetate has formed when equilibrium is reached. You can now summarize the two steps as follows:

Molar concentrations (moles/liter)

	CH_3COOH +	C_2H_5OH	⇄ $CH_3COOC_2H_5$ +	H_2O
Initially	11.70 moles/liter	5.70 moles/liter	0 moles/liter	0 moles/liter
At equilibrium	$(11.70 - y)$ moles/liter	$(5.70 - y)$ moles/liter	y moles/liter	y moles/liter

The concentrations at equilibrium are based on the argument that if y moles/liter of ethyl acetate are present at equilibrium, there must also be y moles/liter of water present. This is because the balanced equation shows that for each mole of ethyl acetate formed, one mole of water must also be formed. It also shows that formation of one mole of ethyl acetate uses up one mole of acetic acid and one mole of ethanol. When y moles of ethyl acetate have been formed, y moles of acetic acid and y of ethanol will have disappeared. Thus the concentrations of them that remain at equilibrium are their initial concentrations minus the concentration of ethyl acetate formed or $(11.70 - y)$ for acetic acid and $(5.70 - y)$ for ethanol. You can now write the *equilibrium* concentrations of the participants in terms of y, the unknown concentration of ethyl acetate at equilibrium:

$$[CH_3COOC_2H_5] = y$$
$$[H_2O] = y$$
$$[CH_3COOH] = 11.70 - y$$
$$[C_2H_5OH] = 5.70 - y$$

You can follow the approach to equilibrium by seeing how the value of the *reaction quotient* (Q) changes. Q has the same form as K but with nonequilibrium concentrations:

$$Q = \frac{(C_{CH_3COOC_2H_5})\,(C_{H_2O})}{(C_{CH_3COOH})\,(C_{C_2H_5OH})}$$

The symbol C is used for concentration in moles/liter to emphasize that the concentration is *not* an equilibrium concentration. Equilibrium concentrations are always enclosed in brackets [ ]. One liter of CH_3COOH ($M = 60$, $d^{20°} = 1.05$ g/ml) weighs 1050 g and contains (1050/60) = 17.50 moles. For C_2H_5OH ($M = 46$, $d^{20°} = 0.79$ g/ml) 0.50 liter weighs 395 g and contains (395/46) = 8.60 moles. The volume of the mixture is (1.00 + 0.50) = 1.50 liters, so C_{CH_3COOH} = (17.5/1.5) = 11.70 moles/liter, and $C_{C_2H_5OH}$ = (8.60/1.50) = 5.70 moles/liter.

Initially the concentrations of ethyl acetate and water (numerator) were zero, so $Q = 0$ at the start of the reaction. As the reaction proceeds, the concentrations in the numerator increase toward the value y moles/liter as products form, while the concentrations in the denominator decrease as reactants are used up. The value of Q begins to grow. Both changes go on until the value of Q becomes equal to the value of the equilibrium constant K (4 in this case). At that time equilibrium will have been reached, and there are no further changes in the concentrations of reactants or products. For the reaction mixture at equilibrium you write, using the equilibrium concentrations of the participants in terms of y, the concentration of ethyl acetate at equilibrium:

$$K = \frac{[CH_3COOC_2H_5]\,[H_2O]}{[CH_3COOH]\,[C_2H_5OH]} = 4 = \frac{(y)\,(y)}{(11.70 - y)\,(5.70 - y)}$$

The right-hand equality in this equation can be rewritten

$$4 = \frac{y^2}{(11.70 - y)\,(5.70 - y)}$$

Doing the indicated algebra gives

$$3y^2 - 69.60y + 266.80 = 0$$

This equation is a *quadratic* equation of the form $ay^2 + by + c = 0$ with $a = 3$, $b = -69.60$, and $c = 266.80$. Using the quadratic formula, we find the value of y that satisfies it to be 4.85 moles/liter. We may now write, for the concentrations at equilibrium:

$$\begin{aligned} y = [CH_3COOC_2H_5] &= 4.85 \text{ moles/liter} \\ [H_2O] = [CH_3COOC_2H_5] &= 4.85 \text{ moles/liter} \\ [CH_3COOH] &= (C_{CH_3COOH} - y) = (11.70 - 4.85) \\ &= 6.85 \text{ moles/liter} \\ [C_2H_5OH] &= (C_{C_2H_5OH} - y) = (5.70 - 4.85) \\ &= 0.85 \text{ mole/liter} \end{aligned}$$

This gives $y = (-b \pm \sqrt{b^2 - 4ac})/2a$ or, in this case,

$$y = \frac{-(-69.60) \pm \sqrt{(-69.60)^2 - (4)\,(3)\,(266.80)}}{(2)\,(3)}$$

Because of the $\pm$ sign, the formula gives two values for y, 4.85 and 18.35. The larger value is clearly meaningless in the context of the problem, because even if *all* the ethyl alcohol, 5.70 moles/liter, were converted to ester, the maximum concentration of ethyl acetate (y) would be 5.70 moles/liter.

To check the arithmetic, put the equilibrium values you have found back into the equation for K to see if they reproduce the value of K. When we do this the value 4 is reproduced, so the arithmetic was error-free.

$$K = \frac{[CH_3COOC_2H_5][H_2O]}{[CH_3COOH][C_2H_5OH]} = \frac{(4.85)(4.85)}{(6.85)(0.85)} = 4.00$$

Because the volume of the solution is 1.50 liters and $[CH_3COOC_2H_5] = 4.85$ moles/liter, the reaction produced 1.50×4.85 moles of ethyl acetate ($M = 88$), or $1.50 \times 4.85 \times 88 = 641$ grams of the substance. Don't let the complexity of the calculations cloud the chemical importance of what has been accomplished. First, we have used the value of an equilibrium constant measured by some other chemist to calculate the results of an experiment we didn't even have to carry out in the laboratory. Second, *we did not start with the* 1-*molar amounts* of acetic acid and ethanol that appear in the chemical equation for the reaction. This is not necessary; we could have taken *any* arbitrary amounts we wanted to and made the same type of calculation, only the numbers would differ. This situation can be described in another way: if C_{CH_3COOH} and $C_{C_2H_5OH}$ are known, *not* necessarily equal, original concentrations of acetic acid and ethanol in moles per liter, and y is the unknown concentration of ethyl acetate (and water) at equilibrium, then the equation

$$\frac{y^2}{(C_{CH_3COOH} - y)(C_{C_2H_5OH} - y)} = 4$$

can be used, along with the quadratic formula, to find the concentrations of all participants at equilibrium. The value $K = 4$ at 20°C clearly *tells the whole quantitative story of the acetic acid–ethanol–ethyl acetate equilibrium.*

Generalizing, we see that the value of K for a reaction

1. Tells you which side of the chemical equation for the reaction is favored at equilibrium.
2. Enables you to calculate the concentrations of participants at equilibrium.
3. Tells you whether useful amounts of energy can be obtained by letting the reaction proceed to equilibrium. If K is large, energy can be obtained. Very little could be obtained from the esterification for which $K = 4$. The value of K must be greater than about 10^4 before useful amounts of energy can be obtained from a reaction.

Exercises

8. For the equilibrium $2SO_2(g) + O_2(g) \rightleftarrows 2SO_3(g)$ one set of equilibrium concentrations at 528°C is $[SO_3] = 0.1000$ mole/liter,

$[O_2] = 2.58 \times 10^{-3}$ mole/liter, and $[SO_2] = 7.74 \times 10^{-3}$ mole/liter,

a. Are the reactants or the product favored at equilibrium?
b. What is the value of the equilibrium constant for the reaction?
c. For the reaction as written $\Delta H = -45$ kcal. Is this reasonable, considering the value of the equilibrium constant? Could energy be obtained by letting the reaction proceed to equilibrium?
d. What are the units of K?

9. For the reaction $2NO_2(g) \rightleftarrows N_2O_4(g)$, we have $K = 171$ liter mole^{-1} at 25°C.

a. N_2O_4 (dinitrogen tetraoxide) is colorless; NO_2 (nitric oxide) is dark orange. If the equilibrium were established in a glass bulb, what color would the gas mixture be?
b. If $[NO_2]$ at equilibrium is 0.020 mole/liter, what is the value of $[N_2O_4]$?

Hydrolysis of esters The reverse of esterification is hydrolysis. (Hydrolysis comes from the Greek words *hydro*, meaning water, and *lysis*, meaning breaking up.) In hydrolysis an ester reacts with water to form an acid and an alcohol. For the hydrolysis of ethyl acetate the chemical equation is

$$CH_3COOC_2H_5 + H_2O \rightleftarrows CH_3COOH + C_2H_5OH$$

and the equilibrium constant equation is

Subscripts are used to identify equilibrium constants when necessary.

$$K_{hyd} = \frac{[CH_3COOH][C_2H_5OH]}{[CH_3COOC_2H_5][H_2O]}$$

The hydrolysis equilibrium is simply the esterification equilibrium looked at from a reverse point of view. Because only the viewpoint has been changed, there is a simple relationship between the equilibrium constants for the two reactions. Letting K_{est} stand for the esterification, we write

$$K_{est} = \frac{[CH_3COOC_2H_5][H_2O]}{[CH_3COOH][C_2H_5OH]}$$

Comparing the expressions for K_{hyd} and K_{est}, we see that they are reciprocally related; that is, $K_{hyd} = 1/K_{est}$:

$$K_{hyd} = \frac{[CH_3COOH][C_2H_5OH]}{[CH_3COOC_2H_5][H_2O]} = \frac{1}{\dfrac{[CH_3COOC_2H_5][H_2O]}{[CH_3COOH][C_2H_5OH]}} = \frac{1}{K_{est}}$$

Since $K_{est} = 4.00$, it follows that $K_{hyd} = 1/4.00 = 0.25$ at 20°C. A general principle is illustrated here:

If K is the value of an equilibrium constant for a reaction, 1/K is the value of the equilibrium constant for the same reaction written in reverse.

EXAMPLE ⊸ Both esterification and hydrolysis are processes vital to life, so the esterification–hydrolysis equilibrium deserves a closer look. Fats are esters of the trihydroxy (3-OH groups) alcohol glycerol ("glycerine") with carboxylic acids of the general formula $R-C(=O)-OH$ called "fatty acids." Tristearin, the main component of beef fat (tallow) has $R = C_{17}H_{35}$, and is hydrolyzed to stearic acid and glycerol:

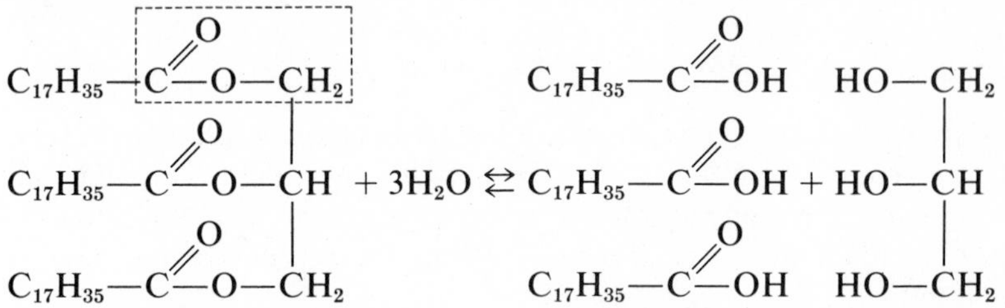

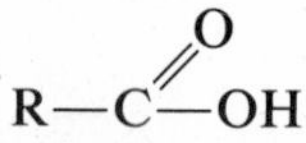

Tristearin (a fat) — 3 moles of stearic acid — 1 mole of glycerol

The reaction involves hydrolysis of three ester groups, one of which is indicated by the dashed-line box. The group $C_{17}H_{35}$ in stearic acid is a long hydrocarbon chain:

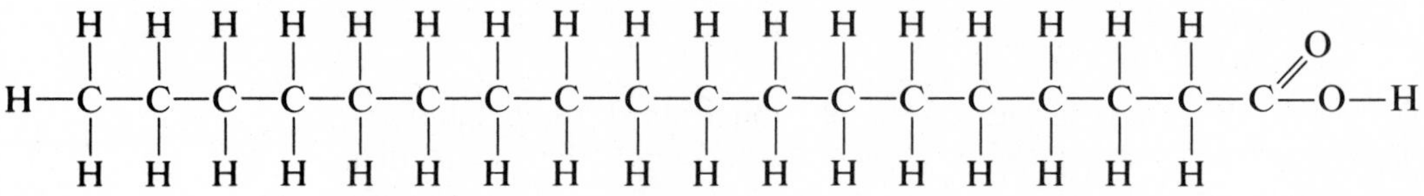

Stearic acid

The digestion of fats occurs in the small intestine and involves hydrolysis to fatty acid and glycerol. Fats are not absorbed through the intestinal wall, although glycerol and fatty acids are. After absorption, esterification converts them back into body fat. The digestion of fat would be inefficient if it depended only on the establishment of the hydrolysis equilibrium, because at equilibrium much unhydrolyzed fat

Stearic acid is called a "saturated" fatty acid because there are no carbon–carbon double bonds in its chain. Fatty acids with two or more double bonds are called "polyunsaturated" and are a necessary part of a sound diet. They are found as esters with glycerol in corn, safflower, and peanut oils.

would be present. How the body achieves complete digestion of fat will give us an example of yet another general property of equilibrium reactions.

For discussing the digestion of tristearin, the reaction may be simplified to

$$\text{Tristearin} + 3H_2O \rightleftarrows 3 \text{ stearic acid} + \text{glycerol}$$

for which

$$K = \frac{[\text{stearic acid}]^3[\text{glycerol}]}{[\text{tristearin}][H_2O]^3} = 0.0156 \qquad \text{at } 20°\text{C}$$

Now suppose that after equilibrium is established the concentration of stearic acid could somehow be reduced below its equilibrium value, as by removing some of it from the equilibrium mixture. How must the equilibrium respond? Removing some of the stearic acid would lower its concentration, and if none of the other concentrations changed, the value of the expression for the reaction quotient Q,

$$Q = \frac{C^3_{\text{stearic acid}}\, C_{\text{glycerol}}}{C_{\text{tristearin}}\, C^3_{H_2O}}$$

would be smaller than the value of K because the stearic acid concentration is in the numerator. To get back to equilibrium (reestablish the required value for K), some tristearin must be hydrolyzed to stearic acid, because this process *decreases* the value of the denominator while *increasing* the value of the numerator. After sufficient reaction, the value of Q reaches K again and a new set of equilibrium concentrations results. Now imagine that additional stearic acid is removed. Again the conversion of tristearin to stearic acid will occur as the reaction proceeds to equilibrium. Thus, if somehow stearic acid could be *continuously* removed from the equilibrium, complete conversion of tristearin to stearic acid and glycerol could be achieved. The same would be true if the glycerol were continuously removed. This is just what happens in the intestine—unlike tristearin, stearic acid and glycerol are absorbed through the bowel wall, continuously removing them from the hydrolysis equilibrium (Figure 7.4). In this way the digestion of fats is driven to completion. The means by which this is achieved—*displacement of equilibrium by removal of a product of reaction*—is an example of a general property of equilibrium systems. ⊸

Exercises

10. The value of K for $H_2(g) + I_2(g) \rightleftarrows 2HI(g)$ is 55 at 425°C. What is the value of the equilibrium constant at the same temperature for the reaction $2HI(g) \rightleftarrows H_2(g) + I_2(g)$?

11. Hydrogen chloride (HCl) gas reacts almost completely with water according to the equation $HCl(g) + H_2O(g) \rightarrow H_3O^+(aq) + Cl^-(aq)$. Predict the effect on the yield of ethyl acetate if HCl gas is passed continuously into a solution of acetic acid and ethanol.

12. The boiling points of stearic acid ($C_{17}H_{35}COOH$), glycerol ($C_3H_8O_3$), and tristearin are all much higher than 200°C. What would be the effect on the amount of tristearin formed if a solution of stearic acid and glycerol were heated to 150°C in an open beaker? (Assume K does not change much with temperature, and remember that water boils at 100°C.)

7.4 Pressure and Temperature Effects on Equilibrium Removing a product from an equilibrium reaction, as in the example of fat digestion, is only one way of driving an equilibrium toward complete reaction. Two other ways are to change the conditions of pressure and temperature. *Pressure effects are important in gas-phase reactions* ***only*** *where there is a change in volume as reactants go to products.* Pressure increase drives the equilibrium toward the side of the reaction having the smaller gas volume. Thus an increase in pressure by com-

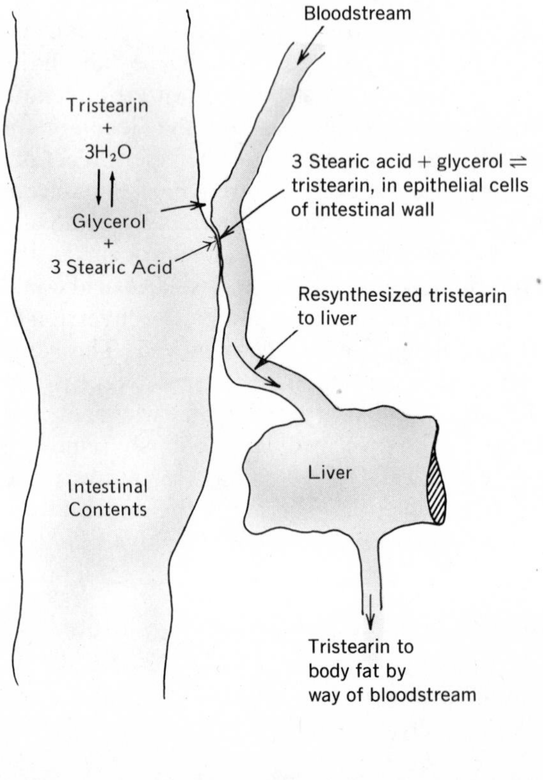

FIG. 7.4 Hydrolysis of a fat (tristearin) in the small intestine, absorption of glycerol and fatty acid through the intestinal wall, followed by resynthesis of stearic acid by esterification.

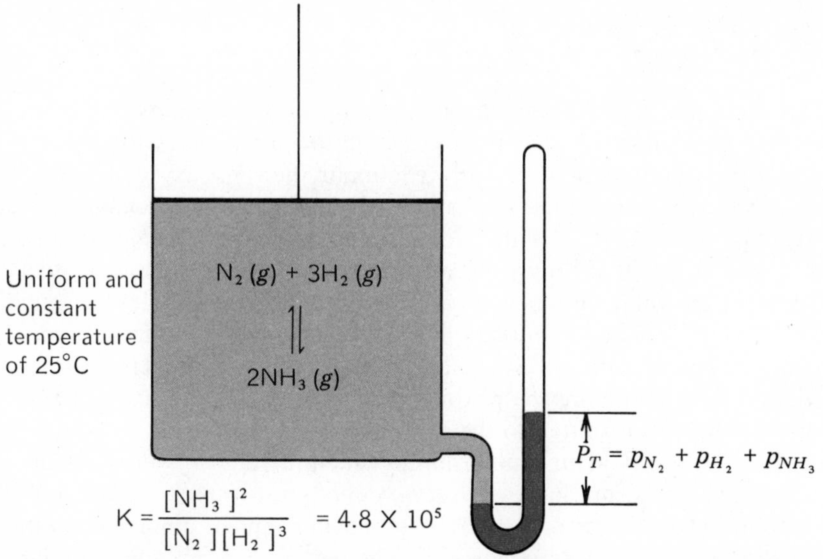

FIG. 7.5 Apparatus for determining the effect of pressure changes on the position of a gas-phase equilibrium.

pression will drive the ammonia equilibrium toward the ammonia side because four volumes of gas on the left go to two volumes on the right.

$$\underset{1\ \text{vol}}{N_2(g)} + \underset{3\ \text{vols}}{3H_2(g)} \rightleftarrows \underset{2\ \text{vols}}{2NH_3(g)}$$

To see why increasing pressure should favor the formation of ammonia we need only consider what the pressure change does to the concentrations of participants appearing in the equilibrium equation and Q[*]:

$$K = \frac{[NH_3]^2}{[N_2][H_2]^3} = 4.8 \times 10^5 \text{ at } 25°C \qquad Q = \frac{C^2_{NH_3}}{C_{N_2}C^3_{H_2}}$$

Let equilibrium be established in some convenient volume in a cylinder under a piston (Figure 7.5). Now holding temperature constant, imagine that you jam the piston abruptly down so that the volume is reduced to one-half the original value. The reduction in volume means that momentarily the concentrations of all participants (in moles/liter) will be doubled. Paying attention to the exponents, we see that the immediate effect is to increase the numerator in the expression for Q by a factor of 4 (since $2^2 = 4$), and the denominator by a factor of 16 (since $2 \times 2^3 = 16$). As a result the value of Q is instantaneously *smaller* than the equilibrium value of 4.8×10^5 by a factor of $\frac{4}{16}$ or $\frac{1}{4}$:

[*] The units of K are liter^2 mole^{-2}, which are omitted for clear presentation. When the complete equilibrium constant equation is given, as here, the units are self-evident from the form of the equation.

$$Q = \frac{C^2_{NH_3}}{C_{N_2}C^3_{H_2}} = \frac{1}{4} \times 4.8 \times 10^5 = 1.2 \times 10^5$$

Remember that the value of K depends only on the temperature. Its value is unaffected by pressure changes.

To return to equilibrium, the momentary concentrations (C) of the participants must be changed by reaction to new values that will restore the value of 4.8×10^5. Reaction in the sense $N_2(g) + 3H_2(g) \rightarrow 2NH_3(g)$ will produce the appropriate concentration changes. As reaction proceeds C_{NH_3} will *increase* because ammonia is being formed; C_{N_2} and C_{H_2} will *decrease* because they are consumed. Both effects result in the value of Q becoming larger, and when it reaches the value of $4.8 \times 10^5 = K$, equilibrium will again exist in the system. In the new state of equilibrium, $[NH_3]$ will be greater and $[N_2]$ and $[H_2]$ smaller than they were before the piston was pushed in. The *position* of equilibrium has been shifted to the NH_3 side.

Now let's look at an equilibrium in which there is *no volume change* to see why pressure increase has *no effect* on its position. Taking the reaction $H_2(g) + Br_2(g) \rightleftarrows 2HBr(g)$ as an example, imagine that the piston-and-cylinder experiment conducted on the ammonia equilibrium is repeated. When the volume of the reaction vessel is reduced to one-half its original value, the concentrations of all participants will be momentarily doubled, just as they were in the ammonia case. The effect of doubling all concentrations is to introduce a factor of $2^2 = 4$ in the numerator and $2 \times 2 = 4$ in the denominator, as the equilibrium constant equation shows:

$$K = \frac{[HBr]^2}{[H_2][Br_2]}$$

Since the same factor 4 appears in both numerator and denominator, it divides out, and no reaction is needed to return the system to equilibrium. This will be the case *whenever there are the same numbers of moles of gas molecules on both sides of the chemical equation*—that is, when no volume change accompanies reaction.

The equilibrium constant in partial pressures—K_p You can write the equilibrium constant for a reaction of gases, using the partial pressures of the participating species at equilibrium instead of their concentrations in moles per liter. For the generalized reaction between gases

$$aA(g) + bB(g) + \cdots + \rightleftarrows cC(g) + dD(g) + \cdots +$$

the equilibrium constant equation, using partial pressures, is

$$K_p = \frac{p^c_C \times p^d_D \times \cdots \times}{p^a_A \times p^b_B \times \cdots \times}$$

As in writing the equation for K, partial pressures of products go in the numerator, partial pressures of reactants go in the denominator, and each partial pressure is raised to the power indicated by the balanced equation for the reaction.

EXAMPLE

For the ammonia equilibrium, vital to the production of ammonia for fertilizing agricultural crops, the equation and expression for K_p are

$$N_2(g) + 3H_2(g) \rightleftarrows 2NH_3(g) \qquad K_p = \frac{p^2_{NH_3}}{p_{N_2} \times p^3_{H_2}} \qquad \text{at } 500°C$$

We saw earlier, using the equation for K, that a decrease in the volume containing the equilibrium mixture of gases shifted the position of equilibrium toward the ammonia side. You can show that the same shift in position of equilibrium must occur when using K_p. Suppose that the volume containing the equilibrium is abruptly halved by pushing in a piston. Remembering Dalton's law that the partial pressure of a gas is the pressure the gas would exert if it alone occupied the volume, we see that halving the volume would have the momentary effect of *doubling* the partial pressures of all the gases present. The instant result is to reduce the value of

$$Q_p = \frac{p^2_{NH_3}}{p_{N_2} \times p^3_{H_2}} \qquad \text{to} \qquad \frac{2^2}{2 \times 2^3} K_p = \frac{1}{4} K_p$$

The only way the partial pressures can adjust to reestablish the value of K_p is for additional NH_3 (numerator) to be formed at the expense of reaction of N_2 with H_2 (denominator). In this way p_{NH_3} will increase and p_{N_2} and p_{H_2} will decrease. The pressure increase drives the position of the equilibrium $N_2(g) + 3H_2(g) \rightleftarrows 2NH_3(g)$ to the right. This is the same result we found for the effect of pressure increase on the position of the equilibrium using K and concentration instead of K_p and partial pressure. As you will see shortly, advantage is taken of the increased pressure on the equilibrium to increase the yield of ammonia from the reaction.

The value of K_p for a reaction, like the value of K, depends only on the temperature. You have seen that in writing the equation for K, the concentrations of solids are omitted. You do the same when setting up the equation for K_p. Take the equilibrium for the preparation of lime (CaO) from limestone ($CaCO_3$):

$$CaCO_3(s) \rightleftarrows CaO(s) + CO_2(g)$$

For this reaction,

$$K_p = p_{CO_2}$$

The value of K_p is 1.06 atm at 900°C. This value for K_p means that if you put some calcium carbonate in a closed container and heat it to 900°C, the pressure of CO_2 in the container will be 1.06 atm or 15.6 psi when equilibrium is reached at 900°C.

The sample of calcium carbonate must be large enough so that both CaO(s) and $CO_2(g)$ are present at equilibrium.

Using Dalton's law of partial pressures, we can show that the relation between the values of K and K_p for an equilibrium is given by the equation

$$K = K_p\left(\frac{1}{RT}\right)^{(c+d+\cdots+)-(a+b+\cdots+)}$$

In this equation $c + d + \cdot\cdot\cdot +$ and $a + b + \cdot\cdot\cdot +$ are the numbers of moles of participants in the chemical equation for the equilibrium:

$$a\text{A} + b\text{B} + \cdot\cdot\cdot + \rightleftarrows c\text{C} + d\text{D} + \cdot\cdot\cdot +$$

For the ammonia equilibrium as written earlier, $a = 1$, $b = 3$, $c = 2$, and $d = 0$. For this equilibrium at 500°C you can write

$$K = K_p\left(\frac{1}{RT}\right)^{(2)-(1+3)} = K_p\left(\frac{1}{RT}\right)^{-2} = K_p \times RT^2$$

Putting in the values of $K_p = 1.5 \times 10^{-5}$ atm^{-1}, $R = 0.08205$ liter atm deg^{-1} mole^{-1}, and $T = (500 + 273) = 773$ K gives the value of K:

$$K = 1.5 \times 10^{-5} \text{ atm}^{-2} \times (0.08205 \text{ liter atm deg}^{-1} \text{ mole}^{-1} \times 773 \text{ deg})^2$$

or, doing the arithmetic, we get

$$K = 6.03 \times 10^{-2} \text{ liter}^2 \text{ mole}^{-2} \quad \text{at } 500°\text{C}$$

An especially simple relationship between K and K_p results when there are equal numbers of moles of gas on both sides of the chemical equation for the equilibrium. In this special case $K = K_p$, for the reason that $(c + d + \cdot\cdot\cdot +) - (a + b + \cdot\cdot\cdot +) = 0$:

$$K = K_p\left(\frac{1}{RT}\right)^{(c+d+\cdots+)-(a+b+\cdots+)} = K_p\left(\frac{1}{RT}\right)^0 = K_p \times 1$$

Any quantity raised to the zero power = 1.

Here are some cases in which $K = K_p$:

$$H_2(g) + I_2(g) \rightleftarrows 2HI(g)$$
1 mole + 1 mole = 2 moles

$$N_2(g) + O_2(g) \rightleftarrows 2NO(g)$$
1 mole + 1 mole = 2 moles

$$2HBr(g) \rightleftarrows H_2(g) + Br_2(g)$$
2 moles = 1 mole + 1 mole

In all of these cases, changing the pressure of the equilibrium produces no change in the position of the equilibrium. This result in the case of the N_2–O_2–$2NO$ equilibrium means that the *quantity* of NO in a particular example of the equilibrium cannot be changed by changes in total pressure. The position of the equilibrium (and therefore, the amount of NO present) *can* be changed by changing the *temperature* of the equilibrium. As you will soon see, this fact about the equilibrium has unpleasant consequences for our environment.

Exercises

13. Much of the $HCl(g)$ for making hydrochloric acid (muriatic acid of commerce) is produced by a combination of elements according to the equation $H_2(g) + Cl_2(g) \rightarrow 2HCl(g)$. Would an increase in pressure have any effect on this equilibrium?

14. What would be the immediate effect on the value of Q for the reaction $2SO_2(g) + O_2(g) \rightleftarrows 2SO_3(g)$ if the pressure were abruptly doubled? Which way would reaction occur to bring the reaction back to equilibrium?

The effect of pressure on the ammonia equilibrium has great practical significance. Millions of tons of ammonia for use as fertilizer are produced each year by the reaction. The favorable effect of increasing pressure on the yield of ammonia results in using pressure as high as 800 atm (11,760 psi). The synthesis of ammonia by this means is called the Haber process, after its discoverer. Figure 7.6 is a schematic diagram illustrating the principles of the process. The diagram may raise some questions in your mind. Does the high temperature of 500°C favor the ammonia side of the equilibrium? That is, does the value of K for the ammonia equilibrium increase with increasing temperature? What is meant by the term "catalyst"? The answers to these questions will greatly increase your understanding of both the theoretical and the practical aspects of equilibrium.

Fritz Haber (1868–1934) was born at Breslau, Germany. His interests in chemistry and physics ranged from the fixing of nitrogen from the air to recovering the traces of gold in seawater. He received the Nobel prize in chemistry in 1918.

Temperature effects on equilibria Temperature changes alter the position of equilibrium by changing *the numerical value of the equilibrium constant* K, or K_p. For the ammonia equilibrium $N_2(g) + 3H_2(g) \rightleftarrows 2NH_3(g)$ the equilibrium constant $K = 4.8 \times 10^5$ at 25°C. In this case, the value of K *decreases* with increasing temperature and thus at 500°C the value has fallen to $K = 0.0603$. Clearly the high temperature at which the Haber process is conducted does *not* favor the am-

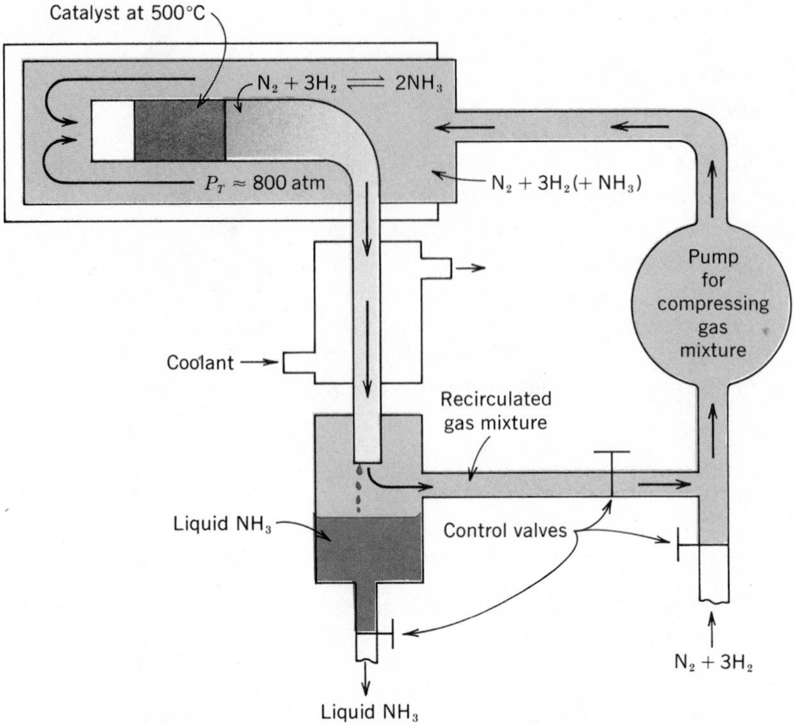

FIG. 7.6 Apparatus showing principles involved in ammonia synthesis by the Haber process.

monia side of the equilibrium. Why not operate at 25°C, where the value of $K = 4.8 \times 10^5$ shows that the ammonia side is strongly favored? The reason is that the *rate* at which equilibrium is established is very, very small at 25°C. Mixtures of nitrogen and hydrogen can be kept at this temperature and at very high pressures indefinitely without any formation of ammonia. Increasing the temperature increases the rates of chemical reactions and hence the rate of attainment of equilibrium. But even at 500°C the reaction of nitrogen with hydrogen is very slow, and this is where the *catalyst* comes into play.

Catalysts are substances (Fe–FeO in the Haber process) that increase the *rate* at which equilibrium becomes established—*they do not affect the position of equilibrium*, once it is reached. Only by the combination of high temperature *and* the catalyst can the Haber process for ammonia synthesis be carried out at a practical rate. The high pressure is used to overcome as much as possible the unfavorable effects of high temperature on the ammonia yield. What the ammonia synthesis needs for improvement is a catalyst active in promoting equilibrium rapidly at 25°C, where the large value of the equilibrium constant favors the ammonia side of the equilibrium. As you might expect, many research chemists are actively trying to invent such a catalyst.

This discussion of the ammonia equilibrium contains two very important and general lessons. First, the value of the equilibrium constant tells you only *where the system will go if it can be brought to equilibrium*. It tells nothing about *how fast* it will get there. And second, it is often necessary to strike a compromise and use the effect of high temperature and pressure on gaseous equilibria, as well as catalysts, to get a practical yield of product at a reasonable rate.

EXAMPLE

Ammonia is not only used as a fertilizer; it is also oxidized to nitric acid (HNO_3) by using a platinum catalyst and oxygen from the air in the Ostwald process:

Wilhelm Ostwald (1853–1932) was a prolific writer on a variety of scientific subjects (over 22 books and 120 papers). He received the Nobel prize in chemistry in 1909 for his work on catalysis.

$$NH_3(g) + 2O_2(g) \xrightarrow{Pt} HNO_3(l) + H_2O(l)$$

Nitric acid is required for making the chemical explosives *trinitrotoluene* (TNT), used in bombs, and *nitroglycerine*, used to make dynamite. Sulfuric acid (H_2SO_4) is used as a catalyst in both reactions:

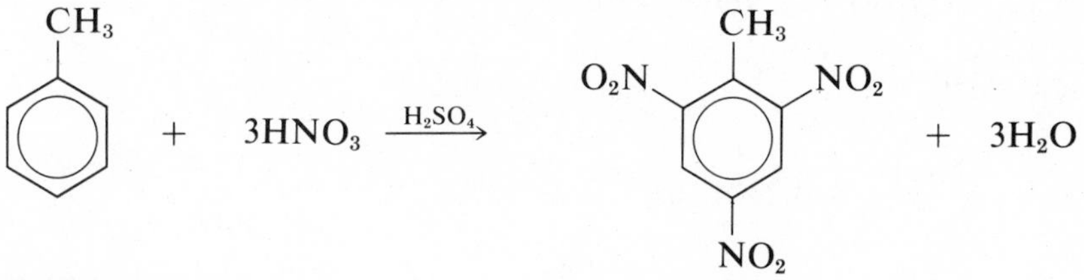

Toluene Trinitrotoluene

$$\begin{array}{l} CH_2-OH \\ | \\ CH-OH \\ | \\ CH_2-OH \end{array} + 3HNO_3 \xrightarrow{H_2SO_4} \begin{array}{l} CH_2-O-NO_2 \\ | \\ CH-O-NO_2 \\ | \\ CH_2-O-NO_2 \end{array} + 3H_2O$$

Glycerol

Nitroglycerine (glycerol trinitrate)

Until the invention of the Haber process in 1913 by the German chemist Fritz Haber, the only plentiful source of nitric acid was the reaction of sulfuric acid with sodium nitrate ($NaNO_3$), which occurs in natural deposits in Chile:

$$2NaNO_3(s) + H_2SO_4(l) \rightarrow 2HNO_3(l) + Na_2SO_4(s)$$

Germans preparing for World War I realized that once the war started the British navy would blockade them to prevent imports by sea of sodium nitrate from Chile. World War I did not start until

Germany was assured of the success of the Haber and Ostwald processes for making the nitric acid needed to produce the necessary explosives.

However, as with all scientific discoveries, there are good and bad sides to the invention of the Haber and Ostwald processes. Almost all of the ammonia and ammonium nitrate ($NH_4^+NO_3^-$) used as fertilizers in agriculture are made by the two processes. Feeding the growing population of the world would be impossible without the fertilizers made in this way. You can see the same problems in the use of nuclear reactions (Chapter 16) either for energy production for the world's developing technical society, or in nuclear bombs that could wipe mankind from the earth in the event of nuclear war.

A less beneficial equilibrium than the ammonia one is the oxygen–nitrogen–nitric oxide reaction mentioned in connection with smog at the end of Section 5.5. You are now ready to understand this equilibrium and its part in smog in more quantitative terms. The reaction is

$$N_2(g) + O_2(g) \rightleftarrows 2NO(g)$$

for which

$$K = \frac{[NO]^2}{[N_2][O_2]} = 4.23 \times 10^{-31} \qquad \text{at } 25°C$$

It is indeed fortunate for life that the value of this equilibrium constant is so tiny at ordinary temperatures. The atmosphere is a mixture mainly of O_2 and N_2. If K were not so very small, there would be significant amounts of the highly poisonous NO in the atmosphere. Life as we know it would be impossible.

Why, then, if the equilibrium concentration of NO is negligible at 25°C, is the nitric oxide equilibrium important to the formation of smog? The reason is that the value of K for this equilibrium *increases* with increasing temperature rather than decreasing as K did for the ammonia equilibrium. At 1730°C (3146°F) the value of K is 2×10^{-4} and thus, if O_2 and N_2 are brought into equilibrium at this temperature, significant amounts of NO are present. It can be shown that the partial pressure of NO (p_{NO}) at equilibrium in air heated to 1730°C is 4.3 torr. This is more than ten times the partial pressure of CO_2 in the atmosphere (Table 5.2) and 100 times the toxic limit for NO – if present,

Nitric oxide reacts with oxygen to give nitrogen dioxide: $2NO(g) + O_2(g) \rightleftarrows 2NO_2(g)$. Nitrogen dioxide is also highly poisonous. This reaction need not concern us here. In discussing smog "NO_X" is used to represent the mixture of NO and NO_2 called "nitrogen oxides." The oxides are toxic at a concentration of 0.00006 g/l of air (25 ppm).

every living thing would die. The point is that if a mixture of O_2 and N_2 is strongly heated, significant amounts of NO are formed. The high temperature of the explosion of the gasoline–air mixture in the cylinders of the automobile engine provides the necessary condition for NO formation. As the spent combustion mixture goes out in the exhaust, it is cooled so quickly that most of the high-temperature equilibrium concentration of NO has no time to go back to N_2 and O_2. That is, the rate of the reaction $2NO(g) \rightarrow N_2(g) + O_2(g)$ is slow at lower temperatures and the high-temperature concentration is "frozen" into the exhaust gas by rapid cooling. The automobile exhaust therefore contains significant amounts of NO. It reacts with unburned hydrocarbons and oxygen in a complex set of reactions to produce ozone (O_3) and the eye-irritating components of smog (Chapter 15). How might we go about solving the smog problem? First, let's see where we are going. For the reaction

$$2NO(g) \rightleftarrows N_2(g) + O_2(g)$$

the equilibrium equation is

$$K' = \frac{[N_2][O_2]}{[NO]^2}$$

This reaction is just the reverse of the reaction for the formation of NO. It follows, then, as you saw earlier, that its equilibrium constant K' is $1/K$, where K is the constant for the formation of NO. That is,

$$K' = \frac{[N_2][O_2]}{[NO]^2} = \frac{1}{K} = \frac{1}{4.23 \times 10^{-31}} = 2.36 \times 10^{30} \qquad \text{at } 25°\text{C}$$

The huge value of K' means that the problem of NO production by cars could be solved if we had a catalyst to speed up its spontaneous decomposition into N_2 and O_2 at ordinary temperatures. Although it is very unstable with respect to reversion to its elements in the equilibrium sense, the reaction is so slow at ordinary temperatures that NO can be kept for years in the absence of catalysts. There are copper–nickel catalysts that will convert up to 90% of the NO in exhaust back to N_2 and O_2. However, there are technical as well as economic difficulties in producing a catalytic converter for installation in the exhaust pipe. Another way to decrease NO production is to lower the temperature of the flame in the cylinders. The value of K for the NO formation reaction decreases rapidly with decreasing temperature, so even a moderate lowering of combustion temperature significantly decreases NO production. At the time of this writing, progress was being made to solve the "oxides of nitrogen" problem, but a complete, practical solution was not in sight. Of course, the problem would disappear if some nonpolluting engine, such as an electrical one, were adapted to the automobile. Photo 22 shows the impact of the automobile and developing industry on air quality in the Los Angeles Basin.

Exercise **15.** Balance the skeleton equations and predict the effect of an increase in pressure on the equilibria that follow:

$$CO(g) + O_2(g) \rightleftharpoons CO_2(g) \quad (1)$$

$$NO(g) \rightleftharpoons N_2(g) + O_2(g) \quad (2)$$

$$NO(g) + O_2(g) \rightleftharpoons NO_2(g) \quad (3)$$

Why is it necessary to balance the equations before you can predict the effect of pressure changes?

Henri Le Chatelier (1850–1936) was a French inorganic chemist and mining engineer. He accomplished the synthesis of ammonia from the elements in 1901, before Fritz Haber, who developed the practical industrial process.

7.5 Le Chatelier's Principle You have seen several examples of the ways in which a system displaced from equilibrium returns to a new position of equilibrium. If the displacement results from altering the concentration of one or more participants, chemical reaction occurs in the proper direction to change participant concentrations, so the numerical value of K is restored when they are put into the equilibrium constant expression. If the displacement results from a change in temperature, reaction occurs to change concentrations so that they give the altered value of K for the new temperature when put into the equilibrium constant expression. There is a general rule called *Le Chatelier's principle* that will tell you qualitatively how *any* equilibrium will

PHOTO 22 Los Angeles on a clear day (left) and the visual effects of Los Angeles smog (right). (*Los Angeles County Air Pollution Control District.*)

respond to a change in conditions. It may be stated "*an attempt to change the conditions of an equilibrium causes a shift in the position of equilibrium in the direction that tends to overcome the attempt*," or, "the system will try its best to get back to where it was." Among the examples discussed in Section 7.4, perhaps the most clear-cut operation of the principle was the shift toward products in the fat-hydrolysis equilibrium. There, when the concentration of one of the products (stearic acid) of the reaction was decreased from its equilibrium value by removal through intestinal absorption, the equilibrium responded by producing more stearic acid to offset the removal. As you saw, the equilibrium could not overcome the imposed change, but it kept trying.

In some cases, an equilibrium may shift to overcome completely an attempt to change its conditions. Ice in equilibrium with water and water vapor at 0°C under the pressure of the atmosphere is an example. Suppose you add a small amount of heat to this equilibrium system in an attempt to raise its temperature. Momentarily the temperature rises above the equilibrium value of 0°C. But very quickly ice melts in just the amount necessary to return the temperature to 0°C. The thermal energy you added is absorbed as heat of fusion of ice, ☛ and the system completely overcomes your attempt to change its temperature. It will do this as long as there is any ice left. The constancy of temperature at 0°C of the ice–water–water vapor equilibrium accounts for its use in calibrating thermometers.

79.9 cal/g or 1439 cal/mole.

The effect of pressure on the ammonia equilibrium is an example in which the equilibrium system cannot completely make up for the change in conditions. A pressure increase results in a shift toward the ammonia side because this change tends to offset the pressure increase. To see why this is so, you can use Dalton's law of partial pressures in the form $P_{total}V = n_{total}RT$. Any decrease in the total number of moles of gas (n_{total}) in a system will have the effect of decreasing the total pressure (P_{total}). Now, according to the equation $N_2(g) + 3H_2(g) \rightarrow 2NH_3(g)$, every time two moles of NH_3 are formed, one mole of N_2 and three moles of H_2 disappear. So the formation of ammonia at the expense of nitrogen and hydrogen has the effect of decreasing the total number of moles of gas present ($4 \rightarrow 2$) and tends to lower the pressure. Another look at the volume-halving experiment with this equilibrium is worthwhile. If temperature is held constant, and *if no reaction occurs*, halving the volume in which the equilibrium was originally present will exactly double the total pressure (Boyle's law, Section 5.3). Momentarily this happens, but as reaction occurs to reestablish equilibrium, the total number of moles of gas present decreases and the pressure is less than doubled when the new position of equilibrium is reached. The change attempted was to double the pressure by halving the volume. The volume was halved, but the pressure less than doubled, because reaction decreased the total number of moles of gas.

Temperature effects on equilibria are also predicted by Le Chatelier's principle. Here the thermal energy change that goes with the reaction is the important thing. Suppose a reaction is *endothermic* like the nitric oxide equilibrium. You can take the thermal energy change into account by writing

$$\text{Heat} + N_2(g) + O_2(g) \rightleftarrows 2NO(g)$$

If you add thermal energy in an attempt to increase the temperature of the equilibrium, it reacts by shifting in the direction that *absorbs* heat, that is, in the direction of NO. It does so because absorbing heat tends to offset the temperature increase. The concentration of NO increases but only at the expense of decreases in the concentrations of N_2 and O_2. The result is that the value of the equilibrium constant given by

$$K = \frac{[NO]^2}{[N_2][O_2]}$$

increases as the temperature is raised. Thus we have the following rules: The value of the equilibrium constant for an *endothermic reaction* increases with increasing temperature. On the other hand, if a reaction is *exothermic*, increasing the temperature drives the equilibrium to the left-hand side, and the value of the equilibrium constant decreases with increasing temperature. The ammonia equilibrium is an example in which K decreases with increasing temperature:

$$N_2(g) + 3H_2(g) \rightleftarrows 2NH_3(g) + \text{heat}$$

$$K = 4.8 \times 10^5 \text{ at } 25°C \qquad K = 0.0603 \text{ at } 500°C$$

These rules also apply in reverse. If you cool an *endothermic* equilibrium, the value of K decreases as the equilibrium shifts in the direction of heat *production* (to the left) in an attempt to overcome the cooling. Likewise, cooling an exothermic equilibrium increases the value of K. (*Note*: these rules are true only for reactions that are more than 15 kcal exothermic or endothermic. See Chapter 10 for more general rules.)

Exercises

16. For the gas-phase equilibrium $CO_2(g) + H_2(g) \rightleftarrows CO(g) + H_2O(g)$ the value of K at 1120°C is 2.0. When the temperature is raised to 1550°C, the value of K increases to 3.5. Is the reaction from left to right endo- or exothermic? Would pressure changes have any effect on the position of this equilibrium?

17. For the reaction $2H_2(g) + O_2(g) \rightleftarrows 2H_2O(l)$ at 25°C, $K = 1.3 \times 10^{83}$, the position of equilibrium lies far to the right. Yet when

$H_2O(g)$ is heated to 3000°C, traces of H_2 and O_2 are produced. Explain. (*Hint*: decide whether the reaction is exothermic or endothermic from the value of K and apply Le Chatelier's principle.)

18. If $K = 4.8 \times 10^5$ at 25°C for the reaction $N_2(g) + 3H_2(g) \rightleftarrows 2NH_3(g)$, what is the value of K at 25°C for the reaction $NH_3(g) \rightleftarrows \frac{1}{2}N_2(g) + \frac{3}{2}H_2(g)$? (*Hint*: $x^{1/2} = \sqrt{x}$.)

The solubility equilibrium We are all familiar with the fact that it is not possible to dissolve an unlimited amount of a solid in a limited amount of a solvent. No matter how long you stir a glass of water to which salt is being added, there comes a time when solid salt remains undissolved and sinks to the bottom when you stop stirring. The solution is said to be *saturated* with salt at this point. A saturated solution in contact with undissolved solute represents yet another case of dynamic equilibrium. Here equilibrium is reached when the rates of dissolving and precipitation (separation of solid) become equal. Figure 7.7 compares the solid $\rightleftarrows$ gas or *sublimation* equilibrium with the solubility equilibrium. In the solubility equilibrium, solvent may be regarded as replacing or "filling up" the free space between the molecules in the sublimation equilibrium.

TABLE 7.1 Range of solubilities[a]

Soluble Substances			*Sparingly Soluble Substances*		
NaI	NaCl	$PbCl_2$	CaF_2	CuS	AgI
159(0°)	35.7(0°)	0.67(0°) 1.0(20°)	2×10^{-3}(20°)	4×10^{-5}(25°)	3×10^{-7}(20°)

[a]Grams of solute per 100 g of water at specified temperature, °C.

The range of solubilities is very great, as shown in Table 7.1. Solubility depends on temperature, usually increasing with increasing temperature, as shown in Figure 7.8. A few cases are known in which

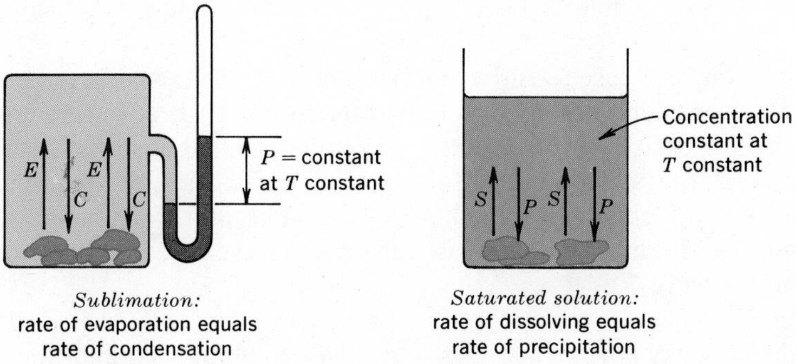

FIG. 7.7 Similarity between the sublimation and saturated solution equilibria.

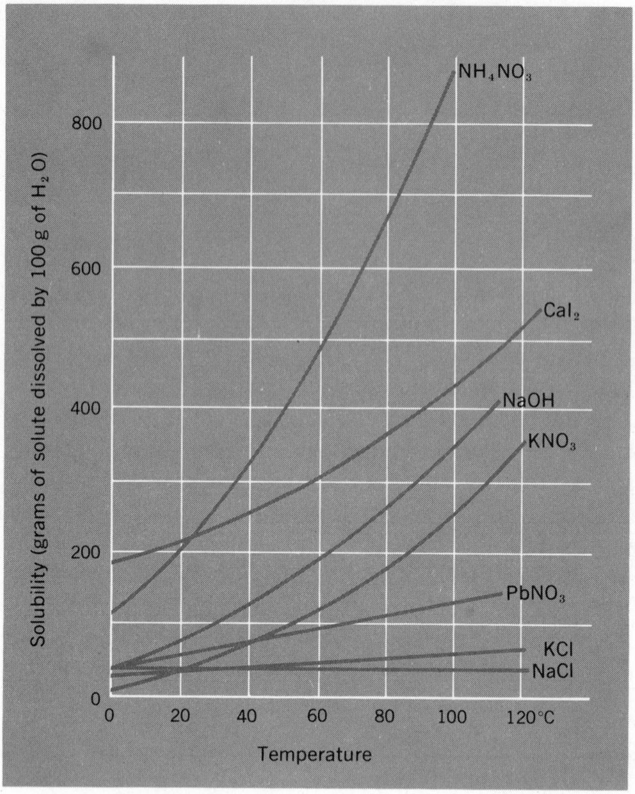

FIG. 7.8 Solubility–temperature curves for some solids (schematic except for the values at 0 and 100°C).

solubility *decreases* with *increasing* temperature. Two examples are shown in Figure 7.9. You can use Le Chatelier's principle to predict temperature effects on solubility if the heat effect accompanying the dissolving of solute in a nearly saturated solution has been measured. Just as with chemical reactions, there are two possibilities:

1. Solute + nearly saturated solution $\rightleftarrows$ saturated solution + heat
2. Heat + solute + nearly saturated solution $\rightleftarrows$ saturated solution

In the first case (positive heat of solution) an increase in temperature decreases solubility as the equilibrium attempts to offset the temperature increase by shifting in the direction (left) that absorbs heat. In the second (negative heat of solution), increasing the temperature produces an increase in solubility as the equilibrium shifts to the right to remove thermal energy. Most substances have negative heats of solution (Figure 7.8).

The solubility product constant, K_{sp} The solubility equilibrium for a slightly water-soluble substance that is ionized in solution is described

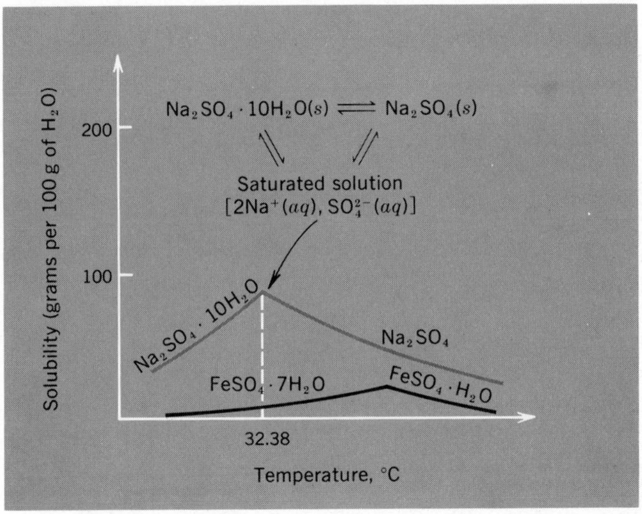

FIG. 7.9 Solubility curves for two substances whose solubilities decrease with increasing temperature.

by the *solubility product* (K_{sp}) *equation*. For calcium fluoride, soluble to the extent of 0.0020 g in 100 g of water, the solubility equilibrium is

$$CaF_2(s) \rightleftarrows Ca^{2+}(aq) + 2F^-(aq)$$

Following the general rule for writing equilibrium constants, we write

$$K = \frac{[Ca^{2+}][F^-]^2}{[CaF_2]}$$

This equation is unnecessarily complicated by the presence in it of $[CaF_2]$, the molar concentration of solid CaF_2. At a given temperature this has a constant value, ☛ so it is omitted in writing the *solubility product constant* K_{sp}. For CaF_2, $K_{sp} = [Ca^{2+}][F^-]^2$. This rule is followed for all solid solutes:

If the density of the solid solute is *d* g/ml at a given temperature, and the molar weight of the solute is M, then the concentration of the solid is (1000 *d*/M) moles/liter.

$$Ca(OH)_2(s) \rightleftarrows Ca^{2+}(aq) + 2OH^-(aq) \qquad K_{sp} = [Ca^{2+}][OH^-]^2$$
$$AgCl(s) \rightleftarrows Ag^+(aq) + Cl^-(aq) \qquad K_{sp} = [Ag^+][Cl^-]$$
$$CaSO_4(s) \rightleftarrows Ca^{2+}(aq) + SO_4^{2-}(aq) \qquad K_{sp} = [Ca^{2+}][SO_4^{2-}]$$
$$Ag_2SO_4(s) \rightleftarrows 2Ag^+(aq) + SO_4^{2-}(aq) \qquad K_{sp} = [Ag]^2[SO_4^{2-}]$$

The value of K_{sp} for a solute at a specified temperature can be readily calculated from its solubility in water at that temperature. For CaF_2, 0.0020 g dissolves in 100 g of H_2O at 20°C. It follows that 1000 g of H_2O will dissolve 10 times as much or 0.020 g. For practical purposes, take 1000 g of water at 20°C as having a volume of 1.00 liter, ☛ so the solution saturated at 20°C contains 0.020 or 2×10^{-2} g/l of dissolved

Actually 1.002 liters since the density of water is 0.998 g/ml at 20°C.

TABLE 7.2 Values of solubility product constants (K_{sp}) for some slightly soluble electrolytes at 25°C

Name of Compound	*Formula*	K_{sp}
Silver chloride	$AgCl$	1.8×10^{-10}
Silver bromide	$AgBr$	4.3×10^{-13}
Silver iodide	AgI	8.3×10^{-17}
Calcium fluoride	CaF_2	4.0×10^{-11}
Silver sulfate	Ag_2SO_4	1.6×10^{-5}
Calcium sulfate	$CaSO_4$	2.4×10^{-5}
Barium sulfate	$BaSO_4$	8.7×10^{-11}
Lead sulfate	$PbSO_4$	1.6×10^{-8}
Calcium hydroxide	$Ca(OH)_2$	7.9×10^{-6}
Barium hydroxide	$Ba(OH)_2$	5.0×10^{-3}
Aluminum hydroxide	$Al(OH)_3$	2.5×10^{-32}
Iron(II) hydroxide	$Fe(OH)_2$	7.9×10^{-15}
Iron(III) hydroxide	$Fe(OH)_3$	6.3×10^{-38}
Copper(II) hydroxide	$Cu(OH)_2$	1.6×10^{-19}
Copper(II) sulfide	CuS	8.0×10^{-36}
Lead sulfide	PbS	1.3×10^{-28}
Tricalcium phosphate	$Ca_3(PO_4)_2$	2.0×10^{-29}

CaF_2. The molar weight of CaF_2 is $40.0 + 2(19.0) = 78.0$, so the concentration of "dissolved CaF_2" is $2 \times 10^{-2}/78 = 2.6 \times 10^{-4}$ mole/liter. But the "dissolved CaF_2" is actually present as Ca^{2+} and $2F^-$ ions, because CaF_2 is ionic and an electrolyte. From the equation for the equilibrium we see that each mole of CaF_2 that dissolves yields one mole of Ca^{2+} but *two* moles of F^-. Therefore, in the solution saturated at 20°C, $[F^-]$ will be $2 \times [Ca^{2+}]$ and

$$[Ca^{2+}] = 2.6 \times 10^{-4} \text{ mole/liter}$$
$$[F^-] = 2 \times 2.6 \times 10^{-4} = 5.2 \times 10^{-4} \text{ mole/liter}$$

Substituting these equilibrium values into the equation for K_{sp} gives its numerical value:

$$K_{sp} = [Ca^{2+}][F^-]^2 = (2.6 \times 10^{-4})\cdot(5.2 \times 10^{-4})^2 = 7.0 \times 10^{-11}$$

The values of K_{sp} for some slightly soluble electrolytes are entered in Table 7.2.

Once the value of K_{sp} at a given temperature is obtained, it can be used to calculate the solubility of the substance in solutions that contain one of its ions from another source. For example, suppose instead of saturating water with CaF_2 at 20°C, solid CaF_2 is added to a 0.5-*M solution* of NaF until no more solid dissolves. From the equilibrium law, $K_{sp} = [Ca^{2+}][F^-]^2$ must equal 7.0×10^{-11} for this solution as well.

In the NaF solution $[F^-] \approx 0.5$ because NaF is a strong electrolyte, and from its formula, for each mole dissolved one mole of F^- is produced. ☛ So we substitute the actual $[F^-]$ in the expression for K_{sp} and write

You can safely ignore the tiny contribution to the F^- concentration made by the dissolving of CaF_2.

$$K_{sp} = [Ca^{2+}][F^-]^2 = [Ca^{2+}][0.5]^2 = 7.0 \times 10^{-11}$$

Solving for $[Ca^{2+}]$ we obtain

$$[Ca^{2+}] = \frac{7.0 \times 10^{-11}}{(0.5)^2} = 2.8 \times 10^{-10} \text{ mole/liter}$$

Since this Ca^{2+} concentration came from the CaF_2 *that dissolved* in the NaF solution, it is the concentration of "dissolved CaF_2"; hence the solubility in moles/liter of CaF_2 in the 0.5-*M* NaF solution.

Comparing this value for $[Ca^{2+}]$ in the NaF solution with the value 2.6×10^{-4} obtained earlier for the saturated solution of CaF_2 in water, we see that *the presence of excess F^- from NaF has tremendously reduced the solubility of CaF_2.* Again you can see Le Chatelier's principle operating. By increasing the concentration of the product F^-, the solubility equilibrium for CaF_2 has been displaced to the left or $CaF_2(s)$ side. This effect of a high concentration of one of its ions from an external source on decreasing the solubility of an electrolyte is called the *common ion effect.* Advantage is taken of it in ensuring more complete precipitation of electrolytes from their solutions in analytical work.

EXAMPLE

Barium sulfate, a chalky white solid, absorbs X rays strongly because of the large mass and electron cloud of the barium atom. Advantage is taken of the strong absorption in X-ray diagnosis of stomach ulcers and other abnormalities in the gastrointestinal system. To get an X-ray picture that shows clearly the shape of your stomach, you are handed a "barium milkshake" to drink. The milkshake contains a lot of $BaSO_4(s)$ and as you are swallowing, the doctor starts taking X-ray pictures of the $BaSO_4$ "on the way down." Regions containing the barium sulfate appear white on the X-ray picture and provide outline shapes for stomach and intestines.

Barium ion is poisonous in relatively small concentrations—ingestion (swallowing) of *soluble* barium compounds [$BaCl_2$, $Ba(OH)_2$, $Ba(NO_3)_2$] causes vomiting, violent abdominal pains, and death by heart failure. Is the use of $BaSO_4$ in X-ray diagnosis safe? The answer is yes, because the $[Ba^{2+}]$ in equilibrium with solid $BaSO_4$ is so tiny. For $BaSO_4$, the solubility product constant $K_{sp} = 8.7 \times 10^{-11}$, so you can write for the saturated solution

$$K_{sp} = [Ba^{2+}][SO_4^{2-}] = 8.7 \times 10^{-11}$$

From the equation for the solubility equilibrium $BaSO_4(s) \rightleftarrows Ba^{2+}(aq) + SO_4^{2-}(aq)$ the ions are produced in equal amounts, so $[Ba^{2+}] = [SO_4^{2-}]$ in the saturated solution, and you can write

$$[Ba^{2+}] \times [SO_4^{2-}] = [Ba^{2+}]^2 = 8.7 \times 10^{-11} \text{ mole}^2 \text{ liter}^{-2}$$

or

$$[Ba^{2+}] = \sqrt{8.7 \times 10^{-11}} \text{ mole/liter} = 0.000009 \text{ mole/liter}$$

$BaSO_4$ can be swallowed safely. The same calculation for a saturated barium hydroxide $(BaOH)_2$ solution gives $[Ba^{2+}] = 0.07$ mole/liter, more than a million times the concentration in the $BaSO_4$ solution. Barium hydroxide is a deadly poison.

Exercises

19. In the solution saturated with silver chloride (AgCl) at 25°C, $[Ag^+] = 1.34 \times 10^{-5}$. What is the value of K_{sp} for silver chloride?

20. A solution saturated with mercury (II) iodide (HgI_2) at 25°C contains 1.5×10^{-4} mole of dissolved solute per liter. What is the value of K_{sp} for HgI_2?

21. For lead sulfide (PbS), $K_{sp} = 1.3 \times 10^{-28}$ at 25°C. What are $[Pb^{2+}]$ and $[S^{2-}]$ in a saturated solution of PbS?

22. For silver sulfate (Ag_2SO_4), $K_{sp} = 1.6 \times 10^{-5}$ at 25°C. Solid Ag_2SO_4 is added to a 1-*M* solution of sodium sulfate, Na_2SO_4. What is $[Ag^+]$ in the solution when equilibrium is reached? Compare this value with that for $[Ag^+]$ in a saturated solution of Ag_2SO_4, and determine whether Le Chatelier's principle applies.

7.6 Summary Guide Chemical equilibrium exists in a closed, isothermal, isopiestic system when all phases possible are present and the rates of a chemical reaction and its reverse are the same. Equilibrium constants K, K_p, and K_{sp} are numerical measures of the position of equilibrium, and the value of the constant for a given reaction depends only on the temperature at which equilibrium is established. In writing the equations for equilibrium constants, the concentrations (or partial pressures of gases) of the products go in the numerator, those of reactants go in the denominator, and each concentration (or partial pressure) is raised to the power of the number before its formula in the balanced chemical equation for the equilibrium.

There is an infinite number of sets of equilibrium participant concentrations (or partial pressures) for any equilibrium, and—provided only that the temperature of the equilibrium system remains unchanged—any set of equilibrium concentrations gives the same value for the equilibrium constant for a reaction when substituted into the equation for the equilibrium constant for the reaction. The equilibrium equations for K and K_{sp} use species molarities and are used for equilibria in solutions. The equilibrium constant equation for K_p uses partial pressures and is used for reactions involving gases.

The reaction quotient Q has the same form as the equilibrium constant equation for a reaction, but its value need not be based on equilibrium concentrations of participants. The reaction quotient is used to predict the reactions that occur on return to equilibrium after the concentrations (or partial pressures) of the participants in an equilibrium are altered from their equilibrium values by adding or removing chemicals from the system, or, with reaction of gases, by changing the volume containing the equilibrium reaction. Removing products from an equilibrium shifts the position of the equilibrium to the product side. Adding reactants has the same effect. A total pressure increase on a gaseous equilibrium produces a shift in position to the side having fewer molecules. If there are the same numbers of molecules on both sides of the equation, total pressure increase is without effect on the position.

If, for an equilibrium, the reaction from left to right is endothermic (heat-absorbing), an increase in the temperature of the equilibrium increases the value of the equilibrium constant and shifts the position of the equilibrium to the right-hand (product) side. If the reaction from left to right in an equilibrium is exothermic (heat-emitting), temperature increase decreases the value of the equilibrium constant and shifts the position to the left-hand (reactant) side.

Equilibria are involved in most of life's reactions, in the production of chemicals for human use, and in reactions that alter our environment. Knowing about the effects of changing conditions on the positions of equilibria makes it possible for us to influence beneficient equilibria to our advantage, and to see what must be done to control equilibria that prove harmful to us.

Exercises

23. Hemoglobin (HGb), the oxygen-transporting protein in red blood cells, reacts with oxygen gas in the lungs to form oxyhemoglobin ($HGbO_2$) according to the equilibrium

$$HGb(aq) + O_2(g) \rightleftarrows HGbO_2(aq)$$

When a solution of hemoglobin in water is exposed to O_2 at a pressure of 37 torr, the concentrations of HGb and $HGbO_2$ at equilibrium are equal. The average value of p_{O_2} in the lungs is 100 torr. Will $[HGbO_2]/[HGb]$ be greater or less than 1 in blood leaving the lungs for the general circulation? (Compare Exercise 20, Chapter 5.)

24. The reaction $C(s) + H_2O(g) \rightleftarrows CO(g) + H_2(g)$ is used industrially to prepare a mixture of carbon monoxide and hydrogen called "producer gas," which—after more hydrogen is added—can be catalytically converted to methanol (methyl alcohol, CH_3OH):

$$CO(g) + 2H_2(g) \xrightarrow{\text{catalyst}} CH_3OH(l)$$

Producer gas can be made by using steam, with coal (even left underground—see Section 13.7) as the carbon source. Methanol made this way is a possible substitute for gasoline as auto fuel when petroleum supplies run out.

For the reaction making producer gas, $K = 1.05 \times 10^{-16}$ at 25°C, and the reaction is strongly endothermic ($\Delta H = 31.4$ kcal) from left to right as written. Give the equilibrium constant equation for K_p for the producer gas equilibrium and predict what conditions of temperature and pressure would favor the generation of producer gas.

For the reaction forming methanol, $K = 1.14 \times 10^5$ at 25°C. Would it be *economically* desirable to carry out the reaction under high pressure?

25. The distribution law is a simple example of an equilibrium law that can be stated "for dilute solutions, the ratio of the molar concentrations of the solute in the two phases of two immiscible liquids in contact with each other has a constant value at a given temperature." For the solute I_2 and the immiscible liquids carbon tetrachloride (CCl_4) and water, the law is

$$K_{\text{dist}} = \frac{[I_2]_{CCl_4} \text{ layer}}{[I_2]_{H_2O} \text{ layer}} = 85 \qquad \text{at } 25°C$$

Assuming that the volumes of the CCl_4 and H_2O layers (phases) are equal in an experiment in which the total amount of I_2 in the system is 0.100 g, calculate the weights of I_2 in the CCl_4 and H_2O layers.

26. In the process of extraction, a solution of a solute in one liquid (A) is shaken with a second, immiscible liquid (B) until equilibrium is established (see Exercise 25). The solution of the solute in liquid B is separated (using a separatory funnel), and the process of adding B, shaking, and separating is repeated several times, pooling the solutions of the solute in solvent *B* obtained. By evaporating the solvent *B* from the pooled solution, the solute is recovered. Suppose for a solute and immiscible solvent system, $K_{\text{dist}} = 1$ and equal volumes of the immiscible liquids are used throughout the extraction. What fraction of the original amount of solute will remain in solution *A* after the first extraction with *B*? After the second? After the *n*th? In principle, how many extractions would be required to remove all of the solute from liquid *A*? Extraction is used extensively to remove biochemicals from the aqueous solutions containing them in living systems.

Key Words

Catalysts substances (or species) that increase the rate at which equilibrium is reached without themselves appearing as reactants or products in the chemical equation for the equilibrium.

Chemical equilibrium the balance or equality of the rates of opposed chemical reactions at the molecular level with relation to a specific chemical equation.

Closed system a system that matter can neither enter nor leave.

Common ion effect the decrease in the solubility of an ionic solute produced by adding a soluble electrolyte that provides a high concentration of one of the ions of the solute.

Equilibrium constant (K) a numerical value calculated by definite rules using the concentrations of the participants of a reaction at equilibrium.

Equilibrium constant in partial pressures (K_p) an equilibrium constant calculated for a reaction of gases but using partial pressures of the participating gases instead of their concentrations.

Esterification a reaction in which the elements of water are eliminated between an acid and an alcohol to form an ester.

Hydrolysis breaking up by reaction with water; usually refers to the reaction of water with an ester to produce the alcohol and acid from which the ester was formed, the reverse of esterification.

Isopiestic system a system in which the pressure is everywhere the same.

Isothermal system a system in which the temperature is everywhere the same.

Le Chatelier's principle an attempt to change the conditions (temperature, pressure, concentrations) of an equilibrium results in a shift in the position of equilibrium in the direction that tends to offset the attempted change.

Reaction quotient (Q) the numerical value that results from putting nonequilibrium values of participant concentrations or partial pressures into the expression for an equilibrium constant.

Saturated solution the solution in equilibrium with excess solute.

Solubility product constant (K_{sp}) the equilibrium constant for the equilibrium of a solute with its saturated solution; when written the concentration of the solid solute is omitted.

Suggested Readings

"Auto Engines of the Future," *Changing Times,* March 1974, pp. 37–40.

Chase, R. L., "Smog and Its Effects," in Wilson, J. D., and Baker, S. D., *Physical Science: Readings on the Environment,* Lexington, Mass., D. C. Heath and Co., 1974, pp. 22–29.

"Cooking to Save Your Heart," Condensed from "The American Heart Association Cookbook," *Reader's Digest,* Sept. 1973, pp. 241–259.

Green, David E., "The Metabolism of Fats," *Bio-Organic Chemistry,* San Francisco, W. H. Freeman and Co., 1968, pp. 242–246.

Klimisch, R. L., and Taylor, K. C., "Exhaust Catalysts: Appropriate Conditions for Comparing Platinum and Base Metal," *Science,* **179**:798–800 (Feb. 23, 1973).

Prud'homme, Robert K., "Automobile Emissions Abatement and Fuels Policy," *Am. Sci.,* **62**:191–199 (March–April 1974).

Tunney, John V., "We Must Have a Smog-Free Car by 1975," *Catalyst,* **1**(2):14–16 (Summer 1970).

chapter 8

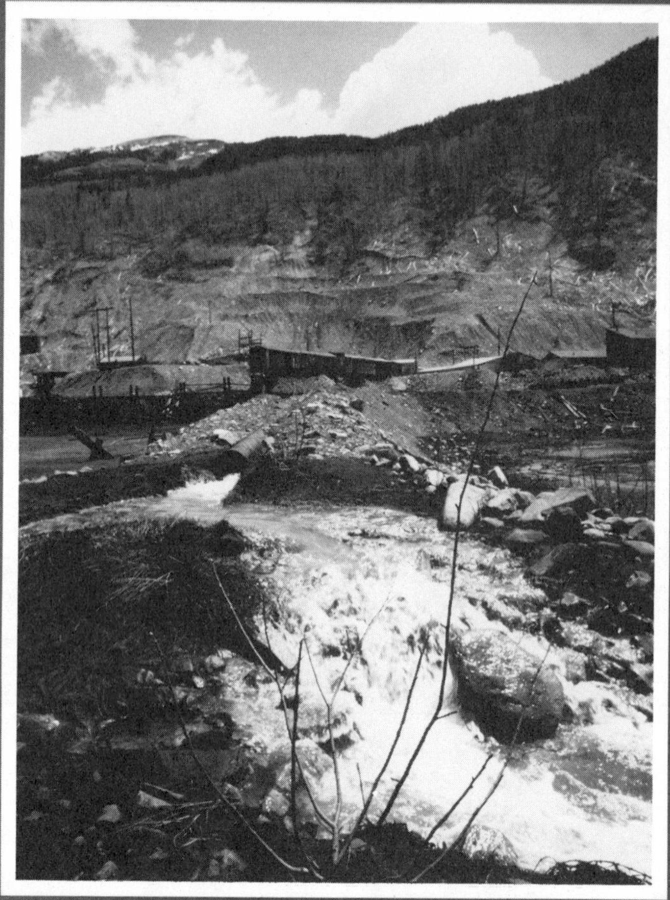

Sulfuric acid waste from a mine in Colorado pours into this mountain stream. *(Photo 23, EPA—DOCUMERICA—Bill Gillette.)*

Solutions of Acids and Bases

8.1 **Introduction** Solutions of acids or bases in water are among the most important types of solutions. Life itself goes on in a water (aqueous) solution of acids, bases, and other chemicals. (The general properties of solutions of acids were discussed in Section 6.4.)

Acids are proton donors; bases are proton acceptors. An acid, which we represent here by the generalized symbol HA, reacts with water to give its proton (H^+) to a water molecule, forming a hydronium ion (H_3O^+) and leaving its anion:

$$HA + H_2O \rightleftarrows H_3O^+ + A^-$$

Bases react with water according to the equation

$$B + H_2O \rightleftharpoons HB^+ + OH^-$$

or, if the base is an anion,

$$A^- + H_2O \rightleftharpoons HA + OH^-$$

These reactions are written as equilibria because they do not go to completion for most acids and bases. The more complete the reactions, the stronger the acid or base.

When an acid reacts with water, a water molecule accepts the proton, so water is acting as a base. In the reaction of a base with water, the water molecule donates a proton to the base (B or A^-), so water is acting as an acid. Water is adaptable—it acts as a base toward acids and as an acid toward bases. Now if water is both an acid and a base, it should react with itself, one molecule donating a proton to another:

$$\underset{\text{Acid}}{H_2O} + \underset{\text{Base}}{H_2O} \rightleftharpoons H_3O^+ + OH^-$$

It does react in this way but to only a very *tiny* extent, because it is a very weak acid and a very weak base. The position of equilibrium is so far to the left at 25°C that the concentrations of hydronium and hydroxide ions in pure water are only 1×10^{-7} mole/liter. This very low concentration of ions is shown by the very weak electrical conductivity of pure water. Nevertheless, this self-ionization or *autoionization* equilibrium of water is very important for the reason *that it exists in all water solutions, including those of other acids, bases, and salts.* And, according to Le Chatelier's principle, if the concentration of H_3O^+ or OH^- is increased by addition of other acids or bases, the position of the autoionization equilibrium will shift to the left. To see this, we write the equilibrium constant equation for the autoionization, $2H_2O \rightleftharpoons H_3O^+ + OH^-$:

$$K = \frac{[H_3O^+][OH^-]}{[H_2O]^2} = 3.26 \times 10^{-18} \qquad \text{at } 25°\text{C}$$

The suffix (aq) for species as in $H_3O^+(aq)$ will be omitted in this chapter to simplify writing equations. It is assumed here, unless otherwise specified, that all species are solvated by water molecules.

Although the *concentrations* of ions are very small, their number per liter is huge. Since one mole represents Avogadro's number (6.023×10^{23}) of particles, there are $1 \times 10^{-7} \times 6.023 \times 10^{23} = 6.023 \times 10^{16}$ of H_3O^+ and OH^- ions per liter of pure water. This number is more than ten million times the population of the earth.

If $[H_3O^+]$ is increased above 1×10^{-7} mole/liter by the addition of some acid such as HCl, $[OH^-]$ must go down to reestablish the constant value of K. Conversely, if you add a base such as Na^+OH^-, then $[OH^-]$ will increase and $[H_3O^+]$ must go down. To look at these changes quantitatively is the first objective in this chapter.

Exercises

1. Using the value $K = 3.26 \times 10^{-18}$ for the autoionization of water and the fact that in pure water $[H_3O^+] = [OH^-] = 1 \times 10^{-7}$, calculate $[H_2O]^2$ and $[H_2O]$ in pure water.

2. Water is not the only liquid that undergoes autoionization. Liquid ammonia (NH_3, bp $-33°C$) and alcohols such as methanol (CH_3—O—H) do also. Write the balanced equations for their autoionization.

3. Write the expression for K for the autoionizations in Exercise 2.

8.2 The Autoionization Constant (K_w) for Water and the pH Scale

Calculations using the autoionization equilibrium for water can be much simpler for you if you see that *however much the position of the equilibrium is shifted by addition of acids or bases, the already large concentration of water in the solution will be changed hardly at all.* In pure water at 25°C, $[H_2O] = 55.4$ moles/liter ☛ so even if 99.9% of the OH^- ions were converted to water molecules by the addition of an acid, the $[H_2O]$ would increase by the relatively tiny amount of 1×10^{-7} mole/liter. The addition of solutes to water will, of course, lower the water concentration from its value in pure water, but if the solution is 1 *M* or less, $[H_2O]$ will remain virtually constant in the range 54.5–55.4 moles/liter. Because the concentration of water does not really change much, we can multiply both sides of the equation for K by $[H_2O]^2$ to obtain a new constant, K_w, the *autoionization constant for water*:

A liter of water at 25°C weighs 997 g, and since the molar weight of water is 18, it follows that $[H_3O^+] = (997/18) = 55.4$ moles/liter.

$$K_w = K[H_2O]^2 = \frac{[H_3O^+][OH^-]}{\cancel{[H_2O]^2}} \times \cancel{[H_2O]^2} = [H_3O^+][OH^-]$$

or

$$K_w = [H_3O^+][OH^-]$$

Since $[H_3O^+] = [OH^-] = 1 \times 10^{-7}$ in pure water at 25°C,

$$K_w = [H_3O^+][OH^-] = (1 \times 10^{-7})(1 \times 10^{-7}) = 1 \times 10^{-14} \quad \text{at } 25°C$$

The importance of $K_w = 1 \times 10^{-14}$ is that *the product of the hydronium and hydroxide ion concentrations in* **any** *dilute water solution is 1×10^{-14} at 25°C.* No matter what acids, bases, or salts are dissolved to give the solution, $[H_3O^+]$ and $[OH^-]$ *must adjust* through equilibrium shifts so that their product keeps the constant value of 1×10^{-14}.

EXAMPLE

The contribution of H_3O^+ from the autoionization of water is negligible compared to that from HCl, and can be safely neglected.

If 0.25 mole of HCl gas is dissolved in 1 liter of water, the $[H_3O^+]$ in the solution will be 0.25 mole/liter because HCl is a *strong* acid (Section 6.4) and for all practical purposes reacts completely according to the equation $HCl(g) + H_2O \rightarrow H_3O^+ + Cl^-$. You can calculate the concentration of hydroxide ion in the solution as follows:

$$K_w = [H_3O^+][OH^-] = 0.25 \times [OH^-] = 1 \times 10^{-14}$$

$$[OH^-] = \frac{1 \times 10^{-14}}{0.25} = 4 \times 10^{-14}$$

Figure 8.1 shows how $[H_3O^+]$ and $[OH^-]$ change in such a way that their product is always 10^{-14}.

If an acid is *weak*, the $[H_3O^+]$ in its solution *will not be equal to the component molarity of the acid.* Acetic acid is a typical weak acid—in 1 *M* acetic acid, $[H_3O^+] = 0.004$, because the reaction $CH_3COOH + H_2O \rightleftarrows H_3O^+ + CH_3COO^-$ is far from complete. In such cases the $[H_3O^+]$ in the solution must be measured. Since measurement shows that $[H_3O^+]$ in 1 *M* acetic acid is 0.004 mole/liter, you calculate $[OH^-]$ in the solution as follows:

$$[H_3O^+][OH^-] = 1 \times 10^{-14}$$

$$0.004[OH^-] = 1 \times 10^{-14}$$

$$[OH^-] = 2.5 \times 10^{-12}$$

A similar situation is seen with *weak* bases. In Section 6.4 we saw that $[OH^-]$ in a 1.00-*M* solution of the weak base ammonia is 0.0042

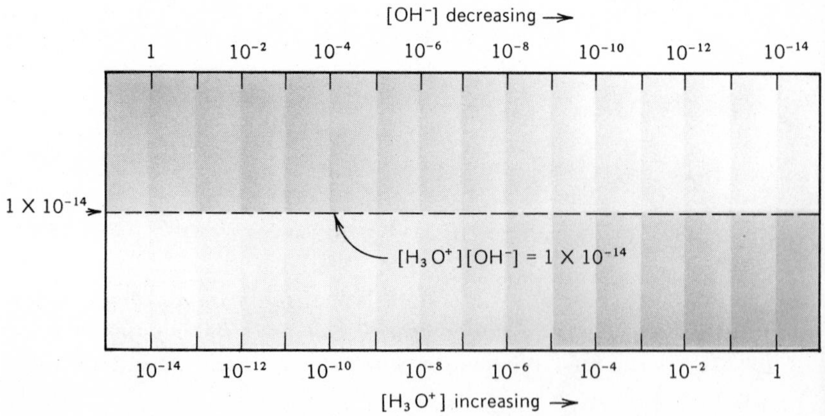

FIG. 8.1 Relation between hydronium and hydroxide ion concentrations in any aqueous solution at 20°C, and the constant value of their product at 1×10^{-14}.

mole/liter as a result of the equilibrium $NH_3 + H_2O \rightleftarrows NH_4^+ + OH^-$. You can calculate $[H_3O^+]$ in the ammonia solution as follows:

$$K_w = [H_3O^+][OH^-] = [H_3O^+][0.0042] = 1 \times 10^{-14}$$

$$[H_3O^+] = \frac{1 \times 10^{-14}}{0.0042} = 2.4 \times 10^{-12}$$

Summarizing, we can say that if the value of either $[H_3O^+]$ or $[OH^-]$ in any aqueous solution is known, we can calculate the value of the other by the equations

$$[H_3O^+] = \frac{K_w}{[OH^-]} = \frac{1 \times 10^{-14}}{[OH^-]}$$

$$[OH^-] = \frac{K_w}{[H_3O^+]} = \frac{1 \times 10^{-14}}{[H_3O^+]} \qquad \text{at } 25°\text{C}$$

You will soon see how valuable it is for understanding what is going on in a solution to be able to do this type of calculation.

Exercises

4. Pure sulfuric acid undergoes autoionization according to the equation $2H_2SO_4 \rightleftarrows H_3SO_4^+ + HSO_4^-$ and $K_{H_2SO_4} = [H_3SO_4^+][HSO_4^-] = 2.5 \times 10^{-4}$ at 25°C. What are the concentrations of $H_3SO_4^+$ and HSO_4^- in pure sulfuric acid? Would you expect pure sulfuric acid to be a better or poorer conductor of electricity than water? How would the conductivity of water change if you added some sulfuric acid to it? Explain your answers. (*Hint*: H_2SO_4 is a strong acid.)

5. The following data show how the value of K_w changes with temperature.

Temperature	K_w
0°C	0.11×10^{-14}
10°C	0.30×10^{-14}
25°C	1.00×10^{-14}
50°C	5.48×10^{-14}

Using Le Chatelier's principle, determine whether the reaction $H_3O^+ + OH^- \rightarrow 2H_2O$ is exothermic or endothermic.

The pH scale The hydronium ion concentration $[H_3O^+]$ of an aqueous solution is one of its most important quantitative properties. For example, in arterial blood $[H_3O^+]$ in health is maintained very close to 3.63×10^{-8} mole/liter. Even small departures from this value result in illness or death. The pH scale is a convenient, shorthand method for describing hydronium ion concentrations. It is used widely in

TABLE 8.1 Relation between hydronium ion concentration and pH

$[H_3O^+]$ *in mole/liter*	pH		*Example Solution*
10^0	0	↑	1.00 *M* HCl
10^{-1}	1		0.10 *M* HCl
10^{-2}	2		0.01 *M* HCl
10^{-3}	3	Acidic	
10^{-4}	4		⋮
10^{-5}	5		
10^{-6}	6	↓	
10^{-7}	7	Neutral	Pure H_2O
10^{-8}	8	↑	
10^{-9}	9		⋮
10^{-10}	10	Basic	
10^{-11}	11		
10^{-12}	12		0.01 *M* NaOH
10^{-13}	13		0.10 *M* NaOH
10^{-14}	14	↓	1.00 *M* NaOH

chemistry, particularly biochemistry, and in medicine. The relation between hydronium ion concentration and pH is shown in Table 8.1. The useful range of pH is from 0 to 14, and the pH is simply the power of ten in the hydronium ion concentration but *taken with positive sign.* Thus we have the defining equation

$$[H_3O^+] = 10^{-pH}$$

and the *larger* the pH the *smaller* the $[H_3O^+]$. For pure water pH = 7, so $[H_3O^+] = 10^{-7}$. Solutions with pH = 7 are said to be *neutral.* Solutions with pH greater than 7 are *basic* or *alkaline*; if the pH is less than 7, the solution is *acidic.* Arterial blood with $[H_3O^+] = 3.63 \times 10^{-8}$ is slightly on the alkaline side, since its pH is between 7 and 8, actually 7.44, as will be calculated shortly.

The relation between $[H_3O^+]$ and pH can be expressed in another way. First you take the logarithm ☛ to base 10 of both sides of the defining equation. This gives

If you're unfamiliar with the use of logarithms, consult Appendix 1.

$$\log_{10} [H_3O^+] = \log_{10} 10^{-pH}$$

The $\log_{10}$ of a number is the power to which 10 must be raised to generate the number, so $\log_{10} 10^a = a$, and $\log_{10} 10^{-pH} = -pH$, giving the equation

$$\log_{10} [H_3O^+] = -pH \qquad \text{or} \qquad pH = -\log_{10} [H_3O^+]$$

EXAMPLE

The subscript 10 in $\log_{10}$ will be omitted from now on with the understanding that "log" means $\log_{10}$.

You can see the convenience of the pH scale by calculating the pH of arterial blood in which $[H_3O^+] = 3.63 \times 10^{-8}$ mole/liter. You write

$$pH = -\log [H_3O^+] = -\log (3.6 \times 10^{-8})$$

Now, the log of the product (3.6×10^{-8}) is the sum of the logs of its factors 3.6 and 10^{-8}; that is, $\log (3.6 \times 10^{-8}) = \log 3.6 + \log 10^{-8}$, so

$$pH = -[\log 3.63 + \log 10^{-8}]$$

The log of 10^{-8} is simply -8, just as $\log 10^2 = 2$, so

$$pH = -[\log 3.63 + (-8)] = 8 - \log 3.63$$

From a logarithm table you find $\log 3.63 = 0.56$, so for arterial blood,

$$pH = 8.00 - 0.56 = 7.44$$

The reverse calculation, determining $[H_3O^+]$ from the value of pH, is most easily done by using the equation $[H_3O^+] = 10^{-pH}$. For example, venous blood has pH 7.37 and is thus slightly *more acidic* (lower pH) than arterial blood with pH = 7.44 (for reasons explained in Section 8.5). Suppose you want to compare $[H_3O^+]$ in arterial and venous bloods. Write

$$[H_3O^+] = 10^{-pH} = 10^{-7.37}$$

The problem is to find the numerical value of 10 raised to the -7.37 power. You can't do this by using log tables directly, because logarithms are given there only for numbers 1 or greater, and $10^{-7.37}$ is between 10^{-7} and 10^{-8}, and so less than 1. You have to separate the exponent into a negative *integer* and a positive decimal fraction by expressing it as $-7.37 = (-8.00 + 0.63)$. Now you have the much more manageable expression

$$[H_3O^+] = 10^{(-8.00+0.63)} = 10^{-8.00} \times 10^{0.63} .$$

Looking in the body of the log table (Appendix 6), you find that 0.63 is the logarithm of 4.27; in other words, $10^{0.63} = 4.27$. For venous blood, then,

$$[H_3O^+] = 4.27 \times 10^{-8} \text{ mole/liter}$$

Comparing this value with the smaller value of 3.63×10^{-8} for $[H_3O^+]$ in arterial blood, you can see that venous blood is indeed slightly more acidic than arterial blood, though both are on the alkaline side of pH = 7.

These sample calculations—pH from $[H_3O^+]$ and $[H_3O^+]$ from pH—are the only kinds of pH calculations that you will encounter. It is important for what follows to be able to do them easily.

Exercises

6. Convert the following values for $[H_3O^+]$ into pH: 10^{-3}, $10^{-5.5}$, 10^{-9}, 1.00, 10.00.

7. Convert the following pH values into $[H_3O^+]$: 1.50, 5.90, 7.20, 0, −1.00.

8. One mole of $Ba(OH)_2$ when dissolved in water yields two moles of OH^-. Calculate $[OH^-]$ and pH for a 0.01-*M* solution of barium hydroxide.

8.3 Strength of Acids—K_a You have had several examples of the way the value of an equilibrium constant measures how far a reaction goes before coming to equilibrium. For the equilibrium of an acid reacting with water according to the general equation

$$HA + H_2O \rightleftarrows H_3O^+ + A^-$$

the equilibrium constant expression is

$$K = \frac{[H_3O^+][A^-]}{[HA][H_2O]}$$

and the value of K is a numerical measure of the *strength* of the acid—that is, of its ability to donate protons to water molecules or other bases. Large values of K mean powerful proton donors (stronger acids) and smaller values of K correspond to weaker acids.

In dilute solutions (1 *M* or less) of weak acids the concentration of water will be changed hardly at all by the small amount of the reaction $HA + H_2O \rightleftarrows H_3O^+ + A^-$. Therefore we can multiply both sides of the equilibrium constant equation for K by the constant concentration of water to obtain K_a, *the dissociation constant for the acid.*

Very strong acids—HCl and $HClO_4$, for example—react so completely with water that $[HA] \approx 0$ at equilibrium. If $[HA] = 0$, then

$$K = \frac{[H_3O^+][A^-]}{0[H_2O]^2} = \infty$$

(infinity because any number divided by zero has the value infinity).

$$K_a = K[H_2O] = \frac{[H_3O^+][A^-]}{[HA]\cancel{[H_2O]}} \times \cancel{[H_2O]} = \frac{[H_3O^+][A^-]}{[HA]}$$

or

$$K_a = \frac{[H_3O^+][A^-]}{[HA]}$$

The numerical value of K_a is a measure of how well HA acts as a donor of protons to water molecules—a measure of the *strength* ☛ of the acid. Table 8.2 contains the K_a values for some typical *weak* acids. Acetic acid (CH_3COOH), which donates only the proton attached to oxygen, is a weak acid with $K_a = 1.8 \times 10^{-5}$. The practical range of K_a values is roughly from 10^{-2} to 10^{-12}; the values for a majority of acids fall within this range. Figure 8.2 compares strong and weak acids. Photo 23 shows an environmental impact of the strong acid H_2SO_4.

Do not confuse the strength of an acid as a proton donor with the *concentration* of its solution. A concentrated solution of acetic acid is still a solution of a weak acid.

The strength of an acid, given by its K_a value, is one of its more important properties. How are these values found? Looking at the defining equation

$$K_a = \frac{[H_3O^+][A^-]}{[HA]}$$

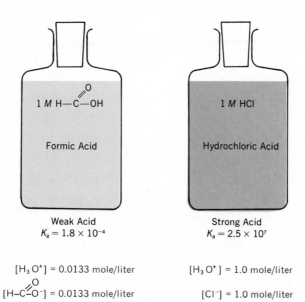

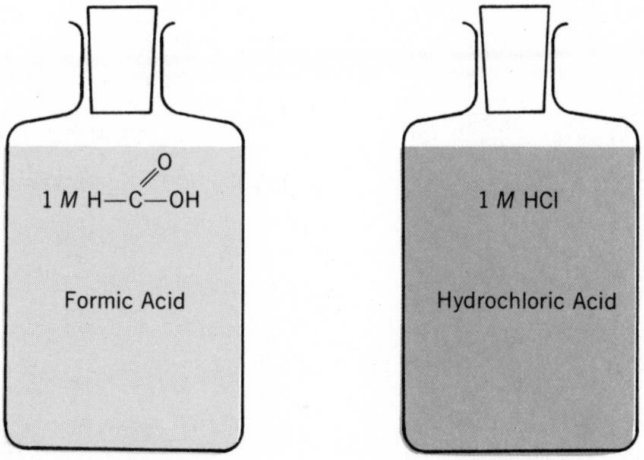

Weak Acid
$K_a = 1.8 \times 10^{-4}$

Strong Acid
$K_a = 2.5 \times 10^{7}$

$[H_3O^+]$ = 0.0133 mole/liter	$[H_3O^+]$ = 1.0 mole/liter
$[H\text{-}C(=O)\text{-}O^-]$ = 0.0133 mole/liter	$[Cl^-]$ = 1.0 mole/liter
$[HC(=O)\text{-}OH]$ = 0.9867 mole/liter	$[HCl]$ = 0.0 mole/liter
$H\text{-}C(=O)\text{-}OH + H_2O \rightleftharpoons H_3O^+ + H\text{-}C(=O)\text{-}O^-$	$HCl + H_2O \rightleftharpoons H_3O^+ + Cl^-$

FIG. 8.2 Comparison of 1 *M* solutions of weak and strong acids.

TABLE 8.2 Equilibrium constants for some acids and their conjugate bases (aqueous solutions, ~0.1 *M*, 25°C)[a]

Strongest acid ↑ Weakest base ↓

Acids		K_a	pK_a	*Conjugate Bases*	K_b
Perchloric	$HClO_4$	—	—	ClO_4^-	—
Hydrochloric	HCl	—	—	Cl^-	—
Sulfuric	H_2SO_4	—	—	HSO_4^-	—
Nitric	HNO_3	$> 10^2$	$<(-2)$	NO_3^-	$< 10^{-16}$
Hydronium ion	H_3O^+	55.6	−1.75	H_2O	1.80×10^{-16}
Sulfurous	$[H_2SO_3]$	1.54×10^{-2}	1.81	HSO_3^-	6.49×10^{-13}
Bisulfate ion	HSO_4^-	1.20×10^{-2}	1.92	SO_4^{2-}	8.33×10^{-13}
Phosphoric	H_3PO_4	7.59×10^{-3}	2.12	$H_2PO_4^-$	1.32×10^{-12}
Nitrous	$[HNO_2]$	5.13×10^{-4}	3.29	NO_2^-	1.95×10^{-11}
Formic	$HCOOH$	1.77×10^{-4}	3.75	$HCOO^-$	5.65×10^{-11}
Acetic	CH_3COOH	1.81×10^{-5}	4.74	CH_3COO^-	5.53×10^{-10}
Methyl red ion	$HMeR^+$	6.31×10^{-6}	5.20	MeR	1.59×10^{-9}
Carbonic	$[H_2CO_3]$	4.27×10^{-7}	6.37	HCO_3^-	2.34×10^{-8}
Bisulfite ion	HSO_3^-	1.23×10^{-7}	6.91	SO_3^{2-}	8.13×10^{-8}
Bromthymol blue	HIn	1.00×10^{-7}	7.00	In^-	1.00×10^{-7}
Hydrosulfuric	H_2S	8.92×10^{-8}	7.05	HS^-	1.12×10^{-7}
Dihydrogen phosphate ion	$H_2PO_4^-$	6.17×10^{-8}	7.21	HPO_4^{2-}	1.62×10^{-7}
Hypochlorous	$[HOCl]$	2.95×10^{-8}	7.53	OCl^-	3.39×10^{-7}
Phenolphthalein	HPhth	7.94×10^{-10}	9.10	$Phth^-$	1.26×10^{-5}
Ammonium ion	NH_4^+	5.63×10^{-10}	9.25	NH_3	1.77×10^{-5}
Hydrocyanic	HCN	4.90×10^{-10}	9.31	CN^-	2.04×10^{-5}
Bicarbonate ion	HCO_3^-	4.79×10^{-11}	10.32	CO_3^{2-}	2.09×10^{-4}
Hydrogen phosphate ion	HPO_4^{2-}	2.14×10^{-13}	12.67	PO_4^{3-}	4.67×10^{-2}
Hydrosulfate ion	SH^-	1.20×10^{-13}	12.92	S^{2-}	8.33×10^{-2}
Water	H_2O	1.80×10^{-16}	15.74	OH^-	55.6
Ammonia	NH_3	10^{-33}	33	NH_2^-	10^{19}

Weakest acid Strongest base

[a] A dash (—) indicates that in water the acid is too strong, and the conjugate base is too weak for K_a and K_b to have defined values. The brackets signify that the acid is unstable. The formulas for indicators are abbreviated $HMeR^+$, HIn, and HPhth; for them $pK_a = pK_{In}$.

you can see that the numerical value of K_a can be calculated if we measure the equilibrium concentrations of H_3O^+, A^-, and HA in a solution of the acid. Consider acetic acid (CH_3COOH). The equilibrium for it is

$$CH_3COOH + H_2O \rightleftarrows H_3O^+ + \underset{\text{Acetate ion}}{CH_3COO^-}$$

and

$$K_a = \frac{[H_3O^+][CH_3COO^-]}{[CH_3COOH]}$$

In 1.00-M CH_3COOH solution the accurate concentrations at equilibrium at 25°C are found to be

$$[H_3O^+] = 0.0042$$
$$[CH_3COO^-] = 0.0042$$
$$[CH_3COOH] = 0.9958$$

Substituting these values into the equation for K_a gives

$$K_a = \frac{[H_3O^+][CH_3COO^-]}{[CH_3COOH]} = \frac{(0.0042)(0.0042)}{0.9958} = 1.8 \times 10^{-5}$$

Exercises

9. Formic acid, $H—\overset{\overset{\displaystyle \ddot{O}:}{\|}}{C}—\ddot{O}—H$, in the sting of certain kinds of ants, has one acidic hydrogen, the one attached to oxygen. Write the balanced equation for the reaction of formic acid with water. The pH of a 1.00-M formic acid solution at 25°C is 1.88. To what $[H_3O^+]$ does this correspond? Is formic acid a weak acid?

10. The value of K_a for an acid HA is 10^{-4}, while that for another acid HB is 10^{-8}. What reaction, if any, would you expect to occur on combining 1.00-M solutions of HB and Na^+A^-, the sodium salt of acid HA? What reaction if any would occur on combining 1.00-M solutions of HA and Na^+B^-? (*Hint*: decide which acid is the better proton donor.)

In practice, to evaluate K_a you don't have to measure *directly* all of the concentrations that appear in the equation for K_a. If the component molarity (C) of the solution of the acid is known from the way it was prepared, and if only $[H_3O^+]$ is measured, the other concentrations can be calculated. To see how this is done, consider a solution of lactic acid (the "sour milk" acid; $CH_3CHOHCOOH$, M = 90.10) made by dissolving 9.01 g of the acid in water and diluting to 1.00 liter. The solution is (9.01/90.10) = 0.10 M lactic acid. The $[H_3O^+]$ in the solution is measured ☛ and found to be 0.00875 mole/liter. Using the "before equilibrium/at equilibrium" argument we used in Section 7.3, we write

Most conveniently with a "pH meter," the principle of which is discussed in Exercise 10.21.

Molar concentration (mole/liter)

	$CH_3CHOHCOOH + H_2O \rightleftarrows$	$H_3O^+ +$	$CH_3CHOHCOO^-$
Before equilibrium	0.100	10^{-7}	0
At equilibrium	0.100 − 0.00875	0.00875	0.00875

The concentrations at equilibrium are obtained as follows: H_3O^+ and $CH_3CHOHCOO^-$ are formed in equal amounts according to the equation, and the $[H_3O^+]$ of 10^{-7} from the autoionization of water is so small compared to 0.00875 from the lactic acid that water's contribution can be neglected. ☛ The $[CH_3CHOHCOOH]$ at equilibrium is the initial concentration of 0.100 *M* minus that which has reacted to form H_3O^+ and lactate ion, $CH_3CHOHCOO^-$. Inserting the equilibrium concentrations into the expression for K_a and doing the arithmetic, you get the value of K_a:

This can be done in any practical case of this type.

$$K_a = \frac{[H_3O^+][CH_3CHOHCOO^-]}{[CH_3CHOHCOOH]} = \frac{(0.00875)(0.00875)}{(0.10000 - 0.00875)}$$

$$K_a = \frac{(0.00875)^2}{0.09125} = 8.4 \times 10^{-4}$$

Comparing this value with $K_a = 1.8 \times 10^{-5}$ found earlier for acetic acid shows that lactic acid is the stronger of the pair by a factor of $(8.4 \times 10^{-4})/(1.8 \times 10^{-5})$ or 47.

Of what use is the value of K_a for an acid, other than for acid strength comparisons? Because it is an equilibrium constant, it provides you with a quantitative description of *any* solution of the acid.

EXAMPLE

When glucose is oxidized in muscle cells to provide energy for muscular activity, lactic acid is an intermediate on the way to the final products CO_2 and H_2O. Suppose that in a leg muscle enough lactic acid accumulated during strenuous exercise—such as a 100-meter dash—to produce a 0.01-*M* solution. What $[H_3O^+]$ would this provide? What would be the local pH? The calculation of $[H_3O^+]$ reverses the pattern just used to obtain the value of K_a. Letting *y* be the unknown $[H_3O^+]$ at equilibrium, we see that the "before and after" pattern becomes ☛

Again you can ignore the $[H_3O^+]$ from the autoionization of water.

Molar concentration (mole/liter)

	$CH_3CHOHCOOH + H_2O \rightleftarrows$	$H_3O^+ +$	$CH_3CHOHCOO^-$
Before equilibrium	0.01	10^{-7}	0
At equilibrium	0.01 − *y*	*y*	*y*

Since the value of K_a for lactic acid is 8.4×10^{-4}, you can write

$$K_a = \frac{[H_3O^+][CH_3CHOHCOO^-]}{[CH_3CHOHCOOH]} = \frac{y^2}{0.01 - y} = 8.4 \times 10^{-4}$$

so

$$y^2 = 8.4 \times 10^{-6} - (8.4 \times 10^{-4})\, y$$

or, rearranging,

$$y^2 + (8.4 \times 10^{-4})\, y - 8.4 \times 10^{-6} = 0$$

This is a quadratic equation that can be solved by using the quadratic formula as we did in Section 7.3. If you do this, you find

$$y = [H_3O^+] = 2.51 \times 10^{-3} \text{ mole/liter}$$

This hydrogen ion concentration is about 100,000 times larger than the value of 3.63×10^{-8} mole/liter for arterial blood. If allowed to persist, this concentration would quickly damage or kill body cells bathed in it. ☛ For practice let's calculate the pH corresponding to $[H_3O^+] = 2.51 \times 10^{-3}$:

How the body prevents this is described in Section 8.5.

$$\text{pH} = -\log[H_3O^+] = -\log(2.51 \times 10^{-3})$$

$$\text{pH} = 3.00 - \log 2.51$$

From the log table, log 2.51 = 0.40, so

$$\text{pH} = 3.00 - 0.40 = 2.60$$

This pH is much *less* than the value of 7.44 for arterial blood. Again note that the *smaller the pH the larger the* $[H_3O^+]$. ⊸

Approximate methods Generalizing from the examples of acetic and lactic acid, and letting C_{HA} be the *component* molarity of the acid HA in the solution, we can write the following equation relating $[H_3O^+]$, $[A^-]$, C_{HA}, and K_a for the acid:

$$K_a = \frac{[H_3O^+][A^-]}{[HA]} = \frac{[H_3O^+][A^-]}{C_{HA} - [H_3O^+]}$$

or, since $[H_3O^+] = [A^-]$,

$$K_a = \frac{[H_3O^+]^2}{C_{HA} - [H_3O^+]}$$

This equation can be used to evaluate K_a if C_{HA} and $[H_3O^+]$ are known, but instead it is most often used in the reverse way to calculate the $[H_3O^+]$ in a solution of known C_{HA} of an acid for which the K_a value has already been found. For another example, suppose you want to calculate $[H_3O^+]$ in a solution of 0.10 *M* HA when K_a for HA is 1×10^{-5}. You write

$$K_a = 1 \times 10^{-5} = \frac{[H_3O^+]^2}{0.10 - [H_3O^+]}$$

which reduces on carrying out the arithmetic to

$$[H_3O^+]^2 + (1 \times 10^{-5}[H_3O^+]) - 1 \times 10^{-6} = 0$$

This quadratic equation could be solved by using the quadratic formula as before, but to simplify the calculations, let's *make the simplifying assumption* that $[H_3O^+]$ is much less than 0.10 mole/liter. This is reasonable if the acid HA is weak, as it is in this case, where the value of K_a is very small. Making the assumption means that $0.10 - [H_3O^+] \approx 0.10$, and you can write the much easier approximate equation

$$K_a = 1 \times 10^{-5} \approx \frac{[H_3O^+]^2}{0.10}$$

or

$$[H_3O^+]^2 \approx 0.10 \times 1 \times 10^{-5} = 1 \times 10^{-6}$$

and

$$[H_3O^+] \approx \sqrt{1 \times 10^{-6}} = 1 \times 10^{-3} \text{ mole/liter} = 0.001 \text{ mole/liter}$$

Checking our assumption, we find that $0.10 - [H_3O^+]$ is $(0.10 - 0.001) = 0.099$, which is very nearly equal to 0.100. This assumption, generally expressed as $(C_{HA} - [H_3O^+]) \approx C_{HA}$, can usually be made to simplify the arithmetic. It always works for solutions of acids with K_a in the range 10^{-4} to 10^{-8}, where C_{HA} is in the range 1 to 10^{-3} mole/liter. Fortunately, most solutions of weak acids that we use meet these requirements, and for them the simple formula

$$[H_3O^+] \approx \sqrt{C_{HA} \times K_a}$$

can be used to calculate hydronium ion concentrations. Suppose you want the value of $[H_3O^+]$ in a 0.01-*M* solution of an acid for which $K_a = 1 \times 10^{-6}$:

$$[H_3O^+] \approx \sqrt{0.01 \times 1 \times 10^{-6}} = \sqrt{1 \times 10^{-8}} = 1 \times 10^{-4} \text{ mole/liter}$$

(Checking the underlying assumption that $C_{HA} - [H_3O^+] \approx C_{HA}$, we see that $0.01 - 1 \times 10^{-4} = 0.01 - 0.0001 \approx 0.01$.)

We have been using the symbol HA to represent an acid or proton donor. Charged species (ions) may also be acids. A simple example is the dihydrogen phosphate ion $H_2PO_4^-$ present in solutions of salts such as $Na^+H_2PO_4^-$ (sodium dihydrogen phosphate). The equilibrium is $H_2PO_4^- + H_2O \rightleftarrows H_3O^+ + HPO_4^{2-}$ and

$$K_a = \frac{[H_3O^+][HPO_4^{2-}]}{[H_2PO_4^-]} = 6.17 \times 10^{-8} \qquad \text{at } 25°C$$

From the value of K_a we see that $H_2PO_4^-$ is a rather weak acid. Other examples of ionic acids and their K_a values may be found in Table 8.2.

Exercises

11. For HSO_3^-, the hydrogen sulf*i*te ion ("bisulf*i*te ion"), $K_a = 1.23 \times 10^{-7}$. Write the equation for the reaction of HSO_3^- with water and calculate the approximate values of $[H_3O^+]$ and pH for a 0.0813-*M* solution of $Na^+HSO_3^-$.

12. For the solution in Exercise 11, what are the approximate values of $[OH^-]$, $[HSO_3^-]$, and $[SO_3^{2-}]$?

13. For HSO_4^-, the hydrogen sulf*a*te ion ("bisulf*a*te ion"), $K_a = 1.20 \times 10^{-2}$. Which solution, 1.00 *M* $Na^+HSO_3^-$ or 1.00 *M* $Na^+HSO_4^-$, would have the lower pH? (Refer to Exercise 11 for data.)

Acid strength—pK_a We have seen the convenience of the pH scale for expressing the $[H_3O^+]$ in aqueous solution. A similar (logarithmic) scale is used, particularly in biochemistry, to give the strengths of acids. It is called the "pK_a scale," and the definition of the pK_a value for an acid is just like the definition of pH except that $[H_3O^+]$ is replaced by the K_a for the acid. The equations are

$$pK_a = -\log K_a \qquad \text{or} \qquad K_a = 10^{-pK_a}$$

just as

$$pH = -\log [H_3O^+] \qquad \text{or} \qquad [H_3O^+] = 10^{-pH}$$

If $K_a = 1 \times 10^{-4}$, then $pK_a = -\log(1 \times 10^{-4}) = -(-4) = 4$. As with $[H_3O^+]$ and pH, the relation between acid strength and pK_a is an *inverse one—the **larger** the* pK_a *value for an acid, the **weaker** the acid.* Table 8.2 lists the pK_a values along with K_a's for some typical acids. In working with the pK_a scale the solvent is assumed to be water, although it can be extended to other solvents. Calculations involving pK_a and K_a values exactly parallel those for pH and $[H_3O^+]$. Two examples will show you how to do them.

EXAMPLE Oxyhemoglobin is the compound responsible for the red color and oxygen-carrying capacity of arterial blood. It is present in the red blood cells (Photo 24) in a concentration of about 5×10^{-3} mole/liter. Although oxyhemoglobin is a complex molecule $(C_{738}H_{1156}FeN_{203}O_{208}S_2)_4$, it is an acid with $pK_a = 6.7$. Suppose in a study of blood chemistry you need the value of $[H_3O^+]$ in a 5×10^{-3} mole/liter solution of oxyhemoglobin. Using $HGbO_2$ to stand for the oxyhemoglobin molecule, we see that the equilibrium producing H_3O^+ ions is

$$HGbO_2(aq) + H_2O(l) \rightleftarrows H_3O^+(aq) + GbO_2^-(aq)$$

and

$$K_a = \frac{[H_3O^+][GbO_2^-]}{[HGbO_2]}$$

To calculate $[H_3O^+]$ you need the value of K_a. To get it you use the equation $K_a = 10^{-pK_a}$. Substituting the value 6.7 for the pK_a of oxyhe-

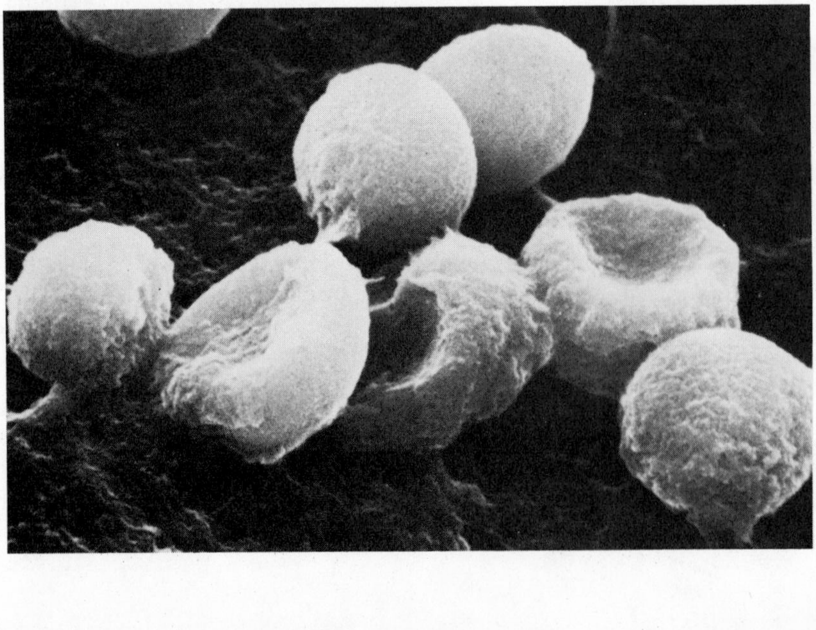

PHOTO 24 A scanning electron micrograph of human red blood cells. Note that the cells are not shaped like spheres but are flat and disc-shaped. This shape may shorten the distance oxygen must diffuse through the cell before it is held by hemoglobin molecules for transport through the body. (Dr. Norman Hodgkin, Micrographics, Newport Beach, California.)

moglobin into the equation and using the same method as for calculating $[H_3O^+]$ from pH gives

$$K_a = 10^{-pK_a} = 10^{-6.7} = 10^{-7.0} \times 10^{0.30}$$

Looking in the body of the log table, you find that 0.30 is the logarithm of 2, that is, $10^{0.30} = 2$, so for oxyhemoglobin

$$K_a = 2 \times 10^{-7}$$

Using the approximate equation for calculating $[H_3O^+]$ in solutions of weak acids gives $[H_3O^+]$ in a 5×10^{-3} mole/liter solution:

$$[H_3O^+] = \sqrt{C \times K_a} = \sqrt{5 \times 10^{-3} \times 2 \times 10^{-7}} = 3.2 \times 10^{-5} \text{ mole/liter}$$

At the body cells, oxyhemoglobin gives up its oxygen for oxidation of glucose to produce the energy for life, becoming converted to hemoglobin (HGb) in the process.

$$HGbO_2(aq) \rightarrow HGb(aq) + O_2(aq) \qquad \text{(to cells)}$$

Hemoglobin is a *weaker* acid than oxyhemoglobin. For hemoglobin,

$$K_a = \frac{[H_3O^+][Gb^-]}{HGb} = 1.2 \times 10^{-8}$$

and $pK_a = 8.00 - \log 1.2 = 7.9$, by the equation $pK_a = -\log K_a$. ⊸

Exercises

14. Ammonium ion reacts with water according to the equation $NH_4^+ + H_2O \rightleftarrows NH_3 + H_3O^+$. For ammonium ion $K_a = 5.63 \times 10^{-10}$. What is the value of pK_a for ammonium ion?

15. Hydrocyanic acid, HCN, has $pK_a = 9.31$. What reaction, if any, would occur on combining 1.00-*M* solutions of sodium cyanide Na^+CN^- and acetic acid CH_3COOH, pK_a 4.74? (*Hint*: which is the stronger base, CH_3COO^- or CN^-?)

8.4 Strengths of Bases—K_b Water acts as a base when it accepts the proton donated by an acid. It can also act as an acid by donating a proton to a base. The equilibrium, with B representing the base, is

$$B + H_2O \rightleftarrows HB^+ + OH^-$$

Taking ammonia as an example, we have

$$NH_3 + H_2O \rightleftarrows NH_4^+ + OH^-$$

In this type of equilibrium, the hydroxide ion concentration is increased from its value of 1×10^{-7} (pH = 7) in pure water to give an alkaline or basic solution. Since in *any* aqueous solution the requirement that $[H_3O^+][OH^-] = 1 \times 10^{-14}$ must be met, $[H_3O^+]$ must be less than 10^{-7} (pH greater than 7) in solutions of bases.

The strength of a base is measured by the position of the equilibrium in its reaction with water. The numerical measure of base strength is the value of K_b, the *basic dissociation constant*, which is defined as follows: ☛

K_h is used instead of K_b by some chemists, and the constant is called the hydrolysis constant of the base; their mathematical forms are the same.

$$K_b = \frac{[HB^+][OH^-]}{[B]} \quad \text{for the reaction} \quad B + H_2O \rightleftarrows HB^+ + OH^-$$

You can see from this equation that the stronger the base is, the larger will be $[HB^+]$ and $[OH^-]$ compared to $[B]$ and the larger will be the value of K_b.

Many bases are negatively charged rather than uncharged. Using A^- to represent an anionic base, we find that the equations become

$$A^- + H_2O \rightleftarrows HA + OH^-$$

and

$$K_b = \frac{[HA][OH^-]}{[A^-]}$$

A typical weak base of this type is acetate ion, CH_3COO^-, present in solutions of salts of acetic acid such as sodium and potassium acetate, $CH_3COO^-Na^+$ and $CH_3COO^-K^+$. For acetate ion as a base the equations are

$$CH_3COO^- + H_2O \rightleftarrows CH_3COOH + OH^-$$

and

$$K_b = \frac{[CH_3COOH][OH^-]}{[CH_3COO^-]}$$

For acetate ion, $K_b = 5.53 \times 10^{-10}$ at 25°C. The basic property of acetate ion means that solutions containing it will be mildly alkaline (pH greater than 7) and also will contain a small concentration of molecules of acetic acid, CH_3COOH. The pH and concentration of acetic acid in these solutions can be calculated from the value of K_b for acetate ion if you know the component molarity of acetate ion in the solution. Suppose you want the pH of a 1.00-*M* sodium acetate solution. Applying the "before" and "at equilibrium" argument used before in talking about K_a values, and letting y be the concentration of $[OH^-]$ at equilibrium, we write

Molar concentration (mole/liter)

	CH_3COO^- + H_2O	⇄ CH_3COOH	+ OH^-
Before equilibrium	1.00	0	1×10^{-7}
At equilibrium	$1.00 - y$	y	$y + 10^{-7}$

If we assume that y is much larger than the 1×10^{-7} mole/liter provided by the autoionization of water, we get the equation

$$K_b = \frac{[CH_3COOH][OH^-]}{[CH_3COO^-]} = \frac{y^2}{1.00 - y} = 5.53 \times 10^{-10}$$

The right-hand equality reduces to a quadratic equation that can be solved exactly by using the quadratic formula. A satisfactory approximate solution is more easily obtained if you see that because acetate ion is a *weak* base y (the $[OH^-]$) is very much less than 1.00. This means that $(1.00 - y) \approx 1.00$ and leads to the simpler equation

$$\frac{y^2}{1.00} = 5.53 \times 10^{-10}$$

from which

$$y = \sqrt{5.53} \times \sqrt{10^{-10}} = 2.35 \times 10^{-5} \text{ mole/liter} = [OH^-]$$

The strategy behind this calculation is like that used earlier to calculate $[H_3O^+]$ in solutions of weak acids, with the important exception that at equilibrium $y = [OH^-]$ instead of $[H_3O^+]$. To get $[H_3O^+]$ you use the equation given earlier,

$$[H_3O^+] = \frac{K_w}{[OH^-]}$$

so

$$[H_3O^+] = \frac{1 \times 10^{-14}}{2.35 \times 10^{-5}} = 4.26 \times 10^{-10} \text{ mole/liter}$$

and

$$\begin{aligned} pH &= -\log [H_3O^+] = -(\log 4.26 \times 10^{-10}) = 10.00 - \log 4.26 \\ &= 10.00 - 0.63 = 9.37 \end{aligned}$$

You can see that a 1.00-M sodium acetate solution is indeed alkaline, since its pH is greater than 7.

The values of K_b for some representative bases are entered in Table 8.2 in the column "Conjugate Bases." The term *conjugate* arises because each base is paired with an acid in the column "Acids." Each base is conjugate to the acid from which it results by removal of a proton:

$$\underset{\text{Acid}}{HA} + H_2O \rightleftarrows H_3O^+ + \underset{\substack{\text{Conjugate}\\ \text{base of the}\\ \text{acid HA}}}{A^-}$$

Comparing K_a values for acids with the K_b values for their conjugate bases shows that *weak bases are conjugate to strong acids and strong bases are conjugate to weak acids.* This is as it must be – if HA is a strong proton donor (strong acid), then its conjugate base A^- must have a weak attraction for protons and be a weak base. On the other hand, if A^- has a large attraction for protons, HA will be a weak proton donor and a weak acid.

The K_b values for bases are determined experimentally by following much the same plan used for K_a values for weak acids. The $[OH^-]$ in a solution of known concentration of base (C_B) is measured, and the values are substituted in the expression

$$K_b = \frac{[OH^-]}{C_B - [OH^-]}$$

In the example of a 0.010000-*M* NH_3 solution ($C_B = 0.010000$), we find $[OH^-]$ to be 0.000412 mole/liter at 25°C. ☚ Substituting these values in the expression for K_b gives

Most conveniently by using a pH meter to determine the $[H_3O^+]$, and the equation

$$[OH^-] = \frac{K_w}{[H_3O^+]} = \frac{10^{-14}}{[H_3O^+]}$$

to calculate $[OH^-]$.

$$K_b = \frac{(0.000412)^2}{0.010000 - 0.000412} = 1.77 \times 10^{-5} \qquad \text{at } 25°\text{C}$$

Comparing this value with the value of $K_a = 1.81 \times 10^{-5}$ for acetic acid shows that ammonia is similar in strength as a base to acetic acid as an acid.

The most important use of K_b values is in calculating the $[H_3O^+]$ or pH at the *equivalence point* in the titration of a weak acid (such as acetic) with a strong base (such as OH^- from Na^+OH^-). Figure 8.3 shows how the pH changes during a titration of this type. At the equivalence point, the number of moles of Na^+OH^- added is equal to the number of moles of weak acid present, because the equation for the reaction is

$$\underset{\text{1 mole}}{HA} + \underset{\text{1 mole}}{Na^+OH^-} = H_2O + Na^+A^-$$

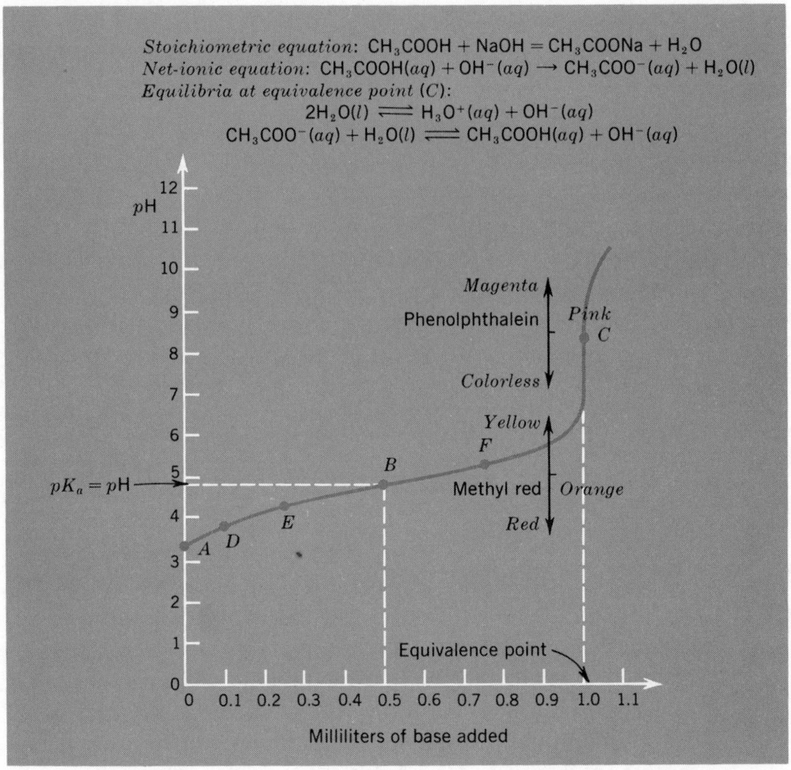

FIG. 8.3 Titration curve for the titration of 100 ml of 0.0100 *M* acetic acid with 1.00 *M* sodium hydroxide.

The number of moles of NaOH added to reach the equivalence point is given by

$$\text{Moles NaOH} = \frac{V_{\text{NaOH}}}{1000} \times C_{\text{NaOH}}$$

where V_{NaOH} is the volume of NaOH added in milliliters and C_{NaOH} is the concentration of the NaOH solution in moles/liter. Measuring V with the buret and knowing C_{NaOH}, we can calculate the number of moles of HA. But how do you know *when* the equivalence point has been reached during the addition of the NaOH solution? At the equivalence point, the solution being titrated will be a solution of the sodium salt (Na^+A^-) of the acid, and if K_b for A^- is known, the $[H_3O^+]$ or pH can be calculated. In most practical cases of this sort the pH at the equivalence point will be in the range 8–9, and it is determined by the change in color of an *indicator* such as phenolphthalein, which is added to the solution being titrated. Indicators are complex organic molecules whose light absorption, and hence color to the eye, changes over a range of about 1 pH unit. Phenolphthalein is colorless in solutions with pH = 8 and bright pink in solutions with pH = 9. Other indicators are methyl orange (red at pH = 3, yellow at pH = 4) and bromothymol blue (yellow at pH = 6, blue at pH = 7).

Exercises

16. For lactate anion ($CH_3CHOHCOO^-$), $K_b = 1.19 \times 10^{-11}$. Write the equation for the reaction of lactate anion as a base with water. Will the pH of a 0.1-*M* solution of sodium lactate ($CH_3CHOHCOO^-Na^+$) be greater or less than 7? Calculate $[H_3O^+]$ and pH in the 0.1-*M* solution. (Use the approximate method.)

17. For hydrocyanic acid, $K_a = 4.90 \times 10^{-10}$ and for cyanide ion (CN^-) $K_b = 2.04 \times 10^{-5}$. Calculate the product K_aK_b of the two constants. Do the same for some other acid-and-conjugate-base pairs found in Table 8.2. Can you prove that K_aK_b for such pairs is always equal to K_w? (*Hint*: use HA to represent the acid and A^- to represent the conjugate base in the expressions for K_a and K_b.)

18. When ammonia (a weak base) is titrated with hydrochloric acid, the net reaction is $NH_3 + H_3O^+ \rightarrow NH_4^+ + H_2O$. For NH_3, $K_b = 1.77 \times 10^{-5}$. Calculate the pH at the equivalence point in the titration if $[NH_4^+]$ is 0.5 *M* at the equivalence point. Which of the indicators mentioned earlier (phenolphthalein, methyl orange, or bromothymol blue) would be most suitable for determining the equivalence point in the titration?

8.5 Buffer Solutions Buffer solutions have the property of maintaining nearly constant pH (or $[H_3O^+]$) when you add strong acids (H_3O^+) or strong bases (OH^-) to them. To grasp the way they work requires understanding of the dissociation constants for acids (K_a) and bases (K_b) discussed in Sections 8.3 and 8.4. As an example of a buffer system we will use one that is important in maintaining the pH of the blood in the range 7.37–7.44, a condition necessary for health.

A buffer solution contains substantial (0.0100–1.000 *M*) *concentrations of a weak acid HA and its conjugate base A^-*. In blood the buffer acid is carbonic acid, H_2CO_3, for which $K_a = 4.27 \times 10^{-7}$:

$$H_2CO_3 + H_2O \rightleftarrows H_3O^+ + HCO_3^-$$

$$K_a = \frac{[H_3O^+][HCO_3^-]}{[H_2CO_3]} = 4.27 \times 10^{-7} \qquad \text{at } 25°\text{C}$$

Although bicarbonate ion HCO_3^- is also an acid ($K_a = 4.79 \times 10^{-11}$ for $HCO_3^- + H_2O \rightleftarrows H_3O^+ + CO_3^{2-}$), it is so weak that it makes no significant contribution to the $[H_3O^+]$ or pH of blood.

The buffer *base* in blood is the bicarbonate ion, HCO_3^-, since it is conjugate to H_2CO_3. ☚ For HCO_3^- the relevant equations are

$$HCO_3^- + H_2O \rightleftarrows H_2CO_3 + OH^-$$

$$K_b = \frac{[H_2CO_3][OH^-]}{[HCO_3^-]} = 2.34 \times 10^{-8} \qquad \text{at } 25°C$$

A typical carbonic acid–bicarbonate buffer solution can be prepared by combining equal volumes of 0.01-M H_2CO_3 and 0.01-M $Na^+HCO_3^-$ solutions. Because the water in each of the solutions dilutes the other when they are poured together, the concentrations of H_2CO_3 and HCO_3^- in the resulting buffer solution will both be very close to one-half their original concentrations or 0.005 mole/liter. This is true because (1) H_2CO_3 is such a weak acid ($K_a = 4.27 \times 10^{-7}$) that the amount lost by the reaction $H_2CO_3 + H_2O \rightleftarrows H_3O^+ + HCO_3^-$ is tiny compared to 0.005; and (2) HCO_3^- is such a weak base ($K_b = 2.34 \times 10^{-8}$) that the amount lost by the reaction $HCO_3^- + H_2O \rightleftarrows H_2CO_3 + OH^-$ can also be ignored. Since the buffer solution contains both H_2CO_3 and HCO_3^-, it follows that K_a for H_2CO_3 must apply to the solution; that is,

$$K_a = \frac{[H_3O^+][HCO_3^-]}{[H_2CO_3]} = 4.27 \times 10^{-7} \qquad \text{at } 25°C$$

Now $[HCO_3^-] = [H_2CO_3] \approx 0.005$ from the method of preparing the buffer solution, so they divide out and we are left with the simple relationship

$$K_a = \frac{[H_3O^+][\cancel{HCO_3^-}]}{[\cancel{H_2CO_3}]} = \frac{[H_3O^+][\cancel{0.005}]}{[\cancel{0.005}]} = 4.27 \times 10^{-7}$$

or

$$[H_3O^+] = K_a = 4.27 \times 10^{-7}$$

Now, since $pK_a = -\log K_a$, and $pH = -\log [H_3O^+]$, you can write

$$pK_a = pH = -(\log 4.27 \times 10^{-10}) = 10.00 - \log 4.27 = 9.37$$

for the buffer solution. ☛ Both equations, that for K_a and that for pH, will be true for any buffer prepared so that $[HA] = [A^-]$.

The first step in understanding how a buffer holds pH nearly constant requires that you (1) write the acid and base reactions *in reverse*, and (2) calculate the values of their equilibrium constants: ☛

$$H_3O^+ + HCO_3^- \rightleftarrows H_2CO_3 + H_2O \qquad K = \frac{1}{K_a} = \frac{1}{4.27 \times 10^{-7}}$$
$$= 2.34 \times 10^6$$

It is not generally true that pK_a = pH. It is true only for those buffer solutions prepared so that the concentrations of the weak acid and its conjugate base are equal.

Remember, if K is the constant for the reaction $aA + bB \rightleftarrows cC + dD$, then the equilibrium constant for the reversed reaction $cC + dD \rightleftarrows aA + bB$ is $1/K$.

$$H_2CO_3 + OH^- \rightleftarrows HCO_3^- + H_2O \qquad K = \frac{1}{K_b} = \frac{1}{2.34 \times 10^{-8}}$$
$$= 4.27 \times 10^7$$

From the large values of their equilibrium constants, the positions of both equilibria must lie far to the right. If some strong acid such as hydrochloric acid is added to the buffer solution, the extra H_3O^+ that results will be almost completely "gobbled up" from the solution by reaction with HCO_3^- according to the first equation. In the same way, if some strong base such as NaOH is added, the OH^- introduced will be removed by reaction with H_2CO_3. Through these two reactions the buffer solution resists the changes in $[H_3O^+]$, or pH, that would occur on adding strong acid or base to water alone.

Figure 8.4 compares the effects on pH of adding strong acid or strong base to pure water, on the one hand, and to a buffer of pH = 7 on the other. The buffer acid, HA ($K_a = 1 \times 10^{-7}$), in this example has $pK_a = 7$ and the concentrations of buffer acid and buffer base, Na^+A^-, are both

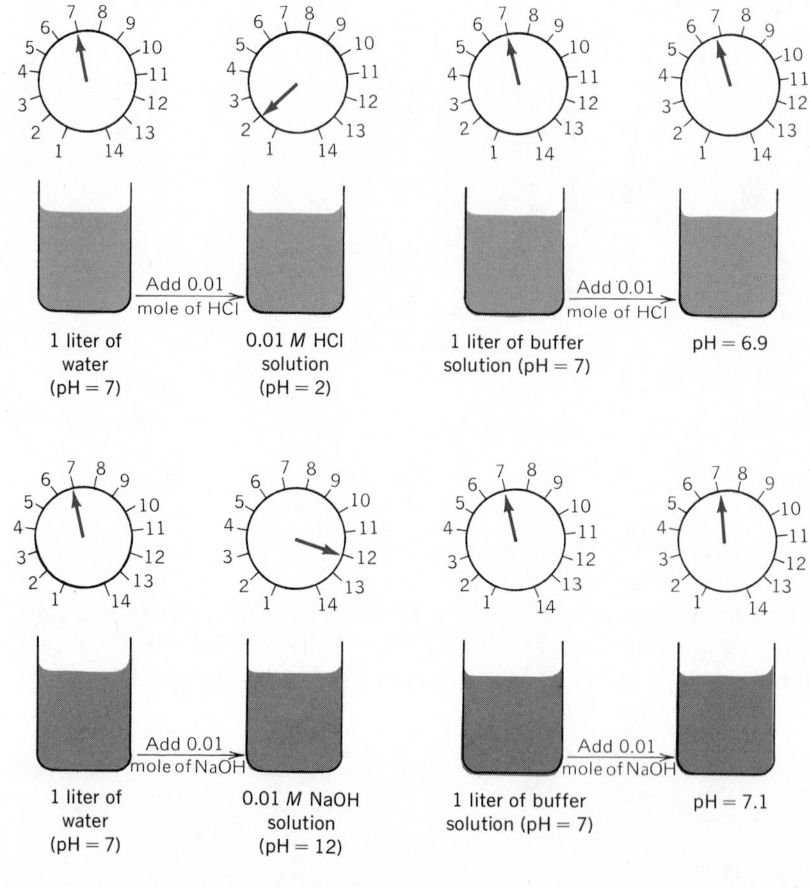

FIG. 8.4 Effects on pH of adding strong acid and strong base to pure water and to a buffer solution at pH = 7.

one mole per liter, so $[HA] = [A^-] = 1$ and $pH = 7$ for the buffer solution. As the pH meters in the figure show, the buffer can't prevent some change in pH on addition of strong acid or base, but it does pretty well – the value $\Delta pH = \pm 0.1$ for the buffer compared to $\Delta pH = \pm 6$ for water shows how well. The small changes away from $pH = 7$ on adding strong acid or base to the buffer solution result from the equilibrium nature of the reactions that absorb the added H_3O^+ or OH^-:

$$H_3O^+ + A^- \rightleftarrows HA + H_2O$$
$$OH^- + HA \rightleftarrows A^- + H_2O$$

Because these reactions are equilibria, they cannot remove every last one of the H_3O^+ or OH^- added. Another important point is that a buffer solution cannot continue indefinitely to absorb added strong acid or base. If so much strong acid is added that nearly all of the buffer base A^- is used up, further additions of acid will drive the $[H_3O^+]$ up (and pH down). Likewise, if too much strong base is added, buffer acid HA will be used up and further additions will cause $[H_3O^+]$ to go down (pH up), according to the equation $[H_3O^+] = K_w/[OH^-]$ derived in Section 8.2.

The quantitative effects on $[H_3O^+]$ and pH of changing the concentration ratio of buffer acid to buffer base, $[HA]/[A^-]$, are given by the equations

$$[H_3O^+] = \frac{K_a[HA]}{[A^-]}$$

$$pH = pK_a + \log \frac{[A^-]}{[HA]}$$

EXAMPLE

For the H_2CO_3–HCO_3^- buffer in blood, the concentrations of buffer acid (H_2CO_3) and buffer base (HCO_3^-) are not equal. Most body reactions produce acids rather than bases – you saw an example in the lactic acid case. For this reason, the H_2CO_3–HCO_3^- buffer is maintained with $[HCO_3^-] \approx 20[H_2CO_3]$ or $[HCO_3^-]/[H_2CO_3] = 20$ so that the buffer has more capacity to absorb acids than bases. You can calculate $[H_3O^+]$ and pH for blood, knowing the $[HCO_3^-]/[H_2CO_3]$ ratio is 20 and $K_a = 4.27 \times 10^{-7}$ ($pK_a = 6.37$) for H_2CO_3, and using the equations just given:

$$[H_3O^+] = K_a\frac{[HA]}{[A^-]} = 4.27 \times 10^{-7} \times \frac{[H_2CO_3]}{[HCO_3^-]}$$

$$= 4.27 \times 10^{-7} \times \frac{1}{20} = 2.14 \times 10^{-8} \text{ mole/liter}$$

$$pH = pK_a + \log \frac{[A^-]}{[HA]} = 6.37 + \log 20 = 7.67$$

Arterial blood has pH $= 7.44$; venous blood pH $= 7.37$. Both of these values are slightly lower than calculated for the H_2CO_3–HCO_3^- buffer because blood is not simply water, and other solutes in blood modify the behavior of the buffer. Venous blood, having the lower pH, is somewhat more acidic than arterial blood, for the reason that venous blood is carrying extra CO_2 (as H_2CO_3), from the metabolism of glucose, back to the lungs to be exhaled:

$$C_6H_{12}O_6 + 6O_2 \rightarrow 6CO_2 + 6H_2O$$

The extra H_2CO_3 in venous blood means that the $[H_2CO_3]/[HCO_3^-]$ ratio in venous blood is larger than in arterial blood, so venous blood is slightly more acidic.

The disease condition called *acidosis* results when for any reason the pH of arterial blood falls below 7.35 (normal range is 7.35–7.45). Acidosis is a medical emergency with symptoms ranging from increasing weakness to unconsciousness, and must be treated immediately. Acidosis appears whenever the $[HCO_3^-]/[H_2CO_3]$ ratio falls below 20 because of an excessive concentration of H_2CO_3 in the blood or a lack of HCO_3^-. Excess H_2CO_3 concentration results when respiration is too weak to allow the CO_2 produced by glucose oxidation to be "blown out" in exhaled air. Diabetes, if uncontrolled by insulin injections, also causes acidosis. With adequate supplies of the hormone insulin from the pancreas (or from injection in diabetes), glucose and fats are oxidized to CO_2 and water. When the supply of insulin is low, acetoacetic and hydroxybutyric acids are produced instead:

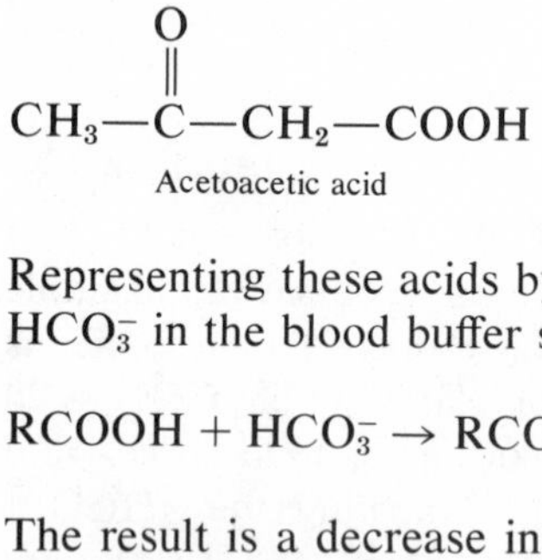

$$\underset{\text{Acetoacetic acid}}{CH_3-\overset{\overset{\textstyle O}{\|}}{C}-CH_2-COOH} \qquad \underset{\text{Hydroxybutyric acid}}{CH_3-\overset{\overset{\textstyle OH}{|}}{CH}-CH_2COOH}$$

Representing these acids by RCOOH, we see that they react with the HCO_3^- in the blood buffer system as follows:

$$RCOOH + HCO_3^- \rightarrow RCOO^- + H_2CO_3$$

The result is a decrease in $[HCO_3^-]$ and an increase in $[H_2CO_3]$ for the blood buffer system, so the $[HCO_3^-]/[H_2CO_3]$ ratio falls below the value of about 20 needed for health and the pH of the blood falls.

You should not confuse acidosis with "acid stomach" due to a temporary excessive acidity in the digestive juice (gastric juice) in the stomach. Acid stomach can be relieved with small amounts of sodium bicarbonate ($\frac{1}{2}$ teaspoon) or any of the numerous "antacid" tablets on the market. Antacids, including sodium bicarbonate, taken by mouth have no effect on the pH of blood or other internal body fluids. They do neutralize excess stomach acid (HCl) by reactions such as

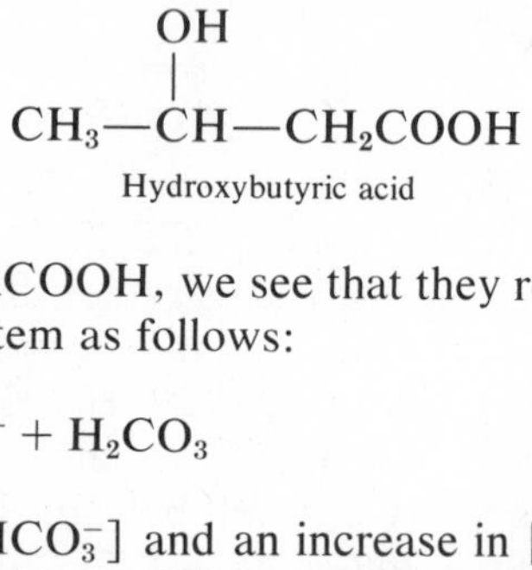

$$NaHCO_3 + HCl \rightarrow NaCl + H_2CO_3$$
$$CaCO_3(s) + 2HCl \rightarrow CaCl_2 + H_2CO_3$$

The carbonic acid decomposes,

$$H_2CO_3 \rightarrow H_2O(l) + CO_2(g)$$

and the CO_2 gas released makes you "burp."

You can see from the H_2CO_3–HCO_3^- example that it is not necessary for [HA] to be equal to [A^-] in a buffer. By changing the ratio of [HA] to [A^-] in the preparation of the buffer solution, it is possible to make buffers that span a range of about 2 pH units centered on the pK_a of the acid used. By choosing acids with the proper pK_a value and adjusting the [HA]/[A^-] ratio, buffers with pH intervals of 0.1 pH unit covering the whole pH range can be made.

Exercises

19. An acetic acid–potassium acetate buffer is prepared by mixing equal volumes of 0.20-*M* CH_3COOH and 0.20-*M* $CH_3COO^-K^+$. Write the equations for the reactions by which this buffer solution scavenges H_3O^+ and OH^- from additions of strong acids and bases.

20. The value of K_a for CH_3COOH is 1.8×10^{-5} at 25°C. Calculate the pH in the buffer described in Exercise 19.

21. In what volume ratio should 0.20-*M* acetic acid and 0.20-*M* sodium acetate ($CH_3COO^-Na^+$) solutions be combined to give a buffer solution having pH = 5.00?

8.6 Dissolving Slightly Soluble Ionic Compounds Most slightly soluble ionic compounds can be dissolved completely by treatment with a reagent that reduces the concentration of one of the ions in the solubility equilibrium for the compound. Hydroxides of the Group 2a metals [$M(OH)_2$] and Group 3a metals [$M(OH)_3$] are sparingly soluble in water but dissolve in solutions of strong acids such as HCl or HNO_3. Using long arrows to indicate the reactions resulting in dissolving, we can write the general equations as follows:

$$M(OH)_2(s) \rightleftharpoons M^{2+} + 2OH^-$$
$$2OH^- + 2H_3O^+ \rightleftharpoons 4H_2O(l)$$

$$M(OH)_2(s) + 2H_3O^+ \longrightarrow M^{2+} + 4H_2O(l)$$

By forming H_2O with the H_3O^+ from the acid, the OH^- in equilibrium with the solid is reduced to such a low level that $Q = C_{M^{2+}} \times C^2_{OH^-}$ becomes much smaller than the value of K_{sp} for $M(OH)_2$. More and more $M(OH)_2$ dissolves in an effort to restore, by increasing $C_{M^{2+}}$, the value of K_{sp} for $C_{M^{2+}} \times C^2_{OH^-}$ (Le Chatelier's principle). This goes on until all of the solid has gone into solution. For calcium hydroxide dissolving in hydrochloric acid the stoichiometric equation is

$$Ca(OH)_2(s) + 2HCl = CaCl_2 + 2H_2O(l)$$

You have to exercise some care in the choice of acid— the cation of the hydroxide may form a slightly soluble compound with the anion of the acid used.

$$Ba(OH)_2(s) + 2H_3O^+ + SO_4^{2-} \rightleftarrows BaSO_4(s) + 4H_2O(l)$$

The position of this equilibrium is far to the right because K_{sp} for $BaSO_4(s)$ is 8.7×10^{-11}, much smaller than the value of 5×10^{-3} for $Ba(OH)_2$ (Table 7.2). Most chlorides and, especially, nitrates are soluble, so this kind of difficulty can usually be avoided by using HCl or HNO_3 as acids instead of H_2SO_4.

The carbonates of the Group 2a metals ($BeCO_3$, $MgCO_3$, $CaCO_3$, $SrCO_3$, $BaCO_3$) dissolve in solutions of nitric or hydrochloric acid according to the general net ionic equation:

$$MCO_3(s) + 2H_3O^+ \rightleftarrows M^{2+} + 2H_2O(l) + H_2CO_3$$

The dissolving is driven to completion in this case because carbonic acid, H_2CO_3, is unstable and decomposes to water and $CO_2(g)$ which bubbles off.

$$H_2CO_3 \rightleftarrows H_2O(l) + CO_2(g)$$

Some sulfides (ZnS, CdS, FeS) are dissolved by acid solutions because the H_3O^+ added ties up the S^{2-} in equilibrium by forming H_2S:

$$Zn(s) \rightleftarrows Zn^{2+} + S^{2-}$$

$$S^{2-} + 2H_3O^+ \rightleftarrows H_2S + 2H_2O(l)$$

$$ZnS(s) + 2H_3O^+ \longrightarrow Zn^{2+} + 2H_2O(l) + H_2S$$

For sulfides with very small solubilities, such as CuS and Ag_2S, addition of strong acid cannot reduce the S^{2-} concentration enough through formation of H_2S to promote dissolving. In these cases the equilibrium $H_2S + 2H_2O(l) \rightleftharpoons 2H_3O^+ + S^{2-}$ provides enough of an S^{2-}

concentration to keep the sulfide from dissolving. "Acid-insoluble" sulfides like CuS and Ag_2S can be dissolved by using concentrated nitric acid and heat. The nitric acid converts (oxidizes) the S^{2-} in the solubility equilibrium to S and the sulfide dissolves completely, giving the metal nitrate.

Many slightly soluble ionic compounds can be dissolved by using reagents that form complex ions with the cation of the compound. Here complexing of the cation reduces the concentration of the cation instead of the anion, and dissolving occurs as the system, following Le Chatelier's principle, tries to reestablish the value of K_{sp} for the solute. Silver chloride ($K_{sp} = 1.8 \times 10^{-10}$) and silver bromide ($K_{sp} = 4.3 \times 10^{-13}$) dissolve in ammonia solutions because of removal of silver ion from the solubility equilibrium through formation of the complex ion $Ag(NH_2)_2^+$:

$$AgCl(s) \rightleftharpoons Ag^+ + Cl^-$$

$$Ag^+ + 2NH_3 \rightleftharpoons Ag(NH_3)_2^+$$

$$AgCl(s) + 2NH_3 \rightleftharpoons Ag(NH_3)_2^+ + Cl^-$$

The reactions are the same for AgBr, but AgI ($K_{sp} = 8.3 \times 10^{-17}$) does not dissolve in ammonia. The $[Ag^+]$ in equilibrium with the complex ion is large enough to prevent dissolving. In this case the equilibria are more balanced:

$$AgI(s) \rightleftharpoons Ag^+ + I^-$$

$$Ag^+ + 2NH_3 \rightleftharpoons Ag(NH_3)_2^+$$

You can quickly satisfy yourself that these explanations are correct by dissolving some silver chloride in ammonia solution, then adding a few drops of sodium or potassium iodide solution. A precipitate of AgI will form immediately.

The silver halides AgCl, AgBr, and AgI make photography possible, and their complex ions play a vital part in the process, as described in Section 11.7. The dissolving of calcium and magnesium carbonates in acidic solution is ruining marble ($CaCO_3$) statues in cities with acidic smog (Photo 12, Chapter 4), and also results in "hard water" (Section 11.3).

8.7 Summary Guide Water (aqueous) solutions are the most important kind of solutions because of the ability of water to dissolve so many different kinds of compounds—electrolytes as well as nonelectrolytes. The water molecule is at the same time a weak acid (proton donor) and a weak base (proton acceptor). Its dual role results in reaction with itself to produce small concentrations of H_3O^+ and OH^- ions in an autoionization equilibrium. This equilibrium exists in all water solutions and requires that for any water solution, whether of acid, base, salt, or covalent molecules, the product of the H_3O^+ and OH^- concentrations in moles per liter must equal K_w or 1×10^{-14} (at 25°C).

Hydrogen ion concentrations in water solutions are conveniently expressed by using the pH scale in which $\text{pH} = -\log [H_3O^+]$ and $[H_3O^+] = 10^{-\text{pH}}$. In neutral solutions the concentrations of H_3O^+ and OH^- ions are equal at 1×10^{-7} mole/liter, and $\text{pH} = 7$. In acidic solutions the concentration of H_3O^+ ions is greater than 10^{-7} mole/liter, and the pH is less than 7. In basic solutions the concentration of H_3O^+ ions is less than 10^{-7}, and the pH is greater than 7.

When acids are dissolved in water they react according to the general equation $HA + H_2O \rightleftarrows H_3O^+ + A^-$, in which water molecules play the part of a base. The strength of an acid is measured by the position of the equilibrium and is expressed numerically by the value of the acid dissociation constant K_a, defined as $[H_3O^+][A^-]/[HA]$. The value of K_a is large for strong acids, small for weak acids. Most acids are weak. Knowing the value of K_a for an acid makes it possible to calculate the concentrations of H_3O^+ and A^- in any water solution in which the component molar concentration of the acid HA is known.

Strengths of acids are often expressed by using the pK_a scale, in which $pK_a = -\log K_a$ or $K_a = 10^{-pK_a}$. The larger the value of pK_a for an acid, the weaker the acid.

When bases (B) are dissolved in water they react according to the general equation $B + H_2O \rightleftarrows HB^+ + OH^-$, with water molecules playing the part of an acid. The strength of a base is given numerically by the value of its basic dissociation constant K_b, defined as $[HB^+][OH^-]/[B]$, or $[HA][OH^-]/[A^-]$ if the base is an anion. The larger the value of K_b, the stronger the base. Knowing the value of K_b for a base and the component molarity of the base in the solution enables calculation of the concentrations of HB^+ and OH^- in a solution. The values of K_b are used to calculate the H_3O^+ concentration or pH of a solution at the equivalence point in a titration. Indicators, complex molecules that change colors over a range of about 1 pH unit, are used to determine the equivalence point in titrations. The equivalence point in an acid–base titration is reached when the number of moles of OH^- added is equal to the number of H_3O^+ provided by complete dissociation of the acid being titrated.

Buffer solutions are made up to contain substantial concentrations of a weak acid HA and its conjugate base A^-. They tend to maintain pH nearly constant in spite of additions of strong acids (H_3O^+) or strong bases (OH^-). Added H_3O^+ ions are removed by the reaction $A^- + H_3O^+ \rightleftarrows HA + H_2O$; added OH^- ions are removed by the reaction $HA + OH^- \rightleftarrows H_2O + A^-$. The pH of a buffer made up so that the concentrations of HA and A^- are the same is equal to the pK_a of the weak acid HA. The pH of a buffer solution can be varied over the range $pH = pK_a \pm 1$ by proper adjustments of the $[HA]/[A^-]$ concentration ratio.

An understanding of pH, K_a, pK_a, K_b, and buffer action is necessary for understanding most biochemical reactions. Many biochemicals are acids or bases, and the reactions of life go on in buffer solutions.

Slightly soluble ionic compounds can often be dissolved by the use of acidic solutions or solutions of reagents that form complex ions with the cation of the ionic compound. In both cases, dissolving results from removal of one of the ions of the ionic compound from its solubility equilibrium. The values of solubility product constants (K_{sp}) enable prediction of the effect of a reagent on the solubility of ionic compounds.

Exercises

22. For FeS(*s*) the solubility product constant $K_{sp} = 5 \times 10^{-18}$. For the reaction $S^{2-} + 2H_3O^+ \rightleftarrows H_2S + 2H_2O(l)$, $K = [H_2S]/[S^{2-}][H_3O^+]^2 = 10^{20}$. In a saturated solution of H_2S in water, $[H_2S] = 0.1$ mole/liter. From these data (all at 25°C) predict whether FeS(*s*) will dissolve appreciably in a 1.00-*M* HCl solution. (*Hints*: calculate the value of $[S^{2-}]$ in a 1.00-*M* HCl solution saturated with H_2S and compare it with the value of $[S^{2-}]$ in a saturated solution of FeS.)

23. For HgS(*s*), the value of $K_{sp} = 1.6 \times 10^{-52}$. Repeat the type of calculation used in Exercise 22 and predict whether HgS(*s*) will dissolve in a 1.00-*M* strong acid solution. Mercury(II) sulfide, HgS (also called "cinnabar" or "vermilion" from its brilliant red color), is an important ore of mercury but FeS(*s*) [iron(II) or ferrous sulfide] is not an ore of iron. Explain. (*Hint*: groundwater is often acidic because of dissolved CO_2.)

24. By excreting urine, our kidneys perform four vital functions: (1) they retain glucose in the blood; (2) they rid the blood of poisonous urea (NH_2CONH_2, about 11 g/l of urine) produced by protein breakdown in the cells; (3) they control the volume and electrolyte concentration (mainly Na^+Cl^-) of blood by passing electrolytes and water into the urine as required; and (4) they control the pH of the blood close to pH = 7.4, slightly on the alkaline side. In kidney disease, one or more of these functions fails in some measure, and if the failure is serious enough, a kidney transplant is necessary to preserve life. The fourth function, that of keeping the blood at about pH = 7.4, is achieved by altering the $[HPO_4^{2-}]/[H_2PO_4^-]$ concentration in blood by excreting $H_2PO_4^-$ at the expense of HPO_4^{2-}. Blood entering the kidneys has a $[H_2PO_4^-]/[HPO_4^{2-}]$ ratio equal to about 1:4. In urine the ratio is about 9:1. Which is more acidic, blood or urine? (*Hint*: think of a buffer system.) Given that, for $H_2PO_4^{2-}$, the value of $K_a = 6.17 \times 10^{-8}$, and neglecting any other acids or bases in urine, calculate the pH of urine.

When oxygen is scarce, as at high altitudes, the concentration of oxyhemoglobin takes over control of breathing.

25. The rate at which you breathe at ordinary altitudes is controlled by the value of the $[H_2CO_3]/[HCO_3^-]$ ratio in your blood, not by the concentration of oxyhemoglobin ($HGbO_2$). Exercise increases $[H_2CO_3]$ through increased oxidation of glucose in muscle cells to produce energy. The equations are

$$C_6H_{12}O_6 + 6O_2 \rightarrow 6CO_2 + 6H_2O + \text{energy}$$

$$CO_2(g) + H_2O(l) \xrightleftharpoons{\text{enzyme}} H_2CO_3$$

A mere 4% increase in the $[H_2CO_3]/[HCO_3^-]$ ratio from exercise signals the respiratory center in the brain to *double* the rate of breathing to "blow off" the extra CO_2 in exhaled breath.

To see just how very sensitive the respiratory center is, calculate the pH change that accompanies the 4% change in the $[H_2CO_3]/[HCO_3^-]$ ratio. (*Hints*: use the buffer equation in the form $\text{pH} = \text{p}K_a + \log \{[HCO_3^-]/[H_2CO_3]\}$; $\text{p}K_a$ for H_2CO_3 is 6.37; use the value of $\frac{10}{1}$ for $[HCO_3^-]/[H_2CO_3]$ before exercise.)

Key Words

Acidic solution an aqueous solution in which the hydronium ion concentration is greater than the hydroxide ion concentration. An aqueous solution having pH less than 7.

Autoionization constant for water (K_w) the product of the hydronium and hydroxide ion concentrations in any aqueous solution, having the value 1×10^{-14} at 25°C.

Autoionization equilibrium an equilibrium established between solvent molecules in which one solvent molecule acts as an acid toward another solvent molecule acting as a base.

Basic or alkaline solution an aqueous solution in which the hydroxide ion concentration is greater than the hydronium ion concentration. An aqueous solution having pH greater than 7.

Buffer solution a solution containing high concentrations of a weak acid and its conjugate base that resists attempts to change its pH by addition of strong acid or base.

Conjugate acid the species that results from adding a proton to a base.

Conjugate base the species that results from the removal of a proton from an acid.

Dissociation constant for an acid (K_a) the value of the equilibrium constant for the reaction of an acid with water, written omitting the concentration of water. A numerical measure of the strength of an acid as a donor of protons.

Dissociation constant for a base (K_b) the value of the equilibrium constant for the reaction of a base with water, written omitting the concentration of water. A numerical measure of the attraction of a base for protons. Also called the hydrolysis constant (K_h) of the base.

Indicator complex molecules or ions whose color changes with changes in the pH of a solution over a range of about 1 pH unit.

Neutral solution an aqueous solution in which the concentrations of hydronium and hydroxide ions are equal at 1×10^{-7} mole/liter. An aqueous solution having pH = 7.

pH scale a convenient, logarithmic method for describing the hydronium ion concentration in an aqueous solution in which larger pH values correspond to smaller hydronium ion concentrations.

pK_a scale a logarithmic method of expressing the strength of an acid in which larger pK_a values correspond to weaker acids.

Titration the technique of measuring the volume of a reagent solution of known concentration required to react completely with a sample of another chemical.

Suggested Readings

Medeiros, Robert W., "Carbon Monoxide: The Invisible Enemy," *Chemistry*, **46**:18–20 (Jan. 1973).

Penman, H. L., "The Water Cycle," *Chemistry in the Environment*, San Francisco, W. H. Freeman and Co., 1973, pp. 23–30.

Perutz, M. F., "The Hemoglobin Molecule," *Bio-Organic Chemistry*, San Francisco, W. H. Freeman and Co., 1968, pp. 42–52.

Platt, Rutherford, *Water: The Wonder of Life*, Englewood Cliffs, N.J., Prentice-Hall, Inc., 1971.

Ratcliff, J. D., "I Am Joe's Bloodstream," *Reader's Digest*, Jan. 1974, pp. 70–74.

Raw, I., and Holleman, G. W., "Water—Energy for Life," *Chemistry*, **46**: 6–11 (May 1973).

VanderWerf, Calvin A., *Acids, Bases, and the Chemistry of the Covalent Bond*, New York, Reinhold Publishing Corp., 1961.

chapter 9

Discarded tires in Lake Tahoe on the California–Nevada border. The half-life of a tire is hundreds of years long. *(Photo 25, EPA-DOCUMERICA—Belinda Rain.)*

Reaction Rates and Pathways

9.1 **Introduction** In Chapters 7 and 8 you learned that the numerical value of the equilibrium constant (K) measures the *amount* of reaction that must take place in a nonequilibrium system before the rates of the forward (→) and reverse (←) reactions become equal and equilibrium is reached. But the value of K does not tell you what the rates of the reactions are, only that they are equal at equilibrium. Equilibrium constants cannot give any information about the rate of approach to equilibrium because an equilibrium situation is *independent of the path by which it is reached.* (You saw this with the H_2–I_2–HI equilibrium in Chapter 7.) In contrast, the speed or *rate of a reaction* carrying a nonequilibrium system to equilibrium is *strongly path-dependent*. We saw an example in the catalysis of the formation of ammonia from its elements. This chapter covers reaction rates and the pathways at the molecular level by which typical reactions proceed.

Reaction rates have practical as well as theoretical importance. The value of the equilibrium constant for the production of a useful chemical may be large, thus favoring the product side of the equilibrium. But a low rate of formation of the product can prevent economical production of the chemical. If a catalyst is invented, the day may be saved.

On the theoretical side, determining the rate of a reaction and how it is affected by changing the concentrations of reactants, the temperature, and other factors is usually the first step in discovering the pathway by which reactant microparticles go to products. Since the objective of chemists is to explain reactions in terms of the behavior of individual atoms, molecules, and ions, reaction rate theory is very important in chemistry. Finding pathways of reactions is one of the great challenges to chemists, and one very difficult to meet successfully. It took many chemists working for more than 50 years to determine the detailed path by which an ester is hydrolyzed to an acid and an alcohol (Section 7.3).

The rates of reactions range widely. The rusting of iron [$4Fe(s) + 3O_2(g) \rightarrow 2Fe_2O_3(s)$] is an example of a slow reaction that takes years for completion. On the other hand, the explosion of a charge of

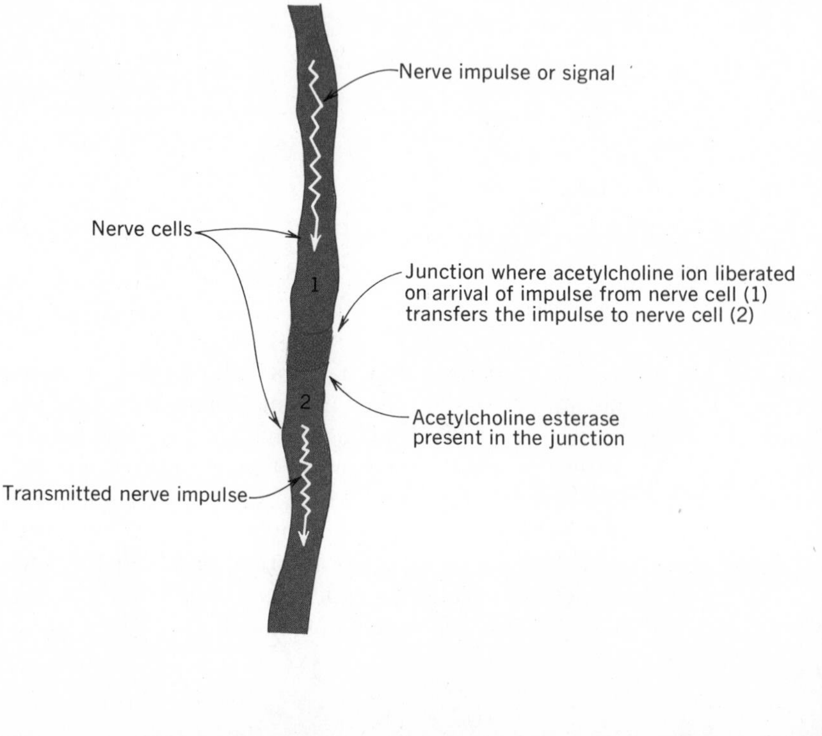

FIG. 9.1 Role of acetylcholine ion and the enzyme acetylcholine esterase in the transmission of nerve impulses.

TNT (trinitrotoluene) is a reaction that is over in a fraction of a second. Reactions in living systems are often hastened by natural catalysts called *enzymes*, leading also to very rapid reactions. The enzyme *acetylcholinesterase* (ACE) catalyzes the hydrolysis of acetylcholine ion, a carrier of nerve impulses across the junctions of nerve cells (Figure 9.1).

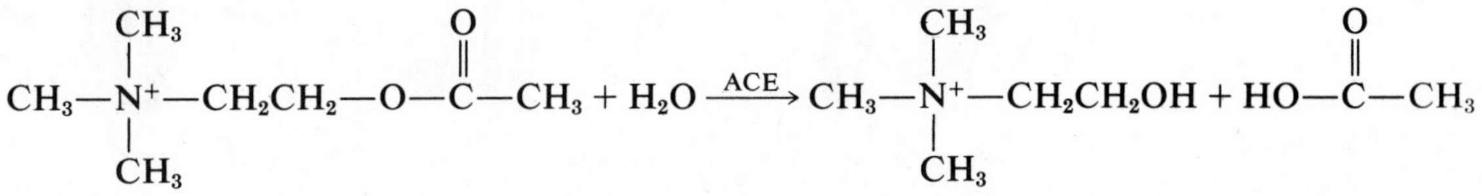

$$
\underset{\text{Acetylcholine ion}}{
\begin{array}{c}
\text{CH}_3 \\
| \\
\text{CH}_3-\text{N}^+-\text{CH}_2\text{CH}_2-\text{O}-\overset{\displaystyle\text{O}}{\overset{\|}{\text{C}}}-\text{CH}_3 \\
| \\
\text{CH}_3
\end{array}
} + \text{H}_2\text{O} \xrightarrow{\text{ACE}}
\underset{\text{Choline ion}}{
\begin{array}{c}
\text{CH}_3 \\
| \\
\text{CH}_3-\text{N}^+-\text{CH}_2\text{CH}_2\text{OH} \\
| \\
\text{CH}_3
\end{array}
} + \underset{\text{Acetic acid}}{
\text{HO}-\overset{\displaystyle\text{O}}{\overset{\|}{\text{C}}}-\text{CH}_3
}
$$

This hydrolysis is needed for transmission of nerve impulses, and it has been found that one acetylcholinesterase molecule catalyzes the hydrolysis of 5000 acetylcholine ions per second. Nerve gases and phosphonate insecticides kill by destroying the activity of acetylcholine-esterase, thus paralyzing the nervous system (Section 9.5).

The early stages in the photosynthesis of glucose from carbon dioxide and water are also very rapid. When a growing plant is exposed to radioactive carbon dioxide ($^{14}_{6}\text{CO}_2$) for only 30 seconds, the labeled ($^{14}_{6}\text{C}$) carbon atoms are built into more than a dozen compounds that are intermediates in the formation of glucose, as well as into the glucose itself.

So you see, reactions may be "fast" or "slow," but these are qualitative terms, and before rate data can be used to find reaction pathways, a quantitative or numerical way of describing the rates of reactions is needed.

9.2 Experimental Rate Laws There are several ways to express the rate of a chemical reaction numerically. The most convenient is usually chosen, but all boil down to measuring the rate at which some property of the system changes with the passage of time as reaction proceeds. For reactions of gases having an increase in the total number of molecules, pressure will increase as reaction proceeds, provided that the reaction is carried out at fixed temperature and in a container of constant volume. An example is the decomposition of hydrogen peroxide conducted at 180°C, where all participants are gases (Figure 9.2).

$$2\text{H}_2\text{O}_2(g) \rightarrow 2\text{H}_2\text{O}(g) + \text{O}_2(g)$$

2 molecules 3 molecules

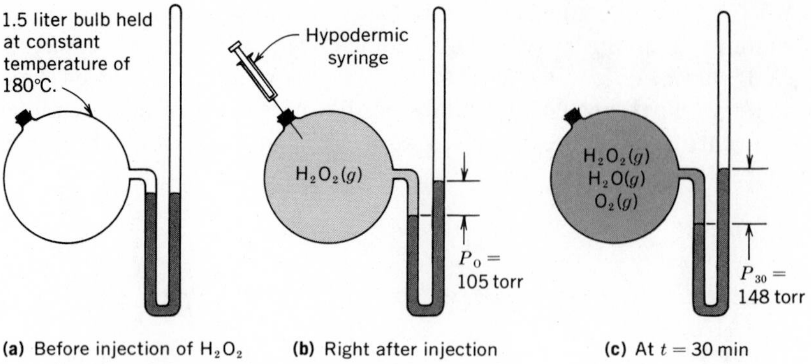

Time (min)	*Pressure (torr)*	*Time (min)*	*Pressure (torr)*
$t_0 \rightarrow$ 0.0	105 $\leftarrow P_0$	20.0	141
2.0	110	25.0	145
4.0	115	30.0	148
6.0	120	40.0	153
8.0	125	50.0	156
10.0	129	60.0	157
12.0	132	70.0	158
14.0	135	80.0	158
16.0	137	$t_\infty \rightarrow$ 90.0	158
18.0	139		

(d) Data table

FIG. 9.2 Apparatus and experimental data for the decomposition of H_2O_2 in the gas phase.

The data for this reaction as followed by measuring the total pressure with the passage of time are shown in Figure 9.2(d). The initial pressure, $p_0 = 105$ torr, is due to H_2O_2 molecules alone. As the decomposition proceeds, the number of H_2O_2 molecules decreases, but the total number of molecules increases, because each time *two* H_2O_2 molecules decompose, *three* molecules—two H_2O plus one O_2—appear. As the number of molecules increases, the pressure increases because the total pressure of a gas at constant volume and temperature is proportional to the total number of molecules (or moles) present $[P_{total} = n_{total}(RT/V)]$. When all of the H_2O_2 has decomposed, the total pressure $p_\infty = 158$ torr. This is 1.5 times the initial pressure $p_0 = 105$ torr, since $105 \times 1.5 = 158$. The factor of 1.5 appears because the total number of molecules at the end of the reaction is 1.5 times the number at the start, $3.0 = 1.5 \times 2$.

When the time-pressure data in Figure 9.2(d) are plotted, you get the graph in Figure 9.3. Three sections of the graph stand out:

1. Between times 0 and 10 minutes, the graph is almost a straight line.
2. In the region 10 minutes to about 50 minutes, the graph is a curve.
3. After about 70 minutes the graph becomes a horizontal straight line.

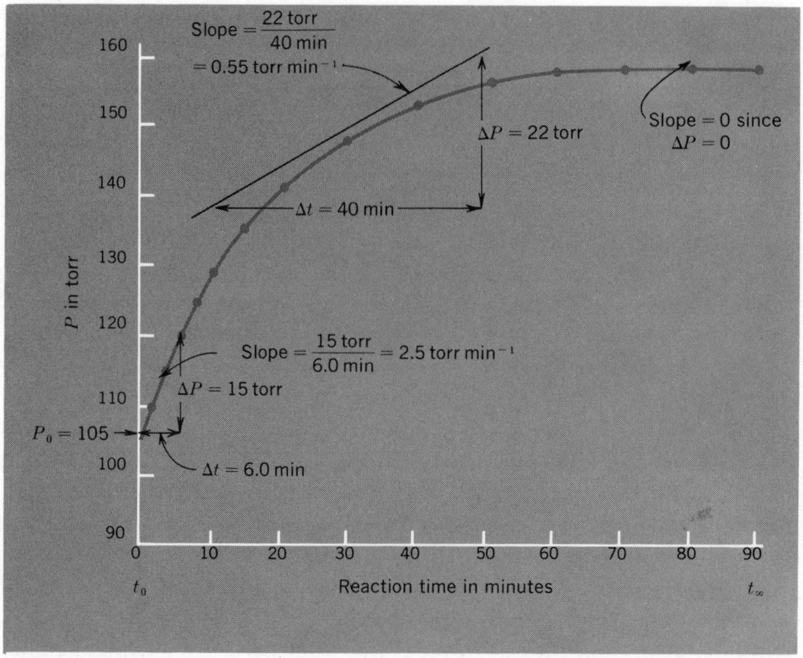

FIG. 9.3 Graphic treatment of a set of data on hydrogen peroxide decomposition.

The different curvatures of the graph in these three regions can be related to the rate of the reaction if *the rate of the reaction is defined as the change* (ΔP) *in pressure divided by the time interval* (Δt) *during which the pressure change occurs*:

$$\text{Rate of reaction} = \frac{\text{pressure change}}{\text{time interval for change}} = \frac{\Delta P}{\Delta t}$$

In the first 10 minutes the straight-line character of the graph means that the rate of reaction during this period doesn't change much. The rate $(\Delta P/\Delta t) = (15 \text{ torr}/6.0 \text{ min}) = 2.5$ torr per min is obtained, as shown in Figure 9.3. The rate is constant in this interval because most of the hydrogen peroxide remains undecomposed, and so is present in a nearly constant amount. As reaction continues, this will no longer be true, and as the amount of H_2O_2 decreases the rate decreases. This decrease in rate produces the curved portion of the graph, where the rate of the reaction as measured by $\Delta P/\Delta t$ is constantly decreasing from the maximum of 2.5 torr per min at the start, toward 0 torr per min at the end. For example, the rate at 30 min is $(22 \text{ torr}/40 \text{ min}) = 0.55$ torr per min, as shown in Figure 9.3.

Beyond 70 min of reaction, pressure change ΔP becomes zero because all of the hydrogen peroxide has decomposed. This means the rate of reaction, $(\Delta P/\Delta t)$, becomes $(0/\Delta t) = 0$. The reaction is finished.

Out of this analysis of the graph comes a very important idea: *the rate of a reaction in a closed system depends on how long the reaction has been going on.* This means that when we speak of the "rate" of a reaction in a closed system we really mean the rate at some particular moment in the "life" of the reaction, or the *instantaneous rate* of the reaction. For the present case, the instantaneous rate at a particular moment is the *slope* $(\Delta P/\Delta t)$ of the line drawn tangent to the curve in Figure 9.3 at the particular moment. Drawing a series of tangents to the curve will quickly show you that the slope (rate) does decrease from 2.50 torr per minute to 0.00 torr per minute.

In *open* systems, such as an oil refinery where crude oil comes in at one end and finished products such as gasoline, kerosene, and lubricating oil go out at the other, the reaction rates do not change with time.

The idea of the "half-life" of a reaction is useful for calculating how much of the reactants remain after a time in the life of a reaction. The half-life of a reaction $(t_{1/2})$ is the time required for one-half of the reactants to go to products. For the decomposition of H_2O_2, the half-life under the conditions giving the graph in Figure 9.3 is 31 min. (The method for calculating half-lives is described in Section 16.2.) This half-life means that after 31 min one-half of the H_2O_2 originally present will have decomposed to H_2O and O_2. Half-lives are particularly useful for understanding how the quantity of reactants decreases in slow or very slow reactions.

EXAMPLE

Nuclear bombs of the fission type (Chapter 16) produce the "unnatural" isotope strontium $^{90}_{38}Sr$ (unnatural because it does not occur in geonormal strontium). The $^{90}_{38}Sr$ comes down to earth in the "fallout" of nuclear explosions. The isotope has a half-life of 28 years, decaying by nuclear reactions to a stable isotope of zirconium $^{90}_{40}Zr$:

$$^{90}_{38}\mathrm{Sr} \rightarrow {}^{90}_{39}\mathrm{Y} + {}^{0}_{-1}e \qquad t_{1/2} = 28 \text{ yr}$$

$$^{90}_{39}\mathrm{Y} \rightarrow {}^{90}_{40}\mathrm{Zr} + {}^{0}_{-1}e \qquad t_{1/2} = 64 \text{ hr}$$

The electrons $(^{0}_{-1}e)$ come out of the nuclei of $^{90}_{38}Sr$ and $^{90}_{39}Y$ at very high speed as β rays (Section 1.4). The value of 28 years for its half-life means that however much $^{90}_{38}Sr$ is produced in a nuclear explosion, one-half will remain undecayed after 28 years. After another 28 years, one-fourth of the original amount will remain, after another 28 years, one-eighth, and so on. The danger of $^{90}_{38}Sr$ pollution is that the isotope replaces calcium in bone, where the β rays, by damaging cells, can

Those who have studied trigonometry will recognize that the slope $(\Delta P/\Delta t)$ is the *tan* of the angle between the line drawn tangent to the curve at any point and the horizontal (t) axis. Those who have studied calculus will recognize that the rate is the value of the first derivative of P with respect to t, or that the instantaneous rate is the value of dP/dt at any point on the curve.

PHOTO 26 Dust, caused by ground collapse following an underground nuclear explosion, rises from the Nevada desert. This dust is *not* radioactive—all radioactivity from such an explosion is trapped within the cavity deep below the surface of the earth. The half-life of the nuclear reaction of the explosion is less than 0.001 second. (*Los Alamos Scientific Laboratory*.)

cause bone cancer and possibly leukemia. Atmospheric testing of nuclear bombs was stopped by the U.S. and the U.S.S.R. several years ago to prevent $^{90}_{38}Sr$ fallout. At the time this book was being written a few other nations were continuing atmospheric tests. Underground tests (Photo 26) do not contaminate the atmosphere.

Chemical compounds released into the environment also can be evaluated with respect to their half-lives. For compounds, in contrast to atoms such as $^{90}_{38}Sr$, half-lives depend on the conditions the compounds are exposed to, so it is not possible to make any specific estimate of the half-lives of compounds. Chlorinated hydrocarbon insecticides have half-lives of at least tens of years. For example, although DDT is rapidly converted to DDE (Section 4.5), DDE appears to last "forever" under the environmental conditions in the fat of birds and animals, including humans. No toxic effects of DDE have appeared in humans as far as is known, but the long-term effects cannot be evaluated. And DDE will be around for a long time because it has an even longer half-life than DDT. Rubber tires also have a long half-life (Photo 25). ⊸

Exercises

1. Draw the tangent to the curve in Figure 9.3 at 40 minutes of reaction time and determine the instantaneous value of the rate of the reaction.

2. For the reaction $H_2(g) + I_2(g) \rightarrow 2HI(g)$ there is no volume change, so pressure change can't be used to follow the rate. How might the rate of this reaction be measured in its early stages? [*Hint*: $I_2(g)$ is deep purple.]

3. Give examples of two slow and two rapid reactions that occur in the home.

Rate equations The rate equation or rate law for a reaction tells how the instantaneous rate of the reaction changes as the reactants are used up. Although you can use pressure changes to follow certain gas reactions, it is usually better to use changes in the molar concentrations of reactants or products. This way, you can handle reactions in solution as well as in the gas phase. Rate equations are usually of either the *first-order* or the *second-order* type:

First order	$\text{Rate} = kC_A$
Second order	$\text{Rate} = kC_A^2$ or $\text{rate} = kC_AC_B$

In these equations C_A and C_B are the *instantaneous* concentrations of the reactants in moles per liter and the k's are proportionality constants called *specific rate constants*. They are numerically equal to the rate of the reaction when the concentrations of all reactants are 1 mole per liter. They differ in value from reaction to reaction, and their value depends on the temperature of the reaction.

For the gas-phase decomposition of hydrogen peroxide the rate expressed in concentration changes is taken as the increase in the molar concentration of O_2 (ΔC_{O_2}) divided by the time interval (Δt) during which the increase occurred or

$$\text{Rate} = \frac{\Delta C_{O_2}}{\Delta t}$$

The way in which C_{O_2} changes during the reaction can be calculated from the pressure data in Figure 9.2(d) by using Dalton's law of partial pressures. Without going into the details of the conversion, we show the results in Figure 9.4. As before, the rate at any instant is the slope of the line drawn tangent to the curve at that instant in time.

The rate law for the decomposition of H_2O_2 is of the first-order type:

$$\text{Rate} = \frac{\Delta C_{O_2}}{\Delta t} = kC_{H_2O_2}$$

At 180°C, the value of the specific rate constant (k) is 0.0225 min^{-1}. We can now write the complete rate law for the reaction at 180°C:

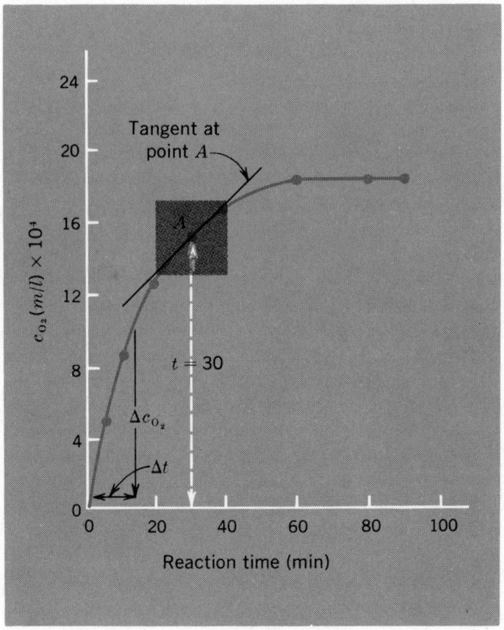

FIG. 9.4 Concentration of oxygen as a function of time in the decomposition of H_2O_2.

$$\text{Rate} = \frac{\Delta C_{O_2}}{\Delta t} = 0.0225\ C_{H_2O_2}\ \text{mole liter}^{-1}\ \text{min}^{-1}$$

EXAMPLE

You can find the value of k for the H_2O_2 decomposition by putting the values of the constant, initial (first 6.00 min) rate found from the slope, and the initial $C_{H_2O_2}$, into the equation

$$\text{Initial rate} = k(\text{initial } C_{H_2O_2})$$

Inserting the values for the left-hand member of the equation gives us

$$\text{Initial rate} = \frac{5.30 \times 10^{-4}}{6.00} = 0.883 \times 10^{-4}\ \text{mole liter}^{-1}\ \text{min}^{-1}$$

The value for initial $C_{H_2O_2}$ (Exercise 4) is

$$\text{Initial } C_{H_2O_2} = 3.71 \times 10^{-3}\ \text{mole liter}^{-1}$$

The entire equation can now be expressed as follows:

$$0.833 \times 10^{-4}\ \text{mole liter}^{-1}\ \text{min}^{-1} = k \times 3.71 \times 10^{-3}\ \text{mole liter}^{-1}$$

from which we obtain

$$k = \frac{0.833 \times 10^{-4} \text{ mole liter}^{-1} \text{ min}^{-1}}{3.71 \times 10^{-3} \text{ mole liter}^{-1}} = 0.0225 \text{ min}^{-1} \; \multimap$$

It is important to realize that we got the rate law for H_2O_2 decomposition by using *only* the *experimental* data in Figure 9.2(d) and Dalton's law. Rate laws obtained this way are called *experimental* or *empirical* rate laws—they are what is found out about the reaction at the macroscopic level in the laboratory. They are the most important clue to the microscopic pathway from reactants to products, as you will soon see. Obtaining the empirical rate law is the first step.

Exercises **4.** Calculate the molar concentration of H_2O_2 in the bulb of Figure 9.1 at 180°C before any has decomposed. Use Dalton's law in the form

$$p_{H2O_2} = \left[\frac{n_{H2O_2}}{V}\right]RT = C_{H2O_2}RT$$

and note that the bracketed factor is really the concentration of H_2O_2 or C_{H2O_2} in moles/liter.

5. Calculate the values of p_{O_2} and p_{H2O} at 180°C when the H_2O_2 decomposition is complete. Use Dalton's law in the form $p_a = X_a P_{total}$, where X_a is the mole fraction of a.

6. What additional information would you need to calculate the amounts in *grams* of H_2O_2 at the beginning of the reaction and H_2O and O_2 at the end?

7. Suppose the time rate of change in the concentration of H_2O_2 is used to express the rate of decomposition of H_2O_2. What relation will exist between the rate expressed in this way and that expressed by using the time rate of change of oxygen concentration?

The hydrolysis of esters in basic solution (saponification) discussed in Chapter 6 is an example of a second-order reaction.

EXAMPLE \multimap For the hydrolysis of ethyl acetate the overall reaction is

$$\underset{\text{Ethyl acetate}}{CH_3-\overset{\overset{\displaystyle O}{\|}}{C}-O-CH_2CH_3} + \underset{\text{Sodium hydroxide}}{Na^+ + OH^-} \rightarrow \underset{\text{Acetate ion}}{CH_3-\overset{\overset{\displaystyle O}{\|}}{C}-O^-} + Na^+ + \underset{\text{Ethanol}}{HO-CH_2CH_3}$$

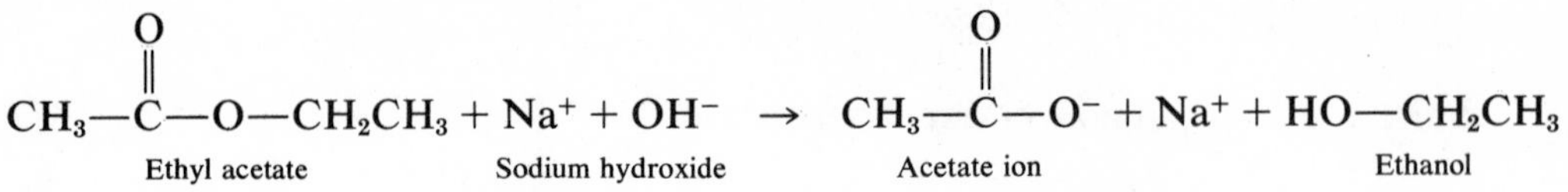

The sodium ion can be omitted in the *net* equation for the reaction because it is only a "bystander" in the reaction:

$$CH_3-\overset{\displaystyle O}{\overset{\|}{C}}-O-CH_2-CH_3 + OH^- \rightarrow CH_3-\overset{\displaystyle O}{\overset{\|}{C}}-O^- + HO-CH_2-CH_3$$

or, in shorthand notation,

$$EtAc + OH^- \rightarrow Ac^- + HOEt$$

The rate equation and order of the reaction are found by measuring its rate in the first few moments of reaction, starting with different concentrations of ethyl acetate and hydroxide ion. Table 9.1 gives typical data from some experiments. From the numbers you can see that doubling the concentration of hydroxide ion doubles the rate of reaction, so C_{OH^-} appears raised to the first power in the rate equation. Doubling the ethyl acetate concentration, C_{EtAc}, while holding C_{OH^-} constant, also doubles the rate of reaction, so C_{OH^-} also appears raised to the first power in the rate equation. The data thus show that the rate law has the second-order form, rate $= kC_AC_B$:

$$\text{Rate} = \frac{\Delta C_{Ac^-}}{\Delta t} = kC_{EtAc} \times C_{OH^-}$$

The value of k from the rate data is 1.06×10^{-1} liter mole^{-1} sec^{-1}, the rate when both reactant concentrations are 1 mole/liter, as shown in the data table. The value can also be obtained by substitution of any other of the data sets into the rate equation. Using the second set, we get

$$2.12 \times 10^{-3} = k \times 0.1 \times 0.2$$

or

$$k = \frac{2.12 \times 10^{-3}}{0.1 \times 0.2} = 1.06 \times 10^{-1} \text{ liter mole}^{-1} \text{ sec}^{-1}$$

TABLE 9.1 Initial rates in the hydrolysis of ethyl acetate at 25°C

Experiment	C_{EtAc} (mole/liter)	C_{OH^-} (mole/liter)	*Initial Rate,* $\Delta C_{HAc}/\Delta t$ (mole liter^{-1} sec^{-1})
1	0.10	0.10	0.00106
2	0.10	0.20	0.00212
3	0.20	0.10	0.00212
4	0.50	0.50	0.0265
5	1.00	1.00	0.106

Exercises

8. Using the data in Table 9.1, calculate the initial rate of hydrolysis of ethyl acetate if the initial concentrations are $C_{EtAc} = 0.15$ mole/liter, and $C_{OH^-} = 0.25$ mole/liter.

9. For a reaction having an equation of the type $2A + B \rightarrow D$ the following data were observed:

Experiment	*Initial* C_A (mole/liter)	*Initial* C_B (mole/liter)	*Initial Reaction Rate,* $\Delta C_D/\Delta t$ (mole liter^{-1} sec^{-1})
1	0.10	0.10	2×10^{-3}
2	0.20	0.20	8×10^{-3}
3	0.10	0.20	4×10^{-3}

Determine the rate law for the reaction and the value of the specific rate constant k. (*Hint*: the order of a reactant is now always apparent from the equation for the reaction.)

10. What is the advantage of measuring the *initial* rate of a reaction in determining its order with respect to reactants?

Once you have found the empirical rate law in the laboratory, the next step is to discover a pathway or *mechanism* for the reaction. This is done by applying reaction rate theory, at the heart of which are the ideas of *simple* or *elementary* reactions and the *fundamental rate laws* for them.

9.3 Reaction Mechanisms—Fundamental Rate Laws Very few of the chemical equations in earlier chapters show the molecular pathway or *mechanism* by which reactant microparticles become product microparticles. All they give is the weight relationships between reactants and products that are "chemicals" in the sense that they can be obtained in bottles or gas cylinders at the chemistry stockroom. Equations of this type are called stoichiometric equations.

Look at the equation for the formation of hydrogen bromide from its elements: $H_2(g) + Br_2(g) = 2HBr(g)$. You can imagine that with some kind of supermicroscope you would see something like the following going on in the reaction:

$$\begin{array}{c} \mathrm{H} \\ | \\ \mathrm{H} \end{array} + \begin{array}{c} \mathrm{Br} \\ | \\ \mathrm{Br} \end{array} \rightarrow \left[\begin{array}{ccc} \mathrm{H} & \cdots & \mathrm{Br} \\ \vdots & & \vdots \\ \mathrm{H} & \cdots & \mathrm{Br} \end{array}\right] \rightarrow \begin{array}{c} \mathrm{H{-}Br} \\ + \\ \mathrm{H{-}Br} \end{array}$$

Molecules of H_2 and Br_2 collide to form a fleeting, half-reacted or *transition* state, in which old bonds are partly broken and new ones partly formed. The transition state then falls apart into products. According to this picture, the only molecules you would see in the reaction mixture would be H_2, Br_2, and HBr molecules. (Transition states are too short-lived to be molecules.) This simple picture is wrong, however. Research on the reaction has shown that it is a *complex* one, and a close look at the research results will teach you a lot about rate theory. While the reaction is going on there are free hydrogen atoms ($H^{\cdot}$) and free bromine atoms ($:\ddot{\underset{..}{Br}}\cdot$) ☛ as well as H_2, Br_2, and HBr molecules flying around in the reaction mixture. Complicated reactions like this can be broken down into a series of *simple* reactions. The *idea of a simple reaction is the starting point for reaction rate theory.*

Hydrogen and bromine *atoms* are not "chemicals" available at the stockroom the way H_2, Br_2, and HBr are.

A simple (or elementary) reaction is a one-step process at the level of atoms, molecules, or ions that is described by a chemical equation in which the particles on the right-hand side (the products) are

1. The *direct* result of the collision of two or more particles on the left-hand side of the equation
2. Or the *direct* result of the breakup of a single particle on the left-hand side of the equation that has been made energy-rich by a molecular collision with an inert molecule (M) or by the absorption of radiation (light, X rays, or the like)

In both cases the particles may be atoms, molecules, or ions, and the reactant particles go through a half-reacted or transition state on the way to products.

If the reaction goes by way (1) of the collision of *two* particles A and B, it is called a *bimolecular reaction* and the theoretical rate law, using Cs for instantaneous concentrations, is

$$\text{Rate} = kC_AC_B \qquad \text{for } A + B \rightarrow \text{products}$$

or, if two identical particles react,

$$\text{Rate} = kC_A^2 \qquad \text{for } 2A \rightarrow \text{products}$$

If the reaction results from the breaking up of a *single* energy-rich particle (way 2), it is called a *unimolecular reaction* and the rate law is

$$\text{Rate} = kC_A \qquad \text{for } A \rightarrow \text{products}$$

Unimolecular reactions are less common than bimolecular ones. Even more rare are *termolecular* (three-molecule) *reactions* of the type

A + B + C → products. The reason is that the collision of three particles *at the same time* is an unlikely event, especially in the gas phase. The termolecular rate law is rate = $kC_AC_BC_C$. In all of these rate equations, the k's are the specific rate constants and the C's are the instantaneous concentrations of reactants. The *molecularity* of a *simple* reaction is the number of molecules in its transition state.

At this point it is vital that you understand the difference between the *order of a reaction* and the *molecularity of a reaction.* Order of reaction is found *experimentally* by methods already described, and it does not take for granted *any mechanism* for the reaction. Molecularity, and the corresponding rate laws, apply *only* to reactions that are *simple*—they are theoretical rather than experimental.

EXAMPLE You may be able to see the difference between molecularity and order from this example: Suppose the hydrogen–bromine reaction was the simple reaction, $H_2 + Br_2 \rightarrow 2HBr$, as we (incorrectly) assumed earlier. It would be bimolecular; that is, the rate law predicted would be the bimolecular one:

$$\text{Rate} = kC_{H_2}C_{Br_2}$$

As a matter of fact the empirical rate law obtained in the laboratory is much more complicated:

$$\text{Rate} = \frac{kC_{H_2}\sqrt{C_{Br_2}}}{1 + k'\ C_{HBr}/C_{Br_2}}$$

If, as with the HBr reaction, the empirical rate law found for a reaction doesn't correspond to any one of the fundamental rate laws for uni-, bi-, or termolecular reactions, then the reaction is complex and is made up of two or more simple reactions. Research has shown that the mechanism of the reaction $H_2(g) + Br_2(g) \rightarrow 2HBr(g)$ is stepwise and consists of the following simple reactions (Br* is an energy-rich bromine molecule produced by collision with M, an inert molecule):

		Type	*Rate Law*
1.	$Br_2 + M \rightarrow Br_2^* + M$	Bimolecular	Rate = $k_1C_{Br_2}C_M$
2.	$Br_2^* \rightarrow 2Br$	Unimolecular	Rate = $k_2C_{Br_2^*}$
3.	$Br + H_2 \rightarrow HBr + H$	Bimolecular	Rate = $k_3C_{Br}C_{H_2}$
4.	$H + Br_2 \rightarrow HBr + Br$	Bimolecular	Rate = $k_4C_HC_{Br_2}$

In step 1, collision of a bromine molecule with an inert molecule (M) makes the bromine molecule energy-rich, so it falls apart into two bromine atoms in the unimolecular step (2). In (bimolecular) step 3 a bromine atom plucks a hydrogen atom out of a hydrogen molecule leaving a hydrogen atom and forming a molecule of HBr. In (bimolecular) step 4 the hydrogen atom from step 3 reacts with a bromine molecule, forming a molecule of HBr and leaving a bromine atom. The bromine atom, as indicated by the dashed line, can repeat step 3. When the product of one reaction is a reactant in another, as here, the pair of simple reactions make up what is called a *reaction chain* or "chain reaction." Steps 3 and 4 make up a chain. Because of this chain you might think that the production of a single bromine atom would result in complete reaction of H_2 and Br_2 to HBr. This doesn't happen, however, because additional simple reactions in the mechanism break the chain by removing hydrogen atoms and bromine atoms from the reaction.

5. $H + H \rightarrow H_2$
6. $Br + Br \rightarrow Br_2$
7. $H + Br \rightarrow HBr$

(Steps 5–7: Chain-terminating steps)

Each of the reactions of any mechanism has its own rate constant and transition state. By a proper combination of the rate laws for the separate simple reactions, you can derive the empirical rate law found for a chain reaction. This has been done for the HBr reaction, but the details would be out of place here.

11. The overall equation for the decomposition of N_2O_5 (nitrogen pentoxide) is $2N_2O_5(g) \rightarrow 4NO_2(g) + O_2(g)$. The reaction is first-order in N_2O_5, and the mechanism is — Exercise

$N_2O_5 \rightarrow NO_3 + NO_2$	Slow	(1)
$NO_3 \rightarrow NO + O_2$	Fast	(2)
$NO + N_2O_5 \rightarrow NO_2 + N_2O_4$	Fast	(3)
$N_2O_4 \rightarrow 2NO_2$	Fast	(4)

a. Which steps are unimolecular? Which bimolecular?

b. Write the rate laws for the steps.

c. The first step is the slowest or *rate-limiting* step in the reaction. Explain how this results in the observed order of the reaction.

9.4 Molecularity, Order, and Mechanisms The molecularity of a reaction is the number of atoms, molecules, or ions on the left-hand side of the equation for *a simple reaction* ***only***. The *order* of a reactant is the power to which the concentration of the reactant is raised in the empirical rate law. Molecularity is a theoretical concept; order comes directly from experiment. The *mechanism* of a reaction is the list of simple reactions through which reactants get to products.

Most equations written for chemical reactions are the overall result or sum of a number of simple reactions. If the empirical rate law for a reaction is *not* one of the simple types—unimolecular (rate $= kC_A$), bimolecular (rate $= kC_AC_B$), or termolecular (rate $= kC_AC_BC_C$), then the mechanism *has to be complex*, as you saw in the example of HBr formation. However, *the opposite is not true*. Just because the empirical rate law takes one of the simple forms *does not mean* that the equation for the reaction must correspond to a simple reaction.

EXAMPLE The oxidation of nitric oxide (NO) to nitrogen dioxide (NO_2), important in the production of nitric acid (HNO_3) and smog, is an example of a complex reaction that has a simple rate law:

$$2NO(g) + O_2(g) \rightarrow 2NO_2(g)$$

$$\text{Empirical rate} = kC_{NO}^2C_{O_2}$$

This empirical rate equation is the one for a termolecular reaction, going through simultaneous collision of two NO molecules and one O_2 molecule. Without going into details, we can say that other data show that the overall reaction is *not* termolecular but is complex, proceeding through two simple reactions:

$$\textbf{1.} \qquad NO(g) + O_2(g) \underset{k_{-1}}{\overset{k_1}{\rightleftharpoons}} OONO(g)$$

$$\textbf{2.} \quad NO(g) + OONO(g) \xrightarrow{k_2} 2NO_2(g)$$

$$\text{Sum} \quad 2NO(g) + O_2(g) \longrightarrow 2NO_2(g)$$

Step 1 is a rapidly established equilibrium. Step 2 is a slower, irreversible reaction. Application of the fundamental rate law to the separate steps of this mechanism gives the empirical rate law found, even though the reaction is not termolecular, but proceeds through the two bimolecular, simple reactions.

Nitric oxide (NO) has a half-life of hours in the low concentrations found in photochemical smog (Section 15.3) for a reason that you can readily see from the mechanisms for the conversion to NO_2. The first step in the mechanism goes very rapidly because the small number of NO molecules are constantly colliding with the large numbers of O_2 molecules present in air. But the second step requires that the relatively few OONO molecules collide with the even rarer NO molecules before NO_2 can be formed. Nitric oxide thus acts as an atmospheric "reservoir" for the production of NO_2. At the high altitudes reached by supersonic aircraft (SST's, Section 15.3), the concentrations of O_2 molecules in the air and of NO from the exhaust of the aircraft are so low that NO is estimated to have a half-life of months, even years, depending on other factors. If high-atmospheric pollution by SST's does have bad effects, those effects could be around for a long time if they result from contamination by NO, even if the SST's were discontinued.

Exercises

12. Step 2 in the mechanism for the oxidation of NO to NO_2 is very much slower than step 1. Which has the larger value, k_2 or k_1?

13. Even though the concentration of the intermediate OONO cannot be measured because it is so small, the fundamental rate law for step 2 in the oxidation of NO can be written. Give the rate law. What is the molecularity of step 2?

14. Give electron-dot structures for NO, ONO, and OONO (note arrangement of kernels). Which would be paramagnetic?

15. Using dashed lines for bonds being formed or broken, sketch the transition states for steps 1 and 2 in the oxidation of NO.

9.5 Conditions Affecting Reaction Rates Reaction rates depend on (1) concentrations of reactants, as we have seen; (2) area of contact between phases if the reaction is heterogeneous, occurring where two phases meet; (3) temperature; and (4) presence of catalysts. We will examine these influences in that order, but before we can even discuss them, it must be known that the reaction of interest "goes" or has the possibility of going.

The *possibility* for a reaction to occur depends on the value of its equilibrium constant. The rate of a reaction cannot be measured unless the value of the equilibrium constant is large enough to require measurable amounts of *products* at equilibrium. For example, the rate of the reaction $CO_2(g) \rightarrow C(s) + O_2(g)$ cannot be measured at 25°C because the value of its equilibrium constant is $K = 10^{-69}$, and the concentrations of products are so tiny that they cannot be measured. On the other hand, the reverse of this reaction, the burning of carbon, $C(s) + O_2(g) \rightarrow CO_2(g)$ has $K = 10^{+69}$. This reaction certainly goes, and has been studied because burning of carbon in the form of coal is one of our important energy sources. The reaction is a *heterogeneous* one, because *solid* carbon reacts with *gaseous* oxygen, and reaction occurs where the solid and gas phases meet. As might be expected, the reaction is much faster in pure oxygen than in air because the higher concentration of O_2 molecules in pure O_2 results in more collisions of them with the surface of the solid carbon where the reaction takes place (Figure 9.5). The rate increases also as the amount of carbon surface exposed to oxygen increases.

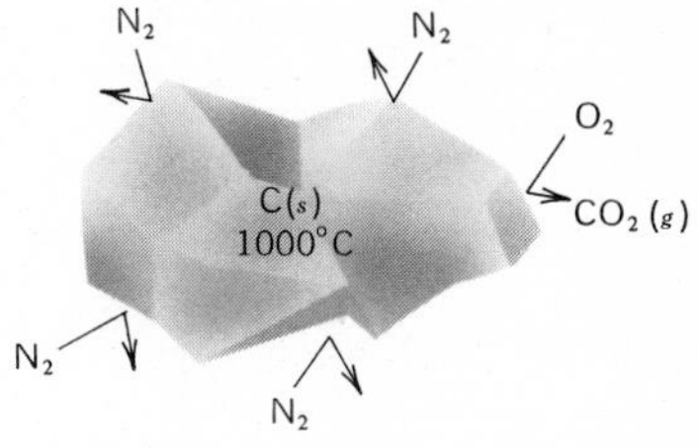

Air: 80% of molecular collisions with the surface are of N_2 molecules, which do not react.

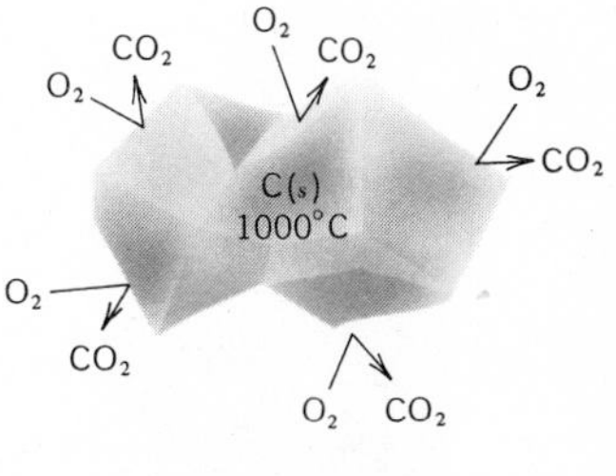

O_2: 100% of collisions with surface are of O_2 molecules, which do react with hot carbon.

FIG. 9.5 Air (20% O_2, 80% N_2) and pure oxygen reacting with solid carbon.

EXAMPLE When a large sample of matter is cut or ground into finer and finer particles, the total amount of surface increases. In very finely divided carbon (lampblack), the total surface of one gram of the powder may be about 1 square meter. Carbon particles this small burn with explosive rapidity when spread out in air and ignited by a spark or flame. Many coal mine disasters have been traced to this kind of "dust explosion," coal dust being a form of finely divided carbon.

It is generally true of heterogeneous reactions that the greater the area of contact between phases is, the faster they go. The reaction $Zn(s) + HCl(aq) \rightarrow ZnCl_2(aq) + H_2(g)$, for example, goes much more rapidly with powdered zinc than with zinc pellets.

Exercises

16. By calculating the area before and after cutting, show that the total surface area increases when a 1.00-cm cube is cut in half.

17. Disastrous dust explosions have also occurred in "grain elevators" (wheat storage warehouses). What combustible dust is responsible? Would you expect flour dust to explode if dispersed in air and ignited?

18. Suppose the *pressure* of O_2 gas over burning carbon is increased. Would the rate of reaction increase or decrease? Explain.

19. At one time the "atmosphere" in the cabin of a manned space vehicle was almost pure O_2 at about the same partial pressure ($p_{O_2} = 159$ torr) as in the atmosphere at sea level. Account for the suddenness and intensity of the fire that killed three astronauts occupying a closed space capsule undergoing ground tests.

Simple reactions in a single phase are called *homogeneous* reactions. The effects of changing the concentrations of reactants on their rates are given by the fundamental rate laws for unimolecular, bimolecular, or termolecular reactions. According to the laws, increasing concentrations of reactants increases rates. This effect is most easily explained with bimolecular reactions. If two molecules are to react, they must first collide with each other. If their concentrations (number of molecules per liter) are increased, there will be more collisions between them and more opportunity for reaction. In the case of unimolecular reactions, increasing the concentration of the reacting particle means that the concentration of "activated" or energy-rich ones will be increased too. The rate therefore increases.

Temperature effects on rates—activation energy For nearly all reactions, only a small fraction of the collisions between reactant particles result in the formation of products. Most of the collisions are of the

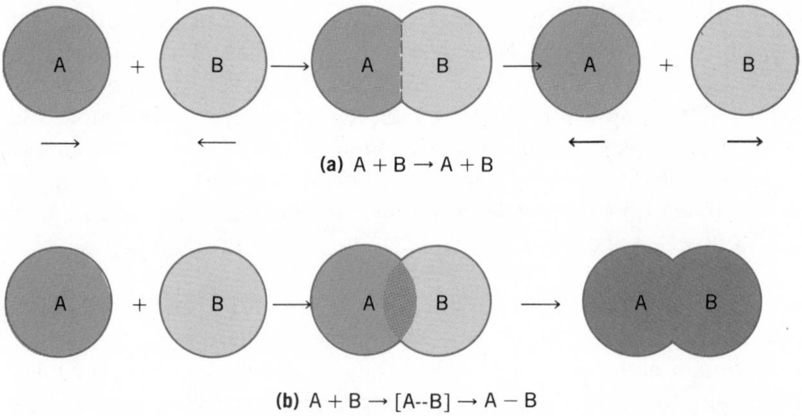

FIG. 9.6 Bimolecular collisions: (a) elastic, no reaction; (b) inelastic or fruitful, leading to an activated complex and products. (Electron clouds are shown with discrete boundaries for convenience in representation.)

elastic or nonfruitful type shown in Figure 9.6(a). Kinetic molecular theory explains why. Using that theory, we can calculate the number of collisions between molecules in 1 liter of a gas at 0°C, 1 atm pressure, to be about 10^{31} per second. Now, one mole is 6.023×10^{23} molecules, so if reaction occurred on every single molecular collision, then $(10^{31}/6.023 \times 10^{23}) = 1.6 \times 10^7$ *moles* (not molecules) would react in 1 liter in 1 second. Very few reactions are anywhere near this fast. Exceptions are reactions between oppositely charged ions, such as $H_3O^+ + OH^- \rightarrow 2H_2O$. This reaction occurs as rapidly as the ions come together, with a rate constant of 10^{11} liter mole^{-1} sec^{-1} at 25°C.

The reason only a fraction of molecular collisions result in reaction is that only the more vigorous collisions between molecules bring their centers close enough together to form new bonds and break old ones. The more vigorous collisions occur between the more rapidly moving molecules, and these are relatively small in number according

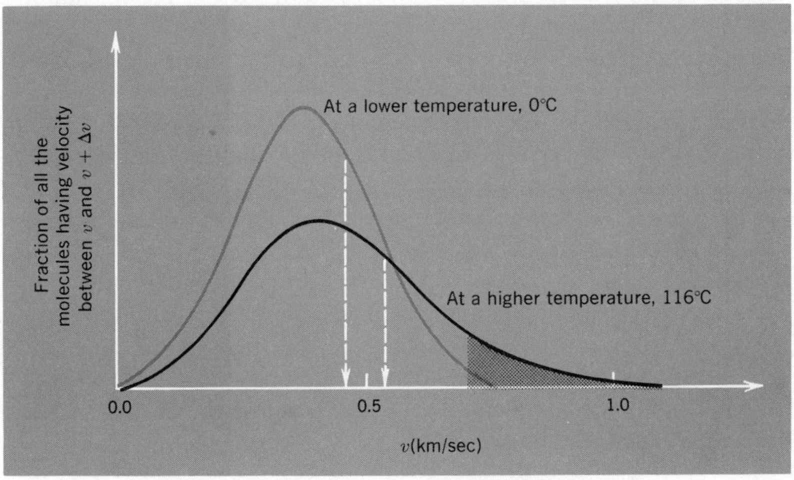

FIG. 9.7 Distribution of molecular velocities in oxygen gas at a lower and higher temperature (schematic; dashed arrows show the value of the average velocities).

to the distribution of molecular velocities shown in Figure 9.7. The collisions that are hard enough to result in reaction are those in the shaded areas at the right-hand "tails" of the curves. As shown in the figure, increasing the temperature increases the fraction of molecules moving rapidly enough to react on collision, so an increase in temperature increases the rates of most reactions. Exceptions are reactions of living things, in which a large enough increase in temperature stops reactions and kills.

EXAMPLE

The normal body temperature of 37°C (98.6°F) is maintained by a sensor in the brain that balances the rate of heat production by the oxidation of glucose and fat against the rate of heat loss from the body. When you feel cold you shiver to warm up by increasing muscle activity and the rate of heat production by the "burning" of glucose and fat. In disease, body temperature may rise to 41°C (106°F) but if it remains that high for very long or goes higher (to 108°F), permanent brain damage or death will occur. The increase in body temperature is the body's way of speeding up the chemical reactions that fight the disease, but if body temperature goes too high, vital reactions stop. Biochemical reactions differ from simple chemical reactions in this way. Their rates increase with increasing temperature up to the point at which the high temperature destroys the delicate balance of reaction rates upon which life depends.

The body can stand lowered temperatures much better than elevated ones. Lowering body temperatures slows down the rate of metabolism, and with the slowdown the need for oxygen decreases rapidly. Advantage is taken of the decrease in need for oxygen during open-heart surgery and other operations in which blood circulation is interrupted. The patient's body temperature is lowered (to as low as 90°F) by circulating his blood through cooling coils or by covering him with hollow blankets through which cold water is pumped. The procedure, called *hypothermia*, reduces oxygen need to the point at which even very slow breathing and slow circulation of the blood bring an adequate amount of oxygen to the cells. The cells of the brain are quickly damaged by lack of oxygen (*anoxia*). After only 5 minutes of nonbreathing (*apnea*) irreparable damage to brain cells occurs at normal body temperature. Under hypothermia, breathing can be interrupted for several minutes without brain damage.

Kidneys for transplant operations are carried in air liners without damage using hypothermia.

Changes in rate accompanying changes in the temperature of a reaction result from changes in the value of the specific rate constant k. This means you can separate concentration and temperature effects. Taking a bimolecular reaction between A and B molecules as an example, we can show the separation as follows:

$$\text{Rate} = kC_{\text{A}}C_{\text{B}}$$

Concentration effects on rate ($C_{\text{A}}C_{\text{B}}$)
Temperature effects on rate (k)

Changing concentrations changes rates by changing the total number of collisions between reacting particles. Temperature changes alter the value of k and the rate by altering the fraction of collisions that lead to reaction.

The Arrhenius equation, named for its developer, Svante Arrhenius, tells how the value of the rate constant k and the rate of a reaction respond to changes in temperature. The equation has many mathematically equivalent forms, and one that would be suitable here is

Other forms are $k = Ae^{-E_a/RT}$ and $2.3 \log k = (-E_a/RT) + C$.

$$k = \frac{B}{10^{(E_a/2.3RT)}}$$

Increase in T makes k larger
Increase in E_a makes k smaller

Svante August Arrhenius (1859–1927) was a Swedish chemist and physicist. His revolutionary electrolytic dissociation theory of acids was gradually accepted, and he received the Nobel prize in chemistry in 1903.

In the equation k is the specific rate constant, B is a constant number different for each reaction but related to the number of collisions per unit time between reacting molecules, E_a is the *activation energy* for the particular reaction, R is the gas constant, and T is the absolute temperature. By measuring the value of k at a series of temperatures the values of B and E_a for a reaction can be obtained, but the details of this procedure need not concern us. What is important to see in the equation is how the values of k and the rate of reaction depend on the temperature (T) of the reaction and on the value of its activation energy, E_a. Because both of these quantities are in the denominator of the equation, it is there we must look. As the value of T increases, the value of the exponent ($E_a/2.3RT$) of 10 in the denominator will get smaller, because E_a will be divided by larger and larger values of $2.3RT$. Decreasing the exponent means that the quantity $10^{(E_a/2.3RT)}$ will also get smaller. With smaller and smaller numbers dividing into B, constant k and the rate become larger. Because the value of k depends as a power of 10 on the value of T, small changes in the temperature of a reaction can produce large changes in the value of k and its rate. For this reason, great care must be taken to maintain constant reaction temperature (±0.01°C) in rate studies. A rule of thumb is that the rate of a reaction that goes at room temperature is approximately doubled by a 10°C increase in temperature.

Activation energy appears in the numerator of the exponent of 10 in the Arrhenius equation, so as E_a *increases* from reaction to reaction *the denominator will increase and the rate constants and rates will decrease*. In other words, slow reactions have large activation energies, and vice versa. As with temperature, the dependence of rate on activation energy is as a power of 10, so small changes in the value of E_a also result in large changes in rate. Doubling E_a decreases the rate a hundredfold.

Exercises

20. The main ingredient of turpentine, α-pinene, rearranges to limonene, a constituent of lemon oil, on being heated to about 200°C.

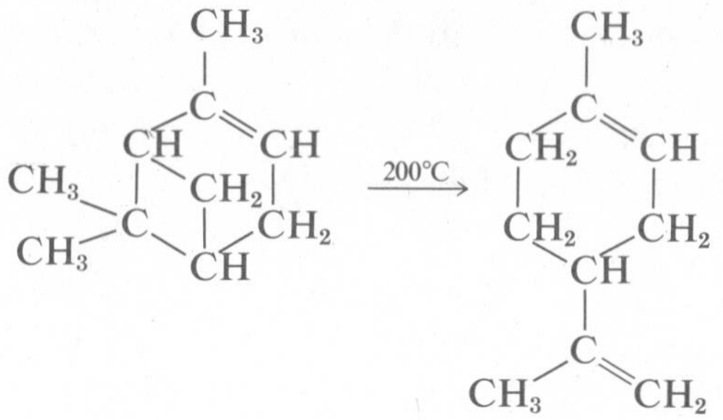

α-Pinene Limonene

The reaction is simple, with $k = 1.5 \times 10^{-5}$ sec^{-1} at 200°C. Give the rate law for the reaction. What are the units of the rate? What is the molecularity of the reaction?

21. The activation energy, E_a, for the pinene–limonene reaction is 43,000 cal $mole^{-1}$. Using the Arrhenius equation with $R = 1.987$ cal $mole^{-1}$ deg^{-1}, calculate the value of the constant B for the reaction at 200°C.

You have seen that the activation energy for a reaction influences its rate by regulating what *fraction* of all collisions are energetic enough to result in reaction. (In reactions having larger activation energies, k is smaller and fewer collisions lead to reaction.) What is activation energy in molecular terms? Formation of the transition state requires the more vigorous molecular collisions, those with energy in excess of the energy of average collisions, and the excess energy above the average is the activation energy, E_a. Activation energies are shown in Figure 9.8 for the two elementary reactions

$$Br(g) + H_2(g) \rightarrow HBr(g) + H(g) \qquad E_a = 18 \text{ kcal/mole}$$
$$H(g) + Br_2(g) \rightarrow HBr(g) + Br(g) \qquad E_a = 1.2 \text{ kcal/mole}$$

In part (a) of the figure, the vertical coordinate measures the energy of collisions between bromine atoms and hydrogen molecules. Part (b) shows the energy for collisions between hydrogen atoms and bromine molecules. When two particles collide, the kinetic energy of their mo-

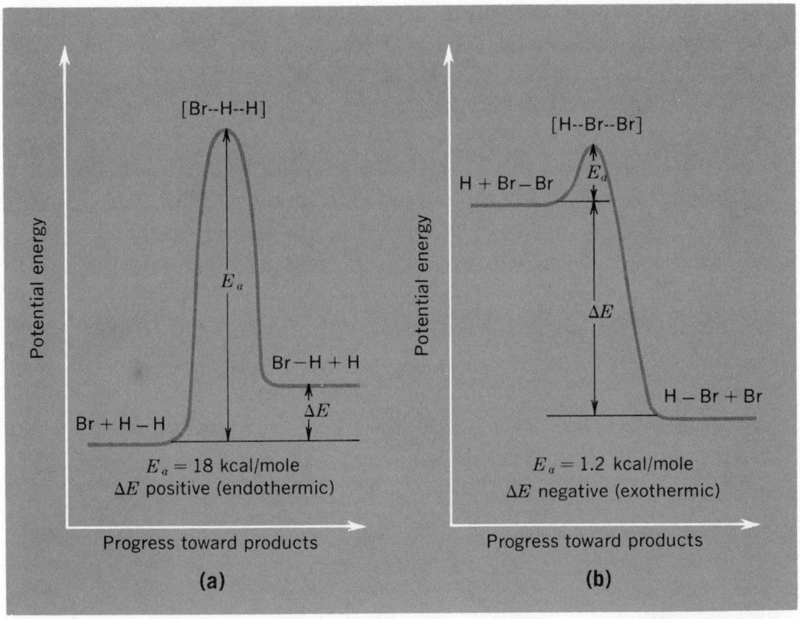

FIG. 9.8 Potential energy diagrams for elementary processes:
(a) Br· + H_2 → HBr + H·
(b) H· + Br_2 → HBr + Br· (not to scale).

tion toward each other is converted into potential energy that reaches its highest value when their motion toward each other stops. (Think of the kinetic-to-potential energy changes in bouncing a rubber ball on a hard surface.) If the collision is elastic (no reaction), the potential energy stored in the distortion of electron clouds at the instant of closest approach of the particles turns back into kinetic energy as they fly apart [Figure 9.6(a)]. The potential energy of average collisions between molecules is about equal to RT per mole of collisions, where R is the gas constant (1.987 cal mole^{-1} deg^{-1}) and T the absolute temperature. Even at 500°C (773 K) the potential energy of *average* collisions is only $1.987 \times 773 = 1540$ cal/mole = 1.54 kcal/mole. This is much less than the activation energy of 18 kcal/mole needed to reach the transition state [Br----H----H] at the top of the "energy hill" in Figure 9.8(a). Most of the collisions between bromine atoms and hydrogen molecules have energies that lie on the lower portion of the left-hand curve in the figure and so are elastic. A few of them are vigorous enough to be inelastic and get the particles over the energy hill and on to products.

A larger activation energy for a reaction means that a higher energy hill must be "climbed" to get to the transition state. The result is a slower reaction. Raising the temperature increases the vigor of collisions and also the fraction having higher energies (Figure 9.7). This is why increasing the temperature of a reaction accelerates it.

The values of activation energies for reactions that take place in the temperature range 0–500°C run from zero to about 50 kcal/mole. The reaction of a hydrogen atom with a bromine molecule has the very low activation energy of only 1.2 kcal/mole [Figure 9.8(b)]. The E_a is so small for this reaction that it occurs with almost every collision between hydrogen atoms and bromine molecules at 300°C, where RT = 1.14 kcal/mole.

Why is there such a large difference in the activation energies for the reactions in Figure 9.8? Answering this question will teach us something of the inner workings of transition states and catalysts.

Look at the transition state [Br----H----H] (Figure 9.9) for the bromine atom–hydrogen molecule reaction (E_a = 18 kcal/mole). It shows that the H—H covalent bond is half-broken and the Br—H bond is half-formed. The H—H bond energy is 104 kcal/mole; to half-break it would require energy input of (104/2) = 52 kcal/mole. The bond energy for H—Br is 88 kcal/mole, so for *forming* half of an H—Br bond we get back (88/2) = 44 kcal/mole. Since we have to put in more energy (52 − 44 = 8 kcal/mole) than we get out in forming the transition state, it is not surprising that the reaction has a considerable activation energy. 🐟

Now compare the reaction of hydrogen atoms with bromine molecules, for which E_a is only 1.2 kcal/mole. The energy required to form the transition state [H----Br----Br], assuming that the H—Br bond (86 kcal/mole) is half-formed and the Br—Br bond (46 kcal/mole) is half-broken, is calculated as follows: the formation of half of a H—Br bond yields (86/2) = 43 kcal/mole; the energy required to half-break the Br—Br bond = (46/2) = 23 kcal/mole. In this case, more energy is released in forming the transition state than is required, 43 − 23 = 20 kcal/mole released. It is not surprising that the activation energy has the small value of 1.20 kcal/mole. You may wonder why there is any activation energy at all for the reaction. Part of the answer is that in coming together, the electron clouds of the H atom and Br$_2$ molecule enter and repel each other (because of their like charge) before their nuclei are close enough for H—Br bond formation to start.

The difference of 8 kcal/mole is less than E_a = 18 kcal/mole for several reasons. One is that the bonds are not exactly "half-formed" and "half-broken," in the transition state.

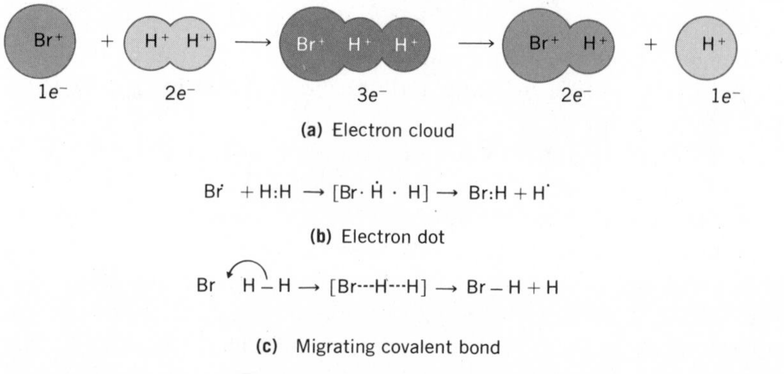

FIG. 9.9 Three representations of the formation of the transition state (Br—H—H) and its transformation to products. [In the electron cloud representation only the electrons directly involved in the bond-making and -breaking processes are shown: The symbol Br$^+$ stands for a bromine nucleus (Z = 35) surrounded by a cloud of 34 electrons which are not directly involved in the reaction; H$^+$ represents the proton.]

(a) Electron cloud

$$Br^+ + H{:}H \longrightarrow [Br\cdot\ \dot{H}\ \cdot\ H] \longrightarrow Br{:}H + H\dot{}$$

(b) Electron dot

$$Br \quad H - H \longrightarrow [Br\text{---}H\text{---}H] \longrightarrow Br - H + H$$

(c) Migrating covalent bond

Transition states are very effective in getting from reactants to products because they make available for breaking bonds the energy being released in new bond formation. Suppose that the Br—H_2 reaction did not go by way of the "helpful" transition state [Br----H----H] but instead went in two simple steps:

1. $H_2 \rightarrow 2H$
2. $Br + H \rightarrow HBr$

Overall $Br + H_2 \rightarrow HBr + H$

Step 1 of this mechanism is endothermic by 104 kcal/mole, the energy needed to break the bond in the hydrogen molecule and form two hydrogen atoms. This much larger energy quantity would be the activation energy for the reaction if it went by the two-step process instead of through the [Br----H----H] transition state, in which bond formation helps bond-breaking.

22. For the light-induced reaction $H_2(g) + Cl_2(g) \rightarrow 2HCl(g)$ the following mechanism has been proposed: Exercise

$$Cl_2 + h\nu \rightarrow Cl_2^* \qquad (1)$$

$$Cl_2^* \rightarrow 2Cl \qquad (2)$$

$$Cl + H_2 \rightarrow HCl + H \qquad E_a = 5.5 \text{ kcal/mole} \qquad (3)$$

$$H + Cl_2 \rightarrow HCl + Cl \qquad E_a = 2.5 \text{ kcal/mole} \qquad (4)$$

$$Cl + Cl \rightarrow Cl_2 \qquad (5)$$

$$H + H \rightarrow H_2 \qquad (6)$$

$$H + Cl \rightarrow HCl \qquad (7)$$

Bond energies in kcal/mole are Cl—Cl, 58.0; H—H, 104; H—Cl, 103.

a. Is the overall reaction $H_2 + Cl_2 \rightarrow 2HCl$ exothermic or endothermic? (*Hint*: use bond energies for H_2, Cl_2, and HCl.)

b. Locate a reaction chain in the mechanism.

c. Account for E_a for step 3 being greater than E_a for step 4.

d. Which steps in the mechanism are unimolecular?

e. Which steps are chain-terminating?

f. Would you expect the empirical rate law for the reaction to be rate $= kC_{H_2}C_{Cl_2}$? Explain.

Catalysts Catalysts provide pathways of low activation energy for reactions and in this regard are like transition states. The effect of a platinum metal catalyst in speeding up the hydrogenation of ethylene

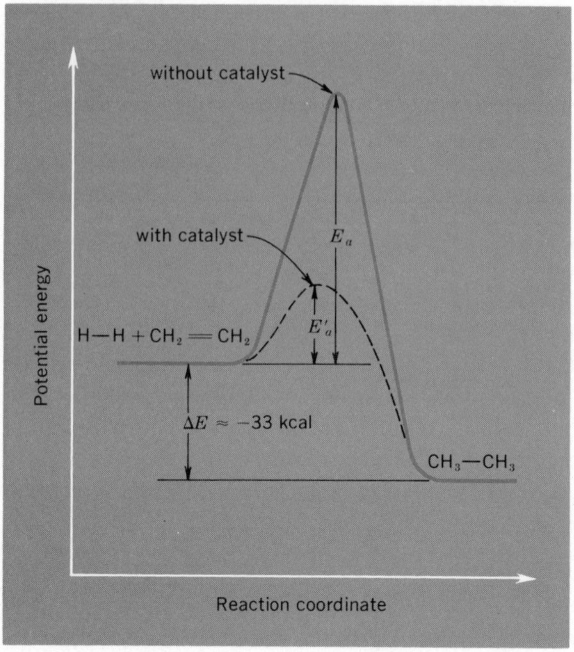

FIG. 9.10 Potential energy relationships in the hydrogenation of ethylene. $E_a \approx 43$ kcal per mole; the value of E'_a depends upon the type, mode of preparation, and history of the catalyst.

to ethane by lowering E_a is shown in Figure 9.10. Catalysts provide a "tunnel" through an activation energy peak. The hydrogenation of ethylene, $CH_2{=}CH_2(g) + H_2(g) \rightarrow CH_3{-}CH_3(g)$, is not a practical process in the absence of a catalyst because very high temperatures must be used to overcome the large activation energy of 43 kcal/mole. With a catalyst the reaction goes smoothly and rapidly at room temperature. Since this is a strongly exothermic ($\Delta E = -33$ kcal/mole) reaction, the ethane ($CH_3{-}CH_3$) side is strongly favored. The way in which the platinum surface assists in the bond-breaking and bond-making processes is shown in Figure 9.11. Catalytic hydrogenation of carbon–carbon double bonds in vegetable oils is used to make oleomargarine. The process is also used in improving the qualities of gasoline and in the manufacture of many medicinal compounds.

Platinum is acting as a *contact* or heterogeneous catalyst where solid and gas phases are brought together in this example. Many catalysts of the contact type are subject to "poisoning" by sulfur and lead compounds that make the catalyst inactive by covering the surface.

Unleaded gasoline must be used in automobiles equipped with catalytic exhaust converters for this reason.

Catalysis is called *homogeneous* when catalyst and reactants are all in the same phase. Hydronium ion is a homogeneous catalyst for the esterification reaction (Section 7.3) between a carboxylic acid and an alcohol. Although catalysts and transition states play similar roles in lowering the activation energy for reactions, they have one important difference: the particles that make up the transition state appear in the

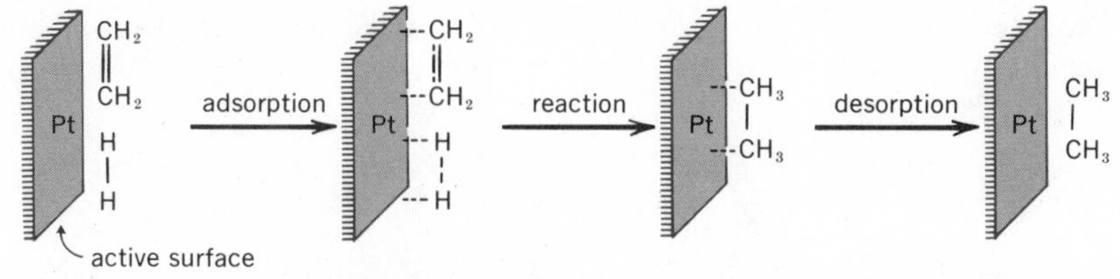

FIG. 9.11 Action of a platinum hydrogenation catalyst. (Dotted lines represent partial bonds.)

products of the reaction, but catalysts do not. They can be recovered after reaction is complete.

EXAMPLE ⚬─ In the "oxo process," the elements of formaldehyde $\overset{H}{\underset{H}{>}}C{=}O$

add to a carbon–carbon double bond under the influence of the catalyst dicobaltoctacarbonyl, $Co_2(CO)_8$, to provide a class of compounds called aldehydes. Carbon monoxide, CO, and H_2 provide the elements of formaldehyde in the reaction. Using propylene, $CH_3{-}CH{=}CH_2$, we see that the reaction is

$$CH_3{-}CH{=}CH_2(g) + CO(g) + H_2(g) \xrightleftharpoons{Co_2(CO)_8} CH_3CH_2CH_2\overset{\overset{\displaystyle H}{|}}{C}{=}O(g)$$

Butyraldehyde

Pressures of 200 atm are employed because of the favorable effect on the equilibrium (3 moles → 1 mole). You can imagine the overall reaction with propylene to go in the following steps:

$$H_2(g) + CO(g) \rightleftarrows \overset{H}{\underset{H}{>}}C{=}O(g)$$

Formaldehyde

$$CH_3{-}\overset{\overset{\displaystyle H}{|}}{C}{=}C\overset{\nearrow H}{\underset{\searrow H}{}} \rightleftarrows CH_3{-}\overset{\overset{\displaystyle H}{|}}{\underset{\underset{\displaystyle H}{|}}{C}}{-}\overset{\overset{\displaystyle H}{|}}{\underset{\underset{\displaystyle H}{|}}{C}}{-}\overset{\overset{\displaystyle H}{|}}{C}{=}O$$
$$\underset{H}{\overset{H}{>}}C{=}O$$

Butyraldehyde

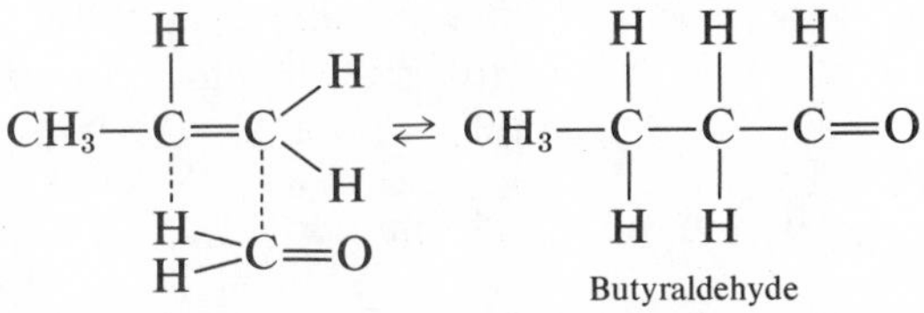

Hydrocarbons with carbon–carbon double bonds such as 1-hexene, $CH_3CH_2CH_2CH_2CH{=}CH_2$, and 1-heptene, $CH_3CH_2CH_2CH_2CH_2CH{=}CH_2$, are by-products of gasoline manufacture and are converted into their corresponding aldehydes by the oxo process. The aldehydes can be catalytically hydrogenated to the corresponding alcohols or oxidized to the corresponding carboxylic acids. Using R to represent the hydrocarbon chain of the aldehydes, we can represent the reactions as follows:

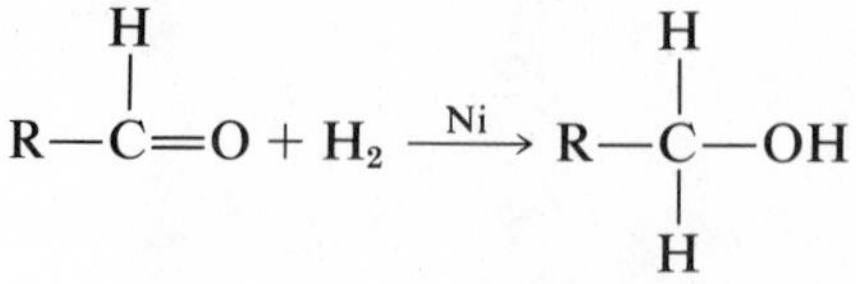

$$R-\overset{\overset{\displaystyle H}{|}}{C}=O + H_2 \xrightarrow{\text{Ni}} R-\overset{\overset{\displaystyle H}{|}}{\underset{\underset{\displaystyle H}{|}}{C}}-OH$$

Alcohols

$$2R-\overset{\overset{\displaystyle H}{|}}{C}=O + O_2 \xrightarrow{\text{ZnO}} 2R-C\overset{\displaystyle O}{\underset{\displaystyle OH}{\diagup}}$$

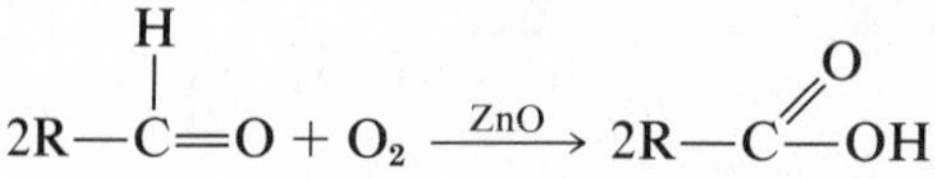

Carboxylic acids

The catalysis probably involves propylene displacing one mole of CO from $Co_2(CO)_8$ to become a ligand of the cobalt atom. Addition of H_2 and CO to the $C=C$ bond gives the aldehyde, which then separates from the cobalt atom and is replaced by CO to regenerate the catalyst. The CO and H_2 needed for the process are produced from coal (C) by the water-gas reaction:

$$C(s) + H_2O(g) \xrightarrow{1100°C} CO(g) + H_2(g)$$

The oxo process provides a way of converting "waste" petroleum refinery gases into useful alcohols and acids by using coal, which is in plentiful supply. The process was developed in the U.S. and used in Germany during World War II, and may come back into wider use with the recognition that petroleum hydrocarbons are too valuable to waste as fuels for heating and running automobile and diesel engines (Section 15.1). ⊶

Enzymes as catalysts If two complicated molecules are to react, their collision (even if energetic enough to form the transition state) must be in the proper *orientation* to bring the reactive sites together. In the saponification of methyl acetate, for example, the OH^- ion must collide with the $-C\overset{\displaystyle O}{\underset{\displaystyle O}{\diagup}}$ carbon in order to form a bond:

$$CH_3-\overset{\overset{\displaystyle O}{\|}}{C}\underset{\underset{\displaystyle -OH}{\uparrow}}{\diagdown}_{O-CH_3} \rightarrow \left[CH_3-\overset{\overset{\displaystyle O^-}{|}}{\underset{\underset{\displaystyle OH}{|}}{C}}-OCH_3 \right] \rightarrow CH_3-\overset{\overset{\displaystyle O}{\|}}{C}\underset{\displaystyle O^-}{\diagdown} + HOCH_3$$

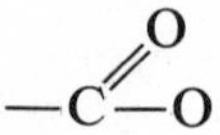

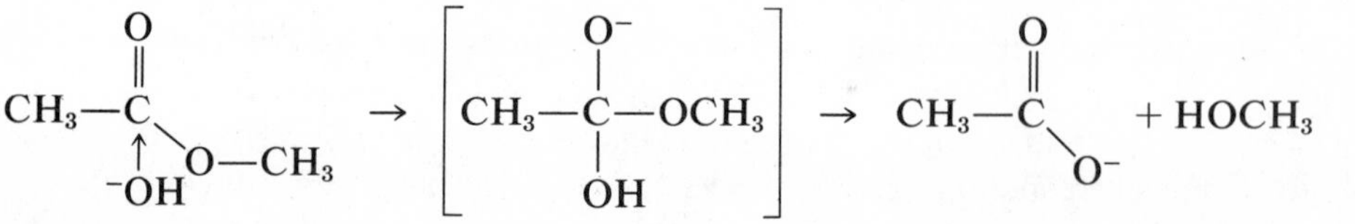

Even the most vigorous collisions with either of the CH_3 groups or the oxygens cannot lead to reaction.

Enzymes are the natural catalysts for the reactions of life. Enzymes are large molecules with an *active site* where high-activation-energy reactions are catalyzed so that they go readily at body temperature (37°C). The active site is of such shape that the reactants fit into it and are put into the right orientation for most efficient catalysis of the reaction. Enzymes are proteins with molecular weights in the tens of thousands and the active site often uses a vitamin molecule attached to the protein chain for the catalysis (Section 14.6). Enzyme action is shown schematically in Figure 9.12. The molecules of nerve gases and phosphonate insecticides mentioned in Section 9.1 destroy the activity of acetylcholinesterase (Figure 9.1) by filling up the active sites. They "poison" the enzyme, in much the same way that sulfur compounds poison a platinum contact catalyst by covering its surface.

9.6 Rates and Equilibrium At equilibrium the rates of the forward and reverse reactions are equal. If the reactions for an equilibrium are *simple reactions*, there is a simple relation between the specific rate constants for the forward and reverse reactions and the equilibrium constant for the reaction. For the gaseous equilibrium $2NO_2(g) \rightleftarrows N_2O_4(g)$ both forward and reverse reactions are simple. With their rate laws, they are

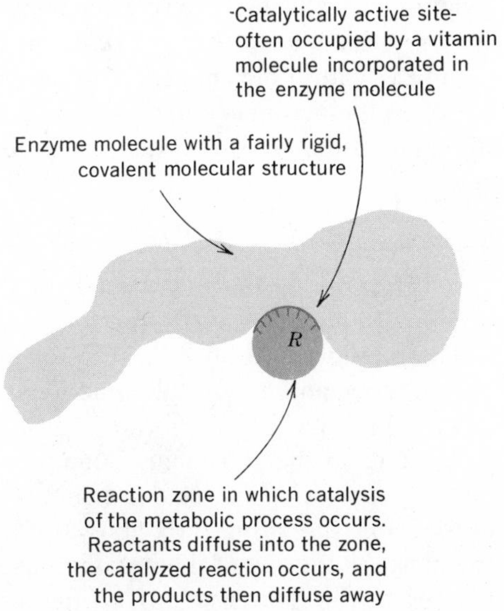
Catalytically active site-often occupied by a vitamin molecule incorporated in the enzyme molecule
Enzyme molecule with a fairly rigid, covalent molecular structure
R
Reaction zone in which catalysis of the metabolic process occurs. Reactants diffuse into the zone, the catalyzed reaction occurs, and the products then diffuse away

FIG. 9.12 Schematic representation of enzyme activity.

$$2NO_2(g) \rightarrow N_2O_4(g) \qquad \text{Rate} = k_1 C^2_{NO_2} \quad \text{(bimolecular)}$$
$$N_2O_4(g) \rightarrow 2NO_2(g) \qquad \text{Rate} = k_{-1} C_{N_2O_4} \quad \text{(unimolecular)}$$

The symbols k_1 and k_{-1} are used for the forward and reverse reactions in an equilibrium.

At equilibrium the two rates must be the same, so you can set them equal to one another:

$$k_1 C^2_{NO_2} = k_{-1} C_{N_2O_4}$$

Rearranging gives

$$\frac{k_1}{k_{-1}} = \frac{C_{N_2O_4}}{C^2_{NO_2}}$$

When equilibrium is reached, the concentrations of participants no longer change with time, so the instantaneous concentrations (C) become equilibrium concentrations ([ ]); that is, $C_{NO_2} = [NO_2]$, and $C_{N_2O_4} = [N_2O_4]$. Making the substitutions, you discover that the ratio of the rate constants is equal to the equilibrium constant for the reaction:

$$\frac{k_1}{k_{-1}} = \frac{C_{N_2O_4}}{C^2_{NO}} = \frac{[N_2O_4]}{[NO_2]^2} = K$$

A derivation of this type can be made for any equilibrium in which the forward and reverse reactions are simple. But *even if they aren't, thermodynamics* tells us that the equilibrium constant equation for a reaction can always be written according to the rules given in Section 7.3. You have seen that the reaction $H_2(g) + Br_2(g) \rightarrow 2HBr(g)$ is complex, with a mechanism of several simple reactions. Even though the overall reaction is complex and its empirical rate law is not simple, the equilibrium constant equation is correctly written according to the rules as

$$K = \frac{[HBr]^2}{[H_2][Br_2]} \qquad \text{or} \qquad K_p = \frac{p^2_{HBr}}{p_{H_2} p_{Br_2}}$$

The relation $(k_1/k_{-1}) = K$ between specific rate constants and equilibrium constants for simple reactions provides an explanation for the change in value of equilibrium constants with temperature changes. According to the Arrhenius equation, the values of both k_1 and k_{-1} must *increase* with *increasing* temperature. Since both change in value in the *same* direction, the only way for the value of their ratio, k_1/k_{-1}, to change is for one of them to increase more rapidly than the other as temperature is increased. If k_1 increases more rapidly than k_{-1} with increasing temperature, then their ratio k_1/k_{-1} will also increase, and so will the equilibrium constant for the reaction, because $K = k_1/k_{-1}$. According to Le Chatelier's principle, this is the case of an *endo*thermic reaction. For an *exo*thermic reaction, K *decreases* in value with increasing temperature. Here k_{-1} must increase more rapidly than k_1.

9.7 Summary Guide The value of the equilibrium constant for a reaction tells what will ultimately happen to reactant concentrations when equilibrium is reached, but gives no hint of the time required to reach equilibrium. The time required is given by the empirical rate law for the reaction. Empirical rate laws are determined by experiment in the laboratory through measuring the effects of changes in concentrations of reactants on the rate of the reaction. The power to which the concentration of a reactant must be raised to give the empirical rate law is the order of the reaction in the reactant. The order of a reaction is the sum of the exponents of the concentrations of reactants in the empirical rate law. Many rate laws are of the first-order or second-order types. Determining the empirical rate law for a reaction is the first step in discovering the pathway or mechanism by which reactants get to products in the reaction.

Rates of reactions are the time rate of change in pressure for gas-phase reactions and the time rate of change of molar concentrations for reactions in solution. The pressures or molar concentrations that appear in the rate law are their values at one instant in the life of the reaction. The time required for one-half of the reactants present at any instant to go to products is called the half-life ($t_{1/2}$) of the reaction. Half-lives are used to calculate how much of a reactant will remain at various times in the life of the reaction.

Most chemical equations do not describe what is going on at the microparticle level of atoms, molecules, or ions as a reaction proceeds. How reactant microparticles get to product microparticles is given by the mechanism of the reaction. The mechanism of a reaction is made up of one or more simple reactions in which the product microparticles are the immediate result of the collision or activation of reactant microparticles. A reaction is called complex if there is more than one simple reaction in its mechanism.

Simple reactions have molecularity, defined as the number of atoms, molecules, or ions in the transition state on the path from reactants to products. The rates of simple reactions are given by the fundamental rate laws for unimolecular, bimolecular, and termolecular reactions. When the mechanism of a reaction is known, the empirical rate law can be derived by using the fundamental rate laws for the simple reactions of the mechanism.

Reactant microparticles rarely react at every collision. Only collisions well above average energy lead to the beginning of new bond making and old bond breaking in the transition state needed for reaction. The activation energy for a reaction is the excess energy needed to produce the transition state. The Arrhenius equation relates the value of the activation energy, the absolute temperature, and the specific rate constant for a reaction. The specific rate constant for a reaction is the proportionality constant (k) in the rate law for the reaction and is equal to the rate of reaction when the concentrations of all reactants are one-molar. For simple reactions the value of the equilibrium constant is the ratio of the values of the specific rate constants for the forward and reverse reactions.

Increasing reactant concentrations increases the rate of reaction through increasing the number of collisions between reactant particles. Reactions with large activation energies are slow. Catalysts speed up slow reactions by providing new pathways of lowered activation energy by which reactants can go to products. Increasing the temperature increases the rate of simple reactions by increasing the number of high-energy collisions between reactant particles. Increasing temperature first accelerates but finally stops the complex reactions of life by destroying the enzymes that catalyze them.

Exercises

23. The hydrolysis of a chlorinated hydrocarbon insecticide (such as DDT) is similar to the hydrolysis of ethyl acetate (Table 9.1) and involves a simple bimolecular reaction that can be abbreviated as follows:

$$\underset{\text{Insecticide}}{R{-}Cl} + OH^- \rightarrow \underset{\text{Harmless hydrolysis products}}{R{-}OH + Cl^-}$$

Write the rate law for this reaction. Suppose the specific rate constant for the hydrolysis of an insecticide like DDT is $k = 1 \times 10^{-3}$ liter mole^{-1} hr^{-1}. Calculate the rates of hydrolysis of the insecticide in terms of C_{R-Cl} at pH = 14 ($C_{OH^-} = 1$ mole/liter), at pH = 8 (a typical alkaline environmental water), and at pH = 5 (a typical acidic environmental water). What do your calculations indicate about the "lifetime" of a chlorinated hydrocarbon insecticide under environmental conditions?

24. To avoid environmental pollution by long-lived chlorinated hydrocarbon insecticides (DDT), insecticides such as Parathion that are quickly hydrolyzed to harmless products are used. Parathion, abbreviated $R{-}OC_2H_5$, is an example. In water, Parathion reacts as follows: $R{-}OC_2H_5 + H_2O \rightarrow ROH + C_2H_5OH$. The rate law for this reaction is of the unimolecular type:

$$\text{Rate} = kC_{R-OC_2H_5}$$

For a unimolecular reaction, the half-life ($t_{1/2}$) of the reaction is given by this equation (Section 16.2): $t_{1/2} = 0.693/k$. Assume that k for the hydrolysis of Parathion under environmental conditions has the value $k = 0.0693\ \text{hr}^{-1}$. Calculate how much of the Parathion originally put on an agricultural field will remain unhydrolyzed 10 hours after application; 20 hours; 10 days.

Key Words

Activation energy (E_a) the energy in excess of the average energy of reactants needed to reach the transition state in a reaction—large values of E_a go with slow reactions.

Bimolecular reaction a simple reaction in which two reactant microparticles go to products after a sufficiently energetic collision.

Catalyst a chemical or species that speeds up a reaction by lowering the activation energy required for reaction without appearing in the stoichiometric equation for the reaction; often written over the arrow for a reaction.

Chain reaction a complex reaction in which two simple reactions are coupled so that a microparticle product of each reaction is a reactant microparticle in the other reaction.

Complex reaction a reaction in which more than one simple reaction is required to go from reactant to product microparticles.

Elementary reaction a simple reaction.

Empirical rate law the relation between the time rate of a reaction, the specific rate constant (k), and the concentrations of reactants, found by experiments in the laboratory.

Enzymes complex, catalytic protein molecules that accelerate reactions in living systems; often have a vitamin as part of their molecular structure.

First-order rate law a rate law (equation) in which the time rate of a reaction is directly proportional to the molar concentration of a single reactant.

Half-life of a reaction ($t_{1/2}$) the time required for one-half of the reactants to go to products.

Heterogeneous or contact catalyst a solid catalyst that accelerates a reaction of liquids or gases.

Heterogeneous reaction a reaction in which the reactants are in two or more different phases.

Homogeneous catalyst a catalyst that works in the same phase as the reactants in a reaction.

Homogeneous reaction a reaction in which the reactants are in the same phase.

Instantaneous rate of reaction the time rate at which reactants are being converted to products at any instant in the life of a reaction.

Mechanism of a reaction the list of simple or elementary reactions that provides the pathway connecting reactant microparticles with product microparticles.

Molecularity of a reaction the number of microparticles needed to create the transition state for a reaction.

Rate of reaction a time measure of how rapidly reactants are converted to products in a chemical reaction, usually expressed as the time rate of increase in the molar concentration of a product of the reaction.

Second-order rate law a rate law (equation) in which the time rate of a reaction is proportional to the product of the molar concentrations of two reactants in a reaction, or the square of a reactant concentration.

Simple reaction a reaction between microparticles (atoms, molecules, ions) described by a chemical equation in which the product microparticles are the immediate result of the collision or activation of the reactant microparticles. Also called an elementary reaction.

Specific rate constant (k) the time rate of a reaction when the concentrations of reactants are all 1 mole/liter, appearing as a proportionality constant in the rate law for the reaction.

Termolecular reaction a simple reaction in which three reactant microparticles collide and go to products.

Transition state a half-reacted state of high energy in which bonds in reactant particles are starting to break while bonds in product particles are starting to form.

Unimolecular reaction a simple reaction in which a single, energy-rich microparticle reactant goes to products.

Suggested Readings

Bach, W., *Atmospheric Pollution,* New York, McGraw-Hill Book Co., Inc., 1972.

Campbell, J. Arthur, *Why Do Chemical Reactions Occur?* Englewood Cliffs, N.J., Prentice-Hall, Inc., 1965.

Gillette, Robert, "DDT: Its Days Are Numbered, Except Perhaps in Pepper Fields," *Science,* **176**:1313–1314 (June 23, 1972).

Kavaler, Lucy, *The Wonders of Cold: Cold Against Disease,* New York, John Day, 1971.

McIntire, G., "Spoiled by Success," *Environment,* **14**:14–29 (July–Aug. 1972).

Newell, Reginald E., "The Global Circulation of Atmospheric Pollutants," *Chemistry in the Environment*, San Francisco, W. H. Freeman and Co., 1973, pp. 255–265.

Tamuru, Kenzi, "New Catalysts for Old Reactions," *Am. Sci.,* **60**:474–479 (July–Aug. 1972).

Voorhoeve, R. J. H., Remeika, J. P., Freeland, P. E., and Matthias, B. T., "Rare-Earth Oxides of Manganese and Cobalt Rival Platinum for the Treatment of Carbon Monoxide in Auto Exhaust," *Science,* **177**:353–354 (July 28, 1972).

Voorhoeve, R. J. H., Remeika, J. P., and Johnson, D. W., Jr., "Rare-Earth Manganites: Catalysts with Low Ammonia Yield in the Reduction of Nitrogen Oxides," *Science,* **180**:62–64 (April 6, 1973).

Wiley, John P., Jr., and Sherman, J. K., "Immortality and the Freezing of Human Bodies," *Nat. Hist.,* **80**:13–22 (Dec. 1971).

chapter 10

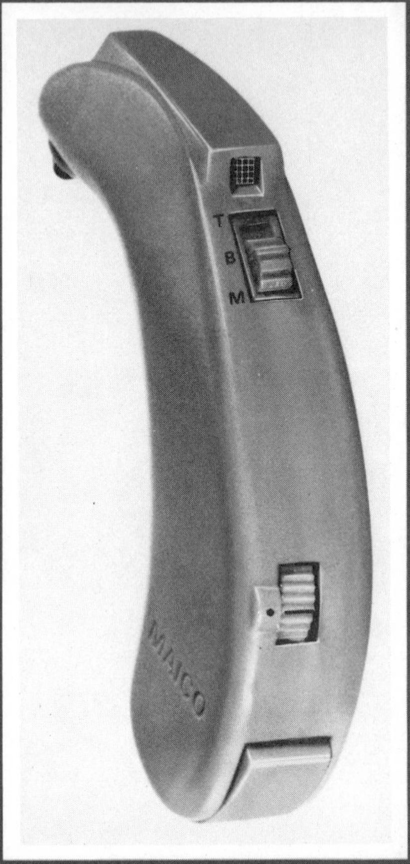

A hearing aid designed to be worn behind the ear. The screened "hatch" near the top leads to the battery compartment, which holds a mercury cell about ¼ inch in diameter. *(Photo 27, Maico Hearing Instruments.)*

Electrochemistry: Oxidation-Reduction Reactions

10.1 Introduction Chapter 8 covered the important proton-transfer or acid–base reactions. This chapter takes up an equally important kind of reaction, electron-transfer or *oxidation–reduction* reactions, sometimes called "redox" reactions. Electrochemistry is the branch of chemistry dealing with oxidation–reduction reactions, particularly those that can be examined by electrical measurements. Unlike acid–base reactions, many oxidation–reduction reactions are a direct and useful source of energy. An electrochemical cell or "battery" can do useful work such as starting a car or running a radio because of an

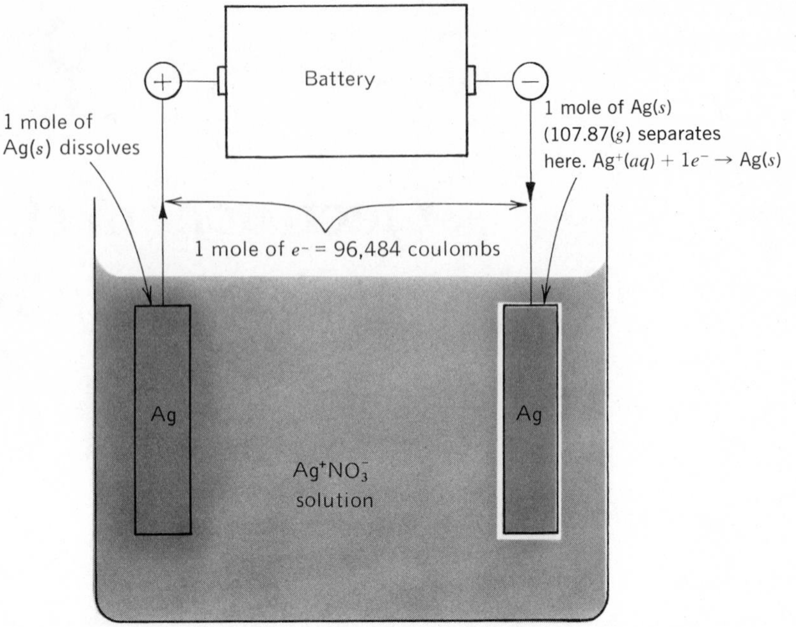

FIG. 10.1 Determination of the value of the faraday of charge.

oxidation–reduction reaction going on inside of it. Batteries can be very small (Photo 27).

A current of electricity in a wire is a flow of electrons somewhat like a flow of water in a pipe. In physics quantities of electricity are expressed in *coulombs*: *one coulomb is the quantity of electricity that deposits* ("plates out") *0.0011180 g of silver from a silver nitrate solution*. A more useful unit for chemists is the *faraday* ($\mathscr{F}$), *defined as the quantity of electricity required to deposit one* ***mole*** *of silver* (107.87 g) *from a silver nitrate solution* (Figure 10.1). A faraday is 96,484 coulombs, calculated this way:

$$1 \text{ faraday } (\mathscr{F}) = \frac{107.87 \text{ g mole}^{-1} \text{ Ag}}{0.0011180 \text{ g Ag coulomb}^{-1}}$$
$$= 96{,}484 \text{ coulomb mole}^{-1}$$

Since silver ions (Ag^+) carry one positive unit of electronic charge, *the faraday is the charge on one mole of* Ag^+ *or, ignoring signs, on one mole of electrons*. The faraday is used *without regard for sign*, being taken as positive in calculations.

The *rate of flow* of water in a pipe is the amount of water that passes a point on the pipe in one second, measured, for example, in "gallons per second." The rate of flow of electricity is measured in *amperes*: *one ampere is a flow of one coulomb per second past a point in a cir-*

cuit. It corresponds to a flow of 6.24×10^{18} electrons past the point each second.

Exercises

1. The electronic charge given earlier was 4.80×10^{-10} in electrostatic units (esu). Calculate the electronic charge in coulombs, using the faraday and the Avogadro number.
2. How many coulombs per second pass a point on a wire that is carrying a current of 5 amps? How many electrons per second?
3. How many seconds must a current of one ampere flow to plate out one mole of Ag in the apparatus in Figure 10.1?
4. How many faradays are required to plate out one mole of copper from a solution of copper nitrate $[Cu^{2+}(NO_3^-)_2]$ if the reaction is $Cu^{2+}(aq) + 2e^- \rightarrow Cu(s)$?
5. How many hours are required to plate out 10 g of Cu, using a current of 10 amps?

Electrons are made to move in a wire carrying a current by an *electromotive force*, abbreviated emf. The emf results from a difference in electric potential or "electron pressure" at the ends of the wire. The electron flow is like the flow of water in a pipe brought about by a pressure difference between the two ends of the pipe. Differences in water pressures are in units of pounds per square inch (psi), *electromotive forces*, $\Delta\mathcal{E}$, *are given in* ***volts*** (Figure 10.2). The emf between the ends (poles) of a flashlight battery is 1.5 V (volts), and can be measured with a voltmeter. If you connect the ends of a flashlight battery to a voltmeter by a wire, electrons flow from the negative pole (−) to the positive pole (+) with an emf of 1.5 V. The electrons flow from (−) to (+) in the wire because of the attraction of their negative charge by the positive charge. A flashlight battery is an example of an *electrochemical cell*.

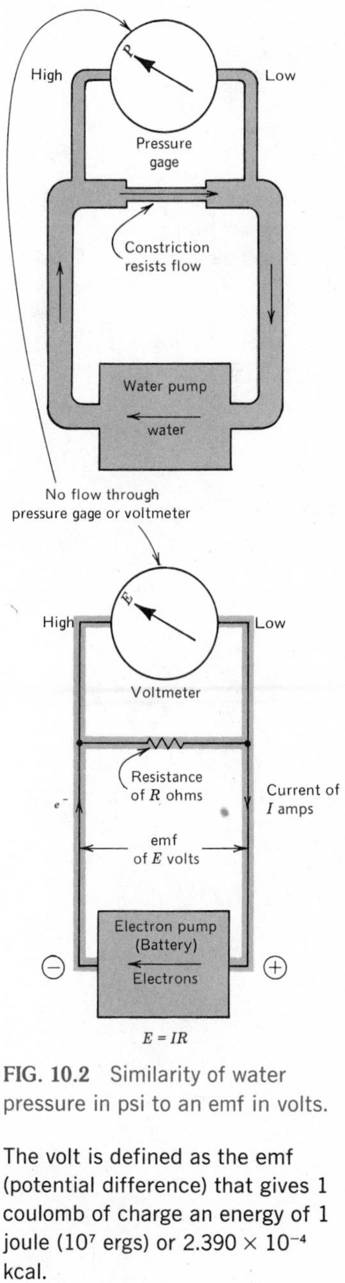

FIG. 10.2 Similarity of water pressure in psi to an emf in volts.

The volt is defined as the emf (potential difference) that gives 1 coulomb of charge an energy of 1 joule (10^7 ergs) or 2.390×10^{-4} kcal.

In the pages that follow we will also be concerned with the amounts of electrical *energy* available from electrochemical cells. According to Figure 1.6, one electron volt of energy is the kinetic energy gained by one electron as it moves from (−) to (+) through a potential difference (emf) of 1 V. One electron volt $= 1.6021 \times 10^{-12}$ ergs $= 3.829 \times 10^{-23}$ kcal. It is far too small an energy quantity to use with electrochemical cells. Instead we will use the *faraday volt*, which is the quantity of energy acquired by *one mole* (6.023×10^{23}) of electrons driven by an emf of 1 V:

$$1 \text{ faraday volt} = 1.6021 \times 10^{-12} \times 6.023 \times 10^{23} \times 1$$
$$= 9.649 \times 10^{11} \text{ erg V}^{-1} \text{ mole}^{-1}$$

Since 1 erg $= 2.390 \times 10^{-11}$ kilocalorie (kcal),

$$1 \text{ faraday volt} = 9.649 \times 10^{11} \times 2.390 \times 10^{-11} \text{ kcal V}^{-1} \text{ mole}^{-1}$$

or

$$1 \text{ faraday volt} = 23.06 \text{ kcal V}^{-1} \text{ mole}^{-1}$$

George Simon Ohm (1787–1854) was a German physicist noted for his law expressing the relationship of voltage, current, and resistance in an electric circuit. The ohm was named after him.

The flow of an electric current in a wire or an electrolyte solution is regulated by the electrical *resistance* of the wire or solution. Resistance is in units of ohms, given by Ohm's law, $R = \Delta\mathscr{E}/I$ or $\Delta\mathscr{E} = IR$, where $\Delta\mathscr{E}$ is the emf in volts that drives a current of I amperes through a resistance of R ohms. According to Ohm's law, the greater the resistance is, the larger the emf has to be to produce the same current. Again, the likeness to water flowing in a pipe is helpful—higher pressure differences are needed to force the same flow of water through smaller pipes because they offer greater resistance.

Exercises

6. Given the definitions 1 volt coulomb = 1 joule = 2.39×10^{-1} cal, and 1 $\mathscr{F}$ = 96,484 coulombs, show that 1 faraday volt = 23.06 kcal V^{-1} mole^{-1}.

7. A portable electric heater draws 12.00 amps at 110 V. Assuming that all of the electrical energy appears as heat, what is the heat output in joules/sec? In cal/sec? In kcal/sec? Calculate the resistance of the heating wire in ohms.

EXAMPLE

A small motor in the guidance mechanism of a rocket has a resistance of 1.50 ohms when running. The motor is to be driven by an emf of 12.60 V from a battery and is expected to be in service for a maximum time of 170 minutes. You need to know how much energy the battery must provide, so that the right size battery can be included in the design of the mechanism.

First you need to calculate the current in amperes that will flow while the motor is doing its work. Using Ohm's law, $\Delta\mathscr{E} = IR$, with $\Delta\mathscr{E} = 12.60$ V and $R = 1.50$ ohms, we obtain

$$12.60 = I \times 1.50$$

or

$$I = \frac{12.60}{1.50} = 8.40 \text{ amps}$$

While running, the motor will draw 8.40 amps at the emf of 12.60 V. Since 1 amp = 1 coulomb sec^{-1}, the motor will require energy at the rate of $8.40 \times 12.60 = 106$ volt coulomb sec^{-1}. Using the definition that 1 volt coulomb $= 2.39 \times 10^{-1}$ cal (Exercise 6), we see that energy will be needed at the rate of 106 volt coulomb $sec^{-1} \times 2.39 \times 10^{-1}$ cal $volt^{-1}$ $coulomb^{-1}$ or at the rate of 25.30 cal sec^{-1}. If the motor is to be operated for 170 min ($170 \times 60 = 10{,}200$ sec), we obtain:

$$\begin{aligned}\text{Total energy needed} &= 25.30 \text{ cal sec}^{-1} \times 10{,}200 \text{ sec} \\ &= 2.58 \times 10^5 \text{ cal} \\ &= 258 \text{ kcal}\end{aligned}$$

This energy quantity is a bit more than 10 faraday volts (1 faraday volt = 23.06 kcal) and is about 30% of the energy stored in one 12-volt automobile storage battery when fully charged, so such a battery could be used with a good safety margin. ⊸

10.2 Oxidation–Reduction Reactions We have seen that proton-transfer reactions occur between proton donors (acids, HA) and proton acceptors (bases, B). In the same way, electron transfers occur between electron donors (reducing agents, red.) and electron acceptors (oxidizing agents, ox.). Using specific examples, we can see that the general forms of the two kinds of transfer are very much alike:

Proton transfer

$$\underset{\text{Acid 1}}{HCl} + \underset{\text{Base 2}}{H_2O} \rightleftarrows \underset{\text{Acid 2}}{H_3O^+} + \underset{\text{Base 1}}{Cl^-}$$

Electron transfer

$$\underset{\text{Red. 1}}{Zn(s)} + \underset{\text{Ox. 2}}{Cu^{2+}(aq)} \rightleftarrows \underset{\text{Red. 2}}{Cu(s)} + \underset{\text{Ox. 1}}{Zn^{2+}(aq)}$$

The position of equilibrium in the proton transfer is decided by the relative attractions of base 2 and base 1 for protons. Equilibrium in the electron transfer is determined by the relative attractions of ox. 2 and ox. 1 for electrons. In spite of these things they have in common, there are two important differences: only *one proton* is transferred in an acid–base reaction but *more than one electron* may be transferred in an oxidation–reduction reaction. Electron-transfer equilibria usually lie very far to one side or the other, while those for acids are more balanced. For typical weak acids the values for K_a are in the range 10^{-2}–10^{-10}. The value of the equilibrium constant for the zinc metal + copper ion reaction is 10^{37} at 25°C and is typical of many oxidation–reduction equilibria.

Knowing that reducing agents donate electrons and oxidizing agents accept them, we can now define the *processes* of oxidation and reduction:

Oxidation is the loss of electrons from an atom, molecule, or ion. Reduction is the addition of electrons to an atom, molecule, or ion.

The two processes are always coupled together—the electrons released by the reducing agent are always transferred to an oxidizing agent. The electrons, like the protons in acid–base equilibria, simply take up a new residence. In talking about oxidation and reduction it is very important to state *exactly* which particles are being oxidized and which reduced. In the zinc–copper ion reaction from left to right as written, zinc *atoms* are *oxidized*, copper *ion* is reduced. (To say simply that "copper" is reduced is wrong.) In the reverse reaction, right to left, zinc ion would be reduced by copper metal:

$$\ddot{Zn}(s) + Cu^{2+}(aq) \rightleftarrows \ddot{Cu}(s) + Zn^{2+}(aq)$$

Forward reaction	Oxidized	Reduced	$\rightarrow$		
Reverse reaction			$\leftarrow$	Oxidized	Reduced

From the value of 10^{37} for the equilibrium constant for the reaction, the forward reaction clearly wins out. Copper ion wins the electrons in the competition because it is a stronger *oxidizing agent* than zinc ion. Put another way, you can say that zinc metal, Zn(*s*), is a stronger *reducing agent* than copper metal, Cu(*s*).

From now on we will assume that ions are solvated by water molecules and omit the "(*aq*)."

The competition between the two metal ions for electrons, as in the $Zn + Cu^{2+}$ reaction, is the type of oxidation–reduction reaction called a *displacement reaction*. In writing equilibrium constant equations for these reactions, the concentrations of the solid metals are omitted. Here are some examples, showing the valence electrons transferred as dots and giving the formula and value of K at 25°C:

We saw in Chapter 7 why concentrations of solids are omitted in writing solubility product constants—the same thinking applies here.

Reducing agents		*Oxidizing agents*		*Oxidizing agents*		*Reducing agents*		
$\ddot{Mg}(s)$	+	Zn^{2+}	$\rightleftarrows$	Mg^{2+}	+	$\ddot{Zn}$	$K_1 = \dfrac{[Mg^{2+}]}{[Zn^{2+}]} = 10^{54}$	(1)
$\ddot{Zn}(s)$	+	Sn^{2+}	$\rightleftarrows$	Zn^{2+}	+	$\ddot{Sn}(s)$	$K_2 = \dfrac{[Zn^{2+}]}{[Sn^{2+}]} = 10^{21}$	(2)
$\ddot{Sn}(s)$	+	Cu^{2+}	$\rightleftarrows$	Sn^{2+}	+	$\ddot{Cu}(s)$	$K_3 = \dfrac{[Sn^{2+}]}{[Cu^{2+}]} = 10^{16}$	(3)
$\ddot{Cu}(s)$	+	$2Ag^{+}$	$\rightleftarrows$	Cu^{2+}	+	$2\dot{Ag}$	$K_4 = \dfrac{[Cu^{2+}]}{[Ag^{+}]^2} = 10^{15}$	(4)

These reactions are all of the type "metal A + metal ion B ⇄ metal ion A + metal B." The large values of the equilibrium constants for the reactions mean that the metal (A) on the left-hand side of each equation will reduce the *ion* (metal ion B) of the metal on the right. By comparing adjacent pairs of reactions, you can put the metals in order according to their strengths as reducing agents. The argument goes like this: according to Equation (1), Mg(*s*) is a stronger reducing agent than Zn(*s*). And according to Equation (2), Zn(*s*) is a stronger reducing agent than Sn(*s*). It follows that Mg(*s*) must be a stronger reducing agent than Sn(*s*). This means that equilibrium in the following reaction must lie to the right:

$$Mg(s) + Sn^{2+} \rightleftharpoons Mg^{2+} + Sn(s) \qquad K = \frac{[Mg^{2+}]}{[Sn^{2+}]} = 10^{76}$$

It does indeed, as shown by the value of its equilibrium constant. Continuing this line of argument gives the order of strengths of the metals as reducing agents as Mg > Zn > Sn > Cu > Ag. By methods you will see later, it is possible to obtain the equilibrium constants for all combinations of the metals and metal ions. Table 10.1 contains the values of the equilibrium constants K in the form of the exponent of 10 for the reaction between the metal (left-hand column) and the metal ion (top row). The values show that Mg(*s*) will effectively reduce all of Zn^{2+}, Sn^{2+}, Cu^{2+}, and Ag^{+} to their metals; that Zn(*s*) will reduce Sn^{2+},

TABLE 10.1 Part of the displacement series of metals[a]

Increasing strength of metal ions as oxidizing agents →

Increasing strength of metals as reducing agents ↑

A ↓ B →	Mg^{2+}	Zn^{2+}	Sn^{2+}	Cu^{2+}	Ag^{+}
Mg		+54	+76	+92	+108
Zn			+21	+37	+53
Sn				+16	+31
Cu					+15
Ag					

[a] numbers are the exponents of 10 in the value of K for the metal A + metal ion B reactions.

Cu^{2+}, and Ag^+ to their metals; that Sn(*s*) will reduce Cu^{2+} and Ag^+; that Cu(*s*) will reduce only Ag^+; and that Ag(*s*) will not reduce any of the other ions.

Table 10.1 shows that the strength of the metals as reducing agents increases upward from silver to magnesium. What about the metal ions? Because they accept the electrons, the metal ions are oxidizing agents. The metal ions are arranged in order of increasing strength as oxidizing agents from left to right across the top of the table. The strongest oxidizing agent in the table is Ag^+ because it oxidizes Mg(*s*), Zn(*s*), Sn(*s*), and Cu(*s*) to their ions. The weakest oxidizing agent is Mg^{2+}. It can oxidize none of the metals. You can see that the strongest reducing metals have ions that are weakest as oxidizing agents. This is like acid–base theory, in which the strongest acids have the weakest conjugate bases (Table 8.2).

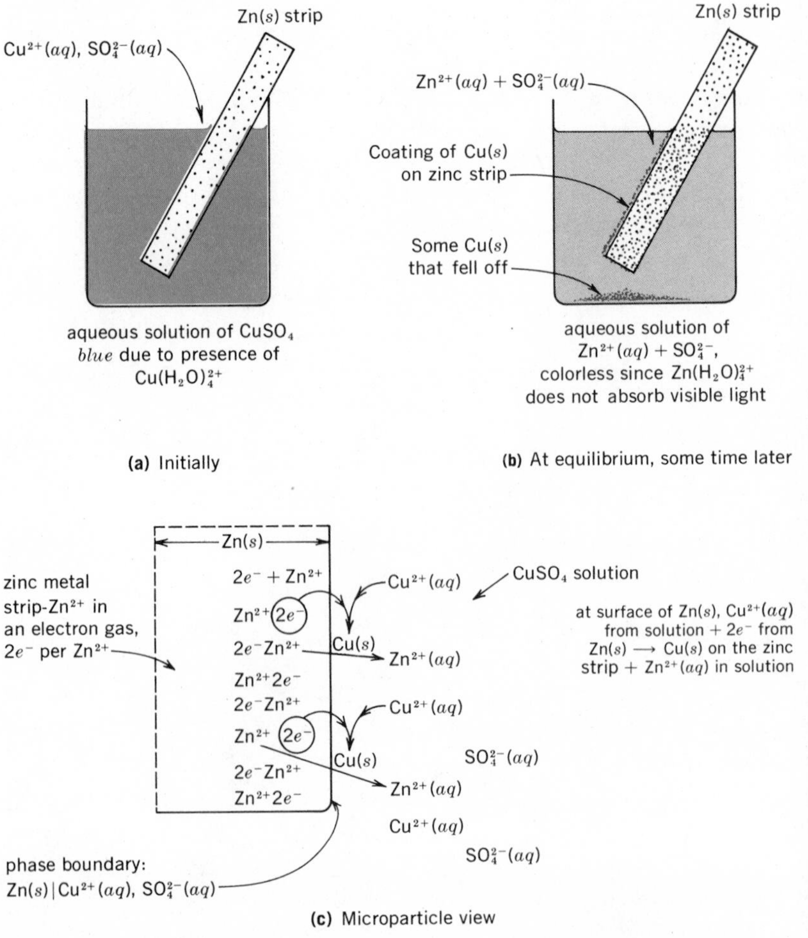

Net Reaction: $Zn(s) + Cu^{2+}(aq) \rightleftharpoons Zn^{2+}(aq) + Cu(s)$

FIG. 10.3 Reduction of $Cu^{2+}(aq)$ by Zn(s): (a) at the instant the zinc strip is placed in the $CuSO_4$ solution; (b) after the system has come to equilibrium; (c) microparticle view of process.

The table of displacements (Table 10.1) can be enlarged to include all of the metals in the Periodic Table, but a more useful way of putting the information together is to separate the overall reactions into their *electron-transfer half-reactions.*

Electron-transfer half-reactions Look at the overall reaction $\ddot{\text{Zn}}(s) + \text{Cu}^{2+} \rightarrow \text{Zn}^{2+} + \ddot{\text{Cu}}(s)$ diagramed in Figure 10.3. The reaction takes place where the solid zinc metal touches the $CuSO_4$ solution. You can think of the reaction occurring in two separate electron-transfer half-reactions:

$$\ddot{\text{Zn}}(s) \rightarrow \text{Zn}^{2+} + 2e^- \qquad \text{Oxidation of Zn}(s)$$

$$\text{Cu}^{2+} + 2e^- \rightarrow \ddot{\text{Cu}}(s) \qquad \text{Reduction of Cu}^{2+}$$

$$\text{Overall} \quad \ddot{\text{Zn}}(s) + \text{Cu}^{2+} \rightarrow \text{Zn}^{2+} + \ddot{\text{Cu}}(s)$$

According to this picture, every time a Cu^{2+} ion picks up two electrons from the zinc strip to form a copper atom, a Zn^{2+} ion goes into solution. It is important that you understand that electrons cannot exist free in water and do not go *into* the copper sulfate *solution*, but jump directly from the zinc strip to the empty valence orbitals of a nearby Cu^{2+} ion. The copper atom produced separates or "plates out" on the zinc strip. Figure 10.3(c) summarizes these ideas.

Each of the reactions in the displacement series of metals in Table 10.1 can be separated into two electron-transfer half-reactions. These are usually written as oxidations (e^- on right) with the strength of the metal as a reducing agent increasing upward.

Oxiding agents—strength increasing downward

$$\ddot{\text{Mg}}(s) \rightleftarrows \text{Mg}^{2+} + 2e^-$$

$$\ddot{\text{Zn}}(s) \rightleftarrows \text{Zn}^{2+} + 2e^-$$

$$\ddot{\text{Sn}}(s) \rightleftarrows \text{Sn}^{2+} + 2e^-$$

$$\ddot{\text{Cu}}(s) \rightleftarrows \text{Cu}^{2+} + 2e^-$$

$$\dot{\text{Ag}}(s) \rightleftarrows \text{Ag}^{+} + e^-$$

Reducing agents—strength increasing upward

In this arrangement, the oxidizing agents are all to the right of the double arrows, and *any oxidizing agent will oxidize any reducing agent above it*. Thus, for example, Cu^{2+} will oxidize $\ddot{\text{Sn}}(s)$, $\ddot{\text{Zn}}(s)$, and $\ddot{\text{Mg}}(s)$, but not $\text{Ag}(s)$. It is important to learn the way this table can be used to predict what will happen, because you can also use the same thinking with the much larger set of standard oxidation potentials in Table 10.2. Even though the predictions you make because of position are *qualitative* rather than numerical (or *quantitative*), being able to make them gets you ready to use the $\mathscr{E}^0$ values in Table 10.2 for numerical calculations of the emf's of a large number of electrochemical cells.

TABLE 10.2 Standard oxidation potentials at 25°C[a]

Reducing Agents		*Oxidizing Agents*	$\mathscr{E}^0$(v)
Strongest		Weakest	
	$K(s) \rightleftharpoons K^+(aq) + e^-$		+2.93
	$Mg(s) \rightleftharpoons Mg^{2+}(aq) + 2e^-$		+2.37
	$Al(s) \rightleftharpoons Al^{3}(aq) + 3e^-$		+1.66
	$Zn(s) \rightleftharpoons Zn^{2+}(aq) + 2e^-$		+0.76
	$Fe(s) \rightleftharpoons Fe^{2+}(aq) + 2e^-$		+0.41
	$Pb(s) + SO_4^{2-}(aq) \rightleftharpoons PbSO_4(aq) + 2e^-$		+0.41
	$Ag(s) + I^-(aq) \rightleftharpoons AgI(s) + e^-$		+0.15
	$Sn(s) \rightleftharpoons Sn^{2+}(aq) + 2e^-$		+0.14
	$Pb(s) \rightleftharpoons Pb^{2+}(aq) + 2e^-$		+0.13
	$H_2(g) \rightleftharpoons 2H^+(aq) + 2e^-$ (*by definition*)		0.00
	$NO_2^-(aq) + 2OH^-(aq) \rightleftharpoons NO_3^-(aq) + H_2O(l) + 2e^-$		0.00
	$Ag(s) + Br^-(aq) \rightleftharpoons AgBr(s) + e^-$		−0.07
	$2S_2O_3^{2-}(aq) \rightleftharpoons S_4O_6^{2-}(aq) + 2e^-$		−0.10
	$H_2S(aq) \rightleftharpoons S(s) + 2H^+(aq) + 2e^-$		−0.14
	$Sn^{2+}(aq) \rightleftharpoons Sn^{4+}(aq) + 2e^-$		−0.15
	$Ag(s) + Cl^-(aq) \rightleftharpoons AgCl(s) + e^-$		−0.22
	$2Hg(l) + 2Cl^-(aq) \rightleftharpoons Hg_2Cl_2(s) + 2e^-$		−0.28
	$Cu(s) \rightleftharpoons Cu^{2+}(aq) + 2e^-$		−0.34
	$4OH^-(aq) \rightleftharpoons O_2(g) + 2H_2O(l) + 4e^-$		−0.40
	$3I^-(aq) \rightleftharpoons I_3^-(aq) + 2e^-$		−0.53
	$2I^-(aq) \rightleftharpoons I_2(s) + 2e^-$		−0.54
	$H_2O_2(aq) \rightleftharpoons O_2(g) + 2H^+(aq) + 2e^-$		−0.68
	$Fe^{2+}(aq) \rightleftharpoons Fe^{3+}(aq) + e^-$		−0.77
	$Ag(s) \rightleftharpoons Ag^+(aq) + e^-$		−0.80
	$Hg(l) \rightleftharpoons Hg^{2+}(aq) + 2e^-$		−0.85
	$NO(g) + 2H_2O(l) \rightleftharpoons NO_3^-(aq) + 4H^+(aq) + 3e^-$		−0.96
	$2Br^-(aq) \rightleftharpoons Br_2(l) + 2e^-$		−1.07
	$2H_2O(l) \rightleftharpoons O_2(g) + 4H^+(aq) + 4e^-$		−1.23
	$2Cr^{3+}(aq) + 7H_2O(l) \rightleftharpoons Cr_2O_7^{2-}(aq) + 14H^+(aq) + 6e^-$		−1.33
	$2Cl^-(aq) \rightleftharpoons Cl_2(g) + 2e^-$		−1.36
	$Mn^{2+}(aq) + 4H_2O(l) \rightleftharpoons MnO_4^-(aq) + 8H^+(aq) + 5e^-$		−1.49
	$Cl_2(g) + 2H_2O(l) \rightleftharpoons 2HOCl(aq) + 2H^+(aq) + 2e^-$		−1.63
	$MnO_2(s) + 2H_2O(l) \rightleftharpoons MnO_4^-(aq) + 4H^+(aq) + 3e^-$		−1.68
	$PbSO_4(s) + 2H_2O(l) \rightleftharpoons PbO_2(s) + SO_4^{2-}(aq) + 4H^+(aq) + 2e^-$		−1.68
	$2H_2O(l) \rightleftharpoons H_2O_2(aq) + 2H^+(aq) + 2e^-$		−1.78
	$2F^-(aq) \rightleftharpoons F_2(g) + 2e^-$		−2.87
Weakest		Strongest	

[a]The dots for the valence electrons of the metals are omitted.

To get an overall reaction that is spontaneous or goes naturally from left to right using two half-reactions in the table, you write the half-reaction that is higher up in the table *as is*, and the one lower down *in reverse*. Electrons are balanced, so the number of electrons donated by the reducing agent equals the number taken up by the oxidizing agent. Then the separate equations are combined. For example, let's predict the reaction between metallic lead, Pb(*s*), and a solution of silver nitrate, $Ag^+NO_3^-$. The half-reaction for lead (Pb) is *above* the one for silver (Ag), so you take the half-reaction for Pb *directly* from Table 10.2, but *you reverse the one for silver*. After multiplying the silver half-reaction through by 2 to balance electrons, you add the half-reactions, so the electrons cancel:

$$Pb(s) \rightleftarrows Pb^{2+} + \cancel{2e^-}$$

$$2Ag^+ + \cancel{2e^-} \rightleftarrows 2Ag(s)$$

$$\text{Overall} \quad Pb(s) + 2Ag^+ \rightleftarrows Pb^{2+} + 2Ag(s)$$

The prediction is that lead metal will displace silver metal from a silver nitrate ($Ag^+NO_3^-$) solution, or put another way, Ag^+ will oxidize Pb(*s*) to Pb^{2+} and be reduced to Ag(*s*). The prediction is correct, as shown by the value of the equilibrium constant for the reaction:

$$K = \frac{[Pb^{2+}]}{[Ag^+]^2} = 10^{32}$$

Look closely at Table 10.2 and you will see that there are a number of half-reactions for the oxidations of molecules (H_2O, NO, H_2O_2) and nonmetal ions (I^-, Cl^-, F^-), as well as for metals. You can use these half-reactions in the same way to make qualitative predictions. For the reaction between silver metal and a solution of fluorine gas in water, notice that the equation for Ag must be multiplied by 2 to balance electrons and write, for the spontaneous reaction,

Oxidation of Ag(*s*)	$2Ag(s) \rightleftarrows 2Ag^+ + \cancel{2e^-}$
Reduction of F_2	$F_2 + \cancel{2e^-} \rightleftarrows 2F^-$
Overall	$2Ag(s) + F_2 \rightleftarrows 2Ag^+ + 2F^-$

8. Using Table 10.2, predict qualitatively what spontaneous reaction (if any) will take place under the circumstances described below. Where reaction does occur, identify the oxidizing and reducing agents and give the equations for the half-reactions and for the overall reaction. Exercise
   a. Mercury (Hg) is poured into a solution of zinc sulfate ($Zn^{2+}SO_4^{2-}$).
   b. Some tin metal is added to a sodium chloride solution.

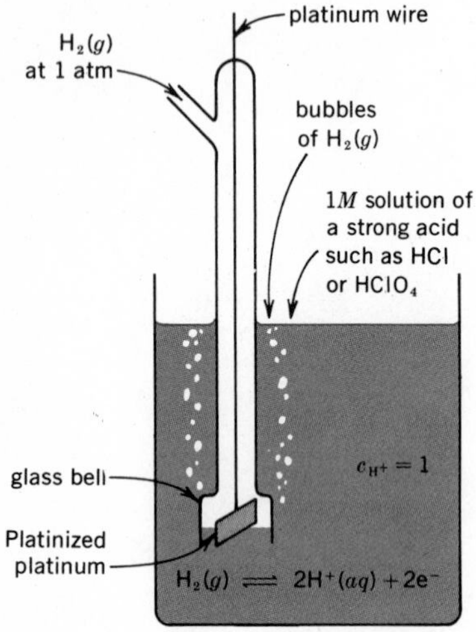

FIG. 10.4 The hydrogen half-cell, Pt(s), platinum black|$H_2(g)$ (1 atm)|$H^+(aq)$(1 *M*).

c. Chlorine gas (Cl_2) is bubbled into a solution of sodium bromide (Na^+Br^-).
d. Some aluminum metal is put in a solution of hydrochloric acid.
e. Solutions of ferric chloride ($Fe^{3+}(Cl^-)_3$) and sodium iodide (Na^+I^-) are combined.

You can use the $\mathscr{E}^0$ values in the table of standard oxidation potentials (Table 10.2) to make *quantitative* calculations of the emf's of electrochemical cells and the values of equilibrium constants for oxidation–reduction reactions. But to be able to do this you first have to know what the $\mathscr{E}^0$ values for the half-reactions mean.

10.3 $\mathscr{E}^0$ Values and Their Uses The $\mathscr{E}^0$ values are called *standard oxidation potentials* and they are measured in *volts*. As their name suggests, they are a *quantitative measure of the ease of oxidation of the species* on the left-hand side of the electron-transfer half-reaction, usually a metal. The more positive the $\mathscr{E}^0$ value, the easier the oxidation (electron removal). In other words, since the strongest reducing agents are the most readily oxidized, the more positive the $\mathscr{E}^0$ value is, the stronger the species on the left of the equation is as a reducing agent. The strongest reducing agent in Table 10.2 is potassium (K); the weakest is fluoride ion (F^-). The strongest oxidizing agent in the table is fluorine gas (F_2); the weakest is potassium ion (K^+).

In this chapter H^+ stands for the hydronium ion H_3O^+, to make balancing equations easier.

Figure 10.5 shows how the scale of $\mathscr{E}^0$ values is a *relative* one based on the hydrogen electrode as reference standard. The equation for the half-reaction for the electrode (Figure 10.4) is $H_2(g) \rightleftharpoons 2H^+ + 2e^-$, and with $p_{H_2} = 1$ atm, $C_{H^+} = 1$ mole/liter, $t = 25°C$, it is taken as the "sea level" of the $\mathscr{E}^0$ scale and assigned the value $\mathscr{E}^0 = 0.000\ldots$ V. The hydrogen electrode makes an electrical connection to H_2 gas and H^+ in solution by bringing them together at a platinum surface.

Figure 10.6 shows how $\mathscr{E}^0$ values are measured in relation to the hydrogen electrode. Notice that the concentrations of all ions are 1 mole/liter and the temperature is 25°C. These conditions, and also the condition that $p_{H_2} = 1$ atm, must be met if the $\mathscr{E}^0$ voltage measured is to be a *standard* oxidation potential, because $\mathscr{E}^0$ values depend on concentration. The dashed lines in the figure stand for "salt bridges," which permit completion of the electrical circuit by the movement of ions. Without this connection the $\mathscr{E}^0$ value cannot be measured. The salt bridge makes it possible to *separate* the two half-reactions making up an overall reaction. And because of this separation we are able to measure the electric potential or emf difference $\Delta\mathscr{E}^0$ for the two reactions.

Data in Table 10.2 show that hydrogen ion will oxidize zinc metal according to the equation

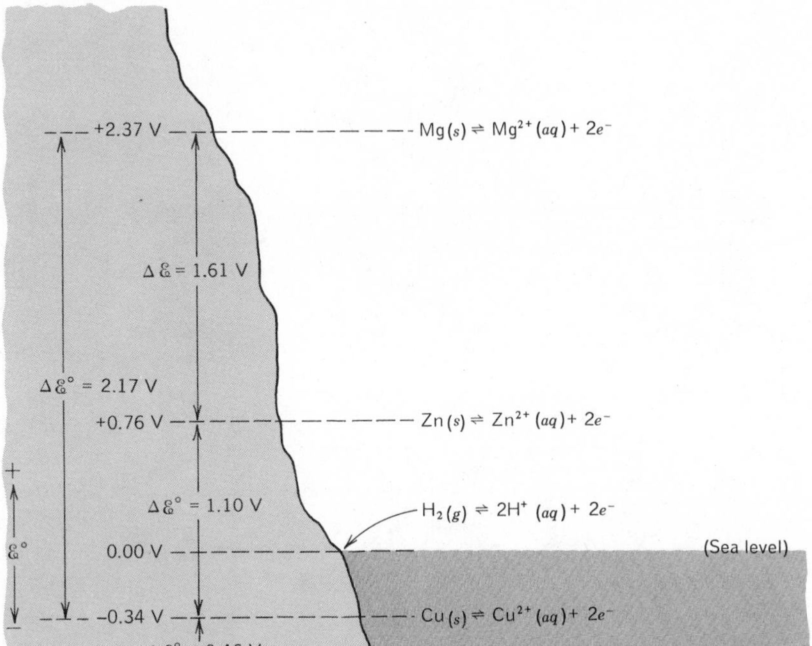

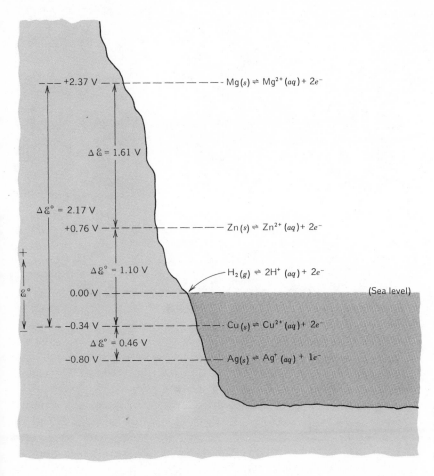

FIG. 10.5 Similarity between $\mathscr{E}^0$ values for electron-transfer half-reactions referred to H_2 electrode and altitudes referred to sea level.

$$Zn(s) + 2H^+ \rightleftarrows Zn^{2+} + H_2(g)$$

This reaction takes place rapidly when zinc comes in contact with an HCl solution, as in "etching" galvanized (zinc-coated) iron before painting it. *In this situation no emf measurements can be made*—the electron transfer occurs right at the zinc surface. By using the salt bridge, $Zn(s)$ and the solution of H^+ don't touch, and the electron transfer is made to occur *through the wire* with the half-reactions going on in separate containers.

Notice in Figure 10.6 that the flow of electrons *in the wire* for the magnesium, zinc, and tin electrodes is *from the metal toward the hydrogen electrode as the cell reaction goes on.* Half-reactions showing this direction of electron flow are said to be "above hydrogen." And if the half-reaction is for a metal, the metal will dissolve in acidic solutions and is called an "active metal." Hydrogen ion oxidizes active metals, so the equilibria that follow have positions very far to the right.

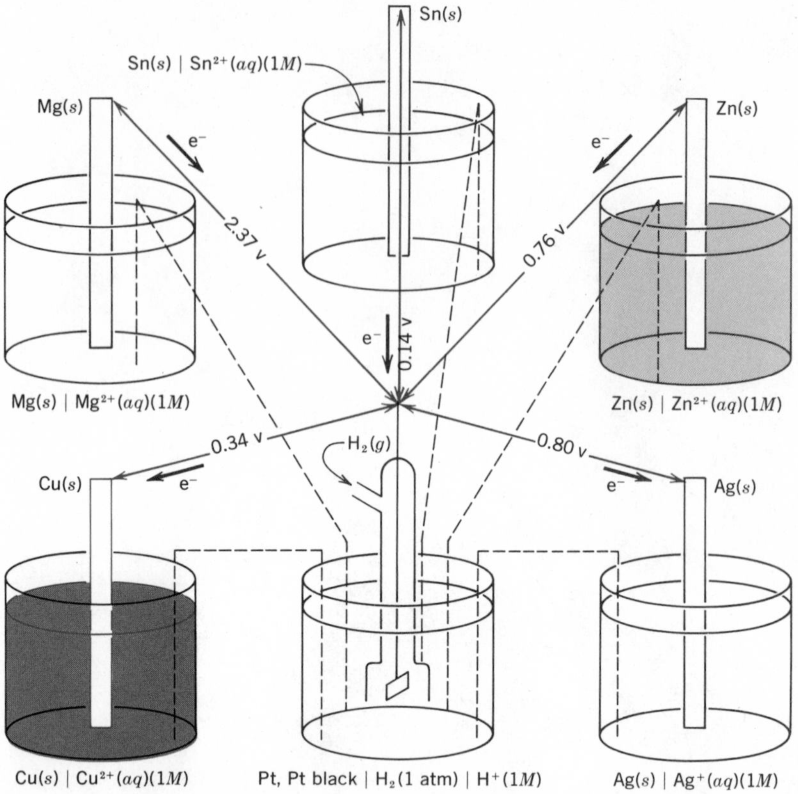

FIG. 10.6 Setup for measuring standard oxidation potentials. The dashed lines stand for salt bridges; arrows show the direction of electron flow in the external circuit.

$$Mg(s) + 2H^+ \rightleftarrows Mg^{2+} + H_2(g) \qquad K = 10^{81}$$
$$Zn(s) + 2H^+ \rightleftarrows Zn^{2+} + H_2(g) \qquad K = 10^{26}$$
$$Sn(s) + 2H^+ \rightleftarrows Sn^{2+} + H_2(g) \qquad K = 10^{5}$$

On the other hand, for the copper and silver electrodes in Figure 10.6, electron flow in the wire is in the opposite direction—*from the hydrogen electrode to the metal electrode*. Copper and silver are "below hydrogen" and are not oxidized to their ions by hydrogen ion. Their $\mathscr{E}^0$ values are given a *negative sign* to show that they are below the "sea level" of the hydrogen electrode (Figure 10.4). Their ions will oxidize hydrogen molecules to hydrogen ion according to the position rule we used earlier. The following equilibria, in which molecular hydrogen acts as reducing agent, lie far to the right:

$$Cu^{2+} + H_2(g) \rightleftarrows Cu(s) + 2H^+ \qquad K = 10^{12}$$
$$2Ag^+ + H_2(g) \rightleftarrows 2Ag^+ + 2H^+ \qquad K = 10^{27}$$

Emfs of electrochemical cells Suppose we put together the cell in Figure 10.7. (Here the purpose of the salt bridge in completing the circuit by allowing ions to migrate is shown.) How could we *calculate*

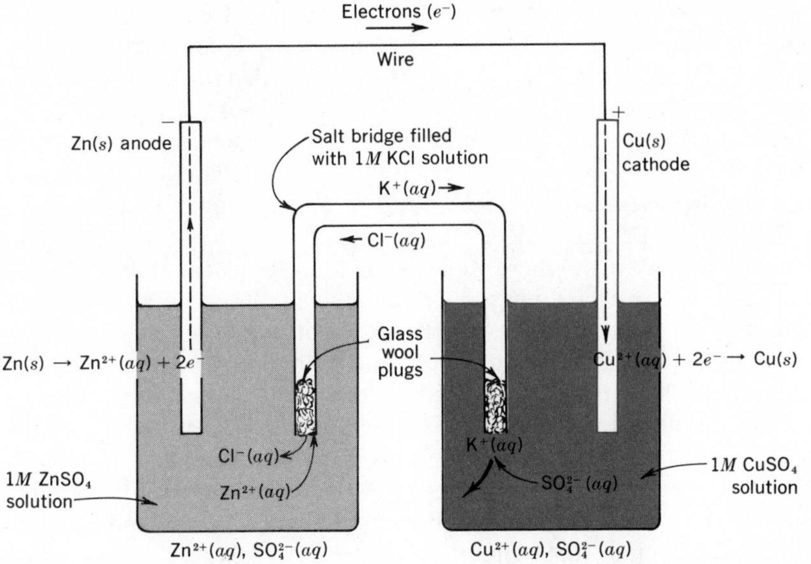

FIG. 10.7 The cell, $Zn(s)|Zn^{2+}(aq)$ (1 *M*)$\vdots K^+Cl^-(aq)$(1 *M*)$\vdots Cu^{2+}(aq)$ (1 *M*)$|Cu(s)$.

the emf of this simple cell or battery? Going back to the "sea-level" picture (Figure 10.5) for $\mathscr{E}^0$ values we find $Zn(s) \rightleftarrows Zn^{2+} + 2e^-$ is 0.76 V "above" the hydrogen electrode at $\mathscr{E}^0 = 0.00$; and the half-reaction $Cu(s) \rightleftarrows Cu^{2+} + 2e^-$ is at -0.34 V "below" the hydrogen electrode. To find $\Delta\mathscr{E}^0$, the standard voltage difference between the zinc electrode and the copper electrode, you make the calculation $0.76 + 0.34 = 1.10$ V $= \Delta\mathscr{E}^0$, as shown by the arrows in the figure. The thinking is exactly like finding the distance Δh between a point 760 ft above sea level and one 340 ft below: $\Delta h = 760 + 340 = 1100$ ft. Now, using the *qualitative* approach described earlier, you can see that the half-reaction for $Cu(s)$ is *below* the one for $Zn(s)$. This means Cu^{2+} will oxidize $Zn(s)$. So, following the rule, you write (and now you put in the $\mathscr{E}^0$ values) the following:

$$Zn(s) \rightarrow Zn^{2+} + 2e^- \qquad \Delta\mathscr{E}^0 = +0.76 \text{ V}$$

$$Cu^{2+} + 2e^- \rightarrow Cu(s) \qquad \Delta\mathscr{E}^0 = +0.34 \text{ V} \quad \text{(Note sign change.)}$$

$$Zn(s) + Cu^{2+} \rightarrow Zn^{2+} + Cu(s) \qquad \Delta\mathscr{E}^0 = 1.10 \text{ V}$$

But $\mathscr{E}^0$ for $Cu(s) \rightleftarrows Cu^{2+} + 2e^-$ is -0.34 V, and we used $+0.34$ V. This is because the half-reaction for Cu was *reversed* by writing it as a *reduction* (electrons on the left). So we have the rule, *if the direction of an electron-transfer half-reaction is reversed*, then the sign of its $\mathscr{E}^0$ value must be reversed. This is true whether the value is positive or negative in the table of $\mathscr{E}^0$ values.

Figure 10.7 also shows the correct arrangement for an electrochemical cell made up of two "half-cells." The electrode at which oxidation occurs as electricity is drawn from the cell is called the *anode*. "Oxidation always occurs at the anode," because electrons leave the cell there. *Reduction* must occur at the *cathode*, where electrons enter the cell. In the shorthand for cells, the anode "half-cell" is always put on the left-hand side, the cathode half-cell on the right, as in

$$\overset{e^-\;\longrightarrow}{Zn|Zn^{2+}\,\vdots\vdots\,Cu^{2+}|Cu}$$

where Zn is the anode, Cu the cathode. The solid vertical bars separate phases (solid|liquid in this case); the dashed vertical lines indicate the salt bridge. Electrons flow from anode to cathode through a connecting wire, in the direction of the arrow.

Suppose the $Zn|Zn^{2+}$ half-cell in Figure 10.7 is replaced by a silver–silver ion half-cell, $Ag|Ag^+$. Is the cell representation $Ag|Ag^+\vdots\vdots Cu^{2+}|Cu$ still correct? It will be *if* the Ag electrode is the *anode*, and this means that the cell reaction must be $2Ag(s) + Cu^{2+} \rightleftarrows 2Ag^+ + Cu(s)$. Reference to the table of $\mathscr{E}^0$ values shows Ag *below* Cu, so the spontaneous reaction is in the *other* direction—that is, Ag^+ oxidizes $Cu(s)$. The correct cell representation is therefore

$$\overset{e^-\;\longrightarrow}{\text{Anode}\quad\text{Cathode}}$$
$$Cu|Cu^{2+}\,\vdots\vdots\,Ag^+|Ag$$

Notice that when a half-cell reaction is multiplied through to balance electrons the $\mathscr{E}^0$ value is *not* multiplied.

And $\Delta\mathscr{E}^0$ is 0.46 V, found as follows:

		$\mathscr{E}^0$
Anode	$Cu(s) \rightleftarrows Cu^{2+} + 2e^-$	−0.34
Cathode	$2Ag^+ + 2e^- \rightleftarrows 2Ag(s)$	+0.80 *(sign changed)*
	$Cu(s) + 2Ag^+ \rightleftarrows Cu^{2+} + Ag(s)$	$\Delta\mathscr{E}^0 =$ 0.46 V

If you make a mistaken prediction for the spontaneous direction of a cell reaction, the error will show up when you calculate the value of $\Delta\mathscr{E}^0$ for the cell—its sign will be negative. You can see this in the following example:

	$\mathscr{E}^0$
$Sn^{2+} + 2e^- \rightleftarrows Sn(s)$	−0.14 *(sign changed)*
$Cu(s) \rightleftarrows Cu^{2+} + 2e^-$	−0.34

$$Cu(s) + Sn^{2+} \rightleftarrows Cu^{2+} + Sn(s)^- \quad \Delta\mathscr{E}^0 = -0.48\ \text{V}$$

The spontaneous reaction is actually from right to left, or $Sn(s) + Cu^{2+} \rightarrow Sn^2 + Cu(s)$, and $\Delta\mathscr{E}^0$ is $(+0.14 + 0.34) = +0.48$ V. The standard voltage difference $\Delta\mathscr{E}^0$ *is always positive in sign if the spontaneous cell reaction is from left to right in the equation as written.*

For the half-reactions in Table 10.2 involving molecules and nonmetal ions, the electrodes are conveniently made of platinum or graphite. The electrode enters into the half-reaction only as source or sink for electrons. The electrode dips into a suitable solution in the half-cell and is bathed in the gas that is a participant. For example, a good O_2 electrode for the half-reaction $4OH^- \rightleftarrows O_2(g) + 2H_2O(l) + 4e^-$ is nickel oxide (NiO) deposited on metallic nickel and bathed in a 1.00-*M* sodium hydroxide solution and oxygen gas at 1 atm pressure.

Exercises

9. Locate the proper half-reactions for the overall reaction $2H_2(g) + O_2(g) \rightleftarrows 2H_2O(l)$ and calculate its $\Delta\mathscr{E}^0$ value. This is the cell reaction in the H_2,O_2 fuel cells used for electrical power in space vehicles. How many such cells would have to be connected in series to give 110 V? (For connection in series, the individual cell emfs add up.)

10. Radio electron tubes require up to 800 V for their operation, while most transistors do the same job using less than 15 V. What effect must the invention of the transistor have had on space exploration?

11. Tin forms ions Sn^{2+} and Sn^{4+}. Which ion will result from dissolving tin metal in hydrochloric acid? Which from reacting tin with chlorine?

12. Metallic copper dissolves rapidly in dilute nitric acid (HNO_3) solution that contains H^+ and NO_3^- ions, but Cu does not dissolve in dilute hydrochloric acid. Explain, identify the oxidizing and reducing agents, and give the balanced equation for the reaction of Cu with HNO_3.

13. Which is the spontaneous direction for the reaction $2Cl^- + Br_2(l) \rightleftarrows Cl_2(g) + 2Br^-$, left to right or right to left? Calculate the value of $\Delta\mathscr{E}^0$ for the spontaneous reaction.

14. Give the shorthand cell representation for the reaction $2Al(s) + 3Cl_2(g) \rightleftarrows 2Al^{3+} + 6Cl^-$, using | for phase separations and ⁞⁞ for the salt bridge, and assuming a carbon (C) rod for the Cl_2 electrode.

15. Give the reaction for the cell $Zn|Zn^{2+}$⁞⁞$I^-|I_2(s)$ Pt. Is this shorthand representation correct? Calculate $\Delta\mathscr{E}^0$ for the cell. What would happen if solid iodine and zinc were ground together in a small amount of water?

Walther Hermann Nernst (1864–1941) was a German physicist and chemist noted for his statement of the third law of thermodynamics. Nernst won the Nobel prize in chemistry in 1920.

10.4 The Nernst Equation—Equilibrium Constants for Oxidation–Reduction Reactions Le Chatelier's principle applies to oxidation–reduction reactions as it does to any reaction. What would it predict for the effect on the emf of the cell $Zn|Zn^{2+} \,\vdots\vdots\, Cu^{2+}|Cu$ in Figure 10.7 if you lowered the copper ion concentration $C_{Cu^{2+}}$ below the value of 1 mole/liter specified for the standard oxidation potentials? You can make a prediction using either the overall cell reaction or the half-reaction at the copper electrode. In the overall cell reaction Cu^{2+} is a *reactant*. Any decrease in its concentration would tend to shift the position of equilibrium to the left, toward the reactant side, according to Le Chatelier. Since the positive emf developed by the cell depends on the reaction going from left to right, a decrease in Cu^{2+} concentration should lower the emf, and it does.

If you apply Le Chatelier's principle to the half-reaction at the copper electrode, the effect on the emf of lowering $C_{Cu^{2+}}$ is even more understandable. When current is not being drawn, think of an equilibrium existing at the Cu electrode, described by the half-reaction $Cu^{2+} + 2e^- \rightleftarrows Cu(s)$. Now the emf of the cell depends on the *difference* in electric potentials or "electron pressures" at the copper and zinc electrodes. Since electrons flow through a wire *from* the zinc *to* the copper electrode as the cell reaction takes place, the electron pressure at the zinc electrode must be higher than at the copper electrode. In the half-reaction for the copper electrode, the two electrons are removed from the electrode by Cu^{2+} ions as the reaction proceeds. If you decrease the Cu^{2+} concentration, there will be fewer Cu^{2+} ions to take electrons from the copper electrode. This will allow the electron pressure in the copper electrode to *increase*, moving it closer to the pressure at the zinc electrode. The emf of the cell should decrease, and it does.

Exercises

16. Predict the effect on cell emf of increasing the zinc ion concentration in the cell of Figure 10.7.

17. What effect (if any) would an increase in C_{H^+} for the hydrogen electrode above 1 mole/liter have on its $\mathscr{E}^0$ value? Would a change in the pressure of hydrogen gas supplied to the electrode alter its $\mathscr{E}^0$ value? If the pressure is increased, will $\mathscr{E}^0$ move up (to positive values) or down (to negative values)? (Use Le Chatelier's principle.)

The numerical effects of concentration changes on cell emfs are given by the Nernst equation. Cell emfs depend slightly on temperature, and the Nernst equation ☛ for use at 25°C has the form

The constant 0.06 is rounded off from the accurate value of 0.05914.

$$\Delta\mathscr{E} = \Delta\mathscr{E}^0 - \frac{0.06}{n} \log Q \qquad \text{at } 25°\text{C}$$

where Q is the reaction quotient (Section 7.4), which has the form of the equilibrium constant expression for the cell reaction, but with non-equilibrium concentrations inserted; n is the number of moles of electrons transferred from reducing to oxidizing agent in the balanced equation for the cell reaction; $\Delta\mathscr{E}^0$ is the standard emf of the cell at 25°C with 1 mole/liter concentrations of all participants, calculated from standard oxidation potentials (Table 10.2); and $\Delta\mathscr{E}$ is the actual emf of the cell at 25°C taking any nonstandard concentrations of participants into account. If all concentrations are 1 mole/liter, then the value of Q also becomes 1, and since $\log 1 = 0$, then $\Delta\mathscr{E} = \Delta\mathscr{E}^0$ under standard conditions.

EXAMPLES

Before the invention of mechanically driven generators for making electricity, the telegraph system for sending messages in Morse code was driven by batteries called Daniel cells, which used the reaction $Zn(s) + Cu^{2+} \rightarrow Zn^{2+} + Cu(s)$, $\Delta\mathscr{E}^0 = 1.10$ V. Suppose after a Daniel cell had operated for a time, $C_{Cu^{2+}}$ in the cell has fallen from 1 mole/liter to 0.50 mole/liter, while $C_{Zn^{2+}}$ has risen to 1.50 mole/liter as a result of the cell reaction. What would be the emf ($\Delta\mathscr{E}$) of the cell with these concentrations according to the Nernst equation? According to the equation for the reaction two moles of electrons are transferred from $Zn(s)$ to Cu^{2+} and $Q = C_{Zn^{2+}}/C_{Cu^{2+}}$, so the Nernst equation takes the form

$$\Delta\mathscr{E} = \Delta\mathscr{E}^0 - \frac{0.06}{2} \log \frac{C_{Zn^{2+}}}{C_{Cu^{2+}}}$$

For the cell reaction $\Delta\mathscr{E}^0 = 1.10$ V, if we use data in Table 10.2 and substitute this value and the values of $C_{Cu^{2+}} = 0.50$, $C_{Zn^{2+}} = 1.50$, we obtain

$$\Delta\mathscr{E} = 1.10 - \frac{0.06}{2} \log \frac{1.5}{0.5}$$

or

$$\Delta\mathscr{E} = 1.10 - 0.03 \log 3$$

The log table gives $\log 3 = 0.477$, so

$$\Delta\mathscr{E} = 1.10 - (0.03 \times 0.477) = 1.09 \text{ V}$$

The effect of decreasing $C_{Cu^{2+}}$ and raising $C_{Zn^{2+}}$ is to lower the emf of the cell by $1.10 - 1.09 = 0.01$ V. If the cell were operated long enough to make $(C_{Zn^{2+}}/C_{Cu^{2+}}) = 10$, then $\Delta\mathscr{E}$ would be

$$\Delta\mathscr{E} = 1.10 - \frac{0.06}{2}\log 10 = 1.10 - 0.03 = 1.07 \text{ V}$$

The lesson here is that cell emfs do not change rapidly as the concentrations of reactants change. From a practical point of view, telegraphers didn't have to worry about losing the emf to drive their messages through the wires, as long as the batteries were replenished with zinc electrodes and copper sulfate from time to time.

Equilibrium constants for oxidation–reduction reactions Hidden in the Nernst equation is a simple relationship between the value of the equilibrium constant for an oxidation–reduction reaction and the standard emf ($\Delta\mathscr{E}^0$) for the corresponding cell. We can find it by looking at what happens as a cell continues to supply electrical energy.

You saw in Chapter 7 that energy can be obtained by letting nonequilibrium systems come to equilibrium. Once there, no more energy can be obtained. Now think of a battery providing the energy to run a radio or an electric motor. So long as it has an emf, it will provide the electricity necessary to do the work, but while it does, the battery is on its way to equilibrium. When it reaches equilibrium, there can be no more changes in the composition of any of its phases, so the overall reaction stops and the emf becomes zero. The battery is discharged or "dead," as the saying goes–its $\Delta\mathscr{E}$ has become zero.

What does $\Delta\mathscr{E} = 0$ mean to the Nernst equation? It turns out to mean that $\Delta\mathscr{E}^0 = (0.06/n) \log K$, where K is the equilibrium constant for the cell reaction, and n is the number of moles of electrons transferred. You can derive this equation as follows:

1. At equilibrium for the cell reaction, $\Delta\mathscr{E} = 0$.
2. The Nernst equation therefore becomes $\Delta\mathscr{E} = 0 = \Delta\mathscr{E}^0 - (0.06/n) \log Q$, or, on rearranging the right-hand equality,

   $$\Delta\mathscr{E}^0 = \frac{0.06}{n}\log Q$$

3. But, whatever the *original* concentrations of participants were in Q, *at equilibrium they become the* ***equilibrium*** *concentrations*, and since K and Q have identical forms, $Q = K$ at equilibrium and we get

   $$\Delta\mathscr{E}^0 = \frac{0.06}{n}\log K \qquad \text{at } 25^\circ\text{C}$$

You can see that this very powerful equation can be used to *calculate* thousands of equilibrium constants for oxidation–reduction reactions using $\mathscr{E}^0$ values. Table 10.2 contains only a small portion of the standard oxidation potentials that have been measured.

The equation can be put into a more convenient form by rearranging it and writing K as a power of 10:

$$\frac{0.06}{n} \log K = \Delta\mathscr{E}^0$$

$$\log K = \frac{n\,\Delta\mathscr{E}^0}{0.06} = 17n\,\Delta\mathscr{E}^0$$

or

$$K = 10^{17n\,\Delta\mathscr{E}^0} \qquad \text{at } 25°\text{C}$$

All of the equilibrium constants for oxidation–reduction reactions given earlier in this chapter were calculated by using this equation and $\mathscr{E}^0$ data from Table 10.2. We have seen for the reaction $Zn(s) + 2Ag^+ \rightleftarrows Zn^{2+} + 2Ag(s)$ that $\Delta\mathscr{E}^0 = 1.56$ V, and $n = 2$, so applying the equation gives

$$K = \frac{[Zn^{2+}]}{[Ag^+]^2} = 10^{17n\,\Delta\mathscr{E}^0} = 10^{17 \times 2 \times 1.56} = 10^{53}$$

This is the value for the reaction entered in Table 10.1. ☛

Some of the values of exponents in the table have been rounded to the nearest whole number.

Exercises

18. The reactions in the methane (CH_4) fuel cell $Pt|CH_4(g)|H^+ \,\vdots\, H^+|O_2|Pt$ are as follows:

Anode	$CH_4(g) + 2H_2O(l) \rightleftarrows CO_2(g) + 8H^+ + 8e^-$
Cathode	$2O_2(g) + 8H^+ + 8e^- \rightleftarrows 4H_2O(l)$
Overall	$CH_4(g) + 2O_2(g) \rightleftarrows CO_2(g) + 2H_2O(l)$

For this reaction $K = 10^{44}$. Using the Nernst equation, calculate $\Delta\mathscr{E}^0$ for the cell.

19. Using necessary data from Table 10.2, calculate the value of K for the reaction $Sn(s) + Fe^{2+} \rightleftarrows Sn^{2+} + Fe(s)$. Predict what reaction, if any, would occur to a significant extent if a piece of iron (Fe) were placed in an $SnCl_2$ solution.

10.5 Standard Free Energy Changes Spontaneous oxidation–reduction reactions going on in electrochemical cells or batteries ☛ are essential sources of electrical energy. The function of the battery is

Technically, a battery is a group or "battery" of electrochemical cells usually connected in series so the emfs of the individual cells add up, but the term is used loosely for even a single cell.

to convert chemical potential energy stored in the reactants for the cell reaction into electrical energy in the form of a flow of electrons through the external circuit. The external circuit may be as simple as a single flashlight bulb or as complex as the ignition, power, and lighting circuits of an airliner. No matter what its complexity, the *maximum* amount of electrical energy the battery powering the external circuit can provide may be calculated by using the idea of the faraday volt discussed in Section 10.1.

A cell will provide its maximum electrical energy only if current is drawn from it by the external circuit at a rate that is slow enough so the emf ($\Delta\mathscr{E}$) of the cell doesn't decrease. This requirement of constant emf or voltage cannot be met by any *real* cell, because *concentration changes that inevitably go along with the cell reaction always cause the emf to decrease.* For calculating the *maximum* energy a cell can provide, imagine one so huge that using a small portion of its energy will cause no measurable change in the concentrations of reactants. Such a cell is often called an "infinite cell." With no changes in concentrations, temperature, or pressure, the emf of the cell will remain constant at the value of $\Delta\mathscr{E}$ during the withdrawal of the energy by the external circuit. How you use the energy (to run a motor, say, or to light a lamp) in the external circuit makes no difference to the cell providing it. Only the quantity is important. However much is used in the external circuit, it must be at the expense of the chemical potential energy of the cell (law of conservation of energy).

The decrease results from an increase in the value of Q in the Nernst equation as product concentrations increase (numerator) and reactant concentrations decrease (denominator).

Electrical energy drawn from cells (or other sources) is in a particularly useful form because it can be converted directly into heat, light, or mechanical work. It's different from thermal energy, which can do mechanical work only on flowing from a source at a high temperature to a sink at a lower temperature (Introduction), because the conversion of electrical energy into work is limited only by the efficiency of the electric motor doing the job. The energy drawn from an electrochemical cell is called "free" or available energy for this reason and given the symbol ΔG. For a process occurring under standard conditions of unit concentrations and 25°C, the symbol ΔG^0 is used. If the emf of a cell is $\Delta\mathscr{E}^0$ and it drives n moles of electrons through the external circuit, then the free energy change *for the cell* is given by the equation

$$\Delta G^0 = -n\mathscr{F}\,\Delta\mathscr{E}^0 = -23.06n\,\Delta\mathscr{E}^0$$

In this equation, $\mathscr{F}$ is the faraday volt with the value 23.06 kcal V^{-1} $mole^{-1}$ (Section 10.1) and n is the number of moles of electrons transferred through the external circuit from the reducing agent to the oxidizing agent according to the overall cell reaction. It has the same value as n in the Nernst equation. The sign of ΔG^0 for the cell is negative because the cell gives up some of its free energy to do the work of the external circuit. *Therefore* ΔG^0 *will be negative in sign for any spontaneous cell reaction* because, as we have seen, $\Delta\mathscr{E}^0$ is always positive in sign for them.

EXAMPLE

The reaction for the Ruben cell (Photo 27) used as the power supply in hearing aids is $Zn(s) + HgO(s) \rightleftarrows ZnO(s) + Hg(l)$, and $\Delta\mathscr{E}^0$ for the cell is 1.33 V. To put the energy used by a hearing aid in scale we can calculate the number of kilowatt hours of electrical energy a hearing-aid battery will provide before it is exhausted. The calculation will show what tiny amounts of energy are used in hearing, even by the hard of hearing, and how tremendously sensitive an "instrument" the human ear is.

Zinc oxide and mercuric oxide, participants in the cell reaction, are ionic compounds, $Zn^{2+}O^{2-}$ and $Hg^{2+}O^{2-}$. Since $Zn(s)$ goes to Zn^{2+} in ZnO, and Hg^{2+} in HgO goes to $Hg(l)$, the cell reaction transfers two moles of electrons from $Zn(s)$ to Hg^{2+}, and $n = 2$ in the equation $\Delta G^0 = -n\mathscr{F}\,\Delta\mathscr{E}^0$ for the free energy change. Putting in the value of 1.33 for $\Delta\mathscr{E}^0$, $n = 2$, and $\mathscr{F} = 23.06$ kcal V^{-1} $mole^{-1}$ gives

$$\Delta G^0 = -2(23.06)(1.33) = -61.3 \text{ kcal}$$

This energy quantity is for a *mole* of Zn being oxidized to ZnO by a mole of HgO. Depending on its size, a hearing-aid battery will contain about 0.001 mole of HgO (mol. wt 216.6) or about 0.2 g of HgO. Reduction of this weight of HgO by zinc will yield 0.001 times ΔG^0 for a mole of HgO, or

$$\Delta G^0 = -0.001 \times 61.3 = -0.0613 \text{ kcal}$$

of energy. One kilowatt hour (kwh) of electricity, the unit used for calculating home electricity bills, is 860 kcal of electrical energy, so the reduction of the HgO in the hearing-aid battery will yield

$$\frac{0.0613 \text{ kcal}}{859 \text{ kcal kwh}^{-1}} = 7.14 \times 10^{-5} \text{ kwh}$$

A hearing-aid battery lasts for about a month, or assuming 12 hr/day of use, about $12 \times 30 = 360$ hr. When turned on, it uses, per hour of operation,

$$\frac{7.14 \times 10^{-5} \text{ kwh}}{30 \text{ hr}} = 2.38 \times 10^{-6} \text{ kwh}$$

About 1/100 of the electricity used by the hearing aid goes into making the sound waves that strike the eardrum, so $2.38 \times 10^{-6} \times 10^{-2} = 2.38$

The resulting sound would probably shatter your eardrums if you were anywhere near the sound.

$\times 10^{-8}$, or 0.0000000238 kwh per hr are used. This is enough to stimulate the auditory nerve through the eardrum. By contrast, a good "hi-fi" amplifier turned up to *maximum volume* would be able to put out about 50 watts of power to drive a speaker, or 0.050 kwh per hour. This is $(0.050/2.38 \times 10^{-8}) = 2{,}100{,}000$ times the energy of the sound produced by the hearing-aid battery during one hour, and heard by the ear. The normal human ear can hear sound in the frequency range 500–5000 cycles per second (hertz), having energy equal to only 10^{-20} kwh per hr.

Free energy changes do not depend on the path or reaction mechanism by which reactants go to products. The ΔG^0 for a cell reaction *is the difference in the free energies of products and reactants*, and these are determined solely by their state, not by how they got there. Using G^0 for free energies *per mole* of reactants and products in their states at standard conditions (25°C, 1 atm), we obtain

$$\Delta G^0 = G^0_{\text{products}} - G^0_{\text{reactants}}$$

Chemists are more interested in *changes* in free energies (ΔGs) than in the free energies (Gs) because *changes in G accompany reactions*, the changes can be readily measured and are easily related to equilibrium constants by the equation $\Delta G^0 = -n\mathscr{F}\,\Delta\mathscr{E}^0$ for cells.

The other quantity for describing chemical systems that is *independent of the path* by which a particular state of the system was reached is the *equilibrium constant*. Since ΔG^0 for a reaction is independent of path and results from progress of a reaction in its spontaneous direction, it would seem reasonable that K and ΔG^0 might be related. They are, and by this equation:

$$\Delta G^0 = -1.364 \log K \qquad \text{at 25°C, 1 atm}$$

The change ΔG^0 is *not* the free energy change of the system going to equilibrium. It is the difference between the free energies of reactants and products, assuming *complete* reaction.

where ΔG^0 is the free energy change of the system according to the equation for the oxidation–reduction reaction, *assuming complete reaction* [*it is in units of kcal* (not kcal/mole, but kcal for the reaction *as written*)], and K is the equilibrium constant at 25°C for the cell reaction *as written*.

The most important aspect of this equation is that *large values* of K *mean large* **negative** values of ΔG^0. Because large values of K correspond to equations for reactions that go *spontaneously* from left to right, *free energy changes are negative for spontaneous reactions.* This is true *for all reactions*, not just oxidation–reduction or cell reactions, provided only that temperature and pressure are held constant. *All spontaneous changes occurring at constant T and P are accompanied by a decrease in the free energy of the system.* With cells, the free energy is used in the work done or heat produced by the flow of electrons in the external circuit. For other reactions it appears as heat.

The equation $\Delta G^0 = -1.364 \log K$ can be derived by combining the equation $\Delta G^0 = -n\mathscr{F}\,\Delta\mathscr{E}^0$ and the Nernst equation for an equilibrium in the following way:

$$\Delta\mathscr{E}^0 = \frac{0.06}{n} \log K \qquad \text{(Nernst equation for equilibrium)}$$

But from $\Delta G^0 = -n\mathscr{F}\,\Delta\mathscr{E}^0$, we have $\Delta\mathscr{E}^0 = -\Delta G^0/n\mathscr{F}$. Substituting this value of $\Delta\mathscr{E}^0$ into the Nernst equation gives

$$\Delta\mathscr{E}^0 = \frac{-\Delta G^0}{n\mathscr{F}} = \frac{0.06}{n} \log K$$

Rearranging the right-hand equality gives you

$$\Delta G^0 = \frac{-\cancel{n}\mathscr{F} \times 0.06}{\cancel{n}} \log K$$

or

$$\Delta G^0 = -0.06\mathscr{F} \log K$$

Inserting 23.06 kcal for the faraday volt ($\mathscr{F}$) gives

$$\Delta G^0 = -0.06 \times 23.06 \log K$$
$$\Delta G^0 = -1.38 \log K$$

The *accurate* value of the constant 0.06 in the Nernst equation is 0.05914. Taking this into account, we see that the accurate form of the equation relating the free energy change and the equilibrium constant for a reaction is

$$\Delta G^0 = -0.05914 \times 23.06 \log K$$
$$\Delta G^0 = -1.364 \log K \qquad \text{at } 25°\text{C}$$

Putting this in exponential form gives a more convenient equation to use:

$$\log K = \frac{-\Delta G^0}{1.364} = -0.733\,\Delta G^0$$

or

$$K = 10^{-0.733\Delta G^0} \quad \text{at } 25°\text{C}$$

The factor 0.733 is used if ΔG^0 is in *kilo*calories. The limitation to 25°C is not as serious as it looks because it is close to room temperature, at which many reactions are studied in the laboratory.

A more important limitation of the equation is the one built into it by the very nature of the equilibrium constant—it tells only *where* a system will go *if a path or mechanism is* available. If $\Delta\mathscr{E}^0$ for an oxidation–reduction reaction is large and positive, K will be a large number and ΔG^0 will be large and negative. According to thermodynamics the reaction is *possible*. But the reaction may not "go" at all.

EXAMPLES The oxidation of iodide ion (I^-) by ferric ion (Fe^{3+}) is predicted to occur and does, rapidly:

		$\mathscr{E}^0$	
$2I^- \rightleftarrows I_2(s) + 2e^-$		-0.54	
$2Fe^{3+} + 2e^- \rightleftarrows 2Fe^{2+}$		$+0.77$	(sign reversed)
$2I^- + 2Fe^{3+} \rightleftarrows I_2(s) + 2Fe^{2+}$	$\Delta\mathscr{E}^0 =$	0.23 V	

and

$$\Delta G^0 = -n\mathscr{F}\,\Delta\mathscr{E}^0 = -2 \times 23.06 \times 0.23 = -10.6 \text{ kcal} \quad \text{at } 25°\text{C}$$

$$K = 10^{(-0.733)(-10.6)} = 10^{7.8} = 5.9 \times 10^7$$

The numbers show that the position of equilibrium is far to the right. In fact, a precipitate of solid iodine separates almost immediately when solutions of ferric chloride ($FeCl_3$) and sodium iodide (NaI) are combined. Because the reaction goes rapidly at 25°C, there must be a *low activation energy path* for Fe^{3+} to oxidize I^-.

The reaction between Fe^{2+} and perchlorate ion is also predicted to occur as the calculations show:

	$\mathscr{E}^0$	
$2Fe^{2+} \rightleftarrows 2Fe^{3+} + 2e^-$	-0.77	
$ClO_4^- + 2H^+ + 2e^- \rightleftarrows ClO_3^- + H_2O(l)$	$+1.19$	*(sign reversed)*
$2Fe^{2+} + ClO_4^- + 2H^+ \rightleftarrows 2Fe^{3+} + ClO_3^- + H_2O(l)$	$\Delta\mathscr{E}^0 = 0.42$ V	

and

$$\Delta G^0 = -n\mathscr{F}\,\Delta\mathscr{E}^0 = -2 \times 23.06 \times 0.42 = -19.4 \text{ kcal} \quad \text{at } 25°\text{C}$$

$$K = 10^{(-0.733)(-19.4)} = 10^{14.2} = 1.6 \times 10^{14}$$

PHOTO 28 This gummy sludge floating on Lake Tahoe will ultimately be oxidized to $CO_2 + H_2O$. (*EPA-DOCUMERICA – Belinda Rain.*)

Perchlorate ion (ClO_4^-) is shown by the calculations to be an oxidizing agent plenty strong enough to oxidize Fe^{2+} to Fe^{3+}. Like ferric iodide, ferrous perchlorate, $Fe(ClO_4)_2$, is a compound whose existence is theoretically impossible. But the experimental facts are that the compound does exist and forms water-soluble white needles that are quite stable. In contrast to the Fe^{3+}—I^- case, there is no path of low activation energy by which ClO_4^- can oxidize Fe^{2+}.

These two examples show that you must take *both* thermodynamics (ΔG^0 and K) *and* kinetic activation energies (E_a) into account when deciding whether a reaction is going to go. Natural pathways for oxidation of the sludge in Photo 28 to carbon dioxide and water exist, but the process is a slow one.

The exercises that follow give you several practical applications of electrochemistry.

Exercises

20. Write equilibrium constant expressions for the reactions

$$2I^- + 2Fe^{3+} \rightleftarrows I_2(s) + 2Fe^{2+}$$

and

$$2Fe^{2+} + ClO_4^- + 2H^+ \rightleftarrows 2Fe^{3+} + ClO_3^- + H_2O(l)$$

Which reaction would you expect to be displaced toward products by increasing C_{H^+}? Explain.

21. Consider the following cell: $Pt|H_2|H^+, C_1 \vdots\vdots H^+, C_2|H_2|Pt$. It is one form of a "pH meter." Diagram the cell, using a salt bridge to connect the two hydrogen-electrode half-cells. (C_1 and C_2 are the H^+ concentrations in the respective half-cells.)

a. What is the $\Delta\mathscr{E}^0$ for the cell? (*Hint*: remember standard conditions for concentrations and pressures.)

b. If the concentrations of H^+ (C_1 and C_2) in the two half-cells are not equal, the cell becomes a *concentration cell*, one whose emf depends only on a *difference in concentrations* of the *same* participants in two otherwise identical half-cells. If C_2 is greater than C_1, the left-hand electrode is the anode, and the overall cell reaction is obtained as follows:

Anode (on left) oxidation of $H_2(g)$	$H_2(g) \rightarrow 2H^+(C_1) + 2e^-$
Cathode (on right), reduction of H^+	$2H^+(C_2) + 2e^- \rightarrow H_2(g)$
Overall	$2H^+(C_2) \rightleftarrows 2H^+(C_1)$

and

$$K = \frac{[H^+(C_1)]^2}{[H^+(C_2)]^2}$$

The cell reaction consists simply of hydrogen ion concentration going from a higher to a lower value as current flows. This is the spontaneous process that follows putting water (or dilute HCl) on top of a layer of concentrated HCl. As time passes, diffusion between the layers results in uniform C_{H^+} and C_{Cl^-} in the solution. In the case of the concentration cell $C_1 = C_2$ at equilibrium. Suppose $C_1 = 0.01$ mole/liter and $C_2 = 1.00$. Using the Nernst equation in the form $\Delta\mathscr{E} = \Delta\mathscr{E}^0 - (0.06/n) \log Q$, calculate $\Delta\mathscr{E}$ for the cell under these conditions. (*Hint*: $\log X^n = n \log X$.) Now suppose the value of C_1 is not known, but $C_2 = 1.00$ mole/liter. What single measurement would be needed to calculate C_1 and the pH of the solution in the anode half-cell?

22. The lead storage cell making up the "6-volt" and "12-volt" storage batteries used in automobiles illustrates the reaction

$$Pb(s) + PbO_2(s) + 4H^+ + 2SO_4^{2-} \rightleftarrows 2PbSO_4(s) + 2H_2O(l)$$

When energy is drawn from the cell as you start the engine, the reaction goes from left to right (spontaneous direction, direction of discharge). In a completely discharged battery both electrodes consist of lead covered with solid lead sulfate ($PbSO_4$). Using Table 10.2 to find the half-reactions for the charged cell, calculate $\Delta\mathscr{E}^0$, K, and ΔG^0 for the reaction *as written*. In 6-V batteries, three lead storage cells are connected in series; in 12-V batteries, six are.

The cell reaction for the lead storage cell is *reversible*, and a discharged battery can be restored to the charged condition by applying a voltage larger than the cell voltage, using an external "battery charger." Give the equation for the charging reaction. [In the fully charged battery one electrode is lead; the other is lead covered with lead dioxide (PbO_2).]

23. The first step in the rusting of iron to Fe_2O_3 ("rust") is the production of ferrous (Fe^{2+}) ions by attack on the iron of H^+ or O_2 acting as oxidizing agents. Using H^+, we can write the reaction as $Fe(s) + 2H^+ \rightleftarrows Fe^{2+} + H_2(g)$. The Fe^{2+} ions are further oxidized by O_2 from the air to Fe^{3+} ions. These then react with water to produce iron rust, Fe_2O_3. Iron is "galvanized" to prevent rusting by coating it with zinc metal. Determine the value of K for the reaction

$$Fe^{2+} + Zn(s) \rightleftarrows Fe(s) + Zn^{2+}$$

Can you explain how the zinc coating protects the iron from rusting *even if* small spots of iron are exposed? Would a coating of tin (as on a tin can) have the same effect? (See Exercise 19.) The zinc on galvanized iron is called a "sacrificial anode." Why? Sacrificial anodes of Mg are often put in home water heaters to protect iron plumbing pipes from rusting.

Standard free energy of formation of compounds The free energy change that accompanies the formation of *one mole* of a compound from its elements under standard thermodynamic conditions ☛ is called the *standard free energy of formation* (ΔG_f^0) of the compound. Standard free energies of formation are important because they tell by their sign (+ or −) whether a compound is stable or not with respect to decomposition into its elements—only if ΔG_f^0 is negative in sign is the compound stable thermodynamically. The formation of one mole of CO_2 from oxygen gas and carbon *releases* 94.3 kcal of free energy so $\Delta G_f^0[CO_2(g)] = -94.3$ kcal, and CO_2 is a stable compound:

Temperature 25°C, pressure 1 atm. Even if the reaction cannot be carried out under standard thermodynamic conditions, the value of ΔG^0 can be calculated from data measured at other pressures and temperatures.

$$C(s) + O_2(g) \rightarrow CO_2(g) \qquad \Delta G^0 = -94.3 \text{ kcal} = \Delta G_f^0(CO_2)$$

In using ΔG_f^0 values you have to remember that they are *molar* quantities. The statement that the standard free energy of formation of liquid water is −56.7 kcal corresponds to the equation forming one mole of liquid water, $H_2O(l)$:

$$H_2(g) + \tfrac{1}{2}O_2(g) \rightarrow H_2O(l) \qquad \Delta G^0 = -56.7 \text{ kcal} = \Delta G_f^0[H_2O(l)]$$

For the reaction as usually written, $\Delta G^0 = 2\Delta G_f^0[H_2O(l)] = 2(-56.7) = -113.4$ kcal, because two moles of H_2O are formed according to the equation

$$2H_2(g) + O_2(g) \rightarrow 2H_2O(l) \qquad \Delta G^0 = -113.4 \text{ kcal}$$

It is also necessary to show the state or condition of reactants and products. For the formation of water *vapor*, $H_2O(g)$, the equation and free energy are

$$H_2(g) + \tfrac{1}{2}O_2(g) \rightarrow H_2O(g) \qquad \Delta G_f^0[H_2O(g)] = -54.6 \text{ kcal}$$

TABLE 10.3 Standard free energies of formation of compounds from their elements

Formula of Compound	*State*	ΔG_f^0 (kcal/mole)
HF	(g)	− 64.7
HCl	(g)	− 22.8
HBr	(g)	− 12.7
HI	(g)	+ 0.3
HI	(aq)	− 12.4
H_2O	(g)	− 54.6
H_2O	(l)	− 56.7
H_2S	(g)	− 7.9
H_2Se	(g)	+ 17.0
H_2Te	(g)	+ 33.1
H_2O_2	(l)	− 28.2
NaCl	(s)	− 91.8
CO_2	(g)	− 94.3
SiO_2	(s)	−192.0
MgO	(s)	−136.1
CaO	(s)	−144.4
GeO_2	(s)	−128.0
SnO_2	(s)	−124.2
PbO_2	(s)	− 52.3
BeO	(s)	−139.0
B_2O_3	(s)	−283.0
NO	(g)	+ 20.7
F_2O	(g)	+ 9.7
CH_4 (methane)	(g)	− 12.1
C_2H_6 (ethane)	(g)	− 7.9
C_2H_2 (acetylene)	(g)	+ 50.0
B_2H_6 (diborane)	(g)	+ 19.8
$C_6H_{12}O_6$ (glucose)	(s)	−220

Table 10.3 lists standard free energy changes for the formation of one mole of some compounds. All compounds in the table are stable at 25°C except HI, H_2Se, H_2Te, NO, F_2O, acetylene, and diborane. There's nothing mysterious about the relation between the sign of a free energy of formation and the stability of a compound. The equation $K = 10^{-0.733\Delta G^0}$ described on p. 388 applies to all equilibria, including those for the formation of compounds from their elements. If ΔG_f^0 is negative, then K must have a large value, and the product is stable. Take the case of $H_2O(g)$ in the last equation, where $\Delta G_f^0 [H_2O(g)] = -54.6$ kcal/mole:

$$K = 10^{-0.733\Delta G_f^0} = 10^{-0.733(-54.6)} = 10^{+40} \quad \text{at } 25°\text{C}$$

Water is indeed stable.

Free energy changes for reactions can always be calculated if the standard free energies of formation, ΔG_f^0, of reactants and products are known. To get ΔG^0 for a reaction you simply subtract the ΔG_f^0 values of reactants from the values for products, *taking into account the fact that ΔG_f^0 values are **molar** values.* To make the calculation you also have to know that *the standard free energies of formation of the elements in their condition or state at 25°C, 1 atm, are zero.* ☛ Suppose you want to know the standard free energy change for the reaction of photosynthesis to produce glucose, $C_6H_{12}O_6$:

Zero because the elements "have already been formed."

$$6CO_2(g) + 6H_2O(l) \rightarrow C_6H_{12}O_6(s) + 6O_2(g)$$

and

$$\Delta G^0 = \{[\Delta G_f^0[C_6H_{12}O_6(s)]\} + \{6\Delta G_f^0[O_2(g)]\} - \{[6\Delta G_f^0[CO_2(g)] + 6\Delta G_f^0[H_2O(l)]\}$$

Inserting ΔG_f^0 values from Table 10.3, remembering that O_2 is an element so $6\Delta G_f^0[O_2(g)] = 6(0) = 0$, gives

$$\Delta G^0 = [(-220) + 0] - [6(-94.3) + 6(-56.7)]$$
$$\Delta G^0 = +686 \text{ kcal/mole of glucose}$$

The large positive free energy change for the photosynthesis reaction shows that the reaction is *not* spontaneous—it requires energy input that in photosynthesis comes from the sun. On the other hand, our lives depend on the *release* of 686 kcal of free energy when a mole of glucose is oxidized to CO_2 and water in our bodies:

$$C_6H_{12}O_6(s) + O_2(g) \rightarrow 6CO_2(g) + 3H_2O(l) \qquad \Delta G^0 = -686 \text{ kcal}$$

10.6 Summary Guide Oxidation–reduction reactions are electron-transfer reactions between electron donors (reducing agents) and electron acceptors (oxidizing agents). Oxidation–reduction reactions can be separated into two electron-transfer half-reactions, one for the reducing agent and one for the oxidizing agent. The half-reaction for the reducing agent has electrons as products; that for the oxidizing agent has electrons as reactants.

The half-reactions for the oxidizing and reducing agents can often be carried out in separate half-cells that together with a salt bridge make up an electrochemical cell. When one half-cell contains a hydrogen electrode, the emf (voltage) of the whole cell is taken as the standard oxidation potential ($\mathscr{E}^0$) for the half-reaction in the other half-cell. The value of $\mathscr{E}^0$ is measured at 25°C with one-molar concentrations for participating atoms, molecules, or ions in the half-cell solutions, and atmospheric pressure for participating gases. The half-reactions are written as oxidations (electrons as products) and arranged in the table of standard oxidation potentials with the half-reaction for the strongest reducing agent at the top of the table and that for the weakest reducing agent at the bottom of the table. Oxidizing agents are products (to the right of the arrows) in the half-reactions in the table, and any oxidizing agent will oxidize any reducing agent above it in the table. The value of $\mathscr{E}^0$ for a half-reaction is a numerical measure of the strength of the reducing agent and is positive in sign for reducing agents that are stronger electron donors than hydrogen, negative in sign for those that are weaker.

When half-reactions are combined to give the equation for an overall reaction, the half-reaction containing the reducing agent is taken directly from the table, the half-reaction containing the oxidizing agent is reversed, and the two half-reactions are added after balancing electrons. When a half-reaction from the table is reversed, the sign of its $\mathscr{E}^0$ value is also reversed. The emf ($\Delta\mathscr{E}^0$) of the whole cell is the algebraic sum of the $\mathscr{E}^0$ values (with signs changed as necessary). The overall reaction will go spontaneously from left to right if in the table the half-reaction for the reducing agent is above the half-reaction containing the oxidizing agent. The $\Delta\mathscr{E}^0$ for spontaneous reactions is positive in sign. The value of the equilibrium constant for a cell reaction at 25°C is given by the equation $K = 10^{17n\Delta\mathscr{E}^0}$. The way the emf ($\Delta\mathscr{E}$) of a cell depends upon the molar concentrations of the reactants and products of the cell reaction is given by the Nernst equation: $\Delta\mathscr{E} = \Delta\mathscr{E}^0 - (0.06/n) \log Q$.

When an electrochemical cell is operated as a battery, the electrical energy provided to the external circuit is given by $n\mathscr{F}\Delta\mathscr{E}^0$, and the free energy change within the cell is $\Delta G^0 = -n\mathscr{F}\,\Delta\mathscr{E}^0$.

The standard free energy of formation of a compound (ΔG_f^0) is the free energy change accompanying the formation of one mole of the compound from its elements in their states at 25°C and 1 atm pressure. A compound is thermodynamically stable with respect to reverting to its elements if its ΔG_f^0 is negative in sign and large in value. The free energy change ΔG^0 for a reaction imagined to proceed at 25°C and 1 atm pressure is calculated by subtracting the standard free energies of formation of the reactants from those for the products, recognizing that ΔG_f^0 values are molar quantities and that ΔG_f^0 for elements are zero.

Negative free energy changes accompany reactions that are thermodynamically spontaneous and have the possibility of providing usable energy. Even though the free energy change for a reaction is large and negative, the reaction cannot go unless a path of accessible activation energy (E_a) connects reactants with products, or can be created by using a catalyst. For any reaction imagined to proceed at 25°C, the free energy change $\Delta G^0 = -1.364 \log K$.

Exercises

24. Chrome (Cr) plating is used extensively for decoration and protection of steel articles. For the half-reaction $Cr(s) \rightleftarrows Cr^{2+} + 2e^-$, the standard oxidation potential $\mathscr{E}^0 = 0.56$ V. Can chromium act as effectively as zinc in the role of sacrificial anode? *Hint*: calculate $\Delta\mathscr{E}^0$ for the reaction $Cr(s) + Fe^{2+} \rightleftarrows Cr^{2+} + Fe(s)$ and compare with the result of Exercise 23. Does something you have seen support your conclusion?

25. At the time of writing this book, a French motor car company announced the development of an electric-powered car driven by a hydrogen–oxygen fuel cell. A top speed of 55 mph and a cruising range of 300 miles on one charge of fuel were claimed. Suppose liquid H_2 (bp −259°C, *d* at bp = 0.07 g/ml) is carried in a 10-gal insulated tank to provide hydrogen for the fuel cell and oxygen is taken from the air. Given that the heat of combustion of $H_2(g)$ is 68.3 kcal/mole, and that of octane (C_8H_{16}, "gasoline," $d^{25} = 0.70$ g/ml) is 1223 kcal/mole, compare the fuel energies in the 10-gal tank filled with liquid H_2 and filled with octane. (One gallon = 3.79 liters.)

26. Lithium hydride (Li^+H^-, mol. wt = 7.95, $d^{20} = 0.82$ g/ml) is a white, crystalline solid (mp 685°C) that gives up its hydrogen on reaction with water: $Li^+H^-(s) + H_2O(l) \rightarrow Li^+OH^-(s) + H_2(g)$. What weight of lithium hydride (in kg, in lb) would yield the weight of hydrogen in 10 gal of liquid hydrogen? (See Exercise 25 for data.) What volume would the lithium hydride occupy in liters? In cu ft? (One cu ft = 28.32 liters.)

27. Sacrificial anodes of zinc or magnesium for protecting the hulls of steel ships can be replaced by a battery- or generator-driven system (cell) that uses an underwater graphite electrode insulated from the hull as anode and the hull of the ship itself as cathode. Calculate the minimum emf that must be applied to the electrodes to prevent the reaction $Fe(s) + 2H^+ \rightarrow Fe^{2+} + H_2(g)$ from occurring under standard conditions of temperature, pressure, and concentrations. Would the emf needed to stop the reaction be greater or less if the water floating the boat had pH = 5?

Key Words

Ampere a measure of the rate of flow of an electric current—a flow of 1 coulomb of charge per second equals 1 ampere.

Anode the electrode at which oxidation occurs in an electrochemical cell—the electrode at which electrons leave the cell.

Cathode the electrode at which reduction occurs in an electrochemical cell—the electrode at which electrons enter the cell.

Coulomb the quantity of electric charge that deposits 0.0011180 g of silver from a silver nitrate ($AgNO_3$) solution; the electric charge on 6.24×10^{18} electrons, disregarding sign.

Displacement reaction an oxidation–reduction reaction in which a metal, A, causes another metal, B, to separate from a solution containing the ions of metal B.

Electrochemical cell a system in which solutions of electrolytes are in contact with metal electrodes and in which an oxidation–reduction reaction can occur, with the electron transfer occurring through an external circuit.

Electromotive force (emf, $\Delta\mathscr{E}$, "volts") the difference in electric potential that causes a current of electrons to flow in a wire or ions to migrate in a solution—measured in units of volts.

Electron volt the kinetic energy of one electron that has been accelerated by an emf of 1 V—equals 1.6022×10^{-12} erg or 3.829×10^{-23} kcal of energy.

Electron-transfer half-reaction the reaction at the electrode in a half-cell.

Equilibrium constant (K) for a cell reaction the equilibrium constant equation for the cell reaction written omitting the concentrations of solids that participate in the cell reaction, $K = 10^{-0.733\Delta G^0}$ or $\Delta G^0 = -1.364 \log K$ (at 25°C) if ΔG^0 for the reaction is in kcal.

Faraday ($\mathscr{F}$) the quantity of electric charge that deposits one mole (107.87 g) of silver from a silver nitrate solution: 96,484 coulombs, the charge on one mole of electrons.

Faraday volt the kinetic energy of one mole of electrons that has been accelerated by an emf of 1 V—equal to 23.06 kcal V^{-1} $mole^{-1}$.

Half-cell a container with a metal electrode that makes contact with a solution containing the species participating at the electrode in either an oxidation or a reduction reaction. Two half-cells connected by a salt bridge make up an electrochemical cell.

Nernst equation an equation that tells how the emf ($\Delta\mathscr{E}$) of a cell changes with changes in the concentration or pressures of participants in the cell reaction—$\Delta\mathscr{E} = \Delta\mathscr{E}^0 - (0.06/n) \log Q$.

Oxidation the removal of electrons from an atom, molecule, or ion.

Oxidation–reduction reaction a reaction in which electrons are transferred from one atom, molecule, or ion to another.

Reduction the addition of electrons to an atom, molecule, or ion.

Resistance (R) the resistance presented by a conductor to a flow of electric current. One ohm is the resistance that allows an emf of 1 V to produce a flow of 1 amp in a conductor that may be a wire or an electrolyte solution.

Salt bridge an electrolyte solution that makes a connection between two electrochemical half-cells and completes the electrical circuit by permitting the movement of ions.

Standard free energy change (ΔG^0) measures the amount of useful energy liberated or consumed in a chemical reaction. For an electrochemical cell, $\Delta G^0 = -n\mathscr{F}\Delta\mathscr{E}^0, = -23.06\ n\Delta\mathscr{E}^0$ kcal.

Standard free energy of formation (ΔG_f^0) the free energy change accompanying the formation of one mole of a compound from its elements at 25°C, 1 atm pressure, with the free energies of formation of the elements taken as zero. A measure of the thermodynamic stability of a compound.

Standard oxidation potential ($\mathscr{E}^0$) a quantitative measure (in volts) of the ease of oxidation of an atom, molecule, or ion, based on the ease of oxidation of hydrogen molecules to hydrogen ions assigned $\mathscr{E}^0 = 0.000\ .\ .\ .$ V.

Volt the difference in electric potential (electron "pressure") required to make a current of 1 ampere flow through an electrical resistance of 1 ohm. The emf that gives one coulomb of charge an energy of one joule.

Suggested Readings

Bauer, P., "Batteries for Space Power Systems," NASA SP-172, Washington, D.C., 1968.

Gough, David A., and Andrade, Joseph D., "Enzyme Electrodes," *Science,* **180**:380–384 (April 27, 1973).

Lyons, Ernest H., Jr., *Introduction to Electrochemistry,* Boston, D. C. Heath and Co., 1967.

Maugh, Thomas H., II, "Fuel Cells: Dispersed Generation of Electricity," *Science*, **178**:1273–1275 (Dec. 22, 1972).

Robinson, Arthur L., "Energy Storage (I): Using Electricity More Efficiently," *Science,* **184**:785–787 (May 17, 1974).

Robinson, Arthur L., "Energy Storage (II): Developing Advanced Technologies," *Science,* **184**:884–887 (May 24, 1974).

Sapio, Joseph P., and Braun, Robert D., "Ion-Selective Electrodes," *Chemistry,* **46**:14–17 (June 1973).

Shelton, William Roy, *Winning the Moon,* Boston, Little, Brown, Inc., 1970.

"Switching on Electric Vehicles," *Env. Sci. and Tech.,* **8**:410–411 (May 1974).

Von Braun, Wernher, *Space Frontier,* Rev. Ed., New York, Holt, Rinehart, and Winston, Inc., 1971.

Weissman, Eugene Y., "Batteries: The Workhorses of Chemical Energy Conversion," *Chemistry,* **45**:6–11 (Nov. 1972).

Wilentz, J. S., *The Senses of Man,* New York, Thomas Y. Crowell Co., 1968.

Winsche, W. E., Hoffman, K. C., and Salzano, F. J., "Hydrogen: Its Future Role in the Nation's Energy Economy," *Science,* **180**:1325–1332 (June 29, 1973).

chapter 11

Mixed nonferrous metals recovered from municipal refuse. *(Photo 29, H. Alter,* Science, *vol. 183, cover, 15 March 1974. Copyright 1974 by the American Association for the Advancement of Science.)*

Metals

11.1 Introduction More than 80% of the 104 elements are *metals*. They differ from the nonmetals and metalloids that make up the rest in having very high electrical conductivity. Table 11.1 shows the conductivities of some typical members of the three classes, compared to the conductivity of silver, which is the best conductor among the elements. As you can see from the table, copper and aluminum are nearly as good conductors and being much cheaper than silver, they are used in electrical wiring. The high conductivity of metals is due to the valence electrons of their atoms occupying completely de-

TABLE 11.1 Relative electrical conductivities of typical metals, nonmetals, and metalloids

Element	*Temperature* (°C)	*Relative Electrical Conductivity* ($Ag = 1.0$)
Metals		
Silver (Ag)	20	1.0 (Reference Standard)[a]
Copper (Cu)	20	0.95
Aluminum (Al)	20	0.60
Sodium (Na)	0	0.38
Magnesium (Mg)	20	0.36
Iron (Fe)	20	0.16
Metalloids		
Antimony (Sb)	0	0.040
Bismuth (Bi)	0	0.020
Carbon (C, graphite)	0	0.0010
Tellurium (Te)	25	0.0000040
Germanium (Ge)	22	0.000000035
Nonmetals		
Iodine (I)	20	1.2×10^{-15}
Phosphorus (P, white)	11	1.6×10^{-17}
Sulfur (S)	20	7.9×10^{-24}

[a]The specific electrical conductivity of Ag is 6.3×10^5 ohm^{-1} cm^{-1}.

localized orbitals (Section 4.5). Because the electrons are not tied down in bonds, they can move freely under an applied emf. Nonmetals are nonconductors or insulators because their valence electrons are firmly held in localized covalent bonds. Diamond (Figure 11.1) is an insulator at ordinary temperatures but conducts weakly at high temperatures. As its temperature is increased, some of the valence electrons get enough energy to move out of the bonds into delocalized orbitals or *conduction bands* (Section 4.5). Graphite is another form of carbon, and it is called a *metalloid* because it conducts weakly. In graphite (Figure 11.1) only three out of four of each carbon atom's valence electrons are localized in covalent bonds. The electrons left over bind the layers together but are still fairly free to move, so graphite is between metals and nonmetals in conductivity (Table 11.1).

Elements can also be classified as metals, nonmetals, and metalloids by their electronegativities (compare Table 3.4):

Metals	0.7–0.8
Metalloids	1.2–2.2
Nonmetals	2.3–4.0

Metals are weakly electronegative or "electropositive," nonmetals are strongly electronegative, and metalloids bridge the gap between them.

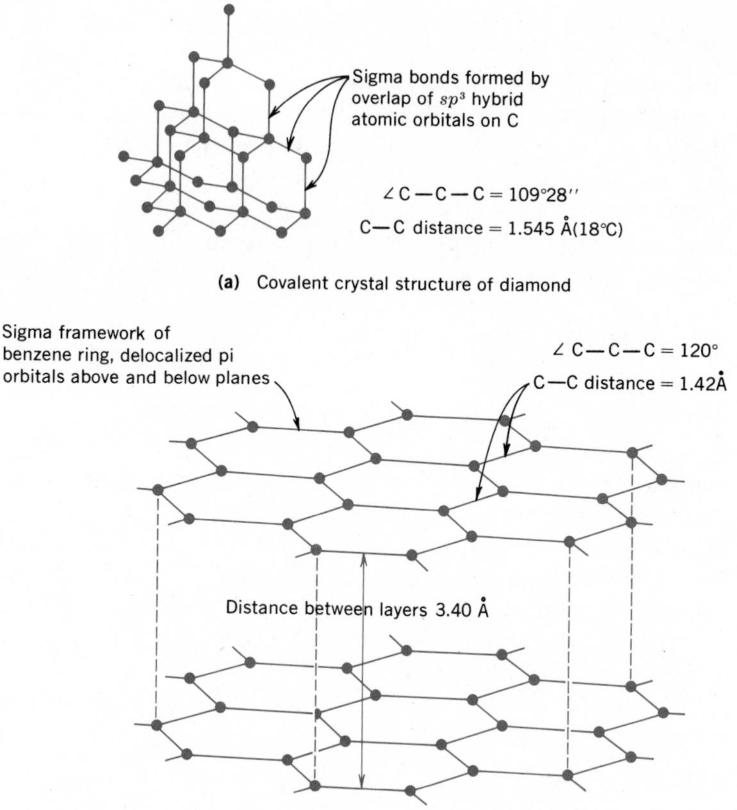

(a) Covalent crystal structure of diamond

(b) Layered structure of graphite

FIG. 11.1 Structures of the allotropes of carbon: (a) diamond; (b) graphite.

Only 19 of the many metals are in the Octet Periodic Table; the remainder are in the various transition metal series (Section 3.7). The octet metals, often called the "representative metals," are strong reducing agents. All of them are oxidized by hydrogen ion, so they dissolve in acid solutions with evolution of hydrogen and the formation of their salts. Examples are

$$2Na + 2HCl \rightarrow 2Na^+Cl^- + H_2$$
$$Mg + HCl \rightarrow Mg^{2+}(Cl^-)_2 + H_2$$
$$Pb + HClO_4 \rightarrow Pb^{2+}(ClO_4^-)_2 + H_2$$

Some of the more active metals (Groups 1a and 2a, excepting Be and Mg) are rapidly oxidized even by water, with formation of hydrogen and the metal hydroxide:

$$2Na + 2H_2O \rightarrow 2Na^+OH^- + H_2$$
$$Ca + 2H_2O \rightarrow Ca^{2+}(OH^-)_2$$

Aluminum, lead, and tin resist the attack of weakly acidic solutions because their surfaces are covered with a protective oxide film. For this reason, aluminum and tin can be used in kitchen ware for cooking acidic fruits and vegetables, but lead is so poisonous that the traces that do dissolve can lead to poisoning even when lead pipes are used to carry drinking water.

Many of the transition metals are also above hydrogen in standard oxidation potential and dissolve in acidic solutions with evolution of hydrogen and the formation of salts.

$$Zn + 2HCl \rightarrow Zn^{2+}(Cl^-)_2$$
$$Fe + H_2SO_4 \rightarrow Fe^{2+}SO_4^{2-}$$
$$Cr + 2HCl \rightarrow Cr^{2+}(Cl^-)_2$$

Only a few transition metals are not oxidized by hydrogen ion. Among the more familiar are copper (Cu), silver (Ag), gold (Au), and platinum (Pt). Copper and silver do, however, dissolve in nitric acid (HNO_3) because it is a much stronger oxidizing agent than hydrogen ion. Gold and platinum do not dissolve even in hot, concentrated nitric acid. The relevant standard oxidation potentials predict this behavior.

	$\mathscr{E}^0$
$Cu(s) \rightleftarrows Cu^{2+} + 2e^-$	−0.34
$Ag(s) \rightleftarrows Ag^+ + e^-$	−0.80
$NO(g) + 2H_2O(l) \rightleftarrows NO_3^- + 4H^+ + 3e^-$	−0.96 (Nitric acid)
$Pt(s) \rightarrow Pt^{2+} + 2e^-$	−1.20
$Au(s) \rightarrow Au^{3+} + 3e^-$	−1.50

The half-reactions for gold and platinum are *below* the nitric acid reaction and are therefore not oxidized.

Oxidation state In talking about the chemistry of metals the idea of *oxidation state* is used. *The oxidation state of a metal in its positive ion is the number of electrons that have been removed from the metal atom to form the ion.* It is equal to the number of electrons in the half-reaction for the standard oxidation potential of the metal. The oxidation state of the metal itself is zero. In naming compounds, roman numerals are used to show oxidation state. Some examples will show you how convenient the idea is compared to the older "ous" and "ic" naming:

$Cu^{2+}SO_4^{2-}$	Copper(II) sulfate	Cupric sulfate
$Pb^{2+}(Cl^-)_2$	Lead(II) chloride	Plumbous chloride

$Sn^{2+}(Cl^-)_2$	Tin(II) chloride	Stannous chloride
$Sn^{4+}(SO_4^{2-})_2$	Tin(IV) sulfate	Stannic sulfate

The roman numerals tell the charge on the ion and make writing formulas easier.

The idea of oxidation state can be used with elements in covalently bonded molecules and ions by applying the following rules:

1. The oxidation state of an element in its elementary state is zero.
2. The oxidation state of hydrogen is +1 in all compounds, except in a few metal hydrides such as lithium hydride, Li^+H^-, in which the oxidation state of H is −1.
3. The oxidation state of oxygen is −2 in all compounds, except in peroxides such as Na_2O_2, in which it is −1. Peroxides have an O—O bond, as in hydrogen peroxide, H—O—O—H, and sodium peroxide, $Na^+[O—O]^{2-}Na^+$.
4. The algebraic total of oxidation states for a compound is zero and for an ion it is equal to the charge on the ion.
5. Oxidation is an increase in oxidation state; reduction is a decrease (Figure 11.2).
6. For half-reactions, the oxidation-state change is the number of moles of electrons in its equation.
7. For an overall reaction the total change in oxidation states is zero.

Here are some examples of these rules:

Formula	*Oxidation States*	*Name of Compound*
Cu	Zero	Copper
Cu_2O	O is −2, Cu is +1 $[-2+2(+1)=0]$	Copper(I) oxide or cuprous oxide
CuO	O is −2, Cu is +2 $(-2+2=0)$	Copper(II) oxide or cupric oxide
HCl	H is +1, Cl is −1 $(+1-1=0)$	Hydrogen chloride
H_3O^+	H is +1, O is −2 $[3(1)-2=+1$ = the charge on the ion]	Hydronium ion
NO_3^-	O is −2, N is +5 $[3(-2)+5=-1$ = the charge on the ion]	Nitrate ion

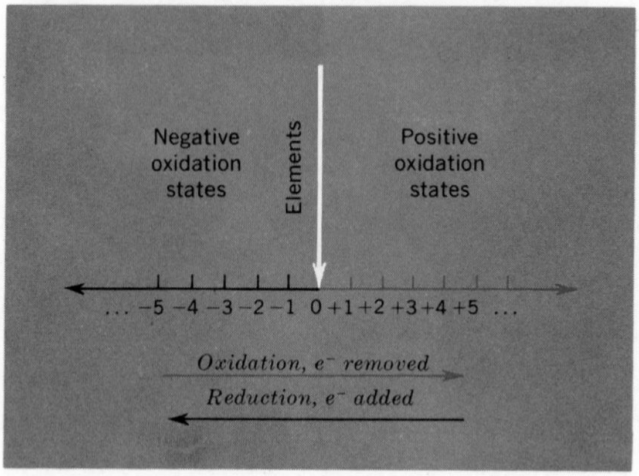

FIG. 11.2 Changes in oxidation state and the processes of oxidation and reduction.

NO	O is −2, N is +2 $[(-2) + 2 = 0]$	Nitrogen(II) oxide or nitric oxide
N_2O	O is −2, N is +1 $[-2 + 2(1) = 0]$	Nitrogen(I) oxide or nitrous oxide

Finding oxidation states makes it possible to see which element is oxidized or reduced in half-reactions for molecules or complex ions, as in the following:

$$NO(g) + 2H_2O(l) \rightleftarrows NO_3^- + 4H^+ + 3e^-$$

The oxidation state of N in NO is +2. The oxidation state of N in NO_3^- is +5. You can see that nitrogen was the element oxidized. The change in oxidation state of N is $+5 - 2 = +3$, and this is equal to the number of electrons in the half-reaction. The oxidation states of H and O do not change.

EXAMPLE The overall reaction for copper dissolving in nitric acid is a good summarizing example, because while copper can be safely used in water pipes, it cannot be used for piping solutions of strong oxidizing agents. The example shows how sometimes both half-reactions must be multiplied through to balance the numbers of electrons:

$$3[Cu(s) \rightarrow Cu^{2+} + 2e^-] \qquad \text{Oxidation of Cu to Cu(II)}$$

$$2[NO_3^- + 4H^+ + 3e^- \rightarrow NO(g) + 2H_2O] \qquad \text{Reduction of } NO_3^- \text{ to NO}$$

$$3Cu(s) + 2NO_3^- + 8H^+ + \cancel{6e^-} \rightarrow 3Cu^{2+} + \cancel{6e^-} + 2NO(g) + 4H_2O(l)$$

Thus

Total change for 3Cu is $3 \times (+2) = +6$
Total change for $2NO_3^-$ is $2 \times (-3) = -6$

Total change for overall reaction = 0

You can calculate $\Delta\mathscr{E}^0$ and K for the reaction, using the $\mathscr{E}^0$ values for the two half-reactions taken from Table 10.2. For the oxidation of copper $\mathscr{E}^0 = -0.34$ V and for the reduction of NO_3^- to NO, $\mathscr{E}^0 = 0.96$ V, so $\Delta\mathscr{E}^0 = -0.34 + 0.96 = 0.62$ V. Using the equation $K = 10^{17n\Delta\mathscr{E}^0}$ with $n = 6$, the number of moles of electrons transferred in the overall reaction, we obtain

$$K = 10^{17 \times 6 \times 0.62} = 1.74 \times 10^{63}$$

The huge value of K shows that the attack of HNO_3 or other strong oxidizing agents on Cu will be very successful.

On the other hand, the attack of $H^+(aq)$ on Cu will most certainly fail. Again taking $\mathscr{E}^0$ values from Table 10.2, for the oxidation of Cu to Cu^{2+} we get $\mathscr{E}^0 = -0.34$ V, and since $\mathscr{E}^0 = 0.00$ for the half-reaction $2H^+(aq) + 2e^- \rightarrow H_2(g)$, we get $\Delta\mathscr{E}^0 = -0.34$ V for the reaction $Cu(s) + 2H^+(aq) \rightleftarrows Cu^{2+}(aq) + H_2(g)$. As before, using the equation $K = 10^{17n\Delta\mathscr{E}^0}$, we get, for the value of K for the attack of $H^+(aq)$ on $Cu(s)$,

$$K = \frac{[Cu^{2+}]}{[H^+]^2} = 10^{17 \times 2 \times (-0.34)} = 2.75 \times 10^{-12}$$

There is no danger of introducing poisonous Cu^{2+} ions in drinking water by using copper pipes for plumbing purposes even if the water they carry is on the acidic side of pH = 7. ⊸

Exercises

1. Separate all of the equations for overall reactions in this section (11.1) into their electron-transfer half-reactions. Identify the oxidizing agent and the reducing agent, and show that the total of oxidation-state changes is zero for the overall reaction.

2. Silver dissolves in nitric acid with the formation of NO. Give the balanced equation for the reaction. Calculate the value of K for the reaction. (*Hints*: compare dissolving of Cu given in the last example; use $K = 10^{17n\Delta\mathscr{E}^0}$.)

All of the octet metals have highest oxidation states equal to their periodic group number, which is the same as the number of valence electrons in their atoms (Table 3.2). Only tin, lead, and thallium also have lower oxidation states: Pb(II) in $PbCl_2$, Sn(II) in $SnCl_2$, Tl(I) in TlCl [thallium(I) chloride, thallous chloride]. Lead (II) and tin(II) are formed by hydrogen ion. The formation of lead(IV), as in PbO_2, and Sn(IV), as in $SnCl_4$, requires stronger oxidizing agents.

In contrast to the small number of oxidation states for the octet metals, transition metals often show several. Manganese is a good example:

$MnCl_2$	Manganese(II) chloride	Manganous chloride
MnF_3	Manganese(III) fluoride	Manganic fluoride
MnO_2	Manganese(IV) oxide	Manganese dioxide
K_2MnO_4	Potassium manganate(MnVI)	
$KMnO_4$	Potassium permanganate(MnVII)	
Mn_2O_7	Manganese(VII) oxide	Manganese heptoxide

The rest of this chapter covers the properties, reactions, and uses of some important octet and transition metals and their compounds.

11.2 The Alkali Metals—Li, Na, K, Rb, Cs, Fr The Group 1a elements are called the alkali metals because their hydroxides, with general formula M^+OH^-, are the best sources of OH^- ions and form strongly alkaline (basic) solutions in water. The gaseous atoms of the elements all have one valence electron in an outer *s* orbital for which the ionization energy lies between the values of 5.39 V (Li) and 3.83 V (Fr). Because of these low values the metals are very willing to donate the electron, even to very weak oxidizing agents. They are therefore among the strongest reducing agents known. Their survival as free metals in the oxidizing environment of the earth is thus impossible, and they are found only as the M^+ ions in combination with the anions of other elements. All of the alkali metals react vigorously with water to give their hydroxides, according to the general equation

Francium occurs as a trace element among the radioactive decay products of uranium. It is so fiercely radioactive and short-lived that there is probably less than 30 g of francium in the earth's crust at any time.

$$2M(s) + 2H_2O(l) \rightarrow 2M^+ + 2OH^- + H_2(g)$$

Cesium is so reactive it combines explosively with ice, even at $-110°C$, to give its hydroxide. These metals are rapidly attacked by the oxygen of the air to give an oxide, a peroxide, or a superoxide, depending on the metal.

$$4Li(s) + O_2(g) \rightarrow 2Li_2O(s) \qquad (Li^+)_2{:}\ddot{\underset{..}{O}}{:}^{2-}$$

Lithium oxide

$$2Na(s) + O_2(g) \rightarrow Na_2O_2 \qquad (Na^+)_2{:}\ddot{\underset{..}{O}}{:}\ddot{\underset{..}{O}}{:}^{2-}$$

Sodium peroxide

$$K(s) + O_2(g) \rightarrow KO_2 \qquad K^+{:}\ddot{\underset{..}{O}}{:}\ddot{\underset{..}{O}}{\cdot}^{-}$$

Potassium superoxide

The lesson in these three reactions is that even though elements are in the same group of the Periodic Table, you must not assume that their reactions will always be the same. Sodium peroxide is used as an oxidizing agent in analytical chemistry. Lithium oxide and potassium superoxide are used to remove moisture and carbon dioxide from the air in the life-support systems for astronauts. For Li_2O the reactions are

$$Li_2O(s) + H_2O(g) \rightarrow 2LiOH(s)$$

$$LiOH(s) + CO_2(g) \rightarrow LiHCO_3(s)$$

The alkali metal elements occur mainly as their chlorides, M^+Cl^-. Sodium chloride makes up 2.7% of seawater, from which the compound is isolated by solar evaporation of the water in large salt ponds, some near San Francisco. It is also obtained, along with potassium chloride, from salt mines and brines in Michigan and Germany. Minerals and mineral waters are sources of the other alkali metal chlorides.

The alkali metals react spontaneously and very vigorously with all of the halogens, according to the general reaction

$$2M(s) + X_2(g) \rightarrow 2M^+X^-(s)$$

This reaction must be reversed to obtain the metals from the chlorides, which are their natural source or "ore." The reversal is brought about by the *electrolysis* (meaning "electrical taking apart") of the molten alkali metal chloride or hydroxide. In electrolysis, an external emf from a battery or generator is used to drive an oxidation–reduction reaction in the reverse of its spontaneous direction. For molten Na^+Cl^- the reactions for the electrolysis and the electric energy input required are

Reduction of Na^+ at cathode	$2Na^+ + 2e^- \rightarrow 2Na(s)$
Oxidation of Cl^- at anode	$2Cl^- \rightarrow Cl_2(g) + 2e^-$
Overall	$2Na^+ + 2Cl^- \rightarrow 2Na(s) + Cl_2(g)$
	$\Delta G^0 = +184$ kcal

Note that in electrolysis, just as in electrochemical cells furnishing electricity, oxidation occurs at the anode, reduction at the cathode.

This process is summarized in detail in Figure 1.12. ☛

Exercises

3. How many faradays ($\mathscr{F}$) are required to produce one mole of sodium by electrolysis of molten NaCl? How long would a current of 10 amps have to flow to do the job? (One amp = 1 coulomb sec^{-1}, 1 $\mathscr{F}$ = 96,484 coulomb $mole^{-1}$.)

4. Using data in Table 10.2, calculate $\Delta\mathscr{E}^0$ for the reaction of sodium with chlorine to produce NaCl. What would be the *minimum* emf

(voltage) necessary for the electrolysis of NaCl, assuming standard conditions? [For $Na(s) \rightleftarrows Na^+ + e^-$, $\mathscr{E}^0 = 2.71$ V.]

5. Lithium is prepared by the electrolysis of molten lithium hydroxide:

$$4Li^+OH^-(l) \rightarrow 4Li(s) + 2H_2O(l) + O_2(g)$$

Separate this overall reaction into its half-reactions and determine $\Delta\mathscr{E}^0$ for the reaction *in the direction given*. Comment on the sign of $\Delta\mathscr{E}^0$ and its meaning. [For $Li(s) \rightleftarrows Li^+ + e^-$, $\mathscr{E}^0 = 3.05$ V.]

The melting points of the alkali metals range downward from 175°C for Li to 28.5°C for Cs, and all are soft enough to be cut with a knife. The freshly cut surface shines like silver, but darkens as it is rapidly corroded by the attack of oxygen and moisture in the air. The metals are good electrical conductors (compare Na and Ag in Table 11.1); and sodium, sealed in copper tubes to protect it from the atmosphere, is used in electrical conductors to transmit very large currents in powerhouses. Molten sodium is also circulated through pipes inside of nuclear reactors to bring the heat outside for steam generation. The hot sodium from the reactor circulates through pipes in boilers to generate the steam, which is then used in turbines to generate electricity.

Compounds of the alkali metals Sodium compounds are the most important compounds of the alkali metals. Among sodium compounds, sodium hydroxide (NaOH), also called "caustic soda" or lye, is the most important, except, of course, sodium chloride (table salt), which is the source of all sodium compounds.

EXAMPLE ⊸ Sodium hydroxide is produced in huge quantities (currently more than 20 billion lb/yr) in the U.S. It is used in the laboratory and is *the* strong base for industrial chemistry. It is produced by the electrolysis of a *solution* of NaCl in water. The cathode and overall reactions for the electrolysis of the solution are quite different from the ones for the electrolysis of *molten* sodium chloride, given earlier. They are as follows:

Oxidation of Cl^- at anode	$2Cl^- \rightarrow Cl_2(g) + 2e^-$
Reduction of H_2O at cathode	$2H_2O(l) + 2e^- \rightarrow H_2(g) + 2OH^-$
Overall	$2Cl^- + 2H_2O(l) \rightarrow Cl_2(g) + H_2(g) + 2OH^-$

The sodium ion is only a "bystander" in the overall reaction. It does not get reduced to the metal at the cathode because $\mathscr{E}^0$ for the half-reaction $H_2(g) + 2OH^- \rightarrow 2H_2O(l) + 2e^-$ is at +0.83 V, *below* $\mathscr{E}^0$ for the half-reaction $Na(s) \rightarrow Na^+ + e^-$ at +2.71 V. In other words, H_2O is a much stronger *oxidizing* agent than Na^+ and is therefore more readily reduced at the cathode. To show how the sodium ion enters into the final product you can write the overall reaction as follows: $2Na^+Cl^-(s) + 2H_2O(l) \rightarrow Cl_2(g) + H_2(g) + 2Na^+OH^-(s)$. The hydrogen produced at the cathode of the cell and the chlorine freed at the anode are combined to make hydrogen chloride and hydrochloric acid. It is necessary in the electrolysis to arrange the cell so that the hydroxide ion produced at the cathode is kept separated from the chlorine freed at the anode. If they get together, they react to form *sodium hypochlorite* (Na^+OCl^-) according to the net equation $2OH^- + Cl_2(g) \rightarrow OCl^- + Cl^- + H_2O(l)$ or, if we put in the bystander Na^+ ion,

$$2Na^+ + 2OH^- + Cl_2(g) \rightarrow 2Na^+ + OCl^- + Cl^- + H_2O(l)$$

Actually, industry takes advantage of this reaction, using a suitable cell setup, to produce huge quantities of sodium hypochlorite (Na^+OCl^-) solution for use in swimming pool sanitation and as a household germicide and laundry bleach (Clorox, Purex).

The main use of the alkali metals is to provide the cation *carriers* of important *anions*. Sodium hydroxide and sodium hypochlorite are not valued for their alkali metal cation, but instead for the properties of the anions. Sodium and potassium compounds are soluble in water (with very few exceptions) and thus quite concentrated solutions of the desired anions are available. Other important compounds of alkali metals and their uses are the following:

$Na_2CO_3 \cdot 10H_2O$	Sodium carbonate decahydrate (soda ash, washing soda, sal soda)	Soap, detergent, and glass manufacture; water softening
NaF	Sodium fluoride	Used in toothpaste to prevent decay
KCN	Potassium cyanide	Electroplating baths
$NaClO_3$	Sodium chlorate	Weed killer, oxidizing agent
$KClO_4$	Potassium perchlorate	Matches, fireworks, oxidizing agent

KNO_3	Potassium nitrate	Black gunpowder
$K_2SO_4 \cdot Al_2(SO_4)_3 \cdot 24H_2O$	Potassium aluminum sulfate (alum)	Astringent, water purification
$LiAlH_4$, $NaBH_4$	Lithium aluminum hydride, sodium borohydride	Reducing agents in organic chemistry
$NaC_{16}H_{17}N_2O_4S$	Sodium penicillin G	Antibiotic

Alkali metals and health Sodium and potassium ions play a key role in maintaining the proper osmotic pressures of the blood and the cells of the body. Very small amounts of sodium and potassium chloride are eliminated in the urine and larger amounts in perspiration, so a continuing supply of them in the diet is vital to life. Because of this, the "electrolyte balance" of sodium and potassium chloride in the blood is carefully watched in the diseased or surgical patient. If it departs very far from healthy values (Table 11.2), additional supplies are given by diet or by injection. *Isotonic saline*, used for intravenous replacement of both electrolyte and water lost by dehydration, contains 0.90% NaCl. Many of you have taken "salt tablets" to offset large losses in sodium chloride through heavy sweating during hard physical exercise.

TABLE 11.2 Range of concentrations of Na^+, Cl^-, K^+ in body fluids in health

Constituent	*Moles of Constituent/Kilogram of Body Weight*	
	Fluid outside cells	*Fluid inside cells*
Na^+	0.024–0.030	0.013–0.016
K^+	0.007–0.009	0.042–0.048
Cl^-	0.020–0.025	0.007–0.009

The data in Table 11.2 show a big difference in the concentrations of Na^+ and K^+ between the fluid inside the cells and that outside. The cell membrane keeps K^+ ions in the cell and Na^+ ions out of the cell in some way that is not yet fully understood. Maintaining the concentration differences for the ions is a vital feature of the transmission of nerve impulses. A nerve impulse is actually a low-voltage electrical signal that passes along the nerve cell at a speed of about 120 m/sec (Figure 11.3). At the point of location of the signal along the nerve cell, the cell membrane becomes "leaky" and momentarily lets Na^+ ions in. After the impulse passes, the membrane somehow repairs itself and "pumps" the Na^+ ions out, restoring the original concentration difference. During this repairing–pumping process, the nerve cannot transmit another impulse. Amazingly, this "refractory period" when the nerve is "turned off" lasts only about 1/1000 of a second. Lithium carbonate ($LiCO_3$) is used to control certain types of mental

It is not like an electric current in a wire through which the current travels at the speed of light, 3×10^8 m/sec.

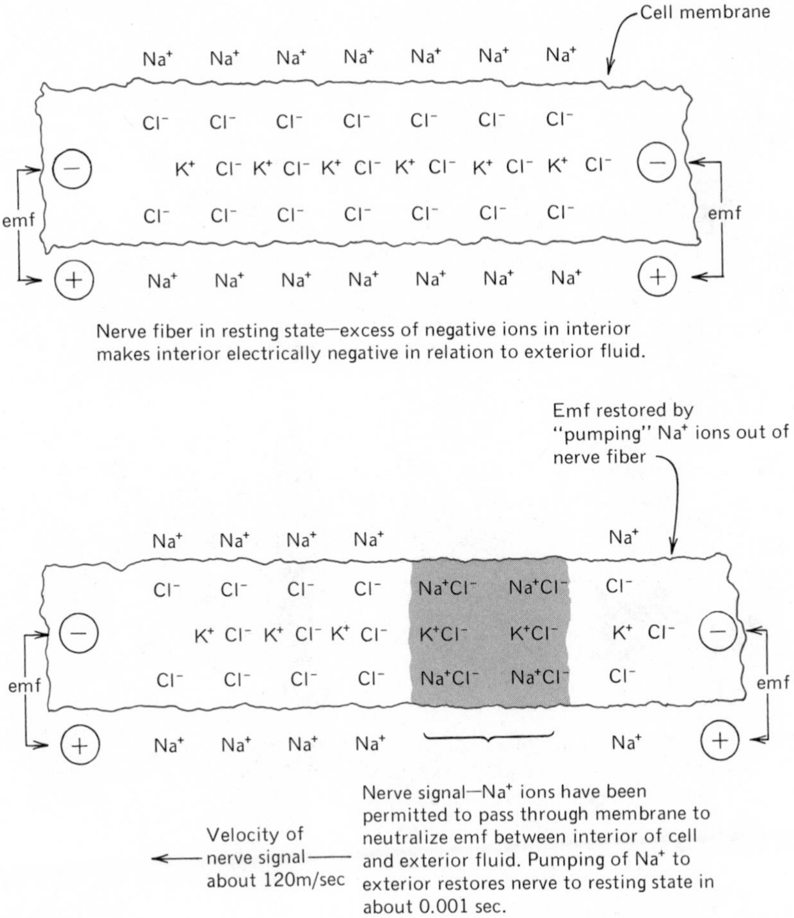

FIG. 11.3 Mode of transmission of a nerve signal along a nerve fiber.

illness because of some as-yet-not-understood effect of Li^+ ions on the cells of the brain.

Exercises

6. Calculate the volume in liters at STP of chlorine produced during the manufacture of 1 lb (454 g) of sodium hydroxide by the electrolysis of NaCl solution.

7. When buying washing soda in the form of $Na_2CO_3 \cdot 10H_2O$, do you pay more for the water of crystallization ($10H_2O$) or for the active ingredient, Na_2CO_3?

11.3 The Alkaline Earth Metals—Be, Mg, Ca, Sr, Ba, Ra The alkaline earth metals (Group 2a) occur on earth mainly as insoluble carbonates having the general formula $M^{2+}CO_3^{2-}$. When heated strongly

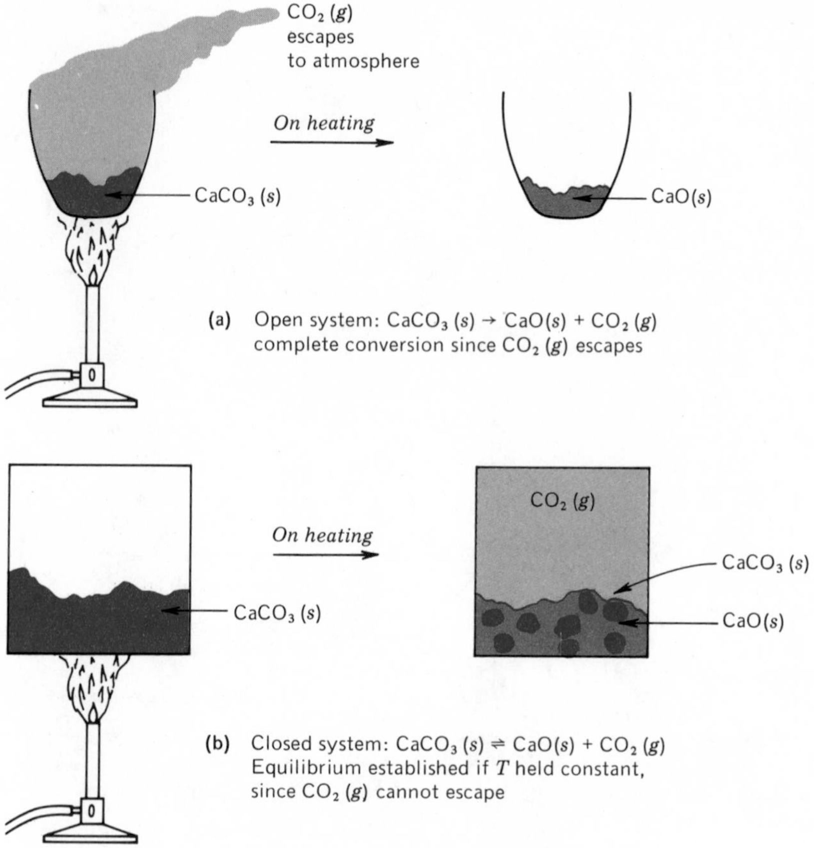
$CO_2\,(g)$ escapes to atmosphere
On heating
$CaCO_3\,(s)$
$CaO(s)$
(a) Open system: $CaCO_3\,(s) \rightarrow CaO(s) + CO_2\,(g)$ complete conversion since $CO_2\,(g)$ escapes
$CO_2\,(g)$
On heating
$CaCO_3\,(s)$
$CaCO_3\,(s)$
$CaO(s)$
(b) Closed system: $CaCO_3\,(s) \rightleftharpoons CaO(s) + CO_2\,(g)$ Equilibrium established if T held constant, since $CO_2\,(g)$ cannot escape

FIG. 11.4 Thermal decomposition of calcium carbonate in open and closed systems. [Both systems, on standing at room temperature for a sufficient time, will revert to $CaCO_3(s)$; the open system gets its $CO_2(g)$ from the atmosphere to form $CaCO_3(s)$.]

in an open system [Figure 11.4(a)] the carbonates decompose into the metal oxide, $M^{2+}O^{2-}$, and CO_2:

$$M^{2+}CO_3^{2-}(s) \xrightarrow{\text{Heat}} M^{2+}O^{2-}(s) + CO_2(g) \qquad \text{(open system)}$$

The oxides react with water to give the hydroxides $M^{2+}(OH^-)_2$, which are alkaline (and so the name "alkaline" earth metals):

$$M^{2+}O^{2-}(s) + H_2O(l) \rightarrow M^{2+}(OH^-)_2(s)$$

Barium hydroxide, $Ba(OH)_2$, is the most soluble, 5 g/100 g of H_2O at 20°C.

Solids are always omitted in equilibrium constant equations.

In contrast to the alkali metal hydroxides, those of the alkaline earths are rather insoluble. ☛ If you heat an alkaline earth carbonate in a system that is *closed* so the CO_2 can't escape, an equilibrium is established among solid carbonate, solid oxide, and CO_2 gas [Figure 11.4(b)]. Using $CaCO_3$ as an example, we can write these equilibrium and equilibrium constant equations: ☛

$$\underset{\text{Limestone}}{CaCO_3(s)} \rightleftarrows \underset{\text{Quicklime}}{CaO(s)} + CO_2(g) \qquad \text{(closed system)}$$

and

$$K_p = p_{CO_2} = 1.04 \text{ atm} \qquad \text{at } 900°C$$

In spite of the small value of K_p, complete conversion to CaO is achieved at lower temperatures (~700°C) by carrying out the reaction in an open system, called a lime kiln, in which the CO_2 is allowed to escape (Figure 11.5). Calcium oxide is called "quicklime" because of the vigor of its exothermic reaction with water to form $Ca(OH)_2$, which is called "slaked lime" because its "thirst" for water has been satisfied.

$$\underset{\text{Quicklime}}{CaO(s)} + H_2O(l) \rightleftarrows \underset{\text{Slaked lime}}{Ca(OH)_2(s)} \qquad \Delta H^0 = -16 \text{ kcal}$$

Ernest Solvay (1838–1922) was born near Brussels, Belgium. The reactions of his process had been known for over 50 years, but all attempts to develop the process commercially had failed.

The reactions have been known from ancient times. Quicklime is used in the preparation of *mortar* for setting bricks; slaked lime is used as fertilizer. Huge quantities of quicklime and slaked lime are also produced and used in the Solvay process for making sodium carbonate.

EXAMPLE

⊸ The Solvay process shows how a spontaneous reaction can be reversed, using equilibrium principles. These are the equations for the process:

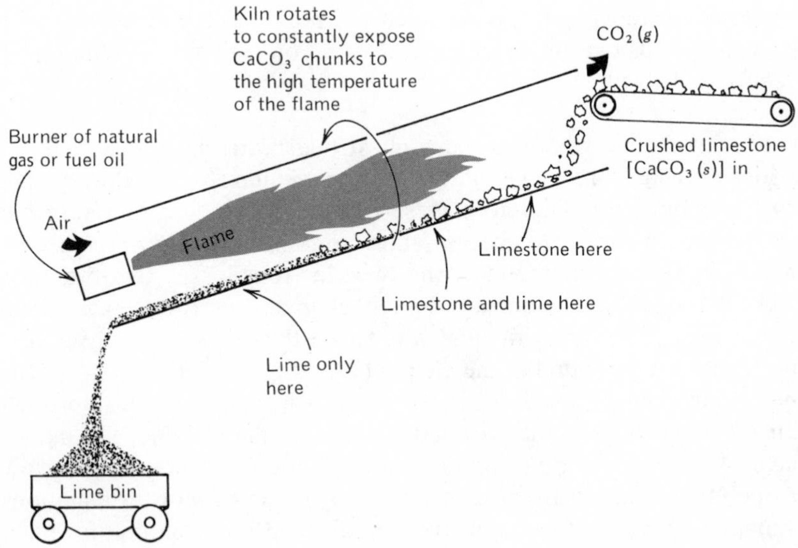

FIG. 11.5 Principles of making lime (CaO) from limestone ($CaCO_3$) in a rotary lime kiln. Kiln about 10 ft in diameter.

$$CaCO_3 \rightleftarrows CaO + CO_2 \quad (1)$$
$$CO_2 + NH_3 + H_2O + NaCl \rightleftarrows NaHCO_3 + NH_4Cl \quad (2)$$
$$2NaHCO_3 \rightleftarrows Na_2CO_3 + CO_2 + H_2O \quad (3)$$
$$CaO + H_2O \rightleftarrows Ca(OH)_2 \quad (4)$$
$$Ca(OH)_2 + 2NH_4Cl \rightleftarrows CaCl_2 + 2H_2O + 2NH_3 \quad (5)$$

This process, which—taken overall—reverses the spontaneous reaction $CaCl_2 + Na_2CO_3 \rightarrow CaCO_3 + 2NaCl$ ($\Delta G^0 = -25$ kcal), succeeds because sodium bicarbonate ($NaHCO_3$, $K_{sp} = 0.62$) is not very soluble in water. The equilibrium in step 2 is pulled to the right by the precipitation of $NaHCO_3$. You can see this better in ionic detail:

$$CO_2(g) + NH_3(g) + H_2O(l) + Na^+ + Cl^- \rightleftarrows NaHCO_3(s) + NH_4^+ + Cl^-$$

The precipitation of $NaHCO_3$ is made as complete as possible by applying Le Chatelier's principle. Ammonia and carbon dioxide gases are passed into a *saturated solution* of sodium chloride. The high sodium ion concentration, $[Na^+]$, in the solution forces the solubility equilibrium for $NaHCO_3$ to the left (common ion effect, Section 7.5):

$$NaHCO_3(s) \rightleftarrows \underset{\uparrow \text{High concentration from } Na^+Cl^-}{Na^+} + HCO_3^-$$

The $NaHCO_3$ precipitate is separated by filtration. Some is purified and sold as "bicarbonate of soda"; the rest goes on to sodium carbonate by step 3. To be economical, the Solvay process has to use the costly ammonia over and over again. The last step (5) recovers the ammonia for recycling, leaving calcium chloride as a by-product. Even the $CaCl_2$ finds use, as an ice-melting agent on highways in winter (see Section 6.3).

The alkaline earth metals are all strong reducing agents with $\mathscr{E}^0$ values ranging from a high of 2.93 V for rubidium (Rb) to a low of 1.85 for beryllium (Be). Therefore these metals are not free in nature but are always in the +2 oxidation state in compounds. To prepare the metals from their compounds, the two electrons lost in the formation of the +2 oxidation state must be put back in the empty *s* atomic orbital of the ions. The only practical way to do this is by electrolysis of a molten ionic compound of the element. Neither the carbonates nor the hydroxides can be used in the electrolysis because they do not melt when heated. Instead they decompose into the metal oxide. The oxides have very high melting points, and only the chlorides melt at low enough temperatures for practical electrolysis. The following equations show the problem of preparing the metals, and its resolution for magnesium.

$$\left.\begin{array}{r}MgCO_3(s) \rightarrow MgO(s) + CO_2 \\ \text{(mp 2800°C)} \\ Mg(OH)_2(s) \rightarrow MgO(s) + CO_2\end{array}\right\} \text{Problem}$$

$$\left.\begin{array}{r}MgCO_3 + 2HCl \rightarrow MgCl_2 + H_2O + CO_2(g) \\ \text{(mp 708°C)} \\ Mg(OH)_2 + 2HCl \rightarrow MgCl_2 + 2H_2O\end{array}\right\} \text{Resolution}$$

$$MgCl_2(l) \xrightarrow{\text{Electrolysis}} Mg(l) + Cl_2(g) \qquad \Delta G0 = +142 \text{ kcal}$$

Most of the magnesium hydroxide used in making the metal is obtained from seawater (0.5% $MgCl_2$) by precipitation with calcium hydroxide:

$$MgCl_2 + Ca(OH)_2 \rightarrow Mg(OH)_2 + CaCl_2$$

The melting point of magnesium oxide is so high that it is used to line the inside of open-hearth furnaces for the production of steel. High melting oxides like MgO are called "refractories"; Al_2O_3 and SiO_2 are other examples.

Magnesium is the only one of the Group 2a elements that is used as the metal in structures. It has a rather high oxidation potential ($\mathscr{E}^0 = 2.37$ V) for this use, but like aluminum, tin, and lead it is protected by a tough surface film of oxide. The largest use of magnesium metal is in *Dow metal*, an alloy of 89% Mg, 9% Al, and 2% Zn. It can be cast, rolled, drawn into wire, and when properly heat-treated is nearly as strong as steel but much lighter. It is used in aircraft construction.

In spite of the protective oxide film, magnesium and its alloys burn spontaneously in air once they are raised to their ignition temperature. The standard free energy change for the oxidation of magnesium is very large and negative, showing the great desire of magnesium to combine with oxygen:

$$2Mg(s) + O_2(g) \rightarrow 2MgO(s)$$
$$\Delta G^0 = -272 \text{ kcal } (-136 \text{ kcal/mole of MgO})$$

The attraction is so great that magnesium fires cannot be put out with water or even a CO_2 fire extinguisher. They fail because both of the following reactions are strongly exothermic:

$$Mg(s) + H_2O(l \text{ or } g) \rightarrow MgO(s) + H_2(g) \qquad \Delta G^0 = -79 \text{ kcal}$$
$$Mg(s) + CO_2(g) \rightarrow MgO(s) + CO(g) \qquad \Delta G^0 = -198 \text{ kcal}$$

The lesson from these equations is this: *don't ever try to put out a magnesium fire with water or a* CO_2 *extinguisher*—they only make matters worse. The only way is to smother the fire with sand or some other noninflammable material that prevents the oxygen in the air from getting to the magnesium.

Softening hard water—ion exchange Water containing dissolved calcium, magnesium, or iron(II) compounds is called "hard" water because it doesn't give you a soapy, cleansing feeling with ordinary amounts of soap. Sodium stearate, $C_{17}H_{35}COO^-Na^+$, a typical soap, behaves this way with hard water.

Soaps are soluble in water, and their cleansing action results from the $C_{17}H_{35}COO^-$ or similar anions present in their solutions (Figure 11.6). The $C_{17}H_{35}$ portion of the soap anion is a long, zigzag hydrocarbon chain that dissolves in oily dirt because oils have a similar hydrocarbon structure and "like dissolves like." The $—COO^-$ part of the soap anion forms strong hydrogen bonds to water molecules (Section 4.6). The result is that the oily material is "dragged into solution" or solubilized in the water. Calcium, magnesium, and iron(II) ions, by forming water-*insoluble* precipitates with the soap anions, remove them from solution and destroy the cleansing power of the soap. Using M^{2+} to represent Ca^{2+}, Mg^{2+}, or Fe^{2+}, we can write the equation as follows:

$$\underset{\text{In solution}}{2C_{17}H_{35}COO^-(aq)} + M^{2+}(aq) \rightarrow \underset{\text{Insoluble}}{(C_{17}H_{35}COO^-)_2M^{2+}(s)}$$

By adding a lot of soap, the M^{2+} ions can be removed through precipitation, and cleansing action restored. But this is unsatisfactory because the precipitate forms a gummy, grimy film or solid curd that clings to clothes or hands. Also, it wastes soap. The wise thing to do would be

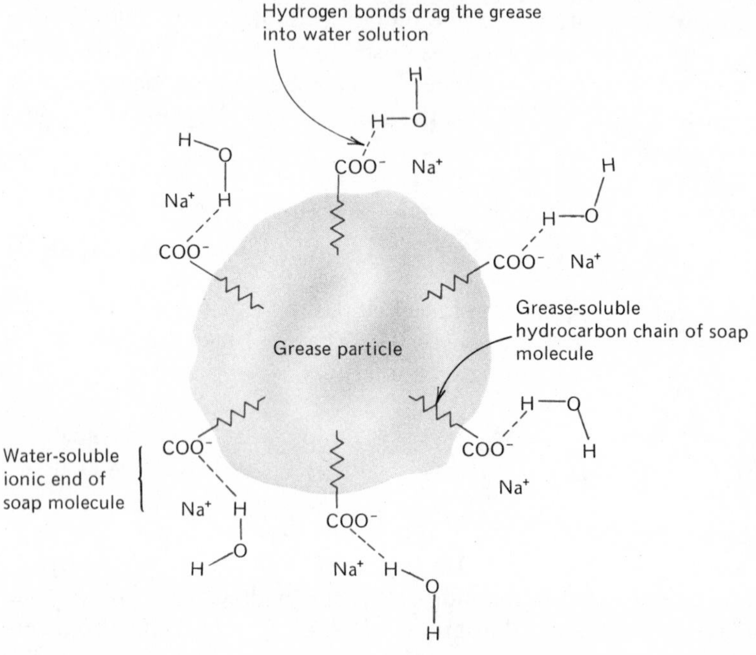

FIG. 11.6 Action of soap in solubilizing grease. With a detergent, an $—\overset{\displaystyle O}{\underset{\displaystyle O}{S}}—O^-Na^+$ group replaces the $—COO^-Na^+$ of the soap.

to remove the troublesome M^{2+} ions somehow, and although this can be done, it turns out to be cheaper to *replace* them with harmless sodium ions instead. This happens in a "water softener" by the process of *ion exchange*, in this case *cation* exchange of $2Na^+$ for M^{2+}.

The zeolite clay or *cation exchange resin* that fills the exchanger in Figure 11.7 is *insoluble in water*, but it has anionic sites (Z^-) on its surface where cations bond. When these sites are occupied by Na^+ ions, the exchange resin or clay is in the "sodium state." If hard water containing Ca^{2+} ions in solution trickles down over the clay or resin in the sodium state, each Ca^{2+} ion pushes two sodium ions off the anionic sites and takes their places. It does this because the ionic bond between $2Z^-$ and Ca^{2+} is stronger than the ionic bond between Z^- and Na^+. Thus if a dilute solution of *calcium* chloride ("hard water") enters the top of the ion exchanger in the sodium state, a dilute solution of *sodium* chloride comes out at the bottom. The result is "soft" water because the Na^+ and Cl^- ions have no effect on the soap anions. Using M^{2+} for Ca^{2+}, Mg^{2+}, or Fe^{2+}, we write the general equation:

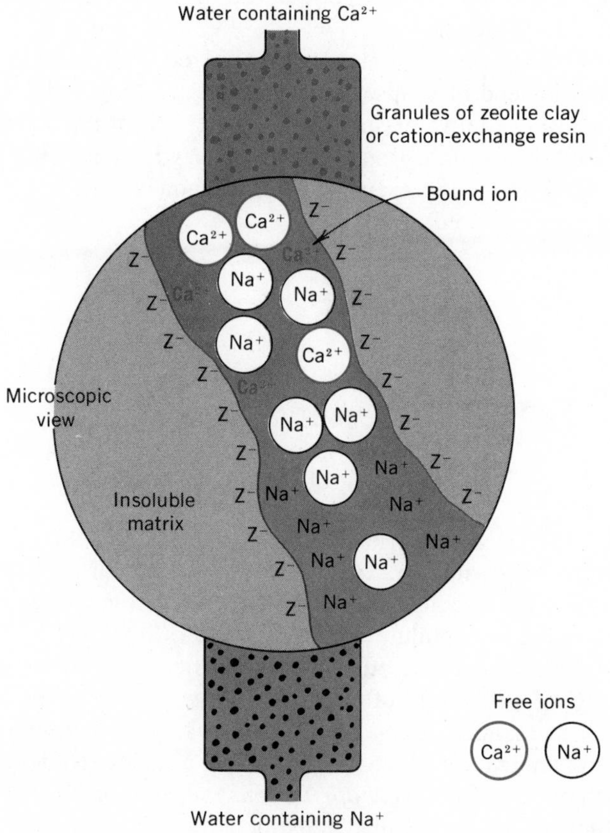

FIG. 11.7 Principle of ion exchange, $2Na^+$ for Ca^{2+} (anions not shown).

$$2Z^-Na^+(\text{on resin}) + M^{2+}(\text{in solution}) \rightleftarrows (Z^-)_2M^{2+}(\text{on resin}) + 2Na^+(\text{in solution})$$

This exchange reaction is written as an equilibrium to show that there is really a competition between the M^{2+} and Na^+ ions for the anionic sites (Z^-) on the insoluble resin or clay. The reversibility of the reaction permits the resin to be *regenerated* from the M^{2+} condition to the Na^+ condition, so that it can be used over and over. Without this feature, the process would be hopelessly expensive – ion exchange resins and clays are far too costly to use once and throw away. To regenerate the resin, a *saturated* solution of sodium chloride is pumped upward through the exchanger, in a process called "backwashing." The high concentration of Na^+ forces the M^{2+} off the anionic sites (Le Chatelier's principle), and the water rich in M^{2+} chloride that comes out the top of the exchanger goes down the drain. When all of the M^{2+} ions have been displaced by Na^+ ions, the resin bed is rinsed with water to remove remaining sodium chloride solution and is then put back into service. Many grocery stores sell bags of inexpensive salt for use in home water softeners that automatically carry out the regenerating steps.

The hard-water problem can be solved in the laundry by using *detergents* instead of soaps. Detergents have a long hydrocarbon chain like soaps, but the water-soluble —COO^- end of the soap molecule is replaced by a sulfate group —O—SO_2O^- which does the same job. Using a zigzag line to represent the hydrocarbon chain of CH_2 groups, the two can be compared:

Soap
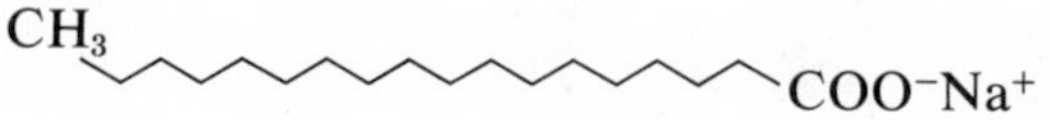

(Sodium stearate)

Detergent
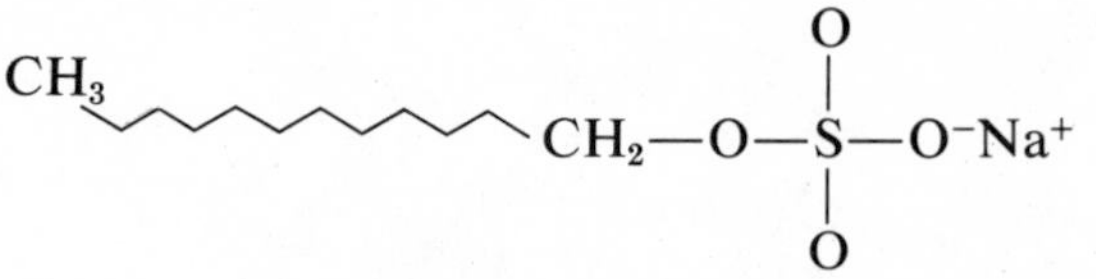

(Sodium dodecylsulfate)

Detergents work in hard water because they do not form insoluble compounds with the M^{2+} ions. Even so, they are not as satisfactory for cleansing the body, they cost more, and some, because they are not biodegradable, pollute the environment (Section 13.8).

Ion exchange has other uses than the cation exchange (such as Na^+ for Ca^{2+}) in water softening. *Anion exchange* resins with R^+ sites to bind exchangeable *anions* are also available. Through the use of both anion and cation exchange resins, it is possible to "demineralize" water – that is, to *remove both cations and anions* from a water supply. Much of what is called "distilled" water in laboratories is actually demineralized water. To remove the foreign anions and cations from water you need a cation exchange resin in the *hydrogen* (H_3O^+) state and an anion exchange resin in the hydroxide (OH^-) state. To see how demineralization works, suppose the water at the faucet contains dissolved $Ca^{2+}SO_4^{2-}$ at a concentration of 0.0020 mole/liter. If you put the water through a cation exchange resin in the hydrogen state, each Ca^{2+} ion from the dissolved calcium sulfate will be replaced by two H_3O^+ ions:

By "foreign" we mean anions and cations other than H_3O^+ and OH^- at concentrations of 10^{-7} mole/liter from the auto-ionization of water: $2H_2O(l) \rightleftarrows H_3O^+(aq) + OH^-(aq)$ (Section 8.1).

$$Ca^{2+}(\text{in solution}) + 2Z^-H_3O^+(\text{on resin}) \rightleftarrows (Z^-)_2Ca^{2+}(\text{on resin}) + 2H_3O^+(\text{in solution})$$

The water coming out of the cation exchanger, called the "effluent," is a dilute solution containing $2H_3O^+ + SO_4^-$ or a solution of sulfuric acid at a concentration of 0.002 mole/liter:

$$H_2SO_4 + 2H_2O \rightleftarrows 2H_3O^+ + SO_4^{2-}$$

If you now put the 0.002-*M* H_2SO_4 effluent from the cation exchanger through the anion exchanger in the OH^- state, each SO_4^{2-} ion will displace $2OH^-$ ions as it becomes bonded to the resin. The equation for this exchange can be written as follows:

$$SO_4^{2-}(\text{in solution}) + 2R^+OH^-(\text{on resin}) \rightleftarrows (R^+)_2SO^{2-}(\text{on resin}) + 2OH^-(\text{in solution})$$

In the cation exchanger, $2H_3O^+$ replaced each Ca^{2+}; in the anion exchanger, $2OH^-$ replaced each SO_4^{2-}. The result is that H_3O^+ and OH^- are in exactly equal numbers to react according to the equation

$$2H_3O^+(aq) + 2OH^-(aq) \rightleftarrows 4H_2O(l)$$

The result of the two exchanges and this last reaction is pure water. The water originally supplied has had all foreign ions ("minerals") removed and has been *demineralized.*

After use in demineralization, the cation exchanger is "revived" or regenerated with a strongly acid solution to provide a high H_3O^+ concentration. A backwash with hydrochloric acid ("muriatic acid") is usually used. For the example of the $Ca^{2+}SO_4^{2-}$ solution the regeneration reaction for the cation exchanger is

$$2H_3O^+(\text{in solution}) + (Z^-)_2Ca^{2+}(\text{on resin}) \rightleftarrows Ca^{2+}(\text{in solution}) + 2Z^-H_3O^+(\text{on resin})$$

To regenerate the anion exchanger, a high concentration of OH^- is needed:

$$(R^+)_2SO_4^{2-}\text{(on resin)} + 2OH^-\text{(in solution)} \rightleftarrows 2R^+OH^-\text{(on resin)} + 2SO_4^{2-}\text{(in solution)}$$

The source of the high OH^- concentration can be a sodium hydroxide or a sodium carbonate solution. Sodium carbonate is less expensive and provides the necessary OH^- concentration because it is hydrolyzed to HCO_3^- and OH^- in water (Table 8.2, Section 8.4):

$$CO_3^{2-}(aq) + H_2O(l) \rightleftarrows HCO_3^-(aq) + OH^-(aq)$$

Ion exchange is used in chemical analysis to separate similar ionic species. For example, with the proper *ion-selective resin*, Mg^{2+} can be removed from a solution of $MgCl_2$ and $CaCl_2$, leaving Ca^{2+} in the effluent. *Ion exchange chromatography*, as this selective ion-exchange process is called, is the best method for separating the different amino acids in the analysis of proteins (Chapter 13).

Alkaline earth elements and health From a health viewpoint, calcium is the most important of the alkaline earths. The others, beryllium, strontium, and barium and their soluble compounds, are poisonous.* Strontium isotope $^{90}_{38}Sr$, a product of nuclear reactions and atomic bomb explosions, is particularly treacherous. It is highly radioactive, and its ions replace calcium ions in bones, where its radiation causes cancer. A small concentration of magnesium (0.012–0.017 mole/liter in cell fluid) is necessary for health, but the chlorophyll of green plants contains magnesium, so almost any diet provides enough of the element.

Calcium occurs in the blood plasma (cell-free part of blood) in amounts of 9–11 mg/100 ml in health. It is required there for proper clotting of blood and elsewhere for the proper functioning of the nervous system. Severe calcium deficiency causes convulsions. But calcium deficiency is rare, because most diets provide an adequate supply and our bones are a large reservoir of the element. Bone is largely tricalcium phosphate, $Ca_3(PO_4)_2$, and with it as a reserve the proper level of Ca^{2+} in the blood is regulated by the parathyroid gland.

Tooth enamel, the hard, protective outer surface of our teeth, requires traces (1 ppm) of fluoride in the diet for its proper formation. Enamel is, like bone, mainly $Ca_3(PO_4)_2$ but with some OH^- and F^- as ligands of the calcium ion. Enamel formed without fluoride is softer and more easily attacked by the bacteria that cause tooth decay.

*But barium *sulfate* is so insoluble ($K_{sp} = 8.7 \times 10^{-11}$ at 25°C) that it is safely used as a contrast medium in X-ray filming of the gastrointestinal tract (Section 7.5).

Milk, cheese, meat, and most foods are sources of calcium in the diet. Fluoride comes in the water supply or in certain toothpastes.

Exercises

8. For magnesium hydroxide, $K_{sp} = 9 \times 10^{-12}$. The $[OH^-]$ in a saturated solution of slaked lime, $Ca(OH)_2$, is 0.04 mole/liter. Calculate the $[Mg^{2+}]$ remaining in seawater after it has been shaken with slaked lime until saturated (all data at 25°C).

9. For the reaction $Ca(s) + O_2(g) \rightarrow 2CaO(s)$, the $\Delta G^0 = -304$ kcal, and for the reaction $Ca(s) + 2H_2O(l) \rightarrow Ca(OH)_2(s) + H_2(g)$, the $\Delta G^0 = -472$ kcal. Could you put out a calcium metal fire with water? Explain.

10. Ordinary soaps do not make suds in seawater. Why?

11. Suppose a 0.10-*M* $FeCl_3$ solution is put through a cation exchange resin in the H_3O^+ state. What would be the chemical nature of the effluent? What would be its concentration?

11.4 Aluminum (Al) and Iron (Fe) You can learn a lot about the chemistry of metals in general by comparing the reactions of aluminum and iron. Aluminum is an *octet* metal with three valence electrons in the arrangement $[Ne]3s^23p^1$. Iron is a transition metal with the arrangement $[Ar]4s^23d^6$. Aluminum uses all three of its valence electrons when forming compounds, so it shows only a +3 oxidation state. Examples are Al_2O_3, Al_2Cl_6, and $Al_2(SO_4)_3$. Iron is like most transition metals in having more than one oxidation state in its compounds. The atom can lose two electrons to form iron(II) (ferrous, Fe^{2+}) compounds or three electrons to form iron(III) (ferric, Fe^{3+}) compounds, depending on the circumstances. ☛

The electron configurations for the gaseous ions are: Fe^{2+}, $[Ar]4s^13d^5$; Fe^{3+}, $[Ar]4s^03d^5$.

Iron and aluminum are both active metals. Aluminum reacts with 1 *M* acids quite rapidly to give the hexaquoaluminum(III) ion: ☛

In this chapter the extent of solvation of ions by water molecules as ligands will be shown where their number is known.

$$2Al(s) + 6H_3O^+ + 6H_2O \rightleftarrows \underset{\text{Hexaquoaluminum ion}}{2Al(H_2O)_6^{3+}} + 3H_2(g) \qquad \Delta\mathscr{E}^0 = +1.66 \text{ V}$$

Iron, on the other hand, will donate only two of its electrons to hydrogen ion, so with acids it yields the hexaquoiron(II) ion:

$$Fe(s) + 2H_3O^+ + 4H_2O \rightleftarrows Fe(H_2O)_6^{2+} + H_2(g) \qquad \Delta\mathscr{E}^0 = +0.41 \text{ V}$$

Oxygen from the air rapidly oxidizes $Fe(H_2O)_6^{2+}$ to the hexaquoiron(III) ion $Fe(H_2O)_6^{3+}$:

$$4Fe(H_2O)_6^{2+} + O_2(g) + 4H_3O^+(aq) \rightarrow 4Fe(H_2O)_6^{3+} + 6H_2O(l) \qquad \Delta\mathscr{E}^0 = +0.46 \text{ V}$$

This ready oxidation means that the overall result of dissolving iron in an acid solution will be the hexaquoiron(III) compound unless precautions are taken to keep oxygen (air) out.

All three of these hexaquoions have the same octahedral arrangement of water molecules as ligands that is found by X-ray crystal analysis for many transition metal ions. Using M^{3+} to stand for Al^{3+} and Fe^{3+}, we have the following octahedral arrangement:

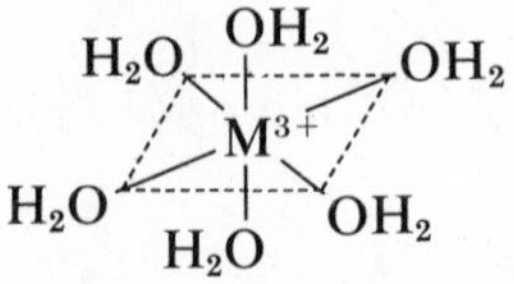

In this arrangement, four of the water molecules lie at the corners of a flat square, one molecule above and one below. Among hundreds of transition metal complex ions that have this octahedral structure are $Cr(H_2O)_6^{3+}$, $Cr(H_2O)_6^{2+}$, $Cr(NH_3)_6^{3+}$, $Ni(NH_3)_6^{2+}$, $Fe(CN)_6^{4+}$, $Co(NH_3)_6^{3+}$.

Metallurgy of iron and aluminum Both iron and aluminum are too easily oxidized to be found as metals on earth, so they occur mainly as their virtually water-insoluble oxides. Iron occurs as Fe_2O_3 (hematite) and Fe_3O_4 (magnetite). Aluminum occurs as Al_2O_3 (bauxite) and in clays as complex compounds with silica, water, and other metal ions. Kaolin, the white clay used to make fine china, has the composition $Al_2O_3 \cdot 2SiO_2 \cdot 2H_2O$; mica is $KAl_3Si_3O_{10}(OH)_2$.

The ores of iron are reduced to the metal by carbon monoxide (CO) in a blast furnace. The operation of a blast furnace for making iron from Fe_2O_3 is shown in Figure 11.8. The standard free energy change for the overall reaction is slightly on the negative (favorable) side:

$$Fe_2O_3(s) + 3CO(g) \rightleftarrows 2Fe(s) + 3CO_2(g) \qquad \Delta G^0 = -7.4 \text{ kcal}$$

Iron from the blast furnace is used directly in making iron castings and, purified further, to make steel and alloys.

Bauxite, a form of Al_2O_3, is the ore for aluminum but it is not reduced by carbon monoxide—the free energy change for that reaction is large and in the wrong direction:

$$Al_2O_3(s) + 3CO(g) \rightleftarrows 2Al + 3CO_2 \qquad \Delta G^0 = +192 \text{ kcal}$$

To put the three electrons back into the valence shell of the Al^{3+} ion requires reduction at the cathode in the electrolysis of a molten ionic aluminum compound. The oxide is ionic but has too high a melting point (2045°C) for practical electrolysis. Fortunately, the oxide dis-

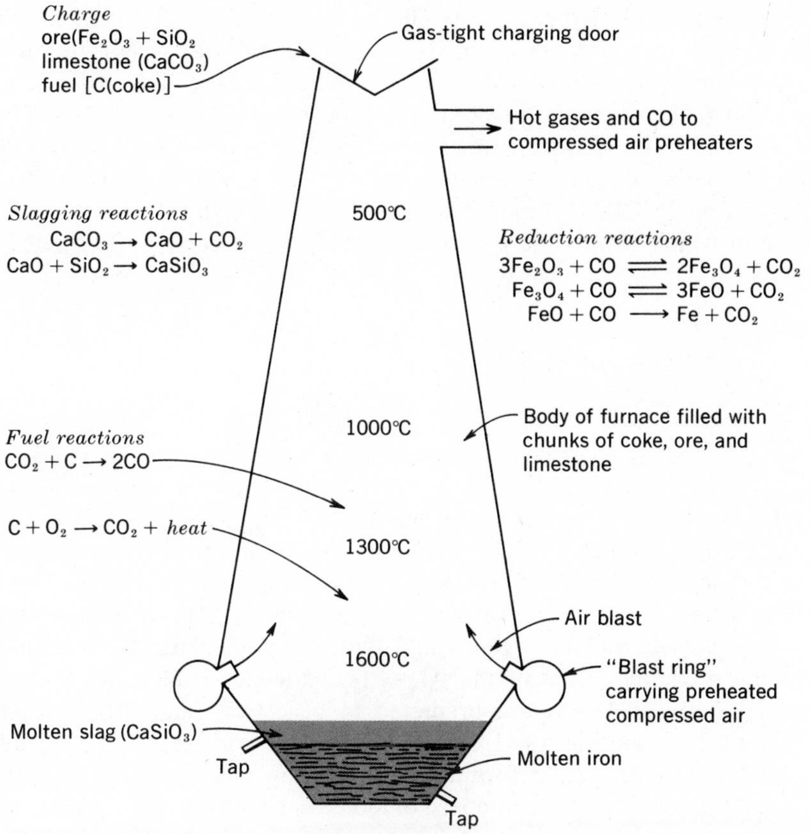

FIG. 11.8 Schematic representation of a blast furnace for the production of pig iron.

solves at 1000°C in the mineral *cryolite*, Na_3AlF_6, which has a much lower melting point, to give a solution that can be electrolyzed to produce aluminum:

$$2Al_2O_3 \text{ (in solution in } Na_3AlF_6) \rightarrow 4Al(s) + 3O_2(g)$$

Pure aluminum is very resistant to corrosion but too weak for use as a structural metal. Its alloy with small amounts of Cu, Mn, and Mg, called *Duralumin* or *Dural*, is much stronger and finds great use in aircraft construction. Dural is much less resistant to corrosion than aluminum, however.

Cryolite is an ionic compound, $(Na^+)_3(AlF_6^{3-})$. The AlF_6^{3-} ion, called the hexafluoroaluminate ion, can be imagined to form by the net reaction $Al^{3+} + 6F^- \rightarrow AlF_6^{3-}$. In this ion the $6F^-$ ligands are arranged octahedrally around the central Al^{3+} ion. Cryolite can be prepared by the reaction $AlF_3 + 3NaF \rightarrow Na_3AlF_6$, but it is also a mineral.

Recalling that the Group 1a and 2a metals are produced by electrolysis of their molten chlorides, you might think that aluminum chloride instead of the oxide would be a good candidate for the electrolysis to produce aluminum. But aluminum chloride is not an ionic compound with formula $Al^{3+}(Cl^-)_3$. When the solid is heated, it doesn't melt; it simply evaporates. The vapor pressure of the solid is 1 atm at 178°C. By measuring the density (d) of the vapor and using the gas laws, we get a molecular weight of 267. This is the value for Al_2Cl_6, not $AlCl_3$. Diffraction studies of the vapor show it has a structure in which two of the chlorines act as bridges between two $AlCl_2$ units.

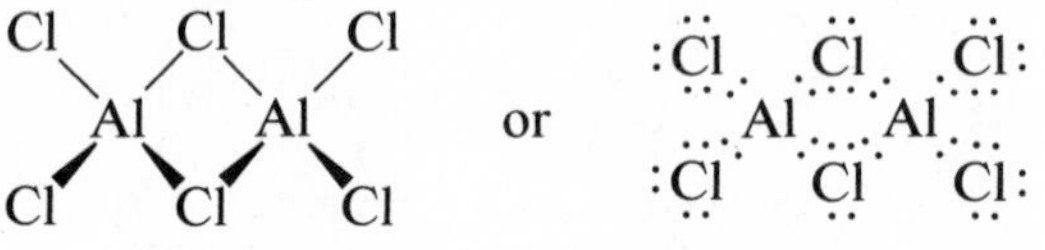

Structure of Al_2Cl_6 molecule

The "wedge" bonds project out of the paper; the ordinary ones lie behind it.

What would the rules for predicting molecular geometry (Section 4.4) say for this molecule? Each aluminum has four chlorines as ligands and no unshared pairs of electrons, so the bonds from each Al kernel should be at the tetrahedral angle of 109.5°. Also, each bridging Cl has two Al kernels as ligands and two unshared pairs of electrons, so the predicted value of the Al—Cl—Al angles is also 109.5°. All angles between bonds are predicted to be 109.5°, and this is what X-ray diffraction shows. Iron(III) chloride (ferric chloride) has the same type of molecular formula, Fe_2Cl_6, it also sublimes (at 300°C), and in the gas state it is composed of molecules with the same geometry as Al_2Cl_6 molecules. By contrast, iron(II) chloride (ferrous chloride, $FeCl_2$) melts at 673°C and is ionic.

EXAMPLE

Aluminum chloride is a colorless solid, but Fe_2Cl_6 with the same bridged structure is deep red. Why? First of all, you must remember that color in a material is due to *absorption* of wavelengths of light in the visible part (4000–8000 Å) of the spectrum (Figure 3.1). If all wavelengths are absorbed, the material is black; if all are reflected, it is silvery like a mirror; if all are transmitted, it is white (colorless, like salt). If *only certain wavelengths are absorbed,* then the material will appear colored to the eye. The light absorbed by Fe_2Cl_6 is in the blue–green region near 5000 Å, so red light is transmitted (or reflected), and the compound is red. To be absorbed, the energy of the photons in the light (Section 3.2) must match some energy-level spacing in the absorbing material. The levels available to the 3d electrons in the iron atoms fit photons in the 5000 Å range, so the photons are absorbed and the red color results. Compounds of transition metals with incompletely filled d orbitals are usually colored. Most of the brilliant colors of minerals and the pigments used in china and in oil painting are due to the transition metal compounds they contain. The octet metals, by contrast, have either completely empty *d* orbitals or completely filled ones. They do not absorb in the visible range, so they do not contribute color to any of their compounds. For example, potassium nitrate, KNO_3, is colorless, but potassium dichromate, $K_2Cr_2O_7$, is bright orange—not because of the K^+ ions but because of *d* electrons in the transition metal chromium atom.

Exercises

12. A 5.00-g sample of iron(III) chloride is evaporated to a gas in a previously pumped out 1.00-liter quartz bulb. At 427°C the pressure of the gas is 0.89 atm. Using the perfect gas law, show that the data indicate that the molecules of the gas are Fe_2Cl_6 rather than $FeCl_3$.

13. When Al_2Cl_6, anhydrous aluminum chloride, is dissolved in water, a violent reaction occurs:

$$Al_2Cl_6(s) + 12H_2O(l) \rightarrow 2Al(H_2O)_6^{3+} + 6Cl^-$$

You can also get aluminum chloride in the form of its hexahydrate, $AlCl_3 \cdot 6H_2O$. Give a detailed structure for the hexahydrate. Would you expect it to make much of a fuss when dissolved in water? Why?

Lewis acids In proton acid–base theory (Section 6.4) an acid HA is a proton donor and a base :B is a proton acceptor. When they react, the proton is transferred from the acid to the base, leaving a new base and forming a new proton acid:

$$\underset{\text{Proton acid}}{H:A} + \underset{\text{Base}}{:B} \rightleftarrows \underset{\text{New base}}{:A^-} + \underset{\text{New proton acid}}{H:B^+}$$

Lewis acid–base theory recognizes that there are many molecules and ions that behave like the proton does in forming bonds with bases. These molecules and ions are called *Lewis acids*. Letting E stand for a Lewis acid, we write its reaction as follows:

$$\underset{\text{Lewis acid}}{E} + \underset{\text{Base}}{:B} \rightleftarrows \underset{\text{Acid–base complex}}{E:B \text{ or } E\text{—}B}$$

The unbonded or "bare" proton H^+ is the simplest and strongest Lewis acid. Because of its tiny size, lack of any electron cloud of its own, and concentrated positive charge, it cannot exist free in the presence of even the weakest bases. It always dives into the electron cloud of a base to form a covalent bond. In the vacuum of the mass spectrometer, H^+ even makes the highly unreactive helium atom behave as a base if the He electron cloud is all that's available for bonding:

$$H^+ + :He \rightarrow [H\!-\!He]^+$$

All other Lewis acids have an electron cloud that makes them weaker acids by shielding the positive charge of their nucleus and by repelling the electron pair of the base. Even the simple Lewis acid Li^+ has a $1s^2$ electron cloud. The proton, however, is a *nuclear*-sized granule with positive charge:

.+

Proton
10^{-13} cm in diameter

Li^+
6×10^{-9} cm in diameter

EXAMPLE

The six nonbonding electrons on each fluorine have been omitted to focus attention on the unshared electron pair of the base.

⌐○ Boron trifluoride, BF_3, is one of the strongest *molecular* Lewis acids and is used as a catalyst in organic chemistry. Here are some examples of its reaction with some bases: ⌐≪

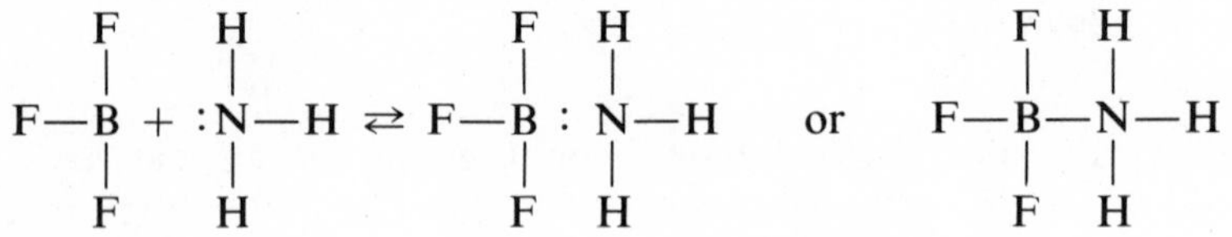

$$
\begin{array}{cccc}
\overset{\textstyle F}{\underset{\textstyle F}{F\!-\!B}} + \overset{\textstyle H}{:N\!-\!H} & \rightleftarrows & \overset{\textstyle F}{\underset{\textstyle F}{F\!-\!B}} : \overset{\textstyle H}{N\!-\!H} & \text{or} \quad \overset{\textstyle F\ \ H}{\underset{\textstyle F\ \ H}{F\!-\!B\!-\!N\!-\!H}}
\end{array}
$$

Boron trifluoride　　　　　　　　　　Boron trifluoride–ammonia complex

$$
\overset{\textstyle F}{\underset{\textstyle F}{F\!-\!B}} + :O\!\!\!\diagup^{CH_3}_{\diagdown CH_3} \quad \rightleftarrows \quad \overset{\textstyle F}{\underset{\textstyle F}{F\!-\!B\!-\!O}}\!\!\diagup^{CH_3}_{\diagdown CH_3}
$$

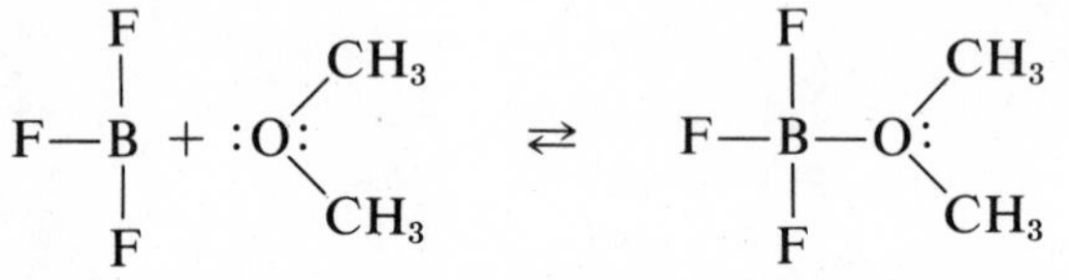

Methyl ether　　　Boron trifluoride–methyl ether complex

An especially interesting reaction involves both Lewis and proton acids. It is the reaction of BF_3 with water to produce the strong trifluoroboric acid:

$$
\overset{\textstyle F}{\underset{\textstyle F}{F\!-\!B}} \quad + \quad 2\,:O\!\!\!\diagup^{H}_{\diagdown H} \quad \rightleftarrows \quad \left[\overset{\textstyle F}{\underset{\textstyle F}{F\!-\!B\!-\!O\!-\!H}} \right]^{-} \ H_3O^+
$$

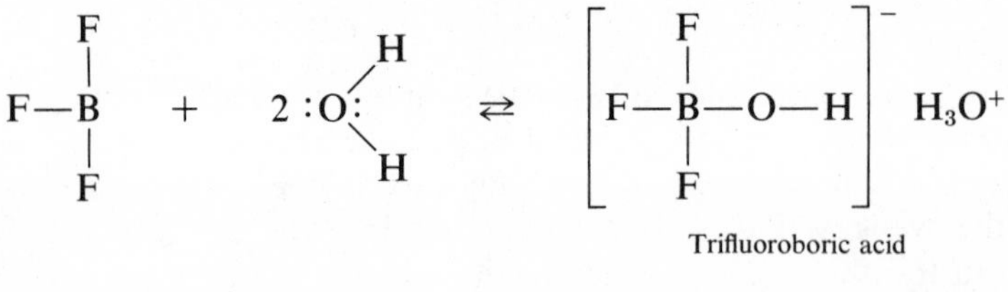

Trifluoroboric acid

In these examples we see that the boron atom achieves an *octet* of electrons by forming the bond to the base. The boron atom in BF_3 is "electron-deficient." Lewis acids usually are electron-deficient—that is why they want to have a share in the electrons of the base.

Exercises

14. For clear presentation, formal charges have been omitted in the Lewis acid–base reactions of boron trifluoride. Calculate the formal charges on the atomic kernels using the equation: formal charge = kernel charge − [no. of unshared e^- + (no. of shared e^-)/2].

15. Lithium ion acts as a Lewis acid in the reaction $Li^+ + H_2O \rightarrow Li(H_2O)^+$. Give the Lewis structure of the product, including formal charges if any. (See Exercise 14.)

16. Using electronegativities (Table 3.4), which would you predict to be the stronger Lewis acid, BF_3 or BCl_3?

17. Would you expect $H—He^+Cl^-$ to be a stable ionic compound? (*Hint*: consider the relative strengths of the bases competing for the proton.)

18. In reacting as a Lewis acid, the boron fluoride molecule undergoes a change in geometry. Predict the change in bond angles, using the electrostatic rules in Section 4.4. Do the molecules of NH_3, H_2O, and $(CH_3)_2O$ undergo any change in geometry when they react with boron trifluoride? Is there any change in oxidation states?

19. NaF and BF_3 react as follows: $Na^+F^- + BF_3 \rightarrow Na^+BF_4^-$. Show that this uses a Lewis acid–base reaction.

The positive ions of metals make up the largest class of Lewis acids. The fundamental reaction of a Lewis acid cation M^{n+} with a base :B is

$$M^{n+} + :B \rightleftarrows (M:B)^{n+}$$

It can also be written showing the covalent bond:

$$M^{n+} + :B \rightleftarrows (M—B)^{n+}$$

Here are some simple examples:

$$Li^+ + H_2O \rightleftarrows (Li:OH_2)^+ \quad \text{or} \quad (Li—OH_2)^+$$

and

$$Mg^{2+} + H_2O \rightleftarrows (Mg:OH_2)^{2+} \quad \text{or} \quad (Mg—OH_2)^{2+}$$

Most reactions of bases with metal cations do not stop at the attachment of only one base molecule as ligand. Attachments of two, four, or six base molecules are more common. In these cases the general form of the equation, where the bases are uncharged, like NH_3 and H_2O, is this: $M^{n+} + lB \rightleftarrows M(B)_l^{n+}$. The number of ligands, l, is called the *ligancy* or coordination number (C.N.) of the complex ion $M(B)_l^{n+}$

Lewis Acid		*Base*		*Formula of Complex Ion*	*Ligancy or C.N.*	*Color*
Ag^+	+	$2NH_3$	$\rightleftarrows$	$Ag(NH_3)_2^+$	2	Colorless
Cu^{2+}	+	$4H_2O$	$\rightleftarrows$	$Cu(H_2O)_4^{2+}$	4	Pale blue
Cu^{2+}	+	$4NH_3$	$\rightleftarrows$	$Cu(NH_3)_4^{2+}$	4	Deep blue
Fe^{2+}	+	$6H_2O$	$\rightleftarrows$	$Fe(H_2O)_6^{2+}$	6	Green
Fe^{3+}	+	$6H_2O$	$\rightleftarrows$	$Fe(H_2O)_6^{3+}$	6	Yellow
Cr^{3+}	+	$6H_2O$	$\rightleftarrows$	$Cr(H_2O)_6^{3+}$	6	Violet
Al^{3+}	+	$6H_2O$	$\rightleftarrows$	$Al(H_2O)_6^{3+}$	6	Colorless

If the base carries a negative charge like F^-, CN^-, OH^-, Cl^-, the charge on the complex ion will be the charge on the Lewis acid cation minus the total charge on the ligands, or what is the same thing, the algebraic sum of all the charges.

$$Al^{3+} + 6F^- \rightleftarrows AlF_6^{3-} \qquad (+3 - 6 = -3)$$

Hexafluoroaluminate ion
(colorless)

$$Fe^{2+} + 6CN \rightleftarrows Fe(CN)_6^{4-} \qquad (+2 - 6 = -4)$$

Ferrocyanide ion
[hexacyanoiron(II) ion]
(yellow)

$$Cu^{2+} + 4Br^- \rightleftarrows CuBr_4^{2-} \qquad (+2 - 4 = -2)$$

Tetrabromocopper(II) ion

Bases compete for *Lewis* acids as they do for the protons of proton acids. Since electrolytes are usually handled in water solutions, *the Lewis acid cations in the solution are usually present as their aquo-complexes* (*aquo* means water), and any added bases must compete with water for the cation. This means that you can't study the equilibrium between a "bare" Lewis acid cation and a base directly. An example will make this clearer.

Exceptions are solutions of compounds in which the metal cation is already more strongly bonded to its ligands than it would be to H_2O. Examples are $K_4Fe(CN)_6$ and $Co(NH_3)_6(NO_3)_3$.

EXAMPLE

Taking silver ion as the Lewis acid, there is no way of measuring the equilibrium constant for the reaction $Ag^+ + 2NH_3 \rightleftarrows Ag(NH_3)_2^+$. What *can* be measured is the equilibrium constant for the reaction

$$Ag(H_2O)_2^+ + 2NH_3 \rightleftarrows Ag(NH_3)_2^+ + 2H_2O$$

in which NH_3 and H_2O compete for the Lewis acid Ag^+. The equilibrium constant for reactions of this type is called the *formation constant* (K_f) for the complex ion on the right. The concentration of water is omitted because it is the solvent. For $Ag(NH_3)_2^+$, the formation constant is

$$K_f = \frac{[Ag(NH_3)_2^+]}{[Ag(H_2O)_2^+][NH_3]^2} = 1 \times 10^8 \qquad \text{at } 25°C$$

Table 11.3 contains K_f values for some typical complex ions. Because of the way K_f is set up, large values mean more stable complex ions.

TABLE 11.3 Formation constants for some complex ions[a]

Complex Ion	K_f	*Complex Ion*	K_f
$Ag(NH_3)_2^+$	1×10^{8}	$HgCl_4^{2-}$	1.6×10^{16}
$Cu(NH_3)_4^{2+}$	1×10^{12}	$Ag(CN)_2^-$	1×10^{21}
$Zn(NH_3)_4^{2+}$	5×10^{8}	$Fe(CN)_6^{3-}$	1×10^{31}
AlF_6^{3-}	7×10^{19}	$Fe(CN)_6^{4-}$	1×10^{28}
CuF^+	10	$Ni(CN)_4^{2-}$	1×10^{30}
$Al(OH)_4^-$	2×10^{28}	$Zn(CN)_4^{2-}$	5×10^{16}
$Zn(OH)_4^{2-}$	5×10^{14}	$Cd(CN)_4^{2-}$	6×10^{18}
$Cr(OH)_4^-$	1×10^{30}	$Hg(CN)_4^{2-}$	4×10^{41}

[a] K_f at 25°C, ligand H_2O molecules omitted.

The formation of complex ions is important in separations of metal ions in analytical chemistry. Hemoglobin, chlorophyll, and vitamin B_{12} are complex ions of Fe^{2+}, Mg^{2+}, and Co^{2+} respectively. Ethylenedithiol ($HS{-}CH_2{-}CH_2{-}SH$) is used in cases of poisoning by arsenic compounds because it forms a nontoxic complex ion with As^{3+} ion, probably as follows:

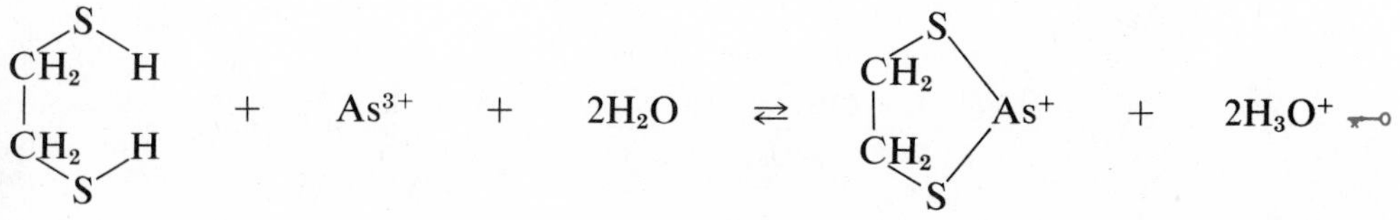

Exercises

20. What reagents could you mix to form a solution containing the complex ion $Fe(CN)_6^{4-}$? (*Hint*: find the oxidation state of Fe first.)

21. What reaction (if any) would you predict for the mixing of a 1.00-*M* $Cu^{2+}SO_4^{2-}$ solution and a 1.00-*M* $Zn(NH_3)_4^{2+}SO_4^{2-}$ solution? (*Hints*: remember that Cu^{2+} is actually $Cu(H_2O)_4^{2+}$ in solution; use data from Table 11.3.)

22. Which of the complex ions in Table 11.3 would you expect to be colored?

11.5 Oxyacids Oxyacids are proton acids in which the proton is linked through an oxygen atom to an atom of some other element. They contain at least one H—O—X group in their structure. The X atom may be a nonmetal, such as S, N, P in H_2SO_4, HNO_3, H_3PO_4, or a metal, such as chromium (Cr) or manganese (Mn):

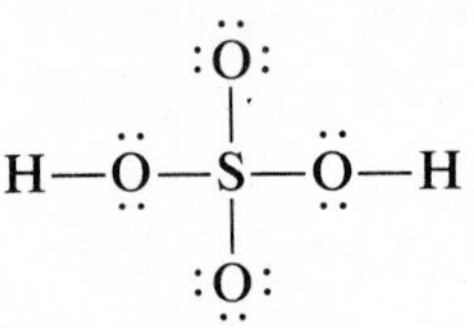
Sulfuric acid

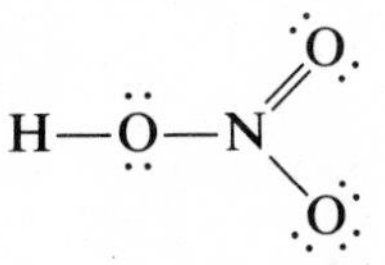
Nitric acid

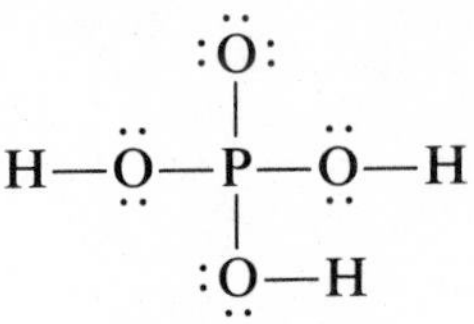
Phosphoric acid

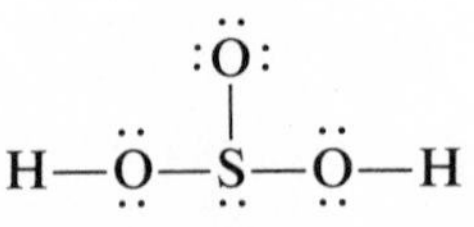
Sulfurous acid

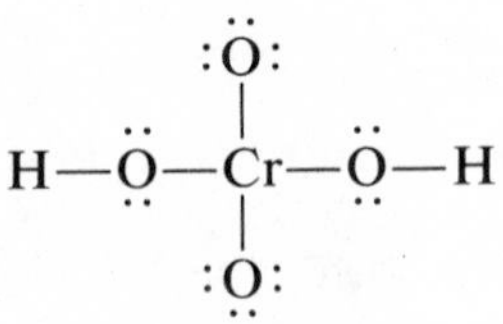
Chromic acid

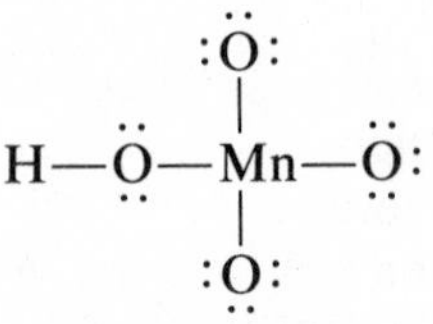
Permanganic acid

Other acids you know, such as $H—\ddot{\underset{..}{Cl}}:$, $H—\ddot{\underset{..}{Br}}:$, $H—\ddot{\underset{..}{I}}:$, are proton acids but not oxyacids.

Aquocomplex ions are frequently weak oxyacids, so their water solutions are acidic. The proton transfer equilibrium and equilibrium constant for the aquocomplex ion $Fe(H_2O)_6^{3+}$ are as follows:

$$Fe(H_2O)_6^{3+} + H_2O \rightleftarrows Fe(H_2O)_5(OH)^{2+} + H_3O^+$$

and

$$K_a = \frac{[Fe(H_2O)_5(OH)^{2+}][H_3O^+]}{[Fe(H_2O)_6^{3+}]} = 9 \times 10^{-4} \qquad \text{at } 25°C$$

You can see that $Fe(H_2O)_6^{3+}$ is acting as an oxyacid if you write its octahedral structure and the reaction in detail. The proton transferred to H_2O is attached to iron through oxygen:

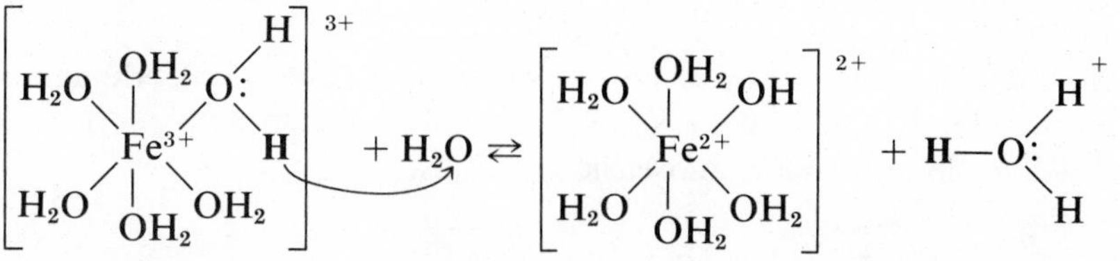

Actually $Fe(H_2O)_6^{3+}$, with $K_a = 9 \times 10^{-4}$, is a stronger acid than acetic acid (CH_3COOH, $K_a = 1.8 \times 10^{-5}$). This means that for solutions of the same component molarity, the aquoiron(III) solution has the higher $[H_3O^+]$. In general, solutions of aquocomplexes of transition metals in oxidation states 2 or higher are acidic rather than neutral. Among the *octet* metal ions, $Al(H_2O)_6^{3+}$ is one of the stronger aquocomplex oxyacids, having $K_a = 1.3 \times 10^{-4}$.

Exercises

23. Using the approximate method (Section 8.3) and with $K_a = 9 \times 10^{-4}$ for $Fe(H_2O)_6^{3+}$, calculate the $[H_3O^+]$ and pH of a 0.01-*M* solution of $Fe(ClO_4)_3$, iron(III) perchlorate.

24. Determine the oxidation states of the central atoms in H_2SO_4, HNO_3, H_3PO_4, H_2SO_3, H_2CrO_4, and $HMnO_4$.

11.6 Amphoteric Hydroxides Compounds that can act either as acids or as bases are called *amphoteric.* Water is the simplest amphoteric compound because it can donate a proton or accept one (Section 8.1). Many slightly soluble metal hydroxides are amphoteric. They act as *bases toward proton acids*, and they act as *Lewis acids toward bases*, particularly toward hydroxide ion. Silver hydroxide, AgOH, is amphoteric, and its reactions with acids and hydroxide ion are a simple example of amphoterism:

Greek, meaning "partly one, partly the other."

As a base $\quad Ag(OH)(s) + H_3O^+ \rightleftarrows Ag(H_2O)_2^+$

As a Lewis acid $\quad Ag(OH)(s) + OH^- \rightarrow Ag(OH)_2^-$

The hydroxides of most of the metallic elements are only very slightly soluble in water; the values of their solubility products (K_{sp}) range from about 10^{-10} to 10^{-60}. Most of them dissolve in acidic solutions, but the amphoteric ones dissolve in basic solutions as well. Zinc hydroxide [$Zn(OH)_2$, $K_{sp} = 10^{-17}$] is amphoteric, and the equations for its dissolving in acid and in base are more complicated than those for AgOH because of the presence of two OH groups and a higher ligancy of the Zn^{2+} ion:

Exceptions are in the Group 1a and 2a hydroxides.

$$Zn(OH)_2(s) + 2H_3O^+ \rightleftarrows Zn(H_2O)_4^{2+}$$
$$Zn(OH)_2(s) + OH^- + H_2O \rightleftarrows Zn(OH)_3(H_2O)^-$$
Zincate ion

In both cases the ligancy of Zn^{2+} is 4 (compare Table 11.3). Aluminum hydroxide, $Al(OH)_3$, with $K_{sp} = 2.5 \times 10^{-32}$ is also amphoteric and dissolves in both acidic and basic solutions.

$$Al(OH)_3(s) + 3H_3O^+ \rightleftarrows Al(H_2O)_6^{3+}$$
$$Al(OH)_3(s) + OH^- + 2H_2O(l) \rightleftarrows Al(OH)_4(H_2O)_2^-$$
Aluminate ion

If $Al(OH)_3$ is treated with hydrochloric acid solution to dissolve it, the result is a solution containing $Al(H_2O)_6^{3+} + 3Cl^-$, a solution of aluminum chloride ("$AlCl_3$"). Dissolving $Al(OH)_3$ in sodium hydroxide solution produces a solution of sodium aluminate, $Na^+ + Al(OH)_4(H_2O)_2^-$.

EXAMPLE The amphoteric properties of metal hydroxides are taken advantage of in separations of metal ions in analytical chemistry and metallurgy. In the production of aluminum from *bauxite* (Al_2O_3 + traces of Fe_2O_3), the amphoterism of $Al(OH)_3$ permits purification of the ore to pure Al_2O_3 before electrolysis to the metal. The bauxite is first hydrated, $Al_2O_3 + 3H_2O \rightleftarrows 2Al(OH)_3$, $Fe_2O_3 + H_2O \rightleftarrows 2Fe(OH)_3$, and the mixture of hydroxides is then treated with sodium hydroxide solution. The $Al(OH)_3$ dissolves according to the equation given earlier but $Fe(OH)_3$ does not dissolve. The solution of sodium aluminate that results is filtered or allowed to settle to remove the solid $Fe(OH)_3$. The solution of pure sodium aluminate obtained is acidified just enough to precipitate $Al(OH)_3$, which is separated and heated to produce pure Al_2O_3 for electrolysis. The equations for the last two steps are as follows:

$$Al(OH)_4(H_2O)_2^- + H_3O^+ \rightleftarrows Al(OH)_3(s) + 4H_2O$$
$$2Al(OH)_3(s) \rightleftarrows Al_2O_3(s) + 3H_2O(g)$$

Exercises

25. Chromium(III) hydroxide, $Cr(OH)_3$, behaves like aluminum hydroxide toward acidic and basic solutions. Give balanced equations for the reactions. What are the oxidation state and the ligancy of Cr^{3+} in the ions? What geometry would you predict for $Cr(H_2O)_6^{3+}$? For $Cr(OH)_4(H_2O)_2^-$?

26. Gold(III) hydroxide, $Au(OH)_3$, is amphoteric with K_f for the formation of $Au(OH)_4^-$ equal to 10^{-3}. Assuming a ligancy of 4 for Au^{3+}, write balanced overall equations for the reaction of $Au(OH)_3$ with nitric acid and sodium hydroxide solutions.

27. When a metal hydroxide shows amphoteric behavior, are there any changes in the oxidation state of the metal?

11.7 Copper (Cu), Silver (Ag), and Gold (Au)—The Coinage Metals

For thousands of years people have used coins made of copper, silver, or gold for money. These metals are among the few that are found uncombined with other elements, or *native*, on earth. ☛ The native deposits have lasted for millions of years because the metals are not oxidized by water, hydrogen ion, or oxygen at ordinary temperatures. They differ from the octet metals in this important way.

Platinum, mercury, and bismuth are others.

Compounds of copper and silver with sulfur, and of gold with tellurium ($AuTe_2$), are very stable and insoluble. For these reasons, sulfides are the principal ores of these metals other than the native deposits. The preparation of copper from its sulfide ores is a particularly important process because deposits of the native metal are small and huge quantities of copper are needed for tubing and electric wire. Copper ore is scarce and becoming scarcer. At the time this book was written it looked as though the value of the copper in a penny might soon become worth more than one cent, making it economically attractive to melt down pennies for their copper. It may become necessary to "mine" city dumps (Photo 29) for the copper and other valuable metals they contain.

EXAMPLE

⊸ Chalcopyrite, $CuFeS_2$, is an important ore of copper. It is crushed and then concentrated ("beneficiated") by flotation to get rid of clay and sand (Figure 11.9). Next, the concentrated ore is roasted in air to eliminate some of the sulfur by the reaction

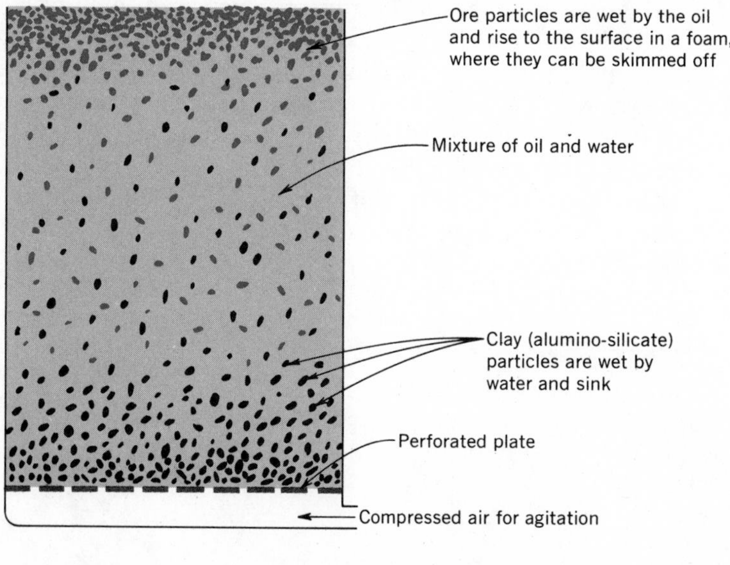
Ore particles are wet by the oil and rise to the surface in a foam, where they can be skimmed off
Mixture of oil and water
Clay (alumino-silicate) particles are wet by water and sink
Perforated plate
Compressed air for agitation

FIG. 11.9 Principle of concentration of a sulfide ore, using flotation.

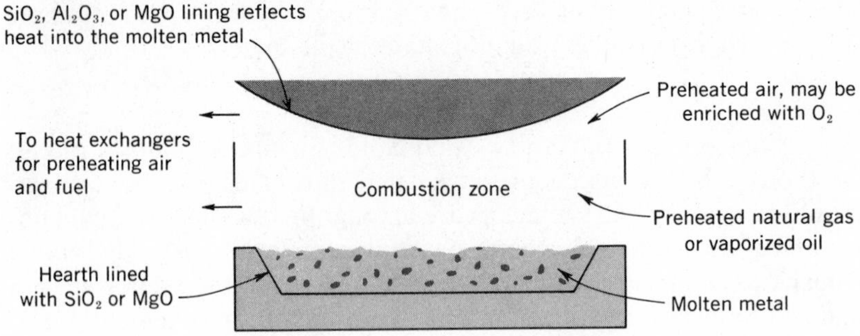

FIG. 11.10 Principle of a reverberatory furnace.

$$2CuFeS_2(s) + 3O_2(g) \rightarrow 2CuS(s) + 2FeO(s) + 2SO_2(g)$$

Roasting is followed by melting in a reverberatory furnace (Figure 11.10) with sand (SiO_2) to remove the iron(II) oxide as a liquid *slag* of iron(II) silicate, $FeSiO_3$:

$$FeO(s) + SiO_2(s) \rightarrow FeSiO_3(l)$$

The slag floats on top of the molten CuS, which is drawn off into a *converter*. Here compressed air is blown through the liquid until all of the sulfur has been removed by the following pair of reactions:

$$2CuS(l) + 3O_2(g) \rightarrow 2CuO(l) + 2SO_2(g)$$
$$CuS(l) + 2CuO(l) \rightarrow 3Cu(l) + SO_2(g)$$

The reactions release huge volumes of the poisonous gas sulfur dioxide, and unless it is reclaimed for the production of sulfuric acid (Section 12.5), it seriously pollutes the air around the refinery.

The copper metal from the converter is impure and cannot be used for electric wire because the impurities, particularly arsenic, greatly reduce its electrical conductivity. The crude copper is purified electrolytically to 99.99+% purity by the process of *electrorefining* described in Figure 11.11. Most of the copper goes into making tubing and electric wire; some goes into alloys such as *brass* (60–90% Cu, 10–40% Zn) and *bronze* (80% Cu, 15% Sn, 5% Zn). The anode sludge

TABLE 11.4 Percentage compositions of some alloys of silver, copper, and gold

Alloy	*Cu*	*Ag*	*Au*
Coin silver	10	90	–
Jewelry silver	20	80	–
Sterling silver	7.5	92.5	–
Coin gold (21.6-carat)	10	–	90
Jewelry gold (18-carat)	25	–	75

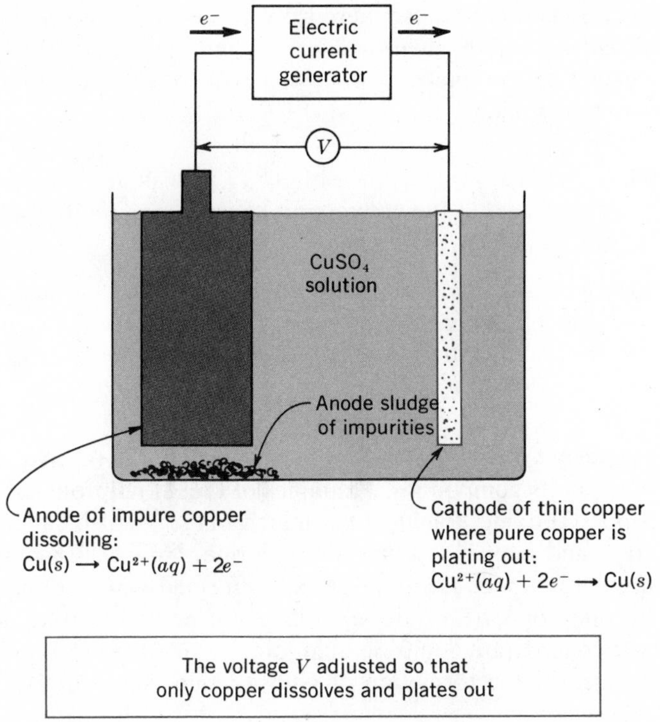

FIG. 11.11 Electrorefining of copper.

from electrorefining is treated by special methods to recover any silver or gold that accompanied copper in the chalcopyrite ore. ⊸

Pure silver and gold are too soft for use in coins or jewelry, so copper is added to produce harder alloys for these purposes. Their percentage compositions are given in Table 11.4. The *carat*, used to give the proportion of gold in a gold alloy, is the number of grams of gold in 24 grams of the alloy—18-carat gold is $(18/24) \times 100 = 75\%$ gold and 25% some other metal, usually copper.

Amalgamation Amalgamation is the dissolving of a metal in liquid mercury, and an *amalgam* is a solution of a metal in mercury. Most metals, including copper, silver, and gold, dissolve in mercury to give amalgams. Prominent exceptions are iron and platinum.

EXAMPLE

⊸ Native silver and gold are often separated by amalgamation from the rock or clay in which they occur. The crushed ore is driven by a stream of water over mercury in the bottom of a "riffle" box. The more dense gold or silver particles sink to the bottom, where they dissolve in the mercury as the insoluble rock and clay are washed away. From time to time the amalgam that has formed is drawn off and freed of mercury by distillation (bp 356.58°C), leaving the gold or silver behind. The mercury is returned to the riffle box. The operation is a hazardous one because mercury vapor is very poisonous, causing loss of hair and teeth, then madness. When mercury is spilled, it must be cleaned up completely, or dusted with powdered sulfur to form the nontoxic sulfide: $Hg(l) + S(s) \rightarrow HgS(s)$.

Exercises

28. Calculate the volume of $SO_2(g)$ measured in liters at STP that is produced in the conversion of 1.00 ton (2000 lb) of chalcopyrite ore, $CuFeS_2$, to copper (1 lb = 454 g).

29. Theoretically, what emf (V) in Figure 11.11 would be required in the electrorefining process for copper? (*Hints*: determine the overall reaction; assume $[Cu^{2+}] = 1$ in the $CuSO_4$ solution used as electrolyte.)

Compounds of Cu, Ag, Au Copper shows both +1 and +2 oxidation states in its compounds. Examples of the +1 (cuprous) state are Cu_2O, copper(I) oxide, found as the mineral *cuprite*, also called "red copper ore," and copper(I) (cuprous) chloride, Cu_2Cl_2. Examples of the +2 state are copper(II) (cupric) oxide, formed when copper is heated in air, and copper(II) (cupric) sulfate, formed by heating the metal with hot, concentrated sulfuric acid. Note that *sulfur* in H_2SO_4, not H^+, is reduced in the formation of $CuSO_4$; thus SO_2, not H_2, is produced:

$$2Cu(s) + O_2(g) \rightleftarrows 2CuO(s)$$

$$Cu(s) + 2H_2SO_4(l) \rightleftarrows CuSO_4(s) + SO_2(g) + 2H_2O(l)$$

Both oxidation states of copper are present in the test for diabetes, using Fehling's solution, by which the amount of glucose ($C_6H_{12}O_6$, blood sugar) in the urine is measured. The solution for the test is prepared just before use by combining copper sulfate solution with a solution of sodium hydroxide and the mixed salt, sodium potassium tartrate ($C_4H_4O_6NaK$, Rochelle salt). An excess of test solution is added to a measured volume of urine and the mixture is boiled. The +2 copper in the solution oxidizes any glucose present to gluconic acid ($C_6H_{12}O_7$) and is itself reduced to copper in the +1 oxidation state. The +1 copper appears as a brick-red precipitate of Cu_2O, copper(I) oxide, at the bottom of the test tube. By weighing it, the amount of sugar in the urine sample can be calculated.

The reaction proceeds beyond oxidation to gluconic acid, depending on the conditions.

The overall equation for the first step in Fehling's test is

$$\underset{\text{Glucose}}{C_6H_{12}O_6} + 2Cu^{2+} + 4OH^- \rightarrow \underset{\text{Gluconic acid}}{C_6H_{12}O_7} + Cu_2O(s) + 2H_2O$$

However, the chemistry of Fehling's test is somewhat more involved than this simple equation suggests. The details are worth following be-

cause they will introduce you to a special kind of Lewis acid–base reaction that forms an important class of complex ions called *chelates*. ☞

Copper in oxidation state +2 oxidizes glucose *only if the solution is
alkaline*. But if a solution of $Cu^{2+}SO_4^{2-}$ and glucose is made alkaline by
adding sodium hydroxide, the Cu^{2+} is almost completely removed
from solution as a precipitate of $Cu(OH)_2$ ($K_{sp} = 1.60 \times 10^{-19}$). As a
result, no oxidation of glucose occurs. The purpose of the sodium
potassium tartrate in Fehling's solution is to keep the copper in oxidation state +2 in a solution that is alkaline. It does this by forming the
cupritartrate complex ion as follows:

"Chela" is Greek for pincers or
crab claws.

$$
\begin{array}{c}
\text{$^-$OOCCH—$\overset{..}{\underset{..}{O}}$H} \\
| \\
\text{$^-$OOCCH—$\overset{..}{\underset{..}{O}}$H}
\end{array}
\quad + \quad Cu(H_2O)_4^{2+} \quad + \quad
\begin{array}{c}
\text{HO—CHCOO$^-$} \\
| \\
\text{HO—CHCOO$^-$}
\end{array}
$$

Tartrate ion Tartrate ion

$$-4H_2\overset{..}{\underset{..}{O}}$$

$$
\left[
\begin{array}{c}
\quad\quad\text{H} \\
\quad\quad| \\
\text{$^-$OOCCH—O^{+1}} \\
| \\
\text{$^-$OOCCH—O^{+1}} \\
\quad\quad| \\
\quad\quad\text{H}
\end{array}
\ \ \text{Cu}^{2-}\ \
\begin{array}{c}
\text{H} \\
| \\
\text{O^{+1}—CHCOO$^-$} \\
\text{O^{+1}—CHCOO$^-$} \\
| \\
\text{H}
\end{array}
\right]^{2-}
$$

$$+4OH^-$$
$$-4H_2\overset{..}{\underset{..}{O}}$$

$$
\left[
\begin{array}{c}
\text{$^-$OOCCH—$\overset{..}{\underset{..}{O}}$} \\
| \\
\text{$^-$OOCCH—$\overset{..}{\underset{..}{O}}$}
\end{array}
\ \ \text{Cu}^{2-}\ \
\begin{array}{c}
\text{$\overset{..}{\underset{..}{O}}$—CHCOO$^-$} \\
| \\
\text{$\overset{..}{\underset{..}{O}}$—CHCOO$^-$}
\end{array}
\right]^{6-}
$$

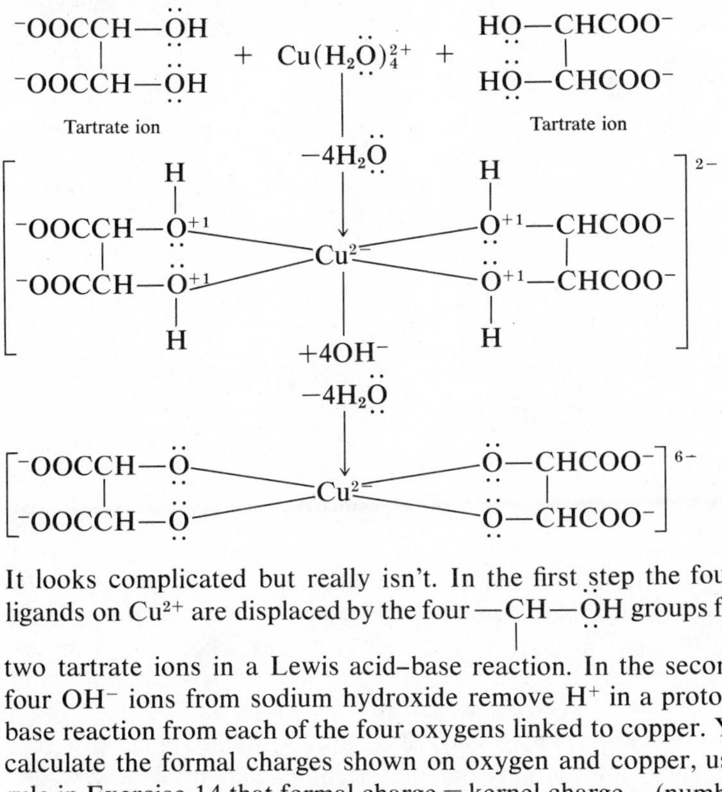

It looks complicated but really isn't. In the first step the four $H_2\overset{..}{\underset{.}{O}}$:
ligands on Cu^{2+} are displaced by the four $—\overset{|}{C}H—\overset{..}{\underset{..}{O}}H$ groups from the

two tartrate ions in a Lewis acid–base reaction. In the second step,
four OH^- ions from sodium hydroxide remove H^+ in a proton acid–
base reaction from each of the four oxygens linked to copper. You can
calculate the formal charges shown on oxygen and copper, using the
rule in Exercise 14 that formal charge = kernel charge − (number of e^-
in unshared pairs + $\frac{1}{2}$ the number of e^- in covalent bonds). The kernel
charge of Cu^{2+} is +2, because two electrons are removed from the
copper atom to make it, so the formal charge on $Cu = +2 - (0 + \frac{8}{2}) = -2$.
The *oxidation state* of Cu in the cupritartrate ion is +2, even though
its formal charge is −2. If the copper were in oxidation state +1, the
kernel charge would be +1 (one e^- removed from Cu), and the formal
charge on Cu would be $+1 - (0 + \frac{8}{2}) = -3$, instead of −2.

Copper in oxidation state +1 does not form a stable complex with
the tartrate anion, so when Cu^{2+} in the cupritartrate ion is reduced to

Cu^+ by glucose in Fehling's test, the complex falls apart, leaving $Cu(H_2O)_4^+$ in the alkaline solution. It reacts immediately with hydroxide ion to give the precipitate of Cu_2O:

$$2Cu(H_2O)_4^+ + 2OH^- \rightarrow Cu_2O(s) + 9H_2O(l)$$

The cupritartrate ion is an example of a *chelate*. Chelates are complexes of metal ions in which two or more of the ligands come from the same molecule or ion. Each tartrate anion provides two —O— ligands for the copper(II) ion. When two ligands come from the *same* molecule or ion, as in cupritartrate ion, the chelating agent is called *bidentate* ("two-tooth"). Tri- and tetradentate chelating agents are also known (three- and four-tooth, respectively). The Fe^{2+} ion in hemoglobin and the Mg^{2+} ion in chlorophyll are both in tetradentate chelates, of the following abbreviated structure:

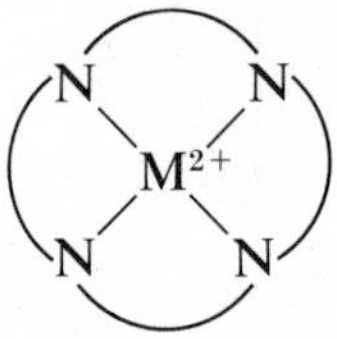

The four nitrogen ligands are all in one very complex molecule, which is different for hemoglobin and chlorophyll.

Exercises

30. If Fehling's test on a 10.00-ml sample of urine yielded 0.0143 g of Cu_2O, what was the molar concentration of glucose in the urine?

31. What bond angles would you predict on electrostatic grounds for the bonds coming out from the Cu(II) ion in the cupritartrate ion? Would the ion be flat? Draw a rough sketch showing its geometry.

The silver halides and photography The three silver halides AgCl, AgBr, and AgI are the most important compounds of silver because they are essential to photography. It is their sensitivity to light that makes photography possible. When light is absorbed by a crystal of a silver halide, the energy of the light (Section 3.2) makes electrons jump from halide ions to silver ions, producing free halogen atoms and free silver atoms. Using $Ag^+ :\ddot{\underset{..}{X}}:^-$ for the silver halide, we can write the reaction as follows: $Ag^+ :\ddot{\underset{..}{X}}: + h\nu \rightarrow \dot{A}g \cdot \ddot{\underset{..}{X}}:$. As more and more light is absorbed, more and more of the lattice points in the silver halide crystal become occupied by silver and bromine atoms instead of their ions. You can see the macroscopic effect of this microscopic process by exposing some freshly prepared silver chloride to sunlight. Originally white, it rapidly turns blue-gray. You can even make a crude photograph by focusing an image on a film covered with a photographic *emulsion* of silver halide grains in gelatin. In the bright places of the image, the film will slowly turn dark bluish-gray; in the unlit places in the image the film remains white. After minutes to hours of exposure to the light image, a *negative* image (light and dark reversed) appears on the film. However, photography as we know it would be impossible if the whole picture had to be produced in this slow way by the absorption of light.

Fortunately, if a grain of silver halide crystal in the film has been exposed to light just long enough to produce a few silver atoms in it, *the whole grain becomes susceptible to* ***chemical*** *reduction* to silver atoms and halide ions. You can think of the silver atoms as catalysts for the reduction. The chemical reduction is called *development* and employs, as "developers," organic reducing agents such as hydroquinone. Development involves the following half-reactions:

Reduction of halide: $Ag^+ :\ddot{\underset{..}{X}}:^- + e^- \rightarrow Ag^{\cdot} + :\ddot{\underset{..}{X}}:^-$ (↑ e^- from developer)

Oxidation of hydroquinone:

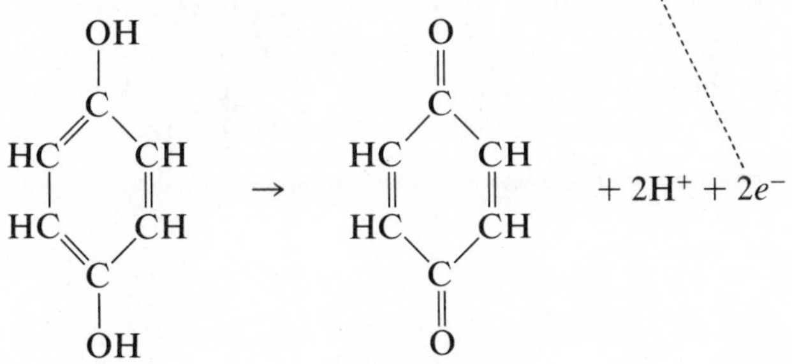

Hydroquinone → Benzoquinone $+ 2H^+ + 2e^-$

The pattern of reduced, partly reduced, and unreduced grains makes up the image on the film.

The microscopic record of the light image that fell on the film (for as little as a thousandth of a second using "fast film" in a good camera) is stored in the relatively few silver atoms in the exposed film and is called the *latent image*. Development even years later produces the macroscopic, *negative* replica of the original image. It is a negative image because *dark* areas due to lots of finely divided silver appear where the light was bright in the image and sensitized a large number of grains of silver halide to development.

After development, the unreduced silver halide in the less exposed areas must be completely removed from the film. Otherwise it turns dark when the film is exposed to light, ruining the picture. The silver halides can't be washed out with water—they are far too insoluble for that (see Table 7.2). Instead, the halide is removed by complexing the silver ion with thiosulfate ion, $S_2O_3^{2-}$. Sodium thiosulfate ($Na_2S_2O_3$, "hypo") provides the ion. As the film lies in the hypo solution, the displacement of the solubility equilibrium shown by the arrows occurs ($X^- = Cl^-, Br^-, I^-$):

$$Ag^+X^-(s) \rightleftarrows Ag^+(aq) + X^-(aq)$$

$$Ag^+(aq) + 2S_2O_3^{2-}(aq) \rightleftarrows Ag(S_2O_3)_2^{3-}$$

and

$$K_f = \frac{[Ag(S_2O_3)_2^{3-}]}{[Ag^+][S_2O_3^{2-}]^2} = 10^{14} \qquad \text{at } 25°C$$

The large value of K_f for the silver ion–thiosulfate ion complex ensures that the $[Ag^+]$ in equilibrium with the complex is smaller than the $[Ag^+]$ in equilibrium with the solid halide, so the halide dissolves (Le Chatelier's principle). The dissolving is like the dissolving of an amphoteric hydroxide in an alkaline solution.

After the hypo treatment, called "fixing," the film is washed thoroughly to remove all chemicals except the metallic silver in the image. This process gives the photographic negative. To get the positive image or "print," light passing through the negative is projected to form an image on a piece of *printing paper*, a paper backing covered with an emulsion of silver halides. The latent *positive* image that results is developed, fixed, washed, and dried in the same way as the negative was. The "print" results.

Exposing the print emulsion to the negative image results in an image that is positive in respect to the image originally photographed.

Color photography also depends on the sensitivity of silver halides to light. The details are very complicated because there are three layers in the emulsion, one for each primary color. But the whole process of developing the image in color starts with exposure of the silver halide in each color layer to light. In color photography, "slide" films like Kodachrome and Agfachrome provide a positive image; color negative films like Kodacolor and Fujicolor provide negatives, in the same manner as black-and-white films.

Magnetic properties of transition metals Matter attracted by a magnet is termed either ferromagnetic or paramagnetic. Ferromagnetism, in which a piece of iron or steel is attracted to a magnet, is more familiar to us. Ferromagnetism and paramagnetism are both due to electrons with unpaired spins in the material. Each spinning electron makes an atomic electromagnet because of its motion. When spins are paired their magnetic effects cancel. In ferromagnetic substances, the atomic magnets due to unpaired electron spins cooperate to produce a strong attractive response to a magnet. Paramagnetism is due to individual atoms, molecules, or ions that have one or more unpaired electron spins. It is weaker than ferromagnetism. Because electrons in the *d* and *f* orbitals of transition and inner transition elements may not all be spin-paired, these elements and their compounds are often ferro- or paramagnetic. Transition metals are valued not only for their magnetic properties but also for making alloys such as manganese steel. This steel has been used to make safes and in the blades of earth-moving equipment because of its extraordinary toughness. Mining the ocean floor for Mn, Ni, and Cu (Photo 30) was planned at the time this book was being written.

EXAMPLES

Any atom, molecule, or ion with an *odd* number of electrons *must* be paramagnetic because it is sure to have one unpaired electron spin. For example, the *gaseous copper atom* with atomic number 29 and electron configuration $[Ar]4s^1 3d^{10}$ is paramagnetic because it has 29 electrons, 18 for [Ar] + 11. *Bulk* copper is *not* paramagnetic, because in the metallic bonding that holds it together, the odd electrons become paired in metallic orbitals. On the other hand, copper(I) *compounds,* for which the electron configuration of Cu^+ is $[Ar]4s^0 3d^{10}$, are not paramagnetic because all spins are paired in the filled 3*d* orbitals on each ion. Copper(II) ions again have an odd number of electrons from the configuration $[Ar]4s^0 3d^9$, which leaves one electron in a 3*d* orbital unpaired. Copper(II) compounds are paramagnetic as a result. Of the aquocomplex ions in the first transition series, those of V^{2+}, Cr^{3+}, Mn^{4+}, Mn^{2+}, Fe^{2+}, and Fe^{3+} are the most strongly paramagnetic—Mn^{2+} and Fe^{3+} have atomic magnetic moments amounting to *five* unpaired electron spins. Measurement of paramagnetism allows assignment of electrons to orbitals. Thus the electron configuration of the Fe^{3+} ion in $Fe(H_2O)_6^{3+}$ is $[Ar]4s^0 3d^5$, and all five of the 3*d* electrons must be in different orbitals and have unpaired spins (Hund's rule, Section 3.4) to account for the paramagnetism of $Fe(H_2O)_6^{3+}$ ions.

The *gaseous atoms* of *all* elements of uneven atomic number (*Z*) are paramagnetic. Because of pairing of spins in the bonding in macroscopic samples of most of them, paramagnetism is not widespread.

Exercises

32. Derive the electron arrangement in the 4*s* and 3*d* orbitals of the Mn^{2+} ion, whose paramagnetism corresponds to five unpaired electron spins. (For Mn, $Z = 25$.)

33. Which would you predict as more likely to be a colored compound, Ag_2SO_4 or $Cr_2(SO_4)_3$? Which would you predict to be paramagnetic? (All alkali metal and alkaline earth sulfates are colorless.)

34. For $Ag(NH_3)_2^+$, the $K_f = 10^8$ and for $Ag(S_2O_3)_2^{3-}$ the $K_f = 10^{14}$, both at 25°C. Which would more effectively dissolve AgCl(*s*), a 1.00-*M* NH_3 or a 1.00-*M* $Na_2S_2O_3$ solution?

PHOTO 30 Manganese ore nodules photographed on the ocean floor thousands of feet below the surface of the Pacific Ocean. The nodules are a likely future source of the transition metals Mn, Ni, and Cu. It is estimated that there are 1.5 trillion tons of manganese nodules on the floor of the Pacific Ocean alone! (*Kennecott Copper Corporation.*)

11.8 Summary Guide Their high electrical conductivity separates the metallic elements from the metalloids, which conduct electricity weakly, and from the nonmetals, which are electrical insulators. Metallic elements are further separated into octet metals that have one, two, or three (rarely four) valence electrons in *s* and *p* orbitals, and transition and inner transition metals that have valence electrons in *d* or *f* orbitals as well as in *s* and *p* orbitals.

The atoms of active metals are oxidized to their ions either by water or by hydronium ions. Most of the metals—octet, transition, and inner transition—are active metals. Metals not oxidized by water or hydronium ions are called noble metals (Ag, Cu, Au, and Pt are examples). Several active metals resist attack by water or dilute acids because of a protective oxide coating (Al, Pb, and Sn are examples). Noble metals are oxidized to their ions by oxidizing agents that are much stronger than hydronium ion.

The oxidation state of a metal in its ion is equal to the number of electrons removed from the metal to form the ion. Oxidation states are useful for balancing oxidation–reduction equations involving metals and can be extended to covalent compounds, using a set of rules based on $+1$ for the oxidation state of H and -2 for the oxidation state of O in covalent compounds and complex ions.

For octet metals, the highest oxidation state of their ions is equal to the number of their valence electrons (the group number of the element). Transition and inner transition metals have multiple oxidation states in most cases. The oxidation state of a transition metal is shown by roman numerals in the name of the compound, as in copper(II) sulfate, $CuSO_4$.

The alkali metals (Group 1a) are the most active metals—all react rapidly with water to produce their hydroxides (M^+OH^-), with oxygen to produce oxides and peroxides, with acids and the halogens to produce salts. Because of the ease of their oxidation, the alkali metals do not occur except as their ions, usually in chlorides (M^+Cl^-). Alkali metals are prepared by electrolysis of their molten chlorides or hydroxides. The alkali metals are soft, of low density, and excellent conductors of heat and electricity. They are so active that they must be protected from air. They are strong reducing agents.

The alkali metal cations (M^+) serve as carriers for important anions. Sodium hydroxide is the most important compound of the alkali metals. It is the strong base in the chemical industry and is prepared by the electrolysis of a solution of sodium chloride. Sodium carbonate, used in making glass, is produced by the Solvay process. Sodium and potassium ions play vital roles in biochemical reactions.

Magnesium and calcium are the most important of the alkaline earth metals (Group 2a). They occur as M^{2+} ions in chlorides and carbonates and are prepared by electrolysis of their molten chlorides. The source of Mg is seawater, in which the element occurs in solution as the chloride. Calcium occurs as the carbonate ($CaCO_3$) in limestone, from which lime is prepared. Limestone is the source of carbon dioxide and lime in the Solvay process for making sodium carbonate and sodium bicarbonate. Only magnesium, among the alkaline earth metals, is used for structural purposes.

Magnesium, calcium, and iron(II) ions (Fe^{2+}) give water a hardness that interferes with the cleansing action of soap. Hard water is softened by using ion exchange to replace Mg^{2+}, Ca^{2+}, and Fe^{2+} ions with Na^+ ions. Water free of all electrolytes (demineralized water) is also made by ion exchange processes, using both anion and cation exchangers. Demineralization of seawater is too expensive to be used for making drinking or irrigation water.

Calcium and magnesium are elements essential for health. Bone and tooth enamel are largely tricalcium phosphate. Magnesium is a part of enzyme systems in the blood.

The hydroxides of the alkali and the alkaline earth metals are all basic oxides. Aluminum hydroxide is amphoteric, showing both acidic and basic properties. Advantage is taken of the amphoteric character of aluminum hydroxide in preparing pure Al_2O_3 for electrolysis to produce aluminum. Iron oxides are reduced to the metal by carbon monoxide in a blast furnace. Aluminum, in the form of its alloys, is widely used as a structural material. Iron is usually made into steel.

Iron(III) chloride and aluminum chloride are covalent molecules of formula type M_2Cl_6. Both chlorides dissolve in water to form octahedral aquocomplex ions of formula type $M(H_2O)_6^{3+}$, which are weak proton oxyacids. Similar oxyacids are found for many transition metals. Metal ions of the M^{+3} type are strong Lewis acids and form covalent bonds to bases such as H_2O to produce the $M(H_2O)_6^{3+}$ ions. Ions of this type for many transition metals are colored. Octet metal ions are colorless.

The coinage metals—copper, silver, and gold—are noble metals and transition metals. They occur to a limited extent in the form of the uncombined (native) metal but mainly as their sulfides. The process of winning copper metal from a sulfide ore releases large volumes of the pollutant sulfur dioxide into the atmosphere unless the SO_2 is used to make sulfuric acid. Crude copper must be purified by electrorefining to get copper useful for electric wiring.

Copper(II) ion plays a central part in Fehling's test for sugar in urine. For this purpose the ion is chelated with tartrate ion. Iron(II) ions are chelated in hemoglobin; magnesium ions are chelated in chlorophyll.

Gold is used mainly for jewelry and as a monetary standard or reserve. Most of the silver produced is used in the form of the silver halides for making photographic film.

Transition metals usually have two or more oxidation states [Cu(I), Cu(II); Fe(II), Fe(III) are examples] and often form compounds that are paramagnetic because of unpaired spins for the electrons in the *d* orbitals of the transition metal ion.

Exercises

35. Predict the effect of increasing $[H_3O^+]$ on the dissolving of $Cu(s)$ by HNO_3.

36. For the half-reaction $Au(s) + 4Cl^- \rightleftarrows AuCl_4^- + 3e^-$, the value of $\mathscr{E}^0 = -1.0$ V. Explain why a mixture of nitric and hydrochloric acids ("aqua regia") dissolves gold while neither acid alone will.

37. Suppose one of your parents is put on a salt-free diet to relieve high blood pressure (hypertension) by decreasing blood volume. Would you advise discontinuing the use of the ion-exchange water softener in the water supply to your home?

38. The K_{sp} for $Na^+HCO_3^-(s) = 0.62$ at 25°C. Calculate $[Na^+]$ in a saturated solution of $Na^+HCO_3^-$ in water, and in a 3.00-*M* $NH_4^+HCO_3^-$ solution. What does your calculation suggest about the pressures of $NH_3(g)$ and $CO_2(g)$ to be used in the Solvay process? $[NH_3(g) + CO_2(g) + H_2O \rightleftarrows NH_4^+HCO_3^-]$

39. In certain types of kidney disease the vital kidney function that excretes urea from blood into the urine is replaced by a "kidney dialysis machine." In the machine, blood from the patient's body is circulated on one side of a membrane permeable to urea, glucose, and electrolytes (Na^+Cl^-, K^+Cl^-, $Na^+HCO_3^-$) but impermeable to blood cells and the large molecules (albumins and globulins) in blood. An aqueous solution circulates past the other side of the membrane to carry away the urea diffusing through the membrane. What should be the composition of the aqueous solution?

Key Words

Alkali metals the Group 1a elements—Li, Na, K, Rb, Cs, Fr.

Alkaline earth metals the Group 2a elements—Be, Mg, Ca, Sr, Ba, Ra.

Amalgamation dissolving a metal in mercury.

Chelate a complex ion in which two or more ligands of the central ion are themselves covalently bonded together.

Electrolysis the reaction that occurs on passing an electric current through a molten ionic compound or a solution that contains ions.

Electrorefining the process of purifying a metal in which the impure metal dissolves at the anode and pure metal separates at the cathode of an electrolysis cell.

Ferromagnetic substance a solid strongly attracted by a magnet in whose atoms, ions, or molecules there are electrons with unpaired spins that cooperate to produce a very strong paramagnetism.

Formation constant (K_f) the equilibrium constant for the reaction forming a complex ion, written omitting the concentration of water.

Hard water water containing M^{2+} metal ions—Ca^{2+}, Mg^{2+}, Fe^{2+}.

Insulator a material that does not conduct electricity—a nonmetal.

Ion exchange replacement of ions in a solution with ions from a zeolite clay or ion exchange resin—used to soften or demineralize water by replacing or completely removing ions in the water, and in ion exchange chromatography for separating amino acids.

Latent image the pattern of silver atoms produced in a photographic film by the light focused on the film by the camera lens during exposure.

Lewis acid any ion or molecule that can utilize the electron pair of a base to form a covalent bond to the base.

Ligand the atoms, ions, or groups of atoms bonded to a single central atom or ion of interest.

Metalloids elements that conduct electricity poorly; they are on the borderline between metals and nonmetals in the Periodic Table, and most have four or five valence electrons in *s* and *p* orbitals.

Metal oxide a compound of a metal with oxygen.

Metals elements that conduct electricity well; they usually have one, two, or three valence electrons in *s* and *p* orbitals.

Nonmetals elements that do not conduct electricity; they usually have five, six, seven, or eight electrons in *s* and *p* orbitals.

Octet metal a metallic element in the Octet Table; they usually have one, two, or three valence electrons (rarely four) in *s* and *p* orbitals.

Oxidation state of a metal ion the positive charge left on the kernel of a metal atom after removal of its valence electrons.

Oxyacids proton acids in which the acidic proton is attached through oxygen to an atom of some other element.

Paramagnetic substance an atom, molecule, or ion that has one or more electrons with unpaired spins and is attracted by a magnet.

Photographic development the process, using reducing agents, that produces the photographic negative from the latent image.

Photographic fixing the process of removing from photographic film any silver halide not reduced to silver during development of a latent photographic image.

Refractories metal and metalloid oxides that have very high melting points.

Transition metal a metallic element having valence electrons in *d* or *f* orbitals.

Suggested Readings

Alter, Harvey, Abert, James G., Bernheisel, J. Frank, "The Economics of Resource Recovery from Municipal Solid Waste," *Science*, **183**:1052–1058 (March 1974).

"Beryllium: After Forty Years, Still a Mystery," *Environment*, **16**(3):35–36 (April 1974).

Brooks, Anita, *The Picture Book of Metals*, New York, John Day, 1972.

Galt, Robert I., *Exploring Minerals and Crystals*, New York, McGraw-Hill Book Company, 1972.

Habashi, Fathi, "Chemistry and Metallurgy," *Chemistry*, **45**:6–10 (Oct. 1972).

Hasewage, R., "Vanadium Concentrations in Atmosphere," *Env. Sci. and Tech.*, **7**:444–448 (May 1973).

Hauser, Ernest O., "Gold—King of Metals," *Reader's Digest*, June 1973, pp. 171–176.

"How Howard Hughes Mines the Ocean Floor," *Bus. Wk.*, Jan. 16, 1973, p. 47.

Kubota, Joe, Mills, Edward L., and Oglesby, Ray T., "Lead, Cd, Zn, Cu, and Co in Streams and Lake Waters of Cayuga Lake Basin, New York," *Env. Sci. and Tech.*, **8**:243–248 (March 1974).

"Manganese Nodules (II): Prospects for Deep Sea Mining," *Science*, **183**: 644–646 (Feb. 15, 1974).

Park, Charles F., Jr., *Affluence in Jeopardy*, San Francisco, Freeman, Cooper and Co., 1968.

Prival, M. J., and Fishe, Farley, "Fluorides in the Air," *Environment*, **15**: 25–32 (April 1973).

Seaborg, Glenn T., *Man-Made Transuranium Elements*, Englewood Cliffs, N.J., Prentice-Hall, Inc., 1963.

"The Electroplaters Are Polishing Up," *Env. Sci. and Tech.*, **8**:406–407 (May 1974).

Wertime, Theodore A., "The Beginnings of Metallurgy: A New Look," *Science*, **182**:875–886 (Nov. 30, 1973).

Zoller, W. H., Gladney, E. S., and Duce, R. A., "Atmospheric Concentrations and Sources of Trace Metals at the South Pole," *Science*, **183**:198–200 (Jan. 18, 1974).

chapter 12

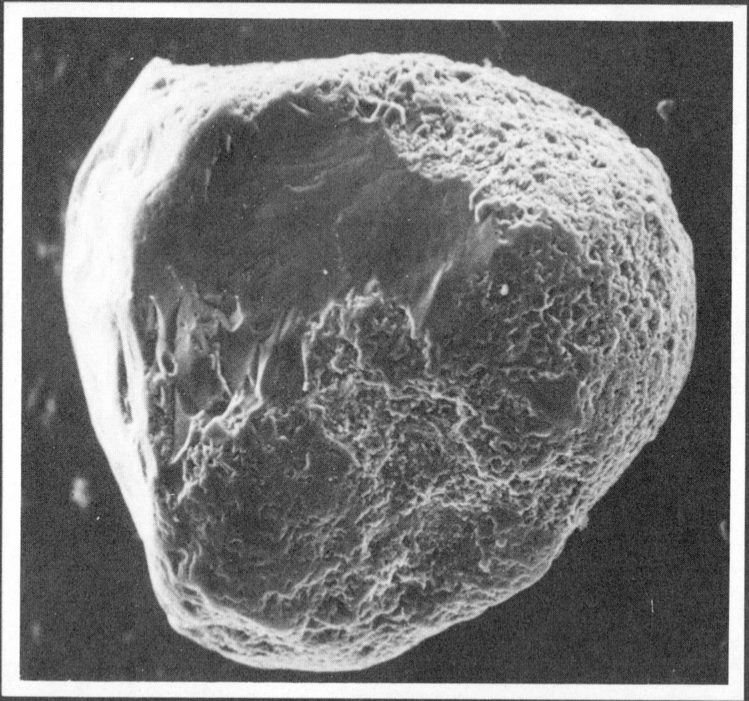

A grain of sand (SiO_2) from a dune on Long Island, New York, as seen by an electron-scanning microscope. The white portions show the scars of wind erosion. The dark region came from a fracture caused by glacier action in the late Pleistocene Epoch (which ended approximately 5000 years ago). *(Photo 31, D. H. Krinsley and I. J. Smalley,* "Sand," American Scientist, *May–June 1972, vol. 60, 286–291.)*

Nonmetals and Metalloids

12.1 Introduction Metallic character or *metallicity* of the elements decreases as you go from left to right across the rows (periods) of the Octet Periodic Table, and metallicity *increases* as you go down columns (groups). For example, period 3 starts with the metal sodium (Na), progresses through metals calcium and aluminum, then with decreasing metallicity goes to the metalloid silicon (Si), and ends with the typical nonmetal chlorine (Cl). Group 5a starts with the nonmetal nitrogen (N) and progresses downward through nonmetal phosphorus (P) and metalloid antimony (Sb) to metal bismuth (Bi). These two trends put all of the nonmetals in the upper-right-hand corner of the Octet Table. The octet metals are on the left side and the metalloids lie in between the two extremes (Table 3.4).

All of the nonmetals and metalloids are in the Octet Table. This means that all *d* and *f* orbitals of their atoms are either completely filled with electrons or completely empty. On this account, you'd predict that paramagnetism, so common among the transition metals with their partly filled *d* orbitals, would be rare with nonmetals and metalloids. This is correct—the *only* nonmetal or metalloid that is paramagnetic in its elementary form is oxygen. The O_2 molecule has two unpaired electron spins (Section 4.3).

There are large differences in molecular structure and chemical properties between second- and third-period nonmetals in the same group—N and P, O and S, F and Cl. For example, nitrogen is a nonreactive gas of triply bonded :N≡N: molecules that make up about $\frac{4}{5}$ of the air we breathe. But phosphorus, just below it, is a waxy solid (mp 44°C) of P_4 molecules [Figure 12.1(a)] that spontaneously burns to P_2O_5 in air and is very poisonous. The differences between third- and fourth-period elements in the same group are not nearly so great—arsenic (As), below phosphorus, is made up of As_4 molecules, is

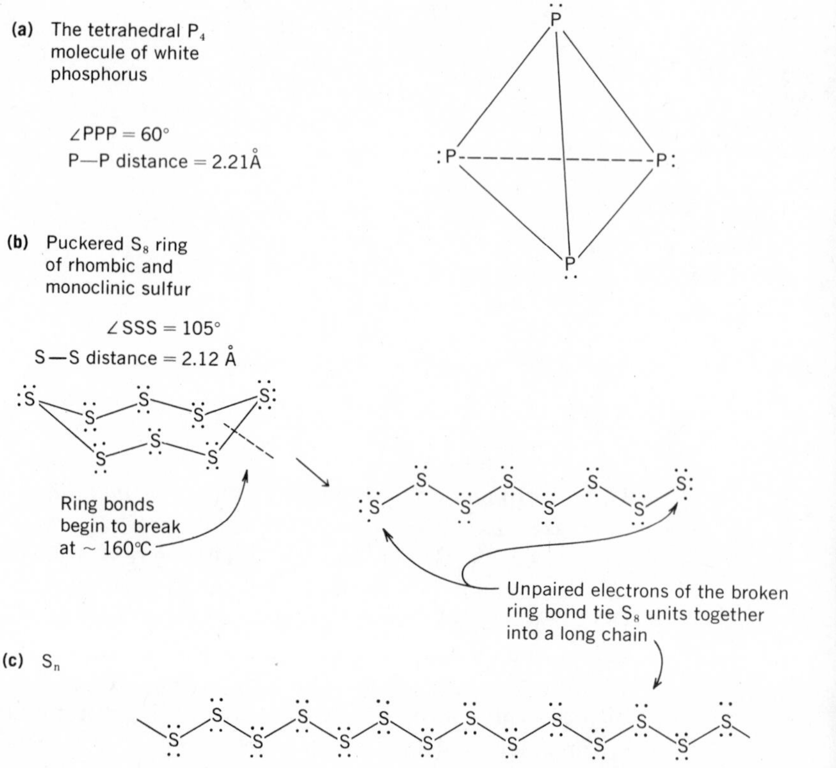

FIG. 12.1 Structures of (a) P_4, (b) S_8, and (c) S_n.

oxidized rapidly (to As_2O_3) when heated in air, and is also very poisonous. In the next group (6a), oxygen (O, period 2) is a gas of paramagnetic O_2 molecules but sulfur (S, period 3), selenium (Se), and tellurium (Te) are nonmagnetic solids in which eight atoms bond together to form a ring [Figure 12.1(b)]. Water is a liquid because of hydrogen bonding (Section 4.6), but H_2S, H_2Se, and H_2Te are all gases at room temperature. Unlike water, they are very poisonous and vile-smelling. Table 3.2 contains additional data for making comparisons of this sort. The sharp break in properties between the second- and third-period elements in the same group is due to the unwillingness of third-period elements to form double or triple bonds.

The bonds between metal and nonmetal atoms are usually ionic like those in Li^+F^-, Na^+Cl^-, $Ca^{2+}O^{2-}$. Bonding between nonmetals (including H) or metalloids is *always* covalent. Electronegativity differences between nonmetals and metalloids are too small to produce the complete transfer of electrons required for ionic bonding. On going from the halogens (Group 7a) toward the center of the table (Group 4a) there is an increasing tendency for nonmetal atoms to bond to each other, either directly or through oxygen. The tendency to bond directly as well as through oxygen and other atoms is highest at carbon. This bonding makes possible the covalent chain, ring, and network molecules that are the chemical basis of life—fats, proteins, carbohydrates, hormones, vitamins, and nucleic acids are examples. Carbon is so many-sided in its bonding that its compounds constitute the whole separate subject of *organic chemistry* (Chapter 13). The tendency to bond *through* oxygen is strong with silicon (Si) and results in the $-\overset{|}{\underset{|}{Si}}-O-\overset{|}{\underset{|}{Si}}-O-$ structure units in many clays, rocks, and silicate minerals.

To trace the increasing tendency of atoms to bond together, you can start with the halogens, Group 7a. They bond to each other to form the molecules F—F, Cl—Cl, Br—Br, I—I, and also interhalogen compounds such as Cl—F, I—Br, and I—Cl. Even *three* iodines are bonded in the triiodide ion $[I—I—I]^-$, but there the tendency to self-link stops. It also stops at three atoms for oxygen, reached in *ozone*, O_3 [Figure 12.2(b)]. But sulfur atoms, just below oxygen in Group 6a, form S_8 rings, as we have seen. And when sulfur is heated beyond its melting point (114°C), the S_8 rings break up into S_8 chains that then join together to form a viscous liquid of infinitely long chains [Figure 12.1(b) and (c)]. When the viscous molten sulfur is cooled, it turns into a solid called "plastic sulfur." This solid is elastic and rubbery because of its long-chain molecular structure. Plastic sulfur would be a useful material of rubberlike properties if it were stable. However, the S_8 ring structure is more stable at room temperature,

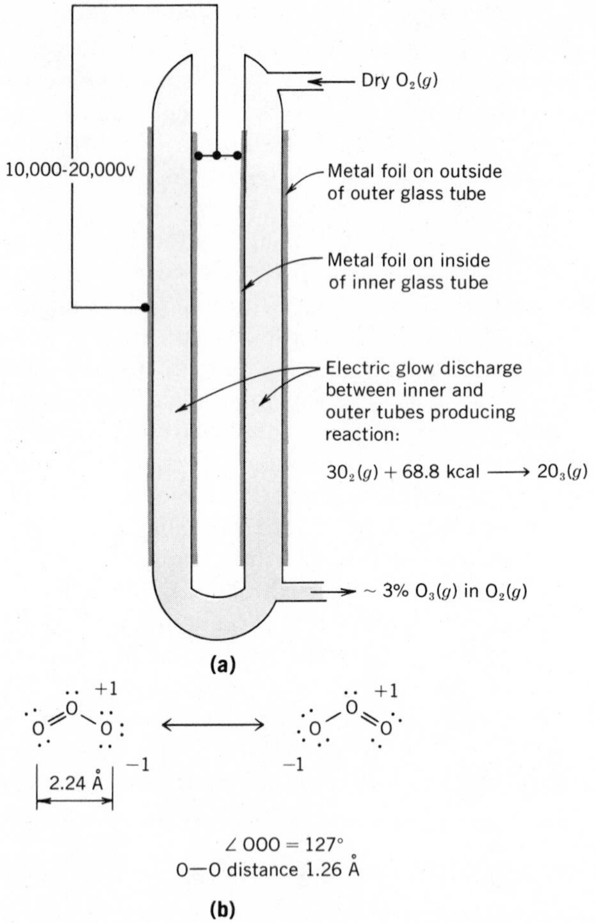

FIG. 12.2 (a) Apparatus for preparation of ozone; (b) ozone as a resonance hybrid.

and the S_n changes back into it within a few hours. The S_8 form is crystalline and brittle.

Linking of nitrogen atoms together stops at N_3^-, the azide ion $[\ddot{N}{=}N{=}\ddot{N}]^-$, but phosphorus atoms form P_4 molecules in the white form of the element [Figure 12.1(a)] and P_n molecules in the red form. The ability of phosphorus to bond to itself *through oxygen*, as seen in the diphosphate and triphosphate units, is of fundamental importance to energy transfer in living cells:

In these formulas and most others that follow, unshared pairs of electrons on O, N, and S are omitted for clarity.

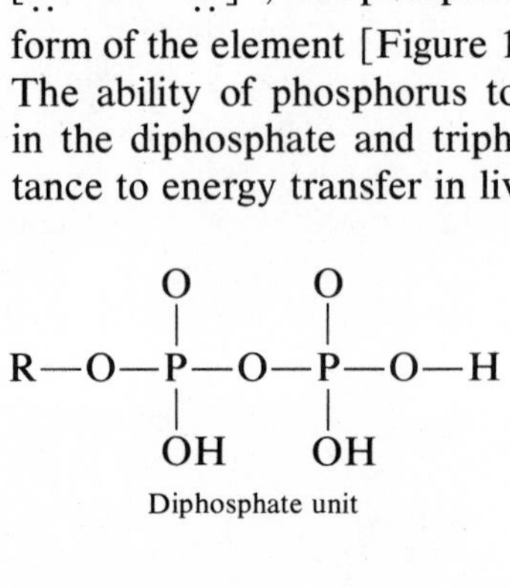

Diphosphate unit

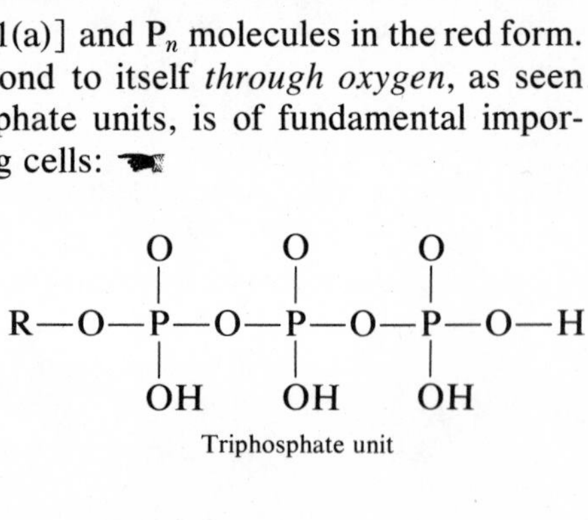

Triphosphate unit

↠ In the biochemicals *adenosine diphosphate* (ADP) and *adenosine triphosphate* (ATP), the R group in the diphosphate and triphosphate units is a complex organic molecule called adenosine (Chapter 13). The enzyme-catalyzed ("enzymatic") conversion of ATP into ADP is the immediate chemical source of muscular energy in our bodies. You can look at the conversion as a breaking of one P—O bond in ATP by hydrolysis, forming the diphosphate (ADP) and phosphoric acid:

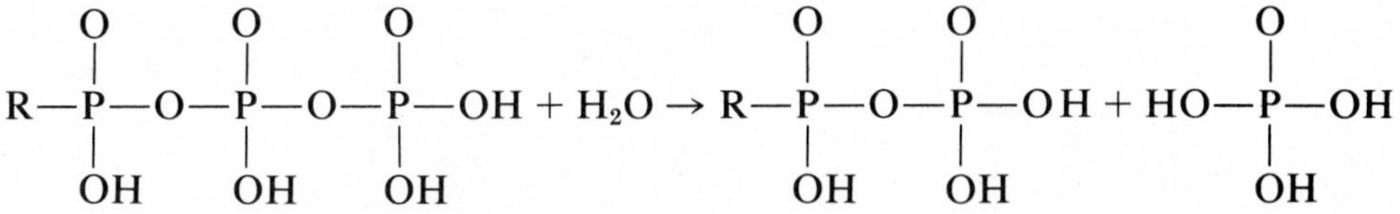

$$
\underset{\text{ATP}}{R-\overset{\overset{O}{\|}}{\underset{\underset{OH}{|}}{P}}-O-\overset{\overset{O}{\|}}{\underset{\underset{OH}{|}}{P}}-O-\overset{\overset{O}{\|}}{\underset{\underset{OH}{|}}{P}}-OH} + H_2O \rightarrow \underset{\text{ADP}}{R-\overset{\overset{O}{\|}}{\underset{\underset{OH}{|}}{P}}-O-\overset{\overset{O}{\|}}{\underset{\underset{OH}{|}}{P}}-OH} + \underset{\text{Phosphoric acid}}{HO-\overset{\overset{O}{\|}}{\underset{\underset{OH}{|}}{P}}-OH}
$$

This reaction is *exo*thermic by 12 kcal/mole. One *mole* of ATP (508 g) going to ADP releases an energy quantity about equal to the energy from burning a *teaspoon* of sugar. In the body, most of the energy released by the oxidation of blood sugar (glucose) is stored in ATP, so ATP serves as the body's "energy bank." ↠

As noted earlier, silicon atoms, like phosphorus atoms, prefer to link through oxygen, and in doing this they provide the

$$
-\overset{|}{\underset{|}{Si}}-O-\overset{|}{\underset{|}{Si}}-O-
$$

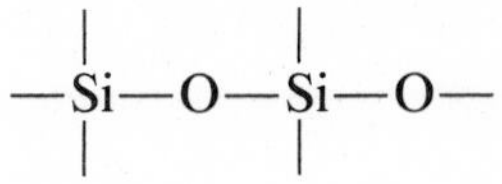

backbone of the *silicates* found in clays, rocks, and many minerals. The silicon atom has four valence electrons in the configuration $[Ne]3s^23p^2$ and forms covalent bonds to four oxygens in the silicate minerals. The ligancy of Si is therefore 4 in these compounds, and the angles between the Si—O bonds have the predicted tetrahedral value, 109.5°. Figure 12.3 shows the symmetry of SiO_4 tetrahedrons and how, by two of them sharing one oxygen, the $(SiO_2)_n$ network of *quartz* (white sand) is produced (Photo 31). Here each Si shares its oxygens with four different Si atoms. Sheets of SiO_4 tetrahedrons in the structure of *mica* result from each silicon sharing *two* oxygens with a second. The structure is an infinite extension of the structure of Al_2Cl_6. and Fe_2Cl_6 discussed in Chapter 11:

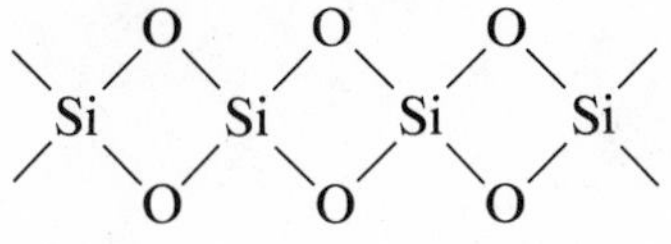

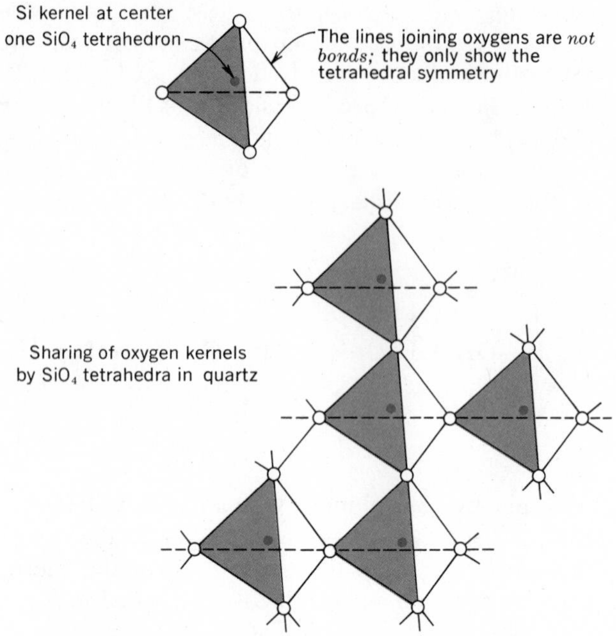

Si kernel at center
one SiO₄ tetrahedron

The lines joining oxygens are *not bonds;* they only show the tetrahedral symmetry

Sharing of oxygen kernels
by SiO₄ tetrahedra in quartz

FIG. 12.3 SiO₄ tetrahedra in quartz (SiO₂).

The silicones, used in waxes, silicone rubber, and cosmetics, also contain repeating Si—O—Si—O units, but the other two bonds of each silicon are to hydrocarbon units. *Methyl* silicones are long-chain molecules with formula $[(CH_3)_2SiO]_n$ in which the repeating structural unit is the one shown here:

$$\left[\begin{array}{cc} CH_3 & CH_3 \\ | & | \\ -Si-O-Si-O- \\ | & | \\ CH_3 & CH_3 \end{array} \right]_n$$

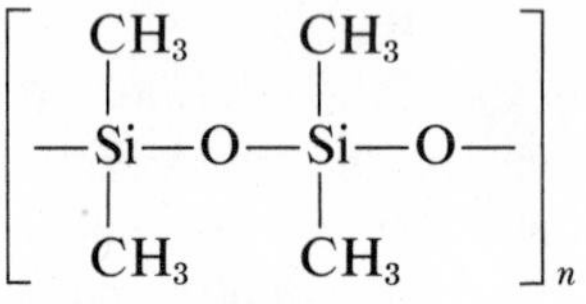

A methyl silicone

The free valences at the ends are joined to H atoms or hydrocarbon groups. Varying the chain length and hydrocarbon groups produces silicone oils and waxes and silicone rubber.

The unwillingness of third-period elements to form double or triple bonds is vividly shown by the failure of attempts to prepare *silicoacetone,* the silicon equivalent of acetone:

$$\begin{array}{c} CH_3 \\ \diagdown \\ \quad Si{=}O \\ \diagup \\ CH_3 \end{array} \qquad \begin{array}{c} CH_3 \\ \diagdown \\ \quad C{=}O \\ \diagup \\ CH_3 \end{array}$$

Silicoacetone Acetone
(unknown)

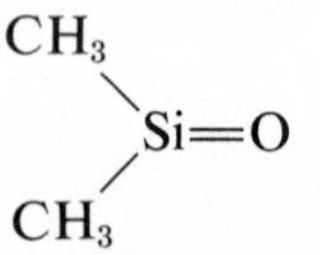

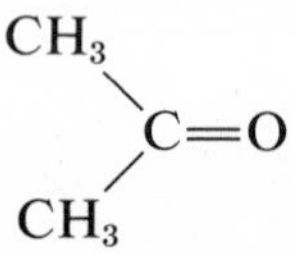

Acetone is a stable liquid (bp 56°C) that shows no tendency to line up (polymerize) to give the following equivalent of a silicone:

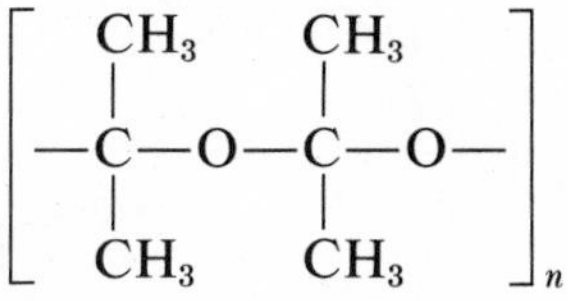

$$\left[\begin{array}{ccc} & CH_3 & CH_3 \\ & | & | \\ -C & -O-C & -O- \\ & | & | \\ & CH_3 & CH_3 \end{array}\right]_n$$

The $\diagdown\!C\!=\!O$ double bond is quite stable. On the other hand, silicoacetone has never been prepared. All efforts to prepare silicoacetone lead to methyl silicones, showing that the $\diagdown\!Si\!=\!O$ unit is quite unstable.

The guides to the molecular architecture expected for compounds of nonmetals and metalloids are covalent bonding directed in space, self-linking of their atoms, and bonding through oxygen, along with the unwillingness of third- and higher-period elements (Si, P, S, Cl, Ge, As, Se, Br) to form double bonds. These guides will help you organize the chemistry of the more important nonmetals and metalloids that follows.

12.2 Oxygen Much of the chemistry of oxygen has been discussed in earlier chapters—its preparation from liquid air by fractional distillation, its part in combustion and smog, its use in steel production, its use as an oxidizing agent in electrochemical fuel cells, and its importance in respiration. Production in the U.S. in 1971 was 26 billion pounds, making it third in the weight of chemicals produced— only ammonia (for fertilizer, mainly) and sulfuric acid (for detergents, among other uses) were made in larger amounts.

Oxygen is the most important nonmetal because it reacts with almost all of the other elements to form their oxides, which are the starting materials for acids, bases, and salts.

Oxides are classified as *acidic, basic,* or *amphoteric.* Acidic oxides react with water to form proton acids and are called *acid anhydrides* for this reason. Most of the nonmetal oxides are of the acidic type. Here are some examples, with their acids:

The word "anhydride" comes from "anhydrous," meaning "without water."

$$SO_3(l) \quad + \quad H_2O(l) \quad \rightarrow \quad H_2SO_4(l)$$

Sulfuric anhydride

Sulfuric acid,
mp 10.4°C,
bp 338°C

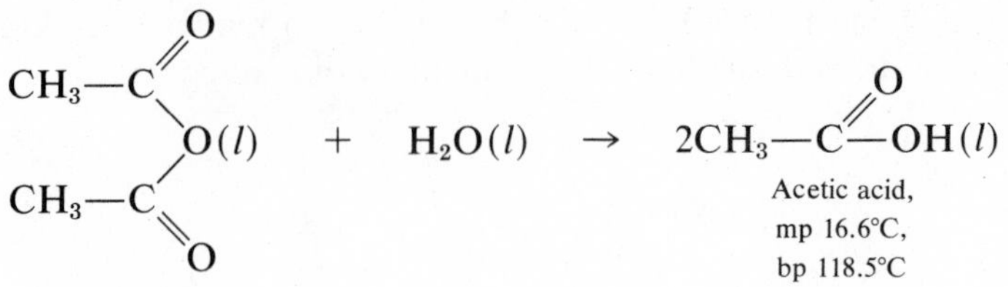

$$CH_3-C(\!\!=\!\!O)(O)\ ; \quad CH_3-C(\!\!=\!\!O)(O)(l) \quad + \quad H_2O(l) \quad \rightarrow \quad 2CH_3-C(\!\!=\!\!O)-OH(l)$$

Acetic acid,
mp 16.6°C,
bp 118.5°C

Acetic anhydride

$$N_2O_5(s) \quad + \quad H_2O(l) \quad \rightarrow \quad 2HNO_3(l)$$

Nitric anhydride
(nitrogen pentoxide)

Nitric acid,
mp −42°C,
bp 83°C

$$P_2O_5(s) \quad + \quad 3H_2O(l) \quad \rightarrow \quad 2H_3PO_4(l)$$

Phosphoric anhydride
(phosphorus pentoxide)

Orthophosphoric acid,
mp 42.4°C,
bp 213°C

The acidic oxides are not as important as the proton acids they produce, and only sulfuric acid and phosphoric acid are actually prepared
from their anhydrides in industry — the others are prepared indirectly
by less expensive means.

EXAMPLE ☞ Nitric acid for making fertilizer, dyes, and explosives is produced
by platinum-catalyzed air oxidation of ammonia in the Ostwald process. The overall equation is

$$NH_3(g) + 2O_2(g) \xrightarrow{Pt} HNO_3(l) + H_2O(l)$$

The ammonia comes from the Haber process (Section 7.4). The nitric
acid is used in the production of chemical explosives such as nitroglycerine, trinitrotoluene (TNT), and nitrocellulose (for smokeless
powder). Much of the HNO_3 is converted to ammonium nitrate for use
as fertilizer and explosive:

$$NH_3(g) + HNO_3(l) \rightarrow NH_4NO_3$$

Acetic anhydride is produced from acetic acid rather than the other
way around. One way is to use the tremendous affinity of P_2O_5 for
water to dehydrate (remove water from) acetic acid.

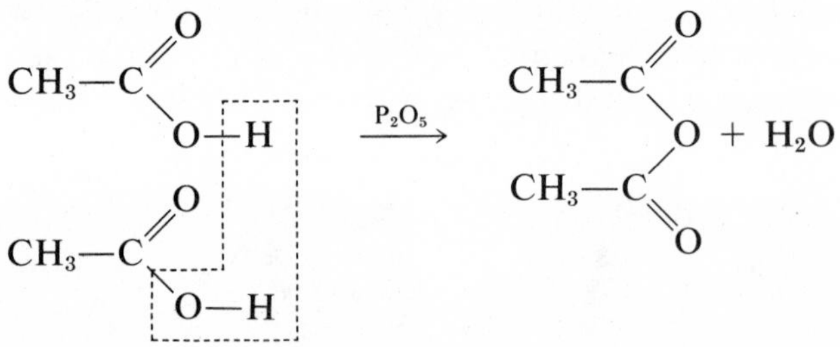

The P_2O_5 is a catalyst but also takes up the water by forming H_3PO_4:

$$P_2O_5(s) + 3H_2O(l) \rightarrow 2H_3PO_4(s)$$

Exercises

1. Give the formulas of the anhydrides that correspond to the following acids:

   Sulfurous acid (H_2SO_3)
   Carbonic acid (H_2CO_3)
   Silicic acid (H_2SiO_3)
   Pyrophosphoric acid ($H_4P_2O_7$)
   Perchloric acid ($HClO_4$)

2. Using the rules in Section 11.1, determine the oxidation states of the nonmetals in the acids and anhydrides in Exercise 1.
3. The annual production of sulfuric acid in 1971 in the U.S. alone was 58.6 billion pounds. Calculate the number of pounds of sulfuric acid per person produced each year in the U.S., assuming a population of 200 million. What would the world production have to be if the whole population of the earth (4 billion) had the same per capita amount of H_2SO_4 per year? How many pounds of sulfur would be needed?

The typical basic and amphoteric oxides of the octet *metals* were covered in Chapter 11. Very few nonmetal or metalloid oxides are either basic or amphoteric; most are acidic. Germanium dioxide [germanium(IV) oxide], GeO_2, a white crystalline solid with mp 1115°C, is amphoteric, as shown by its reactions with acids and bases:

$$GeO_2(s) + 4HCl(aq) \rightarrow GeCl_4(l) + 2H_2O(l)$$
$$GeO_2(s) + 2NaOH(aq) \rightarrow Na_2GeO_3(s) + H_2O(l)$$

The solubility of GeO_2 in water is only 0.5 g/100 ml, but it reacts readily with concentrated hydrochloric acid and dissolves in sodium hydroxide solution according to these equations. Silicon, just above germanium in the Periodic Table, also forms an oxide of the same formula type, SiO_2, with mp 2230°C. But silicon dioxide (quartz, white sand, or silica) is not amphoteric, only acidic. It dissolves slowly in hot sodium hydroxide solution to form a viscous solution of sodium silicate called "water-glass" that is used as an adhesive and for fireproofing wood:

$$SiO_2(s) + 2NaOH(aq) \rightarrow Na_2SiO_3(aq) + H_2O(l)$$

Surprisingly, although SiO_2 doesn't react with acids in general, it does behave like a basic oxide to hydrofluoric acid:

$$SiO_2(s) + 4HF(aq) \rightarrow SiF_4(g) + 2H_2O(l)$$

Silicon tetrafluoride, SiF_4, is covalent, as shown by its remarkably low boiling point of −86°C. Glass is largely SiO_2, and because of this unusual reaction, hydrofluoric acid actually dissolves glass. It is not the acid strength of HF ($K_a = 3.5 \times 10^{-4}$) that makes the reaction go, but instead the unusual thermodynamic stability of the SiF_4 molecule. For its formation reaction, $Si(s) + 2F_2(g) \rightarrow SiF_4(g)$, the value of $\Delta G^0 = -390$ kcal—a huge negative value for a compound of only two elements.

Glass is a complex mixture of Na_2SiO_3 and $CaSiO_3$ prepared by melting together Na_2CO_3, CaO, and SiO_2. In its reaction with HF it can be taken to be SiO_2.

EXAMPLE

Because it dissolves glass, HF is used to etch the graduations into burets, graduated cylinders, and other laboratory glassware. First the glass is covered with paraffin wax (which HF doesn't attack at all), next the desired markings are scratched through the wax to the glass, and then the object is dipped into aqueous hydrofluoric acid for a minute or so. Rinsing and melting off the wax finishes the process. If you should ever try to etch glass, don't get any HF on your skin or eyes. It is a terrible enzyme poison and produces deep burns that take forever to heal. Etching by HF is also used to remove the SiO_2 coating on silicon that goes into transistors and other solid-state electronic devices.

Exercises

4. The equation for the dissolving of $CO_2(g)$ in sodium hydroxide solution is just like the ones for $GeO_2(s)$ and $SiO_2(s)$. Give the equation and name the product. By what process is it produced commercially?

5. Account for the fact that carbon dioxide is a gas at ordinary temperatures while both GeO_2 and SiO_2, from the same group (4a) of elements, are high-melting-point solids.

6. You are given the following thermodynamic data in kcal:

$$Si(s) + O_2(g) \rightarrow SiO_2(s) \qquad \Delta G^0 = -192$$
$$H_2(g) + F_2(g) \rightarrow 2HF(g) \qquad \Delta G^0 = -129$$
$$2H_2(g) + O_2(g) \rightarrow 2H_2O(l) \qquad \Delta G^0 = -113$$
$$Si(s) + 2F_2(g) \rightarrow SiF_4(g) \qquad \Delta G^0 = -360$$

Calculate ΔG^0 for the reaction by which HF dissolves SiO_2. Does it explain why the reaction goes? What is the value of K for the reaction? (*Hints*: arrange the reactions, reversing them if necessary, so that they add up to the glass-dissolving reaction. Remember to change the sign of ΔG^0 if the reaction is reversed.)

12.3 The Hydroxyl Scheme and Oxyacids If you learn what is called the "hydroxyl scheme," you will have much less difficulty remembering or figuring out the formulas of oxyacids. You will also find the scheme useful in organic chemistry. It starts with the positive ion corresponding to the oxidation state of the central kernel in the oxyacid—this is often the group number for the atom in the Octet Periodic Table. You then imagine that a number of OH^- ions equal to the charge on the central kernel are covalently bound to it as ligands. Take the oxyacid of nitrogen with N^{+5} as the central kernel for an example. The first step is to attach five OH^- ions:

$$N^{+5} + 5OH^- \rightarrow \quad \begin{array}{c} HO \quad OH \\ \diagdown \quad | \\ N-OH \\ \diagup \quad | \\ HO \quad OH \end{array}$$

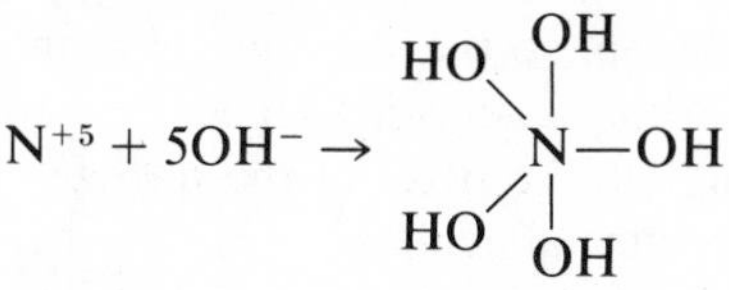

There is no real molecule with this structure, but you can get the correct formula for nitric acid (HNO_3) from it by splitting out two molecules of water as shown in the following by the dashed boxes:

$$\begin{array}{c} HO \quad OH \\ \diagdown \quad | \\ N-OH \\ \diagup \\ HO \quad OH \end{array} \xrightarrow{-2H_2O} \quad \begin{array}{c} O \\ \diagup \\ HO-N \\ \diagdown \\ O \end{array} \quad \text{or} \quad HNO_3$$

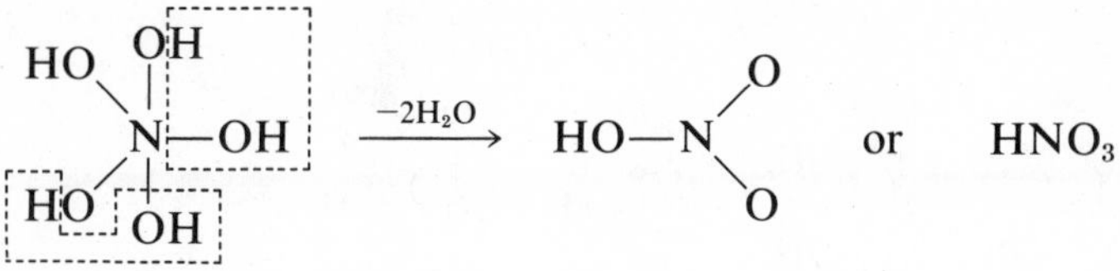

In HNO_3, the hydrogen really is linked through oxygen to the nitrogen as the scheme shows, and according to the rules in Section 11.1, the oxidation state of N is $+5$: three oxygens make $3(-2)$ or -6, one hydrogen is $+1$, so the nitrogen must be $+5$ to make the sum zero as required $(-6+1+5=0)$. The hydroxyl scheme doesn't give you the Lewis dot structure directly but you can usually get it by increasing the bonding to one of the oxygens that carries no hydrogen.

$$\begin{array}{c} O \\ \diagup \\ H-O-N \\ \diagdown \\ O \end{array} \rightarrow \begin{array}{c} O \\ \diagup\!\diagup \\ H-O-N \\ \diagdown \\ O \end{array} \quad \text{or} \quad \begin{array}{c} \ddot{O} \\ \diagup\!\diagup \\ H-\ddot{O}-N \\ \diagdown \\ \ddot{O} \end{array}$$

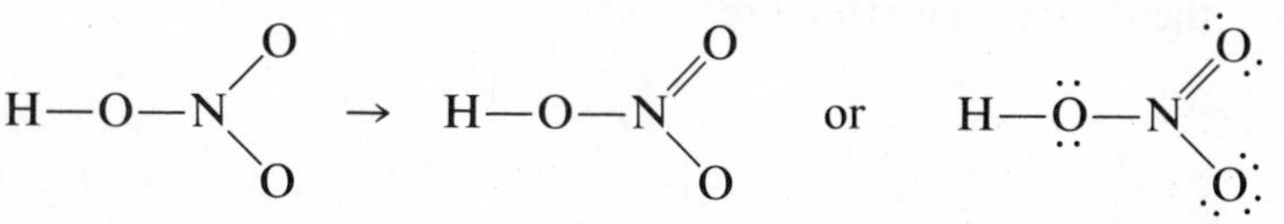

Nitrogen forms another oxyacid, nit*rous* acid, in which the oxidation state of N is $+3$ instead of $+5$. Following the scheme to get its formula, we write

$$N^{+3} + 3OH^- \rightarrow HO-N\begin{matrix} OH \\ \\ OH \end{matrix} \xrightarrow{-H_2O} HO-N\diagup^O \quad \text{or} \quad HNO_2$$

$$H-O-N\diagup^O \rightarrow H-\ddot{O}-N\diagup^{\cdot\ddot{O}\cdot}$$

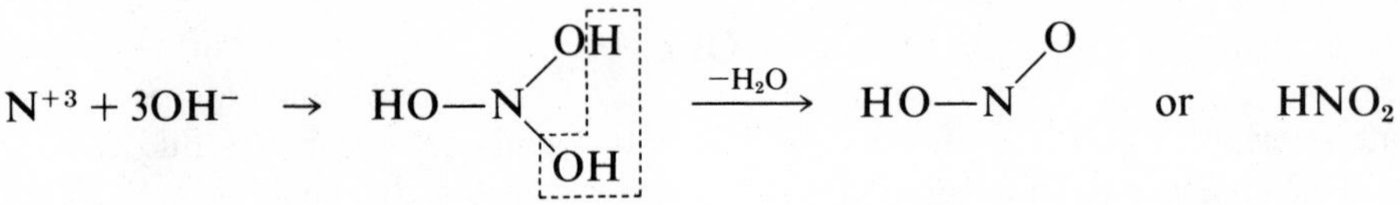

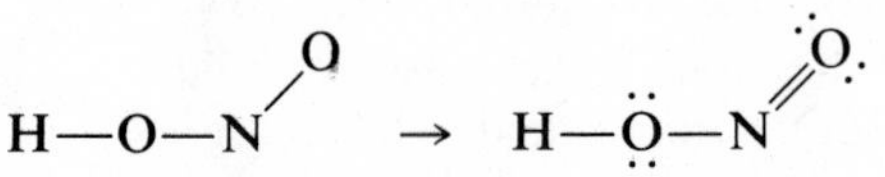

But this last structure doesn't have an octet around nitrogen. Have we shown all of the valence electrons as required? The structure shows eight e^- in bonds and eight e^- in unshared pairs, for a total of sixteen. But H has one valence electron, nitrogen has five, and the two oxygens contribute twelve, making a total of eighteen, so the structure is short two. The only place you can put them without exceeding an octet is on N, and if you put them there they complete the octet of the nitrogen and produce the correct dot formula for nitrous acid:

$$H-\ddot{O}-N\diagup^{\cdot\ddot{O}\cdot} + 2e^- \rightarrow H-\ddot{O}-\ddot{N}\diagup^{\cdot\ddot{O}\cdot}$$

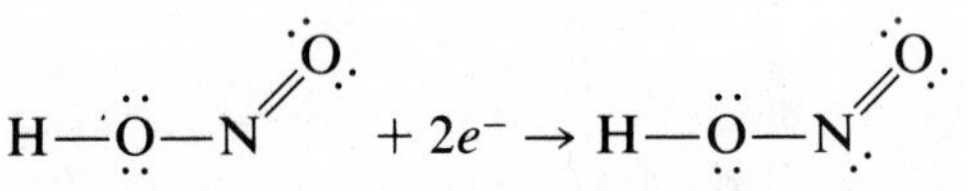

You can get the formulas of sulfuric and sulfurous acids in a similar way:

$$S^{+6} + 6OH^- \rightarrow \begin{matrix} & OH & \\ HO & | & OH \\ & S & \\ HO & | & OH \\ & OH & \end{matrix} \xrightarrow{-2H_2O} \begin{matrix} O \\ \| \\ HO-S-OH \\ | \\ O \end{matrix}$$

Sulfuric acid

$$S^{+4} + 4OH^- \rightarrow \begin{matrix} OH \\ | \\ HO-S-OH \\ | \\ OH \end{matrix} \xrightarrow{-H_2O} HO-S\begin{matrix} OH \\ \diagup \\ \diagdown \\ O \end{matrix}$$

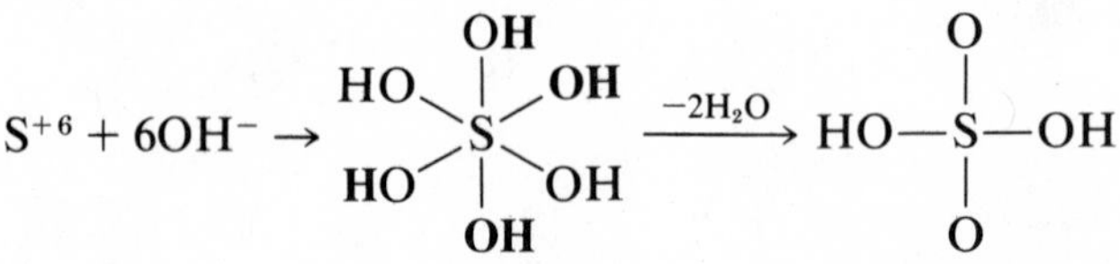

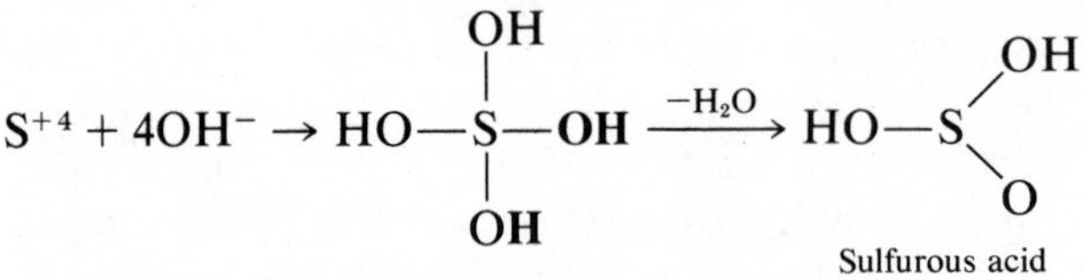

Sulfurous acid

Putting in the valence electrons left over from the bonds to make octets gives the correct dot structures:

$$\begin{matrix} :\ddot{O}: \\ | \\ H-\ddot{O}-S-\ddot{O}-H \\ | \\ :O: \end{matrix} \qquad \begin{matrix} :\ddot{O}: \\ | \\ H-\ddot{O}-\ddot{S}-\ddot{O}-H \end{matrix}$$

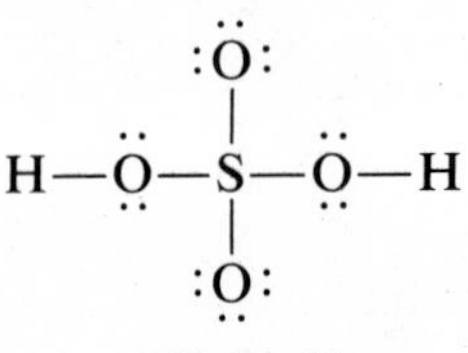

Sulfuric acid Sulfurous acid

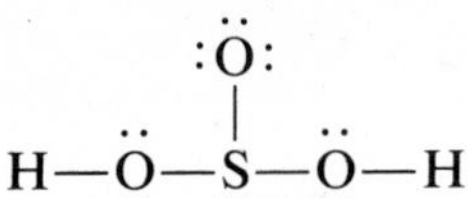

Sulfuric and nitric acids are both strong acids; nitrous and sulfurous acids are weak. For HNO_2, the $pK_a = 3.29$ and for H_2SO_3, the $pK_{a_1} = 1.81$ and $pK_{a_2} = 6.91$. This relationship between the strengths of "ous" and "ic" oxyacids of an element is often true—the acid in which the central atom has the higher oxidation state, the "ic" acid, is stronger. Thus, arsen*ic* acid, H_3AsO_4, with $pK_{a_1} = 2.30$ is stronger than arsen*ous* acid, H_3AsO_3, with $pK_{a_1} = 9.22$.

Sulfuric and nitric acids can be obtained as pure compounds, but nitrous and sulfurous acids are known only in solution. They decompose to their anhydrides if you try to isolate them by boiling off the water. Writing $-H_2O$ over the arrow to show that water must be removed, we have the equations

$$H_2SO_3(aq) \xrightarrow{-H_2O} \underset{\substack{\text{Sulfurous anhydride} \\ \text{(sulfur dioxide)}}}{SO_2(g)}$$

$$2HNO_2(aq) \xrightarrow{-2H_2O} \underset{\substack{\text{Nitrous anhydride} \\ \text{(unstable)}}}{[N_2O_3]} \rightarrow \underset{\text{Nitric oxide}}{NO(g)} + \underset{\text{Nitrogen dioxide}}{NO_2(g)}$$

The *salts* of nitrous and sulfurous acids are well-known, stable compounds. Examples are sodium nitr*ite*, $Na^+NO_2^-$; sodium bisulf*ite* (sodium hydrogen sulf*ite*), $Na^+HSO_3^-$; and sodium sulf*ite*, $(Na^+)_2SO_3^{2-}$. Notice that in naming the salts of these acids the "ite" salt comes from the "ous" acid, the "ate" salt from the "ic" acid. This system is generally used for naming the salts of oxyacids. For example, chlorine forms four oxyacids, and the naming of them and their salts is as follows:

Name of Acid	*Formula of Acid*	*Formula of Salt*	*Name of Salt*
Hypochlor*ous* acid	$HOCl$	Na^+ClO^-	Sodium hypochlor*ite*
Chlor*ous* acid	$HOClO$	$Na^+ClO_2^-$	Sodium chlor*ite*
Chlor*ic* acid	$HOClO_2$	$K^+ClO_3^-$	Potassium chlor*ate*
Perchlor*ic* acid	$HOClO_3$	$K^+ClO_4^-$	Potassium perchlor*ate*

The formulas for the acids are written here to show that H is linked through oxygen to Cl. The formulas are usually written $HClO$, $HClO_2$, $HClO_3$, and $HClO_4$, and their salts written accordingly, as shown. Of the acids, only perchloric acid can be isolated as a pure compound. Salts of the other three are well known, however. Sodium hypochlorite is used for sanitation and as a bleach, sodium chlorite for bleaching wood pulp in paper manufacture, potassium chlorate as a weed killer, and potassium perchlorate as the oxidizing agent in matches and fireworks.

Exercises

7. Is the electron-dot structure obtained for nitric acid the only one possible? What about resonance hybridization?

8. Determine the oxidation states of chlorine in its four oxyacids. Can you relate them to the naming of the acids? (*Hint*: "hypo" means "below," "per" means "above.")

9. Using the electrostatic rules, predict the O—N—O bond angles in nitrate (NO_3^-), nitrite (NO_2^-), and sulfite (SO_3^{2-}) ions.

10. Arrange the oxyacids of chlorine in order of increasing strength.

11. Using the hydroxyl scheme, derive the formulas of the oxyacids of chlorine.

12. Derive the formulas of the anhydrides of the oxyacids of chlorine. (All are known—they are explosive gases.) Which would certainly be paramagnetic?

13. What is the oxidation state of phosphorus in pyrophosphoric acid, $H_4P_2O_7$? There is a —P—O—P— type of bonding in this molecule. Knowing this and using the hydroxyl scheme, obtain the bond structure for pyrophosphoric acid. Where have you seen the —P—O—P— unit before?

14. What is the oxidation state of manganese in the permanganate ion MnO_4^-? Using the hydroxyl scheme, derive the formula of permanganic acid. Permanganic acid exists only in solution, but its anhydride, though explosively unstable, is known. Derive its formula.

12.4 Compounds of Nonmetals and Metalloids with Hydrogen Except for the noble gases (Group 8), all of the nonmetals and metalloids form compounds with hydrogen. The compounds are covalent molecules, and you have seen most of them in earlier chapters. The hydrogen halides (HF, HCl, HBr, HI), water and hydrogen sulfide (H_2S), and ammonia and methane (CH_4) are familiar examples. The compounds with hydrogen fall into groups according to their acidic and basic properties, and the trends in the strength of these properties can be predicted from the positions of the elements in the Periodic Table. In general, acid strength increases from left to right across a period and downward in a group. In this section we'll discuss the preparations and properties of the more important compounds of hydrogen with nonmetals and metalloids.

TABLE 12.1 Properties and preparation of the hydrogen halides

Hydrogen Halide (HX)	*HF*	*HCl*	*HBr*	*HI*
Melting point	−83.1°C	−114.8°C	−88.5°C	−50.8°C
Boiling point	19.5°C	−84.9°C	−67.0°C	−35.4°C
Molar entropy of vaporization at boiling point	25 cal/deg	21 cal/deg	20 cal/deg	20 cal/deg
Bond dissociation energy, $HX(g) \rightarrow H(g) + X(g)$	134 kcal/mole	102 kcal/mole	86 kcal/mole	70 kcal/mole
Dipole moment of gaseous form, in debye units	1.91	1.08	0.80	0.42
Acid strength in water (pK_a)	3.25	−7.4	−9.0	−9.5
Acid strength in acetic acid as solvent	weakest ⟶			⟶ strongest
Preparations	$CaF_2(s) + H_2SO_4(l) = 2HF(g) + CaSO_4(s)$	$2NaCl(s) + H_2SO_4(l) = 2HCl(g) + Na_2SO_4$ $H_2(g) + Cl_2(g) = 2HCl(g)$	$H_2(g) + Br_2(g) = 2HBr(g)$	$2P(s) + 3I_2(s) = 2PI_3(s)$ $PI_3(s) + 3H_2O(l) = H_3PO_3(aq) + 3HI(aq)$

The class of strong acids—the hydrogen halides or HX compounds The preparations and many of the properties of the hydrogen halides are given in Table 12.1. Hydrogen fluoride stands out from the others in two ways—it has the highest boiling point of all, and it is the only weak acid ($pK_a = 3.3$) of the group. The high boiling point (and high molar entropy of evaporation) are both due to strong hydrogen bonding between H—F molecules (Section 4.6). The hydrogen bonding even extends to the gaseous form of the compound. Molecular spectroscopy shows that there are $(H—F)_5$ chains in the gas. Using dashed bonds for the hydrogen bonds, we can draw the structure as follows:

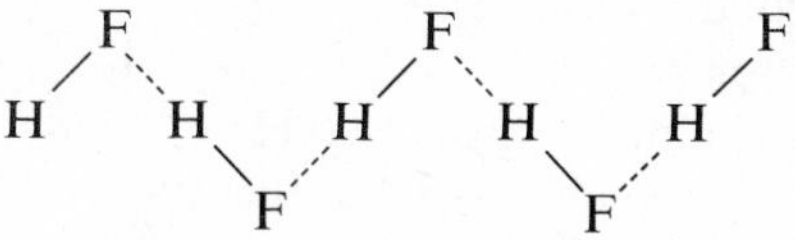

In the chains the H—F distance is 0.92 A, the hydrogen bond distance F----H is 2.6 Å, and the H—F----H bond angle is about 140°. The strong hydrogen bonding makes the reaction of H—F with water different from the reactions of the other hydrogen halides. They react, as you have seen before, in this way:

$$H—X + H_2O \rightarrow H_3O^+ + X^-$$

but the reaction for H—F is more like the following, where $(H—F)_2$ stands for a pair of hydrogen-bonded molecules:

$$(H—F)_2 + H_2O \rightleftarrows H_3O^+ + [F\text{----}H\text{----}F]^-$$

The $[F\text{----}H\text{----}F]^-$ or HF_2^- ion is called the bifluoride ion, and salts like potassium bifluoride $K^+HF_2^-$ are known. The other hydrogen halides do not show this behavior. Acidic strength increases from HF to HI, as the pK_a values in Table 12.1 show. Hydriodic acid (HI in water solution), with p$K_a = -9.5$, is the strongest proton acid made of only two elements.

The electronegativities of Cl, Br, and I are not large enough to produce hydrogen bonding in HCl, HBr, and HI, so their boiling points are lower, and their molar entropies of vaporization agree with Trouton's rule (Section 5.7).

The class of weak acids—the H_2Y compounds The compounds of the Group 6a nonmetals with hydrogen are much weaker acids than the neighboring hydrogen halides in the same period. Water is the weakest of the group, with p$K_a = 15.7$, and acidic strength increases in a regular way as you go down the group: H_2S, pK_a = 7.1; H_2Se (hydrogen selenide), pK_a = 3.80; H_2Te (hydrogen telluride), pK_a = 2.64. Strong hydrogen bonding between water molecules in liquid H_2O is responsible for its unusually high boiling point of 100°C. The other compounds in the group all boil below 0°C, showing that they are not hydrogen-bonded.

Water acts as a base toward strong acids, as we saw in Chapter 8. It is a very weak base, however, because the value of its K_b is only 1.8×10^{-16}. The other members of the group do not show any basic properties.

The class of weak bases—the H_3Z compounds By the time the Group 5a elements (N, P, As, Sb, Bi) are reached, any acidic properties of their hydrogen compounds have virtually disappeared. For NH_3, the p$K_a \approx 33$. However, the value of K_b for NH_3 is 1.8×10^{-5}, so NH_3 is a typical weak base that readily forms ammonium salts ($NH_4^+A^-$) with proton acids (HA) of p$K_a < 7$. You have seen several of these ammonium salts before, NH_4Cl, NH_4NO_3, and $(NH_4)_2SO_4$ among them. Phosphine (PH_3) behaves the same way, forming phosphonium salts like phosphonium iodide, $PH_4^+I^-$, but only with concentrated solutions of proton acids because PH_3 is a weaker base than NH_3. The H_3Z compounds of the lower members of Group 5a don't show significant basic properties. Like phosphine, they have vile odors and are very poisonous.

Ammonia is only weakly hydrogen-bonded in the liquid (bp −33°C). Hydrogen bonding between ammonia molecules and water molecules, $\left(H_3N\cdots H-O\diagdown_H\right)$ is strong, however, and is responsible for the very great solubility of ammonia in water: 47.5 g of NH_3 dissolve in 100 g of H_2O at 0°C. (The solubility of O_2 is 0.007 g/100 g of H_2O at 0°C.) Solutions of ammonia in water are often labeled "ammonium hydroxide" with the formula NH_4OH on the bottle. Although solutions of ammonia in water do contain small concentrations of NH_4^+ and OH^- ions, the ions are a result of the reaction $NH_3(aq) + H_2O(l) \rightleftarrows NH_4^+(aq) + OH^-(aq)$ and do not come from the ionization reaction $NH_4OH(aq) \rightleftarrows NH_4^+(aq) + OH^-(aq)$. There is no NH_4OH molecule.

The class of neutral compounds—the H₄Q compounds The nonmetals of Group 4a—carbon (C), silicon (Si), and germanium (Ge)—all form covalent molecules containing four hydrogens. They are methane (CH_4), silane (SiH_4), and germane (GeH_4), with respective boiling points of −162, −112, and −89°C. The low boiling points show that these compounds are not hydrogen-bonded in the liquid. Unlike all of the other nonmetal hydrides, the H_4Q type has no unshared pairs of valence electrons and so cannot act as a base.

$$H-\overset{..}{\underset{}{X}}: \qquad H-\overset{..}{Y}\diagdown_H \qquad H-\overset{\overset{H}{|}}{\underset{\underset{H}{|}}{Z}}: \qquad H-\overset{\overset{H}{|}}{\underset{\underset{H}{|}}{Q}}-H$$

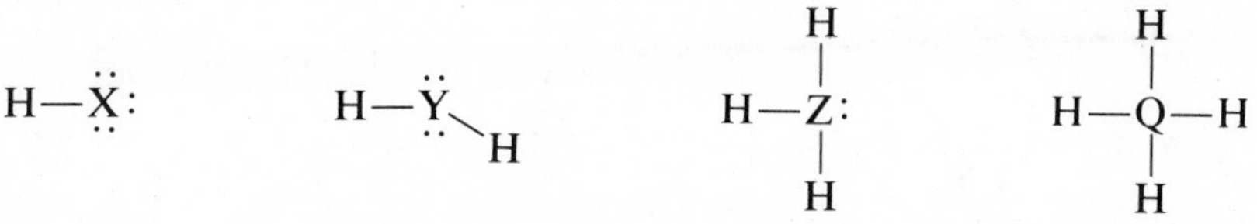

The electronegativities range from 2.00 for Ge to 2.60 for C, and because the value for H is 2.20 (Table 3.4) the H—Q bonds have very little polar character. Although polar character in covalent bonds is not always a good measure of acidic strength, ☛ molecules with nonpolar covalent bonds usually have no significant acidic properties. This is true of methane, silane, and germane. Silane does, however, react with water containing OH^- as a catalyst to produce hydrogen and silicon dioxide:

$$SiH_4(g) + 2H_2O(l) \xrightarrow{OH^-} SiO_2(s) + 4H_2(g)$$

The great stability of the solid SiO_2 formed is the driving force for this reaction. Methane (CH_4) does not react this way.

Hydrofluoric acid (HF), with more polar character (larger dipole moment, Table 12.1) than HCl, is the weaker acid of the two.

Carbon, silicon, and germanium atoms all show a tendency to self-link to produce such molecules as $H_3Q—QH_3$ and $H_3Q—QH_2—QH_3$. With silicon, compounds of this type have been prepared up to Si_6H_{14}, and with germanium up to Ge_3H_8. With carbon, self-linking can go on indefinitely, so hydrocarbons of the type $CH_3—(CH_2)_n—CH_3$, where $n \approx 70{,}000$, are possible. Polyethylene, which you know as the material of the "squeeze-bottle," is an example of virtually indefinite self-linking of carbon atoms, and so are diamond and graphite. Unlike silicon and germanium, carbon also self-links through double and triple bonds and can form covalent bonds to any of the nonmetals (noble gases excepted).

Exercises

15. Give electron-dot structures for phosphine (PH_3), arsine (AsH_3), hydrogen selenide (H_2Se), and hydrogen telluride (H_2Te), indicating the principal quantum number (n) of the valence electrons of the nonmetal atom in each case. (Use Table 3.1.)

16. Tin and lead, though metals, form compounds with hydrogen called stannane and plumbane, respectively. Predict their molecular formulas and geometry from the position of the elements in the Periodic Table. Would you expect them to have boiling points higher or lower than that for germane? Why?

17. Why don't the compounds of the H_4Q group hydrogen-bond?

18. For the reaction $CH_4(g) + 2H_2O(l) \rightarrow CO_2(g) + 4H_2(g)$, the $\Delta G^0 = +31.2$ kcal, while for the similar reaction $SiH_4(g) + 2H_2O(l) \rightarrow SiO_2(s) + 4H_2(g)$, the $\Delta G^0 = -69$ kcal. Which reaction is thermodynamically spontaneous? Use the following data to explain why:

$$C(s) + O_2(g) \rightarrow CO_2(g) \qquad \Delta G^0 = -94.3 \text{ kcal}$$

$$Si(s) + O_2(g) \rightarrow SiO_2(s) \qquad \Delta G^0 = -192 \text{ kcal}$$

12.5 The Nonmetallic Elements The sources and methods for preparing some of the more important nonmetallic elements have been given before—O_2 and N_2 and the noble gases are produced by fractional distillation of liquid air; Cl_2 is a by-product of the electrolysis of sodium chloride in the production of sodium and sodium hydroxide. Now let's look at phosphorus, sulfur, and the rest of the halogens, the remaining important nonmetals.

Phosphorus Phosphorus occurs naturally in the form of tricalcium phosphate [$Ca_3(PO_4)_2$] deposits. It is prepared by heating tricalcium phosphate with sand (SiO_2) and carbon (coke) at 1500°C in an electric furnace. The reaction goes in two steps:

$$2Ca_3(PO_4)_2(s) + 6SiO_2(s) \rightarrow P_4O_{10}(g) + 6CaSiO_3(s)$$
$$P_4O_{10}(g) + 10C(s) \rightarrow P_4(g) + 10CO(g)$$

The gas of P_4 molecules is led out of the furnace and collected under water, where it condenses to a yellowish, waxy solid called white phosphorus. White phosphorus bursts into flame on contact with air (spontaneous combustion) at about 35°C and burns to the pentoxide, ☛ so it is stored under water. When touched to the skin it produces terrible burns, so it must be treated with great caution—never touched with the fingers. When white phosphorus (mp 44°C) is heated at 250°C in the absence of air, it becomes *red* phosphorus, which is much less reactive. Red and white phosphorus are electrically nonconducting, but when the white form is subjected to very high pressure, it changes to *black* phosphorus, which has a layered structure like graphite (Figure 11.1) and conducts electricity like graphite. These forms of phosphorus—white, red, black—are called the *allotropes* of the element. *Different molecular forms of any element are called its allotropes.* The allotropes of oxygen are O_2 and O_3.

The gaseous form of phosphorus pentoxide, P_2O_5, is P_4O_{10} molecules, as is one crystalline form.

Most of the white phosphorus prepared is converted to the pentoxide and then to phosphoric acid (H_3PO_4) by reactions described earlier. Red phosphorus mixed with glue, sand, and potassium chlorate is used to make matchheads. When the match is struck, the frictional heat sets off rapid oxidation of the phosphorus by the potassium chlorate, and the heat produced ignites the match.

Phosphorus reacts spontaneously with the halogens to form penta- and trihalides—PCl_5 and PI_3 are examples that are used in organic chemistry. The structures and relationships between some oxides and oxyacids of phosphorus are shown in Figure 12.4.

Growing plants need a supply of phosphorus in the form of phosphates, so they are put in fertilizers such as "bone meal," largely $Ca_3(PO_4)_2$, and other combinations, often with NH_4NO_3 to provide nitrogen as well.

Water-soluble phosphates, particularly sodium trimetaphosphate, $Na_3(PO_3)_3$, are mixed with soaps and detergents to make washing powders for dishes and clothes. The $(PO_3)_3^{3-}$ ion hydrolyzes in water to give an alkaline solution that improves the cleansing action of soaps and detergents:

$$(PO_3)_3^{3-}(aq) + H_2O(l) \rightleftarrows H(PO_3)_3^{2-}(aq) + OH^-(aq)$$

The trimetaphosphate ion, by forming soluble complexes with the Ca^{2+}, Mg^{2+}, Fe^{2+} ions in hard water, softens it to further improve clean-

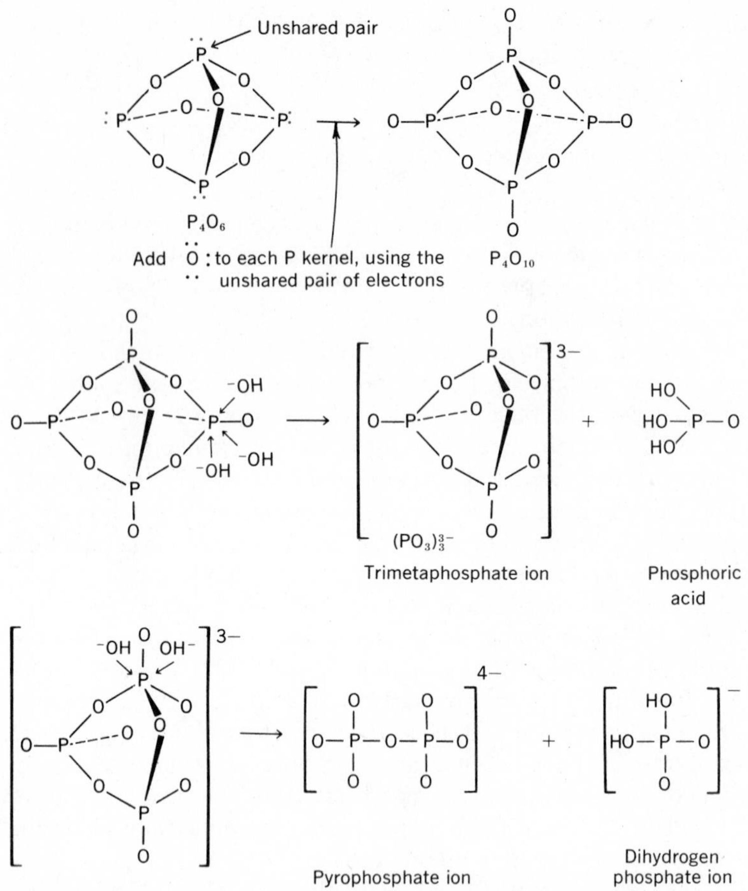

FIG. 12.4 Structures of some oxides and oxyanions of phosphorus.

ing. Helpful as they are in cleaning, soluble phosphates pollute lakes and streams into which waste wash waters flow (Photo 32).

Sulfur Sulfur occurs as the free element in underground deposits in Louisiana and Texas. To get it to the surface, water is heated to 160°C under pressure and pumped down pipes that extend into the deposit. Sulfur melts at 114°C, and the sulfur is melted by the hot water. Other pipes carry it back to the surface. The molten sulfur is run into huge wooden bins (1200 ft × 200 ft × 50 ft), where it solidifies to the yellow solid form, as much as 600,000 tons of sulfur in one chunk. This "mining" method is called the Frasch process after its inventor.

Herman Frasch (1851–1914) was born in Germany and came to the U.S. when he was 17. The sulfur in Louisiana had long been inaccessible because of quicksand when he performed his experimental mining method in a 10-in. tube 625 ft long.

The most important use for the element is in the *contact process* for the production of sulfuric acid (59 billion pounds in the U.S. in 1971). In this process the sulfur is melted, sprayed into a chamber, and burned in air to $SO_2(g)$ much like oil is burned in an oil furnace. After thorough purification to remove catalyst poisons such as As_2O_3, the SO_2

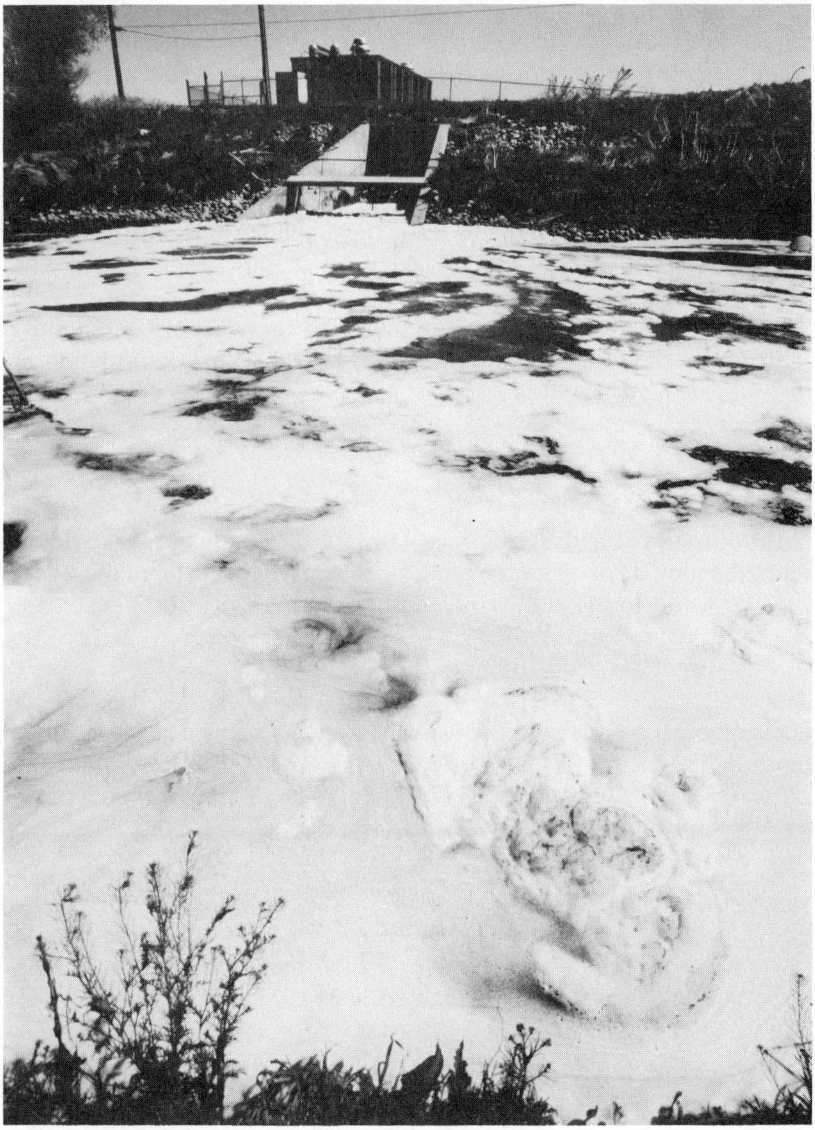

PHOTO 32 Discharge from a sewage plant on the South Platte River in Denver. The incompletely treated sewage contains phosphate and, as you can plainly see, detergent. (*EPA-DOCUMERICA—Bruce McAlister.*)

gas produced in the chamber is mixed with air and led over a catalyst of platinum metal or vanadium pentoxide (V_2O_5) spread out on silica (SiO_2). Oxidation to the trioxide goes rapidly at 500°C in the presence of the catalyst:

$$S(l) + O_2(g) \rightarrow SO_2(g) \qquad K = 10^{53} \text{ at } 25°C$$

$$2SO_2(g) + O_2(g) \xrightarrow{V_2O_5} 2SO_3(g) \qquad K = 10^{14} \text{ at } 25°C$$

Enough air is mixed with the SO_2 to give a large excess of O_2, so the conversion to SO_3 in the second step is more rapid and complete (Le Chatelier's principle). You may find it strange that excess O_2 is necessary when the value of K for the SO_2–O_2–SO_3 equilibrium is so large (10^{14} at 25°C). However, even with a catalyst, equilibrium is not established at all rapidly at 25°C. Instead, temperatures near 500°C are necessary, even with a catalyst, to get to equilibrium rapidly, and at the higher temperature the value of K for the equilibrium is only 6.5×10^4. For this reason, excess O_2 is used to push the equilibrium to the SO_3 side and speed up the reaction.

Some of the SO_3 is condensed to the liquid form (bp 45°C), but most of it is absorbed in concentrated sulfuric acid with the formation of *pyrosulfuric acid*, $H_2S_2O_7$:

$$SO_3(g) + H_2SO_4(l) \rightarrow \underset{\text{Pyrosulfuric acid}}{H_2S_2O_7(l)}$$

Pyrosulfuric acid and SO_3 are used in large quantities to make detergents (Section 11.3). The remainder of the pyroacid is diluted by *adding it to water* to make the concentrated (95%) H_2SO_4 of industry:

$$H_2S_2O_7(l) + H_2O(l) \rightarrow 2H_2SO_4(l)$$

The acid must be added to the water to prevent the heat of the strongly exothermic reaction $H_2S_2O_7 + H_2O \rightarrow H_3O^+ + HS_2O_7^-$ from making the water boil and spattering acid all over. The three "A's" of chemistry are "*Always Add Acid*" *to water* when preparing dilute solutions.

Exercises

19. Phosphorus pentoxide sublimes at 300°C. What method could you use to show that its molecular formula is P_4O_{10} rather than P_2O_5 in the gas state? (*Hint*: recall Al_2Cl_6 and Fe_2Cl_6.)

20. A compound of phosphorus and oxygen has the composition 56.34% P, 43.66% O; when a 6.35-g sample was vaporized in a 2.24-liter bulb at 200°C, it exerted a pressure of 0.50 atm. What are the empirical and molecular formulas of the compound? (Refer to Figure 12.4 to check.) What is the oxidation state of P in the compound?

21. Phosphorus pentachloride vapor consists of covalent PCl_5 molecules, but when condensed to the solid form the bonding becomes the ionic type between the complex PCl_4^+ and PCl_6^- ions. Write an equation for the reaction, give dot structures, and predict the geometry for these two ions. (*Hint*: phosphorus does not always bond according to octet theory.)

22. The vapor pressure of water at 160°C is 4636 torr. Calculate the minimum pressure in atm and psi that must be exerted on the water in the Frasch process to keep it from boiling (1 atm = 14.70 psi).

23. Give another example of heterogeneous catalysis like that in the contact process for making H_2SO_4.

24. For the reaction $2SO_2(g) + O_2(g) \rightleftarrows 2SO_3(g)$, the value of $K = 10^{14}$ at 25°C and 6.50×10^4 at 500°C. Is the reaction exothermic or endothermic? What principle is involved?

25. What is the oxidation state of S in pyrosulfuric acid? Derive the structure of the acid, using the hydroxyl scheme. (*Hint*: recall

—P—O—P— bonding in pyrophosphoric acid.) What is the

formula of the anhydride of pyrosulfuric acid?

Sulfuric acid is also made by the *lead chamber process*, so named because the reaction is a gas-phase one carried out in huge lead-walled chambers (as large as 50,000 cu ft). The process is used to recover SO_2 from the roasting of sulfide ores such as $CuFeS_2$ (Section 11.7). Like the contact process, this process is catalytic, but the catalysis is *homogeneous* (one-phase) rather than heterogeneous, and poisoning of the catalyst cannot occur. The otherwise very slow oxidation of SO_2 to SO_3 at ordinary temperatures is catalyzed by nitrogen dioxide, NO_2. When NO_2, SO_2, air, and a misty spray of water are led into the lead chambers, the reactions are

$$SO_2(g) + H_2O(l) + NO_2(g) \rightarrow H_2SO_4(l) + NO(g)$$

$$2NO(g) + O_2(g) \rightarrow 2NO_2(g)$$

Rapid oxidation of nitric oxide (NO) by air renews the nitrogen dioxide, as shown in the second equation. The oxides of nitrogen are known to act as oxygen "carriers" in this way, but the detailed mechanism of the catalytic reaction at the molecular level is not well understood. Unlike the contact process, the lead chamber process does not produce SO_3, $H_2S_2O_7$, or 95% H_2SO_4, but instead produces only 66% H_2SO_4 (in water solution). This is a major shortcoming, but the process has the advantage of not requiring purified SO_2 and is being used more and more to prevent air pollution by SO_2 from sulfide ore refining.

The largest quantities of sulfuric acid go into petroleum refining, detergent manufacture, and the production of ammonium sulfate for fertilizer:

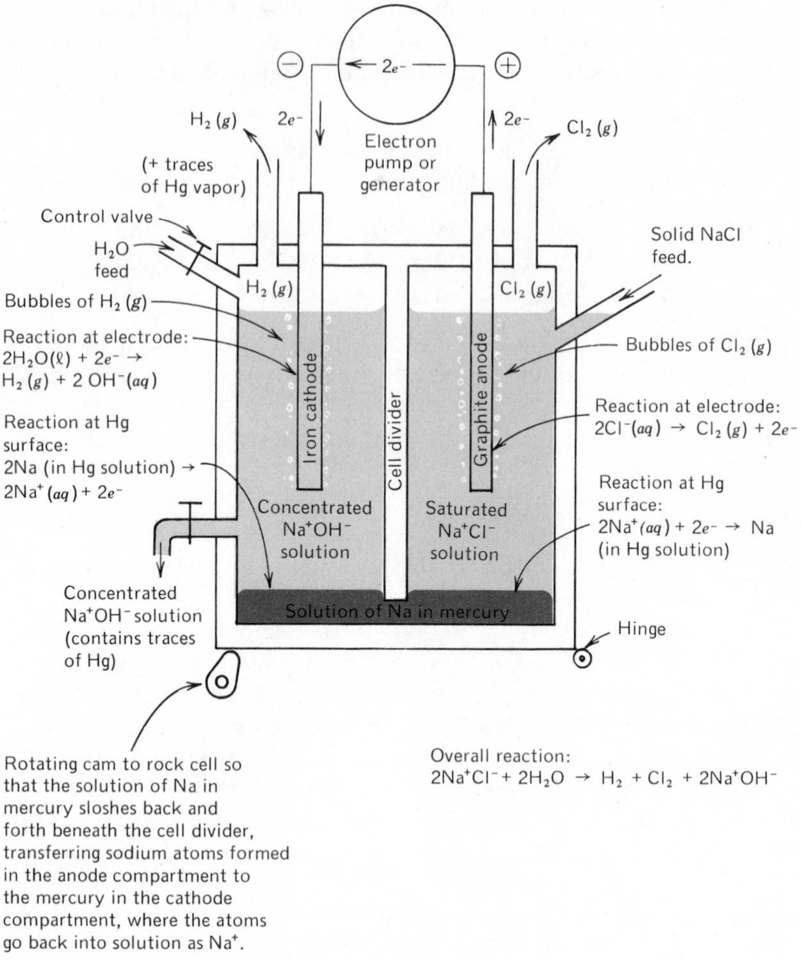

FIG.12.5 Principles of operation of a mercury electrolysis cell making Na^+OH^- from Na^+Cl^-. (Cell produces Na^+OH^- free from Na^+Cl^- by separating anode and cathode compartments, using Hg.)

$$2NH_3(g) + H_2SO_4(l) \rightarrow (NH_4)_2SO_4(s)$$

Large amounts are also used in the automotive and steel industries to remove rust and scale before painting automobile bodies and structural steel.

The amount of H_2SO_4 produced by a country can be used as a "barometer" of industrial activity because H_2SO_4 is used in nearly every industry. To show how industry has grown in the U.S. since 1945 you can compare the production then of 20 billion lb with the more than 60 billion lb produced today.

Astatine (At) is a fiercely radioactive element produced in only trace amounts by nuclear reactions.

Halogens—Group 7a The word "halogen" means "salt former" and comes from the formation of salts such as NaCl, NaBr, and KI between metals and the Group 7a elements (F, Cl, Br, I, At). ☛ You have already seen that electrolysis of "salt" (NaCl) is used to prepare chlorine and sodium hydroxide. This electrolysis (Section 11.2) is an important source of contamination of the environment by mercury when the electrolysis is carried out in a *mercury cell* of the type shown schematically in Figure 12.5. Although the vapor pressure (0.0012 torr at 20°C) and solubility of mercury are small, some of it inevitably gets into the environment from the operation of these cells. It has been estimated that about 0.20 lb of mercury is lost to the environment for every ton (2000 lb) of chlorine produced. This doesn't sound like much mercury until you learn that chlorine production exceeds 10 million *tons* per year in the U.S.—this means as much as 2 million lb of mercury can go into our environment from this source. With a population

CLEAR ZONE

A clean stream has plenty of oxygen, a diversity of organisms and ample food for all of them.

When leaves fall into a stream, that's natural pollution. The leaves and tiny plants such as diatoms are also food substances for animals which eat them and, in so doing, help keep the stream clean.

These animals are food for others called predators.

A clean stream is nature's own waste disposal plant—though poets and romantics might never admit it.

SEPTIC ZONE

Untreated: A discharge into a stream itself is not necessarily harmful. In fact, it can be beneficial by providing nutrients for plant and marine growth.

Stream pollutants include untreated municipal or industrial wastes, runoffs from roads and farms, litter and debris. Too many pollutants upset the natural cycles that work to keep water clean. An abundance of organic material breeds bacteria that use up oxygen. Clean water organisms can't live in this environment.

Barring further pollution, these diseased waters may recover, but they will have to travel many miles before nature makes them healthy again.

Treated: To prevent stream pollution, many industries are investing huge sums in sedimentation tanks, trickling filters, aeration tanks and other pollution abatement facilities.

Proper treatment removes harmful organic and chemical matter so the wastes can safely be discharged into streams. Bacteria in the sewage treatment facilities turn the wastes into nutrients or other substances that will not harm the stream's animal and plant life.

RECOVERY ZONE

One of the wonders of nature is the ability of a stream to cleanse itself, even after it has been polluted. This happens in the Recovery Zone which is downstream—but how far downstream depends on the extent and severity of the pollution.

As it meanders along, the stream is diluted, oxygen is continuously added, and a diversity of plant and animal life begins to reappear. Eventually, the stream reaches a point where the organisms once again can become the stream's own waste disposal plant.

PHOTO 33 The ecology of a stream. The microscopic views (not drawn to scale) shown here depict the busy "world" of a stream. Left alone, a stream maintains itself. We can help keep the stream healthy, or we can cause widespread degradation. The choice is ours. The middle section of the chart (the Septic Zone) explains the alternatives. (*DuPont Context, copyright © 1972 by E. I. du Pont de Nemours and Company; and the Academy of Natural Sciences.*)

of 200 million this amount of Hg comes to 0.01 lb or about 4.5 g per person per year. Liquid mercury is poisonous in a cumulative way, but it is not as hazardous as its water-soluble compounds. In lake and stream bottoms, where the dense liquid mercury accumulates from electrolytic cell wastes, there are bacteria that convert it to methyl mercury (CH_3HgCH_3). Methyl mercury is a far more dangerous form of the element from a health standpoint, because it dissolves in water and is taken up by the tiny organisms that fish feed on. The result is an accumulation of mercury in the tissues of the fish people eat. On several occasions, mercury content exceeding safe standards set by the government has been found in fish caught in the vicinity of chlorine-alkali plants using mercury cells. At the time of writing this book vigorous efforts were being made to reduce this type of pollution. Pollution by streams of organic wastes can be overcome by the natural process of aeration and bacterial action (Photo 33), but mercury cannot be made harmless by these actions.

One simple solution would be to cut down on the production of chlorine and sodium hydroxide. This is unworkable because NaOH is *the* strong alkali of industry (10 million tons used in 1971), and chlorine is essential to the manufacture of plastics (polyvinylchloride or PVC), insecticides (DDT, dieldrin), laundry bleach, and many other chemicals. Another solution would be to use other electrolytic methods, but they aren't as efficient and don't provide NaOH of as high purity as the mercury cells.

Until recently, bromine was needed in very large amounts for the manufacture of tetraethyl lead and 1,2-dibromoethane, the chemicals in ethyl fluid, which has been used to improve the antiknock properties of gasoline and increase the efficiency of car motors.

$$CH_3-CH_2-\overset{\displaystyle CH_2-CH_3}{\underset{\displaystyle CH_2-CH_3}{\overset{|}{\underset{|}{Pb}}}}-CH_2-CH_3 \qquad\qquad \overset{\displaystyle Br \quad Br}{\overset{|\qquad|}{CH_2-CH_2}}$$

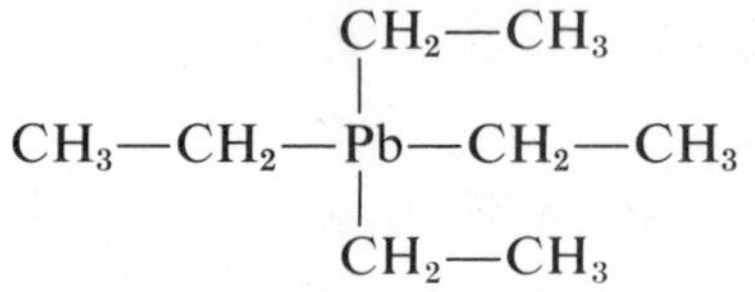

Tetraethyl lead 1,2-dibromoethane
 ("ethylene dibromide")

The use of ethyl fluid is being phased out because of the way its use poisons the environment with lead (Section 2.7). Nevertheless, bromine is still an important element and the method developed for obtaining it in quantity from seawater is a good example of how an element present in very small amounts in some natural source can be concentrated, using chemical principles you have learned.

EXAMPLE ⊶ Bromide ion (from Na^+Br^-) is present to the extent of 65 ppm (parts per million) in seawater. This means that 1 million lb of seawater contains only 65 lb of bromine (in the form of bromide ion), and the low

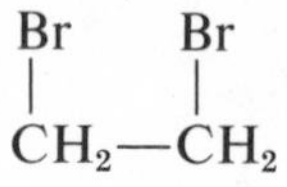

concentration presents a substantial problem for isolation of the bromine. Any attempt to crystallize sodium bromide directly from seawater by evaporation is bound to fail because of the much larger proportion of NaCl present (19,000 ppm as Cl^-). However, the solution left after salt has been prepared by evaporation of seawater (Section 11.2) is far richer in NaBr and can be used as a source of the element.

The first step in the isolation is to treat seawater (concentrated by evaporation or not) with chlorine gas. From Table 10.2 the oxidation potentials of bromide and chloride ions and their half-reactions are

	$\mathscr{E}^0$
$2Br^-(aq) \rightleftarrows Br_2(l) + 2e^-$	-1.07 V
$2Cl^-(aq) \rightleftarrows Cl_2(g) + 2e^-$	-1.36 V

Following the rules in Chapter 10, we see that these values mean that for the reaction $2Br^-(aq) + Cl_2(g) \rightleftarrows Br_2(l) + 2Cl^-(aq)$, the $\Delta\mathscr{E}^0 = +0.29$ V and that the equilibrium constant for the reaction is $K = 10^{17n\,\Delta\mathscr{E}^0} = 10^{17\times2\times0.29} = 10^{9.86} = 7.24 \times 10^9$. This means Cl_2 oxidizes Br^- very efficiently. Thus addition of Cl_2 to seawater converts almost all of the bromide ion present to bromine (Br_2). Pure bromine at room temperature is a deep red liquid with red-orange vapor above it (the boiling point of Br_2 is 59°C). The solubility of bromine in water is so large (36 g/liter at 20°C) that it does not separate from the "chlorinated seawater" as a liquid phase that could be drained off. However, if air bubbles are blown through the seawater solution of Br_2, an equilibrium between dissolved Br_2 in the water and gaseous Br_2 in the air bubbles is set up. The very dilute gaseous solution of bromine in air obtained is continuously bubbled through an alkaline solution (NaOH or Na_2CO_3), where the Br_2 is absorbed by the net reaction

$$3Br_2(g) + 6OH^-(aq) \rightarrow 5Br^-(aq) + BrO_3^-(aq) + 3H_2O(l)$$

The result, after much of this air "extraction" of Br_2 from the seawater, is a concentrated solution of NaBr and $NaBrO_3$ (sodium bromate). This solution, on acidification (H_2SO_4), yields $Br_2(l)$ in sufficient quantity to separate from the solution as a liquid phase that can be drawn off. The net reaction liberating the Br_2 is

$$5Br^-(aq) + BrO_3^-(aq) + 6H^+(aq) \rightarrow 3Br_2(l) + 3H_2O(l)$$

The development of this process required a lot of knowledge about solubilities, oxidation–reduction reactions, and equilibria between phases. Bromine is still an important element in spite of the approaching end of the use of ethyl fluid. It is an essential element, in the form of AgBr, in photography (Section 11.7).

TABLE 12.2 Properties and preparations of the halogens

Halogen (X)	*F*	*Cl*	*Br*	*I*
State at 20°C, 1 atm pressure	F_2 molecules, yellow gas	Cl_2 molecules, greenish-yellow gas	Br_2 molecules, deep-red liquid	I_2 molecules, purplish-black solid
Molecular weight, X_2	38.00	70.91	159.82	253.81
Melting point	−223°C	−102°C	−7.3°C	114°C
Boiling point	−187°C	−33.7°C	58.8°C	183°C (sublimes)
Electron configuration by principal shells	2,7	2,8,7	2,8,18,7	2,8,18,18,7
	increasing numbers of electrons → increasing mp and bp			
Bond dissociation energy, $X_2(g) \rightarrow 2X(g)$	36 kcal/mole	57 kcal/mole	46 kcal/mole	36 kcal/mole
Electronegativity	4.0	3.2	3.0	2.7
$\mathscr{E}^0$ for $2X^-(aq) \rightleftharpoons X_2(g) + 2e^-$	−2.87 V	−1.36 V	−1.07 V	−0.54 V
Natural occurrence	fluorite, CaF_2 cryolite, Na_3AlF_6	rock salt in mines, seawater (19,000 ppm as Cl^-)	salt mines (NaBr), seawater (65 ppm as Br^-)	kelp, $NaIO_3$ (Chile) brine wells, as NaI (Texas)
Preparations	(1) electrolysis of HF dissolved in KHF_2 $2HF(l) = H_2(g) + F_2(g)$	(1) electrolysis of NaCl solution: $2NaCl(aq) + 2H_2O(l) = 2NaOH(aq) + H_2(g) + Cl_2(g)$	(1) $2Br^-(aq) + Cl_2(g) = Br_2(l) + 2Cl^-(aq)$	(1) $2I^-(aq) + 2NO_2^-(aq) + 4H^+(aq) = I_2(s) + 2NO(g) + 2H_2O(l)$
		(2) $MnO_2(s) + 2NaX(s) + 3H_2SO_4(l) = 2NaHSO_4(s) + MnSO_4(s) + 2H_2O(l) + X_2(g, l, s)$		

Some of the other chemistry of the halogens is summarized in Table 12.2. The trends in such quantities as ionization potential, electronegativity, oxidation potential ($\mathscr{E}^0$) (for the ions), and boiling points are fairly typical of any group of octet elements.

Noble gas compounds—Group 8a Until about 15 years ago, all efforts to prepare compounds of the noble gases (He, Ne, Ar, Kr, Xe, Rn) had failed. They were even called the "inert gases" until then because it was believed they wouldn't react with any other element. Their unreactivity was part of the foundation of the octet theory of valence—because they already had the ns^2np^6 octet ($1s^2$ duet for He), it was thought that they couldn't react. Then in 1962 Neil Bartlett of the University of British Columbia brought platinum(VI) fluoride (PtF_6) and xenon (Xe) together for the first time and was happy to see them react to form the *first noble gas compound,* $XePtF_6$. Now why did he try this particular experiment? Applying only principles that you have learned and knowing what he knew about ionization potentials, you might have tried the same experiment—unless, of course, you accepted octet *theory* as absolute truth, which he clearly didn't. What he had observed earlier is that molecular oxygen (O_2) reacts with PtF_6 to give a compound having the formula O_2PtF_6. To form this compound the O_2 molecule is called on to *donate* electrons to PtF_6—the high electronegativity of the fluorine makes it most unlikely that Pt has any electrons left to give to O_2. Bartlett knew that the ionization potential of the Xe *atom* was *less than* the ionization potential of the O_2 *molecule*—that it is easier to remove an electron from a xenon atom to give Xe^+ than it is to remove an electron from O_2 to give O_2^+. The data are as follows:

$$O_2(g) + 12.5\text{ eV} \rightarrow O_2^+(g) + e^-$$
$$Xe(g) + 12.1\text{ eV} \rightarrow Xe^+(g) + e^-$$

So he reasoned that if O_2 reacts with PtF_6 by donating electrons, so should Xe. He tried the experiment, it worked, the "inert gases" became the "noble gases," and noble gas chemistry began.

Since that time, other compounds of Xe and compounds of Kr and Rn have also been prepared. Now noble gas chemistry is expanding to such an extent that whole books are being written about it. Here we can only touch on some interesting highlights of xenon chemistry. (One purpose is to show you how old principles can be applied to new chemistry.)

It was soon found that when mixtures of xenon and fluorine were heated together at 300–400°C an equilibrium with xenon fluorides XeF_2, XeF_4, and XeF_6 was established in only a few hours. ☛ The reaction does not go at all at room temperature, and it is believed that the high temperature starts the reaction by dissociating the F_2 molecule into F atoms as shown in the following mechanism:

Years before, Linus Pauling had suggested on theoretical grounds that xenon and fluorine might form compounds because of the very high electronegativity and electron-attracting power of the fluorine atom.

$$F_2(g) \rightleftarrows 2F(g)$$
$$F(g) + Xe(g) \rightleftarrows XeF(g)$$
$$XeF(g) + F(g) \rightleftarrows XeF_2(g)$$

and so on to XeF_6.

The value of only 36 kcal/mole for the bond dissociation energy of F_2 means that at 400–500°C ample numbers of F atoms are present in the mixture. By applying Le Chatelier's principle, the formation of XeF_6 was made practically complete. What conditions would you use to favor the product side of the following reaction?

$$Xe(g) + 3F_2(g) \rightleftarrows XeF_6(g)$$

The answer is "high pressure, because there is a volume decrease (4 moles of gas go to 1 mole), and a high concentration of F_2, because it is a reactant." Pressures of 100 atm and F_2/Xe molar ratios of 20:1 do indeed lead to good yields of XeF_6.

Xenon hexafluoride [xenon(VI) fluoride] is a white crystalline compound (mp 50°C) that is stable in absolutely dry air. It reacts rapidly with water and with glass (SiO_2), which, like HF, it etches:

$$2XeF_6(s) + SiO_2(g) \rightarrow SiF_4(g) + 2XeOF_4$$
$$XeF_6(s) + H_2O(l) \rightarrow 2HF(aq) + XeOF_4$$

The product of these reactions, $XeOF_4$, is a liquid (mp −30°C) that reacts further with water to give HF and XeO_3, xenon trioxide [xenon(VI) oxide]. The trioxide has a formula like SO_3 and CrO_3, which are the anhydrides of the strong acids H_2SO_4 and H_2CrO_4 (Section 12.2). Water solutions of XeO_3 are only weakly acidic, indicating the possible presence of some xenic acid, H_2XeO_4, but spectroscopy shows that there are mainly XeO_3 molecules in the solution, rather than H_2XeO_4 molecules.

Although XeO_3 is stable in solution, when dry it is a violent explosive that must be handled with extreme caution. The explosion results from its going to its gaseous elements, which expand suddenly because of heating by the large amount of thermal energy released. The equation for the explosion is

$$2XeO_3(s) \rightarrow 2Xe(g) + 3O_2(g) + 96 \text{ kcal}$$

By contrast, SO_3 and CrO_3 are both very stable and *require* large amounts of energy for decomposition:

$$2SO_3(l) + 189 \text{ kcal} \rightarrow 2S(s) + 3O_2(g)$$
$$2CrO_3(s) + 277 \text{ kcal} \rightarrow 2Cr(s) + 3O_2(g)$$

The difference is due to the fact that the outer octet ($5s^25p^6$) of electrons for xenon does represent great stability.

This brief introduction to xenon chemistry should show you that even though new chemistry is constantly being discovered, the older principles still work, and often suggest new experiments and ways to carry them out.

Exercises

26. Balance the skeleton equation $XeOF_4 + H_2O \nrightarrow XeO_3 + HF$.

27. The melting point of XeF_6 is 50°C, and its vapor pressure is 29 torr at 25°C. What kind of bonding is present in XeF_6 and what kind of forces are responsible for the solid? What geometry would you predict for the molecule?

28. When Xe is ionized to Xe^+, which electron is removed? When O_2 is ionized to O_2^+, which electron is removed? (*Hint*: see Table 4.2.)

29. Draw an electron-dot octet structure for XeO_3. According to X-ray data there is only one Xe—O distance of 1.76 Å and the O—Xe—O angle is 103°. Does this geometry fit your structure?

30. In SO_3, the S—O bond length is 1.43 Å, and the O—S—O angle is 120°. Why are there these differences between SO_3 and XeO_3 (Exercise 29)? (*Hints*: draw the dot structure and remember resonance theory.)

31. Is the difference in Xe—O and S—O bond lengths in XeO_3 and SO_3 in line with their stabilities? (*Hint*: remember that short bonds are strong bonds.)

32. XeO_3 is a strong oxidizing agent. Balance its electron-transfer half-reaction, for which the skeleton is

$$XeO_3(aq) + H^+ + e^- \nrightarrow Xe(g) + H_2O(l)$$

12.6 Summary Guide The elements H, F, Cl, Br, I, O, S, N, P and the noble gases are the nonmetals and have in common the property of being nonconductors of electricity. Metalloids are elements that are weak conductors of electricity and are found on the borderline between the metals and the nonmetals in the Octet Periodic Table.

Bonding between metals and nonmetals is usually ionic; bonding between nonmetals and between nonmetals and metalloids is always covalent. Because of the reluctance of the nonmetallic elements in period 3 (and higher) to form double or triple bonds, a third-period nonmetal has very different properties from those of the nonmetal above it in period 2 of the Octet Table.

On going from the Group 7a nonmetals toward the center of the Octet Table, there is an increasing tendency of nonmetal atoms to bond to each other, either directly or through oxygen. Bonding of silicon atoms through oxygen gives rise to the largest class of minerals, the silicates. Linking of phosphorus atoms through oxygen atoms permits the formation of adenosine di- and triphosphates. Hydrolysis of a P—O—P link in adenosine triphosphate releases the free energy for muscular action. The carbon atom displays the largest number of ways of forming bonds to itself and to the atoms of other nonmetals. This ability allows the formation of the complex molecules needed for life.

Oxygen is the most important nonmetal because it reacts with most metals and nonmetals to form oxides. Oxides of the nonmetals react with water to form proton acids, called oxyacids. The formula of an oxyacid can be derived by attaching to the central atom a number of OH groups equal to the oxidation state of the central atom, then eliminating water either from within or between the OH structures that result. Between the oxyacids of an element, the oxyacid in which the element has the higher oxidation state will be the stronger acid.

The compounds of Periodic Groups 7a, 6a, 5a, and 4a with hydrogen have molecular formulas HX, H_2Y, H_3Z, and H_4Q, respectively, where X, Y, Z and Q are elements in the groups. The HX compounds, the hydrogen halides, are strong proton acids with the exception of HF, which because of hydrogen bonding is weak. Acid strengths of the HX compounds increase from HF to HI. The H_2Y compounds, with H_2O leading them, are both weak acids and weak bases at the same time, with acid strengths increasing and basic strengths decreasing on going from H_2O to H_2Te. The H_3Z compounds are weak bases and very weak acids, and the H_4Q compounds are without significant strength as acids or bases.

The nonmetals in Groups 6a, 5a, and 4a, with the exception of N, show allotropy by having more than one covalent molecular structure that can exist temporarily under ordinary conditions of temperature and pressure (O_2 and O_3, S_8 and S_n, P_4 and P_n are examples).

Salts of the phosphoric acids are used as water-softening agents in laundry soaps and detergents, and they pollute streams and lakes into which sewage is discharged.

Sulfur is a necessary element in a technological society, where it appears as sulfuric acid, the strong acid of industry. Chlorine, produced by the electrolysis of sodium chloride solutions, is the most important halogen in industry, appearing in insecticides of the chlorinated hydrocarbon type (which are pollutants) and in plastics such as polyvinylchloride (PVC). Production of chlorine by electrolysis, using mercury cells, discharges polluting quantities of mercury into the environment. Bromine is isolated from the bromides present in seawater and is an essential ingredient, in the form of silver bromide, in photographic film.

The first compounds of the noble gases (XeF_6 and XeO_3 are examples) were prepared in the last 15 years. Their existence and structure are in line with general chemical principles.

Exercises

33. For most crops, the pH of the soil should be in the range 5.8–6.8 for maximum growth. Ammonium nitrate, $NH_4^+NO_3^-$, is the fertilizer of choice for soils that are too alkaline (pH > 7). Why is this so? (*Hint*: using data from Table 8.2, calculate the pH of a 0.01-*M* solution of $NH_4^+NO_3^-$.)

34. Table 12.2 shows that $I_2(s)$ is obtained from I^- (in brine from wells in Texas) by oxidation, using an acidic solution of nitrite ion (NO_2^- from $Na^+NO_3^-$). For the half-reaction $NO(g) + H_2O(l) \rightleftharpoons 2H^+ + NO_2^- + e^-$, the $\mathscr{E}^0 = -1.19$ V. Using data from Table 10.2, calculate $\Delta\mathscr{E}^0$ and K for the reaction $2I^- + 2NO_2^- + 4H^+ \rightleftharpoons I_2(s) + 2NO(g) + 2H_2O(l)$ to see how efficient the oxidation is.

35. Trisodium phosphate ("TSP," Na_3PO_4) is used in water solution to clean painted surfaces prior to repainting. The solution works to clean because it is mildly alkaline owing to the presence of the moderately strong base PO_4^{3-} ($K_b = 4.67 \times 10^{-2}$, Table 8.2). Calculate the pH of a 0.10-*M* solution of Na_3PO_4, taking only the reaction $PO_4^{3-} + H_2O(l) \rightleftharpoons HPO_4^{2-} + OH^-$ into account. Is the solution alkaline?

36. What is the ground-state electron configuration of the gaseous atoms of the element of atomic number (Z) 32? (Do not use any outside data.) Is the element a metal, nonmetal, or metalloid?

37. Using data in Table 5.3, calculate the partial pressure of argon (p_{Ar}) in the atmosphere when the barometric pressure is 740 torr.

38. In thionyl chloride ($SOCl_2$, bp 79°C) the oxygen and both chlorines are bonded directly to sulfur. Give the Lewis structure for $SOCl_2$. One of the products of its (vigorous) reaction with water is $SO_2(g)$. What is the other? Thionyl chloride is used in organic chemistry to convert alcohols (R—OH compounds such as CH_3—CH_2—OH) into their chlorides R—Cl. Write the balanced equation for the reaction of R—OH with $SOCl_2$.

39. Referring to Table 10.3, what generalization can you make about the stability of the compounds of hydrogen with periodic groups of nonmetals and metalloids as the atomic number of the nonmetal or metalloid increases?

Key Words

Acid anhydride a compound that reacts with water to form a proton acid.

Allotropes different molecular forms of an element.

Halogens the Group 7a elements.

Hydrogen halides HF, HCl, HBr, HI.

Hydroxyl scheme a method using oxidation state, hydroxyl groups, and elimination of water to derive the formulas of oxyacids.

H_4Q compounds compounds of hydrogen with the Group 4a elements.

HX compounds compounds of hydrogen with the Group 7a elements—also called the hydrogen halides.

H_2Y compounds compounds of hydrogen with the Group 6a elements.

H_3Z compounds compounds of hydrogen with the Group 5a elements.

Lead chamber process a method for making sulfuric acid, using nitrogen oxides as catalysts.

Mercury cell an electrolytic cell, using mercury, for making sodium hydroxide and chlorine from sodium chloride.

Metallicity metallic character.

Noble gases the Group 8a elements He, Ne, Ar, Kr, Xe, and Rn.

Suggested Readings

Bruty, D., Chester, R., and Padgham, R. C., "Mercury in Lake Sediments: A Possible Indicator of Technological Growth," *Nature,* **24**:450–451 (Feb. 16, 1973).

Clapp, Leallyn B., *The Chemistry of the OH Group,* Englewood Cliffs, N.J., Prentice-Hall, Inc., 1967.

Frieden, Earl, "The Chemical Elements of Life," *Sci. Am.,* **227**:52–60 (July 1972).

Hall, Stephen K., "Sulfur Compounds in the Atmosphere," *Chemistry,* **45**: 16–18 (March 1972).

Krinsley, D. H., and Smalley, I. J., "Sand," *Am. Sci.,* **60**:286–291 (May-June 1972).

"Ozone Technology Makes Strides in Water and Waste Water Cleanup," *Env. Sci. and Tech.,* **8**:108–109 (Feb. 1974).

Schofield, Maurice, "Early Days of Sulfuric Acid," *Chemistry,* **45**:11–13 (Oct. 1972).

Slack, A. V., "Removing SO_2 from Stack Gases," *Env. Sci. and Tech.,* **7**: 110–119 (Feb. 1973).

Sutherland, Earl W., "Studies on the Mechanism of Hormone Action," *Science,* **177**:401–407 (Aug. 4, 1972).

"Those Nasty Phosphatic Clay Ponds," *Env. Sci. and Tech.*, **8**:106–107 (April 1974).

VanderWerf, Calvin A., *Acids, Bases, and the Chemistry of the Covalent Bond,* New York, Reinhold Publishing Corp., 1961.

chapter 13

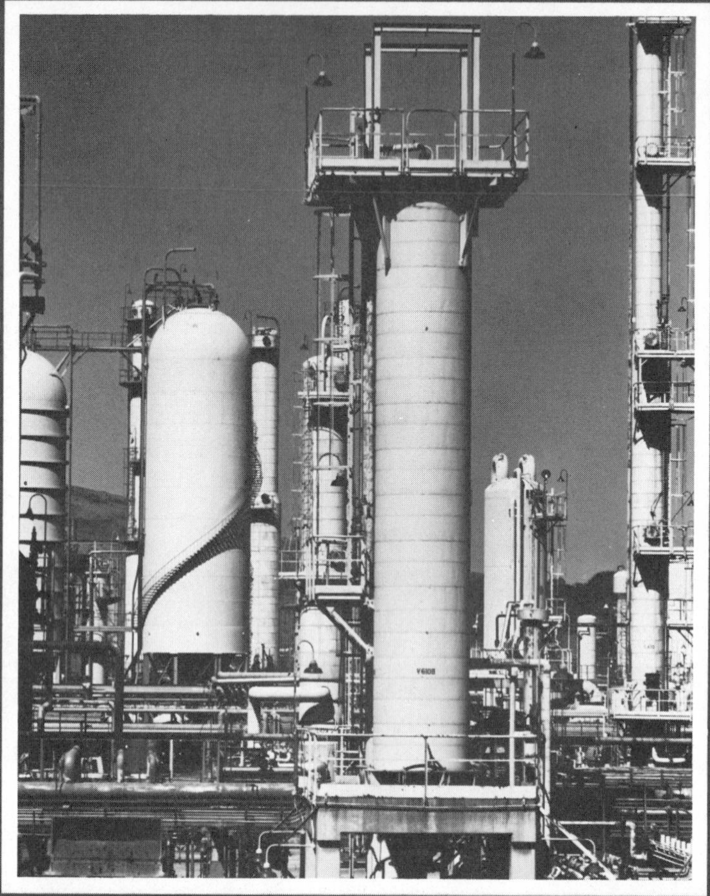

A catalytic hydrocarbon cracking installation ("cat-cracker") for increasing the yield of high-octane gasoline from crude oil. (*Photo 34, Standard Oil Company of California.*)

Organic Chemistry

13.1 Introduction The chemistry of the compounds of carbon makes up the huge subject called organic chemistry—there are more than a million carbon compounds known already, and each day hundreds of new ones are either discovered in nature for the first time or made by the methods of synthetic organic chemistry. The name "organic" comes from history—the first compounds of carbon, hydrogen, and oxygen discovered were produced by *organ*isms. Alcohol (ethanol) and acetic acid, both produced by bacterial fermentation of sugar, and sugar itself, produced by plants, were among the first organic compounds studied by chemists. Now the word covers the whole range of carbon chemistry, whether or not the compounds are formed by living things.

Organic chemistry is *molecular* chemistry and it is *three-dimensional* chemistry. It's molecular because carbon forms only covalent bonds to other atoms, making ionic bonding and "infinite" ionic crystals impossible. It's three-dimensional because the covalent bonds from carbon to carbon atoms and other atoms have definite directions in space. Each of these aspects of carbon bonding has its own effects on the molecular architecture of carbon compounds, which will be discussed in Sections 13.2, 13.3, and 13.4. The ability of a carbon atom to form covalent bonds to itself and to practically any other atom results in *molecular skeletons* and *functional groups*. The bonds that hold the atoms together in the skeleton are of the sigma (σ) type (Section 4.3), but may be assisted by π bonds. Understanding these structural features of organic molecules will make it possible for you to appreciate what a wonderfully exciting subject organic chemistry is.

13.2 Molecular Frameworks and Functional Groups The framework or skeleton of a molecule contains *all* of the atoms in the molecule and shows the ways in which they are connected by single bonds (—). Double and triple bonds, if present, are put in later to give the *bond* structure of the molecule. These multiple bonds are of the pi (π) type (Figure 4.6). Molecular skeletons may be simple like those for ethanol and dimethyl ether (Section 4.2):

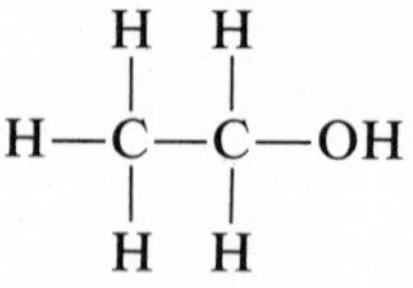

Ethanol

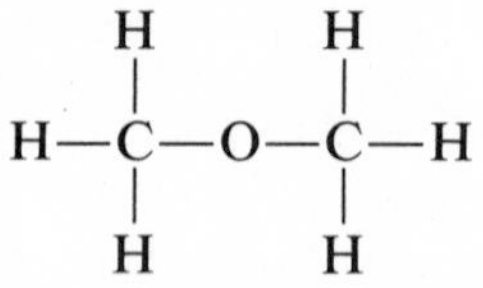

Dimethyl ether

The skeletons of ethanol and dimethyl ether are the *complete* bond structures. For an example in which the skeleton is not the complete bond structure, the skeleton of penicillin is a good one:

The skeleton is that of penicillin G. In other penicillins, of which there are many, the benzene ring is replaced by a different group.

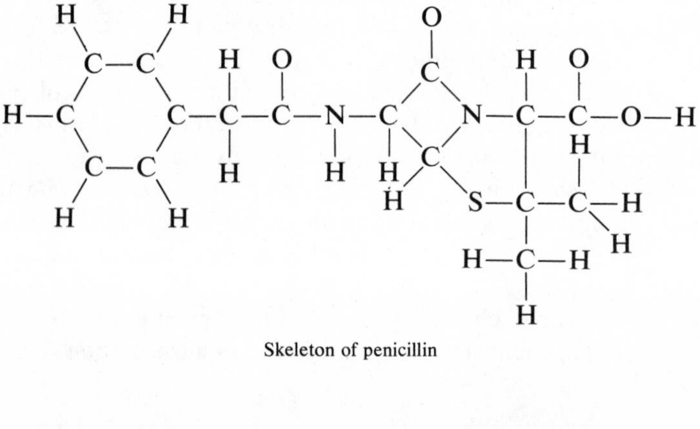

Skeleton of penicillin

Penicillin has a very complicated skeleton indeed, but when the double bonds are put in, you will recognize several structural units. Some are indicated in the following bond structure for the molecule:

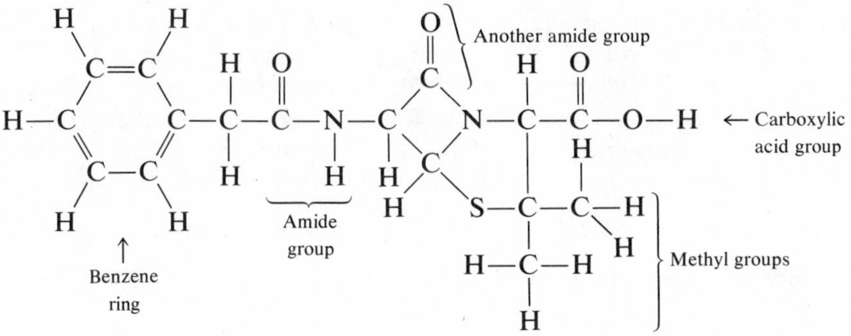

Bond structure of penicillin

Penicillin kills bacterial cells while leaving body cells unharmed. For such *selective* action it is no surprise that a complicated molecule is needed. As you can see in the *bond* structure for penicillin, every carbon atom has four bonds, oxygens have two, sulfur has two, nitrogens have three, and hydrogens have only one. These are almost always the numbers of bonds for C, O, S, N, and H in organic molecules.

When you look at the bond structure of a molecule, you should first find its skeleton. Then you should look for *reactive centers*. These are places in the molecule where it can be attacked by acids, bases, oxidizing and reducing agents, and other reagents. *All carbon–carbon double or triple bonds and all atoms other than carbon and hydrogen are reactive centers.* You can count many reactive centers in the penicillin molecule, and you know some of the reactions or "functions" of one of them already: the $-\mathrm{C}(=\mathrm{O})-\mathrm{O}-\mathrm{H}$ or carboxylic acid group. It reacts with sodium hydroxide to give the sodium salt of penicillin and can be esterified by reaction with an alcohol (Section 7.3). Using R— to represent the rest of the penicillin molecule, we can write the reactions as follows:

$$\underset{\text{Penicillin}}{\mathrm{R-C(=O)-O-H}} + \mathrm{Na^+OH^-} \rightarrow \underset{\text{"Penicillin sodium"}}{\mathrm{R-C(=O)-O^-Na^+}} + \mathrm{H_2O}$$

$$\mathrm{R-C(=O)-O-H} + \mathrm{CH_3-CH_2-OH} \rightarrow \underset{\text{Penicillin ethyl ester}}{\mathrm{R-C(=O)-O-CH_2-CH_3}} + \mathrm{H_2O}$$

These reactions are typical of the carboxylic acid group whether R is only a methyl group (CH_3—) as in acetic acid, $CH_3-\overset{\overset{\displaystyle O}{\|}}{C}-O-H$, or all the rest of the penicillin molecule. If you've ever had a penicillin shot," it was probably the sodium salt of penicillin you got.

The reactive centers in molecules are found in what are called *functional groups*. "Functional group" has a larger meaning than "reactive center" because the group of atoms making up a functional group may have more than one reactive center. For example, the functional group called the *carbonyl group*, $\diagdown C = \ddot{O}$, has two reactive centers, one toward acids and one toward bases:

$$\overset{\diagdown}{\underset{\diagup}{\underset{\uparrow}{C}}} = \ddot{O} \quad \leftarrow \text{Acids} \qquad \overset{\diagdown}{\underset{\diagup}{C}} = O + OH^- \rightleftarrows \overset{\diagdown}{\underset{\diagup}{\underset{|}{C}}} - \ddot{\ddot{O}}{:}^- \qquad \overset{\diagdown}{\underset{\diagup}{C}} = \ddot{O} + H_3O^+ \rightleftarrows \overset{\diagdown}{\underset{\diagup}{C}} = \overset{+}{\ddot{O}} - H + H_2O$$

Bases Base OH Acid

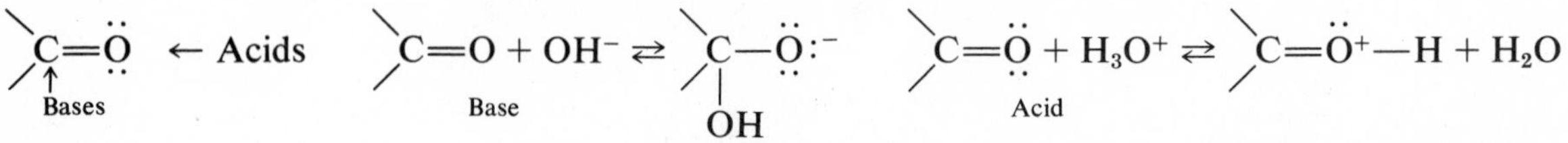

Examples of the important functional groups are listed in Table 13.1. You should learn their names and structures, and train yourself to recognize them even if they are written in different ways. All of the following are ways of writing the same amide group with the same bonding between atoms: ▬◖

Unshared pairs of electrons are usually omitted in structural formulas for organic molecules.

$$-\overset{\overset{\displaystyle O}{\|}}{\underset{\underset{\diagdown H}{|}}{C}}-\overset{\diagup H}{N} \qquad -\overset{\overset{\displaystyle O}{\|}}{C}-NH_2 \qquad -CONH_2 \qquad H_2N-\overset{\overset{\displaystyle O}{\|}}{C}- \qquad H_2NCO-$$

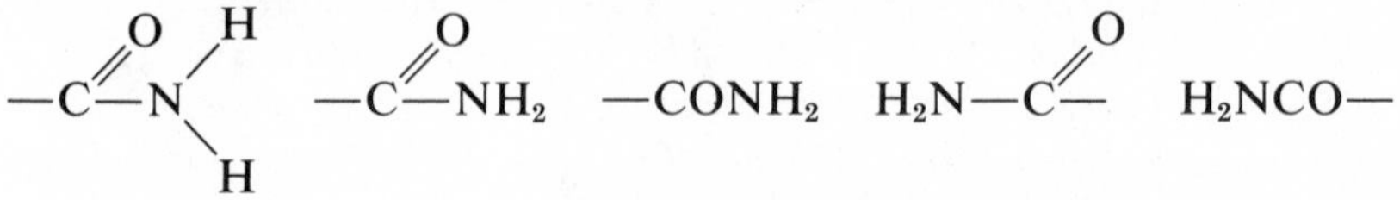

Many functional groups can be derived by using the hydroxyl scheme for oxyacids (Section 12.3) with one slight change. Instead of using the charge on carbon to give you the number of —OH groups to attach, you just replace each hydrogen with an OH group. To get the carboxyl group $-\overset{\overset{\displaystyle O}{\diagup\!\!\!\|}}{C}-OH$, you start with a —$CH_3$ group, replace 3Hs by 3OHs to get —$C(OH)_3$, and split out water as shown at (a) in Table 13.2. The carboxyl group is written in all of the following ways, depending on what is being looked at:

$$-CO_2H, \quad -COOH, \quad -\overset{\overset{\displaystyle O}{\diagup\!\!\!\|}}{C}-OH, \text{ and } -\overset{\overset{\displaystyle O}{\diagup\!\!\!\|}}{C}-O-H.$$

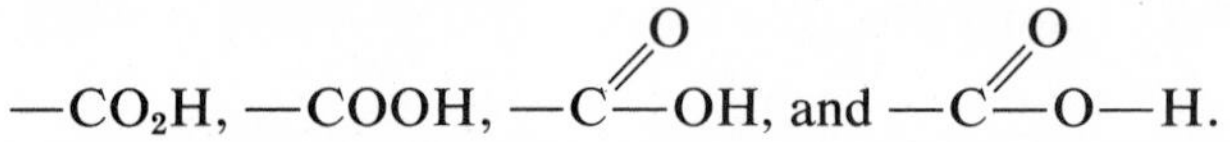

Other examples are in Table 13.2, and the 2Hs and O split out as H_2O are indicated by boldface type.

TABLE 13.1 Functional groups

Structure	*Suffix*	*Corresponding Class of Compounds*	*Example*	*Example Name*
$>C=C<$ or[a] $—CR=CR—$	-ene	olefins	$CH_3—C(H)=CH_2$	propene
$—C\equiv C—$	-yne	acetylenes	$CH_3—C\equiv C—H$	propyne
$—OH$	-ol (or "hydroxy" as prefix)	alcohols	$CH_3—CH_2—OH$	ethanol[b]
$—C(H)=O$ or $—CH=O$	-al	aldehydes	$CH_3—C(H)=O$	ethanal[b]
$>C=O$ or $—CO—$	-one	ketones	$(CH_3)_2C=O$	propanone[b]
$—C(=O)—OH$ or $—COOH$	-oic acid	carboxylic acids	$CH_3C(=O)—OH$	ethanoic[b] acid
$—NH_2$	amine	amines	$CH_3—NH_2$	methylamine
$—C(=O)—NH_2$ or $—CONH_2$	-oic acid amide	amides	$CH_3—CH_2—C(=O)—NH_2$	propanoic acid amide
$—C(=O)—O—C\lt$ or $—COOC\lt$	-oic acid ester	esters	$CH_3CH_2C(=O)—O—CH_3$	propanoic acid methyl ester
$—O—$	ether	ethers	$CH_3—O—CH_3$	dimethyl ether

[a]The alternatives are condensed forms; R = H or an alkyl radical, such as $CH_3—$, $CH_3—CH_2—$ (Section 13.4).
[b]Trivial names are often used for the early members of a homologous series. Thus ethanol, ethanal, propanone, and ethanoic acid are commonly called ethyl alcohol, acetaldehyde, acetone, and acetic acid, respectively.

TABLE 13.2 Functional groups and the hydroxyl convention

(a)	$\underset{\displaystyle -\overset{\displaystyle \text{HO}\quad\text{HO}}{\underset{\displaystyle \big\backslash \quad\big/}{\text{C}}}-\text{OH}}{}$	\longrightarrow $-\overset{\text{O}}{\underset{}{\text{C}}}-\text{OH} + \text{H}_2\text{O}$
(b)	$-\overset{\text{HO}\quad\text{HO}}{\underset{}{\text{C}}}-$	\longrightarrow $-\overset{\text{O}}{\underset{}{\text{C}}}- + \text{H}_2\text{O}$
(c)	$\text{CH}_3-\text{O}\boxed{\text{H}\quad\text{HO}}-\text{CH}_3$	\longrightarrow $\text{CH}_3-\text{O}-\text{CH}_3 + \text{H}_2\text{O}$
(d)	$-\overset{\text{O}}{\underset{}{\text{C}}}\boxed{\text{OH}\quad\text{H}}\text{O}-\text{CH}_3$	\longrightarrow $-\overset{\text{O}}{\underset{}{\text{C}}}-\text{O}-\text{CH}_3 + \text{H}_2\text{O}$
(e)	$-\overset{\text{O}}{\underset{}{\text{C}}}\boxed{\text{OH} + \text{H}}\text{NH}_2$	\longrightarrow $-\overset{\text{O}}{\underset{}{\text{C}}}-\text{NH}_2 + \text{H}_2\text{O}$
(f)	$-\overset{\displaystyle}{\underset{\boxed{\text{H}\ \text{HO}}}{\text{C}}}-\text{C}\diagup$	\longrightarrow $\diagdown\!\!\text{C}\!=\!\text{C}\!\diagup + \text{H}_2\text{O}$

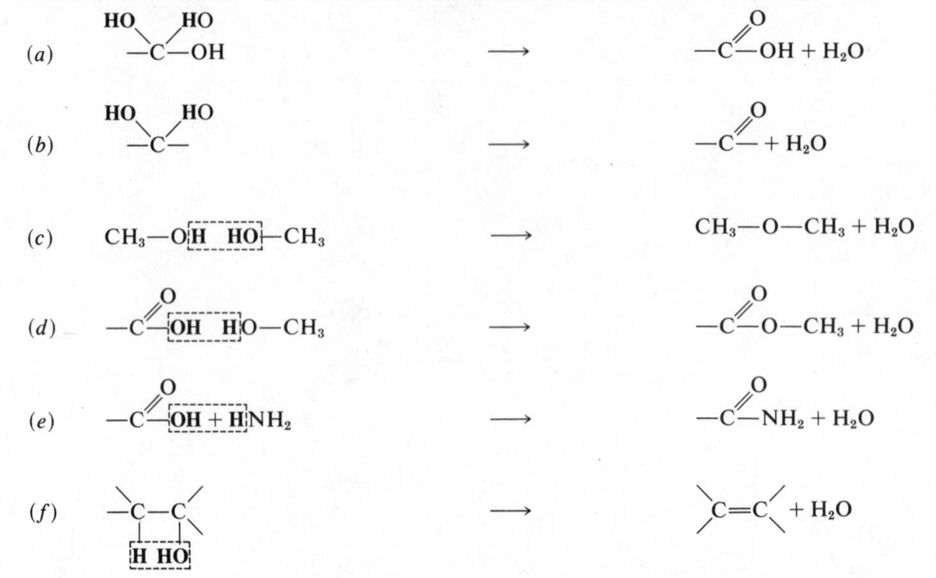

Building up organic molecules can be compared to decorating a Christmas tree — the tree is the molecular skeleton, the ornaments and lights are the functional groups. In synthetic organic chemistry, in which organic chemists build up skeletons and functional groups, methods are available for making almost any stable molecule you could imagine, and some examples will come later. Needless to say, organic syntheses require hard, imaginative thinking and much work in the laboratory (where reactions don't always go as smoothly as they look on paper).

Exercises **1.** The skeleton of vitamin C (ascorbic acid) is as follows:

$$
\begin{array}{c}
\text{O} \\
| \\
\text{C} \\
\diagup \quad \diagdown \\
\text{H}-\text{O}-\text{C} \qquad \text{O} \\
| \qquad\qquad \diagup \\
\text{H}-\text{O}-\text{C} \diagdown \quad \\
\text{H}-\text{C} \\
| \\
\text{H}-\text{O}-\text{C}-\text{H} \\
| \\
\text{H}-\text{C}-\text{O}-\text{H} \\
| \\
\text{H}
\end{array}
$$

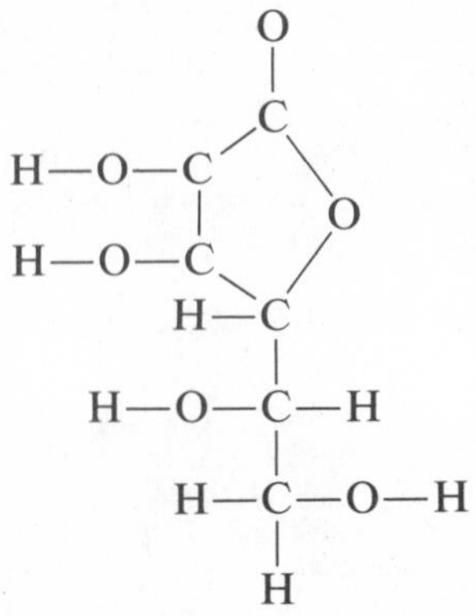

By putting in the necessary bonds, give the bond structure for the vitamin. How many functional groups does it have?

2. In Table 13.2 the formation of an ester is shown at (d) to go as follows:

$$-C\overset{O}{\diagup}\overset{\Vert}{\underset{}{}}\vdash\text{OH}\quad \text{H}\dashv\text{O}-\text{CH}_3 \rightarrow -C\overset{O}{\diagup}\overset{\Vert}{\underset{}{}}\text{O}-\text{CH}_3 + \text{H}_2\text{O}$$

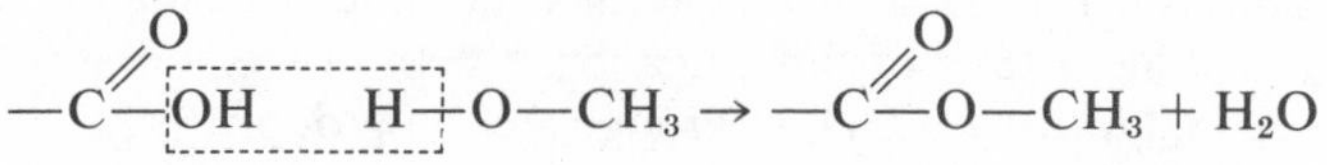

It could go by splitting out water the other way:

$$-C\overset{O}{\diagup}\overset{\Vert}{\underset{}{}}\text{O}\vdash\text{H}\quad \text{HO}\dashv\text{CH}_3 \rightarrow -C\overset{O}{\diagup}\overset{\Vert}{\underset{}{}}\text{O}-\text{CH}_3 + \text{H}_2\text{O}$$

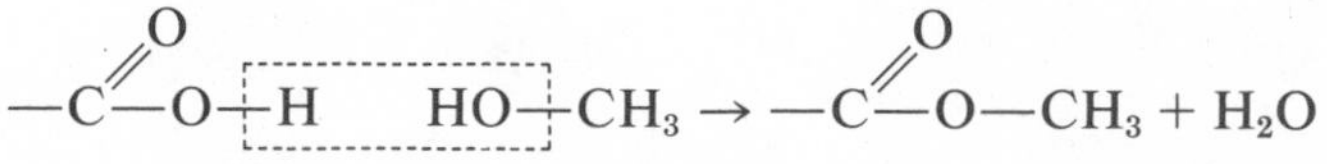

Suppose you had isotopically labeled methanol $\text{CH}_3\text{—}^{18}\text{O—H}$. What experiment could you do to find out by which path the reaction goes? (*Hint*: what would mass spectroscopic analysis of the *products* of esterification reveal?)

3. The thioether functional group $\diagdown\!\!\diagup C\text{—S—}C\diagup\!\!\diagdown$ is like the ether group but with —S— replacing —O—. Is there a thioether group in penicillin?

13.3 Bond Angles—Geometry of Molecules

You can predict bond angles and the geometry of nearly all organic molecules quite well by using the electron repulsion rules in Section 4.4. The carbon atom has four valence electrons, so when it covalently bonds to four other atoms it will have four ligands (L) and no unshared pairs. For this case, the predicted L—C—L angles all have the tetrahedral value of 109.5° (Figure 13.1). To show tetrahedral geometry you should use "wedge" bonds to connect ligands that lie above the plane of the paper and dashed (\vdots) bonds for the ligands that are behind. This method is shown in the following examples, in which two ligands are behind and two in front of the plane of the paper: ☛

Work with some kind of a molecular model set is really necessary to understand organic chemistry. The set can be as simple as one using different colored gumdrops for atoms and toothpicks for bonds.

L \vdots L►C◄L \vdots L	H \vdots H►C◄H \vdots H	Cl \vdots Cl►C◄Cl \vdots Cl	H \vdots H►C◄OH \vdots H
4-ligand carbon	Methane	Tetrachloromethane (carbon tetrachloride)	Methanol (methyl alcohol)

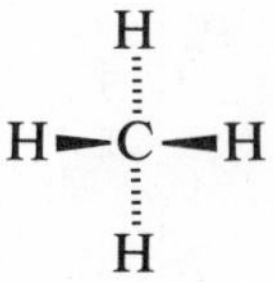

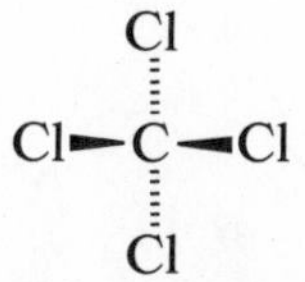

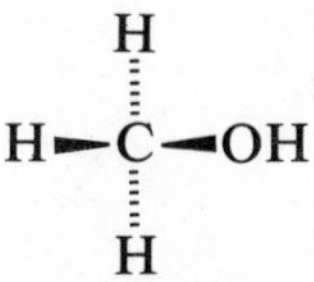

Small differences from tetrahedral angles are found if bigger ligands crowd smaller ones, but the angles are usually within a degree or so of 109.5° for four-ligand carbon.

The rules for naming organic compounds are internationally agreed on. In this chapter the naming according to the rules comes first; other commonly used names are in parentheses.

Remember that when counting the ligands of an atom it makes no difference whether they are singly, doubly, or triply bonded to it (Section 4.4).

The electron repulsion rules predict L—C—L angles of 120° for three-ligand carbon atoms. Ethene (ethylene) ☚ is a simple example of this geometry. Each carbon in ethene has the other carbon and two hydrogens, making a total of three ligands. ☚ In propene one of the hydrogens is replaced by a CH_3— group (methyl group):

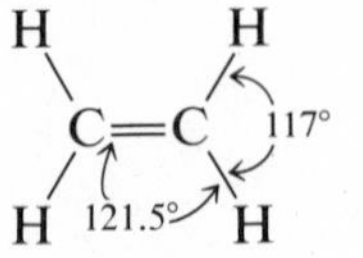

Ethene
(ethylene)

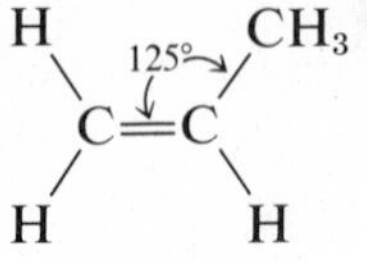

Propene
(propylene)

The angles shown for ethene and propene are measured by spectroscopy and are not exactly the 120° predicted. However, the predictions from the rules are close enough to give you an excellent idea of the shapes of the molecules. Other examples of three-ligand carbon in which the angles indicated by the arrows are very close to 120° are the following:

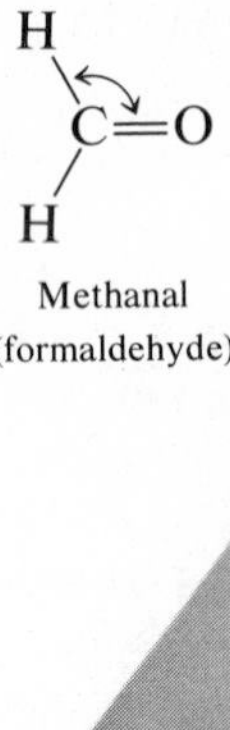

Methanal
(formaldehyde)

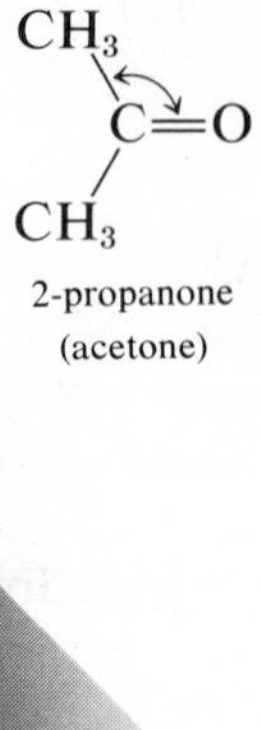

2-propanone
(acetone)

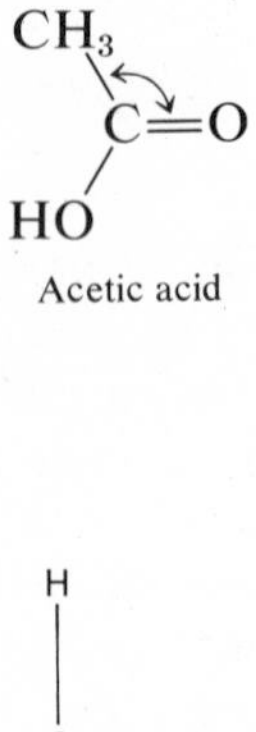

Acetic acid

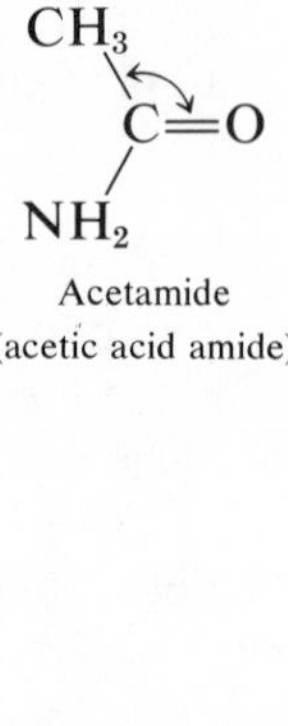

Acetamide
(acetic acid amide)

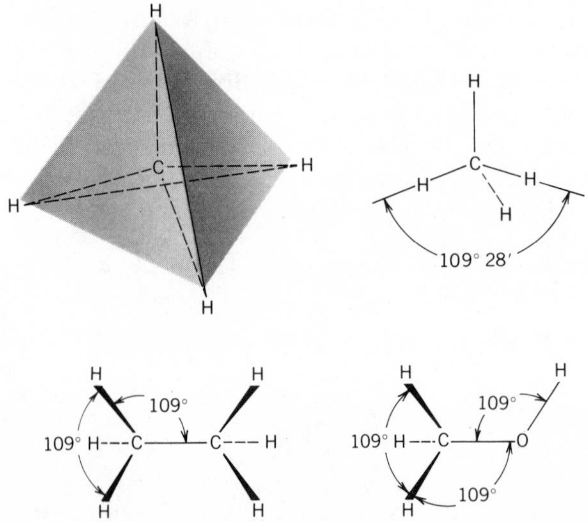

FIG. 13.1 Tetrahedral symmetry of the methane molecule, and time-average values of bond angles to ±1° in the ethane and methyl alcohol molecules (values from spectroscopic data).

In these and any other cases of three-ligand carbon it is important to realize that *the centers of the carbon atom and its three ligands all lie in the same plane*—three ligands result in "flat" molecules. Of course, it is only the *nuclei* of the carbon atoms and the central atoms of the three ligands that really lie in the same plane; any ligands on the other atoms and the electron clouds of all the atoms are above and below the plane occupied by the carbon atom and its three ligands. In the example of propene, the CH_3— group itself has tetrahedral H—C—H angles because carbon has four ligands, 3H and C. Three-ligand carbon is often called "trigonal carbon" because the three ligands are at the corners of a triangle (trigon) with the carbon atom at the center. You can show this arrangement for formaldehyde by drawing dashed lines connecting the three ligands O and 2H:

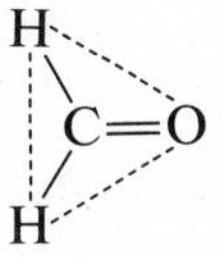

In two-ligand carbon compounds, the centers of the carbon atom and its ligands lie on a straight line, making a *linear* compound. Here are some examples:

H—C≡C—H	CH_3—C≡C—H	CH_3—C≡N
Acetylene	Propyne (methyl acetylene)	Acetonitrile

Tetrahedral (4—L) and trigonal (3—L) carbon are much more important and frequent than the linear (2—L) type.

Most often organic structures are written by using ordinary solid bonds (—). But when the direction of the bonds in space is the important consideration, you have to use the solid, dashed, and wedge bonds to show ligands that are *on* (—), *behind* (⋯⋯), or *in front of* (◀) the plane of the page.

Exercises

4. Draw Lewis electron-dot structures for all the molecules in this section. (Include all valence electrons either in bonds or as dots for unshared pairs.)

5. Predict the bond angles for the bonds to the carbon atoms in boldface type in the following compounds. Using solid, dashed, and wedge bonds, show their geometry, and don't forget unshared pairs on O and N. (You need to show the geometry of the methyl group only once, then write it CH_3—.)

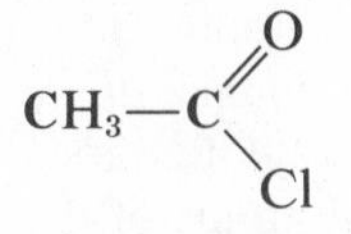

Acetic acid chloride
(acetyl chloride)

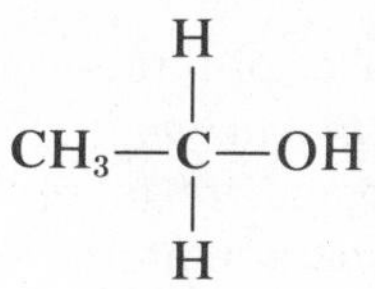

Ethanol
(ethyl alcohol)

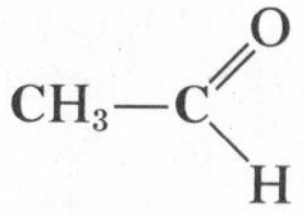

Ethanal
(acetaldehyde)

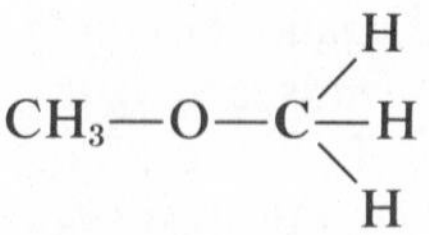

Dimethyl ether

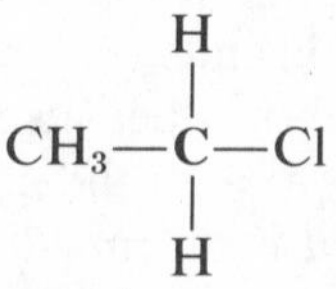

Chloroethane
(ethyl chloride)

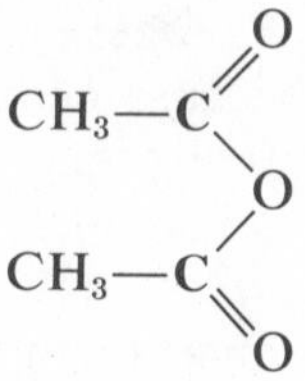

Acetic anhydride

$$CH_3—C\equiv C—MgBr$$

1-propynemagnesium
bromide

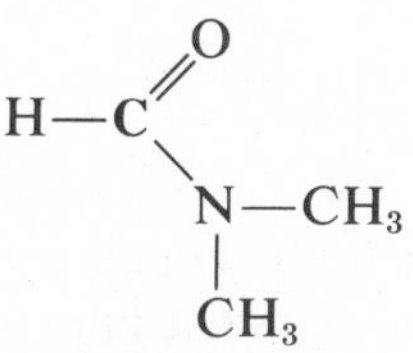

Dimethyl formamide

6. The I—C—I angle in iodoform (HCI_3, used as an antiseptic) is 113°. Why is it larger than the 109.5° predicted?

7. In phosgene, $COCl_2$, used in protein chemistry and as a war gas, the oxygen and both chlorines are bonded directly to carbon. Give the Lewis structure of phosgene and predict the value of the Cl—C—O angle.

13.4 Naming Organic Compounds—Skeletal and Functional Isomers If chemists, pharmacists, doctors, and nurses are to be able to communicate clearly in words as well as by drawing structures, they have to agree about the names they use. Therefore this section introduces you to the rules for naming organic compounds, but only very briefly, because the internationally agreed on or IUPAC naming rules cover more than 50 pages of fine print.

IUPAC—International Union of Pure and Applied Chemistry. The rules are in any recent edition of the *Handbook of Chemistry and Physics* published by the Chemical Rubber Company.

The first step in naming an organic compound is to find the carbon skeletons in it. Carbon skeletons, unlike the *complete skeleton* just explained for penicillin, consist of carbon atoms *only*. For example, ethyl propyl ether has the following complete skeleton:

$$H-\underset{\underset{H}{|}}{\overset{\overset{H}{|}}{C}}-\underset{\underset{H}{|}}{\overset{\overset{H}{|}}{C}}-O-\underset{\underset{H}{|}}{\overset{\overset{H}{|}}{C}}-\underset{\underset{H}{|}}{\overset{\overset{H}{|}}{C}}-\underset{\underset{H}{|}}{\overset{\overset{H}{|}}{C}}-H$$

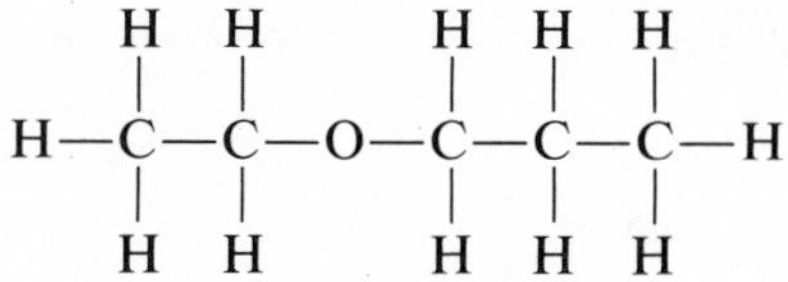

but the molecule has *two* carbon skeletons in it, one of two and one of three carbon atoms:

C—C C—C—C

Carbon skeletons take their names from the straight-chain or "normal" hydrocarbon that has the same number of carbon atoms. The names, structures, and skeletons of a few of these hydrocarbons are as follows:

Methane	CH_4	C
Ethane	$CH_3—CH_3$	C—C
Propane	$CH_3—CH_2—CH_3$	C—C—C
Butane	$CH_3—CH_2—CH_2—CH_3$	C—C—C—C
Pentane	$CH_3—CH_2—CH_2—CH_2—CH_3$	C—C—C—C—C
Hexane	$CH_3—CH_2—CH_2—CH_2—CH_2—CH_3$	C—C—C—C—C—C

Starting at pentane, the names of the hydrocarbons consist of "-ane" preceded by the Greek word for the number of carbon atoms: *hex*ane for six carbon atoms, *hept*ane for seven, *oct*ane for eight, *non*ane for nine, *dec*ane (pronounced "deck-ane") for ten, and so on. Each hydrocarbon name matches the name of a *group* (or radical) obtained from the hydrocarbon by removing one hydrogen to permit attachment of the hydrocarbon group to the rest of the molecule. Groups are named by replacing "-ane" in the name of the hydrocarbon by "-yl." Thus $CH_3—$ from meth*ane* is the meth*yl* group, $CH_3—CH_2—$ from eth*ane* becomes the eth*yl* group, and so on. Looking at the structure of ethyl propyl ether, you can see that it results from joining an ethyl group and a propyl group to an oxygen atom, and is named according to the C—C and C—C—C carbon skeletons.

$CH_3—CH_3$ $CH_3—CH_2—CH_3$
 Ethane Propane

$CH_3—CH_2—$ $CH_3—CH_2—CH_2—$ $CH_3—CH_2—O—CH_2—CH_2—CH_3$
 Ethyl group Propyl group Ethyl propyl ether

However, there is another way that you can make a propyl group, namely, by removing a hydrogen from the $—CH_2—$ instead of from one of the $—CH_3—$ groups in propane. To indicate this, the group is called

2-propyl because the attachment is at the second carbon atom in the chain,

$$CH_3-CH_2-CH_2-$$

Propyl group

$$CH_3-\overset{|}{\underset{3}{C}}H-\underset{1}{CH_3}$$
$$2$$

2-propyl group

rather than on the end. There should then be two ethyl propyl ethers, the one just discussed and one named ethyl 2-propyl ether:

$$CH_3CH_2-O-\overset{\overset{\displaystyle CH_3}{|}}{\underset{\underset{\displaystyle CH_3}{|}}{C}}-H$$

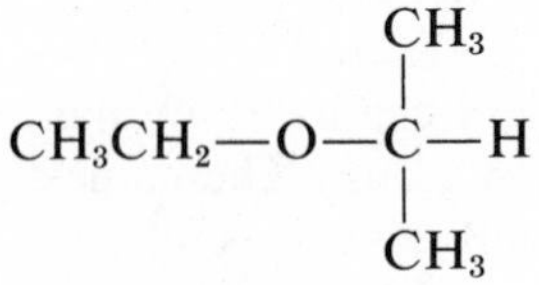

Ethyl 2-propyl ether

These two ethers are *different* compounds because they have *different skeletons*: ethyl propyl ether boils at 64°C; ethyl 2-propyl ether boils at 53°C. Different compounds that have the same molecular formula ($C_5H_{12}O$ in this case) are called *isomers*. Isomers of this type are called skeletal or positional isomers because the isomers have the same *functional group* (ether group in this example) but their skeletons are different. Other examples of this type of isomerism using the functional group names given in Table 13.1 follow. In these the carbon skeletons are numbered to show where the functional group is attached:

1. $CH_3-CH_2-CH_2-OH$　　and　　$CH_3-\overset{\overset{\displaystyle OH}{|}}{\underset{2}{C}}H-\underset{1}{CH_3}$
$$　　　　　　　　　　　　　　　　　　　3

1-propan*ol*　　　　　　　　　　　　　　　2-propan*ol*
(propyl alcohol)　　　　　　　　　　　(isopropyl alcohol)

2. $\underset{4}{CH_3}-\underset{3}{CH_2}-\underset{2}{CH_2}-\underset{1}{CH_2}-Br$　　and　　$\underset{1}{CH_3}-\overset{\overset{\displaystyle Br}{|}}{\underset{2}{C}}H-\underset{3}{CH_2}-\underset{4}{CH_3}$
$$　　　　　　　　　　　　　　　　　　　　$(4)\quad(3)\qquad(2)\quad(1) \leftarrow$ Incorrect

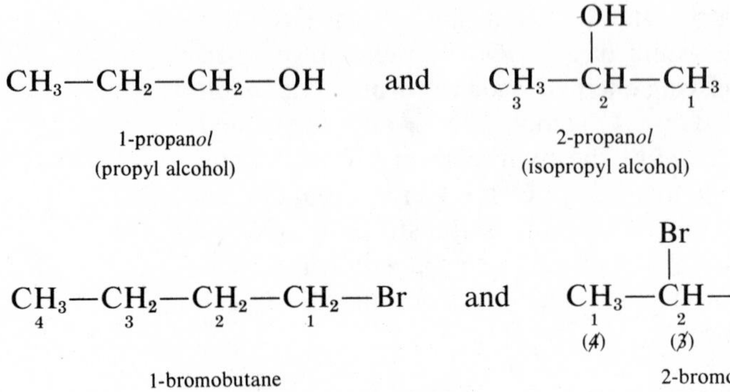

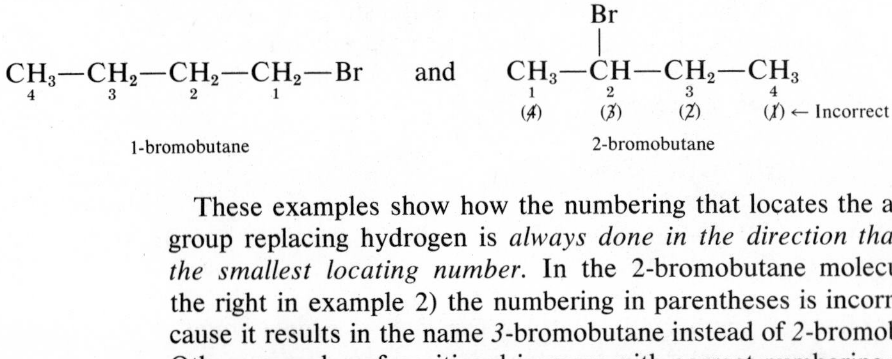

1-bromobutane　　　　　　　　　　　　2-bromobutane

These examples show how the numbering that locates the atom or group replacing hydrogen is *always done in the direction that gives the smallest locating number*. In the 2-bromobutane molecule (on the right in example 2) the numbering in parentheses is incorrect because it results in the name *3*-bromobutane instead of *2*-bromobutane. Other examples of positional isomers with correct numbering follow:

3. $CH_3-CH_2-\underset{||}{\overset{O}{C}}-CH\underset{CH_3}{\overset{CH_3}{<}}$ and $CH_3-CH_2-\underset{||}{\overset{O}{C}}-CH_2-CH_2-CH_3$

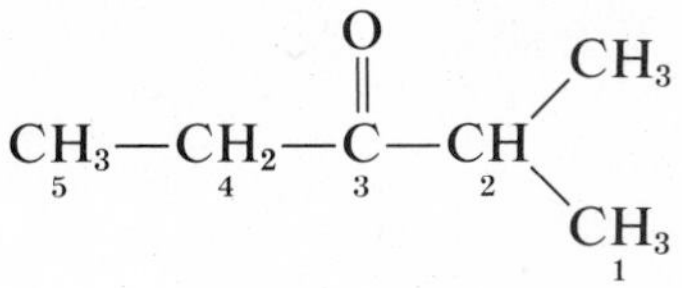

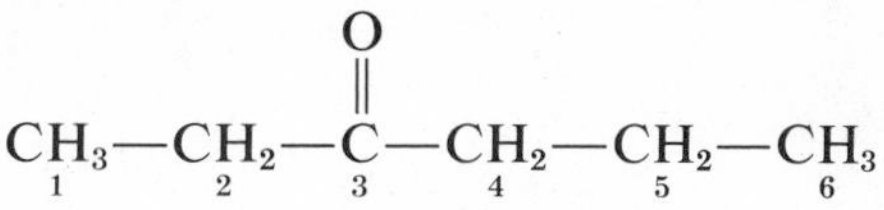

2-methyl-3-propanone
(ethyl isopropyl ketone)

3-hexanone
(ethyl propyl ketone)

4. $CH_3-CH_2-\underset{||}{\overset{O}{C}}-CH\underset{CH_3}{\overset{CH_2CH_2CH_3}{<}}$ and $\underset{CH_3}{\overset{CH_3}{>}}CH-\underset{||}{\overset{O}{C}}-CH_2-CH\underset{CH_3}{\overset{CH_3}{<}}$

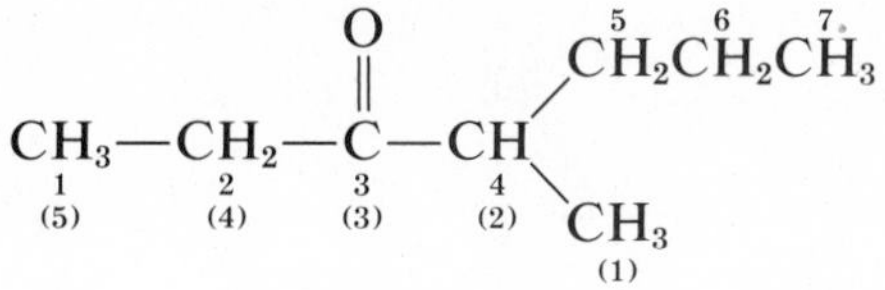

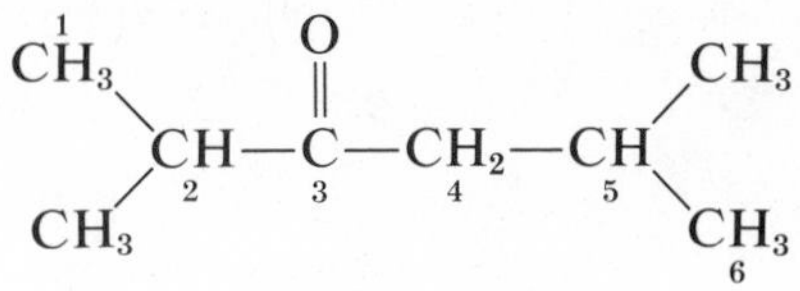

4-methyl-3-heptanone

2,5-dimethyl-3-hexanone
(isobutyl isopropyl ketone)

Naming the ketone on the left in example 4 illustrates the rule that *the longest chain of self-linked carbons is numbered*. It is wrong to name the ketone 2-propyl-3-pentanone according to the numbers in parentheses because the shorter five-carbon chain instead of the longer seven-carbon chain was numbered.

Molecules also may be isomeric because they have different functional groups rather than just different skeletons. Dimethyl ether and ethanol (Section 4.2), both with molecular formulas C_2H_6O, are *functional isomers*:

CH_3-O-CH_3 and CH_3-CH_2-OH
Dimethyl ether Ethanol

If isomers are of the functional type, their names will show it:

5. $CH_3-CH_2-\underset{||}{\overset{O}{C}}-CH_3$ and $CH_3-CH_2-CH_2-\underset{|}{\overset{H}{C}}=O$

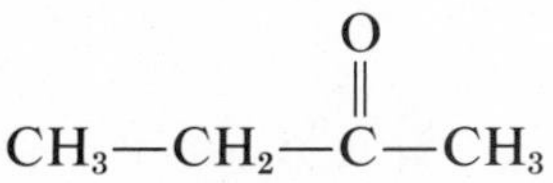

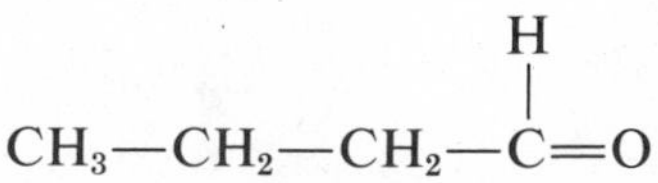

2-butan*one*
(methyl ethyl ket*one*)

Butan*al* ☞
(butyr*aldehyde*)

Naming a structure you are looking at is always more difficult than writing the structure if you have its name. You can't expect to be able to name complicated molecules from this brief set of rules, but the rules and the functional groups in Table 13.1 should help you to translate names of compounds into bond structures.

It is not necessary to locate the —CHO (aldehyde) group with a number (for example, "1-butanal") because the aldehyde function has to be at the end of the chain. Notice also that the carbon atom in the aldehyde group is counted in finding the chain length.

8. Give the bond structures for the following compounds:
 a. 3-chloropentane

Exercises

b. 2-hexanone
 c. Octanal
 d. 2,2,4-trimethylpentane ("isooctane")
 e. 3-methyl-1-decanol
 f. 2-bromo-2-methylpropane
 g. Decanoic acid (refer to Table 13.1 and be sure to count the carboxyl carbon)
 h. 2-heptanol

9. Name the following structures, using Table 13.1 as needed.

a.
$$\underset{\text{a.}}{}\quad CH_3-CH_2-\overset{\displaystyle OH}{\underset{\displaystyle |}{CH}}-CH_3$$

b. $CH_3-CH=CH-CH_2-CH_3$

c.
$$CH_3-CH_2-\overset{\displaystyle O}{\overset{\displaystyle ||}{C}}-CH\overset{\displaystyle CH_2-CH_2-CH_2-CH_3}{\underset{\displaystyle CH_2-CH_3}{<}}$$

d. $CH_3-CH_2-CH_2-\overset{\displaystyle O}{\overset{\displaystyle \nearrow\!\!\!\!\!\diagup}{C}}-OH$

e.
$$CH_3-CH_2-\overset{\displaystyle CH_3}{\underset{\displaystyle |}{CH}}-\overset{\displaystyle O}{\overset{\nearrow}{C}}-OH$$

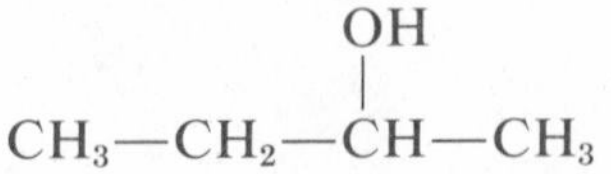

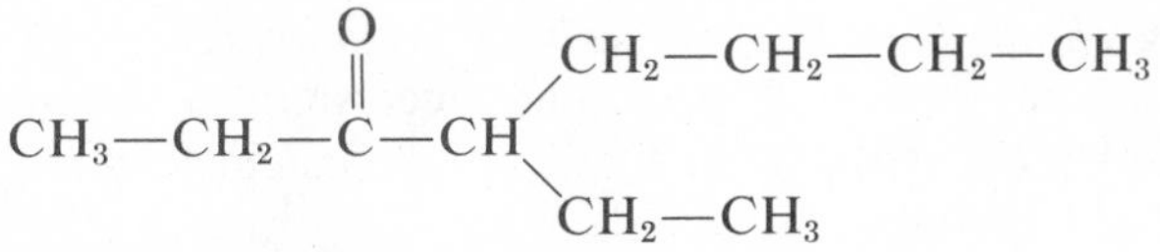

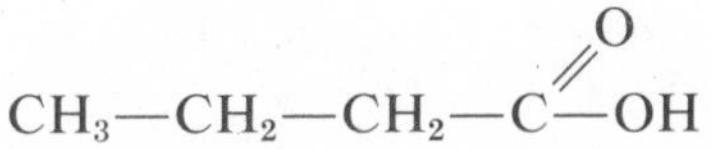

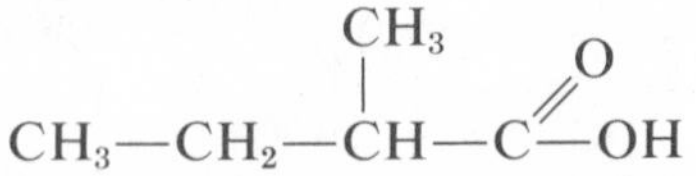

f. $CH_3-CH_2-CH_2-C\equiv C-CH_3$

g. HCl_3

10. Give structural formulas and names for the isomers of C_5H_{12}. What kind of isomers are these?

11. Give structural formulas and names for all the isomers, skeletal and functional, having the molecular formula $C_4H_{10}O$.

12. Freons 20 and 22, used in home refrigerators and air conditioners (Section 5.7), are, respectively, difluorodichloromethane and difluorochloromethane. Give their bond structures.

13.5 Stereoisomerism — *Cis–Trans* and Optical Isomers Two molecules that have the *same molecular formula* are *isomers* of one kind or another if their molecules cannot be superposed so that their atoms all coincide when one is placed on top of the other. Butane and isobutane molecules, isomers having formula C_4H_{10}, cannot be superposed because their carbon skeletons are different—they are *skeletal isomers.*

$$CH_3{-}CH_2{-}CH_2{-}CH_3$$

Butane

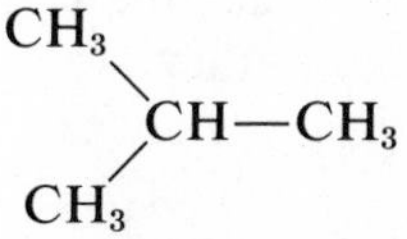

Isobutane
(2-methylpropane)

In testing for identity or isomerism there are some operations you can do and some you can't. Free rotation around the *single* bonds in a chain is permitted, but breaking and remaking bonds is not. Free rotation means that the following structures all stand for the *same* molecule, pentane—the bent-chain formulas can be turned into the straight-chain one by rotating about the single bonds:

$$CH_3{-}CH_2{-}CH_2{-}CH_2{-}CH_3$$

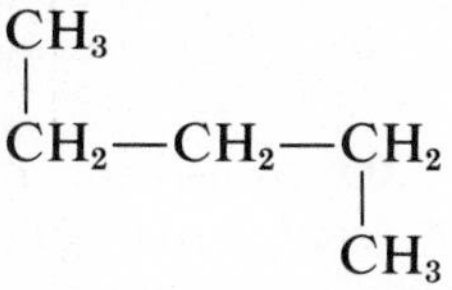

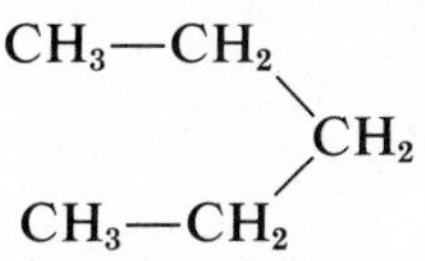

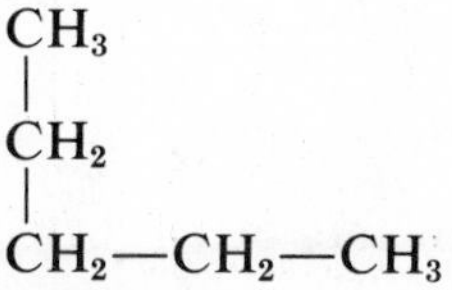

You can quickly convince yourself by making a model of pentane that these are just different ways of drawing the same molecule. You will find that you can make arrangements like those given (and many more) by rotations about single bonds, and without breaking any bonds. At ordinary temperatures rotation about single bonds in a chain is going on all the time, so the different structures for pentane are not isomers.

On the other hand, *for a single bond in a ring, free rotation is impossible.* The ring compound *cis*-1,2-dihydroxycyclopentane ☛ cannot be converted to the *trans*-isomer without breaking and remaking bonds:

Carbon ring compounds are named by putting "cyclo" before the name of the straight-chain hydrocarbon that has the same number of carbon atoms.

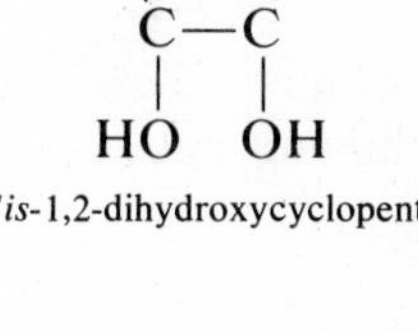

Cis-1,2-dihydroxycyclopentane

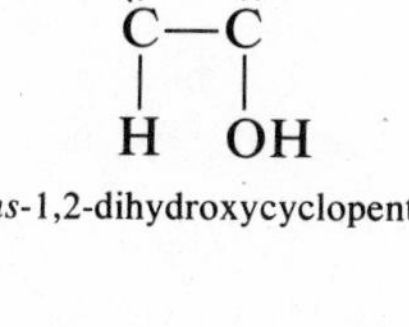

Trans-1,2-dihydroxycyclopentane

Molecular models will quickly convince you that these *cis*- and *trans*-isomers cannot be superposed, that changing one to the other requires bond breaking and remaking. These two compounds are an example of *cis–trans* or *geometric* isomers in a cyclic compound. In naming isomers of this type, *cis* means the OH groups are on the same side of the ring, *trans* means that they are across from each other.

Molecules that contain a carbon–carbon double bond can also show *cis–trans* isomerism. Rotation of carbon atoms joined by a double bond is not permitted when trying to superpose two molecules. The *cis*- and *trans*-isomers of 1,2-dichloroethene are an example of double-bond geometric isomerism:

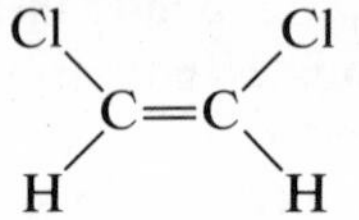

Cis-1,2-dichloroethene (bp 60°C, dipole moment 1.9 debye)

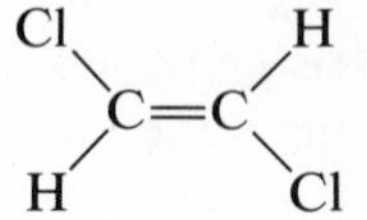

Trans-1,2-dichloroethene (bp 47°C, dipole moment 0)

The dipole of the *trans*-isomer is zero because the carbon–chlorine bond moments cancel each other as a result of the geometry of the molecule (Section 4.3 – Polar Molecules). The boiling point of the *cis*-isomer is higher because of dipole–dipole attractions (Section 4.6).

Exercises

13. Draw the structures of *cis*- and *trans*-1,2-dibromocyclopentane.

14. The structure of 1,1-dichloroethene is

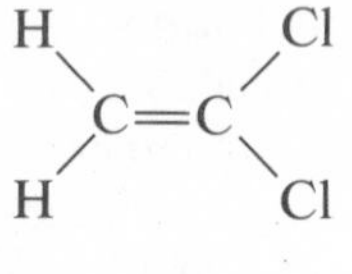

Is it capable of *cis–trans* isomerism? What kind of isomer of the 1,2-dichloroethenes is it, functional or skeletal?

15. If *cis*-1,2-dichloroethene is heated strongly in the presence of a catalyst, it is converted to the *trans*-isomer. Give *two* bond-breaking, bond-making mechanisms for the reaction. From an energy point of view, which reaction is more likely? (*Hint*: breaking and remaking two bonds is harder than breaking and remaking one.)

16. Using molecular models if necessary, find out which, if any, of the following structures are incorrect for the ethanol molecule:

```
   H  H              H  H              H  H
   |  |              |  |              |  |
H—C—C—OH         H—C—C—H          H—C—C—H
   |  |              |  |              |  |
   H  H              OH H              H  OH

  HO  H              H  H              H  OH
   |  |              |  |              |  |
H—C—C—H          HO—C—C—H         H—C—C—H
   |  |              |  |              |  |
   H  H              H  H              H  H
```

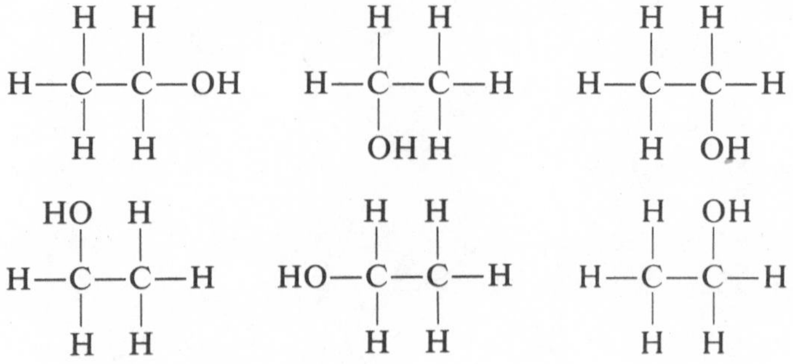

17. What kind of isomers are methyl acetate, CH_3COOCH_3, and propionic acid, CH_3CH_2COOH? Can you write another isomer having their molecular formula, $C_3H_6O_2$? (*Hint*: the simplest carboxylic acid is formic acid, $H—C\overset{O}{\big\Vert}—OH$.)

18. Suppose the angles between carbon atoms were 90° instead of 109.5°. This would mean that a carbon atom and all four of its ligands would lie in the same plane. How many *skeletal* isomers of chlorobromoiodomethane would there be?

Optical isomerism Optical isomerism has to do with handedness or *chirality* (from the Greek word *cheir,* meaning "hand") in the structure of molecules. Optical isomers have the same molecular formula, the same skeleton, and the same functional groups, but they cannot be superposed so that their atoms coincide, just as you cannot superpose your own hands ☞ so that your fingers coincide. Like our hands, optical isomers come in pairs, and each isomer is the mirror image of the other because of the way their atoms are arranged in space. Figure 13.2(a) (p. 504) compares a *left* hand with its mirror image. The image in the mirror is superposable on a *right* hand and is therefore identical with a right hand. A molecule or object that doesn't have chirality is superposable on its *own* mirror image. ☞

Optical isomerism results whenever a carbon atom has *four different ligands*. Figure 13.2(b) shows the pair of optical isomers for the compound chlorobromoiodomethane, a carbon atom with H, Cl, Br, and I as ligands. The formulas for the two optical isomers are written by using the dashed-line and wedge bonds described in Section 13.1 to show roughly how the ligands are arranged (tetrahedrally) in space.

Placing your hands palm-to-palm is *not* superposition. Superpose means "put on top of," and you can't get your hands to coincide by putting one on top of the other.

These objects or molecules are termed "achiral," meaning "without chirality."

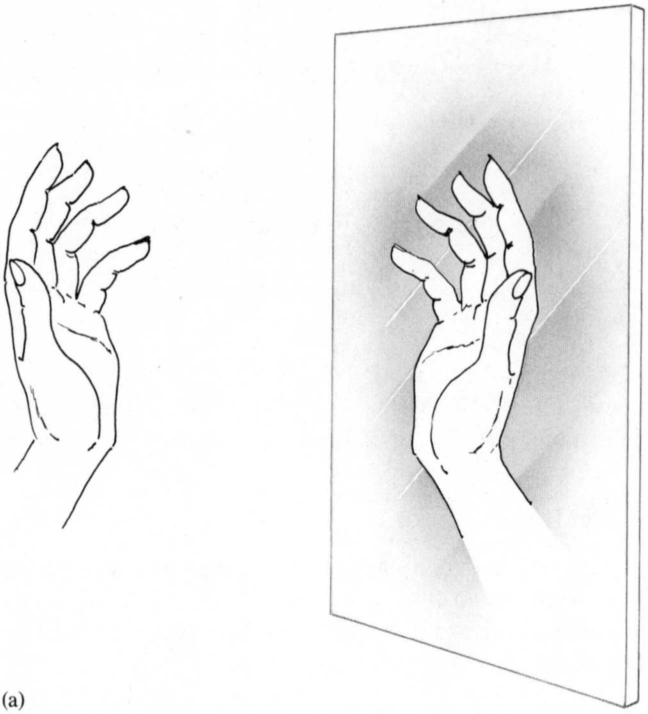

FIG. 13.2 (a) A left hand and its mirror image, a right hand.

(a)

From these structures you can see that the isomers are not superposable. If you rotate the right-hand structure (b) in the plane of the paper as indicated by arrow 1 until the Cl and Br ligands coincide, you get (c), which is not identical with (a) because the H and I ligands do not coincide. If you flip (b) over around the H······C······I axis as shown by arrow 2 in (b) to put the chlorine on the left as in (a), you get (d).

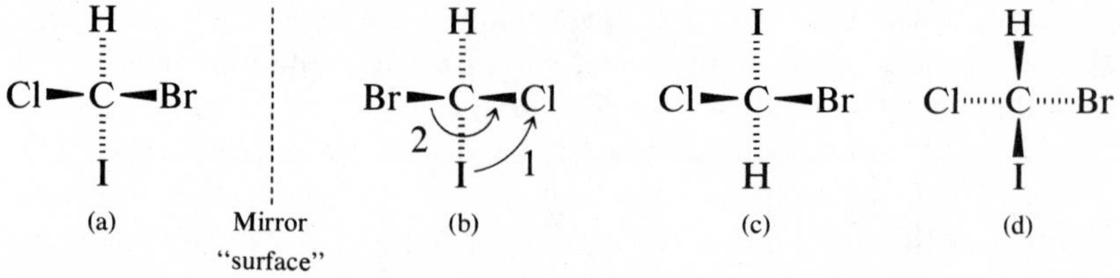

Optical isomers of chlorobromoiodomethane

The flipping puts Cl and Br *behind* the plane of the paper and brings H and I forward out of the plane, so (d) is not superposable on (a) either. Work with toothpick–gumdrop or other models will quickly remove any uncertainty in your mind about the truth of these statements.

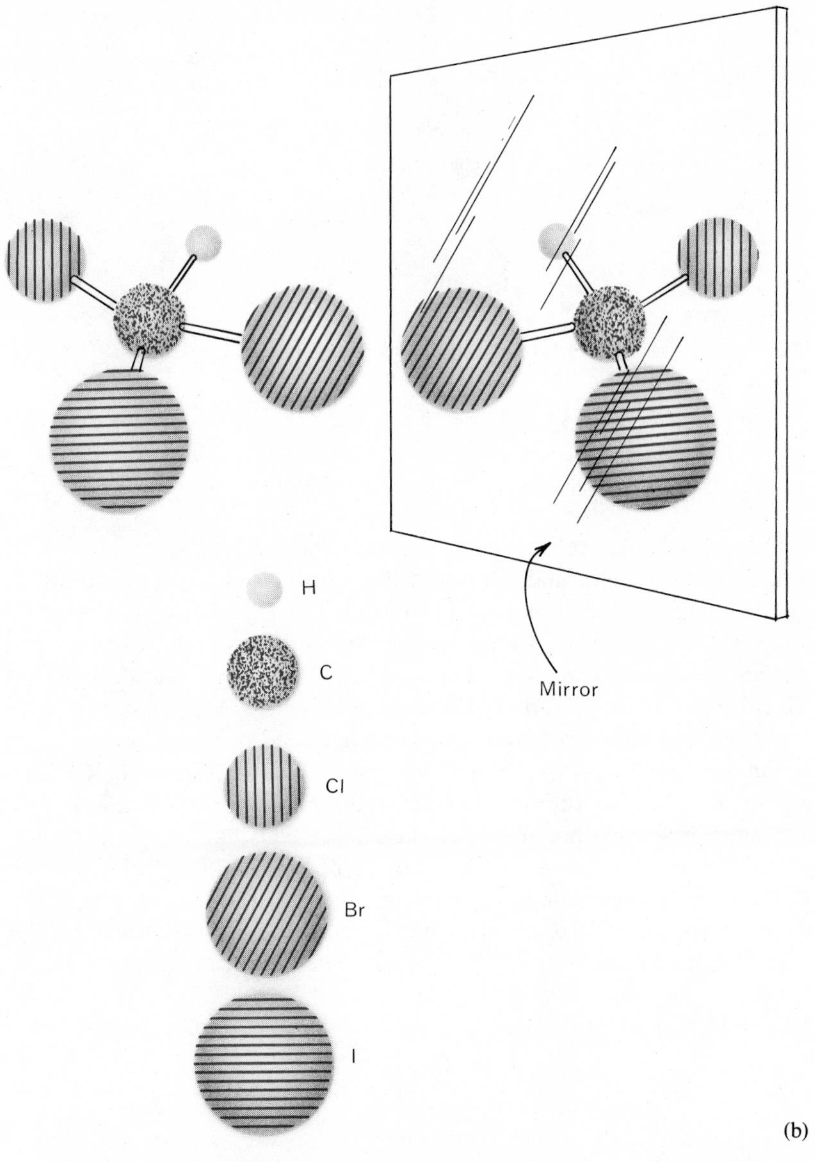

FIG. 13.2 (b) the optical isomers of chlorobromoiodomethane.

Optical isomers always result when a carbon atom has four different ligands. Carbon atoms of this type are *chiral centers* (also called asymmetric or dissymmetric carbon atoms) and are often marked with an asterisk to identify them in more complicated molecules.

EXAMPLE

Some biochemically important molecules that have two optical isomers are the following:

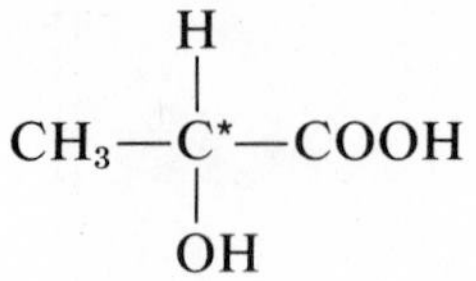

Lactic acid

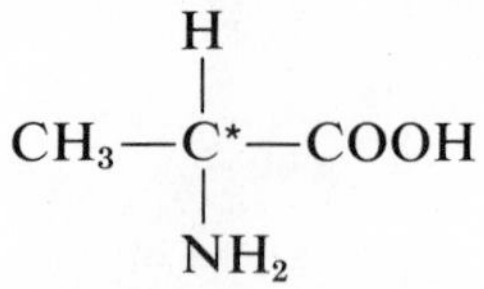

2-aminopropionic acid
(alanine)

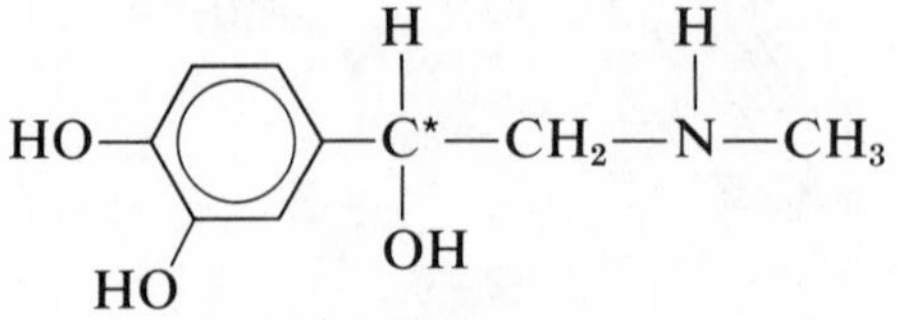

Adrenalin
(epinephrine)

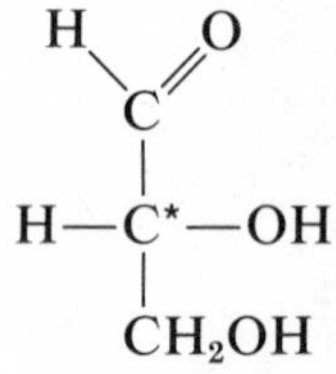

Glyceraldehyde
(2,3-dihydroxypropanal)

One optical isomer of lactic acid is produced by fermentation of milk sugar (lactose) and is responsible for the taste of sour milk; its mirror-image molecule is found in muscle tissue. Alanine is one of 19 aminocarboxylic acids found in protein (meat, muscle, and so on), where the aminocarboxylic acids are joined through peptide linkages (Section 14.2). Your adrenal glands release the hormone adrenalin into your bloodstream when you are frightened or under pressure. In seconds it prepares your body for maximum performance. Glyceraldehyde serves as one of the intermediate compounds in the photosynthetic reaction that forms glucose from CO_2 and water in plants. All of these examples have one pair of optical isomers, which is the rule for molecules that have only one chiral carbon atom. However, if there are n chiral carbon atoms in a molecule, there will be $2n^2$ optical isomers.

The *n* chiral carbons must be what are called "different;" that is, none of them can have the same set of four ligands for the rule to work.

Exercises

19. Which of the following structures will have optical isomers?

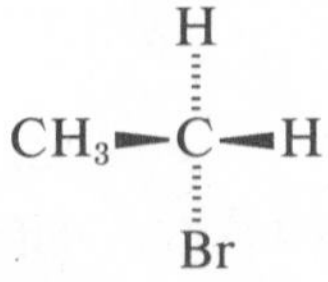

(a)

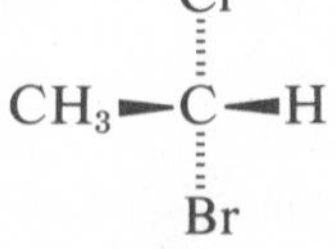

(b)

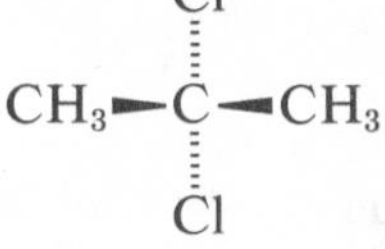

(c)

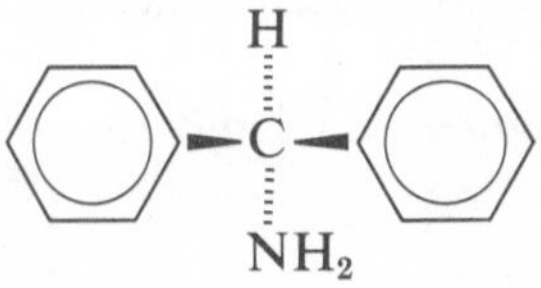

(d)

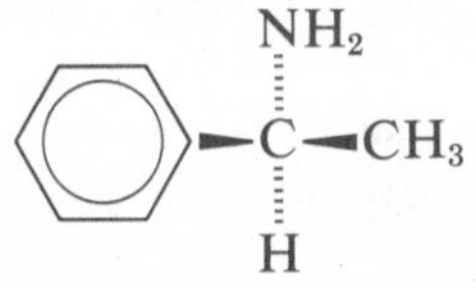

(e)

For those that do, give the structure of the other isomer, using dashed and wedge bonds.

20. The open-chain form of glucose (blood sugar) has the following structure, in which the carbon chain is seen as lying on the plane of the paper with H and OH groups projecting toward the reader:

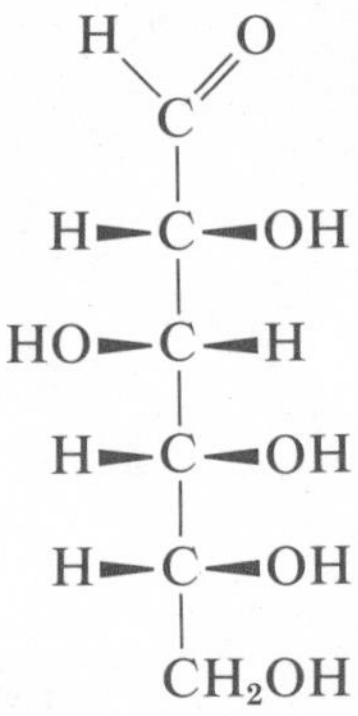

How many chiral carbon atoms are there in the glucose molecule? How many optical isomers are there? The glucose isomer given is called D-glucose. Draw its mirror image.

21. Suppose, as we did in Exercise 18, that the carbon atom had its four bonds in the *same plane*. How many, if any, *optical* isomers of chlorobromoiodomethane would there be? How important to the existence of optical isomers is the tetrahedral direction of the four bonds from carbon?

Optical isomers have exactly the same molecular formulas, functional groups, and boiling and melting points. How can we tell them apart? How is it known that they even exist? *All* of the physical properties of a pair of optical isomers are alike except *one*. That one property is the direction in which the isomers rotate the plane of polarization of light (Figure 13.3). One of the isomers will rotate the plane of polarization a certain number of degrees in the clockwise direction, the other an exactly equal number of degrees but in a counterclockwise direction, ¶ provided only that the thicknesses of the sample are equal. The clockwise rotating isomer is called the dextrorotatory (right-rotating) isomer and is indicated by a (+) sign as in (+) lactic acid. The counterclockwise rotating isomer is termed levorotatory (left-rotating) and is identified with a (−) sign as for (−) lactic acid in Figure 13.3. Pairs of optical isomers are called enantiomorphic forms of the com-

¶ If the isomers are not liquid, then their rotations must be measured, using solutions of them, and not only the lengths of the light paths but also the concentrations of the isomers in their solutions must be equal.

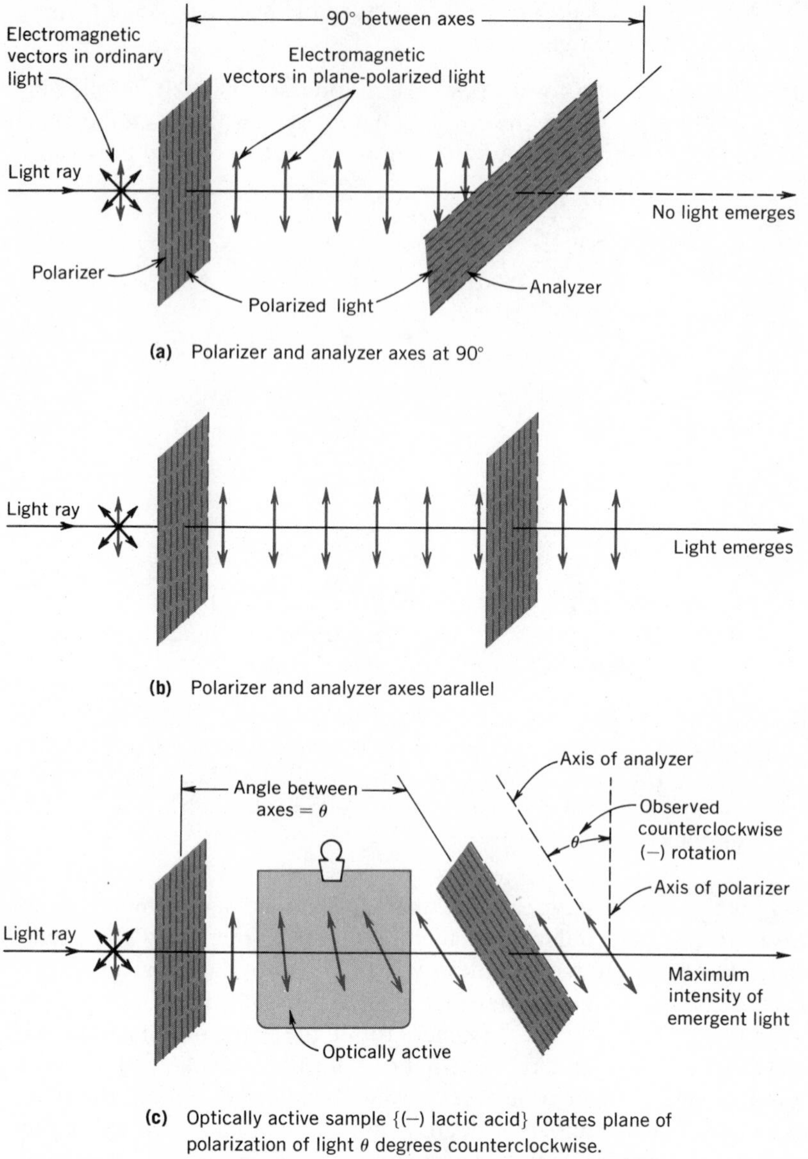

FIG. 13.3 Polarization of light and optical activity.

pound or simply "enantiomorphs." Enantiomorphs frequently behave differently in the biochemical reactions in our bodies and other living systems. This remarkable property is discussed in Chapter 14.

At this point the foundation for understanding the structures and uses of many organic compounds has been laid. In the rest of this chapter, "case histories" of some important organic compounds will show you that many of the principles you have learned have very broad application in organic chemistry.

Exercises

22. One of the optical isomers of benzedrine ("speed," a potent, useful, but dangerous drug) has the following structure:

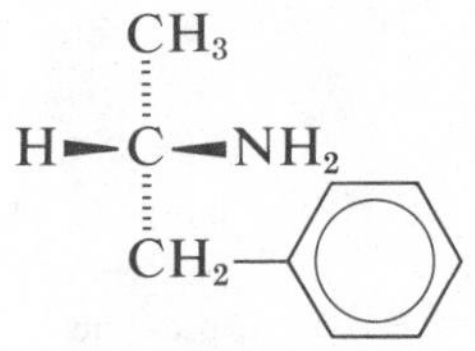

Benzedrine
(amphetamine)

Give the structure of the other optical isomer.

23. The amino acid valine is 2-amino-3-methylbutanoic acid. How many chiral carbon atoms are there in the molecule? Draw the optical isomers of valine, using dashed and wedge bonds.

13.6 Hydrocarbons Compounds made up of carbon and hydrogen only are called hydrocarbons and are divided into the classes shown in the table on the next page.

The main sources of alkanes, alkenes, and cycloalkanes are natural gas and crude oil (petroleum), from which they are separated by fractional distillation (Section 6.3). Arenes are found in petroleum and coal tar. Natural gas is mostly methane (CH_4, bp −163°C) plus small amounts of ethane (C_2H_6, bp −89°C), propane (C_3H_8, bp −45°C), and butane (C_4H_{10}, bp −0.5°C). Natural gas is liquefied by compression and cooling to produce liquid natural gas or "LNG." Propane and butane from the distillation of petroleum are compressed into tanks to form liquid petroleum gas or "LPG." This LPG is used for heating and cooking in regions where natural gas is not available. Oceangoing tankers transport huge tonnages of LPG and LNG from foreign oil and gas fields to the U.S.

A developing use for LPG and LNG is as the motor fuel for buses and trucks, replacing diesel oil and gasoline. The C_1—C_4 alkanes they contain burn cleanly to CO_2 and water, giving an almost odorless exhaust that contributes much less unburned hydrocarbon to smog than the usual fuels. By installing a special carburetor you can even convert your automobile to use LPG. The only disadvantage is that LPG must be stored under pressure in steel tanks because the vapor pressure over the liquid is quite high—at 25°C the vapor pressure of propane is 10 atm (147 psi) and that of butane is about 3 atm.

Huge quantities of natural gas are burned to heat buildings, to provide steam for generating electricity, and to provide thermal energy for countless industrial processes. For the main component, the reaction and heat of combustion are as follows:

Class	*General Formula*	*Examples*	
Alkanes (aliphatic hydrocarbons) (saturated hydrocarbons)	C_nH_{2n+2}	CH_4	methane
		C_6H_{14}	hexane(s)
		C_8H_{18}	octane(s)
Alkenes (olefinic hydrocarbons) (unsaturated hydrocarbons)	C_nH_{2n}	$CH_2{=}CH_2$	ethene
		$CH_3{-}CH{=}CH_2$	propene
Alkynes (acetylenic hydrocarbons)	C_nH_{2n-2}	$HC{\equiv}HC$	acetylene
		$CH_3{-}C{\equiv}CH$	propyne
Cycloalkanes (alicyclic hydrocarbons)	C_nH_{2n}	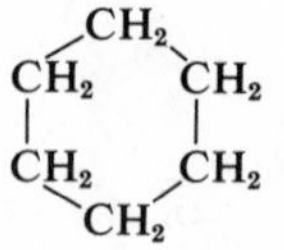	cyclohexane
Arenes (aromatic hydrocarbons)	none	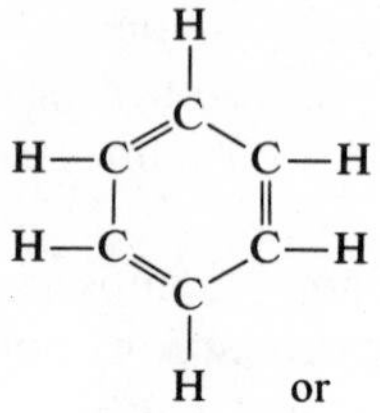 or 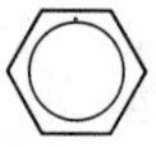	benzene
		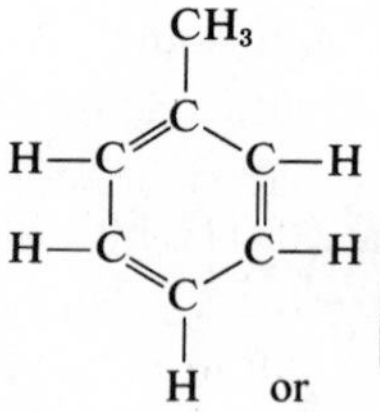 or 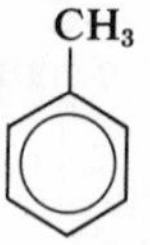	toluene (methylbenzene)

$$CH_4(g) + 2O_2(g) \rightarrow CO_2(g) + 2H_2O(g) \qquad \Delta H = -192 \text{ kcal}$$

Like LPG, methane (LNG) is a clean fuel, contributing virtually nothing to air pollution. On the other hand, coal and fuel oils from many sources contain sulfur compounds that pollute the air with SO_2 when they are burned. For this reason, many electric generating and industrial plants have switched to natural gas to cut down air pollution. The U.S. has large proven reserves of natural gas, and there are large

A tiny trace of methyl mercaptan, CH_3—SH, which stinks, is added to natural gas used in houses to help detect leaks. Methane itself has very little odor and forms terribly explosives mixtures with air.

Isomerism of the skeletal type first appears at C_4H_{10} in butane $CH_3—CH_2—CH_2—CH_3$, and isobutane (2-methylpropane) $\begin{matrix}CH_3\\ CH_3\end{matrix}\!\!>\!CH—CH_3$. It increases rapidly with the number of carbon atoms in the hydrocarbon. Theoretically there are 366,319 skeletal isomers of the alkane eicosane, $C_{20}H_{42}$. There are so many isomers of $C_{30}H_{62}$ that it would take 130 years to draw all their formulas at the rate of one per second.

sources overseas, but if the rate of energy production from its burning continues to increase as it has in the last few years, the supply will be used up in the foreseeable future. Coal and shale-oil resources in the U.S. are large enough to provide energy for hundreds of years, but much of the coal is high in sulfur, and cheap ways for extracting petroleum from shale oil have yet to be found. As these "fossil fuels"—gas, coal, and oil—are used up nuclear energy will have to take their places. Estimates are that within 40 years, other methods of electric power generation will have to be found.

Exercises

24. Make a graph of the boiling points given earlier for the first four alkanes against the number of carbon atoms they contain. By drawing the best straight line through the points, predict the boiling point for pentane. Does the boiling point of methane indicate anything unusual about this hydrocarbon? (Compare with bp of the noble gas of about the same molecular weight.)

25. If you were planning to use LPG as the fuel for a "rush to the North Pole" in snowmobiles equipped for a gaseous fuel, would you want LPG that was mostly butane or mostly propane? Explain. (*Hint*: think of arctic temperatures.)

26. Suppose the source for our bodies' energy was methane. What volume in liters at standard conditions (0°C, 1 atm) would correspond to a typical 2400-kcal diet? What volume of O_2 and of air (0°C, 1 atm) would we have to breathe? (Table 5.3.)

27. Give balanced equations for the combustion (burning) of propane and butane.

Gasoline Crude oil (petroleum) is a very complex mixture of organic compounds, mainly hydrocarbons in the C_4 to C_{40} range. It contains alkanes, alkenes, cycloalkenes, and arenes as well as small amounts of nitrogen and sulfur compounds, depending on the source. When you realize that there are already nine skeletal isomers of formula C_7H_{16} (heptane), you can easily see that the number of hydrocarbons in petroleum can be very large even if only a few of the isomers of each hydrocarbon are present. ☛ Petroleum is separated into mixtures of hydrocarbons by fractional distillation (Section 6.3). Gasoline is the fraction of petroleum boiling at temperatures between about room temperature and 200°C and is mainly made up of C_5 to C_{11} hydrocarbons, with C_8 predominating. It has been estimated that there are as many as 500 different hydrocarbons in gasoline alone. About 150 of them have been separated and identified—a very difficult task because the closeness of their boiling points makes separation by fractional distillation and other methods very difficult. Other fractions obtained on distillation of crude oil are *jet fuel* (kerosene, bp 200–

The number of skeletal isomers with molecular formula $C_{30}H_{62}$ is 4,111,846,763, as calculated by a complicated equation.

250°C), *fuel oil,* and *diesel oil* (bp 250–400°C), and lubricating oils and greases of even higher boiling point.

Catalytic cracking of the hydrocarbons in the kerosene, diesel oil, and higher boiling fractions of petroleum breaks the long-chain hydrocarbons they contain into smaller "gasoline-size" molecules and increases the yield of gasoline (Photo 34). Cracking also produces substantial amounts of ethene and propene, which are used to make plastics and alcohols by methods to be described shortly.

If you have ever "gassed up" a car, you may know that the most important property of a gasoline is its "octane" rating or number. You have to decide between "regular" and "premium"—between gasolines of lower and higher octane rating—to get good performance from your car's engine. The octane rating requirements of car engines vary according to the compression ratio. (This ratio is the volume of the cylinder with the piston at the bottom of its stroke divided by the volume of the cylinder with the piston at the top of its stroke.) Compression ratios range from 6:1 to a high of 9:1 in some engines—the higher the compression ratio, the higher the octane number required for the gasoline if the engine is to run smoothly without "knocking" or "pinging" on acceleration or when pulling up a hill. "Pinging" results from too rapid combustion (detonation) of the air–gasoline mixture in the cylinders and can definitely damage the engine if permitted to go on very long.

Octane rating is based on the knocking performance of the gasoline compared to that of a mixture of heptane and 2,2,4-trimethylpentane in a standard test engine. Heptane, like all straight-chain hydrocarbons, knocks or pings very badly, 2,2,4-trimethylpentane not at all.

$$CH_3{-}CH_2{-}CH_2{-}CH_2{-}CH_2{-}CH_2{-}CH_3 \qquad CH_3{-}\underset{\underset{\displaystyle CH_3}{|}}{\overset{\overset{\displaystyle CH_3}{|}}{C}}{-}CH_2{-}\underset{}{\overset{\overset{\displaystyle CH_3}{|}}{CH}}{-}CH_3$$

Heptane	2,2,4-trimethylpentane
(octane rating = 0)	(octane rating = 100)

The octane rating of a gasoline is the percentage of 2,2,4-trimethylpentane in a mixture with heptane that shows the same knocking behavior. Ratings above 90 are required by most high-performance car engines. Regular gasoline has a range from 80 to 90 octane, premium from 90 to 100. However, you gain nothing in performance by using a gasoline with higher octane rating than your car's engine is designed for.

The octane rating of gasoline distilled directly from petroleum is low (50–70) but can be increased in several ways. The more highly branched hydrocarbons have higher octane numbers, and *isomeriza-*

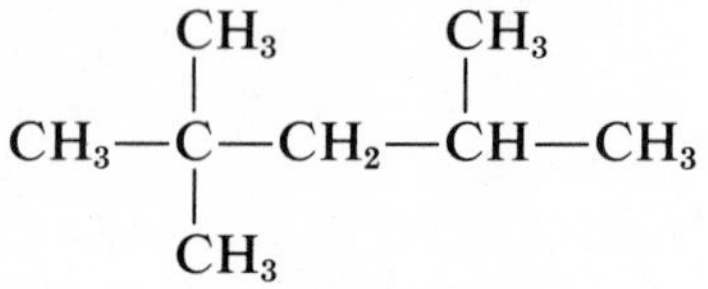

tion catalysts are used (Al_2Cl_6 mainly) to convert chains to branched chains. Platinum catalysts cause *cyclization* and *dehydrogenation* with the production of arenes, which also have high octane rating.

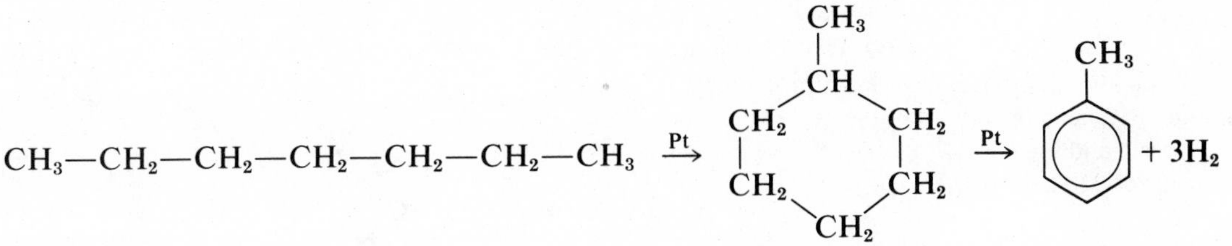

Methylcyclohexane Toluene

Finally, ethyl fluid (tetraethyl lead + 1,2-dibromoethane) can be added, although this practice is being phased out because the poisonous lead compounds in the exhaust pollute the air and the earth.

Octane rating has had important historical impact. Because the fighter planes of the British Royal Air Force had higher octane gasoline, they were able to fly higher than their enemies in the "Battle of Britain" in World War II.

Alkenes Alkenes are hydrocarbons whose molecules contain one or more carbon–carbon double bonds. Alkenes are also called *olefins* or *unsaturated hydrocarbons*. Alkenes are unsaturated because they undergo addition reactions that remove the double bond. They add hydrogen (Pt or Ni catalyst), halogens, water (acid catalyst), halogen acids, and many other reagents, becoming saturated in the process:

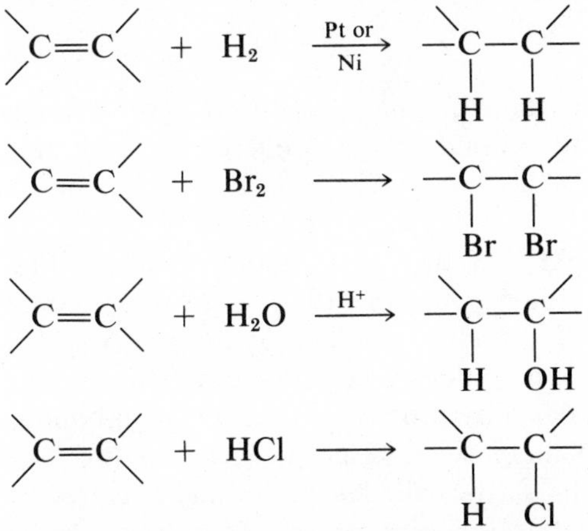

All of these reactions as well as the polymerization of alkenes to be described shortly are strongly exothermic and proceed virtually to completion.

EXAMPLE ⟜∘ You can use bond energies to calculate the energy changes accompanying addition reactions of alkenes. Taking the addition of hydrogen to ethene as an example, the following thought experiment gives the enthalpy change (ΔH) for the reaction:

Step 1. Breaking the π bond in ethene requires an energy input of 63 kcal/mole: 🖅

The value is the difference between the C=C bond energy of 146 kcal/mole and the C—C bond energy of 83 kcal/mole.

$$\begin{array}{ccc} \text{H} & & \text{H} \\ \diagdown & & \diagup \\ & \text{C}=\text{C} & \\ \diagup & & \diagdown \\ \text{H} & & \text{H} \end{array} \quad + \; 63 \text{ kcal} \rightarrow \quad \begin{array}{ccc} \text{H} & & \text{H} \\ \diagdown & & \diagup \\ & \overset{\cdot}{\text{C}}-\overset{\cdot}{\text{C}} & \\ \diagup & & \diagdown \\ \text{H} & & \text{H} \end{array}$$

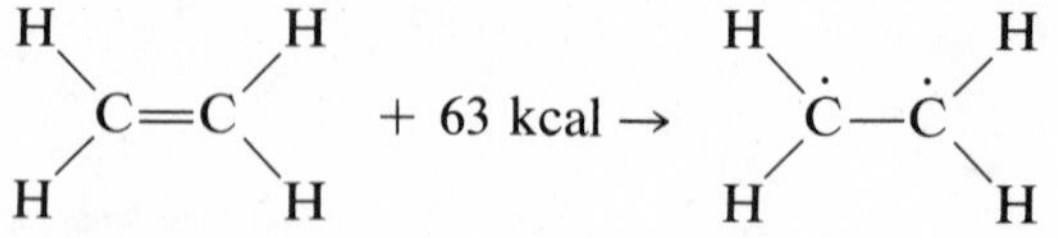

Step 2. Breaking the σ bond in the H_2 molecule requires energy input of 104 kcal/mole:

$$H_2 + 104 \text{ kcal} \rightarrow 2H\cdot$$

Step 3. Adding the $2H\cdot$ to the product of step 1 forms ethane and *releases* the bond energy of two C—H bonds or $2 \times 99 = 198$ kcal. Striking an energy balance, we see that $63 + 104$ kcal $= 167$ kcal was put in to break bonds in steps 1 and 2, and 198 kcal was released in bond formation in step 3. The difference ($198 - 167 = 31$ kcal/mole) is called the heat of hydrogenation of ethene to ethane: 🖅

The experimentally measured value is 32.8 kcal/mole. The difference of 1.8 kcal from the calculated value is typical of the accuracy of calculations using bond energies.

$$\begin{array}{ccc} \text{H} & & \text{H} \\ \diagdown & & \diagup \\ & \text{C}=\text{C} & \\ \diagup & & \diagdown \\ \text{H} & & \text{H} \end{array} \text{(g)} \; + \; H_2\text{(g)} \xrightarrow{\text{Pt}} \begin{array}{ccc} & \text{H} & \text{H} \\ & | & | \\ \text{H}-&\text{C}-\text{C}&-\text{H} \\ & | & | \\ & \text{H} & \text{H} \end{array} + \; 31 \text{ kcal}$$

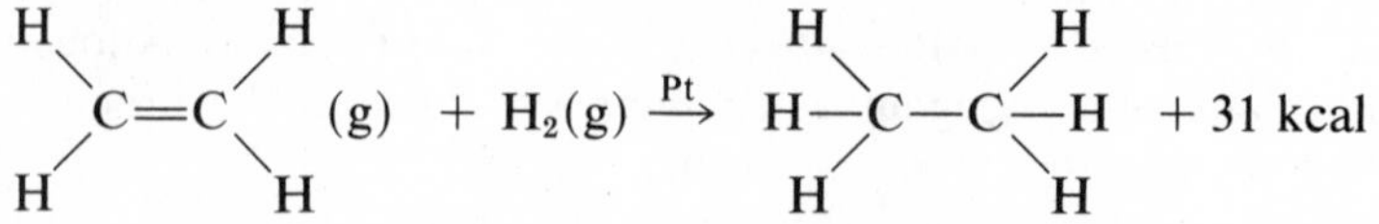

A very large number of organic chemical reactions "go" because the conversion of π bonds to σ bonds releases energy, as in the hydrogenation of ethene. ⟜∘

Ethene (ethylene, CH_2=CH_2) is the most important alkene. Production in the U.S. was 19 billion pounds in 1971, and was exceeded only by the productions of chlorine, sodium hydroxide, oxygen, ammonia, and sulfuric acid. Where does ethene come from and where does it go? Most of it comes as a by-product of cracking high-boiling-point crude oil fractions to increase the yield of gasoline. 🖅 And much of it goes into making the plastic called *polyethylene*. The process is called *polymerization* of the *monomer* ethene. To start the polymerization a catalyst or *chain initiator,* which we will represent by R—, is needed:

Until recently most of the ethene from cracking operations was simply burned at the end of a tall pipe producing a refinery "flare." Now ethene is being recovered.

$$R— + CH_2{=}CH_2 \rightarrow R—CH_2—CH_2—$$
$$R—CH_2—CH_2— + CH_2{=}CH_2 \rightarrow R—CH_2CH_2CH_2CH_2—$$

and so on. Hundreds of thousands of ethene molecules "zip up" to form a chain of CH_2 groups. The polymerization can be represented by using arrows to show bond formation:

$$R \quad CH_2{=}CH_2 \quad CH_2{=}CH_2 \quad CH_2{=}CH_2 \rightarrow RCH_2CH_2CH_2CH_2CH_2CH_2— \cdots$$

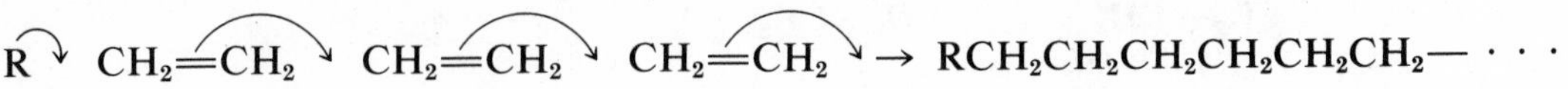

The chain stops building when the free valence at the end of one chain bonds to the free end of another or when some other chain-ending process occurs. The driving force for polymerization is conversion of (weaker) π bonds to (stronger) σ bonds.

Polymerization of ethene is the simplest example of the process called *vinyl polymerization* by which substituted ethenes in which one of the H's has been replaced are converted into a number of very useful plastics. Letting X represent the substituent taking the place of H and R— be the initiator, we can write the reaction as follows:

$$R \quad CH_2{=}CH \quad CH_2{=}CH \quad CH_2{=}CH \rightarrow R—CH_2—CH—CH_2—CH—CH_2—CH \cdots$$
$$\qquad \quad X \qquad\quad X \qquad\quad X \qquad\qquad\quad X \qquad\quad X \qquad\quad X$$

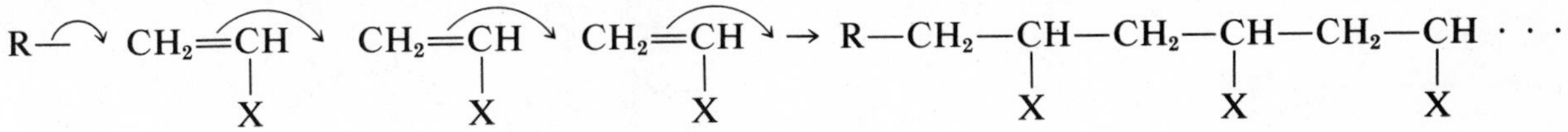

If X = Cl, then polyvinylchloride (PVC), used for shower curtains and raincoats, tubing (Tygon), and "vinyl" upholstery, results. ¶ If X = CN, polyacrylonitrile (Orlon) is produced, and if a mixture of acrylonitrile $CH_2{=}CH—CN$ and *butadiene* $CH_2{=}CH—CH{=}CH_2$ is polymerized, *nitrile rubber* is formed. Unlike natural rubber, nitrile rubber is unaffected by oils and solvents and is used to line fuel tanks and hoses. If X is a benzene ring ($C_6H_5—$), the monomer, $CH_2{=}CH—C_6H_5$, called styrene, polymerizes to *polystyrene*, familiar in a large number of plastic products from toys to red tail-light lenses. Polymerization of butadiene ($CH_2{=}CH—CH{=}CH_2$) with some styrene added gives GRS or "cold rubber," used extensively in tires. Polymerization of *isoprene* (2-methyl-1,3-butadiene) gives a synthetic rubber identical with the natural product. Natural rubber is cheaper, but the process does provide insurance against a cutoff in rubber from overseas:

The $CH_2{=}CH—$ group is called a vinyl group. Vinyl chloride, $CH_2{=}CH—Cl$ (bp $-14°C$), but not its polymer, has been found to cause liver cancer when inhaled in tiny amounts. Its use as a propellant in aerosol sprays has been stopped.

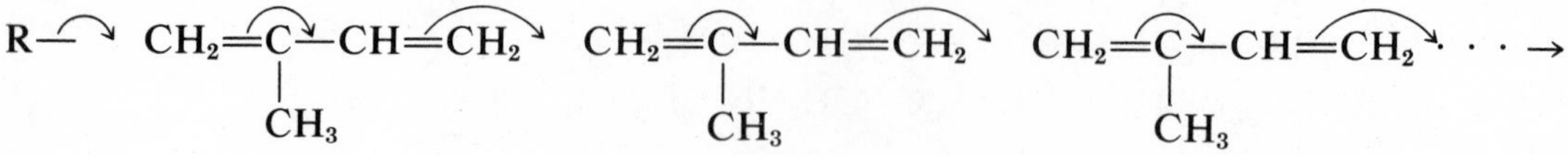

Isoprene

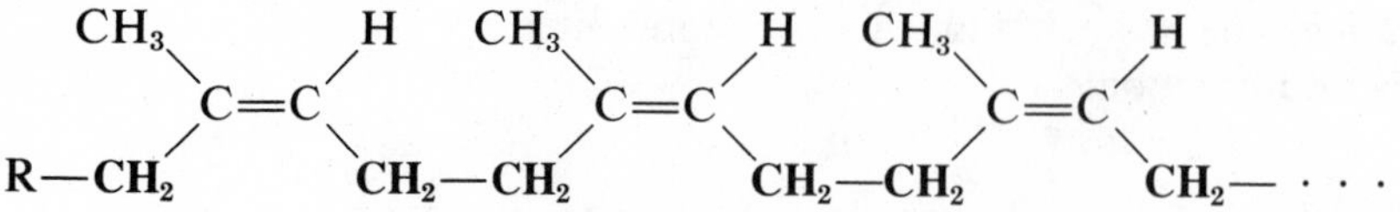

Natural rubber

Very long chains are formed with molecular weights in the millions. In natural rubber the polymerization is head-to-tail, and the double bonds are all *cis*—that is, the CH_2 groups (in boldface type in the formula) are on the same side of the double bond. The natural product, in which the double bonds are all *trans*, is called gutta-percha and is brittle rather than elastic. When rubber is *vulcanized* by heating with as little as 0.30% of sulfur, adjacent chains are linked together by reaction of S with the double bonds, as follows:

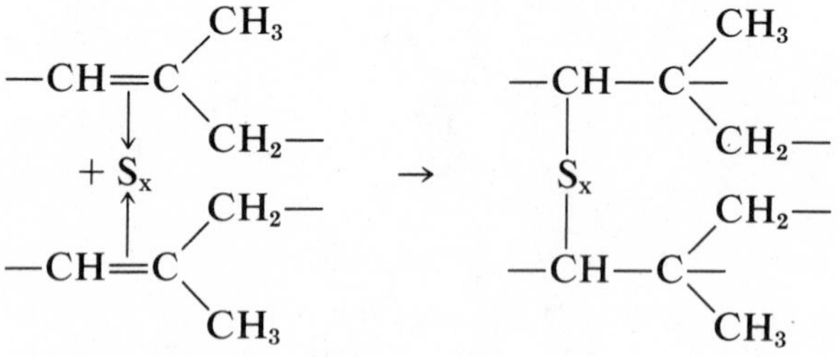

The *cross-linking* through one or more sulfur atoms ($X = 1 - 3$) turns the rubber from a soft, sticky material to the familiar elastic form.

If the methyl group in isoprene is replaced by chlorine, chloroprene (2-chloro-1,3-butadiene) results, and its polymerization gives the synthetic rubber called Neoprene, which is very resistant to attack by solvents, acids, bases, and other corrosive reagents. Laboratory "rubber" stoppers are usually made of Neoprene.

The most chemically unreactive vinyl plastic is polytetrafluoroethylene (Teflon), made by polymerizing tetrafluoroethylene (tetrafluoroethene):

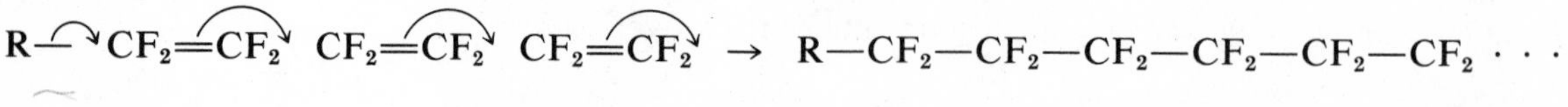

Teflon is polyethylene in which all hydrogens have been replaced by fluorines—it is resistant to all reagents except molten alkali metals (Group 1a) and stands up well at temperatures up to 250°C. At 600–800°C it depolymerizes into tetrafluoroethene without charring.

EXAMPLE

Few plastics are as stable as Teflon. For one thing, most plastics burn freely once ignited. Plastics made of monomers composed only of carbon, hydrogen, and oxygen burn to water and carbon dioxide, which is toxic only at high concentrations. Plastics made from monomers that contain nitrogen or chlorine in addition to carbon and hydrogen produce very toxic substances when burned. Chlorinated plastics such as polyvinylchloride produce the toxic gases HCl and Cl_2 when they burn. Burning of vinyl upholstery makes an especially dangerous fire. Under certain conditions, nitrogen-containing plastics give off the very toxic gases hydrogen cyanide, HCN, and cyanogen, $(CN)_2$, when burned. Although Teflon does not burn, if it is heated to a high temperature in a fire being fought with water the very poisonous gas hydrogen fluoride, HF, can be produced. The lesson here is that special gas masks or a separate oxygen supply *must be used* in fighting fires involving most plastics.

Exercises

28. Polypropylene is made by polymerizing propene,

$$CH_3{-}CH{=}CH_2$$

Show a portion of the polymer chain.

29. The xylenes are dimethylbenzenes. Using superposition to determine isomerism, how many isomeric xylenes do you find?

30. Draw the carbon skeletons of the nine isomers of heptane.

31. Show a portion of the following polymer chains:
a. Gutta-percha (like natural rubber but all *trans* double bonds)
b. Polyvinyl alcohol (X = OH)
c. Polystyrene
d. Lucite or Plexiglas, for which the monomer is methyl methacrylate,

$$CH_2{=}\underset{\displaystyle CH_3}{\underset{|}{C}}{-}COOCH_3$$

32. Hydrocarbons are very poisonous and, of course, cannot be used for food. Nevertheless, it is interesting to calculate the amount of gasoline (octane, C_8H_{18}) needed to fulfill the daily energy requirement of the average body for 2400 kcal. The heat of combustion of octane is 1223 kcal/mole, and its density at 20°C is 0.70 g/ml. What volume of liquid octane ("gasoline") would provide 2400 kcal on combustion? Taking 1 gal equal to 3.79 liters, how many kcal of energy are used when a car engine uses 10 gal of gasoline (taken as octane) to drive 150 mi?

13.7 Alcohols, Aldehydes, and Ketones The alcohols used in largest amounts are methanol (methyl alcohol, wood alcohol), ethanol (ethyl alcohol, beverage alcohol, alcohol), and 2-propanol (isopropyl alcohol, rubbing alcohol). All three are liquids that boil below 100°C and dissolve completely in water in any proportions. Methyl alcohol is particularly poisonous, causing blindness and death if ingested in even small amounts. All three of these alcohols are used as solvents and as starting materials for making many other chemicals.

Methanol was once obtained by heating wood strongly in the absence of air (destructive distillation of wood) but is now made by starting with methane from natural gas. Partial oxidation of methane, using oxygen from liquid air, gives a mixture of carbon monoxide and hydrogen:

$$2CH_4(g) + O_2(g) \rightarrow 2CO(g) + 4H_2(g)$$

When the mixture is passed over a zinc oxide catalyst, methanol results:

$$CO(g) + 2H_2(g) \rightarrow CH_3OH(l)$$

Methanol is also prepared using the mixture of CO and H_2 produced by the water gas reaction, in which steam is passed over red hot coal (carbon):

$$C(s) + H_2O(g) \xrightarrow{1100°C} \underbrace{CO(g) + H_2(g)}_{\text{Water gas}}$$

The water gas reaction is an important industrial source of hydrogen, but may soon provide a way of making methanol for use as a fuel from coal without disfiguring the environment with mines (Photo 35).

EXAMPLE The water gas reaction is strongly endothermic (by 31 kcal), so the coal has to be heated to keep the reaction going. The necessary high temperature is maintained by mixing some oxygen (from liquid air) with the steam. The whole process can be carried out on coal that is left underground. Closely spaced pairs of wells are drilled into the coal seam, and explosives set off at the bottoms of the wells break up the coal. The coal is ignited, and then a mixture of steam and oxygen is pumped down one member of each pair of wells and the $CO + H_2$ produced underground by the water gas reaction is collected from the other wells in each pair.

The water gas reaction does not yield CO and H_2 in the correct molar ratio for conversion to methanol. This difficulty is overcome by sacrificing some of the CO to make H_2, using the following reaction:

$$CO(g) + H_2O(g) \xrightarrow[\text{catalyst}]{500°C} CO_2(g) + H_2(g)$$

At the time of writing this book methanol was seen as a substitute fuel for dwindling supplies of petroleum and natural gas. Someday, methanol might even be the fuel for fuel cells that provide electricity for homes. The cell reaction and ΔG^0 value are as follows:

$$CH_3OH(l) + \tfrac{3}{2}O_2(g) \rightleftarrows CO_2(g) + 2H_2O(l) \qquad \Delta G^0 = -168 \text{ kcal}$$

Fuel cells utilizing methanol to generate electricity have already been constructed. Methanol makes an attractive fuel because it burns cleanly to CO_2 and water in a furnace, an internal combustion engine, or a fuel cell, and does not pollute the environment.

Most of the methanol produced is converted to formaldehyde by catalytic oxidation (reaction with oxygen), using platinum or silver:

$$2 \; H{-}\underset{\underset{\textstyle H}{|}}{\overset{\overset{\textstyle H}{|}}{C}}{-}O{-}H + O_2 \xrightarrow{\text{catalyst}} 2 \; \underset{H}{\overset{H}{>}}C{=}O + {}_2H_2O$$

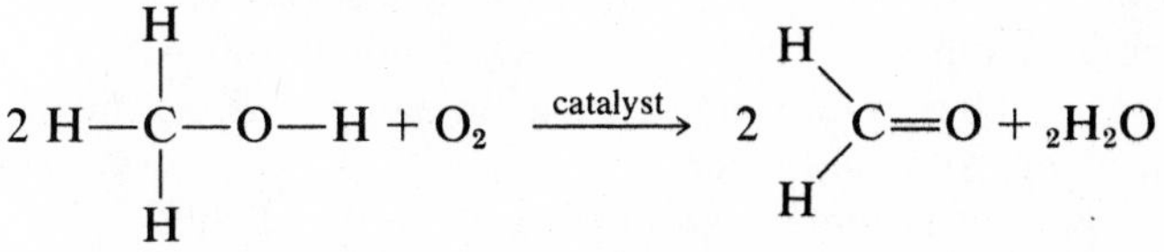

Formaldehyde is a gas at ordinary temperatures, and the liquid (bp $-21°C$, mp $-92°C$) spontaneously polymerizes in a reaction that may be explosive because it is so fast. The polymerization is like that of ethylene but requires no more than the trace of OH^- ions in water to make it go:

$$HO^- \; C{=}O \; C{=}O \; C{=}O \; C{=}O \qquad H{-}O{-}H$$

$$HO{-}C{-}O{-}C{-}O{-}C{-}O{-}C{-}O{-} \cdots \qquad C{-}O{-}C{-}OH + OH^-$$

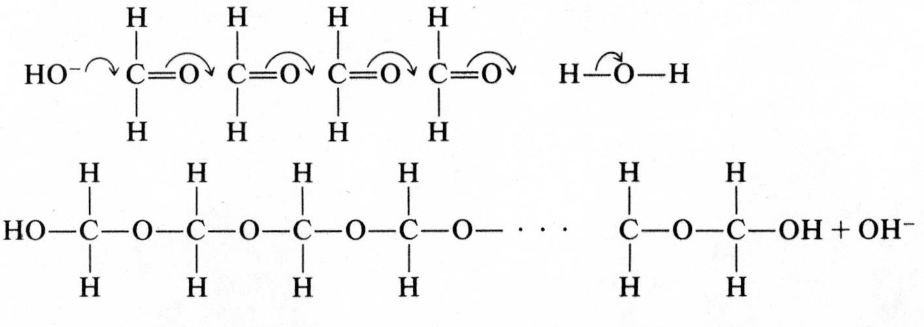

Unlike polyethylene, polymeric formaldehyde (polyoxymethylene) readily depolymerizes or "unzips" to formaldehyde when heated. However, if the OH *end groups* are protected (blocked) by esterification, the unzipping can't occur and tough useful plastics (Delrin) result:

$$\mathrm{HO{-}CH_2{-}O{-}CH_2{-}(O{-}CH_2)_n{-}O{-}CH_2{-}OH}$$

$$\downarrow \longleftarrow \text{Esterification using acetic anhydride}$$

$$\mathrm{CH_3{-}\overset{\overset{\displaystyle O}{\|}}{C}{-}O{-}CH_2{-}O{-}CH_2{-}(O{-}CH_2)_n{-}O{-}CH_2{-}O{-}\overset{\overset{\displaystyle O}{\|}}{C}{-}CH_3}$$

EXAMPLE The difference in the ease of depolymerizing polyoxymethylene (to formaldehyde) and polyethylene (to ethene) is readily explained by using bond energies. The π bond energy for ethene is 63 kcal/mole, and the energy for a C—C σ bond is 83 kcal/mole. In the depolymerization of polyethylene, $83 - 63 = 20$ kcal must be *provided* for each mole of σ bonds converted to π bonds, so the depolymerization is strongly endothermic:

$$\mathrm{R{-}CH_2{-}(CH_2)_n{-}CH_2{-}CH_2{-}CH_2{-}} + 20\ \text{kcal} \rightarrow \mathrm{R{-}CH_2{-}(CH_2)_n{-}CH_2{-}} + \mathrm{CH_2{=}CH_2}$$

For the π bond in formaldehyde the bond energy is 80.5 kcal/mole and the C—O σ bond energy is only slightly larger at 85.5 kcal/mole. The energy input required to convert C—O σ bonds into C═O π bonds is only $85.5 - 80.5 = 5.0$ kcal for each mole of bonds, so the depolymerization of polyoxymethylene is not nearly as endothermic as the depolymerization of polymethylene and goes more readily:

PHOTO 35 Strip mining for coal on old Indian burial grounds in Arizona. (*EPA-DOCUMERICA—Lyntha Scott Eiler.*)

$$R-\underset{\underset{H}{|}}{\overset{\overset{H}{|}}{C}}-O-\underset{\underset{H}{|}}{\overset{\overset{H}{|}}{C}}-O-\underset{\underset{H}{|}}{\overset{\overset{H}{|}}{C}}-O- + 5 \text{ kcal} \rightarrow R-\underset{\underset{H}{|}}{\overset{\overset{H}{|}}{C}}-O-\underset{\underset{H}{|}}{\overset{\overset{H}{|}}{C}}-O- + \underset{H}{\overset{H}{\diagdown}}C=O$$

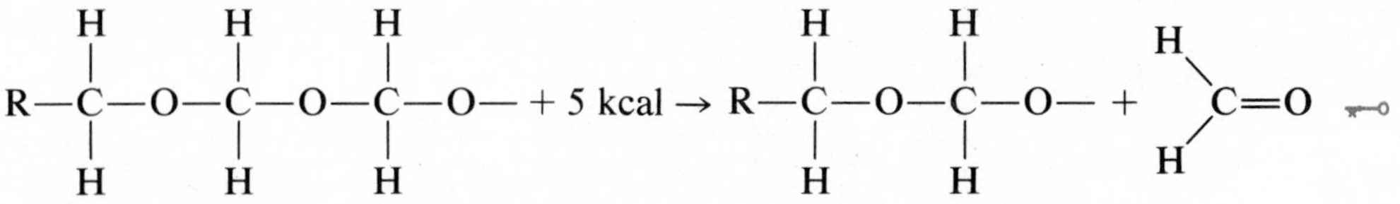

Ethanol and 2-propanol are both made by the acid-catalyzed addition of water to hydrocarbon by-products of catalytic cracking to make gasoline. Addition of water (hydration) to ethene gives ethanol, and to propene gives 2-propanol (isopropyl alcohol):

$$CH_2{=}CH_2 + H_2O \xrightarrow{\;H^+\;} CH_3CH_2OH$$
Ethanol

$$CH_3-CH{=}CH_2 + H_2O \xrightarrow{\;H^+\;} \underset{CH_3}{\overset{CH_3}{\diagup}}C\underset{OH}{\overset{H}{\diagdown}}$$
2-propanol

This kind of hydration reaction goes well with most alkenes (but not with arenes), and the OH group of the water added becomes attached to the carbon in the alkene that has the smallest number of hydrogens — just as in the formation of 2-propanol instead of 1-propanol from propene. The reactions can be carried out in the gas phase by using steam and a solid acidic catalyst such as clay that has been coated with phosphoric acid (H_3PO_4) or hydrogen fluoride. Under different conditions (H_2SO_4 catalysis) ethanol is converted to diethyl ether, which is used extensively as an anesthetic ☞ in surgery:

From *an*, meaning "not"; *esthetic*, meaning "feeling."

$$2CH_3CH_2OH \xrightarrow{\;H_2SO_4\;} CH_3CH_2-O-CH_2CH_3 + H_2O$$
Diethyl ether

Ether is one of the safest anesthetics in the body, but it has the disadvantages of being extremely flammable and forming very explosive mixtures with air.

Ethanol is also prepared from the biopolymer starch (corn, wheat) by fermentation. The steps, with the necessary enzymes and their sources, are as follows:

$$2(C_6H_{10}O_5)_n + nH_2O \xrightarrow[\text{(from malt)}]{\text{Diastase}} nC_{12}H_{22}O_{11}$$
Starch Maltose

Malt is grain, usually barley, that is softened in water, allowed to sprout, and then dried and ground. Malt is rich in proteins and carbohydrates, and is an essential ingredient in brewing beer and ale. The finest Scotch whiskies ("all-malt" Scotch) are made from malted barley that is fermented, distilled, and then filtered through peat charcoal.

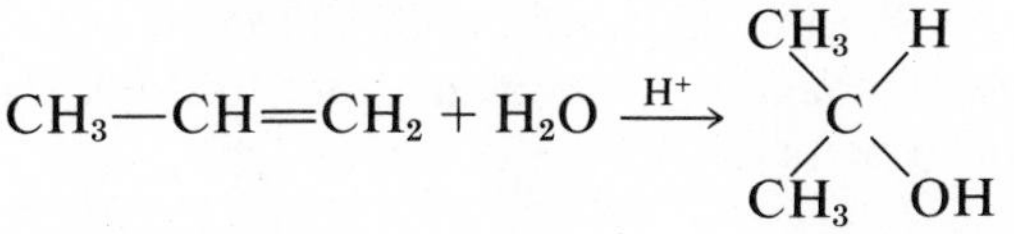

$$\underset{\text{Maltose}}{C_{12}H_{22}O_{11}} + H_2O \xrightarrow[\text{(from yeast)}]{\text{Maltase}} \underset{\text{Glucose}}{2C_6H_{12}O_6}$$

$$C_6H_{12}O_6 \xrightarrow[\text{(from yeast)}]{\text{Zymase}} 2CO_2 + 2CH_3CH_2OH$$

The first two reactions are hydrolysis reactions (bond-breaking by addition of water) that don't involve large energies, but the last reaction is strongly exothermic ($\Delta H = -26$ kcal), and the fermenting glucose solution must be kept cool to prevent the heat from killing the yeast and stopping the fermentation. Alcohol produced from grain (starch) is called beverage alcohol; that produced by hydration of ethene is industrial alcohol. Chemically the same, their prices are very different because beverage alcohol is heavily taxed. A 55-gal drum of alcohol contains alcohol worth about $37 but carries a tax of $1135. Alcohol for industrial purposes must be *denatured* or made unfit to drink by adding chemicals if it is not to be taxed.

EXAMPLE

Alcohol is a drug because it influences the central nervous system. Small quantities of alcohol stimulate, larger quantities depress, and unconsciousness and even death can result from an overdose of alcohol. Blood levels up to 100 mg per 100 ml of blood produce exhilaration and are reached in a 160-pound person after about three $1\frac{1}{2}$-ounce drinks of whiskey or its alcoholic equivalent. In legal language, if the blood concentration is 0.10 percent, the person is "under the influence of alcohol" and, if operating an automobile, is liable to prosecution for drunken driving. The National Council on Alcoholism considers a person *intoxicated* if his blood alcohol is 150 mg per 100 ml of blood or 0.15 percent. Blood levels of 0.50 percent are likely to result in death. Beyond a blood level of about 0.10 percent muscular coordination and judgment become progressively impaired, making a driver dangerous on the highway. In Sweden a person is considered unable to drive a car properly after even a single drink. Bicycles are very popular there.

Three bottles of beer or two glasses of wine, for instance.

Ethanol and water cannot be completely separated by fractional distillation because they form a mixture (called an azeotrope) that has a lower boiling point than either one by itself. This mixture is 95.6% ethanol by weight, and distills off first because it boils at 78.13°C while pure ethanol boils slightly higher at 78.3°C. Special methods are used to remove the water from "95%" ethanol to give "absolute" ethanol (99.9%), which is needed for some industrial and laboratory reactions. (Esterifications usually use absolute ethanol.) The "proof" of an ethanol-water solution is based on "100 proof spirit," defined as a mixture of equal volumes of ethanol and water. A 100 proof solution contains 42.5% ethanol by weight. The azeotropic mixture "95%" ethanol is 190 proof; absolute alcohol is 200 proof.

Most of the 2-propanol from hydration of propene is converted to acetone, the most important ketone in industry. One conversion method produces hydrogen peroxide (H_2O_2) as a valuable by-product:

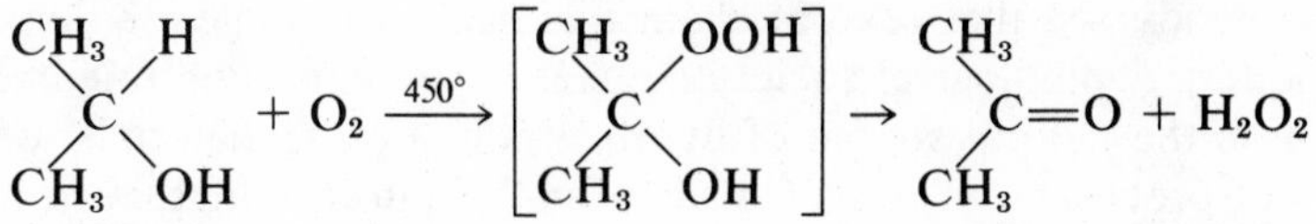

$$\underset{\text{2-propanol}}{\overset{\text{CH}_3\quad\text{H}}{\underset{\text{CH}_3\quad\text{OH}}{\text{C}}}} + O_2 \xrightarrow{450°} \left[\underset{\text{CH}_3\quad\text{OH}}{\overset{\text{CH}_3\quad\text{OOH}}{\text{C}}}\right] \rightarrow \underset{\text{CH}_3}{\overset{\text{CH}_3}{\text{C}}}{=}O + H_2O_2$$

Acetone (propanone)

Acetone is also produced in a similar oxidation process that converts isopropyl benzene (cumene, 2-phenylpropane) into phenol (hydroxy benzene) and acetone:

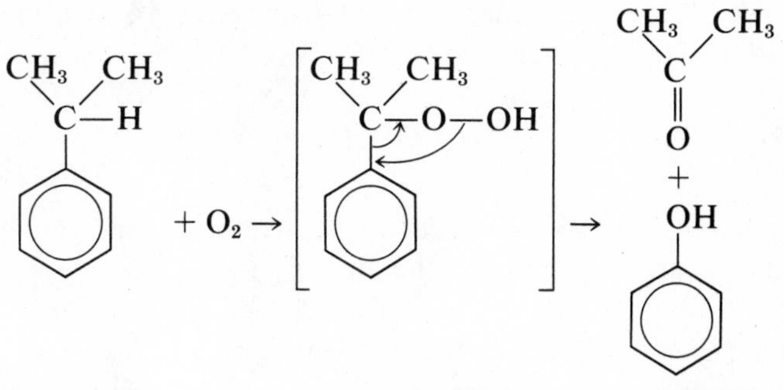

Cumene
(2-propylbenzene) Cumyl hydroperoxide Phenol

Acetone is used as a solvent for plastics, in making photographic film, and as a starting material for other chemicals. Phenol ("carbolic acid") is used in countless ways, one being to make phenol-formaldehyde plastics (Bakelite).

33. The reaction forming methanol from carbon monoxide and hydrogen is carried out at gas pressures as high as 300 atm (4410 psi). Why?

34. Give the structure and name of the alcohol formed by acid-catalyzed hydration of 1-butene.

Exercises

The word *proof* came from a way of testing the alcohol content of liquor (rum or whiskey) in colonial times. The liquor was poured over a pile of gunpowder and ignited. Burning of the gunpowder showed that the percentage of water in the liquor was so low that the gunpowder didn't get wet and fail to go off. Liquor passing the test was "100 proof."

35. What advantage is there to using absolute ethanol instead of 95% ethanol in an esterification reaction? (*Hint*: recall the equilibrium nature of the esterification reaction.)

This discussion of organic chemicals based on petroleum barely scratches the surface of what is called the *petrochemicals industry*. You can already see that several chemicals needed in huge amounts by our modern technological society come ultimately from petroleum. Petroleum is the natural source of hydrocarbon starting materials for hundreds of processes. It is so valuable for this purpose that using it for fuel must be discontinued soon as being too wasteful of limited natural resources.

13.8 Carboxylic Acids *Acetic acid* is the most important carboxylic acid industrially, and is made in a number of ways. One way is by the enzyme-catalyzed oxidation of ethanol to make vinegar, a 5% solution of acetic acid in water plus some flavorings:

$$CH_3CH_2{-}OH + O_2(\text{air}) \xrightarrow[\text{enzyme}]{\text{bacterial}} CH_3COOH + H_2O$$

Ethanol can be oxidized in stages to acetaldehyde (ethanal) and acetic acid, using catalysts and oxygen from the air.

$$2CH_3CH_2OH + O_2 \xrightarrow{\text{catalyst}} \underset{\text{Acetaldehyde}}{2CH_3CHO} + 2H_2O$$

$$2CH_3CHO + O_2 \xrightarrow[Co^{2+}]{Mn^{2+}} 2CH_3COOH$$

In this last reaction, the transition metal cations act as catalysts, and at the time of writing this book a transition-metal-catalyzed addition of carbon monoxide to methanol to give acetic acid had just been developed:

$$CH_3OH + CO \xrightarrow{\text{catalyst}} CH_3COOH$$

This is the simplest reaction producing acetic acid. The transition metal catalyst just inserts CO between the carbon and oxygen atoms of methanol:

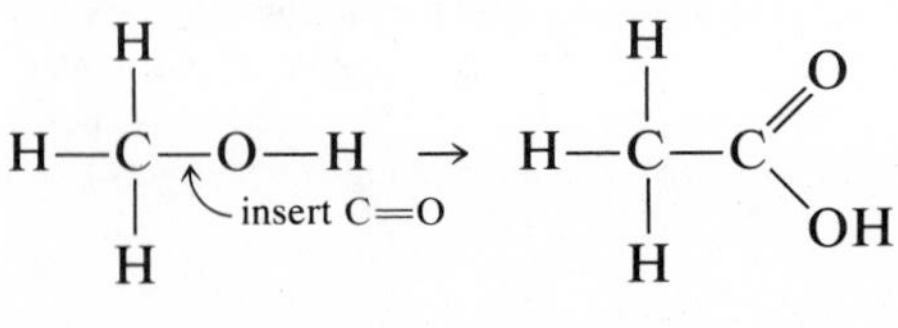

It seems likely that the transition metal ion has methanol and carbon monoxide as two of its ligands. Bonding to the metal ion activates them energetically and brings them close enough together so that the "insertion reaction" can take place. With methanol and carbon monoxide available from coal, acetic acid can be made from coal, air, and water.

Large amounts of acetic acid are made from butane by transition metal ion (Mn^{2+}, Co^{2+}) catalyzed air oxidation in the liquid phase. Methyl ethyl ketone (MEK) is the first product, and some of it is diverted for use as a solvent for paints and varnishes:

$$CH_3CH_2CH_2CH_3 + O_2 \xrightarrow{\text{catalyst}} CH_3CH_2{-}\overset{\displaystyle O}{\overset{\|}{C}}{-}CH_3 + H_2O$$

Methyl ethyl ketone

$$2CH_3CH_2{-}\overset{\displaystyle O}{\overset{\|}{C}}{-}CH_3 + 3O_2 \xrightarrow{\text{catalyst}} 4CH_3COOH$$

The largest single use of acetic acid is to make *acetic anhydride*—2 billion lbs of the acid and 1.6 billion of the anhydride were made in 1973 in the U.S. Acetic anhydride is used to make acetate esters, and the largest tonnage goes into the production of aspirin, cellulose acetate for photographic film, and acetate rayon fabrics.

Acetic anhydride is made by reaction of ketene (pronounced "key-teen"), $CH_2{=}C{=}O$, with acetic acid.

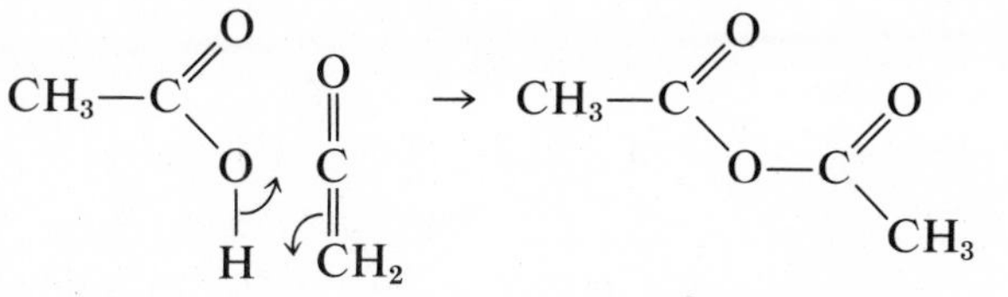

Acetic anhydride

The ketene is in turn made by decomposition (dehydration) of acetic acid at 700°C:

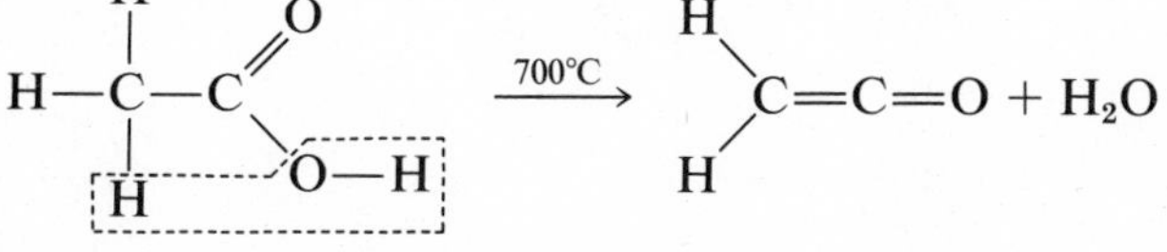

Exercises

36. Give the IUPAC names for acetone, methyl ethyl ketone, and ketene.

37. Using the example of the reaction of ketene with the OH group of acetic acid, predict the product of the reaction between ketene and methanol.

38. As we saw earlier, acetic anhydride reacts with water to give two moles of acetic acid. When acetic anhydride reacts with an alcohol R—OH, one mole of acetic acid is formed. What type of compound is the other product of the reaction?

39. What product would you predict for the reaction of ketene with water?

"Dodec" means 2 (do) + 10 (dec) or 12, the number of carbon atoms in the molecule.

Detergents Carboxylic acids in the C_{10}–C_{12} range, prepared from coconut oil, are starting materials for detergents such as sodium dodecylsulfate (Section 11.3 – Ion Exchange). To make sodium dodecylsulfate, dodecanoic acid ($C_{11}H_{23}COOH$) is esterified with methanol, and the methyl ester function is then catalytically reduced with hydrogen to give dodecanol, $C_{11}H_{23}—CH_2—OH$. Sulfation (formation of a sulfate ester with sulfuric acid) followed by formation of the sodium salt using sodium carbonate from the Solvay process (Section 11.2) gives the detergent. Letting $C_{11}H_{23}—$ stand for the $CH_3(CH_2)_{10}—$ hydrocarbon chain of dodecanoic acid, we can write the reactions as follows:

$$\underset{\text{Dodecanoic acid}}{C_{11}H_{23}\overset{O}{\overset{/\!/}{C}}—OH} + CH_3OH \xrightleftharpoons{H^+} \underset{\text{Methyl dodecanoate}}{C_{11}H_{23}\overset{O}{\overset{/\!/}{C}}—OCH_3} + H_2O$$

$$C_{11}H_{23}\overset{O}{\overset{/\!/}{C}}—OCH_3 + 2H_2 \xrightarrow[(CuCrO_3)]{\text{catalyst}} \underset{\text{Dodecanol}}{C_{11}H_{23}CH_2OH} + CH_3OH$$

$$C_{11}H_{23}CH_2—OH + H—O—\underset{\underset{O}{|}}{\overset{\overset{O}{|}}{S}}—OH \rightarrow C_{11}H_{23}CH_2—O—\underset{\underset{O}{|}}{\overset{\overset{O}{|}}{S}}—OH + H_2O$$

$$2C_{11}H_{23}CH_2—O—\underset{\underset{O}{|}}{\overset{\overset{O}{|}}{S}}—OH + Na_2CO_3 \rightarrow \underset{\text{Sodium dodecylsulfate}}{2C_{11}H_{23}CH_2—O—\underset{\underset{O}{|}}{\overset{\overset{O}{|}}{S}}—O^- + 2Na^+} + H_2O + CO_2$$

This set of reactions shows quite clearly what "functional group" means. The $C_{11}H_{23}—CH_2—$ hydrocarbon chain does not participate in any of the reactions – it is a mere bystander for what goes on at the functional group carbon atom at the end of the molecule. The hydrocarbon chain is necessary for detergent action (Figure 11.6), however.

Detergents like sodium dodecylsulfate that have a *straight* hydrocarbon chain can serve as food for microorganisms in soil, and for this property are called "biodegradable" to CO_2 and water. Branches in the carbon chain or benzene rings stop the action of the microorganisms and result in "nonbiodegradable" detergents. Because they collect in and pollute water supplies, the use of nonbiodegradable detergents in laundry and dishwashing powders should be discontinued.

Exercises

40. The straight-chain hydrocarbon with 18 carbon atoms is named octadecane. What is the systematic (IUPAC) name for stearic acid ($C_{17}H_{35}COOH$)?

41. Ethylene glycol (1,2-ethanediol), used as car radiator antifreeze, is a close but very poisonous relative ☛ of glycerol, 1,2,3-propantriol. Give their structural formulas. Ethylene glycol is made from a hydrocarbon. Which hydrocarbon do you think is the most likely candidate? Would the reaction forming ethylene glycol be a reduction or an oxidation?

Several people died when an uninformed pharmacist used "glycol" (another name for ethylene glycol) instead of "glycerol" in a cough syrup. This incident points up the importance of naming compounds correctly.

13.9 Arenes and Aromatic Compounds Arenes are ring hydrocarbons with 3, 5, 7, 9, . . . , double bonds alternating with single bonds around the ring in their bond structures. Bond structures and other ways of writing formulas ☛ of the three most important arenes are as follows:

The type of formula used for an arene depends on the subject of the discussion.

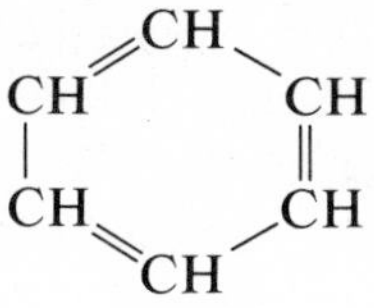

or

or

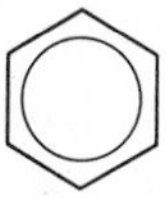

Benzene

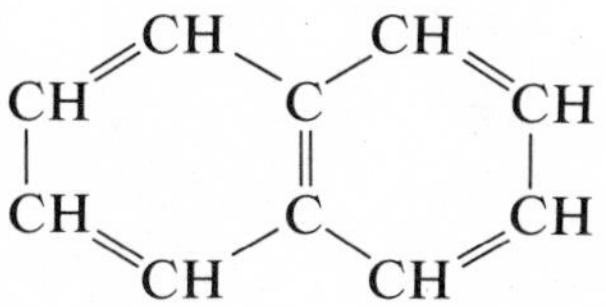

or

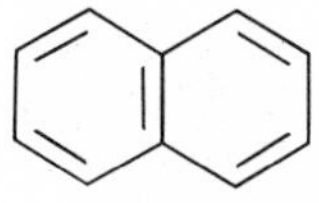

or

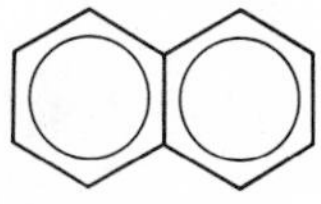

Napthalene

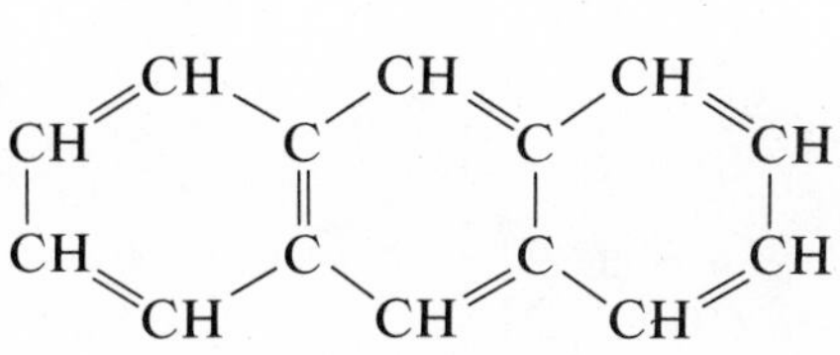

or

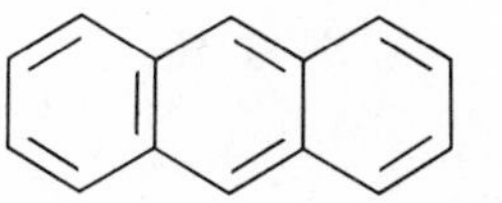

or

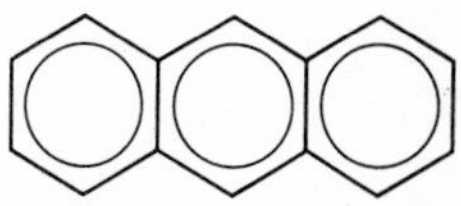

Anthracene

In Chapter 4 we saw that the bond lengths in benzene could not be explained by using only one Lewis structure. Resonance theory using two structures, and molecular orbitals for π bonding, both explained why all of the C—C bonds in benzene are the same length. One way to show that the bond lengths are all equal is to use a circle to stand for the six electrons in the three double bonds in the benzene ring, as shown. Resonance and molecular orbital theory also explain the bond lengths and angles in other arenes such as napthalene and anthracene.

The double bonds in the bond structure of an arene do not behave chemically like the double bonds in alkenes. Alkenes readily add water in the presence of an acid catalyst to form alcohols:

$$\begin{array}{c} \diagup\!\!\!\!\diagdown \\ C \\ \| \\ C \\ \diagup\!\!\!\!\diagdown \end{array} + H_2O \xrightarrow{\ H^+\ } \begin{array}{c} \diagup\!\!\diagdown \\ C\!-\!H \\ | \\ C\!-\!OH \\ \diagup\!\!\diagdown \end{array}$$

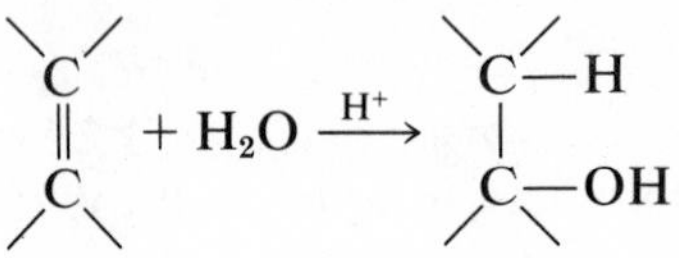

Alkenes also add hydrogen halides:

$$\begin{array}{c} \diagup\!\!\!\!\diagdown \\ C \\ \| \\ C \\ \diagup\!\!\!\!\diagdown \end{array} + HCl \rightarrow \begin{array}{c} \diagup\!\!\diagdown \\ C\!-\!H \\ | \\ C\!-\!Cl \\ \diagup\!\!\diagdown \end{array}$$

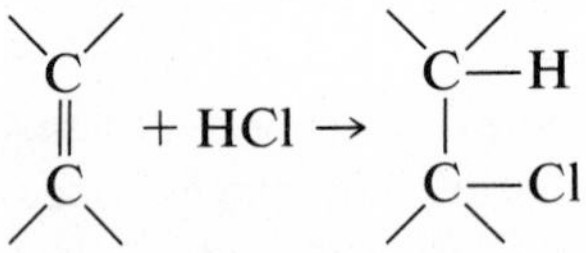

Benzene and other arenes don't show either of these reactions, so their double bonds must be unreactive for some reason. The reason is not hard to find if you consider what would happen to the resonance energy of an arene in additions of the type that go with alkenes. Taking the case of benzene, you can see that the resonance energy of 40 kcal/mole is lost if addition of water or a hydrogen halide takes place:

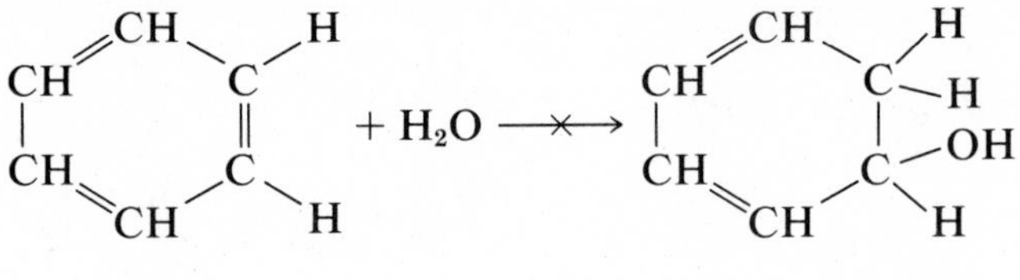

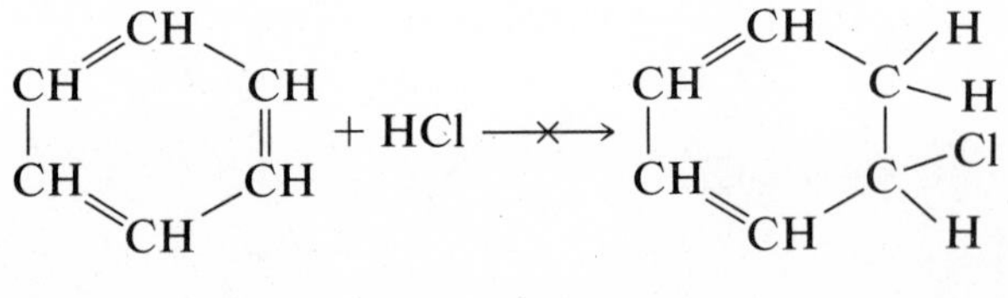

You cannot write a second bond structure for the product of addition of water (or HCl), so the resonance energy lost makes the addition *endo*thermic by about 40 kcal/mole. The same conclusion results if molecular orbital theory is used—the addition breaks up the stable six-electron π bonding (Section 4.5) by pulling two of the π electrons out to form σ bonds to H and OH (or Cl). Other arenes show the same lack of reactivity in these addition reactions.

On the other hand, arenes *are* reactive in a way that alkenes are not. Arenes readily undergo *substitution reactions* ☛ in which a hydrogen atom of one of the CH groups in the ring is replaced by a functional group. Substitution of a nitro group ($-NO_2$) for hydrogen on benzene is one of the best-understood arene substitution reactions. In this reaction benzene is treated with a mixture of concentrated nitric (HNO_3) and sulfuric (H_2SO_4) acids. The acids react as follows to produce the nitronium ion, NO_2^+:

Also called "aromatic substitution" reactions and "electrophilic substitution" reactions, the latter because NO_2^+ is "looking for electrons" to bond with. Electrophilic means "electron-loving."

$$HO-\overset{\overset{O}{|}}{\underset{\underset{O}{|}}{S}}-OH + HO-N\overset{O}{\underset{O}{\diagup}} \rightarrow HO-\overset{\overset{O}{|}}{\underset{\underset{O}{|}}{S}}-O^- + H_2O + NO_2^+$$

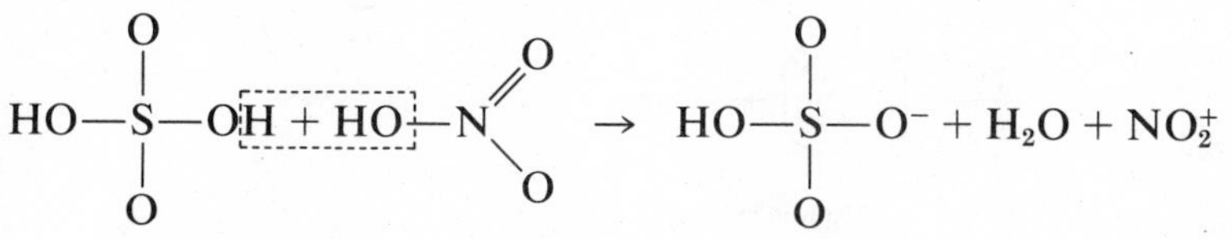

The nitronium ion then attacks the π electron system of benzene to form a $C-NO_2$ bond:

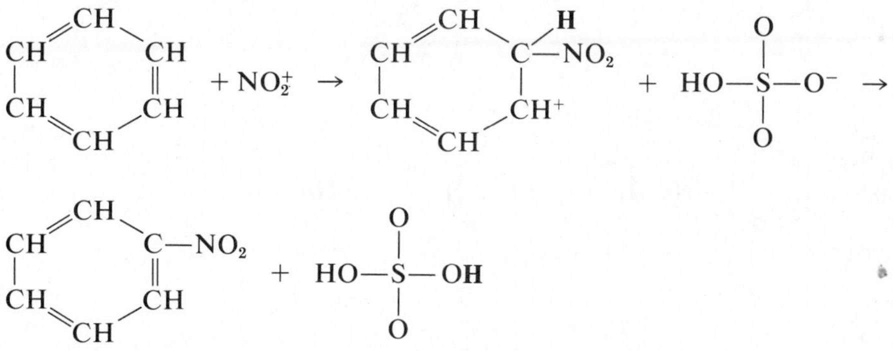

Nitrobenzene

The bisulfate ion (HSO_4^-), acting as a base, then removes the ring proton (indicated by boldface type), and the resonance energy of the benzene ring is restored. All arenes undergo this type of substitution reaction to give nitro compounds. An important practical example is the preparation of the high explosive trinitrotoluene (TNT) from the hydrocarbon toluene. Here three ring hydrogens are replaced by nitro groups.

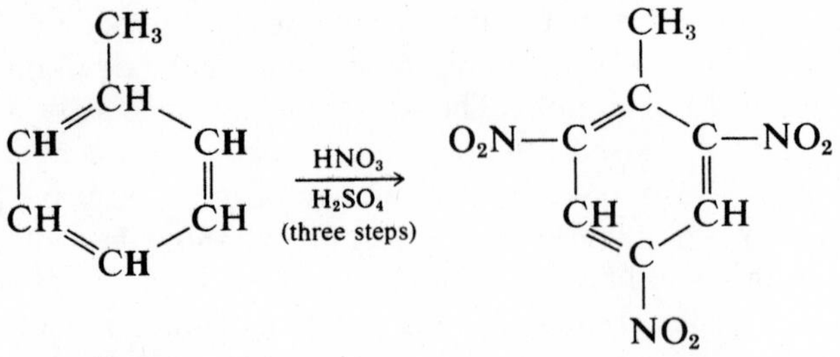

Toluene (methylbenzene)

Trinitrotoluene

Benzene (and other arenes) undergo a large number of substitution reactions. Some of the more important ones for benzene are the following:

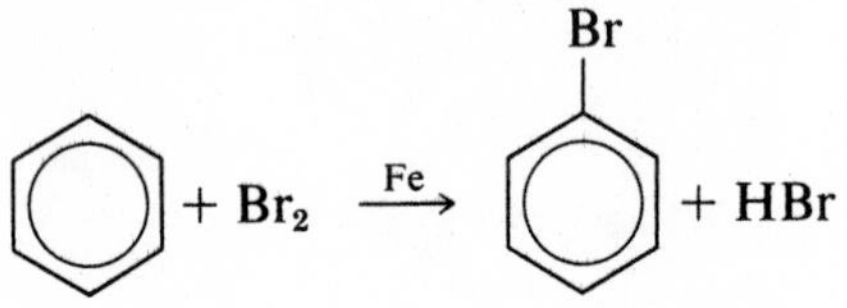

Bromobenzene

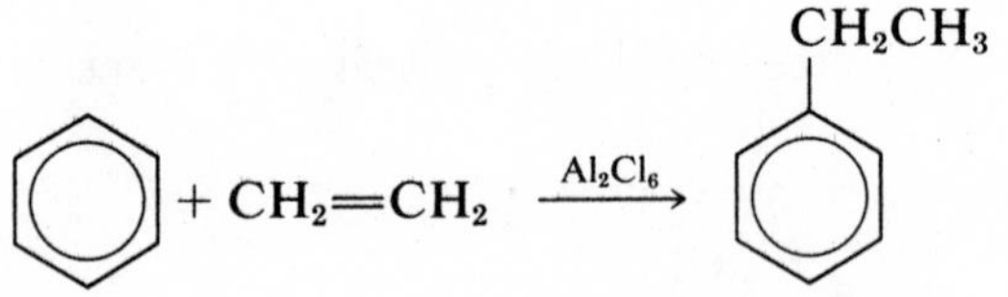

Ethyl benzene

Charles Friedel (1832–1899) was a Professor of Chemistry at the Sorbonne in France. He was the first to produce isopropyl alcohol by the reduction of acetone.

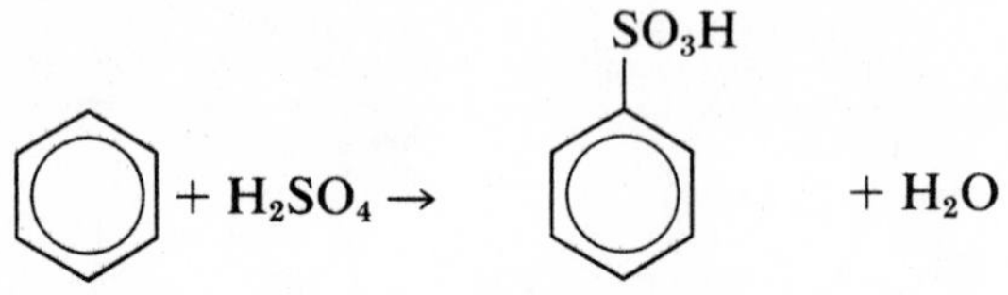

Benzenesulfonic acid

James Mason Crafts (1839–1917) collaborated with Friedel at the Sorbonne for 17 years. He returned to the Massachusetts Institute of Technology in 1891 as Professor of Chemistry and became President of the Institute in 1897.

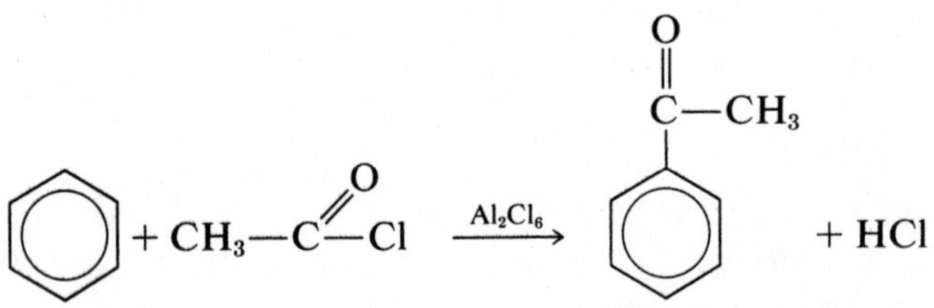

Acetyl chloride

Acetophenone (acetylbenzene)

The $CH_3—C(=O)—$ group is called the acetyl group because it is present in *acetic* acid $CH_3—C(=O)—OH$.

Of these reactions, those that use aluminum chloride (Al_2Cl_6) as a catalyst are called Friedel-Crafts reactions in honor of the chemists who discovered them.

At this point it is important not to make the mistake of thinking that bond structures of the following type, where A is a substituent, are *isomers* because of the different positions of the double bonds:

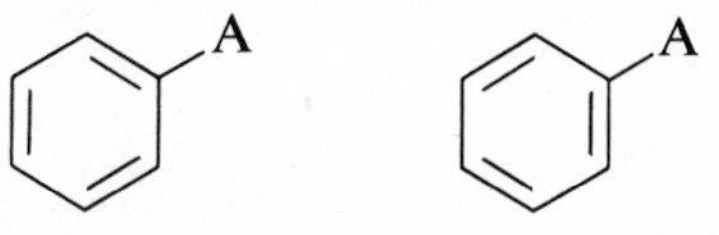

The molecule is a hybrid of these two structures, and you can't assume that the double-bond positions are fixed. One way of showing this clearly is to use the circle in the ring as we have done to represent the six electrons in the three double bonds of the benzene ring. Or you can use the double-headed arrow (Section 4.2) to indicate resonance hybridization of the two bond structures:

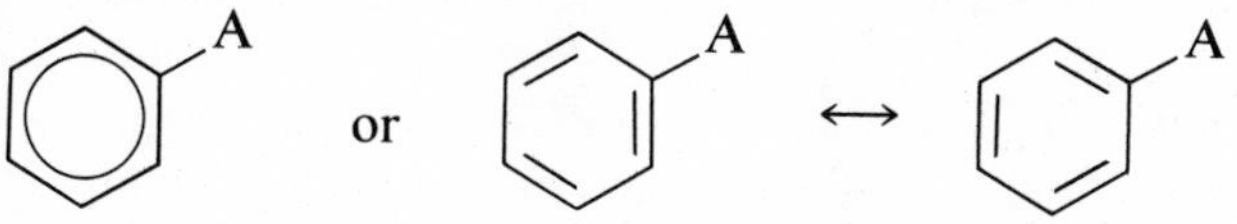

Aromatic compounds Aromatic compounds are functional derivatives of benzene and other arenes. The term aromatic arose because many of these compounds have pleasant odors. Among those found in the natural oils indicated are the following:

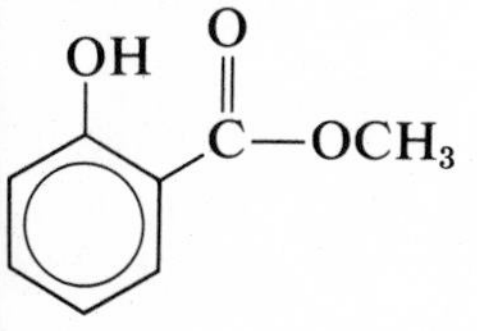

Methyl salicylate
(wintergreen)

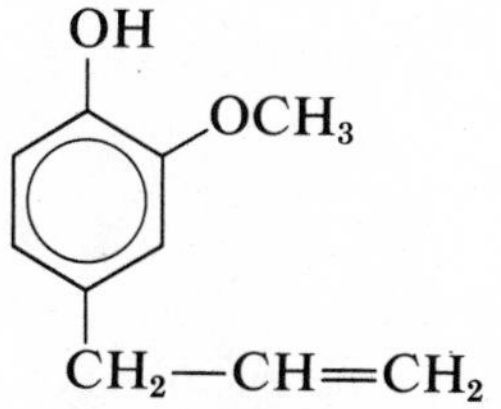

Eugenol
(clove)

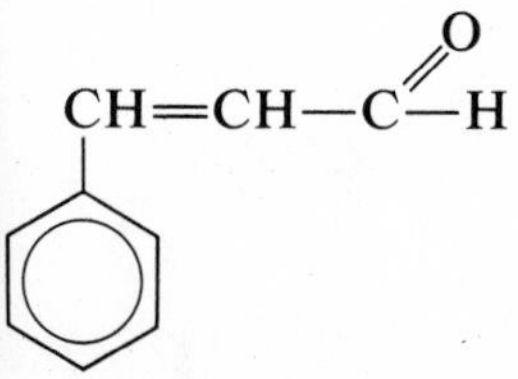

Cinnamaldehyde
(cinnamon)

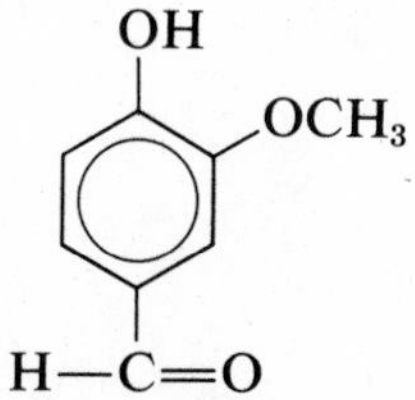

Vanillin
(vanilla)

Once a substituent is on the benzene ring, reactions can usually be carried out on the substituent without changing the ring. For example, eugenol is converted to the more valuable vanillin by shifting the position of the double bond with a base, followed by oxidation, using potassium permanganate solution to cleave the double bond and produce the aldehyde functional group:

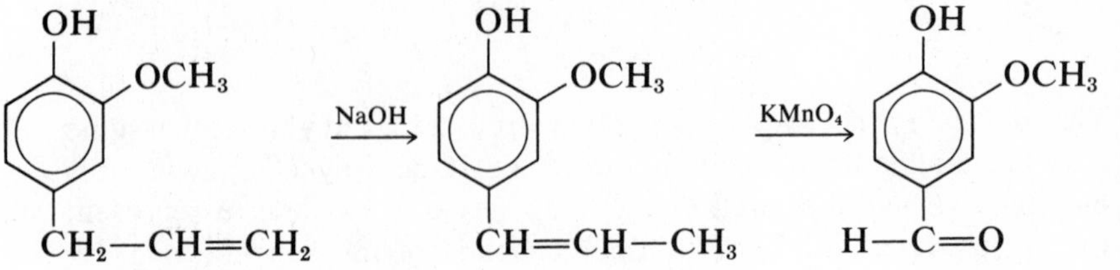

Eugenol Isoeugenol Vanillin

The benzene ring is often called the *benzene nucleus* because it tends to survive substitution reactions and reactions of its substituents.

Naming aromatic compounds Naming substitution products of benzene and other arenes according to IUPAC rules can become very complicated. Limiting ourselves to benzene derivatives makes naming simpler. Take the case of trinitrotoluene mentioned earlier—it has the carbon skeleton of toluene (methyl benzene), so the ring is numbered starting at the point of attachment of the methyl group:

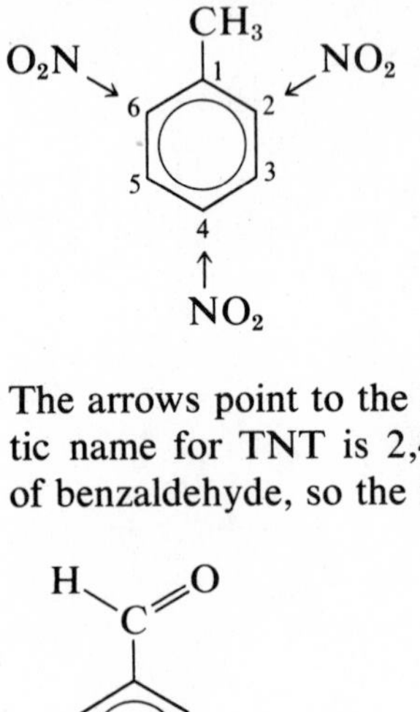

The arrows point to the locations of the nitro groups, so the systematic name for TNT is 2,4,6-trinitrotoluene. Vanillin has the skeleton of benzaldehyde, so the numbering and naming are as shown:

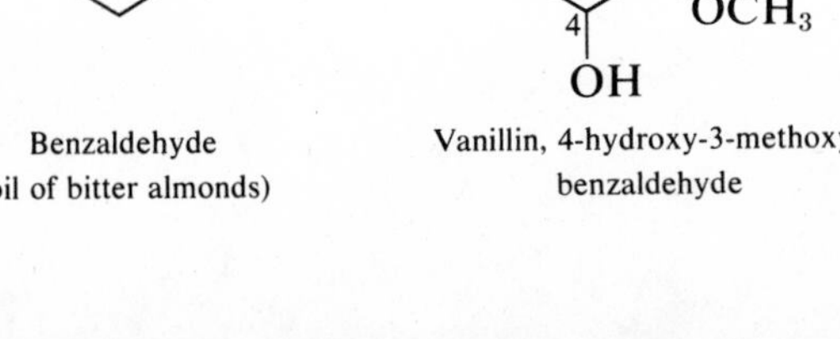

Benzaldehyde (oil of bitter almonds)

Vanillin, 4-hydroxy-3-methoxybenzaldehyde

If there are only two substituents attached to the benzene ring, the words *ortho, meta,* and *para* (abbreviated to *o, m,* and *p*) are used to indicate substituent positions in the following way:

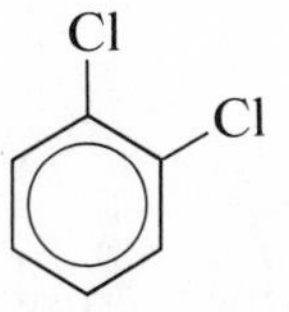

o-dichlorobenzene
(1,2-dichlorobenzene)

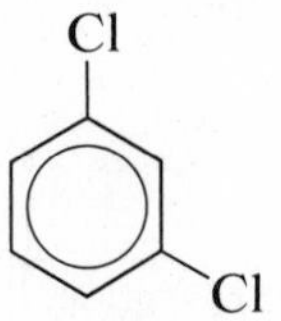

m-dichlorobenzene
(1,3-dichlorobenzene)

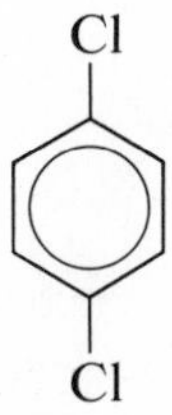

p-dichlorobenzene
(1,4-dichlorobenzene)
(moth crystals)

The substituents do not have to be the same to use this system:

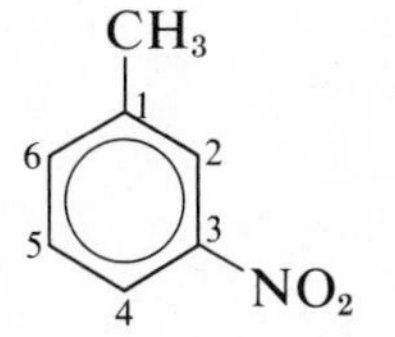

m-nitrotoluene
(3-nitrotoluene)

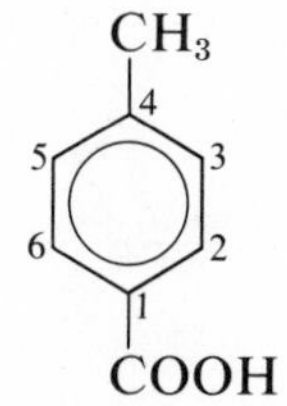

p-toluic acid
(4-methylbenzoic acid)

Here are some other important skeletons that serve as the basic structure in naming:

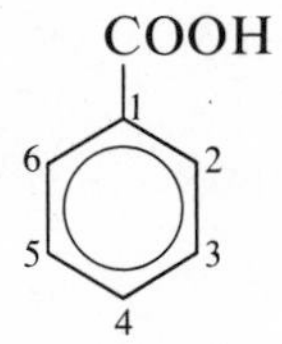

Benzoic acid

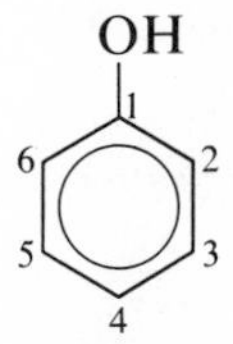

Phenol

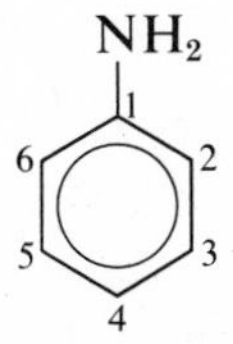

Aniline

In naming substitution products of these compounds, the numbering of the ring starts at the group already present as shown on the formulas, and is done in the direction that gives the *lowest* numbers to the other groups on the ring. The following structure is named 2-chloro-3-hydroxy-4-nitrobenzoic acid (and *not* 6-chloro-5-hydroxy-4-nitrobenzoic acid):

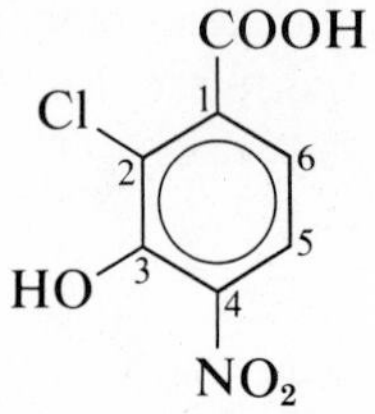

(Correct numbering)

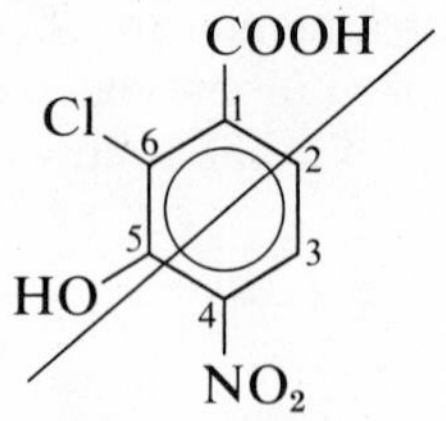

(Incorrect numbering)

Sometimes the benzene ring is treated as a substituent (for hydrogen) on other molecules. In this case the C_6H_5— group of benzene is called "phenyl":

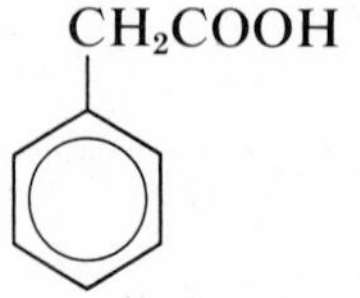

Phenylacetic acid

or $C_6H_5CH_2COOH$

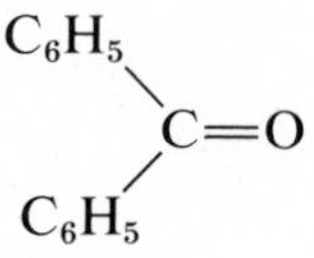

Diphenyl ketone (benzophenone)

Exercises

42. Give the bond structures for the following names (other names are in parentheses).

a. 1,3-dimethylbenzene (*m*-xylene)
b. 1,4-dimethylbenzene (*p*-xylene)
c. 1,3,5-trinitrobenzene
d. *p*-aminophenol (the amino group is NH_2—)
e. 3,5-dinitrobenzoic acid
f. diphenylmethane
g. phenylacetaldehyde (phenylethanal)
h. *p*-nitrophenyl acetic acid

43. How many isomeric tribromobenzenes are there? (Give structures.)

44. How many isomeric methylnapthalenes are there? (Give structures.)

45. Name the following compounds, using either numbers or the *o, m, p* system:

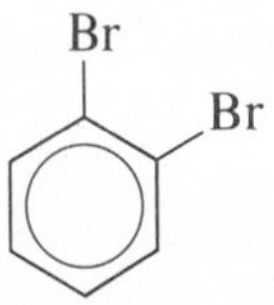

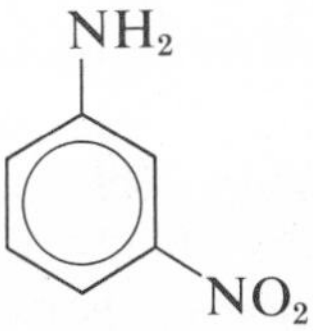

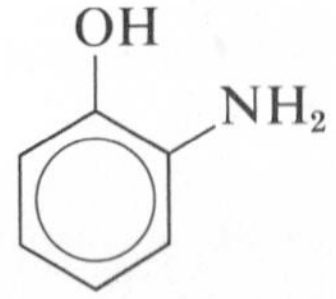

OH

O_2N — (ring with NO_2 at top right, NO_2 at bottom)

$$\begin{array}{c} C_6H_5 \\ \diagdown \\ \quad CHCOOH \\ \diagup \\ C_6H_5 \end{array}$$

$$\begin{array}{c} C_6H_5 \\ \diagdown \\ \quad C{=}CHCOOH \\ \diagup \\ CH_3 \end{array}$$

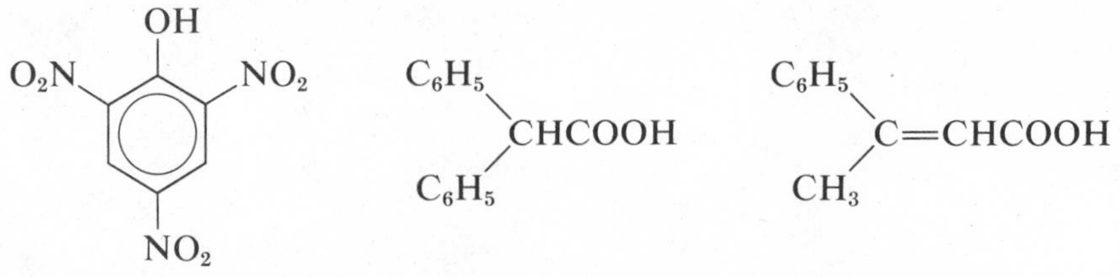

46. The trinitrophenol in Exercise 45 is also called picric acid. When dry, picric acid is an explosive like TNT. What do the two explosives have in common?

Sources and uses of arenes All of the arene hydrocarbons — benzene, naphthalene, anthracene, and others — come from petroleum and coal tar. Heating coal to a high temperature in the absence of air produces coke (largely carbon), used in iron and steel manufacture (Section 11.4), and volatile organic compounds that make up coal tar when condensed. Naphthalene, aniline, and anthracene are used mainly in making dyes. Benzene, toluene, and the dimethylbenzenes (xylenes) are starting materials for drugs, plastics, and many other chemicals.

An example of the use of benzene as a starting material for plastics is the preparation of styrene for use in polystyrene plastic. Benzene is first converted to ethyl benzene by reaction with ethene from petroleum cracking gases, using aluminum chloride as a catalyst. Dehydrogenation (removal of H_2) of ethyl benzene, using ZnO or Fe_2O_3 as catalyst, gives styrene:

$$\bigcirc + CH_2{=}CH_2 \xrightarrow{Al_2Cl_6} \underset{\text{Ethyl benzene}}{\overset{CH_2CH_3}{\bigcirc}} \xrightarrow[\text{heat}]{ZnO} \underset{\text{Styrene}}{\overset{CH{=}CH_2}{\bigcirc}} + H_2$$

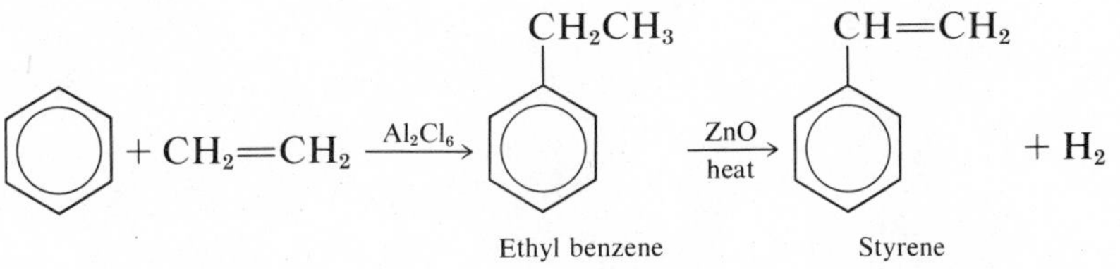

The synthesis of the antibacterial sulfanilimide using benzene as the starting material shows one of the roles of benzene in medicine:

$$\bigcirc \xrightarrow[H_2SO_4]{HNO_3} \overset{NO_2}{\bigcirc} \xrightarrow[HCl]{Fe} \overset{NH_2}{\bigcirc} \xrightarrow{CH_3COCl} \overset{NH{-}\overset{\overset{\displaystyle O}{\|}}{C}{-}CH_3}{\bigcirc} \xrightarrow[Al_2Cl_6]{ClSO_3H}$$

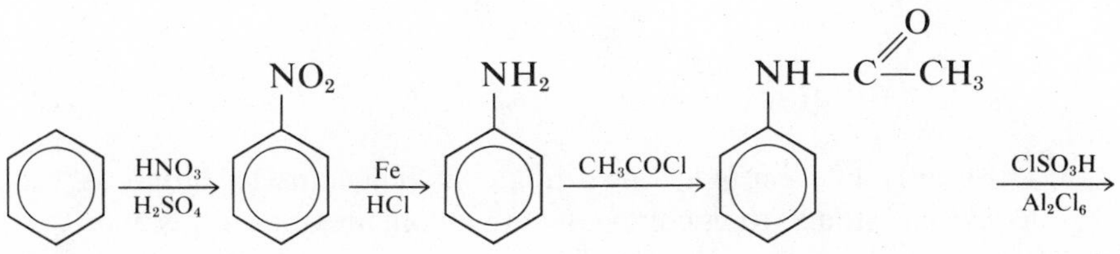

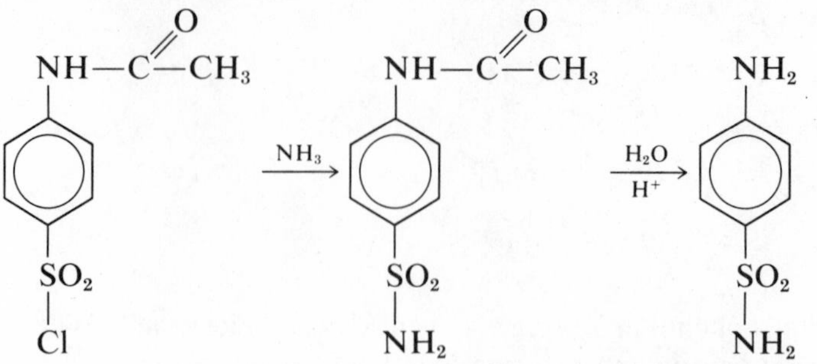

Sulfanilimide
(p-aminobenzenesulfonamide)

This set of reactions makes two points: the poisonous hydrocarbon benzene can be converted to the potent, life-saving antibacterial compound sulfanilimide, and the benzene ring or nucleus comes through all of the reactions without falling apart. Benzene is the starting material for countless medicines and drugs—examples are aspirin (sodium acetylsalicylate), the local anesthetic Novocain, and Dilantin (diphenylhydantoin sodium), used to prevent epileptic seizures. Even penicillin, discussed at the beginning of Section 13.2, contains a benzene ring:

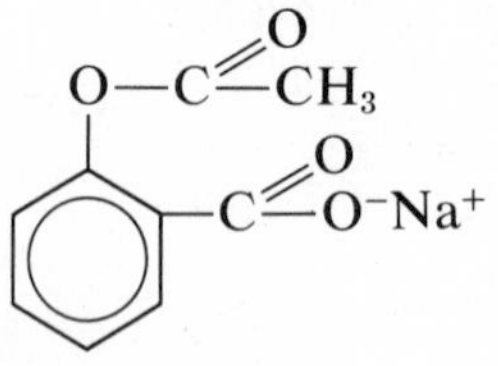

Aspirin
(analgesic)

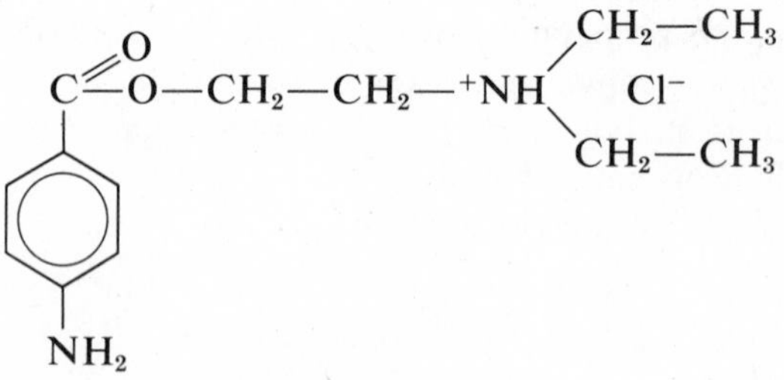

Novocaine
(anesthetic)

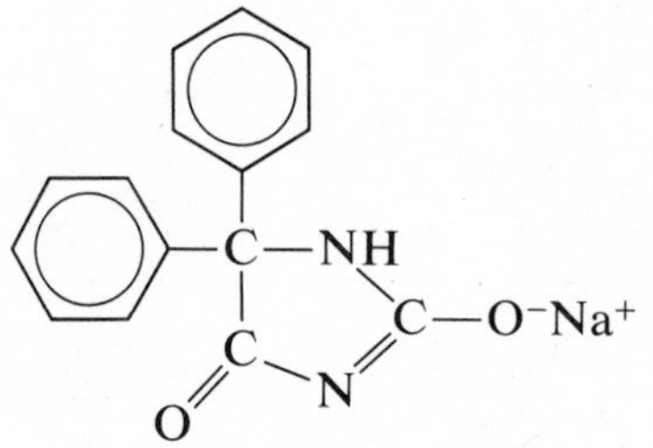

Dilantin
(antispasmodic)

Dilantin is different in having a ring containing nitrogen atoms as well as carbon atoms. Ring compounds that contain atoms other than carbon are called *heterocyclic* compounds. Many vital biochemicals are of this type (Chapter 14).

Polyester fabrics and plastics The ortho and para xylenes (1,2- and 1,4-dimethylbenzenes) from coal tar and petroleum are starting materials for important plastics. Dacron textile fiber and Mylar film begin with *p*-xylene. In a series of steps the methyl groups are converted to carboxylic acid groups by oxidation, giving the benzene dicarboxylic acid called terephthalic acid, which is then esterified with methanol:

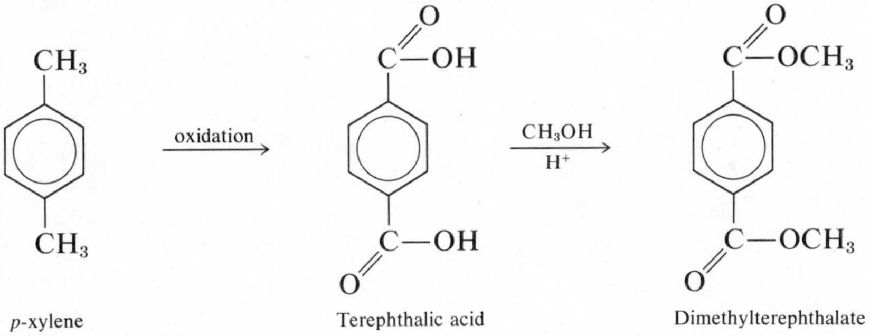

The polyester chain forms when dimethyl terephthalate is heated with ethylene glycol:

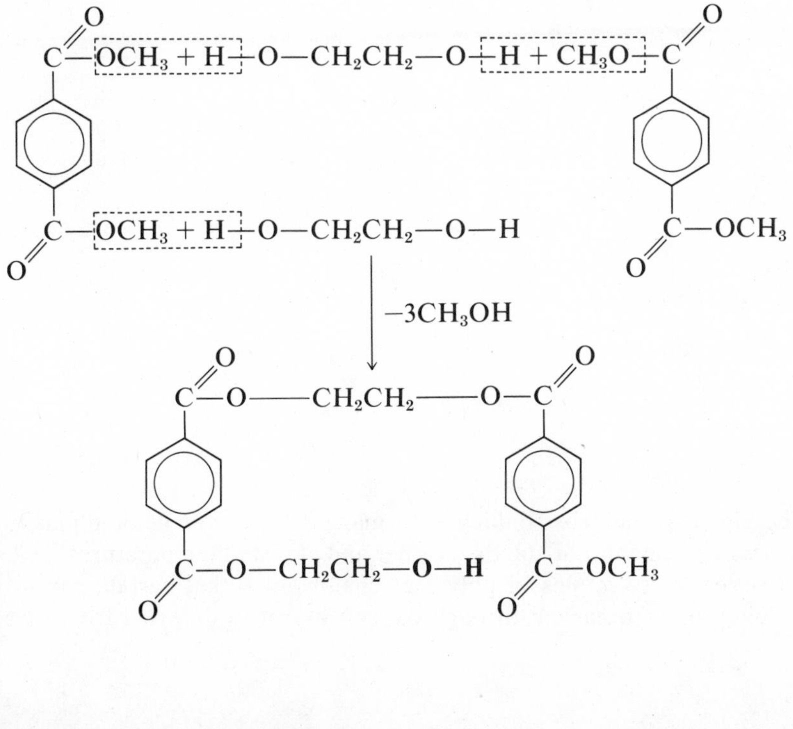

The growing end of the polymer chain is the OH group printed in boldface type, and the repeating unit in the polyester chain is

$$\left(-\overset{\overset{\displaystyle O}{\parallel}}{C}-\!\!\!\bigcirc\!\!\!-\overset{\overset{\displaystyle O}{\parallel}}{C}-O-CH_2-CH_2-O-\right)$$

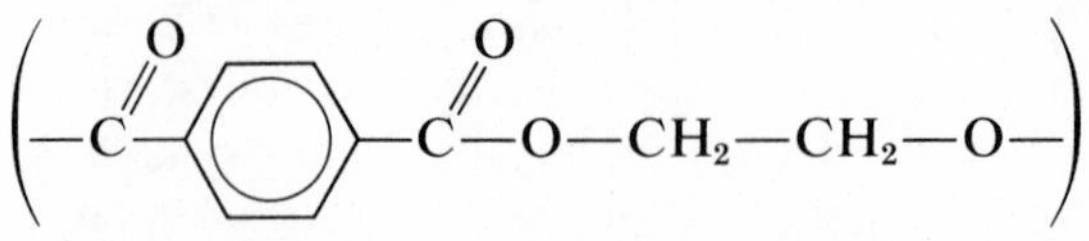

There are from 100 to 130 units in each chain.

The *alkyd resins* and Glyptal used in paint and other protective coatings start with *o*-xylene. Oxidation of *o*-xylene yields phthalic acid, which on heating goes to *phthalic anhydride* with loss of water:

| *o*-Xylene | Phthalic acid | Phthalic anhydride |

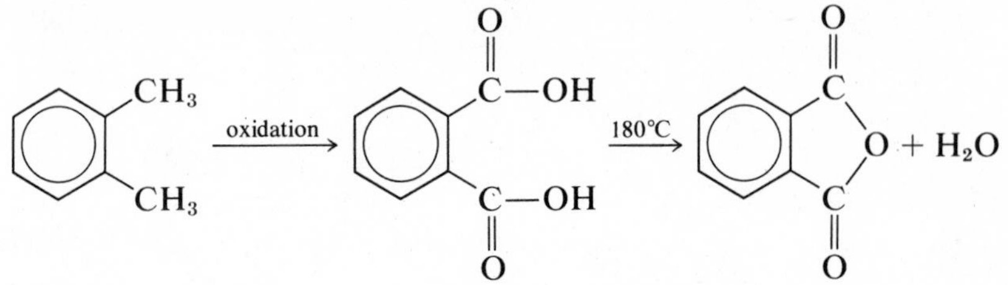

Glyptal, a covering and cementing plastic, is formed when phthalic anhydride is heated with glycerol (1,2,3-propanetriol). Because glycerol has *three* OH groups instead of two like ethylene glycol, a *cross-linked* polyester resin is formed. The repeating unit is

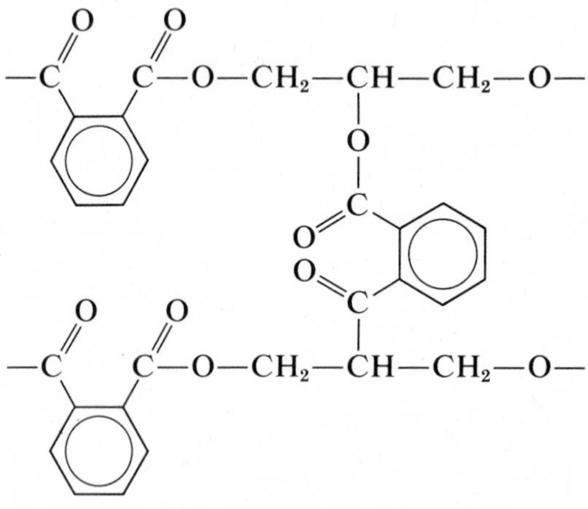

The effect of the cross-linking is to make a tough yet flexible plastic coating or cement that stands weather and elevated temperatures well.

These two examples of polyester chains show that carbon can indefinitely link to carbon through oxygen in a way different from that in polymeric formaldehyde. In Chapter 14 we will find that cellulose (wood), starch, and other carbohydrates are natural polymers in which carbon is linked to carbon through oxygen.

Exercises

47. The reactions in the synthesis of sulfanilimide involve aromatic substitution, reduction, hydrolysis, and amide formation. Using these terms, identify each of the steps in the synthesis of sulfanilimide. What are the other products of each of the reactions?

48. Sometimes styrene is polymerized with a small amount of *p*-divinylbenzene added:

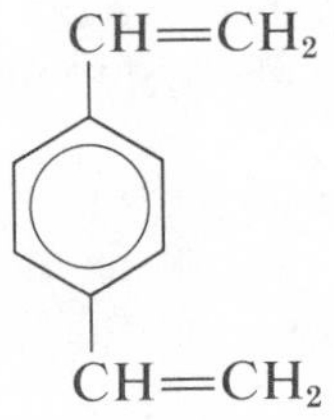

What effect would the addition have on the structure of the polymer?

13.10 Summary Guide Organic chemistry is the chemistry of the compounds of carbon. The carbon atom can form single, double, and triple bonds to other carbons and to atoms of other elements. The bonds are covalent and have definite directions in space. The network of single (sigma, σ) bonds that connects the atoms in an organic molecule is called the skeleton of the molecule. A molecular skeleton may consist of a single carbon skeleton or of two or more carbon skeletons connected through atoms other than carbon, such as O, N, P, and S. Bond angles in molecules can be estimated accurately by using electrostatic repulsion rules and Lewis structures.

A functional group in an organic molecule is a region in the molecule where reactions can take place. Double and triple bonds and atoms other than carbon make up functional groups. Organic compounds are named according to their carbon skeletons, with numbers on the skeleton to locate functional groups if needed. Carbon skeletons are in turn named from the straight-chain alkane or cycloalkane that has the same carbon skeleton.

Isomers are different compounds that have the same molecular formula but different structural formulas. Skeletal isomers have the same functional groups but different skeletons. Functional isomers have different functional groups. Geometric or *cis–trans* isomers differ in the order in which groups are attached to a double bond or a ring compound. Optical isomers are mirror-image molecules that cannot be superposed. They arise when a molecule has chirality or handedness, most frequently due to the presence of a carbon atom (or atoms) carrying four different ligands in the molecule. Optical isomers have identical properties excepting equal but opposite effects in rotating the plane of polarization of light, and their reactions in living systems.

Hydrocarbons are classified as alkanes, alkenes, cycloalkanes, cycloalkenes, acetylenes, and arenes. Petroleum and natural gas are the most important sources of alkanes, alkenes, and cycloalkanes. Coal tar is an important source of arenes. High-molecular-weight alkanes are broken down (cracked) catalytically to increase the yield of gasoline from petroleum. Ethene is an important by-product of cracking and is used to make plastics and ethanol.

Oxygen-containing functional groups are found in alcohols (—OH), aldehydes (—CHO), and ketones (—CO—), as well as in carboxylic acids (—COOH) and their esters (—COOR) and anhydrides ($-\overset{\overset{\displaystyle O}{/\!/}}{C}-O-\overset{\overset{\displaystyle O}{/\!/}}{C}-$). Methanol and ethanol are the most important industrial alcohols. Methanol is made by catalytic hydrogenation of carbon monoxide and may one day replace gasoline and natural gas as a fuel because it can be made from coal. Ethanol is made by fermentation of starch and by hydration of ethene. Formaldehyde is made by catalytic oxidation of methanol and is used in phenol-formaldehyde and polyoxymethylene plastics. Phenol and acetone are made by a process that starts with benzene and propene. Acetone is an important industrial solvent.

Acetic acid is the most important carboxylic acid in industry and is made by fermenting ethanol and by oxidation. Long-chain carboxylic acids occur as esters of glycerol in fats and oils. Soaps are sodium or potassium salts of fat acids. Detergents are long-chain hydrocarbons containing a terminal sulfate group.

Arenes are cyclic, planar molecules written with single and double bonds alternating around the ring. Benzene is the most important arene, and benzene rings are found in several plastics, many medicines, amino acids, dyes, flavors, and other kinds of compounds.

Benzene and other arenes do not undergo the addition reactions shown by alkenes and alkynes, but do undergo aromatic substitution reactions in which an atom or group of atoms replaces one of the hydrogens of the ring. Aromatic substitution reactions on benzene produce a wide variety of useful products. The benzene ring system is particularly stable because its three bonding π molecular orbitals are filled with electrons. This stability makes it possible for the benzene ring to persist intact through many reactions.

Exercises

49. For safety reasons LNG (liquid CH_4) is preferred to LPG (liquid C_2H_6, C_3H_8, C_4H_{10} mixture) as a fuel in the stoves of sailing yachts and cruisers. Why? (*Hint*: think of mol. wt.)

50. Methane has a higher energy as a fuel than methanol. Why is this so?

51. By considering resonance, explain why the π bond of the $\rangle C{=}O$ group is just about as strong as the σ bond in the group.

52. Explain why the bp of acetic acid (CH_3COOH, mol. wt 60) is 119°C while that of butane (C_4H_{10}, mol. wt 58) is −138°C.

53. Alkyl derivatives of ammonia, such as methylamine, $CH_3{-}NH_2$ (bp −7°C), are weak bases in water solution. Write the equation for the reaction of methylamine with water. For methylamine $K_b = 8 \times 10^{-4}$. Calculate the pH of a 0.01-*M* solution of methylamine. What is the value of pK_a for the $CH_3NH_3^+$ ion?

54. The boiling points for methylamine (mol. wt 31), dimethylamine (mol. wt 44), and trimethylamine (mol. wt 59) are, respectively, −7°C, +7°C, +4°C. Using hydrogen bonding, explain why the boiling points do not increase with increasing molecular weight in the usual way.

55. Sodium benzoate, $C_6H_5{-}COO^-Na^+$, is used as a food preservative. What weight of benzoic acid, C_6H_5COOH, in kg would be required for making a ton (2000 lb) of sodium benzoate? (One ton = 907 kg.)

Key Words

Arenes (aromatic hydrocarbons) hydrocarbons written with double and single bonds alternating around rings, as in benzene.

Aromatic substitution reaction a reaction in which a hydrogen atom in an arene molecule is replaced by some other atom or group of atoms.

Carbon skeleton the part or parts of a molecule made up of interconnected carbon atoms only, named after the hydrocarbon having the same skeleton.

Chiral center or asymmetric carbon carbon atom having four different ligands that confer handedness (chirality) on a molecule.

***Cis-trans* isomers** isomers having different arrangements in space of groups attached to double bonds or ring compounds—also called geometric isomers. *Cis*-isomers have groups on the same side of the double bond or ring; *trans*-isomers have the groups on opposite sides.

Enantiomorphs pairs of optical isomers.

Functional group one or more atoms connected together in a molecule that make up a region subject to attack by external reagent species.

Functional isomers molecules with the same molecular formula but different functional groups.

Group or radical what is left after removing one hydrogen from a hydrocarbon—named by replacing -ane, -ene, or -yne in the name of the hydrocarbon by -yl.

Isomers different compounds that have the same molecular formula but different structural formulas.

Linear carbon a carbon atom having two ligands in a molecule—bond angles at 180°.

Molecular skeleton the complete framework of the atoms connected by sigma (σ) bonds in a molecule.

Monomers the small molecular units that are linked together in polymers and plastics.

Optical isomers two isomers that are not superposable because each is the mirror image of the other.

Organic chemistry the chemistry of carbon compounds.

Petrochemicals organic compounds prepared from the hydrocarbons found in petroleum.

Plastics (polymers) giant molecules produced by linking together of small molecules.

Polymerization the process of tying small molecules together by covalent bonds to make giant molecules.

Reactive centers locations subject to reaction in a functional group.

Saturated hydrocarbon or molecule a molecule that has no double or triple bonds, only single bonds.

Skeletal isomers molecules with the same molecular formula and functional groups but differing in carbon skeletons.

Tetrahedral carbon a carbon atom having four ligands in a molecule—bond angles at 109.5°.

Trigonal carbon a carbon atom having three ligands in a molecule—bond angles at 120°.

Suggested Readings

"Coal Gasification Plant Begins Operation," Technology Report, *Chem. and Eng. News,* Nov. 5, 1973, p. 21.

Cook, Paul L., and Crump, John W., *Organic Chemistry: A Contemporary View,* Lexington, Mass., D. C. Heath and Co., 1969.

"Energy: How Do We Go from Here?" *DuPont Context,* **3**(2):3–23 (1974).

Friggens, Paul, "The Great Western Coal Rush," *Reader's Digest*, May 1974, pp. 104–108.

Hart, Donald F., "The Role of Natural Gas in the Energy Crisis," *Catalyst,* **3**(4):20–24 (1973).

Lambert, J. B., "The Shapes of Organic Molecules," *Sci. Am.,* **222**:58–70 (Jan. 1970).

Metz, William D., "Power Gas and Combined Cycles: Clean Power from Fossil Fuels," *Science,* **179**:54–56 (Jan. 5, 1973).

Morton, Rogers B., "Coal: Our Ace in the Hole to Meet the Energy Crisis," *Reader's Digest,* March 1974, pp. 87–90.

Reed, R. B., and Lerner, R. M., "Methanol: A Versatile Fuel for Immediate Use," *Science,* **182**:1299–1304 (Dec. 28, 1973).

Squires, Arthur M., "Clean Fuels from Coal Gasification," *Science,* **184**: 340–346 (April 19, 1974).

"This 'Energy Crisis'—Is It Real?" *Changing Times,* **26**:43–46 (Oct. 1972).

"Western Coal—An Important Element in National Energy Outlook," *Coal Age,* Western Coal Edition, **78**:57–66 (April 1973).

chapter 14

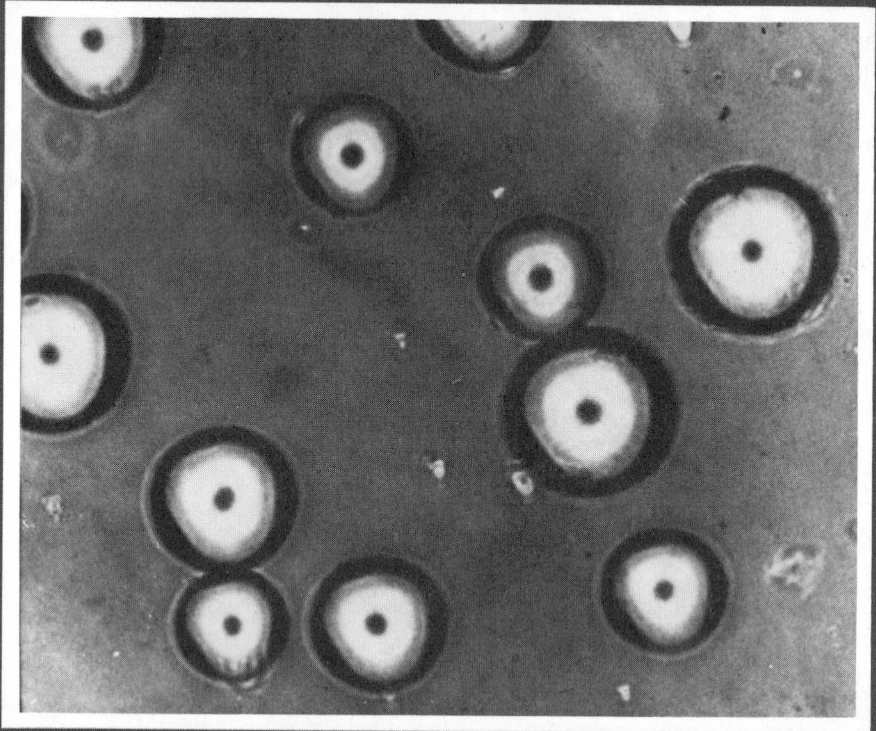

White fat cells obtained by enzymatic digestion of parametrial adipose tissue. (*Photo 36, as used in a study of membrane-mediated responses by M. P. Czech and W. S. Lynn, Department of Biochemistry, Duke University Medical Center.*)

Bioorganic Chemistry

14.1 Introduction Biochemists try to understand the reactions that go on in living things. The molecules taking part in those reactions fall into several classes and range in size from CO_2 and H_2O to huge *biopolymers* with molecular weights in the millions.

Nucleic acids are the giant molecules at the very foundation of life. Located in the genes of living cells, nucleic acids by their structures completely determine whether a single fertilized egg will grow and multiply to become a woman, man, fish, tree, or some other living thing. The nucleic acids direct the putting together of amino acids into the *proteins* that make up most of the structures in which the chemical reactions of life go on. Cell membranes, brain, skin, muscle, hair, and connective tissues are among the many proteins whose biosynthesis is directed by nucleic acids.

Life processes require the release of the chemical potential energy stored in green plants by photosynthesis. Most people get their energy from *carbohydrates* (starch and sugar) and *fats* (triglycerides), although protein (meat) can also serve. Energy that is quickly available is stored in the muscles and liver in the form of the carbohydrate biopolymer called *glycogen* [$(C_6H_{10}O_5)_n$, $n \approx 600{,}000$]. On demand, the *hormones* insulin and adrenalin by adding water to glycogen quickly "unzip" it to *glucose* (blood sugar), the basic energy source (food) for body cells.

$$\underset{\text{Glycogen}}{(C_6H_{10}O_5)_n} + nH_2O \rightarrow \underset{\text{Glucose}}{nC_6H_{12}O_6}$$

Energy for long-term use is stored in body fat, which is formed whenever the potential energy in the food you eat exceeds the amount of energy you spend in exercise, thought, and keeping your body warm and repaired (Photo 36).

Many other hormones secreted into the bloodstream by glands (pancreas, thyroid, pituitary, adrenal, and so on) are like insulin in being complex protein molecules. But some hormones have fairly simple structures. The male sex hormones, androsterone and testosterone, and the female hormones, estrone and progesterone, belong to the class of compounds called *steroids*. Their molecular structures are not much more complicated than some of the molecules you saw in Chapter 13.

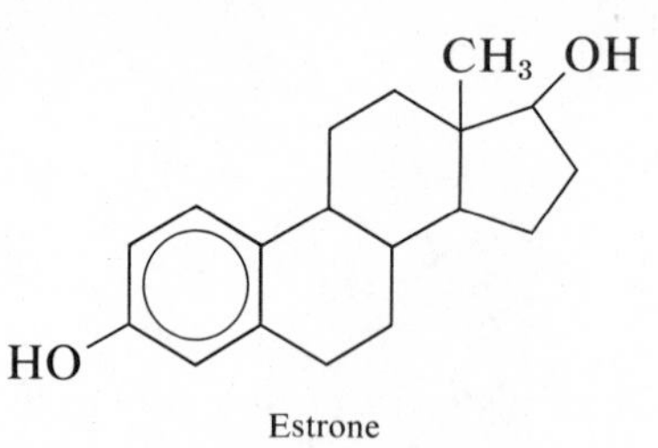

Estrone

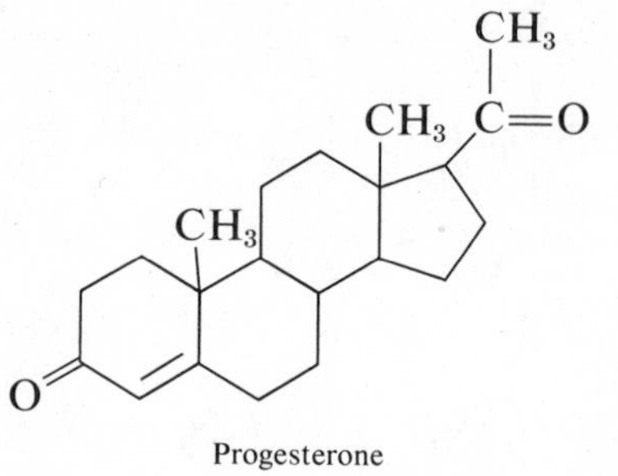

Progesterone

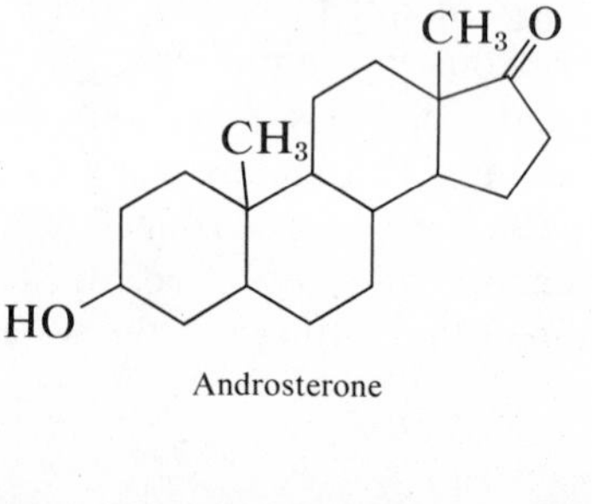

Androsterone

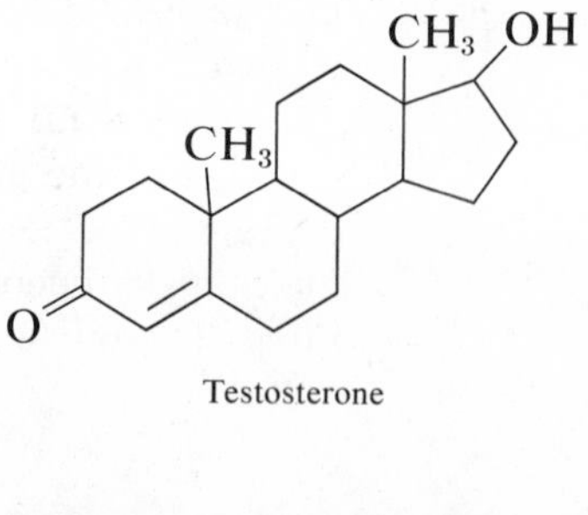

Testosterone

The hormones are the *regulators* of the reactions of life — they tell reactions when to start and when to stop.

Enzymes are protein molecules that catalyze the reactions of life so that they can proceed rapidly at ordinary temperature. *Vitamins* are often located in the catalytically active sites of enzymes.

In summary, nucleic acids, proteins, fats, carbohydrates, hormones, enzymes, and vitamins are the classes of organic chemicals for life. In this chapter we give examples of the structure and function of a few of these chemicals. The purpose is to give you some feeling for the wonderfully interesting and challenging field called biochemistry, or more descriptively, "bioorganic chemistry."

Many of the compounds in living things are electrolytes and are more or less ionized in cell fluids. Ionization will be neglected in this chapter, so you will find carboxylic and phosphoric acids in their molecular form HA instead of as H_3O^+ and A^-.

Exercises

1. Calculate the approximate molecular weight of glycogen $(C_6H_{10}O_5)_{600,000}$. The liver is about 3% of body weight, and may contain as much as 15% glycogen by weight. Calculate the number of glycogen molecules in the liver of a 75-kg person from these figures.

2. How many chiral carbon atoms are there in the testosterone molecule? Using the 2^n rule, determine how many optical isomers are possible for testosterone. (Only one of the isomers is made and used by the body.)

14.2 2-Amino Carboxylic Acids and Proteins Cells, whether of human, animal, plant, or microorganism, produce 2-amino carboxylic acids to serve as building blocks for constructing the high-molecular-weight biopolymers called *proteins*. There are 21 of these *natural* amino acids, ☛ and they all have the same arrangement of ligands at carbon atom number 2. Letting R stand for the rest of the amino acid molecule, we have this arrangement:

From now on "amino acid" will mean 2-amino carboxylic acid. The acids are called α-amino acids when another set of naming rules is used.

$$
\begin{array}{c}
\overset{\displaystyle O}{\overset{\displaystyle \|}{^1C}}\!\!-\!OH \\
H_2N\!-\!^2C\!-\!H \\
|\\
R
\end{array}
$$

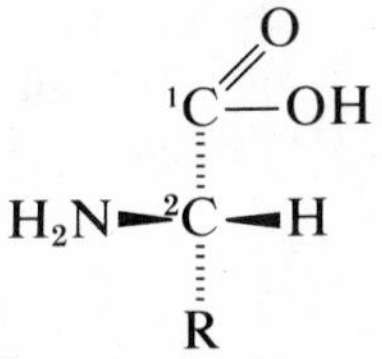

Amino acids are amphoteric (Section 11.6) — their carboxyl group behaves as an acid toward bases, their amino group as a base toward acids. You might guess that they would react with themselves by transferring the carboxyl proton to the unshared electron pair on the amino group. This they do, and the double ion formed is called the *dipolar ion form*:

$$
\begin{array}{ccc}
\overset{\displaystyle O}{\underset{\displaystyle}{\overset{\|}{C}}} \!\!-\!\mathrm{OH} & & \overset{\displaystyle O}{\underset{\displaystyle}{\overset{\|}{C}}} \!\!-\!\mathrm{O}^{-} \\
H_2N \!-\! \underset{\displaystyle R}{\overset{\displaystyle|}{C}} \!-\! H & \rightleftarrows & H_3^+N \!-\! \underset{\displaystyle R}{\overset{\displaystyle|}{C}} \!-\! H
\end{array}
$$

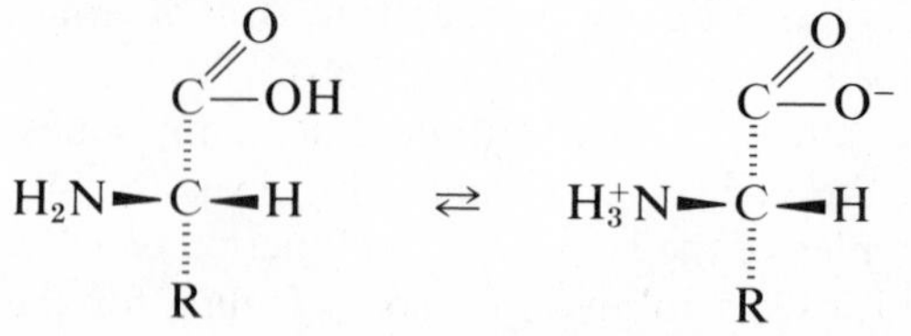

Dipolar ion form

Solid amino acids and solutions of them in water contain mainly the dipolar ion form. ¶ In the following sections the un-ionized formulas are used because they make the formation of proteins easier to understand.

Proteins are giant molecules formed by removal of water from amino acids, using a hydrogen from the amino (NH_2) group of one amino acid and the OH from the carboxyl group of another. Two steps in the polymerization using dashed boxes to show the elimination of water can be outlined as follows:

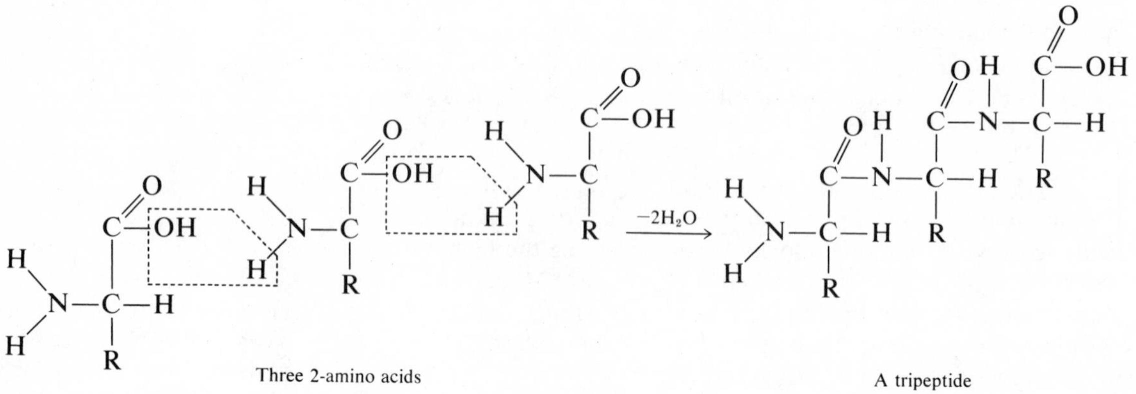

Three 2-amino acids A tripeptide

Continued building of the polymer in this way produces the covalent bond or *primary structure* of the protein. This is a zigzag chain held together by carbon-to-carbon bonds alternating with *peptide*

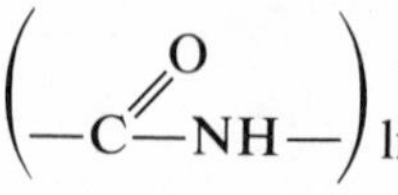

$$
\left(-\overset{\displaystyle O}{\overset{\|}{C}}-NH-\right) \text{linkages:}
$$

The amount of dipolar ion form present depends on the pH of the solution. The pH at which the concentration of dipolar ion is largest is called the "isoelectric point" for the amino acid. Solubility of the amino acid in water is usually smallest at the isoelectric point.

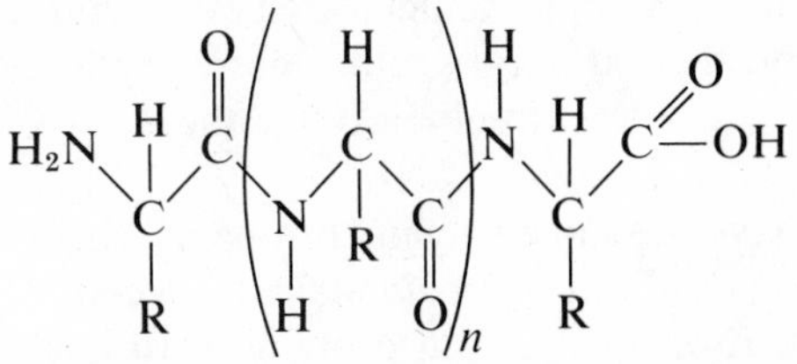

Proteins range in molecular weight from tens of thousands to tens of millions and show great variety because nature provides the 21 *natural* amino acids through differences in the part of the structure represented by R in the general formula. Most proteins contain several different amino acids in their primary structure, and you can readily see that with so many different building blocks available there is no limit to the number of protein molecules possible.

The simplest amino acid is *glycine* (amino acetic acid), in which R = H:

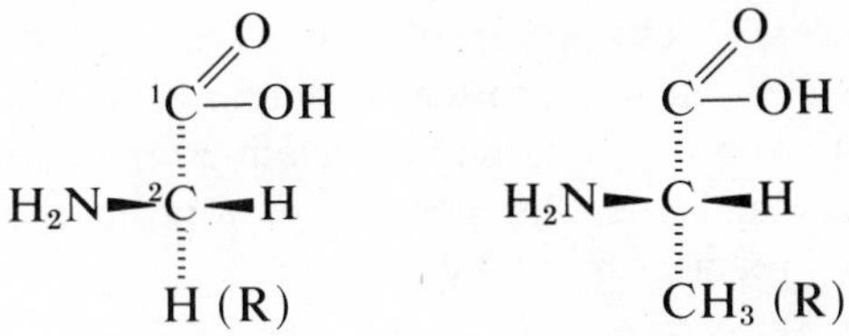

Glycine
(aminoacetic acid)

Alanine
(2-aminopropionic acid)

Next comes *alanine*, in which R is a methyl group. For the other 19 amino acids R becomes more and more complicated. Except for glycine, all the amino acids have four *different ligands* on the number 2 carbon atom, making it a *chiral* carbon atom. This means all of the amino acids except glycine have optical isomers. Most remarkable is the fact that all of the natural amino acids have the *same* arrangement of groups around the chiral carbon atom C-2. This arrangement of the groups in space is called the *natural* or L *configuration*.

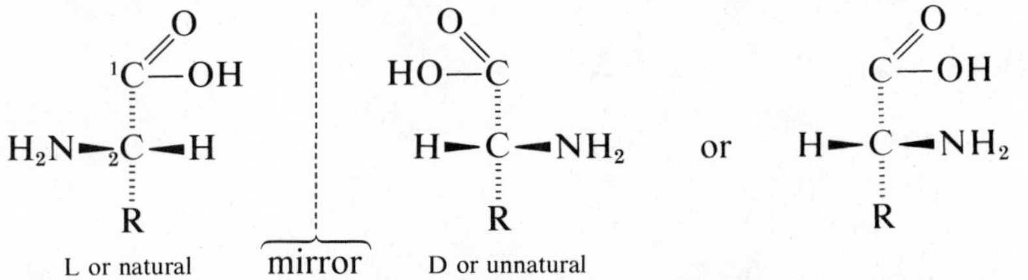

L or natural configuration

D or unnatural configuration

The letters D and L used to show configuration have nothing to do with the direction of rotation of the plane of polarized light—L-alanine ($R = CH_3$) is actually the (+) or dextrorotatory isomer (Section 13.5). The D- and L-isomers *are* always mirror images of each other, however.

The mirror image or enantiomorph of the L configuration is called the D configuration. (How you write the carboxyl group doesn't actually make any difference because it rotates freely around the C_1—C_2 bond.)

The L- and D-amino acids behave the same way in "ordinary" chemical reactions, but they behave very differently in *biochemical reactions* that are catalyzed by enzymes. Enzymes have a protein chain backbone and therefore also have "handedness" or chirality. They will catalyze a reaction of one optical isomer but not the other. The molecule that is to react must have the correct chirality to fit into the chiral active site on the enzyme. (You can't put a right shoe on a left foot.) Body enzymes can use L-alanine in building up the proteins of skin, hair, brain, muscle, and the like, but they reject the D-isomer. If you eat a mixture of D- and L-isomers, only the L will find its way into your proteins. All of the amino acids that come from our food have the L or natural configuration.

Unless an optical isomer is used as a reagent, laboratory preparations of compounds that possess a chiral carbon atom will always contain equal amounts of the (+) and (−) optical isomers. Mixtures like this are called *racemic mixtures*. For example, racemic mixtures of the 2-bromopropionic acids and the alanines (2-aminopropionic acids) result from the following reaction sequences:

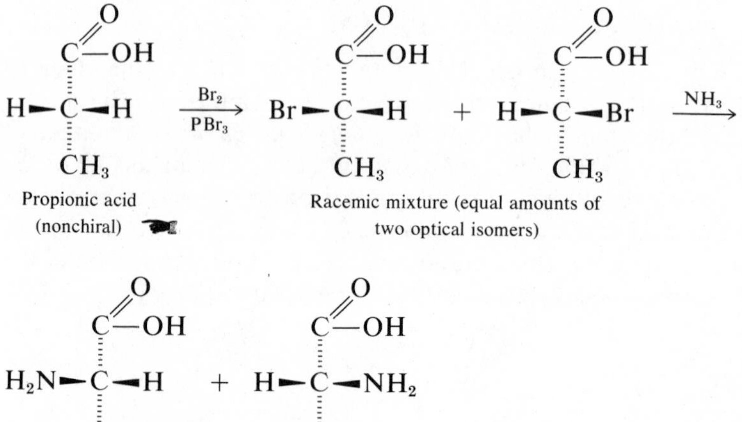

Propionic acid (nonchiral)

Racemic mixture (equal amounts of two optical isomers)

The word *achiral*, meaning "without chirality," is also used.

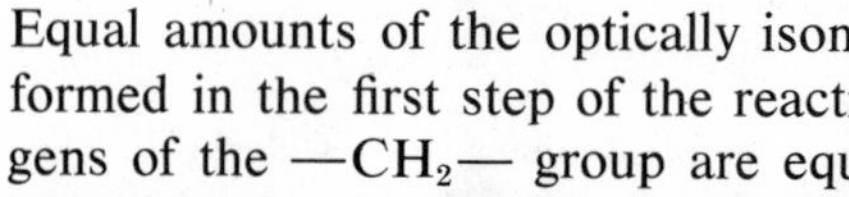

Racemic mixture

Equal amounts of the optically isomeric 2-bromopropionic acids are formed in the first step of the reaction sequence because the hydrogens of the —CH_2— group are equivalent and equally liable to replacement by Br. When NH_2 replaces Br in the second step, equal amounts of the (+) and (−) or D- and L-isomers of alanine will also be formed. On the other hand, if you have just one of the optical isomers of 2-bromopropionic acid, it will give just *one* of the alanine optical isomers on treatment with ammonia.

A racemic mixture of optical isomers doesn't rotate the plane of polarization of light because the mixture contains equal numbers of (+)- and (−)-isomer molecules and their equal but opposite rotations cancel each other. Racemic mixtures can be separated into their (+) and (−) or D- and L-isomers by the process called *resolution*. Any description of this process would be out of place here; you will learn about it if you take a course in organic chemistry.

Organic molecules having a chiral carbon atom and found in plants and animals almost always consist of either one or the other of the two optical isomers of the compound. For example, when milk sours, the milk sugar (lactose) it contains is converted by the enzymes of a microorganism into (−)-lactic acid, but when we exercise our muscles, (+)-lactic acid is produced from glucose. Both of the isomers of lactic acid are "natural," but they occur separately, not as the racemic mixture.

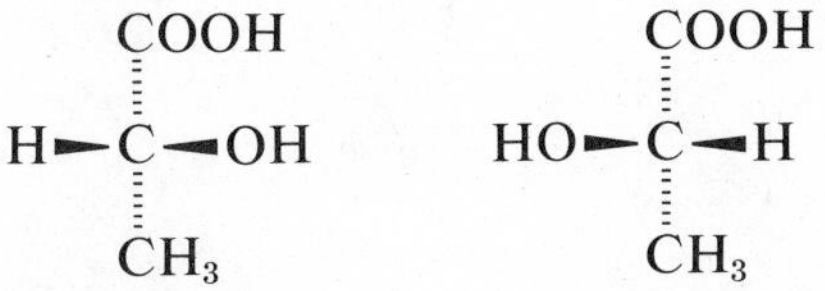

$$\begin{array}{cc} \text{COOH} & \text{COOH} \\ | & | \\ \text{H—C—OH} & \text{HO—C—H} \\ | & | \\ \text{CH}_3 & \text{CH}_3 \end{array}$$

(−)-lactic acid (+)-lactic acid
(D-lactic acid) (L-lactic acid)

One question that still has scientists puzzled is how in the evolution of life more of one optical isomer than of the other (as in the case of L- over D-amino acids) was produced. There is still no firm answer to this question, but you can be sure of one thing—the tetrahedral valence bond directions in four-ligand carbon that make chirality possible were the starting point.

Shapes of protein molecules In talking about the structures and shapes of protein molecules, biochemists use the idea of an *amino acid residue*. To form one, you remove the elements of water from the amino acid in a thought experiment:

$$\begin{array}{ccc} & \text{O} & & & \text{O} \\ & \| & & & \| \\ \boxed{\text{H}} & \text{C}{-}\boxed{\text{OH}} & & & \text{C}{-} \\ \backslash & | & & & | \\ \text{N—C—H} & \longrightarrow & \text{N—C—H} + \text{H}_2\text{O} \\ / & | & & / & | \\ \text{H} & \text{R} & & \text{H} & \text{R} \end{array}$$

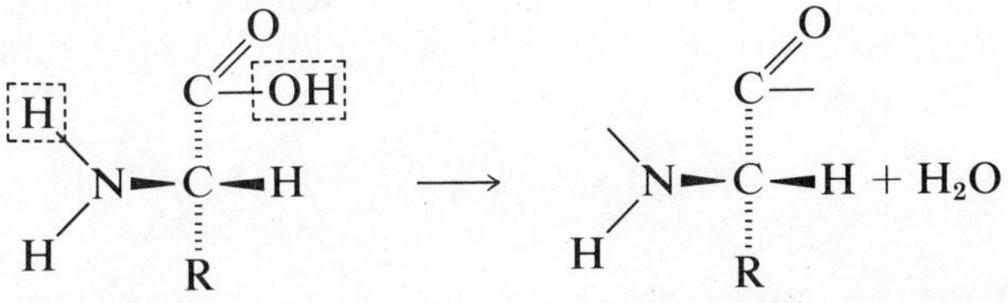

Amino acid Amino acid residue

There are 21 amino acid residues corresponding to the 21 different amino acids. Residues are identified by three- or four-letter abbreviations of the name of the amino acid. Some examples are these:

	Amino acid	*Residue*	*Abbreviation*

Glycine

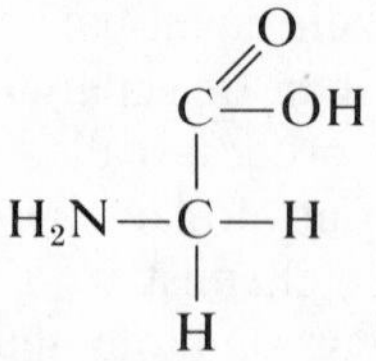

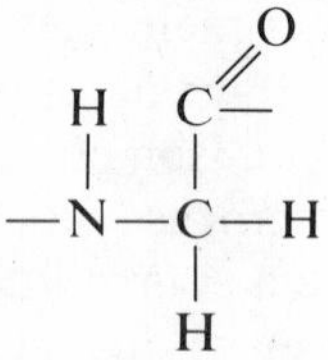

$$\begin{array}{c} O \\ \parallel \\ C-OH \\ | \\ H_2N-C-H \\ | \\ H \end{array}$$

$$\begin{array}{c} \quad O \\ \quad \parallel \\ H \quad C- \\ | \quad | \\ -N-C-H \\ \quad | \\ \quad H \end{array}$$

Gly

Alanine

$$\begin{array}{c} O \\ \parallel \\ C-OH \\ | \\ H_2N-C-H \\ | \\ CH_3 \end{array}$$

$$\begin{array}{c} \quad O \\ \quad \parallel \\ H \quad C- \\ | \quad | \\ -N-C-H \\ \quad | \\ \quad CH_3 \end{array}$$

Ala

Cysteine

$$\begin{array}{c} O \\ \parallel \\ C-OH \\ | \\ H_2N-C-H \\ | \\ CH_2SH \end{array}$$

$$\begin{array}{c} \quad O \\ \quad \parallel \\ H \quad C- \\ | \quad | \\ -N-C-H \\ \quad | \\ \quad CH_2SH \end{array}$$

Cys

Tyrosine

$$\begin{array}{c} O \\ \parallel \\ C-OH \\ | \\ H_2N-C-H \\ | \\ CH_2 \end{array}$$
(benzene ring)
$$OH$$

$$\begin{array}{c} \quad O \\ \quad \parallel \\ H \quad C- \\ | \quad | \\ -N-C-H \\ \quad | \\ \quad CH_2 \end{array}$$
(benzene ring)
$$OH$$

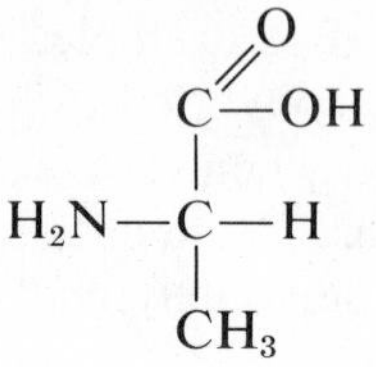

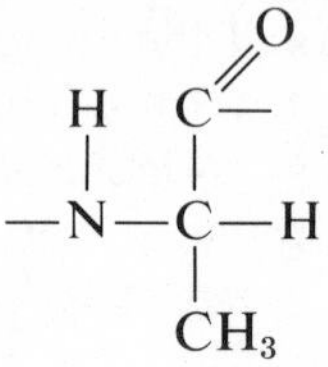

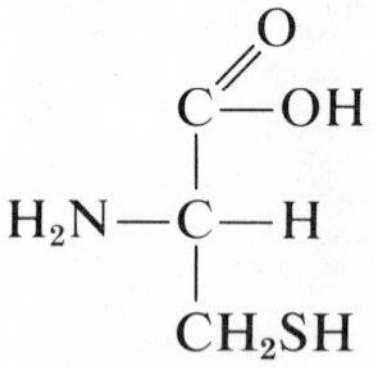

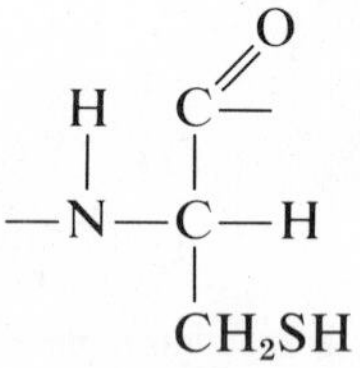

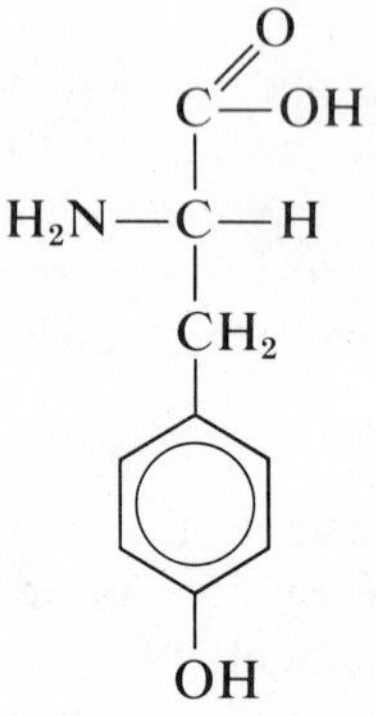

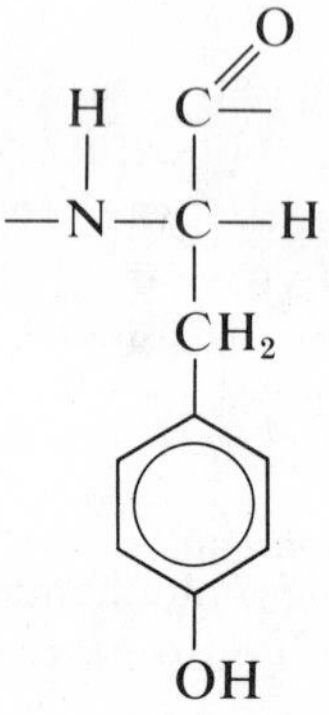

Tyr

The abbreviations for residues at the ends of polypeptide chains are completed by adding an H to show an H_2N—*terminal* group or an

$$OH \text{ to show a } -\overset{\displaystyle O}{\overset{\displaystyle \parallel}{C}}-OH \text{ terminal group.} \quad \P \text{ Part of the "A" chain,}$$

Amino acids with two carboxyl groups (glutamic acid, aspartic acid) or with two amino groups (lysine, arginine) are among the 21, so depending on how the polypeptide chain is put together, it can be terminated with Hs or OHs at *both ends*.

one of the two polypeptide chains in the *insulin* molecule, shows how terminal groups and amino acid residues are used to show the structure of proteins:

H—Gly┼Ileu–Val–Glu ← 15 more residues → Cys┼Asn—OH

H—NH—CH$_2$—C$\overset{O}{\diagdown}$—

$\overset{\overset{O}{\diagdown}}{C}$—OH

—NH—C—H

CH$_2$—CO—NH$_2$
$\overset{O}{\diagdown}$

Terminal NH$_2$ group
(glycine residue)

Terminal C—OH group
(asparagine residue) ☛

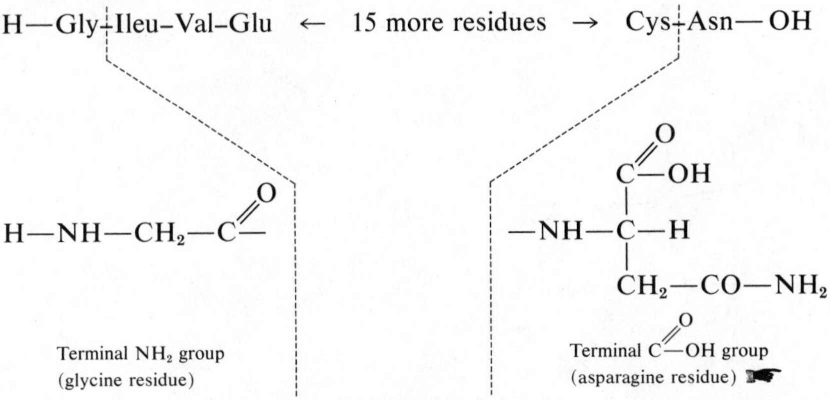

Part of A chain of insulin

The —CO—NH$_2$ amino group is part of the R group in asparagine and does not count as a terminal group.

The amino acid cysteine (3-mercapto-2-aminopropionic acid) stands out from the rest because two cysteine residues in *different* polypeptide chains can tie the two chains together, using a *disulfide link* (—S—S—). This link is formed between the —SH (thiol or mercaptan) groups of the two cysteine residues by oxidation:

$$\xrightarrow[\text{agent}]{\text{Oxidizing}} \; [O]$$

H$_2$O +

Disulfide link

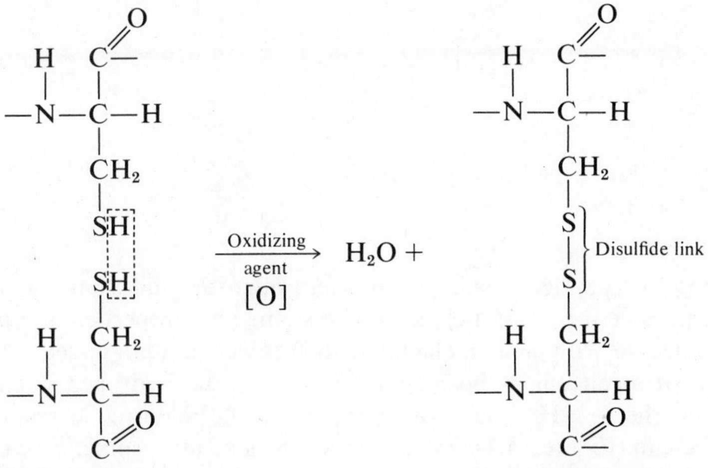

Cystein residues in two
different polypeptide chains

Two chains linked together

The link can also be formed between two cysteine residues at different locations in the same polypeptide chain. The process is similar to cross-linking of hydrocarbon chains in rubber during vulcanization (Section 13.6).

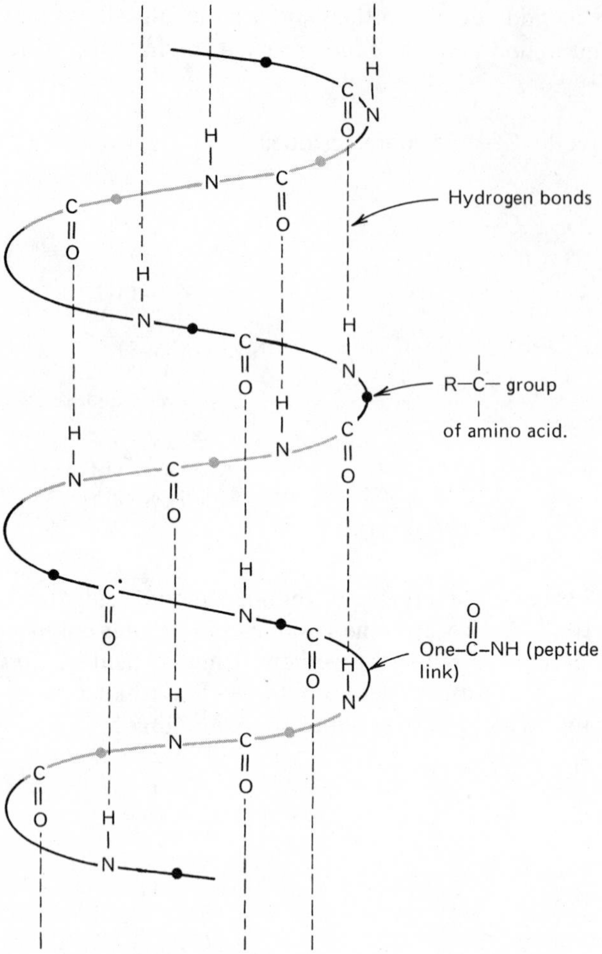

FIG. 14.1 The helical secondary structure imposed on the polypeptide chain of a protein by hydrogen bonding (not to scale).

Tying polypeptide chains together through disulfide links in this way can have a powerful effect on the shape and properties of a protein molecule. Polypeptide chains by themselves prefer a coiled spring or *helical* arrangement because it maximizes the hydrogen bonding between the —NH— and >CO groups in the amino acid residues of the chain (Figure 14.1). The hydrogen-bonded arrangement is called the *secondary structure*. By tying chains together, disulfide links can make globe-shaped proteins out of helical, rod-shaped ones. The A and B chains of insulin are tied together by two disulfide links to give the protein molecule a more spherical shape. Without going into the detailed structure of the polypeptide chains, you can see how two long chains could be put together to form a more "round" molecule.

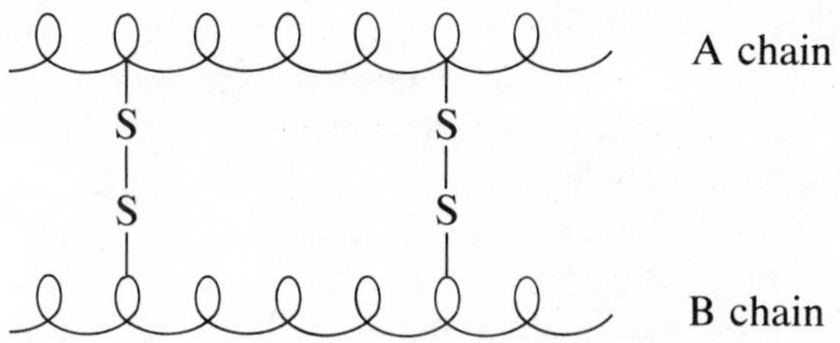

Disulfide links can be broken by reducing agents. Permanent waving of hair results from breaking disulfide links in hair protein (keratin), changing the shape of the hair fiber by curling or waving, then re-forming the disulfide links to hold the new (waved) position, using an oxidizing agent. Letting R—S—S—R stand for the disulfide link between two polypeptide chains, we see that the breaking and re-forming reactions are as follows:

$$R{-}S{-}S{-}R + \underset{\text{Ammonium mercaptoacetate}}{HS{-}CH_2{-}COO^-NH_4^+} \rightarrow 2R{-}SH + \begin{array}{l} S{-}CH_2COO^-NH_4^+ \\ | \\ S{-}CH_2COO^-NH_4^+ \end{array}$$

$$6R{-}SH + \underset{\text{Potassium bromate}}{KBrO_3} \rightarrow 3R{-}S{-}S{-}R + KBr + 3H_2O$$

Breaking disulfide links with reducing agents is one way of changing the natural, covalent-bond, or primary structures of proteins. Heat, strong acids, and strong bases have the same effect. The secondary, spiral structure, held together by hydrogen bonds between the $\rangle C{=}O$ and —NH— groups in the polypeptide chain, can be broken up using solutions of urea, $H_2N{-}\overset{O}{\overset{\|}{C}}{-}NH_2$, and other solutes. Urea forms hydrogen bonds with the $\rangle C{=}O$ and —NH— groups that are stronger than the natural ones, and causes the helical secondary structure to unwind into a random coil or zigzag chain.

Folding of the spiral coils back on themselves gives the tertiary structure of the protein. The tertiary structure of a protein is the most sensitive to change and it is—at the same time—necessary if the protein is to do the job the nucleic acids designed it for. Changes in electrolyte concentrations in the solution of a protein can destroy its tertiary structure. Any treatment that destroys protein structure is said to *denature* the protein. If the denaturation is "gentle" enough, changing the conditions may let the structure re-form. This often happens when only the tertiary structure is destroyed.

Proteins are very "particular" in their reactions because of their delicate tertiary and secondary structure. Ordinary temperatures (20–37°C) provide just enough thermal energy to give their delicate structures flexibility without destroying them.

Proteins carry on many reactions vital to life. Enzymes and viruses are part protein. In enzymes, the protein provides the chiral backbone structure that carries the catalytically active center. Located in the active center is the *prosthetic group* (often a vitamin), built into the protein molecule by covalent bonds. The prosthetic group is the actual catalyst, but it cannot function without the supporting protein structure and its chirality.

The complete structures of several proteins are already known, and the use of new X-ray crystallographic techniques and new ways of determining the sequence of amino acid residues in the polypeptide chains promise rapid progress in this vital area of biochemistry.

EXAMPLE

After purifying a protein and determining its molecular weight, the next step toward finding out its structure is to hydrolyze the protein to its constituent amino acids and find out the kinds, sequence, and amounts of the various amino acid residues in its molecules. This kind of analysis presented almost impossible difficulties until the invention of ion exchange chromatography.

Chromatography in general is a process for separating molecules or ions, using a column of finely divided particles of an absorbent that selectively holds back solute species contained in a solution that percolates through the column. Ion exchange chromatography uses a column of an ion exchange resin in the way described earlier (Section 11.3) for softening or demineralizing hard water. However, the exchange resin is highly selective as to the ease with which the ions are exchanged. The exchange resins are called ion-selective resins, and using them makes it possible to separate different ions of even the same charge type. For example, it is possible to separate Na^+ and K^+ ions from a solution containing K^+Cl^- and Na^+Cl^-. The resin exchanges H_3O^+ ions for both ions but does so much more readily for K^+ ions than for Na^+ ions. The result is that a solution of Na^+Cl^- in $H_3O^+Cl^-$ appears first in the effluent solution at the bottom of the column, as the sample containing Na^+, K^+, and Cl^- ions is followed by water. Then by pouring an acid solution on at the top of the column, the K^+ ions are dislodged by H_3O^+ ions from the sites on the exchange resin, and they too come out (but separately) in the effluent.

When the solution is a gaseous one, the technique is called liquid-vapor partition chromatography or simply "gas" chromatography.

Amino acids are electrolytes and ionize in a way determined by the pH of the solution containing them. Exchange resins that pass or hold up ions derived from the amino acids in a protein hydrolysis mixture in a selective way separate the amino acid mixture into its component ions. The process is automated and determines not only which amino acids are present but their amounts and molar ratios as well. Knowing the molecular weight of the protein and the molar ratios of the amino acids enables calculation of the number of amino acid residues of each type in the original protein.

Using enzymes that bring about hydrolysis of peptide links between specific amino acid residues, ion exchange chromatography makes it possible to find the sequence of amino acid residues in the protein and its primary (covalent bond) structure. Secondary and tertiary structures are then determined by X-ray crystallographic analysis, using a crystal of the protein. If the protein cannot be obtained in crystalline form, at least the amino acid sequence can be found and other, spectrometric methods can be applied to get at the secondary and tertiary structures.

In diagnosing gene-linked diseases, ion exchange chromatography can be applied to the proteins in blood to see if there are any of abnormal structure.

Exercises

3. Using the part of the A chain of insulin given in this section, give the complete formula of the amino acid asparagine.

4. Using dashed lines for hydrogen bonds, show how urea bonds to both the >C=O and —NH— groups in a polypeptide chain.

5. The pituitary gland hormone *oxytocin,* responsible for uterine contractions and the start of milk flow at childbirth, has the following amino acid sequence: H—Gly-Leu-Pro-Cys-Asn-Gln-Ileu-Tyr-Cys—OH. The primary (covalent bond) structure is not a zigzag chain but looks more like this:

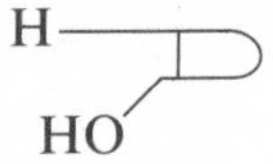

Account for the bent structure. Oxytocin has been synthesized from its separate amino acids in the laboratory.

Essential amino acids Proteins in the diet are the source of the constant supply of amino acids our bodies need for growth and repair. During digestion in the stomach the enzyme *pepsin,* itself a protein, catalyzes the hydrolysis of the protein chain into smaller polypeptides called *proteoses* and *peptones.*

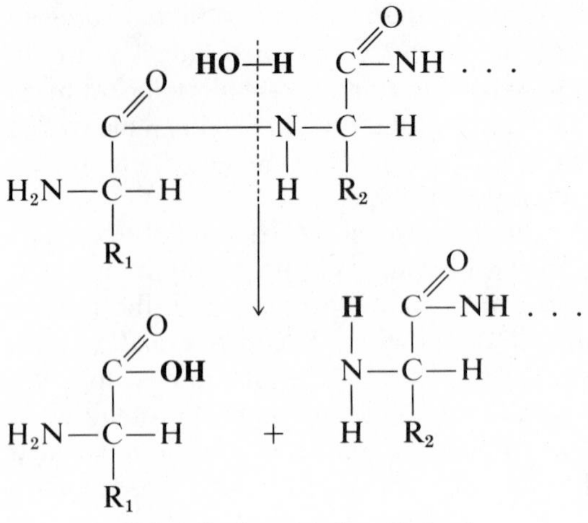

Hydrolysis of a terminal peptide bond

Further hydrolysis to the separate amino acids occurs in the small intestine with the pancreatic enzyme *trypsin* as catalyst. The amino acids are absorbed through the intestinal wall into the bloodstream to provide the raw material for protein synthesis in the liver and other organs. The Food and Nutrition Board of the National Research Council recommends a protein intake of 1 g (dry weight) per kg of body weight each day. This amount is provided by $4\frac{1}{2}$ oz of meat, fish, or poultry plus an egg and a pint of milk (or 2 oz of cheese) for a person of average weight. Smaller amounts of protein come from breads and vegetables.

If, for one reason or another (crash diet to lose weight?), you take in less protein than your body needs, you go into what is called *negative nitrogen balance*—the condition in which you excrete more nitrogen, in the form of waste products in urine and feces, than you take in in the form of protein in your diet. If negative nitrogen balance continues, muscles waste away as their protein is hydrolyzed to provide amino acids for the synthesis of the more vital proteins of the blood (albumin and globulin). When the protein reserves in the body are exhausted, the concentration of blood proteins has to fall for lack of amino acid building blocks. The result is a lowering of the osmotic pressure of the blood. Just as in the case of diseased kidneys (Section 6.3, Osmotic Pressure) leaking blood protein into the urine, lowered osmotic pressure of the blood results in movement of water from the blood to the tissues, producing swelling (edema), particularly of the abdomen.

Proteins of blood, brain, and heart are last to go.

Thus people suffering from extreme protein deficiency look "fat."

The structure of α-chymotrypsin (one of the trypsins) contains 221 amino acid residues in three polypeptide chains that are tied together by disulfide links between cysteine (Cys) residues. It has been synthesized in the laboratory.

Humans and animals are able to synthesize many amino acids from other amino acids by the process called *transamination*. For example, if your body has plenty of alanine for protein synthesis but lacks glutamic acid that it needs for a particular protein it's putting together, the following transfer of an —NH_2 group takes place:

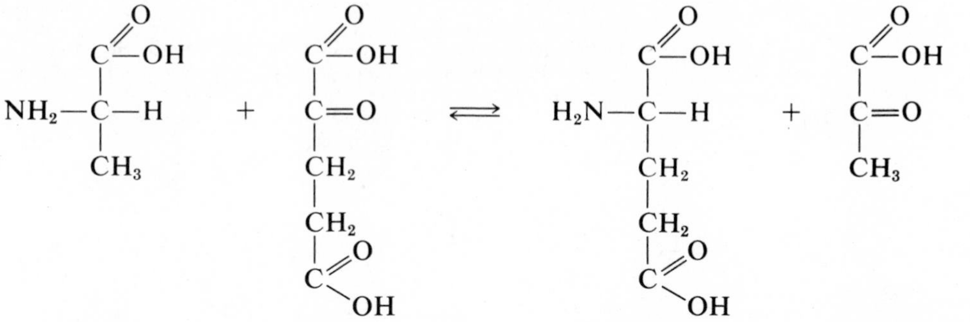

$$
\begin{array}{ccccc}
\underset{\text{Alanine}}{\begin{array}{c} \text{O} \\ \text{C}-\text{OH} \\ | \\ \text{NH}_2-\text{C}-\text{H} \\ | \\ \text{CH}_3 \end{array}}
& + &
\underset{\text{2-ketoglutaric acid}}{\begin{array}{c} \text{O} \\ \text{C}-\text{OH} \\ | \\ \text{C}=\text{O} \\ | \\ \text{CH}_2 \\ | \\ \text{CH}_2 \\ | \\ \text{C} \overset{\text{O}}{\underset{\text{OH}}{}} \end{array}}
& \rightleftharpoons &
\underset{\text{Glutamic acid}}{\begin{array}{c} \text{O} \\ \text{C}-\text{OH} \\ | \\ \text{H}_2\text{N}-\text{C}-\text{H} \\ | \\ \text{CH}_2 \\ | \\ \text{CH}_2 \\ | \\ \text{C} \overset{\text{O}}{\underset{\text{OH}}{}} \end{array}}
& + &
\underset{\begin{array}{c}\text{Pyruvic acid}\\\text{(2-ketopropionic}\\\text{acid)}\end{array}}{\begin{array}{c} \text{O} \\ \text{C}-\text{OH} \\ | \\ \text{C}=\text{O} \\ | \\ \text{CH}_3 \end{array}}
\end{array}
$$

Also named "α-ketoglutaric acid."

There is a pool of various 2-ketocarboxylic acids in the carbohydrate and fat reaction systems in the body that participate in transamination. Transaminations are reversible so the body can make many amino acids not in dietary protein. But there are some amino acids, called the *essential amino acids*, which the body cannot make at all—they must be in the diet, just as vitamins must be.

There are eight essential amino acids for humans and other mammals: isoleucine, leucine, lysine, methionine, phenylalanine, threonine, tryptophan, and valine. A protein is called "complete" if its molecules contain the residues of the essential amino acids among others. The proteins of milk, cheese, eggs, meat, soybeans, and whole wheat are complete. Corn and gelatin lack tryptophan and are very low in cysteine—the "cross-linking amino acid."

Children whose diet consists purely of carbohydrates do not get enough protein or enough complete protein and usually not enough vitamins. These children become ill with a disease called "Kwashiorkor." The symptoms are lack of growth, lack of interest in surroundings, low levels of blood protein (albumin) resulting in swollen bodies (edema), anemia, and general weakness. If the disease is treated early enough—all that's needed is enough milk in the diet—recovery can be complete. There is growing evidence that protein-deficient diets during growing years produce mental retardation.

The World Health Organization and the Food and Agriculture Organization of the United Nations are both actively pursuing education and agricultural research programs to rid the world of protein-defi-

FIG. 14.2 Some of the vitamins with their biochemical effects.

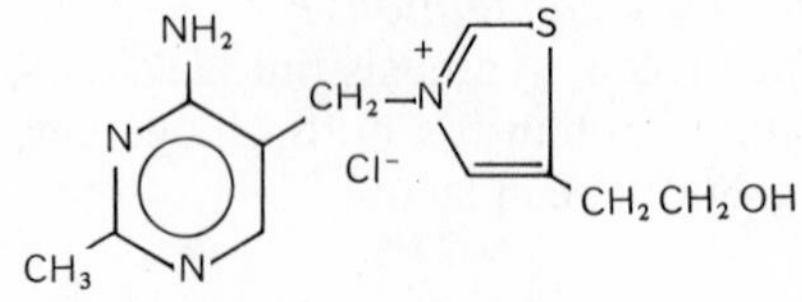

Vitamin B$_1$ (thiamine). Deficiency diseases: decreased growth, improper carbohydrate metabolism, nerve paralysis, mental depression.

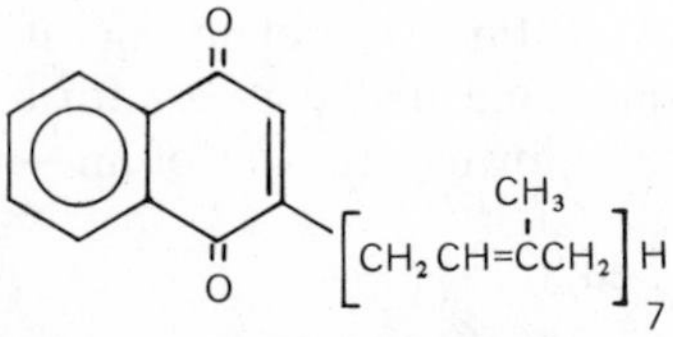

Vitamin K. Antihemorrhagic vitamin. Deficiency diseases: hemorrhages resulting from insufficient formation of prothrombin, the protein that forms blood clots.

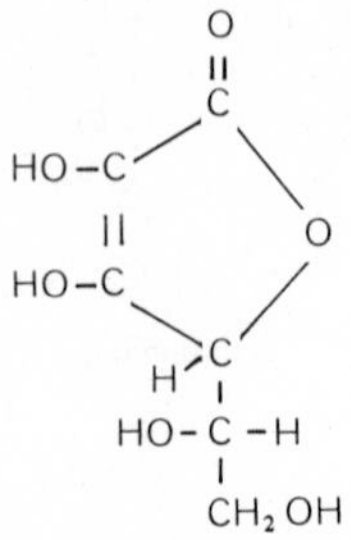

Vitamin C (ascorbic acid). Antiscurvy vitamin. Deficiency diseases: bleeding due to weak walls of blood vessels, retarded growth of bones and teeth.

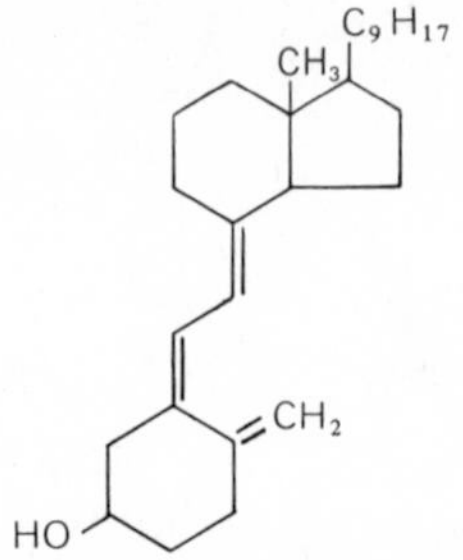

Vitamin D (calciferol). Deficiency diseases: rickets, softening of bones, poor utilization of Ca^{2+} and PO_4^{3-} to form $Ca_3(PO_4)_2$ of bone.

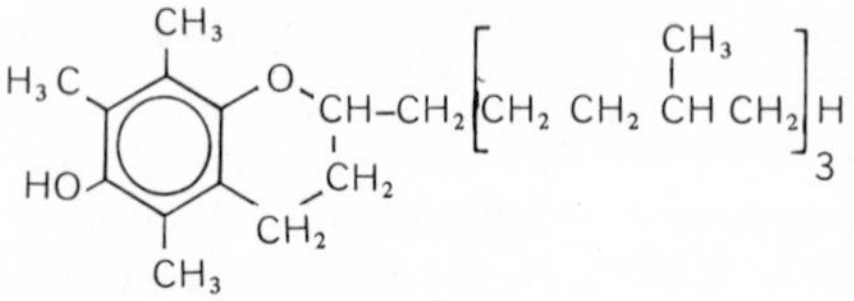

Vitamin E (α-tocopherol). Deficiency diseases (in animal, presumably in man): sterility of males, abortion by females.

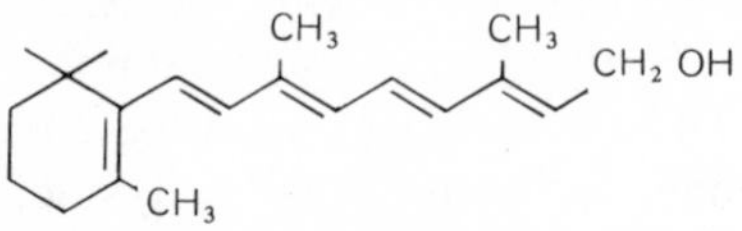

Vitamin A (Retinol). Deficiency diseases: retarded growth, poor vision in dim light ("night blindness"), nerve degeneration.

ciency disease. There have been breakthroughs by plant geneticists in producing protein-rich grain species that will grow almost anywhere, and a process has been developed for converting fish protein to an odorless, nearly tasteless, powder that can be mixed with milk to make a complete, high-protein beverage.

Exercises

6. Transamination of 2-keto-3-phenylpropionic acid (2-keto-3-phenylpropanoic acid) with glycine would give the amino acid phenylalanine. Give the structure of phenylalanine and the other product of the transamination.

7. Phenylalanine is one of the essential amino acids. Does the transamination reaction of Exercise 6 go on in the body?

14.3 Vitamins Vitamins, which our bodies cannot make, are a group of compounds vital to life. Like the essential amino acids just described, they must be in your diet. ☛ The word "vitamin" comes from "vital amine" because vitamin B_1, the first vitamin to be identified, contains an amino (—NH_2) group. The structures of some of the vitamins and the deficiency diseases they prevent are given in Figure 14.2. Whole books have been written on the structures of vitamins and their biochemical roles. Here we will take a closer look at vitamins D and A and one of the B vitamins in relation to life processes.

Vitamins A and D can be produced in the body provided that their *precursors* (compounds of structure close to that of the vitamin) are in the diet. More on this shortly.

Vitamin D controls the amounts of calcium and phosphate ions in your blood, keeping them at the proper level for growing and maintaining bone (largely Ca_3PO_4). In the absence of the vitamin, bones in older people become porous and easily broken. In children, vitamin D deficiency results in soft bones that bend permanently when a child starts to walk. The deficiency disease is called "rickets."

Cod and halibut liver oils contain vitamin D itself. Many other foods contain precursors of the vitamin, the sterols ergosterol ☛ and 7-dehydrocholesterol. When the ultraviolet rays in sunlight penetrate the skin, the precursors are converted to D vitamins. Because people are less exposed to the sun in winter, rickets develop most often in that season. In most countries vitamin D is added to milk, either by adding the vitamin itself or by irradiating the milk with ultraviolet light to convert the sterol precursors in it into the vitamin. The conversion of ergosterol to vitamin D starts with a photochemical reaction in which light energy ($h\nu$) breaks bonds to produce a previtamin, then involves an intramolecular hydrogen transfer (shown by arrows) to give the vitamin itself:

The vitamin from ergosterol is called vitamin D_2 and differs from the one (D_3) from 7-dehydrocholesterol only in having a double bond and an extra methyl group in the hydrocarbon side chain. Vitamin D_3 is more active as a D vitamin than D_2.

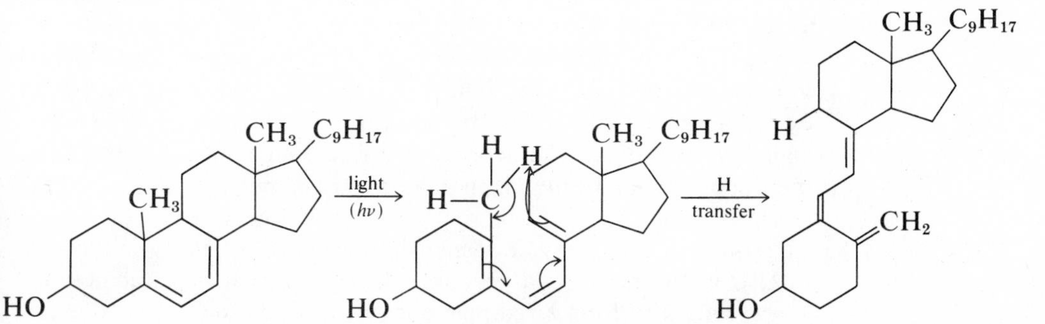

Ergosterol Previtamin D_2 Vitamin D_2 (calciferol)

Vitamin D is a second example of nature's use of the steroid ring system in compounds for controlling life processes. (The sex hormones discussed earlier are also steroids.)

Vitamin A is necessary for maintaining the skin, eyes, and nervous system. It is found in fish liver oils and in the livers of mammals, in which enzymes convert β-carotene (a conjugated polyene hydrocarbon) to vitamin A by cleavage and oxidation of the double bond in the center of the molecule (arrow):

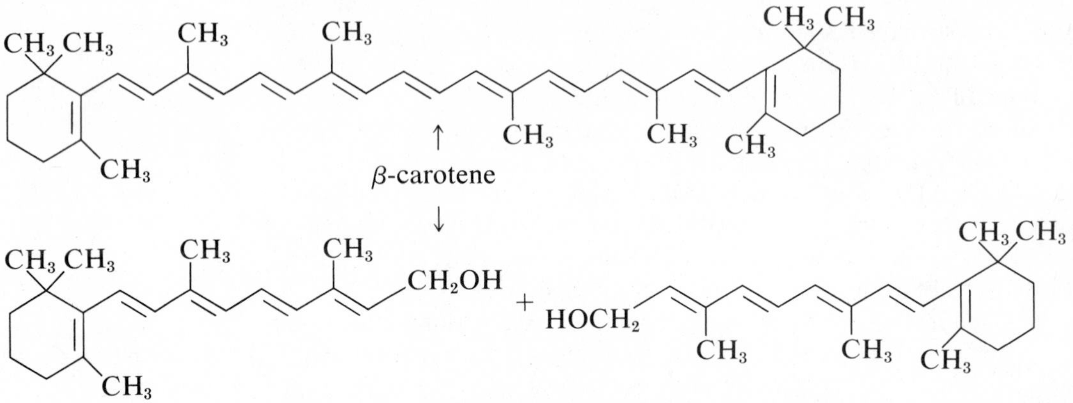

Our bodies can't make the vitamin "from scratch" but can make it from β-carotene, which is plentiful in a balanced diet, because β-carotene is responsible for the color of carrots and is present in corn, leaf vegetables, and many fruits. Butter and milk contain the vitamin itself, the amount depending on the amount of β-carotene in the cow's diet. β-carotene absorbs light of wavelengths in the region around 4500 Å, which accounts for its orange color. The colors of tomatoes and other vegetables are due to conjugated polyenes of structure close to that of β-carotene. They are called *carotenoids*, and some are converted to vitamin A in the body.

The yellow color of oleomargarine is due to the addition of β-carotene.

Vitamin A is also called *trans-retinol* for the reason that it plays a vital part in the *retina*, the light-sensitive membrane at the back of the eye. There the vitamin is oxidized to the corresponding aldehyde called *trans*-retin*al*. The "*trans*" in these names indicates that all of the double bonds in the chain have *trans* geometry. *Trans* → *cis* isomerization of *trans*-retinal at C-11 then converts it to the isomer 11-*cis*-retinal. The enzyme responsible is called retinal isomerase.

Also in the retina is a protein called *opsin*, which combines with 11-*cis*-retinal to form a red pigment called *rhodopsin*. Opsin does not combine with *trans*-retinal or with any other of the *cis–trans* isomers possible for retinal. Rhodopsin is the key substance in "night vision"

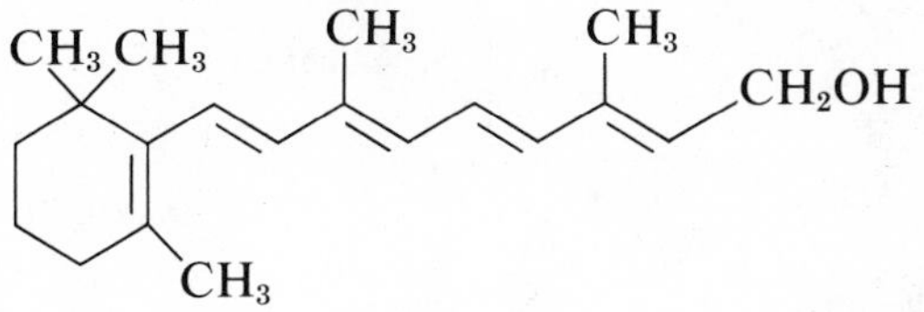

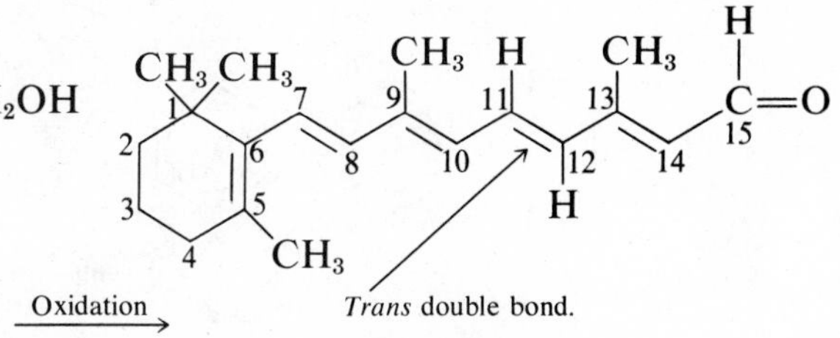

Vitamin A (*trans*-retinol)

Trans-retinal (vitamin A aldehyde, retinene)

↓

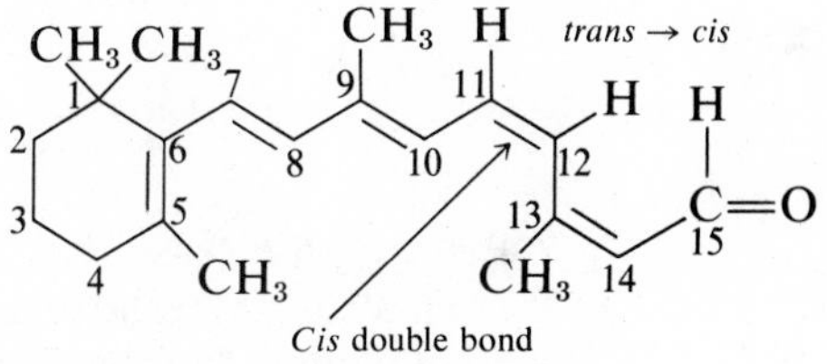

11-*cis*-retinal (neoretinene)

or vision in dim light. It is located in the rod cells of the eye, which are mainly in the region of the retina outside of the center of vision. When even very faint light of any color but red falls on rhodopsin, the light energy is absorbed and converts the 11-*cis*-retinal in rhodopsin to *trans*-retinal, at the same time generating the nerve impulse for vision. Red light does not cause this change because rhodopsin is itself red and reflects (does not absorb) red light. The combination of opsin with *trans*-retinal formed by the light absorption is unstable and separates into opsin and *trans*-retinal. Isomerization of *trans*-retinal to 11-*cis*-retinal and re-forming of rhodopsin completes the cycle. No reactions go perfectly, even biochemical ones, so there is some natural loss of the retinals. For this reason you must have a regular source of vitamin A for replacement or you will become "night-blind," as well as suffer other deficiency diseases.

Nicotinic acid amide or nicotinamide (niacin), one of the B vitamins, is directly involved in the oxidation of vitamin A to *trans*-retinal. Nicotinic acid got its name because it was first identified as an oxidation product of the tobacco alkaloid nicotine. ☛ The alkaloid and acid both contain a pyridine ring, which—like benzene—has a resonance hybrid structure:

Alkaloids are nitrogen-containing compounds, usually poisonous, found in plants. Their name comes from their being bases (alkaline) because of amino groups in their molecules. Morphine, quinine, heroin, and cocaine are examples of alkaloids.

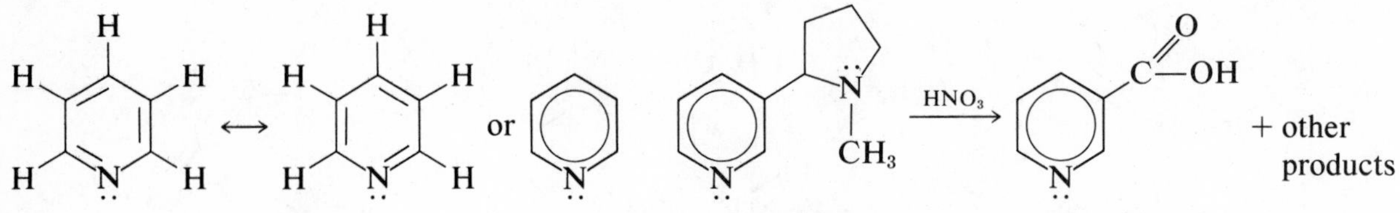

Pyridine ring — Nicotine — Nicotinic acid

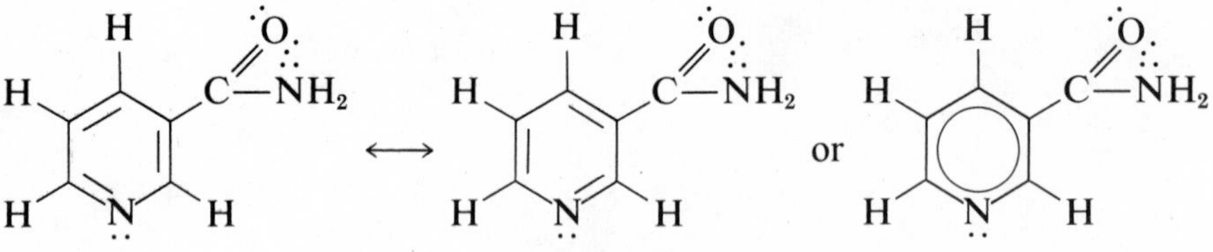

Nicotinamide

Pellagra, a disease that involves failure of the digestive, nervous, and brain systems, results from nicotinamide deficiency. There is plenty of nicotinamide to prevent pellagra in a balanced diet.

NAD$^+$ ≡ nicotinamide adenine dinucleotide.

Nicotinamide is the working part of the biochemical oxidation–reduction enzyme system abbreviated NADH ⇄ NAD$^+$ ☛ that is used to catalyze many biochemical reactions in both plants and animals. This enzyme system has a large, complicated covalent structure attached to the nitrogen of the pyridine ring. Letting R stand for this complex part of the molecule in its oxidized (NAD$^+$) and reduced form (NADH), the electron-transfer half-reaction for the oxidation of NADH to NAD$^+$ can be written as follows:

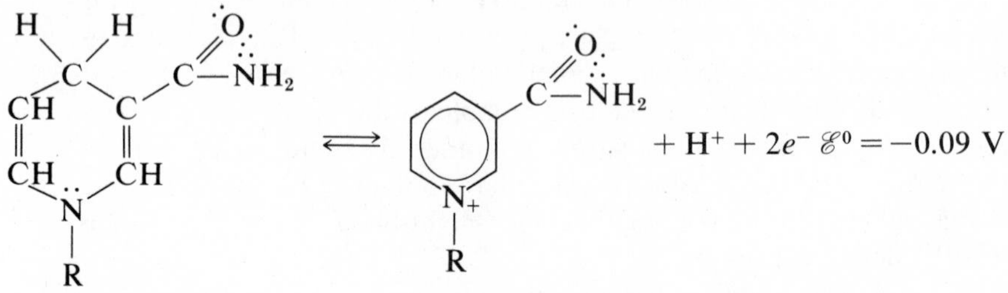

NADH reducing agent — NAD$^+$ oxidizing agent ☛

NAD$^+$ is a cation and has one anion (Cl^-) associated with it to balance the + charge.

The value of $\mathscr{E}^0 = -0.09$ V is for standard electrochemical cell conditions, where $C_{H^+} = 1$ mole/liter. Correcting to pH = 7 (the situation in the body) gives $\mathscr{E}^0_7 = 0.32$ V, a value that means NADH in the body is a reducing agent about as strong as iron metal ($\mathscr{E}^0 = 0.41$ V, Table 10.2). The reason NADH is such a strong reducing agent lies in the release of the resonance energy of the pyridine ring (21 kcal/mole) when NADH is oxidized to NAD$^+$. One function of NAD$^+$ in the body is to oxidize vitamin A (retinol) to the corresponding aldehyde retinal needed for night vision.

EXAMPLE

The electron-transfer half-reaction for the oxidation of an alcohol to an aldehyde, using A to represent the rest of the alcohol molecule, is a two-electron reaction. For the retinol–retinal oxidation A is $C_{20}H_{27}$— and the half-reaction is as follows:

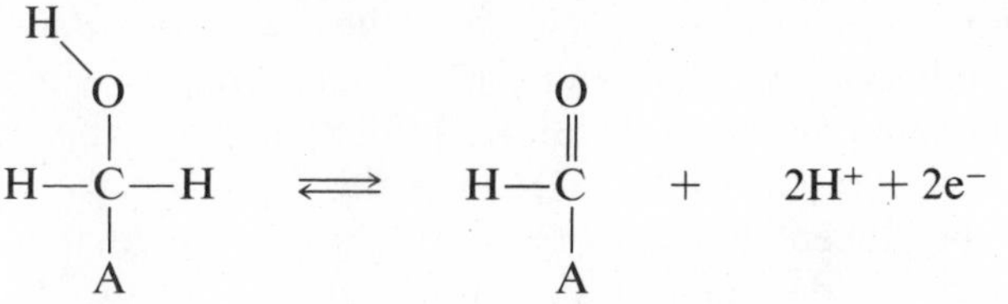

Alcohol (retinol) Aldehyde (retinal)

Adding this half-reaction to the one for NADH $\rightleftarrows$ $NAD^+ + H^+ + 2e^-$ but written in reverse gives the overall equation for the oxidation of vitamin A (retinol) to its aldehyde (retinal):

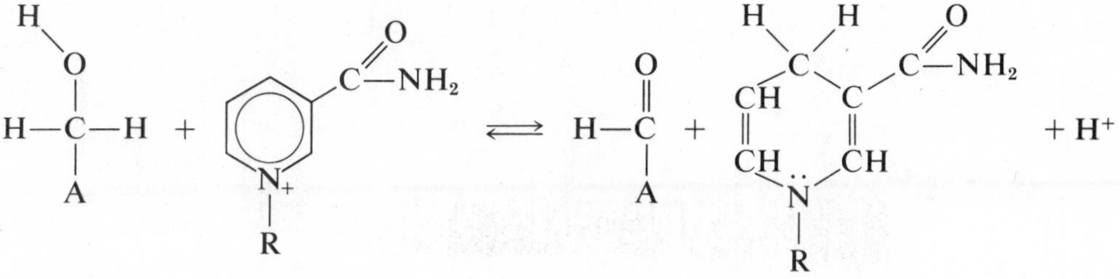

Retinol NAD$^+$ Retinal NADH

The position of this oxidation–reduction equilibrium is far to the *left* because NADH is a strong reducing agent and the reduction of an aldehyde to an alcohol is an exoenergetic reaction ($\Delta G^0 \approx -20$ kcal/mole). How then can the NAD$^+$–NADH system *oxidize* vitamin A to its aldehyde in the body? It can for the reason that the NAD$^+$–NADH system does not work all by itself in the body—it is *coupled* through a chain of six separate electron-transfer systems to the final oxidizing agent in the body, O_2. Moreover, the overall reaction for the chain is spontaneous and strongly *exo*energetic:

Complex compounds called ubiquinones and cytochromes make up the system.

$$\tfrac{1}{2}O_2 + NADH + H^+ \rightleftarrows NAD^+ + H_2O \qquad \Delta G^0 = -52.4 \text{ kcal}$$

The effects of the chain and this overall reaction are continuously to remove from the equilibrium even the tiny amounts of NADH in equilibrium with retinol, retinal, and NAD^+. The removal of the *product* NADH forces the oxidation of retinol to retinal to completion. The chain nature of the reaction breaks the overall energy release of 52.4 kcal/mole up into smaller pieces (12–24 kcal). This is a vital function of the chain because the body is not equipped to handle large chunks of energy all at once. This kind of stepwise energy chain is found in the detailed mechanisms of all biochemical reactions that yield (or require) large quantities of energy.

Using electron-transfer half-reactions for the retinol–retinal–NADH-NAD^+ equilibrium may mislead you by suggesting that the protons needed can come from anywhere (H_2O, for example). This is not true. Using deuterium labeled NADH (actually NAD**D** because of the deuterium label), acetaldehyde is reduced to 1-deuterioethanol, showing that "outside" hydrogen atoms or protons do not get into the action. Using arrows to show electron shifts, we can show the hydrogen (deuterium) transfer, which must be something like this:

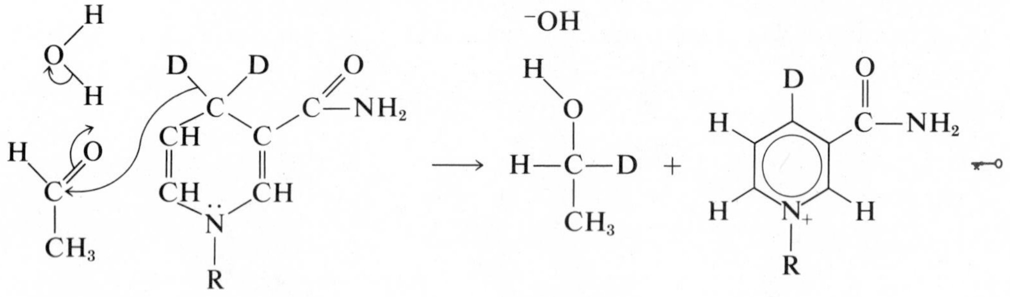

Acetaldehyde NADH 1-deuterioethanol

Exercises

8. How many double-bond geometric (*cis–trans*) isomers of the vitamin D_2 molecule are there?

9. Are optical isomers possible for vitamin D_2? For β-carotene? For vitamin A? For nicotinamide?

10. Why is the oxidation potential for NADH larger at pH = 7 than at pH = 0 (C_{H^+} = 1 mole/liter)? (*Hint*: consider the position of H^+ in the electron-transfer half-reaction for NADH.)

14.4 Fatty Acids—Triglycerides

Carboxylic acids with a long hydrocarbon chain attached to the carboxyl group are called *fatty acids* because they occur as esters of glycerol (1,2,3-propanetriol) in animal fats (tallow, lard) and vegetable oils such as corn oil and safflower oil. The esters are called *triglycerides*. Using R— to stand for the long hydrocarbon chain of the fatty acid, we can write the general formula of a triglyceride and the equation for its hydrolysis to fatty acid and glycerol as follows:

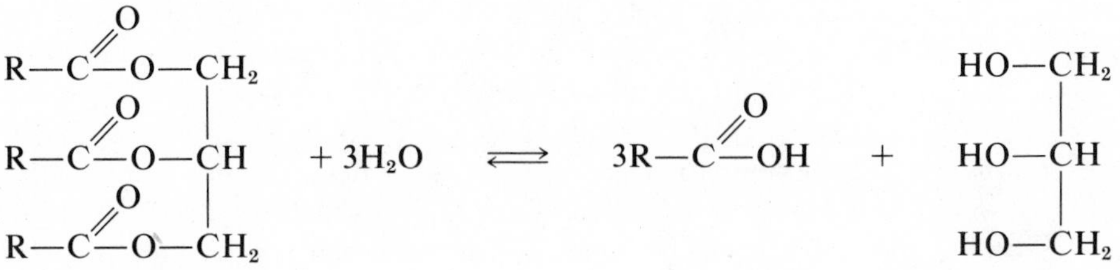

Triglyceride Fatty acid Glycerol

The equilibrium nature of this reaction was described in Section 7.3 for tristearin, the stearic acid triglyceride that makes up beef tallow. For tristearin, R— stands for the straight-chain hydrocarbon group $C_{17}H_{35}$—, so stearic acid has the formula $C_{17}H_{35}COOH$ or, in more detail, $CH_3(CH_2)_{16}COOH$.

Other important fatty acids have one or more carbon–carbon double bonds. They are termed "unsaturated" if they have one double bond, "polyunsaturated" if they have two or more. Double bonds lower the melting points of triglycerides, so the triglycerides of unsaturated and polyunsaturated fatty acids are usually oils instead of solids. Oleic acid, $CH_3(CH_2)_7CH{=}CH(CH_2)_7COOH$, occurs in the triglycerides of corn and olive oils, and linoleic acid, $CH_3(CH_2)_4CH{=}CHCH_2CH{=}CH(CH_2)_7COOH$, occurs in the triglycerides of safflower oil.

Most triglycerides from plants (vegetable oils) are liquids at room temperature because there are one or more double bonds in the hydrocarbon chain. The double bonds can be removed by catalyzed addition of hydrogen to them (Section 9.5), thus raising the melting point of the triglyceride. Oleomargarine, the butter substitute, is made from corn, peanut, safflower, and other vegetable oils in this way. In the past no thought was given to how completely the double bonds were removed by hydrogenation–the only objective was to get a product that spread like butter. Then a few years ago medical research showed that triglycerides of polyunsaturated fatty acids in the diet help prevent hardening of the arteries (arteriosclerosis), particularly the arteries of the heart, where hardening or plugging leads to heart attack by closing off the blood supply to the heart muscle. Now in the production of oleomargarine, hydrogenation is stopped before all of the polyunsaturated triglycerides are hydrogenated. The result is a softer but more healthful spread. Butter contains very little polyunsaturated triglycerides, only about 5%; safflower oil contains about 75%.

As we saw in Section 14.15 fat is the form in which your body stores energy for the "long haul." It is very well suited to this purpose because oxidation of one gram of fat produces 9.3 kcal of energy, while oxidation of one gram of carbohydrate (starch, glycogen) yields only 4.2 kcal. Furthermore, fats are not water soluble to any real extent, so they do not act as *solutes* in blood. As a result, they have no effect on the osmotic pressure of blood and they can be stored in any amounts in the body without upsetting the delicate balance of osmotic pressure between blood and tissue fluid.

The breakdown of fatty acids to CO_2 and water in the body is a complex but well-understood process. It goes through many, many steps so that the energy is delivered in small pieces instead of all at once. It takes five steps, three of which are energy-releasing, to chop just the last two carbon atoms of the chain off the fatty acid. However, the breakdown is done two carbon atoms at a time:

$$\text{R—}\overset{3}{\text{C}}\text{H}_2\overset{2}{\text{C}}\text{H}_2\overset{1}{\text{C}}\text{OOH} \xrightarrow{\text{5 steps}} \text{R—}\overset{3}{\text{C}}\text{OOH} + \overset{2}{\text{C}}\text{H}_3\overset{1}{\text{C}}\text{OOH}$$

Fat biosynthesis is the reverse of the breakdown—starting with acetic acid, the chains of the fatty acids are built up, using the two carbon atoms from acetic acid at a time. This two-carbon pattern explains why fatty acids with an even number of carbon atoms (C_{20}, C_{18}, C_{16}, C_{14}, C_{12}, . . .) are more common in nature.

EXAMPLE

The straight-chain alkane $C_{20}H_{42}$ is named eicosane.

The polyunsaturated C_{20} carboxylic acid, eicosa-8,11,14-trienoic acid, has recently been shown to be the precursor of an important regulatory hormone called prostaglandin E_1 (PGE_1).

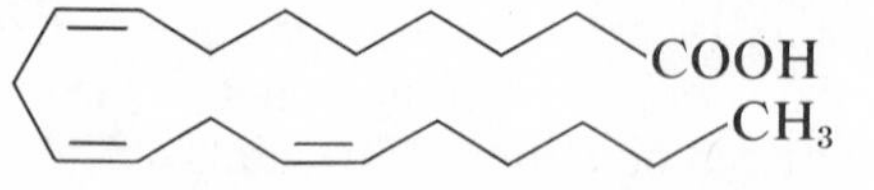

Eicosa-8,11,14-trienoic acid

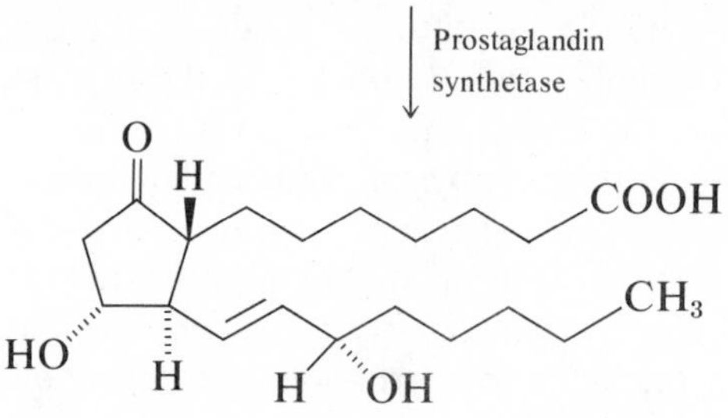

PGE_1

Prostaglandin E_1 is formed from the polyunsaturated acid through the action of the enzyme prostaglandin synthetase, and is one of several prostaglandins formed in a similar way from other C_{20} polyunsaturated carboxylic acids. Prostaglandins are widespread, though in tiny amounts, in the cells of the body. They have regulating effects on many body systems including the reproductive and circulatory systems.

They control several aspects of reproduction including stimulating the maturation of the ovum, ovulation, and the growth of the "yellow body" (corpus luteum) which replaces the ovum after ovulation. Progesterone, the female sex hormone that stops menstruation during pregnancy, is synthesized in the corpus luteum.

One step in the breakdown or *catabolism* of a fatty acid chain uses the same NAD^+–NADH oxidation–reduction enzyme that oxidizes vitamin A to its aldehyde as described in Section 14.3. The step in the catabolism of a fatty acid requires the oxidation of a 3-hydroxy carboxylic acid to the corresponding ketone:

It is actually the ester of the acid with acetyl-coenzyme A,

$$R{-}CH_2{-}\overset{\overset{\displaystyle OH}{|}}{C}H{-}CH_2{-}\overset{\overset{\displaystyle O}{\|}}{C}{-}S{-}CoA$$

that undergoes the oxidation (Section 14.6).

$$R{-}CH_2{-}\overset{\overset{\displaystyle OH}{|}}{\overset{3}{C}H}{-}\overset{2}{C}H_2\overset{1}{\overset{\overset{\displaystyle O}{\|}}{C}}{-}OH + NAD^+ \rightarrow R{-}CH_2{-}\overset{3}{\overset{\overset{\displaystyle O}{\|}}{C}}{-}\overset{2}{C}H_2{-}\overset{1}{\overset{\overset{\displaystyle O}{\|}}{C}}{-}OH + NADH + H^+$$

As we saw with vitamin A, the NADH will be oxidized back to NAD^+ with a large energy release occurring through a chain of many steps, leading finally to O_2.

Exercises

11. Burning 1.00 g of tristearin (mol. wt 891.5) produces 9.3 kcal, while burning 1.00 g of glucose (mol. wt 180.16) produces 3.7 kcal. Which of these foods is richer in energy on a *mole* basis?

12. Referring to Exercise 11, write the balanced equations for the burning (combustion) of tristearin and glucose. Why, on a gram basis, is tristearin a richer energy source than glucose?

13. Most of the naturally occurring fatty acids have the *cis* arrangement at their double bonds. Give the detailed structures for the *cis*- and *trans*-isomers of oleic acid.

14. Natural waxes are esters between fatty acids and straight-chain alcohols. One of the esters in *carnauba* wax, prized for its hardness, is between a C_{24} fatty acid and a C_{32} straight-chain alcohol. Give its structure. The synthetic waxes called "Carbowaxes," used in water-based floor waxes, are high-molecular-weight (10^3 to 10^4) polymers made from ethylene oxide, $\overset{\displaystyle O}{CH_2{-}CH_2}$ (ring), and have the structure $HO{-}CH_2CH_2{-}(O{-}CH_2CH_2)_n{-}OCH_2CH_2{-}OH$. Carbowaxes are water soluble but natural waxes are not. Explain the difference in solubility.

*Cat*abolism means the breaking down of a fat, protein, or carbohydrate. Its opposite is *an*abolism, meaning their building up or synthesis. *Met*abolism is the whole picture, anabolism and catabolism taken together.

15. Can you see any relation in meaning between the pairs of terms *an*ode and *ca*thode, *an*abolism and *cat*abolism? (*Hint*: consider the changes in oxidation number that occur in the half-reactions at the anode and cathode of an electrochemical cell.)

14.5 Carbohydrates and Sugars The term carbohydrate means "compound of carbon and water" and arose because a large class of natural products have formulas of the type $C_xH_{2n}O_n$ in which there is a 2:1 ratio of hydrogen to oxygen atoms, as in H_2O (Greek, *hydros*). Examples of carbohydrates are glucose $C_6H_{12}O_6$, sucrose (table sugar) $C_{12}H_{22}O_{11}$, ribose $C_5H_{10}O_5$ (in nucleic acids), and the biopolymers starch $(C_6H_{10}O_5)_n$, glycogen $(C_6H_{10}O_5)_m$, and cellulose (cotton, wood) $(C_6H_{10}O_5)_x$.

Starch, glycogen, and cellulose are all high-molecular-weight (one million to ten million) biopolymers of glucose, making the glucose structural unit the most abundant organic chemical group on earth. The biopolymers of glucose are formed by the elimination of water between glucose molecules, using the H of one glucose molecule and the OH from another. Using HO—Gluc—OH [where Gluc stands for the rest of the glucose molecule $(C_6H_{10}O_4)$], we see that the thought experiment of forming carbohydrate biopolymers goes like this:

$$\text{H—O—Gluc}\boxed{\text{—O—H}\quad\text{H—}}\text{O—Gluc}\boxed{\text{—O—H}\quad\text{H—}}\text{O—Gluc—O— }\cdots$$

$$\Big\downarrow -2H_2O$$

$$\text{H—O—Gluc—O}\!\left(\text{Gluc—O}\right)_{\!n}\text{Gluc—O— }\cdots$$

The repeating unit in the polymer chain, —Gluc—O—, has the formula $C_6H_{10}O_5$. Knowing that proteins can be hydrolyzed to their amino acids, you might guess that the biopolymers of glucose can be hydrolyzed back to glucose. This can be done—the cellulose of wood or cotton can be hydrolyzed to glucose but only under vigorous conditions —0.5% H_2SO_4 at 130°C in a pressure vessel (autoclave): ☛

The vapor pressure of H_2O at 130°C is 2026 torr or 39.20 psi.

$$\underset{\text{Cellulose}}{(C_6H_{10}O_5)_n} + nH_2O \xrightarrow[130°C]{H^+} \underset{\text{Glucose}}{nC_6H_{12}O_6}$$

Plant-eating animals have hydrolytic enzymes that digest cellulose to glucose—humans lack the enzymes for digesting cellulose, and must depend on starch or sugar for his carbohydrate food. Large quantities of corn starch are hydrolyzed to glucose to make corn syrup and crystalline glucose, used in candy manufacture. The hydrolysis of cellulose to glucose doesn't compete economically with starch hydrolysis, however.

In carbohydrates the monomer units (called monosaccharides) are joined through an oxygen atom in a special kind of ether linkage called a *glycosidic link*. This linkage does for carbohydrates what the peptide link does for proteins. To understand the power of glycosidic linking to provide variety in the structures of carbohydrates, you first have to master some of the stereoisomeric details of glucose. In solution and in the solid, D-glucose ☛ is present in two stereoisomeric *ring* forms called α-D-glucose and β-D glucose. ☛

The D in D-glucose indicates the optical isomer of glucose that occurs in nature.

Or α-D-glucopyranose and β-D-glucopyranose to show they contain the six-membered, oxygen-containing *pyran* ring.

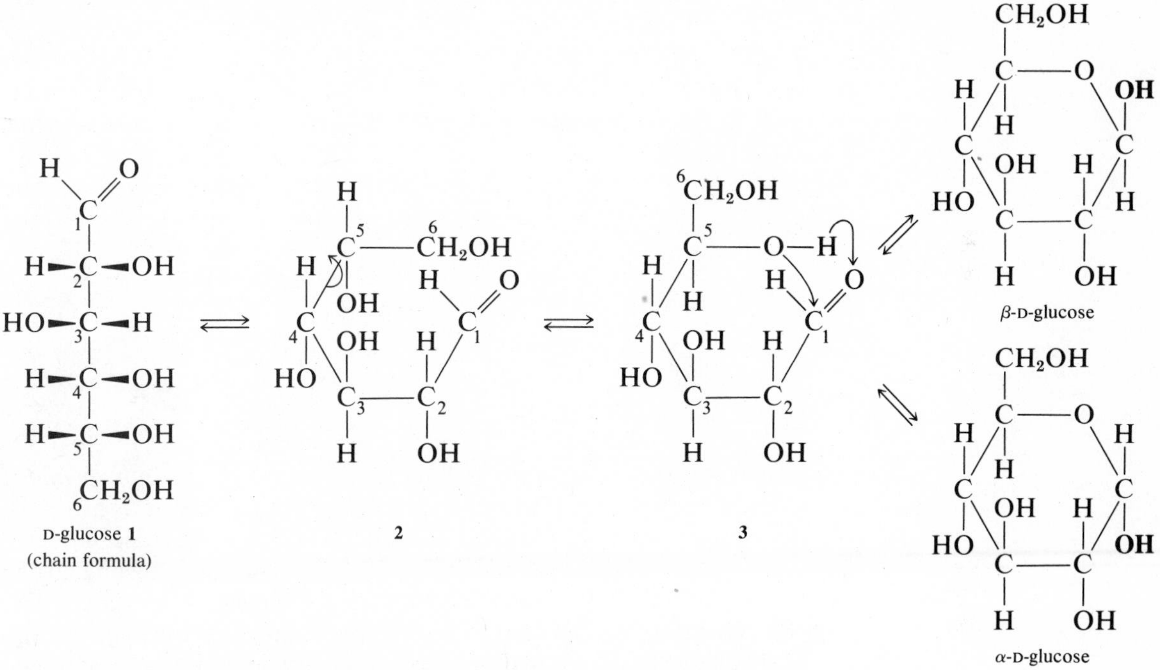

D-glucose **1** (chain formula) **2** **3** β-D-glucose α-D-glucose

Formation of the ring from the chain form is most easily understood if you do it in steps, preferably using a molecular model. First, you bend the chain form of D-glucose (**1**) around into the open ring (**2**). Rotation about the C-4 to C-5 bond, shown by the arrow in **2**, then puts the C-5 hydroxyl group in the position shown in (**3**), where it can add to the aldehyde group at C-1 to close the ring. The process is reversible, but the position of equilibrium is so far on the side of the ring compounds that the chain form is virtually undetectable in water solutions of glucose. The chain form of glucose has four chiral carbon atoms (C_2, C_3, C_4, and C_5). As the ring closes, an additional *chiral carbon atom is generated at C-1*, the original location of the aldehyde group. Using A—O—H for the C-5 hydroxyl group and B—C(=O)—H for the aldehyde group, you can see more clearly how the new chiral carbon atom is formed. Attack on one side of the aldehyde carbon gives one arrangement of ligands:

$$A-O-H \quad + \quad \underset{B}{\overset{H}{\underset{|}{C}}}{=}O \quad \longrightarrow \quad A-O-\underset{B}{\overset{H}{\underset{|}{C}}}-OH$$

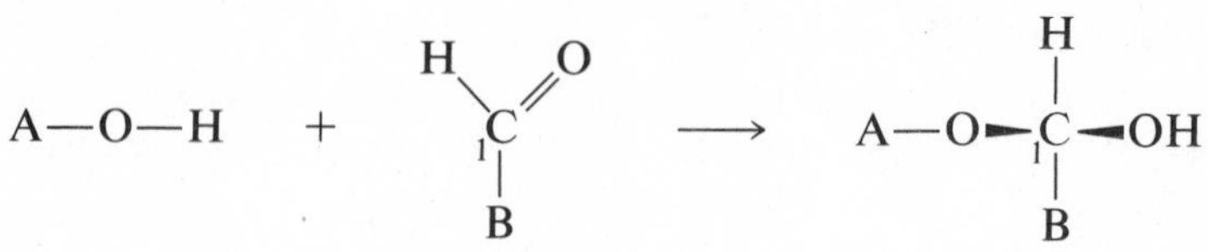

But, because the aldehyde group is flat, attack in the opposite way to give the mirror-image arrangement of ligands at *C*-1 is also possible:

$$O{=}\underset{B}{\overset{H}{\underset{|}{C}}} \quad + H-O-A \quad \longrightarrow \quad HO-\underset{B}{\overset{H}{\underset{|}{C}}}-O-A$$

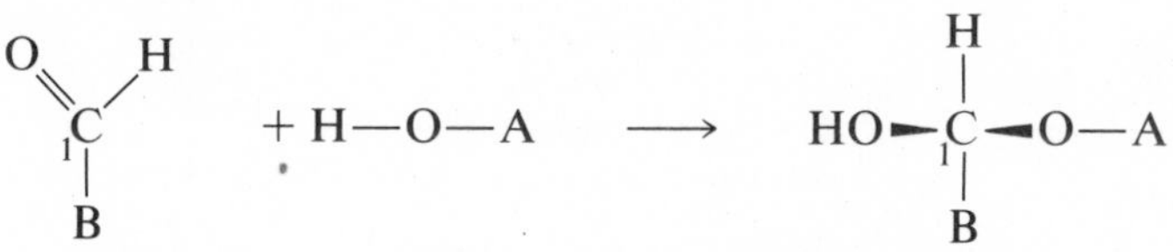

The key idea here is "mirror-image arrangements at C-1." It is important to see that α-D-glucose and β-D-glucose are *not mirror-image molecules*. Only the arrangements of ligands at C-1 are mirror images; the stereochemistry of the rest of both molecules is the same. The two ring isomers of glucose are *different compounds* and belong to the larger class of molecules called *diastereoisomers* or *diastereomers*. Diastereomers are molecules having the same skeleton that are optical isomers but not mirror images of each other. In the case of carbohydrates and sugars the diastereoisomers resulting from ring closure are called the α- and β-*anomers*, and the C-1 carbon is called the anomeric carbon atom. In the α-anomer (as in α-D-glucose) the OH group on the anomeric carbon is on the side of the ring *opposite* the —CH$_2$OH group—a *trans* arrangement of the two groups. In the β-anomer (β-D-glucose) the two groups are on the *same* side of the ring or *cis* to each other.

When two glucose molecules (C$_6$H$_{12}$O$_6$) are joined in a disaccharide (C$_{12}$H$_{22}$O$_{11}$) by eliminating water between them to form a *glycosidic link*, the OH group on C-4 of the ring of one molecule and the OH group on the anomeric carbon (C-1) of the other are used. The glycosidic link can form from the α-anomer or from the β-anomer, so *two isomeric disaccharides are possible*—using outline formulas for the carbon part of the rings, except for the anomeric carbons, the α-glycosidic link is formed as follows:

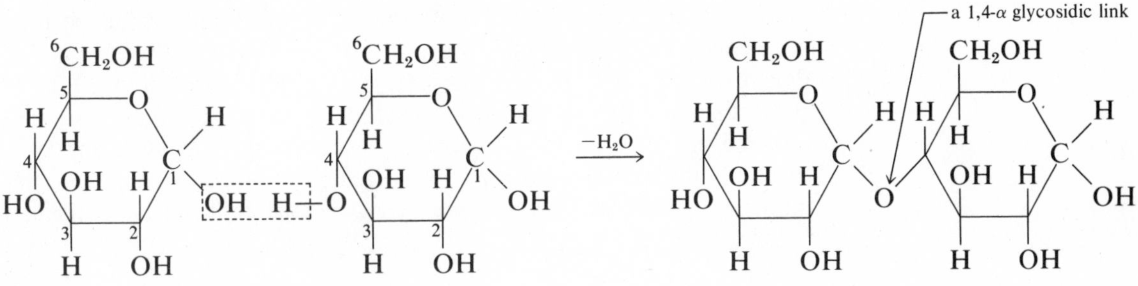

Two molecules of α-D-glucose

Maltose (mp 160°C)

To show the 1,4-β-glycosidic link clearly, it helps to flip one of the glucose units over:

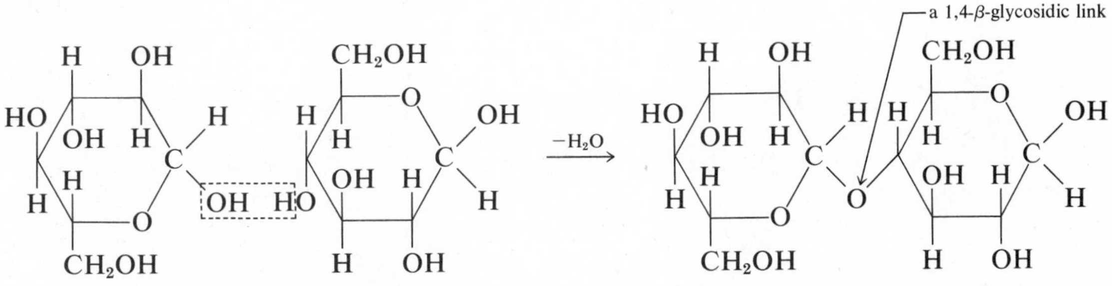

(Flipped over)

Two β-D-glucose molecules

Cellobiose (mp 255°C)

There is still an "unused" anomeric carbon in the right-hand rings of both maltose and cellobiose, so α- and β-anomers of each of these sugars are possible—the structures given are those of the α-anomer of maltose and the β-anomer of cellobiose.

Maltose is the first step on the way to the biopolymer *starch*, and cellobiose is the first step toward *cellulose*. Starch is built up from maltose by building on successive glucose molecules, using mainly 1,4-α-glycosidic links; ☛ cellulose results from cellobiose in the same way but using 1,4-β-glycosidic links instead. The repeating units in the two polymers are as follows:

There are a few β-glycosidic links in starch.

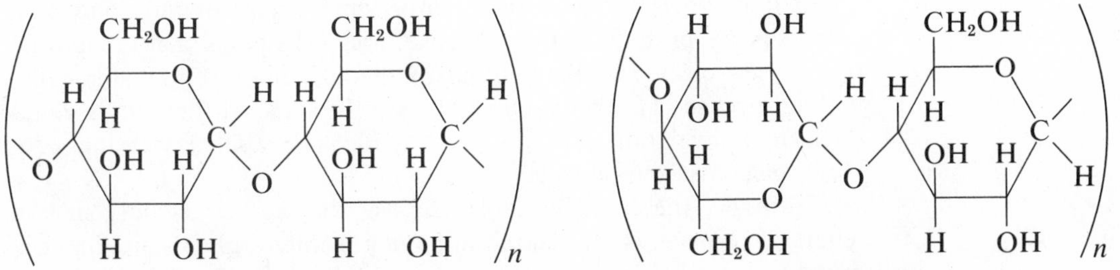

α-glycosidic link in starch

β-glycosidic link in cellulose

The difference between the structures of starch and cellulose may seem small but it explains why cows can live on hay and we can't. Our digestive systems lack the enzyme that catalyzes the hydrolysis of β-glycosidic linkages to convert cellulose to glucose. The missing enzyme is called β-glucosidase and is among the digestive enzymes of all plant-eating animals. We can digest starch because we have the enzyme α-glucosidase, which cleaves α-glycosidic links. Animals also have α-glucosidase.

EXAMPLES ⊸ Milk sugar (lactose), present to about 5% in mammalian milk, is a β-linked disaccharide made up from D-galactose and D-glucose:

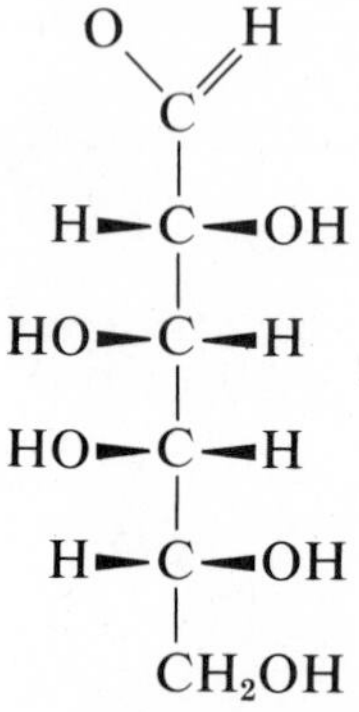

D-galactose

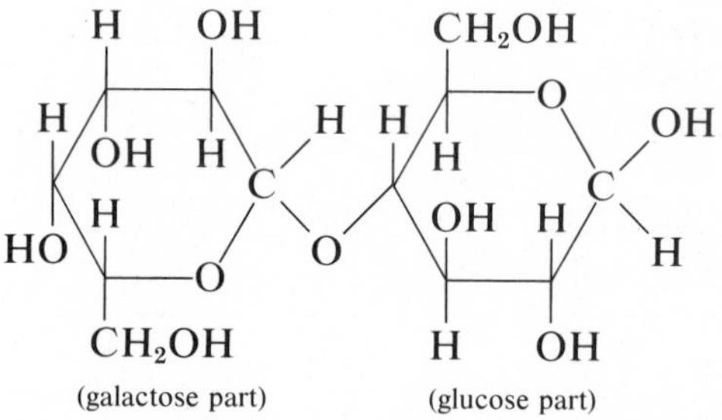

Lactose

Galactosemia ("galactose in the blood") is a disease of nursing infants caused by their lack of the enzyme (called *p*-galactose-uridyl-transferase) for converting galactose into glucose. Digestion hydrolyzes lactose in the normal way to glucose and galactose, both of which are absorbed into the baby's bloodstream. But cells cannot "eat" (catabolize) galactose, so it accumulates as a poison in the infant's blood, causing vomiting and lack of growth after a few days of feeding. Death may result from malnutrition and wasting. Victims of galactosemia who recover without treatment suffer mental retardation and are stunted in growth. Early treatment—replacement of milk and milk products by proteins, fruits, glucose, and vegetables that contain no galactose—clears up all symptoms. Galactosemia is a typical genetic-defect disease. The deoxynucleic acids (DNA's) in the baby's genes lack the information needed for the synthesis of the enzyme that converts galactose into glucose.

Glycogen, the biopolymer of glucose that serves as our "instant energy" reserve, is like starch in having mainly α-glycosidic linkages but has some branches using β-glycosidic links. Molecular weights as high as 100 million have been measured for glycogen. Glycogen storage disease results from the body's storing glycogen in excessive amounts in various organs. It is due to a failure at some point in the enzyme system that converts glycogen to glucose. It is very rare but usually fatal. ━○

16. Give the structure of the β-anomer of maltose (β-maltose).

17. Give the open-chain structure of L-glucose, the enantiomer (mirror image) of D-glucose.

18. Give the structure of the cyclic form of β-L-glucose (L-glucose is not found in nature, but has been made in the laboratory).

Exercises

Other sugars Other sugars important in biological reactions are the monosaccharides *fructose* and *ribose* and the disaccharide *sucrose* (table sugar). Glucose, which has been discussed in detail, belongs to the class of monosaccharides called the *aldohexoses* — "ald" because of the aldehyde group in its chain form, "hex" because it has six carbon atoms like hexane, and "ose" because it is a sugar. *Fructose* (fruit sugar) belongs to the class of *ketohexose* in which the aldehyde group of the aldohexose is replaced by a CH_2OH group and a ketone group appears at carbon atom number 2 of the chain. The natural isomer is called D-fructose because it has the same arrangement of ligands (configuration) about C-5 in the chain. ━▶ In fructose C-2 is the anomeric carbon atom in the five-membered ring or *furanoside* form produced by adding the C-5 hydroxyl group to the keto group:

All sugars with the arrangement

$$H - C - OH$$
$$CH_2OH$$

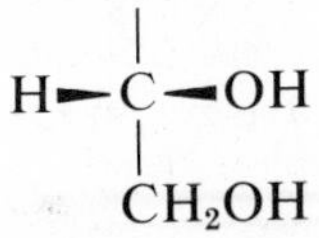

for the last two carbon atoms in the chain belong to the D- or natural sugar series.

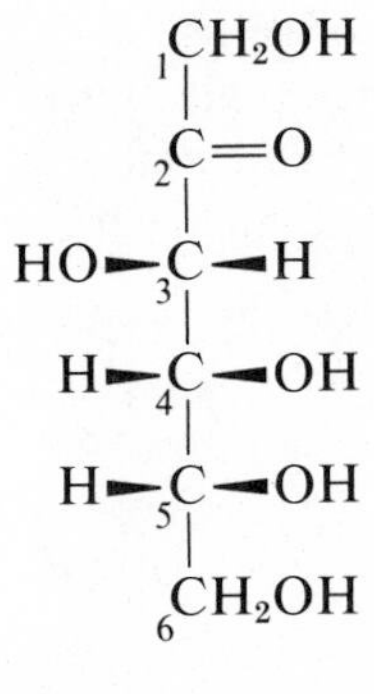

D-fructose

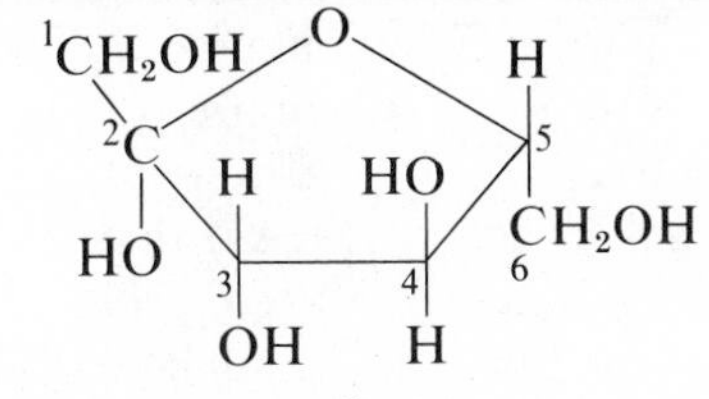

β-D-fructose

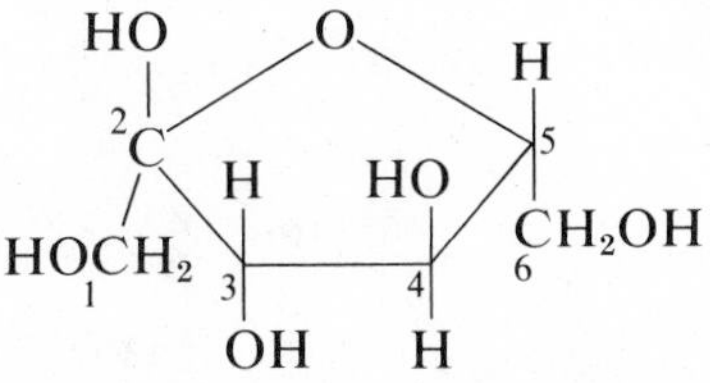

α-D-fructose

Fructose is an important intermediate in the metabolism of glucose (Section 14.7). It also forms the disaccharide *sucrose* ("sugar") by linking up to a glucose molecule. The linkage is not of the 1,4-glycosidic type seen in maltose, cellobiose, starch, glycogen, and cellulose, but instead is an ether link between the anomeric carbon atoms, C-1 for glucose, C-2 for fructose:

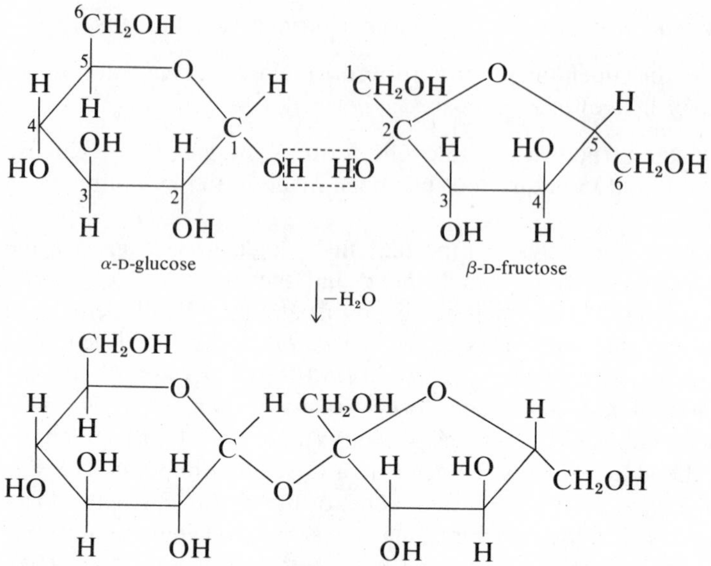

Sucrose

The systematic name for sucrose is α-D-glucopyranosyl-β-D-fructofuranose, because the oxygen link is α for glucose and β for fructose, and both monosaccharide units are in their cyclic forms.

Ribose is a five-carbon, aldehyde sugar—an *aldopentose*—that is an essential part of many enzymes and of the compounds called ribonucleic acids (RNAs), which are present in all parts of living cells. A close derivative of ribose, *deoxyribose* plays the part of ribose in the *deoxyribonucleic* acids (DNAs) that make up the genes that control inheritance. Both ribose and deoxyribose have chain and five-membered ring (furanoside) forms, and both belong to the natural or D-sugar series:

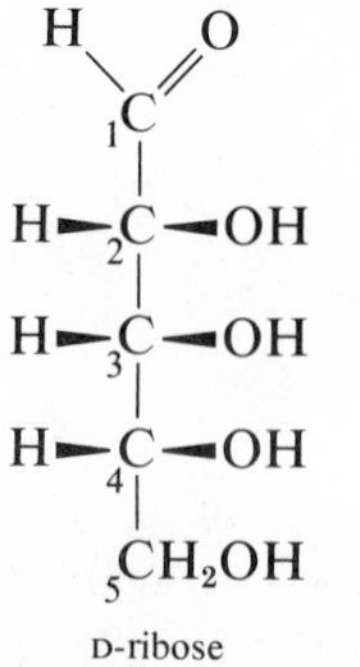

D-ribose

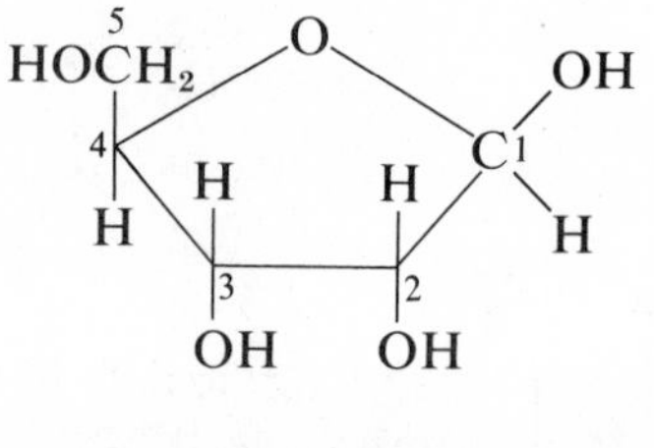

β-D-ribose

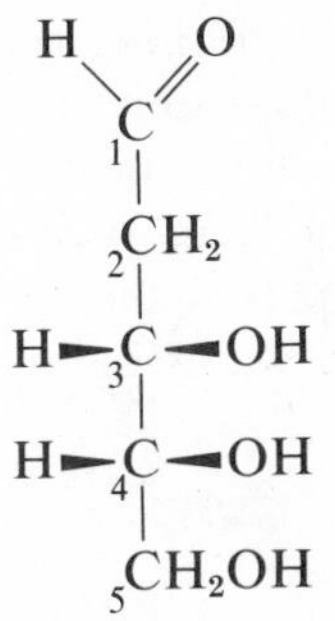
2-Deoxy-D-ribose

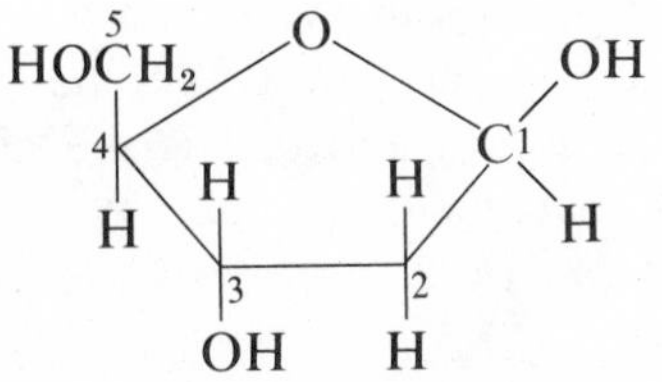
β-2-Deoxy-D-ribose

Exercises

19. Give the structure for the chain form of L-galactose and for α-D-ribose.

20. The three-carbon compounds glyceric acid (2,3-dihydroxypropanoic acid) and glyceraldehyde ("glycerose" or 2,3-dihydroxypropanal) are important intermediates in the metabolism of D-glucose. Give the structures of their D and their L optical isomers, assuming that C-2 in their structures corresponds to C-5 in the hexoses.

21. Are D-galactose and D-glucose diastereoisomers? Are diastereoisomers possible for D-glyceraldehyde?

14.6 Enzymes and Coenzymes A complete enzyme system is made up of a *protein* and a *cofactor* or *coenzyme*, which often contains a built-in vitamin structure. The coenzyme is the main working part of the enzyme molecule and carries out only one particular *type* of chemical reaction, such as hydrogen transfer, peptide link hydrolysis, dehydrogenation, or the like. The coenzyme often does its particular job in a number of metabolic reactions. The oxidation–reduction enzyme system NAD^+–NADH, described earlier in explaining the role of vitamin A in night vision, is used in the photosynthesis reactions forming glucose from CO_2 and water and also in several steps in the catabolism of glucose that yields the energy for life. Which reaction a particular enzyme will catalyze is determined by the protein part of the enzyme molecule. It is because enzymes are *catalysts* and therefore needed only in tiny amounts that our daily requirements for many vitamins are in the milligram to microgram range.

A great deal is known about the structure and chemical reactions of many coenzymes, particularly those built around the B vitamins. The complete structure of NAD^+ (nicotinamide adenine dinucleotide), earlier given in abbreviated form, using R to represent everything but the nicotinamide part of the molecule, has five structural units covalently bonded. One is derived from a two-ring (bicyclic), aromatic compound called adenine, two come from ribose molecules, one is a diphosphate group, and one is the nicotinamide part. You can imagine NAD^+ being put together by eliminations of water in the following ways:

In adenine, as in pyridine, —CH= groups are replaced by —N= groups, which are equivalent as far as bonding is concerned.

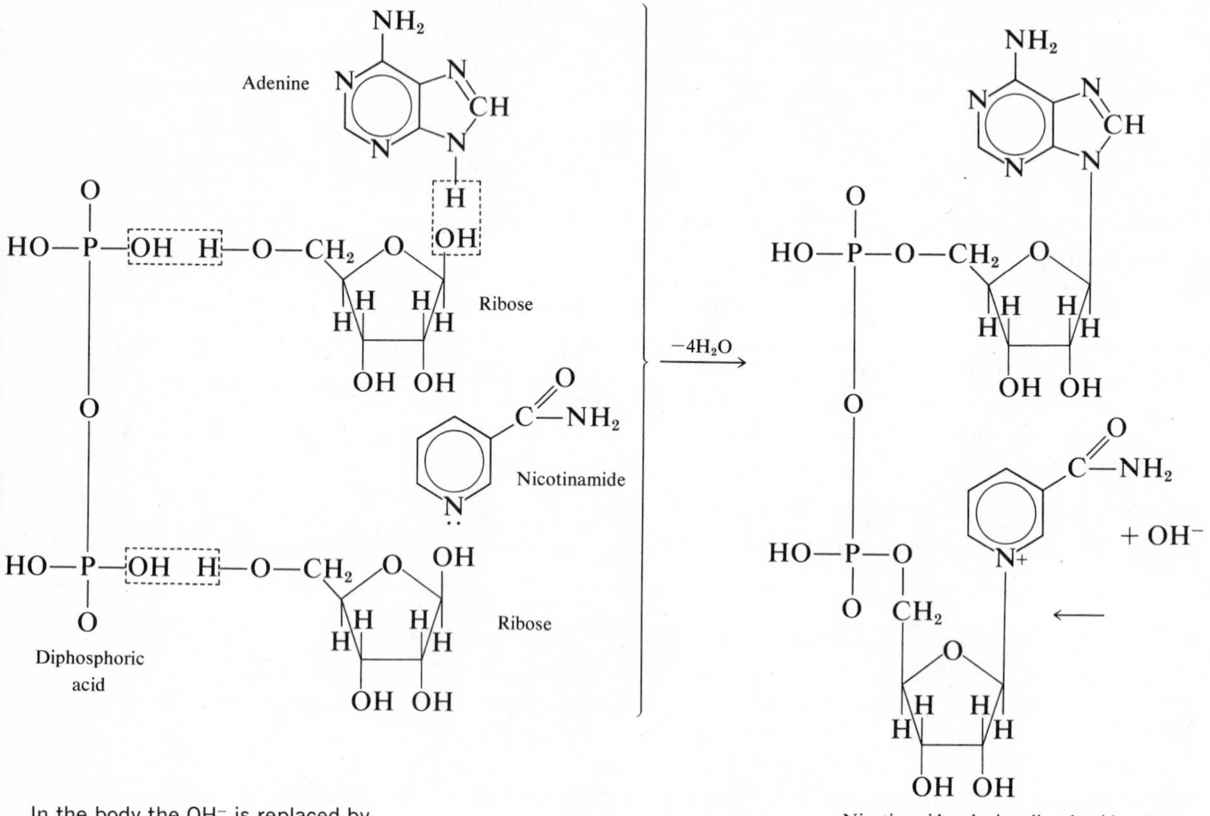

Nicotinamide adenine dinucleotide

In the body the OH^- is replaced by Cl^- and the HO— groups attached to phosphorus are ionized to ^-O— groups. Both of these changes are due to the pH of about 7 in body fluid.

The formation of the bond (arrow) between the nicotinamide nitrogen and the lower ribose molecule is not as mysterious as it looks. The unshared pair of electrons on the nitrogen displaces an OH^- from the anomeric carbon of the ribose molecule, forming the covalent bond:

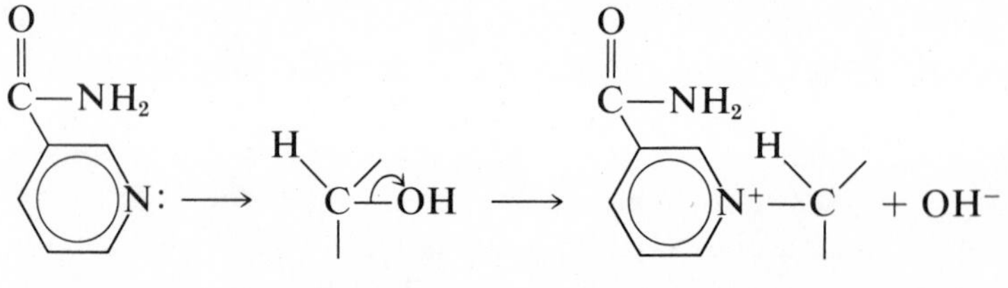

The (+) charge appears on the nitrogen because it donates *both* electrons to the new N—C bond. NAD^+ is most certainly *not* put together in this simple "spontaneous" way—you can't just mix solutions of adenosine, ribose, diphosphoric acid, and nicotinamide and expect NAD^+ to form and crystallize out. (There are five OH groups in the ribose molecule, and any one of them can react with diphosphoric acid to give a diphosphate ester.) Cofactors themselves are put together by enzymes in living systems, although many of them have been synthesized in the laboratory.

EXAMPLE

The history of the study of the NAD^+–NADH enzyme system is both interesting and typical of many enzyme systems. In 1906 biochemists studying alcoholic fermentation found that a water-soluble, low-molecular-weight, heat-stable (nonprotein) substance was required for the reaction. But the active substance was not isolated as a crystalline compound until 1963. Until then it was not possible to determine its structure, although what the cofactor did in biochemical systems was known. In many cases the discovery of the existence of a cofactor came long before its structure was known. Biochemists have methods for characterizing a cofactor and what it does without ever having the pure compound in hand.

Proteins are often water soluble but they are denatured by heat (Section 14.2—Shapes of Protein Molecules), and they are *never* of low molecular weight.

Coenzyme A Coenzyme A or "CoA" is probably the most versatile and important of the coenzymes in cell metabolism. It is *the* "two-carbon transfer reagent" in the breakdown of glucose to water and CO_2, and is the key substance in the biosynthesis of aromatic compounds, fatty acids, sterols, carotenoids, and terpenes. For simplicity in describing the two-carbon transfers, the abbreviation HS-CoA is used for the coenzyme. The *thiol* or *mercapto* group HS— binds the two carbons being transferred in the form of a *thiol acetate ester*. Thiol esters are like ordinary esters but with sulfur replacing oxygen:

$$CH_3C(=O){-}OH + HS{-}CH_3 \rightarrow CH_3{-}C(=O){-}S{-}CH_3 + H_2O$$

Acetic acid; Methylthiol (methyl mercaptan); Thiomethyl acetate

$$CH_3C(=O){-}OH + HS{-}CoA \rightarrow CH_3C(=O){-}S{-}CoA$$

Acetyl Coenzyme A ("Acetyl CoA")

Terpenes are hydrocarbons based on the formula $(C_5H_8)_n$, where C_5H_8 stands for isoprene (Section 13.6) and the *n* range is from 2 to 10—except for rubber and gutta percha, which are giant molecules. Turpentine is a mixture of $C_{10}H_{16}$ terpenes.

It is the two carbons in the acetyl($CH_3\overset{O}{\overset{\|}{C}}$—) group of acetyl-CoA that are transferred. For example, in the biosynthesis of fatty acids the first

step is the attachment or transfer of the $CH_3\overset{O}{\overset{\|}{C}}$— group of acetyl-CoA to a carbon dioxide molecule to form *malonyl-CoA*, the half thioester between malonic acid (methane dicarboxylic acid) and HS-CoA:

$$CH_3\overset{O}{\overset{\|}{C}}-S-CoA + CO_2 \longrightarrow$$

Acetyl CoA

$$HO-\overset{O}{\overset{\|}{C}}-CH_2-\overset{O}{\overset{\|}{C}}-S-CoA \qquad \left(HO-\overset{O}{\overset{\|}{C}}-CH_2-\overset{O}{\overset{\|}{C}}-OH\right)$$

Malonyl-Co-A Malonic acid

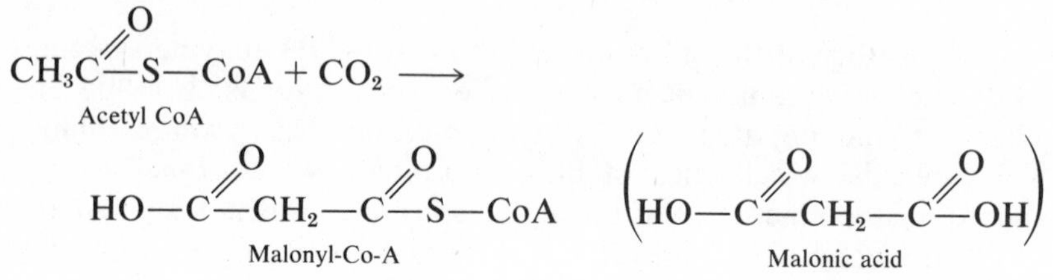

Transfer of a second $CH_3\overset{O}{\overset{\|}{C}}$— group to malonyl-CoA accompanied by loss of CO_2 and HS-CoA gives *acetoacetyl-CoA*:

$$CH_3\overset{O}{\overset{\|}{C}}-S-CoA + HO-\overset{O}{\overset{\|}{C}}-CH_2-\overset{O}{\overset{\|}{C}}-S-CoA \xrightarrow[-HS-CoA]{-CO_2}$$

Acetyl CoA Malonyl CoA

$$CH_3\overset{O}{\overset{\|}{C}}CH_2\overset{O}{\overset{\|}{C}}-SCoA$$

Acetoacetyl-CoA

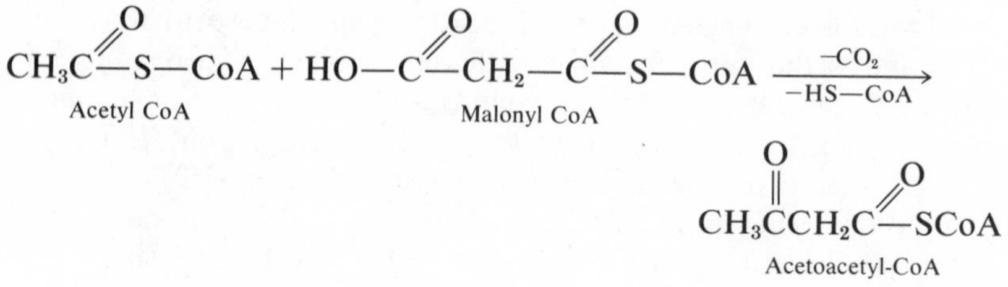

Enzymatic reduction of the keto group in acetoacetyl-CoA gives $CH_3CH_2CH_2\overset{O}{\overset{\|}{C}}-S-CoA,$ (butyryl-CoA), and the fatty acid chain is launched. Repetition of the cycle builds up the fatty acid chain, two carbons at a time. It's no accident that fatty acids with even numbers of carbon atoms are far more plentiful in nature than those with odd numbers.

 The complete structure of coenzyme A is worth looking at because of similarities to the structure of NAD^+. It contains one of the B vitamins (pantothenic acid), an adenine ring system, ribose, a diphosphate linkage, and two amide linkages, which you have seen in proteins.

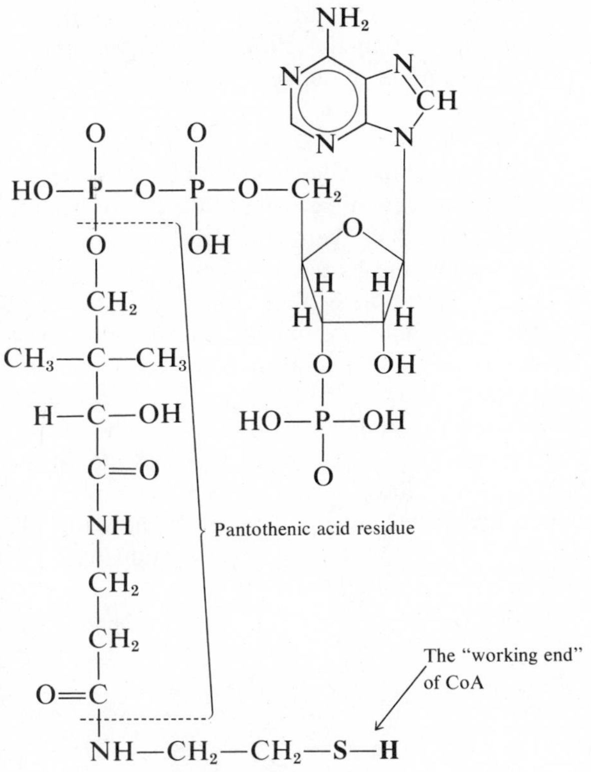

Coenzyme A

Exercises

22. Plants use the nicotinamide–adenine–dinucleotide enzyme system mainly in the reductive direction $NADH \rightarrow NAD^+ + H^+ + 2e^-$, but in animals the direction is reversed to the oxidizing direction, to $NAD^+ + H^+ + 2e^- \rightarrow NADH$. Why should this be so? (*Hint*: compare the overall reactions for the photosynthesis of glucose and for its breakdown, as oxidation–reduction reactions.)

23. In fatty acid biosynthesis, the first step in turning the keto group in acetoacetyl-CoA into a $—CH_2—$ group is reduction to the corresponding alcohol by NADH. Give the equation for the reaction. Would you be willing to bet that the alcohol produced is optically active? Why? (The further steps in the conversion $—\overset{\overset{\displaystyle O}{\|}}{C}— \rightarrow —CH_2—$ are dehydration of the alcohol and hydrogenation of the $\gt C{=}C \lt$ that is formed. Two other enzyme systems are involved.)

24. By hydrolyzing the linkages indicated by the dashed lines in the formula for coenzyme A, obtain the structure for pantothenic acid (vitamin B_3). Can the molecule have optical isomers?

14.7 Bioenergetics—ADP, ATP, and Glucose Bioenergetics deals with the release and use of energy by living things. The energy source is chemical potential energy and it is converted into thermal, mechanical, and electrical energy by chemical reactions going on in cells—whether your muscle cells in a 100-meter race, yeast cells making a fine wine, or cells in a growing plant. The oxidation or catabolism of glucose serves as the main source of chemical potential energy released in animal cells, and *adenosine triphosphate* (ATP) stores much of the energy. Adenosine itself is a molecule you have seen before—if you look back at the structures of NAD^+ and coenzyme A, you'll see that both have in them a cyclic ribose sugar unit linked through its anomeric carbon atom to adenine. This combination is the molecule called *adenosine*, and its ester at the C-5 hydroxyl group with triphosphoric acid is *adenosine triphosphate* (ATP):

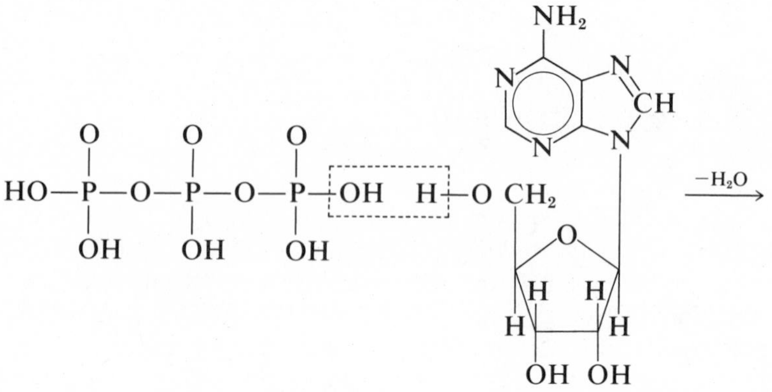

Triphosphoric acid   Adenosine

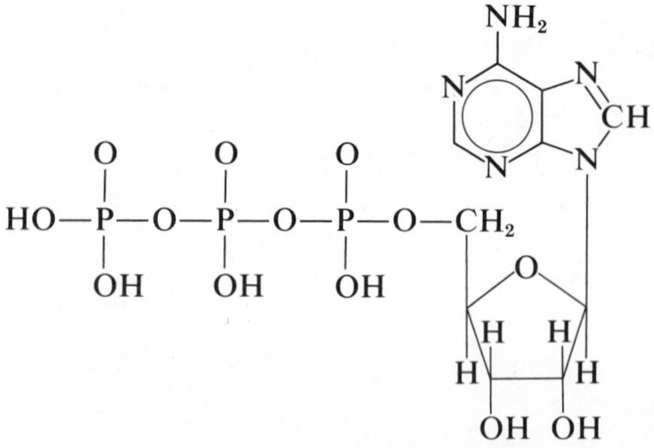

Adenosine triphosphate (ATP)

These phosphoric acids and phosphates are strong multiproton acids and are present at pH ≈ 7 in body fluids as their multiple anions.

Some of the free energy released in the catabolism of glucose and fats is used in the synthesis of ATP from *adenosine diphosphate* (ADP) in a reaction that is endothermic by 9 kcal/mole. Letting A stand for the adenosine group, we show the reaction as follows:

$$
\begin{array}{c}
\text{O} \\ \| \\
\text{HO—P—OH} \\ | \\ \text{OH}
\end{array}
\;+\;
\begin{array}{c}
\text{O} \qquad \text{O} \\ \| \qquad \| \\
\text{HO—P—O—P—A} \\ | \qquad | \\ \text{OH} \qquad \text{O}
\end{array}
\;+\;9\text{ kcal}\;\rightarrow\;
\begin{array}{c}
\text{O} \qquad \text{O} \qquad \text{O} \\ \| \qquad \| \qquad \| \\
\text{HO—P—O—P—O—P—A} \\ | \qquad | \qquad | \\ \text{OH} \qquad \text{OH} \qquad \text{OH}
\end{array}
\;+\;\text{H}_2\text{O}
$$

Phosphoric acid Adenosine diphosphate Adenosine triphosphate

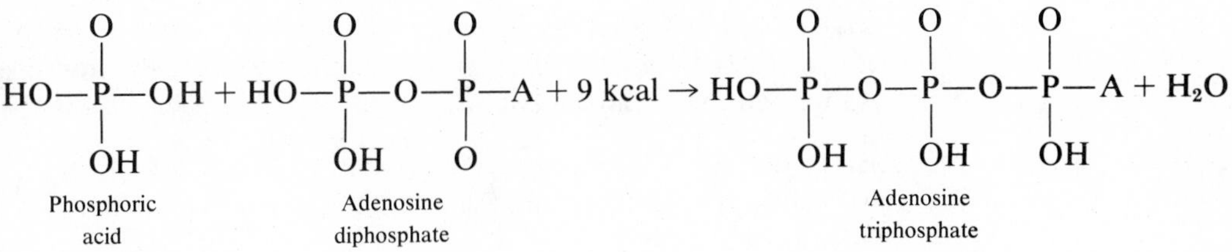

The reaction does not go by this simple mechanism, but the point is that the *reverse reaction, hydrolysis of ATP to ADP and phosphoric acid,* ✋ *yields 9 kcal/mole of free energy.* This energy, coming originally from glucose, makes your muscles work, your eyes blink, and your brain think. The value of ΔG^0 for the conversion of one mole of glucose to CO_2 and water is -686 kcal. During the conversion in living cells 36 molecules of ADP are converted to ATP, using the energy released, so of the 686 kcal total available, $9 \times 36 = 324$ kcal or 47% is stored as chemical potential energy in ATP, where it is "instantly available" for life processes. The catabolism of glucose and fatty acids is broken down into many, many steps, most of them *slightly* energy-releasing (exoenergetic). The energy released by catabolism of a mole of glucose is about what you'd get from burning $\frac{1}{3}$ lb of sugar—the body simply couldn't take that much energy all at once.

A special kind of arrow arrangement is used to show that a biochemical reaction uses or produces a mole of ATP, or other reagents:

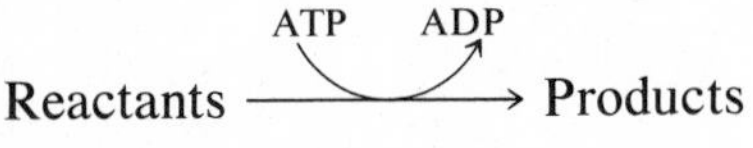

 ATP ADP

Reactants ⟶ Products 1 mole of ATP used

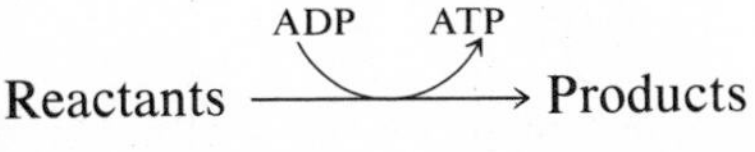

 ADP ATP

Reactants ⟶ Products 1 mole of ATP produced

Understanding energy storage by ATP will allow us shortly to take a look at some of the details of the oxidation of glucose in cells.

25. Determine the oxidation states of phosphorus in phosphoric acid (H_3PO_4), diphosphoric acid ($H_4P_2O_7$), and triphosphoric acid

Exercises

Phosphoric acid is called "inorganic phosphate" in biochemistry. The extra P—O—P link in ATP compared to ADP is sometimes called a "high-energy phosphate bond." The 9 kcal/mole is the free energy released by hydrolysis under the concentration conditions in living cells. The ΔG^0 for the hydrolysis of ATP to ADP under standard thermodynamic conditions is -7 kcal.

($H_5P_3O_{10}$). Is the conversion of ATP to ADP an oxidation–reduction reaction?

26. Derive the structure of triphosphoric acid, using the hydroxyl scheme (Section 12.3).

27. Give the complete Lewis structure of ATP, showing all electrons either in bonds or as unshared pairs. Is there any reason to think ATP might be paramagnetic?

Oxidation of glucose The complete oxidation of glucose to CO_2 and water in cells goes through more than 25 intermediate compounds and involves an even larger number of enzyme systems. Biochemists have worked out the mechanisms of the steps. Here we will take a look at a few of the important reactions and an overall look at the process.

The first big part of the program that starts glucose on its way to CO_2 and water cleaves the glucose molecule between carbon atoms number 3 and 4 to give two molecules of *pyruvic acid* (2-ketopropionic acid), $CH_3COCOOH$. In the overall reaction six moles of ADP are converted into six moles of ATP:

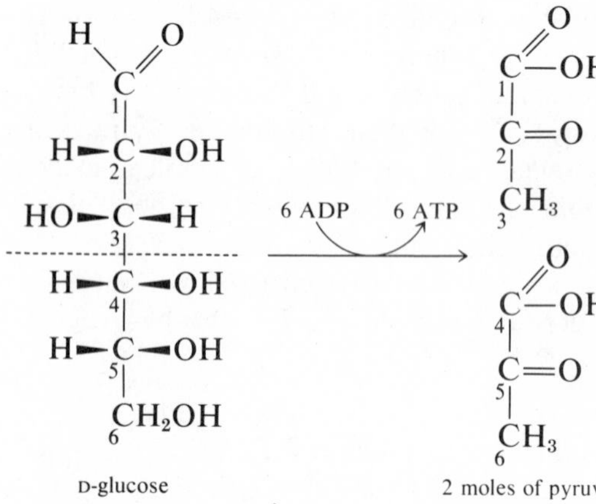

D-glucose — 2 moles of pyruvic acid

$C_6H_{12}O_6 + O_2 \rightarrow 2C_3H_4O_3 + 2H_2O \qquad \Delta G^0 = -140$ kcal

Quite a bit of rearranging of hydrogens and the complete removal of two of them are needed to go from reactants to products—how this all occurs is shown by the steps in the detailed program in Figure 14.3. *Phosphorylation*, the formation of phosphate esters by ATP, is an essential reaction in the program and one that is irreversible (shown by the single arrows in the reaction scheme). For clarity and saving space, we let P stand for the $-\overset{\displaystyle O}{\underset{\displaystyle OH}{P}}-OH$ part of phosphate esters, so

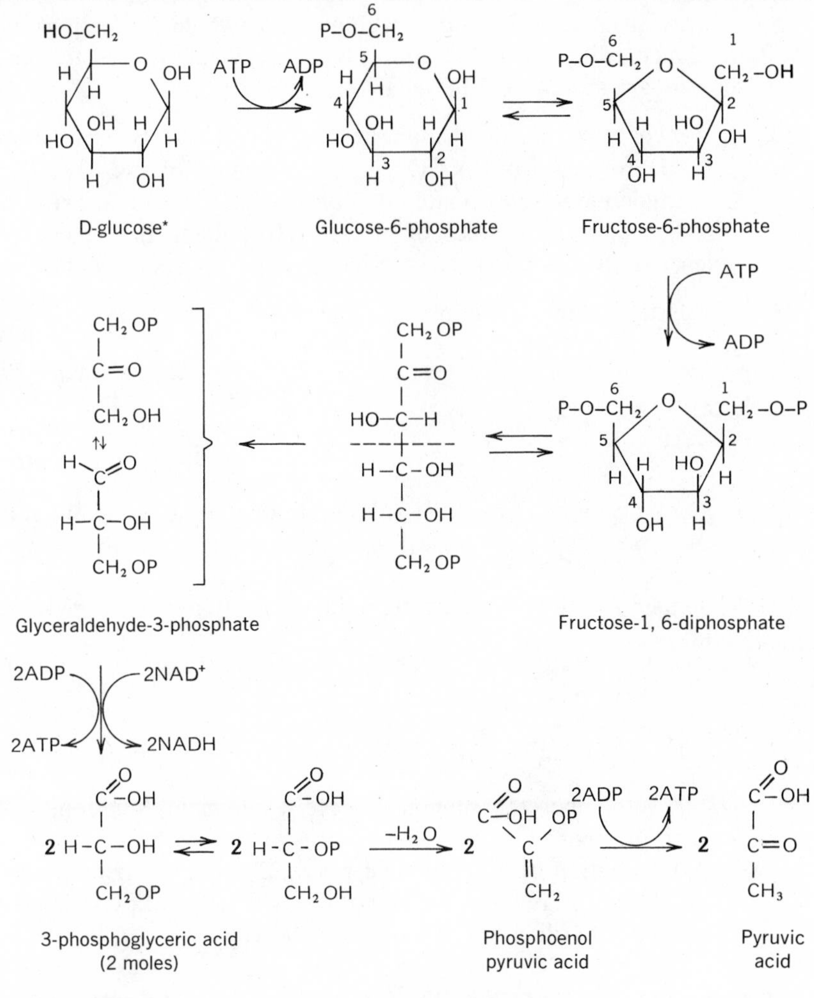

FIG. 14.3 Catabolism of glucose to pyruvic acid. P stands for —P(OH)—OH with OH (i.e., $-P(OH)_2$ drawn as OH above, —P—OH, OH below)

*β-D-glucose is shown but since it is in equilibrium with the α-anomer through equilibrium with the open chain form, either anomer will work.

R—O—P(=O)(OH)—OH (O above P, OH below P) becomes ROP in the set of reactions. Each of the steps is catalyzed by an enzyme.

Exercises

28. Locate the compound 2-phosphoglyceric acid in the scheme for the oxidation of glucose. Are optical isomers possible for the acid?

29. What is the *net* number of ADP molecules converted to ATP by energy from the oxidation of glucose to two moles of pyruvic acid according to the scheme given? How many kcal of free energy are stored in ATP as a result of the process?

30. Somewhere in the cell catabolizing glucose, the two moles of NADH formed from NAD^+ in the oxidation of two moles of glyceraldehyde-3-phosphate to 3-phosphoglyceric acid are oxidized back to two moles of NAD^+. How much free energy is released? (See Section 14.3 – Nicotinamide for energy data.)

31. Given the following thermodynamic data,

$$\underset{\text{(Glucose)}}{C_6H_{12}O_6} + 6O_2 \rightarrow 6CO_2 + 12H_2O \qquad \Delta G^0 = -686 \text{ kcal}$$

$$\underset{\text{(Glucose)}}{C_6H_{12}O_6} + O_2 \rightarrow \underset{\text{(Pyruvic acid)}}{2CH_3\overset{O}{\overset{\|}{C}}-\overset{O}{\overset{\|}{C}}-OH} + 2H_2O \qquad \Delta G^0 = -140 \text{ kcal}$$

Calculate ΔG^0 for the complete oxidation of two moles of pyruvic acid:

$$2CH_3\overset{O}{\overset{\|}{C}}-\overset{O}{\overset{\|}{C}}-OH + 5O_2 \rightarrow 6CO_2 + 4H_2O$$

(*Hint*: remember that if you reverse a chemical equation, you change the sign of its ΔG^0 value.)

What fraction of the chemical potential energy originally in a mole of glucose remains unused in the two moles of pyruvic acid produced according to the oxidation shown?

Summary on acetyl-coenzyme A The partial oxidation of a mole of glucose to two moles of pyruvic acid yields only about $\frac{1}{5}$ of the chemical potential energy stored by photosynthesis in glucose for life processes. Most of the potential energy remains tied up in the pyruvic acid molecules. At the expense of one mole of NAD^+ going to NADH, and helped by the formation of a mole of CO_2, the pyruvic acid is converted to the high-energy compound acetyl-coenzyme A, "acetyl-CoA." This conversion is the next of many steps remaining in the complete oxidation of glucose to CO_2 and water:

$$\underset{\text{(Pyruvic acid)}}{CH_3\overset{O}{\overset{\|}{C}}-\overset{O}{\overset{\|}{C}}-OH} + HS-CoA + NAD^+ \rightarrow \underset{\text{(Acetyl-CoA)}}{CH_3\overset{O}{\overset{\|}{C}}-S-CoA} + CO_2 + NADH + H^+$$

This reaction carried out on the two moles of pyruvic acid coming from one mole of glucose yields six moles of ATP, bringing to twelve moles the total of ATP made in going from one mole of glucose to two moles of acetyl-CoA.

Because of its high energy, acetyl-CoA is a very reactive compound, and it takes part in hundreds of two-carbon transfers both of biosynthesis and of energy release. The following flow diagram shows the final results of some of the many reactions of acetyl-CoA.

$$
\begin{array}{c}
\text{O} \\
\parallel \\
\text{CH}_3\text{C}-\text{S}-\text{CoA}
\end{array}
$$

Biosynthesis of terpenes, steroids, carotenoids

Biosynthesis of fatty acids, aromatic compounds

Amino acids ⟵ Citric acid cycle ⟶ Naturally occurring *di* and *tri* carboxylic acids — citric acid (oranges, lemons), malic acid (apples), oxaloacetic acid, succinic acid

Energy
+
CO_2 and H_2O

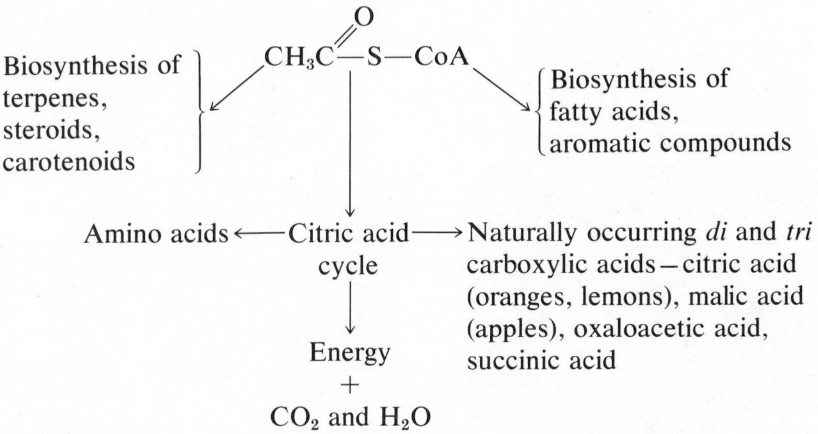

The citric acid cycle (Figure 14.4), also called the tricarboxylic acid cycle or the Krebs cycle, occupies a central position in the use of energy stored in acetyl-CoA because the cycle produces much ATP. There are nine carboxylic acids in the cycle and many more enzymes are used to convert them from one to the next. Acetyl-CoA from pyruvic acid enters the cycle by reacting with oxaloacetic acid to form citric acid:

Sir Hans Adolf Krebs (b. 1900) was born in Germany but became a British physiologist. He won the Nobel prize for physiological medicine (with Lippman) for research into metabolic processes. He was knighted in 1958.

$$
\begin{array}{c}
\text{O} \quad \text{O} \\
\parallel \quad \parallel \\
\text{C}-\text{C}-\text{OH} \\
\mid \\
\text{CH}_2 \\
\mid \quad \text{O} \\
\quad \parallel \\
\text{C}-\text{OH}
\end{array}
+ \text{CH}_3\text{C}\overset{\text{O}}{-}\text{S}-\text{CoA} + \text{H}_2\text{O} \rightarrow
\begin{array}{c}
\text{O} \\
\parallel \\
\text{C}-\text{OH} \\
\mid \quad \quad \text{O} \\
\text{HO}-\text{C}-\text{CH}_2\text{C}\overset{}{-}\text{OH} \\
\mid \quad \quad \parallel \\
\text{CH}_2\text{C}-\text{OH} \\
\quad \quad \parallel \\
\text{O}
\end{array}
+ \text{HSCoA}
$$

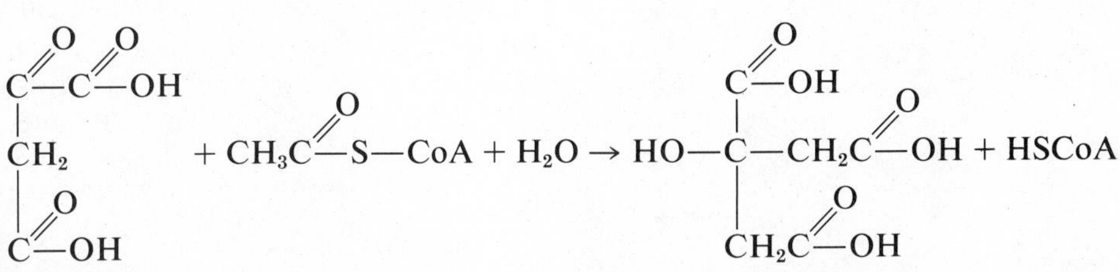

Oxaloacetic acid Citric acid

Citric acid starts the series of eight consecutive carboxylic acids that finally ends with oxaloacetic acid to start the cycle over again. The net result of going around the cycle once is the oxidation of one molecule of acetic acid to carbon dioxide and water:

$$CH_3\overset{O}{\overset{\|}{C}}—OH(l) + 2O_2(g) \rightarrow 2CO_2(g) + 2H_2O(l) \qquad \Delta G^0 = -186 \text{ kcal}$$

The free energy originally in acetyl-CoA that is given up by the reactions of the citric acid cycle is stored in 12 moles of ATP according to the summarizing equation:

The symbol P_i stands for "inorganic phosphate" H_3PO_4 or its ions, $H_2PO_4^-$ and HPO_4^{2-}.

$$CH_3CO—SCoA + 2O_2 + 12ADP + 12P_i \rightarrow 2CO_2 + HS—CoA + 12\ ATP$$

Each molecule of glucose goes to *two* of pyruvic acid and from there, on to *two* of acetyl-CoA, so 1 mole of glucose yields 24 moles of ATP

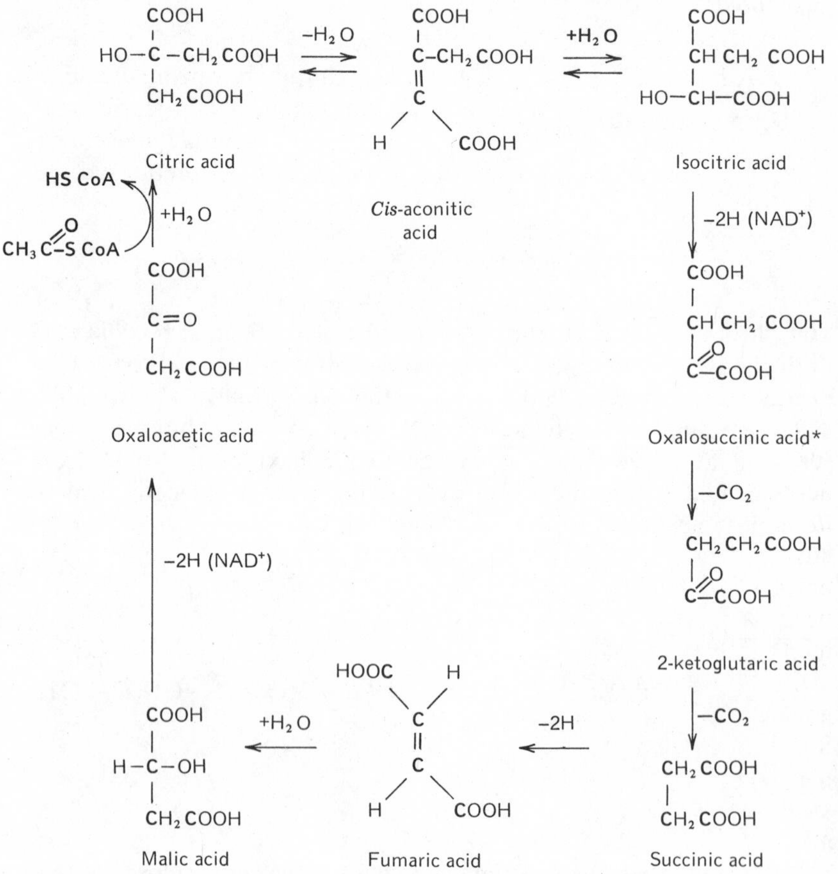

* The two carbon dicarboxylic acid HOOC–COOH is called *oxalic* acid, so $HOOC—\overset{O}{\overset{\|}{C}}—$ is the "oxalo" group. The five carbon acid $HOOC—CH_2\ CH_2\ CH_2—COOH$ is called *glutaric* acid.

FIG. 14.4 Tricarboxylic acid or Krebs cycle.

in one turn of the cycle. Because $\Delta G^0 = -9$ kcal for the reaction ATP $+ H_2O \rightarrow ADP + P_i$, there are $9 \times 24 = 216$ kcal stored in ATP by the citric acid cycle alone for each mole of glucose oxidized. The ATP is made from ADP in enzyme reactions we will not go into. Adding the 24 moles of ATP from the citric acid cycle to the 12 formed by glucose going to $2CH_3C(=O)—S—CoA$ gives a total of 36 moles of ATP formed per molecule of glucose oxidized to CO_2 and water.

Acetyl-CoA and ATP are very powerful reagents, as you have seen – ATP for energy and acetyl-CoA for biosynthesis as well as energy production. As you will see in Section 14.8, adenosine, which is part of the NAD^+, ATP, and CoA molecules, also plays an important part in the "recipes" that our genes use for the synthesis of the proteins that make us what we are. It seems that nature is very economical – one structural unit is designed for many uses.

Exercises

32. Give the structure of *trans*-aconitic acid (refer to the citric acid cycle in Figure 14.4).

33. Can optical isomers of malic acid exist?

34. The *cis*-isomer of fumaric acid is called *maleic* acid. Give its structure and the structure of its anhydride.

14.8 Nucleic Acids You are what your proteins and enzymes make you. Different proteins and enzymes mean a different living thing – a different person or even a different species: an animal, a plant, or a bacterial cell. But men and women produce human babies; bulls and cows produce calves. Somewhere in the cells is stored a set of rules for putting together proteins of different kinds from the natural amino acids and, through the proteins, building the whole organism. The *deoxynucleic acids* (DNAs) are biopolymers or codes in genes that store in their molecular structures the instructions for protein and enzyme biosynthesis. In this way, genes determine heredity. They can be thought of as an information storage and retrieval system, like a magnetic tape in a computer.

Cell division to make new cells for growth and repair is constantly going on in living systems, and *each cell* must have the *full* instructions for protein and enzyme synthesis. This requirement for all cell division means that DNAs must have the power of duplicating (or replicating) ☛ their own structure. How they do this is rather well understood, although fundamental mysteries remain to be worked out. As you might expect, the detailed molecular structure of DNA's has to be complex to do so many things. You may find it helpful to think of your own DNA, half of which came from your mother, half from your father, as a master sound tape of a symphony. A master tape can be electronically copied or replicated indefinitely.

Replicating means duplicating over and over again.

The DNAs making up the genes are a very special kind of mixed biopolymer. There are four *different* monomer molecules or basic structural units for building DNA molecules. The monomers have structures something like that of adenosine, which you've already seen in NAD^+ and CoA. The monomers are *cytosine deoxyribonucleotide* (abbreviated C), *thymine deoxyribonucleotide* (T), *adenine deoxyribonucleotide* (A), and *guanine deoxyribonucleotide* (G). Letting DN stand for "deoxyribonucleotide," we can draw the four monomer structures as follows:

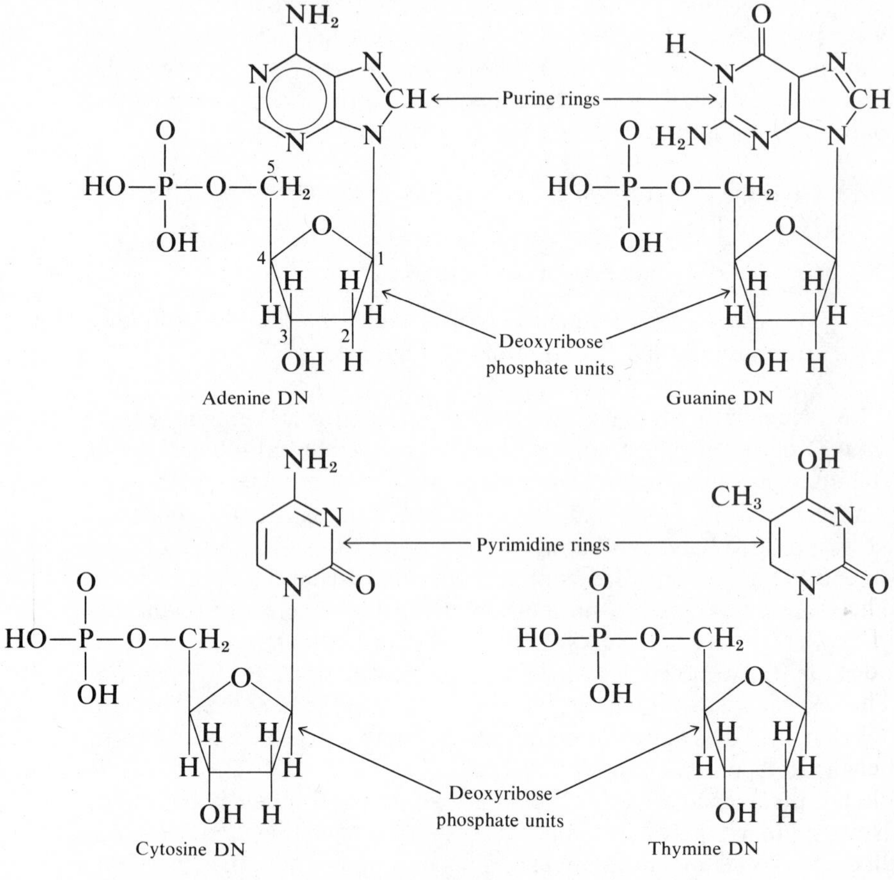

The primary or covalent bond structure of a DNA chain or *strand*, as it is called, is built up using phosphate ester linkages between the phosphoric acid group at C-5 of one deoxyribose ring and the hydroxyl group at C-3 of another. The building of a strand can be imagined to start in the following way, with R standing for the purine or pyrimidine ring of any one of the nucleotides:

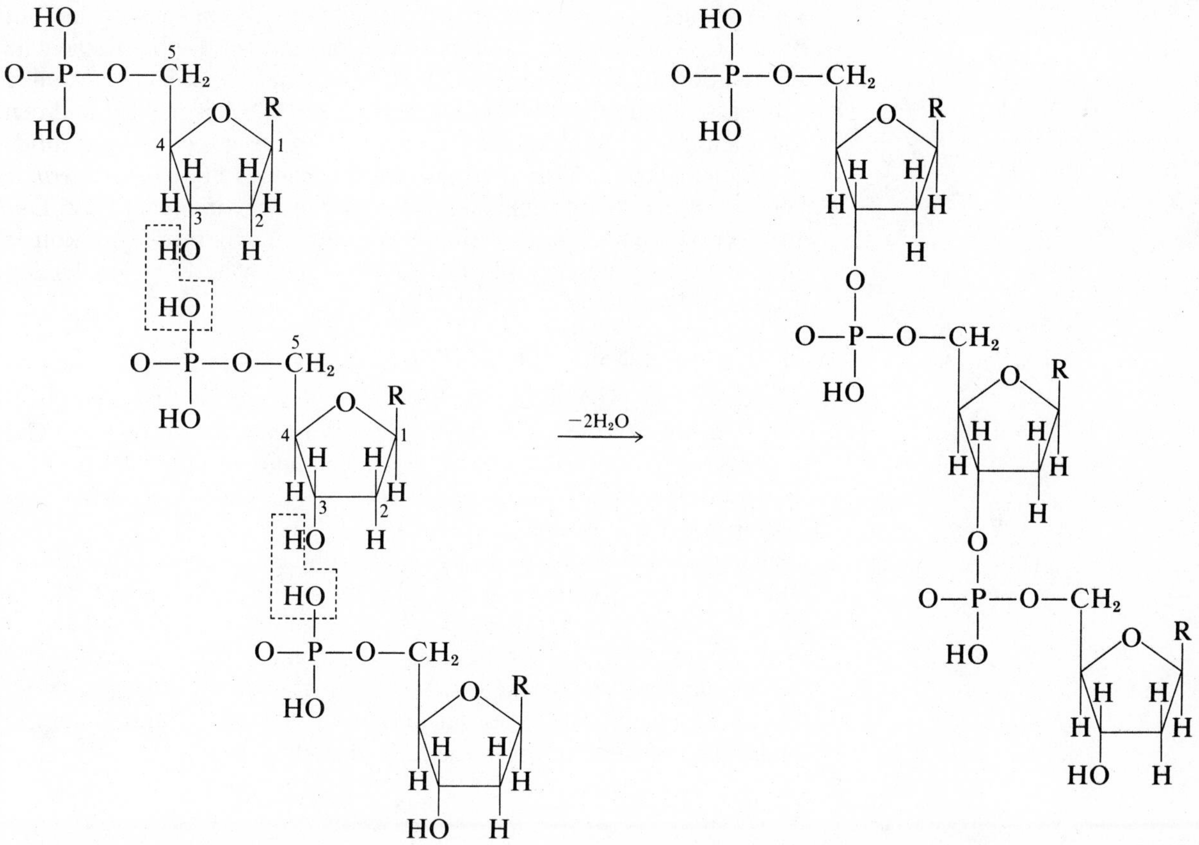

Three mononucleotides

A trinucleotide

The adding on of nucleotides continues to produce a giant molecule of DNA. Notice that the purine or pyrimidine ring parts of the nucleotides don't take part in forming the primary structure of a DNA strand. They have an equally important job, to be described shortly.

Chains of DNA as long as 30,000 Å have been isolated from cells but are probably longer in the cell, and are broken by isolation procedures.

With only the *four* nucleotides available as building blocks, the number of structures possible for a DNA strand is smaller than the possible number of protein structures, where 21 amino acids are available. Nevertheless, the sequence of nucleotides in a DNA strand carries all of the information needed for the complete development of the cell, and through the cells the growth of the individual organism—person, animal, plant, or bacterial cell. If put together in one strand, the DNA in a *single* human cell would be close to a meter (about 1 yard) in length. This length of DNA is put together when sperm fertilizes ovum and it has to be duplicated every time the cells, starting with the first, divide until the whole organism is produced. Taking 10 Å (10×10^{-8} cm) as the length of one nucleotide residue in the chain, a meter of chain would contain 10^9 nucleotides. With any sequencing of the four nucleotides allowed, the number of arrangements to provide genetic information is really without limit. It has been shown that a particular sequence of only three nucleotide residues in DNA is needed to direct the addition of one particular amino acid to the growing polypeptide chain of a protein. The three nucleotide residues are called a triplet, and one of the great triumphs of molecular biology in recent years was discovering the nucleotide residues in triplets that put each amino acid onto a growing polypeptide chain. The relation between nucleotide triplets and amino acids is called the *genetic code*.

The double-strand structure of DNA A molecule of DNA contains *two* strands of the primary, covalently bonded structure just described. The two strands are held together by hydrogen bonds between the purine and pyrimidine parts of the nucleotides in each strand. The hydrogen bonding is *very selective*—an *adenine* residue (A) can hydrogen-bond effectively *only* to a *thymine* residue (T) and a *guanine* residue (G) bonds only to a *cytosine* residue (C). Bonding between A and T uses two hydrogen bonds, shown by A⋯T; bonding between G and C uses three hydrogen bonds, shown by G⋯C. The selectivity in hydrogen bonding comes from the geometry of the purine and pyrimidine rings in the nucleotide. There's a sort of "lock-and-key" relationship in the pairing of nucleotides in the two strands. The permitted pairings are shown in the following structures:

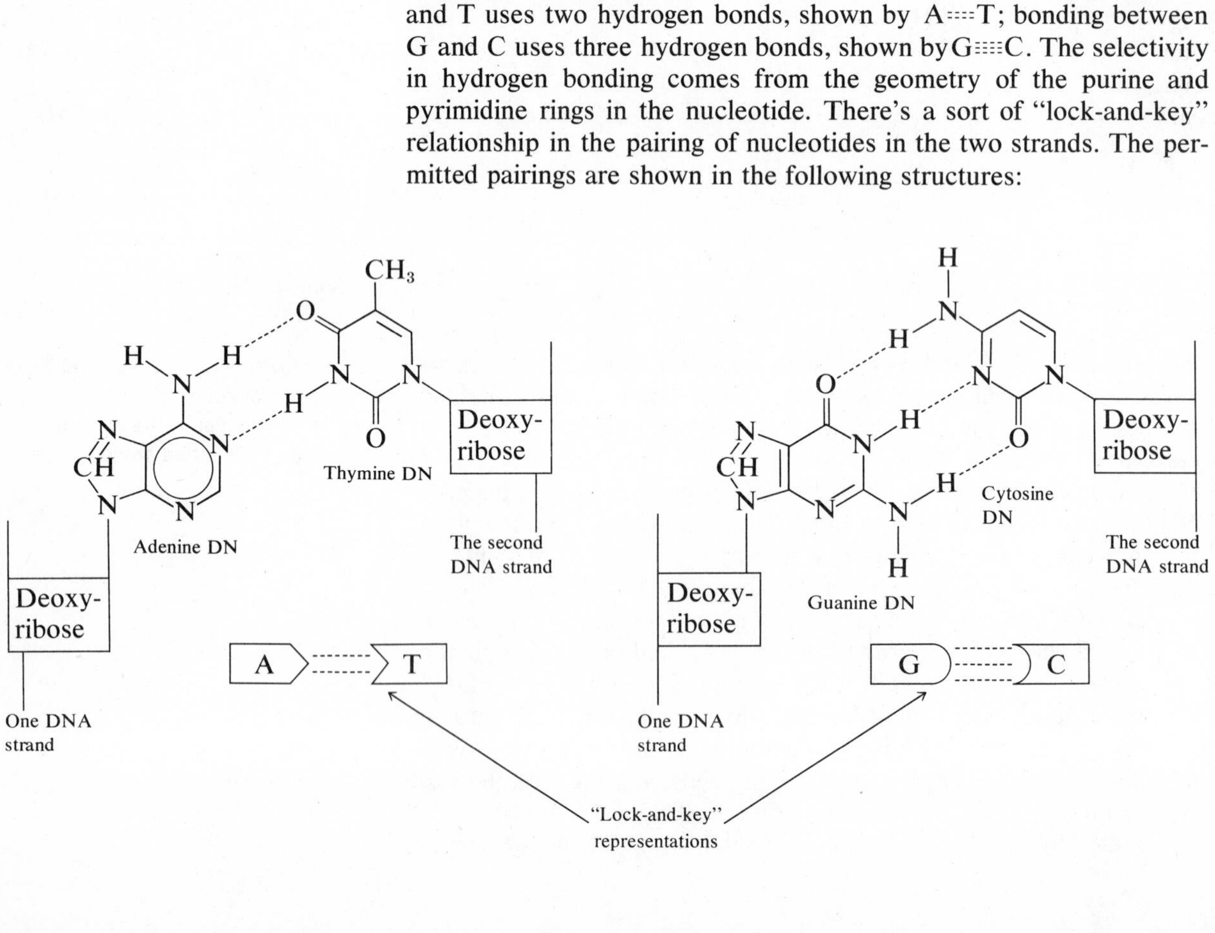

Using abbreviations for the nucleotide units, we can show the permitted pairings between the two DNA strands in a DNA molecule as follows:

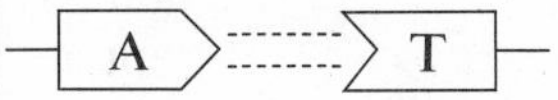

and

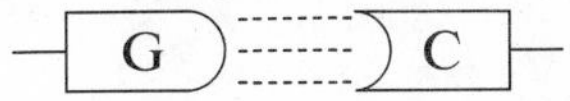

The other combinations are not permitted because the hydrogen bonds don't fit together properly:

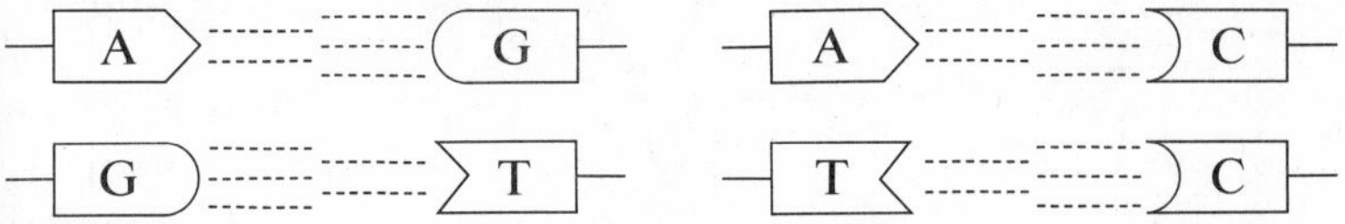

The need for only A····T or G····C pairing of nucleotide residues may seem unimportant, but it is not—*the pairing makes it possible for* ***either strand*** *of nucleotides to serve as the "master tape" for duplication of the DNA molecule of which they are parts.* To see why, let's first look at two strands in a tiny part of a possible DNA molecule, using the "box" abbreviations for the nucleotide residue as shown in Figure 14.5.

When a cell divides to make two cells, the A and B strands begin to separate or "unzip" at one end as hydrogen bonds are broken. ☛ If

As pointed out in Section 4.6, energies available in collisions between molecules at room temperature can break hydrogen bonds. Exactly what chemical reaction starts the separation of the two strands is not yet understood.

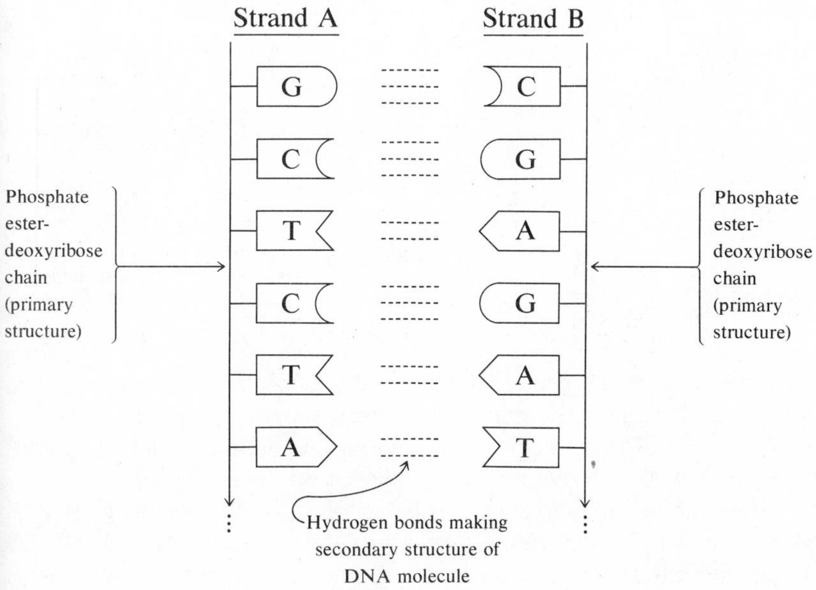

FIG. 14.5 Hydrogen bonding between nucleotide pairs in two strands of DNA.

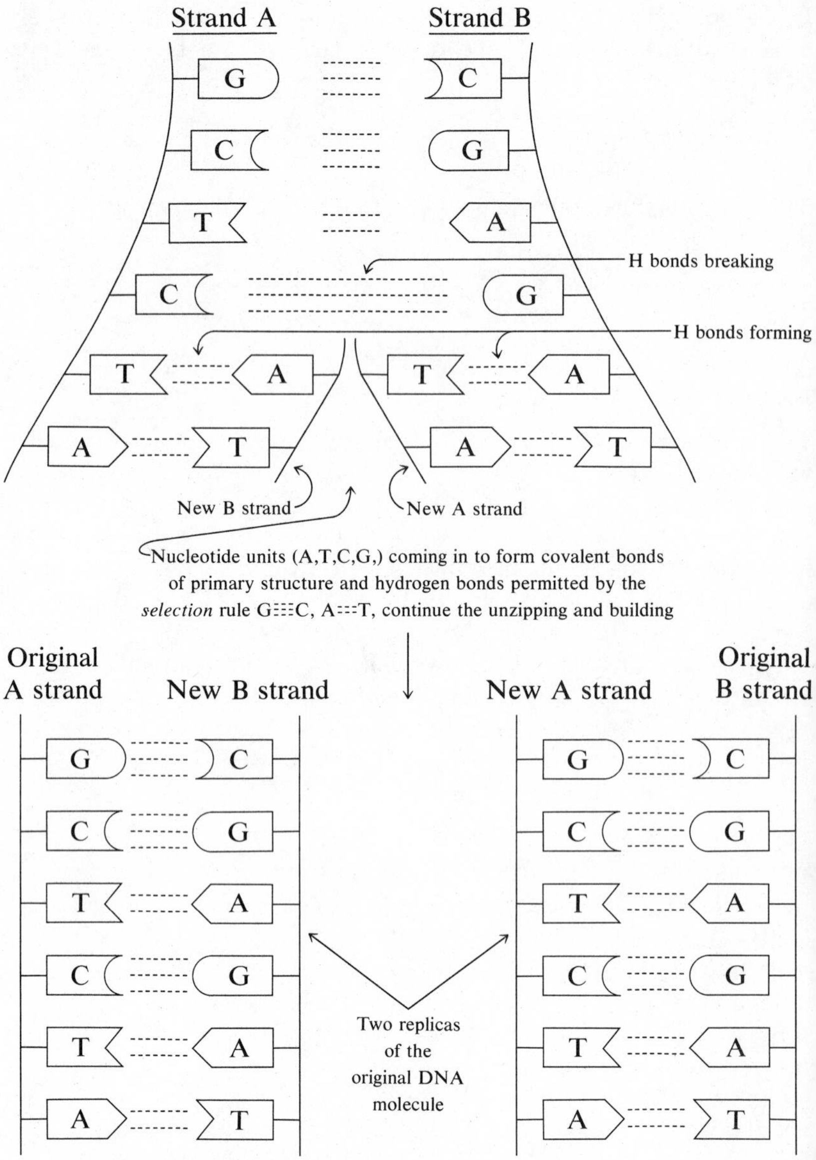

FIG. 14.6 Replication of DNA strands during cell division.

you could catch the gene in the act of duplicating its DNA molecules, you would see something like the pattern shown in Figure 14.6.

The result of the breaking–building process is to produce *two* molecules of DNA with the *same* sequence of nucleotide residues.

When the cell divides into two parts, one DNA molecule goes with each that was in the original DNA molecule. Each of the new DNA molecules can then go on to duplicate itself in the same way when its

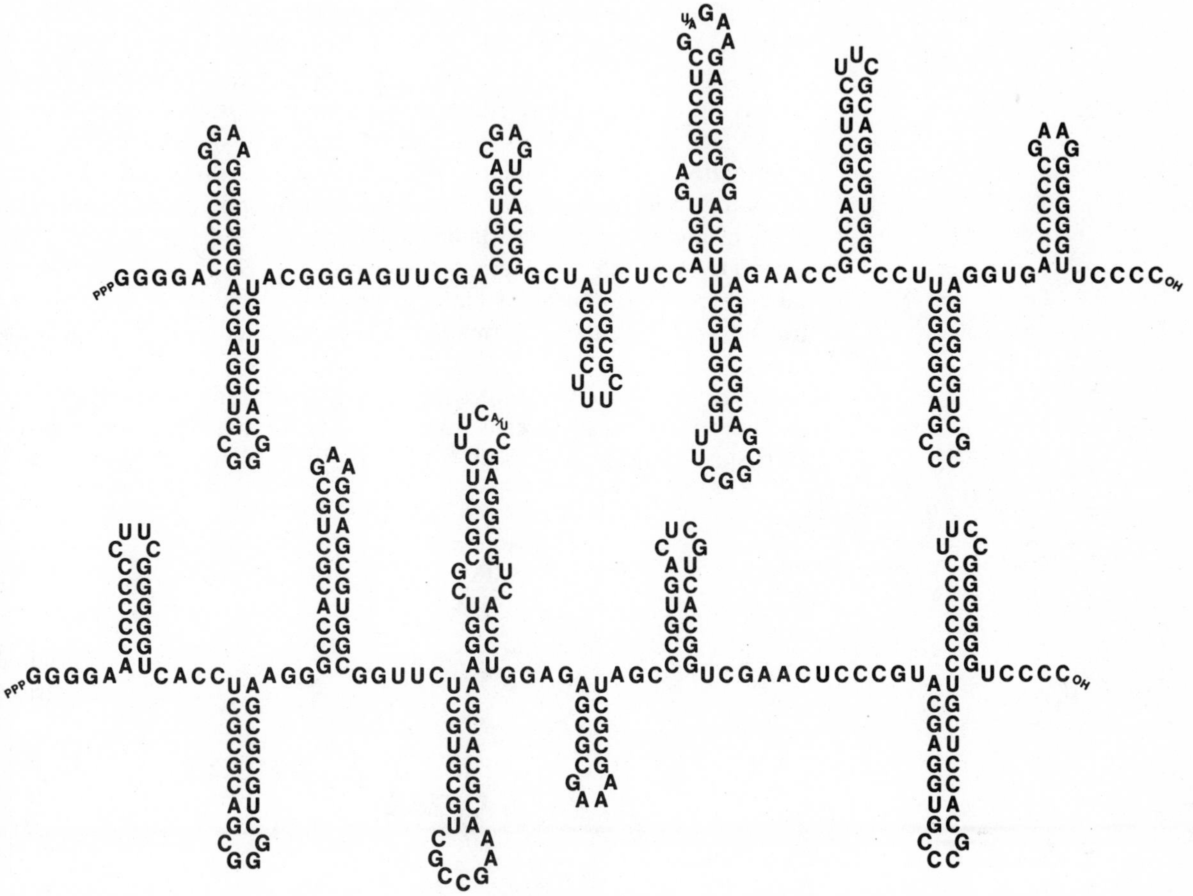

PHOTO 37 The first complete nucleotide sequence of a molecule capable of extracellular replication. *(D. R. Mills and others, Science, vol. 180, 1 June 1973. Copyright 1973 by the American Association for the Advancement of Science.)*

cell is called on to divide. This is how DNA is replicated with each cell division. It occurs trillions of times in your development, starting after conception, when sperm and ovum get together to make your DNA molecules. And it continues all your life as new cells are made to replace old ones.

Photo 37 shows the first RNA molecular nucleotide sequence made synthetically (1973) that is capable of replication outside cells. In a test tube, provided with the necessary ribonucleotides as food, one molecule of the sequence will generate 10^{12} copies in 20 minutes by replication. Recently biochemists have agreed to strict controls on such experimentation because of possible production of "unnatural" nucleotide sequences that could spawn a whole new set of viruses harmful to life. ☛

See *The Adromeda Strain* by Michael Crichton (Dell Publishing Company, New York, 1969).

The DNAs are in the genes of cells, and genes are in the *nucleus* or "intelligence center" of cells. The instructions the DNA molecules have for protein synthesis are carried by *ribonucleic acids*

(RNAs) from the nucleus to the rest of the cell, where the synthesis of proteins from amino acids goes on. RNA molecules are of several types—they all differ from DNA molecules in being single strands and by having ribose sugar residues rather than deoxyribose units in their phosphate ester primary structure. "Messenger RNA" carries the information stored in the nucleotide residues in DNA to tell the amino acids the order in which to join in the growing polypeptide chains of proteins and enzymes. The details of the transfer of genetic information from DNA to RNA are complex but rather well understood. If you take a course in molecular biology, you will learn about the way the transfer of the genetic information is accomplished using the "genetic code."

Electron-microscope photographs of DNA strands show that they are like unbranched threads, and whole DNA molecules are rodlike in shape. X-ray analysis shows that the two strands of the molecule

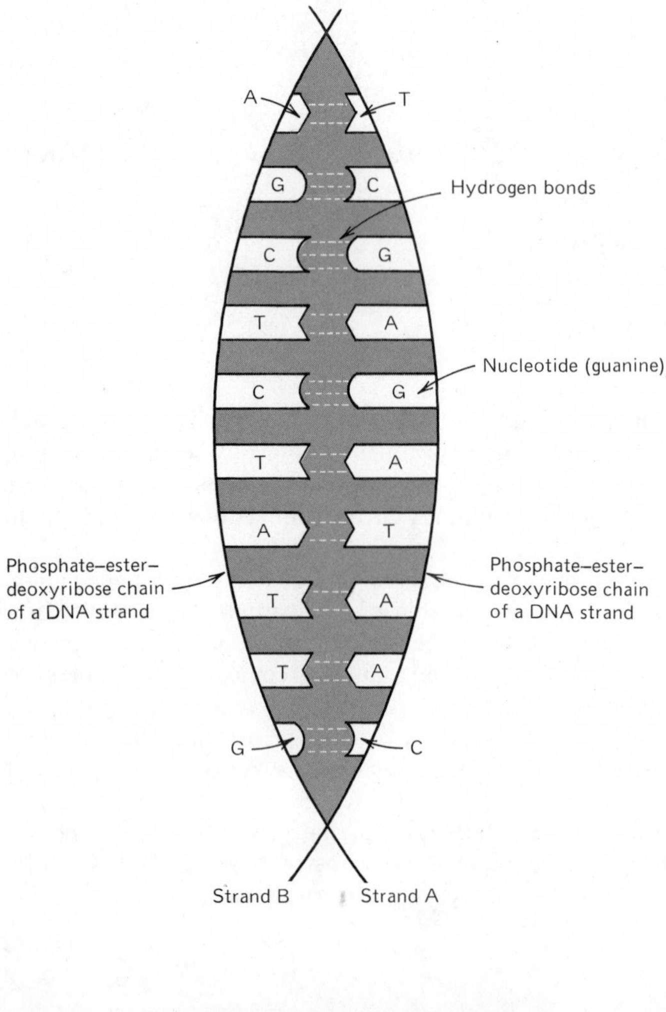

FIG. 14.7 Double helix formed from two strands of DNA by hydrogen bonding between nucleotide pairs.

are twisted around each other in the rods to give a *double-helix arrangement* (Figure 14.7). In discussing the process of DNA duplication, we used straight chains for the covalent, phosphate–ribose primary structure because it makes the process easier to see. The separation of the strands is actually an untwisting that begins at one end of the double-stranded coil.

Exercise

35. Hydrolysis of DNA molecules from any organism gives the separated nucleotides. No matter what the source of the DNA, the number of moles of adenine equals the number of moles of thymine, and the number of moles of guanine equals the number of moles of cytosine. Why should this be so? Would you expect the ratio of (adenine + thymine) to (guanine + cytosine) to be the same for all RNA molecules, no matter what organism the DNA came from?

Genetic mutations Nucleotide sequences in DNA molecules ordinarily go unchanged through unlimited numbers of duplications. But certain chemicals and radiation (ultraviolet rays in sunshine, X rays and α, β, and γ rays from radioactivity) can interrupt the nucleotide sequence by damaging the structure of one or more nucleotides. These ordinarily rare events are called *genetic mutation,* and they result in changes in the structure of the proteins put together according to the changed information in the DNA sequences. Not long ago, dozens of infants were born with undeveloped arms and legs that looked more like seal flippers than human limbs. The cause was traced to a mutation-causing (teratogenic) drug named "thalidomide" taken by the mothers during pregnancy. Now all new drugs are carefully tested in animals before being made available for medical use. Rubella (German measles) during pregnancy causes mutations that lead to serious birth defects, including mental retardation, blindness, and deafness. Apparently the rubella virus interferes at some stage in the transfer of genetic information. (Rubella can be prevented by vaccination.) Mutations also appear to cause cancer—extensive heavy exposure to the sun's rays over many years can cause skin cancer in cattle and in humans. The testing of nuclear bombs in the atmosphere was almost entirely stopped by international agreements when it was realized that the radioactive isotopes in the "fallout" produced by the explosions would greatly increase the chance of cancer-producing mutations in people.

Most mutations are lethal or at least harmful, but not all. The evolution of species could only occur through many mutations that resulted in more complex organisms, right up to humans.

14.9 Summary Guide Proteins, the main structural material for cells, are made up of amino acid (2-amino acid) residues joined by peptide linkages. The amino acid sequence of a protein is determined by the sequence of nucleotides in the deoxynucleic acids (DNA) in the genes of cells. Ribonucleic acids (RNA) transfer the recipes for proteins from DNA to the sites of protein synthesis from amino acids in the cell.

There are 21 natural amino acids, all of which except glycine are optical isomers having the same (L) configuration at C-2. This configuration is passed on to the proteins, making them chiral as well. A protein has a primary covalent bond structure upon which hydrogen bonding between —NH— and —C(=O)— groups imposes a secondary coiled (helical) structure. Residues of the amino acid cysteine can greatly change the shape of a protein molecule by connecting two polypeptide chains through disulfide (—S—S—) links. Eight of the 21 natural amino acids are an essential part of the diet of mammals, and an adequate supply of all amino acids is necessary for growth and health.

Vitamins are organic compounds necessary for life; yet they cannot be made by animal cells and so must be in the diet. A vitamin often occupies the active or prosthetic center in the cofactor of an enzyme that catalyzes a biochemical reaction. The B vitamin nicotinamide is part of the nicotinamide–adenine–dinucleotide enzyme system (NAD^+–NADH) that catalyzes many oxidation–reduction reactions in cells.

Hormones are compounds synthesized in the glands of the organism and carried by its circulatory system to regulate the activity of other cells. A hormone may be a rather simple steroid molecule (sex hormones) or a complex protein (insulin).

Fats are the esters of long-chain fatty acids with glycerol that store any energy in the diet that is in excess of the energy expended by the body. Fatty acids with two or more carbon–carbon double bonds (polyunsaturated fatty acids) are essential for health and must be in the fat of the diet.

Carbohydrates are the major energy source in the body; glucose is the source for cells, and the biopolymer glycogen stores glucose for rapid recall. The carbohydrate cellulose is a biopolymer of glucose that serves as the main structural material in plants. Grazing animals digest cellulose to glucose but people cannot. Starch is another biopolymer of glucose, found in cereal grains and digestible by man. Glycogen, cellulose, and starch are all biopolymers of glucose but differ in the stereochemical way the glucose units are connected by glycosidic links. Sucrose is a disaccharide made from a glucose residue linked to a fructose residue.

Ribose is a five-carbon sugar that plays an essential part in RNA. Deoxyribose, a closely related molecule, plays a similar role in DNA. Ribose is part of adenosine di- and triphosphates (ADP, ATP), the latter of which stores most of the energy released in the catabolism of glucose for use by the body. The catabolism of glucose, whose overall reaction is $C_6H_{12}O_6 + 6O_2 \rightarrow 6CO_2 + 6H_2O$, proceeds through many enzyme-catalyzed steps that deliver the energy released in the small amounts compatible with life.

Acetyl-coenzyme A (acetyl-CoA) is the most important of the enzyme cofactors in cell metabolism. It participates in the catabolism of glucose by way of the Krebs (tricarboxylic acid) cycle and is the intermediate in the synthesis of fatty acids, steroids, carotenoids, and amino acids.

The deoxyribonucleotides (DN) contain a purine (adenine DN, guanine DN) or pyrimidine (cytosine DN, thymine DN) ring joined through a ribose residue to a phosphoric acid unit. They are the materials that make up the biopolymer DNA in the genes that carries the genetic code for protein synthesis. The covalent backbone of a nucleic acid strand consists of deoxyribose units linked through phosphate ester groups. Two nucleic acid strands are held together in DNA in a helical coil by hydrogen bonds between adenine DN and thymine DN, and between guanine DN and cytosine DN. These two are the only pairings of deoxynucleotide residues possible, and the requirement for this pairing makes it possible for each DNA strand to make a duplicate of itself at the time of cell division. Duplication (or replication) of DNA strands permits a gene to pass on its genetic code at every cell division, and is the basis of inherited characteristics.

A genetic mutation occurs when a nucleotide residue in a DNA strand suffers a change in structure from the action of a foreign chemical or radiation. Genetic mutations are often fatal to the organism but nevertheless are the basis of the evolution of species when they produce a more viable organism.

Exercises

36. The monoanion of adenosine monophosphate, an intermediate in phosphate metabolism, can be represented as

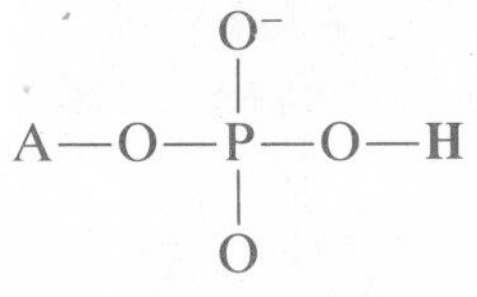

or AMP—OH⁻

where A stands for adenosine. If $pK_a = 7.21$ for the hydrogen (in boldface type) in the anion, what is the value of the concentration ratio $[AMPO^{2-}]/[AMPOH^-]$ in blood of $pH = 7.40$? (*Hint*: use the buffer equation pH $=$ pK_a + log $[A^-]/[HA]$ from Section 8.5.)

37. Cellulose $(C_6H_{10}O_5)_n$ does not dissolve appreciably in any solvent, including water. However, when the hydrogens of the —OH groups in cellulose (Section 14.5) are replaced by $-\overset{O}{\overset{\|}{C}}-CH_3$ groups using acetic anhydride in the "acetylation" reaction

$$R{-}OH + CH_3\overset{O}{\overset{\|}{C}}{-}O{-}\overset{O}{\overset{\|}{C}}{-}CH_3 \xrightarrow{H_2SO_4} R{-}O{-}\overset{O}{\overset{\|}{C}}{-}CH_3 + CH_3\overset{O}{\overset{\|}{C}}{-}OH$$

the product (cellulose acetate) becomes soluble in solvents such as dichloromethane (CH_2Cl_2). Why should the solubility characteristics change? (Cellulose acetate is the material of photographic film, made by spreading a solution of the acetate in CH_2Cl_2 out into a thin layer and letting the solvent evaporate.)

38. Does the triglyceride tristearin (Section 14.4) have optical isomers?

39. Taking the affinity of Pb^{2+} for S^{2-} ions as represented by the value of 1.3×10^{-28} for the K_{sp} of PbS, can you suggest a way in which poisoning by lead compounds might result? (*Hint*: consider the possibility of reaction between coenzyme A and Pb^{2+}.)

40. Vitamin C (ascorbic acid, white crystals, mp 190°C) is made commercially from L-sorbose by a series of steps:

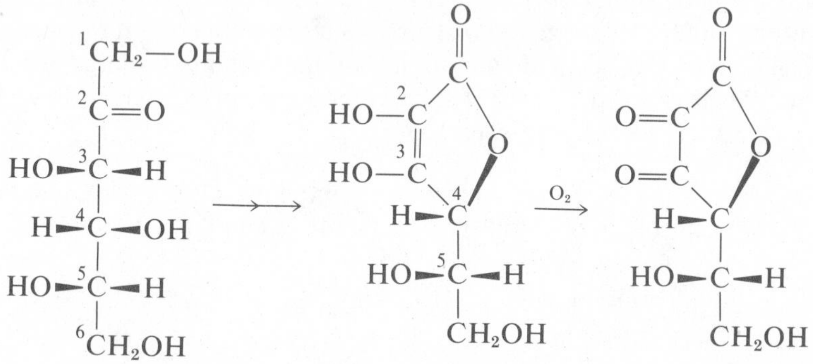

L-sorbose | Vitamin C | Vitamin C quinone

Is L-sorbose the enantiomer of D-fructose (Section 14.5) or is it a diastereoisomer of D-fructose? Does the configuration at C-5 in vitamin C correspond to that in the "natural" D-sugars? [Note that L-sorbose does not occur naturally—it is made from sorbitol, a naturally occurring hexahydroxyhexane, by bacterial oxidation at C-2 by the organism *Acetobacter* ("mother of vinegar").]

Vitamin C is rapidly oxidized by oxygen in air to vitamin C quinone as shown, and fruit juices and vegetables containing the vitamin rapidly lose it by this route when exposed to air. Write the balanced equation for the oxidation, using $C_6H_8O_6$ for the vitamin, and separate the equation into the balanced electron-transfer half-reactions.

Key Words

ADP adenosine diphosphate, the molecule produced with energy release from hydrolysis of ATP to ADP + phosphoric acid.

Aldohexose a six-carbon sugar with an aldehyde group at C-1—glucose is an example.

Amino acid residue what is left of an amino acid when it has been built into a protein—amino acid minus the elements of H_2O.

Amino acids (2-amino acids, α-amino acids) compounds of the general formula $RCHNH_2COOH$ that are building blocks for proteins.

Anabolism the building up of glucose or other metabolic intermediates, mainly by photosynthesis.

ATP adenosine triphosphate, the molecule that stores energy released in the oxidation of glucose or fats.

Bioenergetics deals with the production or utilization of energy by living systems.

Biopolymer giant molecule of a protein, carbohydrate, or nucleic acid that performs a vital function in a living organism.

Carbohydrates compounds having the general formula $C_n(H_2O)_m$.

Catabolism the oxidation process leading from glucose or metabolic intermediates to CO_2 and water.

Cofactor nonprotein part of an enzyme that contains the prosthetic center.

Denaturation alteration of the secondary structure of a protein by heat or chemicals.

Diastereoisomers molecules having the same molecular skeleton that are optical isomers but not mirror images (enantiomorphs) of each other—also called diastereomers.

Dipolar ion form (of an amino acid) a double ion having the general formula $RCHNH_3^+COO^-$ formed from an amino acid by proton transfer.

DNA deoxyribonucleic acid, the double-strand, nucleotide biopolymer in the genes that carries the information for the synthesis of the proteins making up an organism – the storehouse for inherited characteristics of an organism that are duplicated during cell division.

Double helix the intertwined springlike coil of two nucleotide chains to make a DNA molecule.

Enzymes protein molecules containing a prosthetic group that catalyze biochemical reactions.

Essential amino acids amino acids necessary for life that must be in the diet of an organism because they cannot be synthesized by the organism.

Fats esters formed between glycerol and long-chain carboxylic acids – also called triglycerides.

Genetic code the information carried by the nucleotide sequence in DNA and transferred to RNA to direct protein synthesis in a cell.

Genetic mutation alteration of the nucleotide sequence in DNA by chemicals or radiation.

Glucose blood sugar, having molecular formula $C_6H_{12}O_6$.

Glycogen a carbohydrate biopolymer of formula $(C_6H_{10}O_5)_n$ that stores glucose in mammals.

Glycosidic link an ether (—O—) linkage connecting C-1 of an aldose sugar (C-2 of a ketose) with C-4 of another aldose or ketose sugar – may be either α or β, depending on stereochemistry at C-1 in an aldose, C-2 in a ketose.

Helix the coiled-spring arrangement produced from the primary structure of a protein or nucleic acid by hydrogen bonding.

Hormones the regulators of the reactions of life.

Ketohexose a six-carbon sugar having a keto group at C-2 – fructose is an example.

Krebs (citric acid) cycle the path by which pyruvic acid is converted through acetyl-CoA to CO_2 and water in the catabolism of glucose.

NADH $\rightleftarrows$ NAD$^+$ the nicotinic acid–adenine–dinucleotide oxidation-reduction system found in cells.

Nucleic acids biopolymers having a ribose (RNA) or deoxyribose (DNA) phosphate ester chain that holds or transmits information for building the proteins of an organism.

Peptide linkage the amide linkage ($\text{—}\overset{\text{O}}{\overset{\|}{\text{C}}}\text{—NH—}$) that links amino acid residues in proteins.

Polyunsaturated fatty acid a long hydrocarbon chain (C_{16}–C_{20}) carboxylic acid having two or more carbon–carbon double bonds in the chain – necessary in diet for proper nutrition.

Primary structure the covalently bonded structure or backbone of a protein or nucleic acid molecule.

Prosthetic center region in an enzyme molecule where the reaction catalyzed by the enzyme goes on—often occupied by a vitamin.

Proteins biopolymers built of amino acid residues connected by peptide linkages according to instructions contained in the DNA of the genes of a cell.

Racemic mixture a mixture containing equal amounts of the two optical isomers of a compound having one or more chiral carbon atoms.

Retinal vitamin A aldehyde, necessary for vision in dim light, found in the rod cells of the retina of the eye.

Retinol vitamin A.

RNA ribonucleic acid, the single-strand biopolymer of nucleotides that carries information for amino acid residue sequences from DNA to the site of protein synthesis in a cell.

Secondary structure the arrangement of amino acids or nucleotide residues in a protein or nucleic acid produced by hydrogen bonding.

Steroids compounds having the ring structure of cholesterol—often they are hormones or vitamin D precursors.

Strand the primary bond structure of a DNA or RNA chain.

Terminal groups groups ($—NH_2$ or —COOH) at the ends of a polypeptide chain.

Transamination synthesis (in a cell) of an amino acid from an α-keto acid by transfer of an $—NH_2$ group from another amino acid.

Vitamins a group of compounds, often containing nitrogen atoms in combination, that are necessary for life yet cannot be made by the cells of an animal organism, and are therefore a necessary part of the diet.

Suggested Readings

Altschul, Aaron M., "Fortification of Foods with Amino Acids," *Nature,* **248**:643–646 (April 19, 1974).

Anfinsen, Christian B., "Principles That Govern the Folding of Protein Chains," *Science,* **181**:223–229 (July 20, 1973).

Edson, Lee, "Those Enterprising Enzymes," *Reader's Digest,* Feb. 1972, pp. 19–24.

Fernstrom, John D., and Wurtman, Richard J., "Nutrition and the Brain," *Sci. Am.,* **230**(2):84–91 (Feb. 1974).

Gustafson, Tryggve, and Toneby, Mark I., "How Genes Control Morphogenesis," *Am. Sci.,* **59**:452–462 (July–Aug. 1971).

Koshland, Daniel E., Jr., "Protein Shape and Biological Control," *Sci. Am.,* **229**:52–64 (Oct. 1973).

Lappé, Frances Moore, "Protein from Plants," *Chemistry,* **46**:10–13 (Oct. 1973).

Lowe, James N., and Ingraham, Loyd L., *An Introduction to Biochemical Reaction Mechanisms,* Englewood Cliffs, N.J.: Prentice-Hall, 1974.

Majtenyi, Joan Z., "Food Additives—Food for Thought," *Chemistry,* **47**: 6–13 (May 1974).

Majumder, Sanat K., "Vegetarianism: Fad, Faith, or Fact?" *Am. Sci.,* **60**: 175–179 (March–April 1972).

Olsen, I. D., *Metabolism,* Indianapolis, Bobbs-Merrill Co., Inc., 1973.

Ratcliff, J. D., "I Am Joe's Adrenal Gland," *Reader's Digest,* May 1971, pp. 127–130.

Ratcliff, J. D., "I Am Joe's Cell," *Reader's Digest,* Dec. 1973, pp. 120–123.

Routh, Joseph I., *Introduction to Biochemistry,* Philadelphia, W. B. Saunders Co., 1971.

Stadtman, Thressa C., "Selenium Biochemistry," *Science,* **183**:915–921 (March 8, 1974).

Stent, Gunther S., "Cellular Communication," *Sci. Am.,* **227**:43–51 (Sept. 1972).

Watson, J. D., *The Double Helix,* New York, Signet (New American Library), 1968.

chapter 15

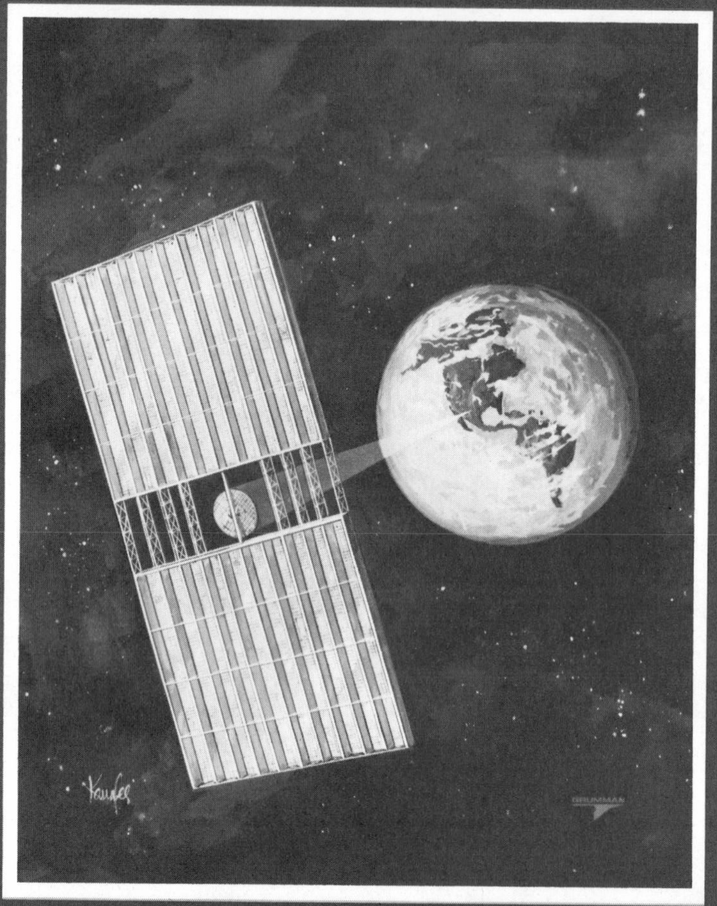

Artist's conception of the proposed Satellite Solar Power Station. The satellite would convert sunlight into electricity and transmit the electricity to a receiver on the ground. (*Photo 38, Arthur D. Little, Inc.*)

The Biosphere, Energy, and Man

15.1 Introduction The parts of planet earth that most affect us and other living things are the *atmosphere* (air), the *hydrosphere* (water in oceans, lakes, ice, and rivers), and the *lithosphere* (the solid outer crust of earth). Most of our activities go on in a layer about two miles thick and including parts of all three of these "spheres." This rather thin layer is the *biosphere* or "life region" of earth. In this chapter we take a look at the chemistry of air, water, and the earth's crust, applying the essential chemical principles of earlier chapters to understand the effects of human activities on the biosphere.

The development of science seen in modern industrial society has made possible vast improvements in the conditions people live under. Over the last century, scientific discoveries have led ever more rapidly to practical applications. You have only to realize that most of the medicines, plastics, and synthetic fibers we take for granted today were unknown even 50 years ago, to see major changes resulting from chemical discoveries. In 1860, the automobile had not yet been invented and the first oil well had just been drilled. Today, the rate of consumption of petroleum (in transportation and in generating electricity) is challenging the rate at which new supplies are discovered, and very soon petroleum will become too valuable to burn just for heat or transportation—it will have to be reserved as a source of organic chemicals for the petrochemicals industry if high standards of living are to be maintained and extended to all peoples.

We have gone wildly ahead in using scientific discoveries for material gain without much thought for the effects of our actions on the biosphere. It was not until 1947 that any action was taken to try to end smog and air pollution in Los Angeles. And we are still far from a satisfactory solution to the smog problem today. At best, controls keep up with the increase in the rate at which polluting chemicals are being added to the atmosphere, hydrosphere, and lithosphere by increasing population, auto use, and industrial production. We are beginning to understand that the atmosphere, oceans, lakes, and rivers are not places in which we can continue dumping vast quantities of waste gases, liquids, and particulates without paying any penalties. Increased burning of fossil fuels (coal, oil, gas) is estimated to have increased the concentration of CO_2 in the atmosphere by 20 parts per million (ppm) (Section 5.6) since 1860. ☚ And even in the much shorter period 1959–1969, atmospheric CO_2 increased from 314 to 321 ppm. There may be other explanations for the increase, but our use of fire to produce energy is a prime suspect.

See "greenhouse effect" of CO_2, Section 5.7.

The quality of water, both fresh and salt, has probably suffered more than earth or air at the hands of humans. In about 40 years Lake Erie has changed from a beautiful place for fishing, swimming, and boating to a lifeless, smelly cesspool. But as this book is being written, rigid control of the dumping of industrial wastes by the cities of Detroit, Cleveland, Erie, and Toledo seems to be giving Lake Erie a chance to recover.

Agricultural wastes (both fertilizers and insecticides) entering rivers, lakes, and estuaries of the ocean have also taken their toll of water quality and aquatic life. The oceans are becoming contaminated, not only with insecticides and other chemicals entering from the rivers that feed them, but also from oil spills. And at locations on the U.S. East Coast, tiny, sandlike grains of polystyrene have been discovered trapped in the gills of fish. Much of the marine life in the oceans is in the shallows over the continental shelves close to shore, and this is just where pollutants from rivers are concentrated and can do the most damage.

Finally there is the earth's crust, that part of the planet people mine for metal ores, coal, and other chemicals. Some metal ores are already in short supply and with increasing use in modern technology may soon be available only according to realistic priority and in conjunction with recycling programs. Furthermore, we have to find ways to dispose of the dangerously radioactive isotopic wastes from the "burning" of nuclear fuels in atomic energy power generating plants. These wastes can't be dispersed in the air, nor can they be dumped in the oceans. One plan is to bury them deep underground in abandoned salt mines where their activity will gradually decay to safe levels, though only after thousands of years. It has even been suggested that the wastes be loaded into rockets and shot into the sun.

One way you can find out what people are doing to their own immediate environment, and the whole biosphere as well, is to follow an element or a compound in its travels from earth's crust to the hydrosphere, the atmosphere, and back to the crust. Many vital elements and compounds go through this kind of circular path or "cycle." In this chapter we will examine the impact of human activities on some of the more important cycles—water, carbon and oxygen, and nitrogen.

15.2 The Carbon–Oxygen Cycle At the heart of the carbon–oxygen cycle, shown in Figure 15.1, are photosynthesis of glucose in plants and catabolism of glucose in animals:

Photosynthesis $$686 \text{ kcal} + 6CO_2 + 6H_2O \rightarrow C_6H_{12}O_6 + 6O_2$$

Catabolism $$C_6H_{12}O_6 + 6O_2 \rightarrow 6CO_2 + 6H_2O + 686 \text{ kcal}$$

These two equations can be written as one, using double arrows, but don't assume that a state of equilibrium exists for the two processes —*the biosphere is not a closed system* and is not at equilibrium, because energy in the form of sunlight is constantly coming in.

$$C_6H_{12}O_6 + 6O_2 \underset{\text{photosynthesis}}{\overset{\text{catabolism}}{\rightleftharpoons}} 6CO_2 + 6H_2O + \text{energy}$$

You saw some of the details of the catabolism of glucose in Chapter 14. Much is also known about the detailed steps in photosynthesis. The overall changes occurring can be summarized, using a set of equations that show only the oxidations and reductions that go on. The

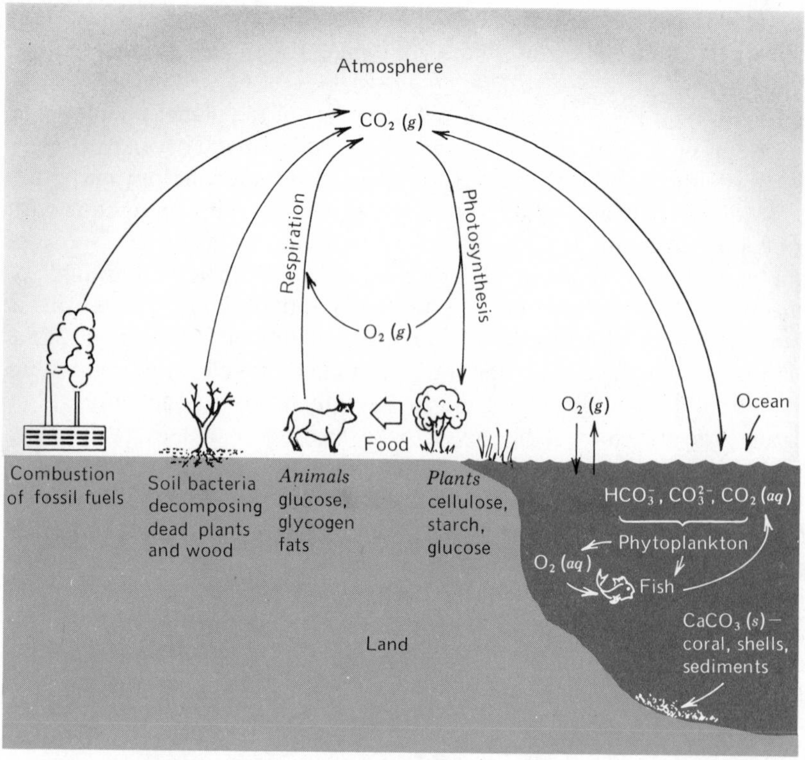

FIG. 15.1 Carbon-oxygen cycle.

enzyme cofactor abbreviated $NADP^+$ serves as an intermediate to transfer electrons and protons from water to CO_2.

Oxidation of H_2O	$12H_2O \rightarrow 6O_2 + 24H^+ + 24e^-$
Reduction of $NADP^+$	$12NADP^+ + 24e^- + 12H^+ \rightarrow 12NADPH$
Reduction of CO_2	$12H^+ + 12NADPH + 6CO_2 \rightarrow 12NADP^+ + 6H_2O + C_6H_{12}O_6$
Sum	$6CO_2 + 6H_2O \rightarrow C_6H_{12}O_6 + 6O_2$

Actually chlorophyll *a*, chlorophyll *b*, and β-carotene, when it is present in a green leaf.

The 686 kcal of energy required for the photosynthesis of the sugar glucose is absorbed by chlorophyll from sunlight and passed over to a long series of reaction steps that lead from CO_2 and H_2O to glucose. None of these steps goes directly in the simple way shown in the reaction scheme.

$NADP^+$, an abbreviation for *nicotinamide adenine dinucleotide phosphate*, which differs from NAD^+, described earlier, by an extra phosphoric ester group $\mathrm{H{-}\overset{|}{\underset{|}{C}}{-}O{-}\overset{O}{\overset{\|}{\underset{\underset{OH}{|}}{P}}}{-}OH}$ on one of the ribose units. Otherwise, NAD^+ and $NADP^+$ work in the same way to transfer hydrogens and electrons, using the nicotinamide parts of their molecules.

Photosynthesis produces oxygen (O_2), and it is believed that the oxygen now in the earth's atmosphere came originally from this source, beginning about 2 billion years ago when the first primitive plants (phytoplankton in the oceans) appeared. Since our life depends on O_2 for respiration, we owe those primitive plants and those that evolved from them a great debt. The oxygen concentration in the atmosphere is believed to have reached its present level (20%) about 20 million years ago.

EXAMPLE

In these days of growing world population it's instructive to estimate the area of photosynthesizing green leaf each of us needs for food. To simplify things, assume that the diet is based on starch only and that we each need about 3000 kcal of energy per day. Only about 1.2% of the light falling on leaves is used in photosynthesis. A square meter of leaf area in a temperate climate absorbs a total of about 4700 kcal per day in the form of light energy. The portion of absorbed energy going into photosynthesis is then 1.2% of 4700 kcal or $0.012 \times 4700 = 56.4$ kcal per square meter each day. For a 3000-kcal daily diet, $3000/56.4 = 53.2$ m^2 of leaf are needed. A "very closely planted year-round garden" at least 11 m by 5 m (36 ft by 16 ft) is needed to feed one person. This calculation assumes that humans eat the starch produced by photosynthesis directly as bread, rice, corn, or the like.

If the biodynamic/French-intensive method of gardening is used, the same amount of food can be grown in a garden of about 18 ft by 8 ft.

Now suppose a person wants to live on meat alone. Beef cattle are very poor converters of energy in the cellulose and starch they eat to energy in protein (meat). Only about 10% of the carbohydrate a steer eats goes into meat. The remaining 90% keeps his body warm and is burned away in exercise. These are very rough estimates, but they indicate that about 10% of the energy from photosynthesis is available in food produced by the food chain plant → animal → human. Instead of the 53 m^2 of leaf area needed for the 3000-kcal diet based on starch, about 10 times that area, or 530 m^2 (0.13 *acre*) of leaf area, is required for the all-meat diet.

Put another way, the average American consumes the *equivalent* of 1600 lb of cereal grain (wheat, rice, corn, oats) per year. Of the 1600 lb only 160 lb or 10% is consumed *directly* in the form of bread, breakfast cereals, and other products made directly from grain. The remaining 1440 lb (90%) of the grain goes to feed animals and is consumed *indirectly* in the form of meat, eggs, and milk products. By contrast people living in pretechnological societies consume an average of about 360 lb of grain per year (1750 kcal per day), nearly all of it directly in the form of bread and rice. These are rough figures, but they do show that getting your calories from beef is wasteful of the energy fixed by photosynthesis in grains and other feed stocks. The efficiency with which poultry convert feed stocks to meat is about three times that of cattle. The increased efficiency is a result of advances in breeding and feeding chickens and turkeys. Many of the savings have been passed along to consumers in the form of lower prices.

Figure 15.1 shows the molecular and ionic forms in which carbon cycles through the atmosphere, hydrosphere, and lithosphere. Most of the carbon (~99%) on earth is in the form of calcium carbonate ($CaCO_3$) in limestone, coral, and other sedimentary rocks. The 320 ppm of CO_2 in the atmosphere is only 0.0035% of the total, and all the rest of the biosphere contains about 0.2% in the form of plants (mainly trees), living and dead.

The total amount of carbon in the atmosphere, hydrosphere, and lithosphere is estimated to be 20,000,000 billion or 2.0×10^{16} metric tons (1 metric ton = 1000 kg = 2205 lb).

By burning fossil fuels (coal, gas, and oil) our activities are adding about 5 billion tons of CO_2 per year to the atmosphere. The atmosphere contains about 700 billion tons. If all the CO_2 from burning fuels stayed in the atmosphere, the CO_2 content would increase by about 2 ppm per year. The current annual concentration increase is only about 0.70 ppm or $\frac{1}{3}$ that expected, so the other $\frac{2}{3}$ of the expected increase must be stored through increased photosynthesis or in the oceans.

The carbon–oxygen cycle in the lakes and oceans can pretty much run by itself because through photosynthesis the algae and phytoplankton produce enough dissolved oxygen and food for all fish to live on. The fish feed on algae, plankton, or other fish and catabolize glucose, thus giving CO_2 that the plankton need to grow. This kind of healthy balance can be destroyed by polluting chemicals and sewage that are discharged into rivers that flow into lakes and the ocean.

Pollutants in water are of two types—*true poisons* that kill marine animals and plants, and *nutrients* in the form of organic matter in sewage, phosphates, and fixed-nitrogen compounds (nitrates and ammonia). Nutrients accelerate the natural aging process (eutrophication) of a lake by causing overpopulation of algae. The overgrowth of algae proceeds until the dissolved CO_2, upon which the algae depend for photosynthesis, is used up. The algae and other marine plants then die and sink to the bottom of the lake. There bacterial decomposition of their organic matter to CO_2 and water begins. The bacteria need oxygen to carry out the decomposition, so the concentration of dissolved oxygen decreases. Soon so little oxygen is available that the fish die and add their bodies to the organic matter at the bottom. These are among the first steps in the process of eutrophication. In the normal course of things eutrophication occurs naturally over thousands of years because even unpolluted streams continuously bring in small amounts of phosphates and nitrogen compounds that have been leached from the soil through which they flow. Eutrophication is thus a natural process, but the process is speeded up when a lake or pond receives an oversupply of nutrients or when untreated sewage is dumped into the water. The sewage contains organic matter, which consumes oxygen during de-

Any process that converts N_2 into nitrogen compounds is said to "fix" the nitrogen. Most organisms need fixed nitrogen for growth.

composition, and nitrates and phosphates that stimulate algal growth. Lake Erie went through the first stages of eutrophication in about only 50 years because of human activities.

Although total agreement has not been reached, it seems most likely that raw sewage as well as ammonia and phosphates from the use of fertilizers in farming play important parts in accelerating eutrophication. Also involved, in the case of Lake Erie, was the dumping of sulfite liquors (Na_2SO_3) from paper manufacture. The sulfite ion consumes dissolved oxygen when it is oxidized to sulfate ion: $2SO_3^{2-}(aq) + O_2(aq) \rightarrow 2SO_4^{2-}(aq)$. The dumping of sulfites was stopped several years ago.

Since the water from lakes finally reaches the ocean, pollution in streams ultimately pollutes the ocean as well. The ocean is so vast that the poisons work slowly. Nevertheless, in bays and estuaries, pollution that contaminates marine life has already appeared. For years, much of the waste produced by the millions of people in New York City has been dumped into the New York bight—about a dozen miles from the mouth of the Hudson River. Now scientists have discovered that the gummy acidic sludge which has collected on the sandy ocean floor is creeping toward land, and New York's beaches, including Fire Island National Seashore, and shellfish breeding areas are threatened. Scientists are monitoring the "crawl" of the acid waste, but no one knows how to contain or control the sludge.

Exercises

1. Give the structure changes in the *nicotinamide part* of NADPH that occur as CO_2 is reduced to $(CH_2O)_n$ and H_2O during photosynthesis.
2. Show which species are oxidized and which are reduced in the set of equations given for the photosynthesis of glucose. Write the electron-transfer half-reactions for each species oxidized and for each reduced, in their *simplest* forms.
3. Give the balanced electron-transfer half-reactions for the oxidation of sulfite ion to sulfate ion by molecular oxygen.

15.3 The Nitrogen Cycle The main parts of the nitrogen cycle are shown in Figure 15.2. This cycle is to proteins and nucleic acids what the carbon–oxygen cycle is to carbohydrates. All organisms need some form of fixed nitrogen (NH_3, NH_4^+, NO_2^-, NO_3^-) to make the proteins for their growth. A few microorganisms and the blue-green algae can convert the molecular nitrogen of the atmosphere into usable, fixed forms. Certain bacteria (*Rhizobium*) that grow in swellings on the roots of legumes (beans, peas, alfalfa) synthesize amino acids, using N_2 from the air. The bacteria and the plant roots have a *symbiotic* ("live together by helping each other") relationship—the bacteria give amino acids to the plant for its growth and the plant donates carbohydrate to feed the bacteria. Other soil bacteria (*Clostridium* and *Azotobacter*) can reduce nitrogen to ammonia if they have a source of carbohydrate, such as dead wood.

Lightning bolts also fix N_2 in the form of nitric oxide (NO), which goes on through nitrogen dioxide (NO_2) to nitric acid and nitrates. The lightning spark produces the very high temperature and energy needed for the reaction $N_2(g) + O_2(g) \rightarrow 2NO(g)$, $\Delta G^0 = +41.4$ kcal (Section 7.5). It is estimated that 7–8 million metric tons of nitrogen are fixed by lightning each year. But the majority of fixed nitrogen entering the biosphere is put there by our use of nitrogen fertilizers based on the Haber process for the synthesis of ammonia (Section 7.4). More than 30 million metric tons (66 billion lb) of fixed nitrogen go into the biosphere per year at the present time in the form of fertilizer to increase yields of farm crops. Much of the fertilizer is in the form of am-

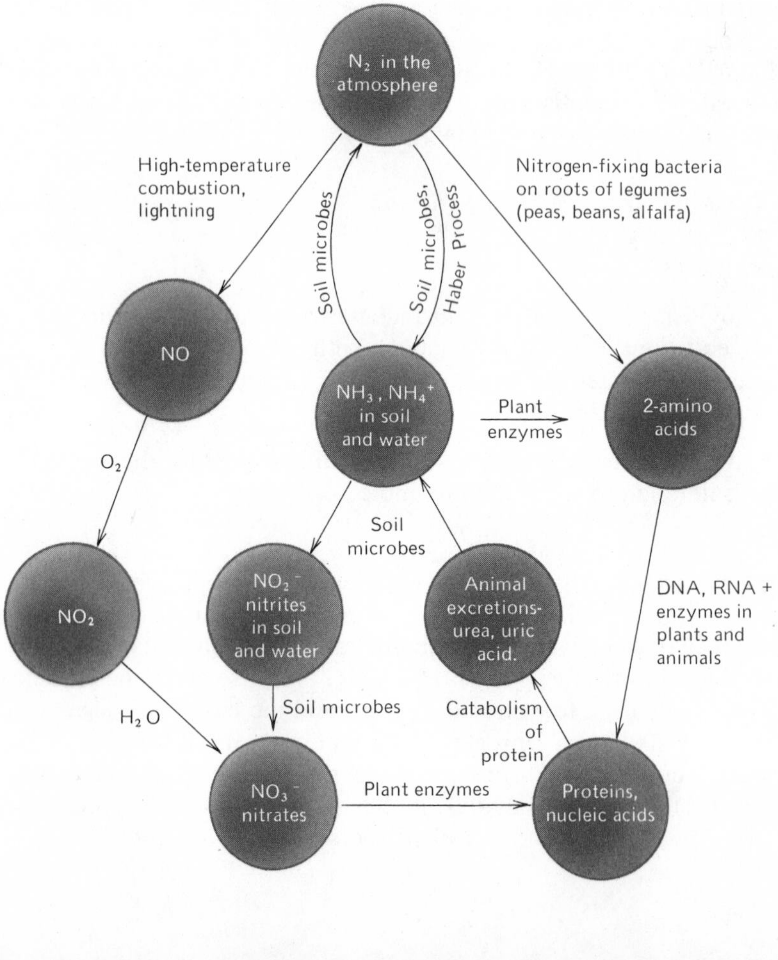

FIG. 15.2 Nitrogen cycle.

monium nitrate ($NH_4^+NO_3^-$), a rich source of fixed nitrogen because both ammonium ion and nitrate ion are in key positions of the nitrogen cycle (Figure 15.2). To make ammonium nitrate, ammonia from the Haber process is oxidized to produce nitric acid, using a platinum catalyst (Ostwald process) and air as the oxygen source. The nitric acid is then neutralized with additional ammonia to give $NH_4^+NO_3^-$.

The heavy use of fertilizers raises a question—where does the fixed nitrogen from fertilizers finally go in the environment? As shown in the nitrogen cycle, pathways are provided through the activities of soil bacteria to return fixed nitrogen to the atmosphere as N_2. But these natural pathways were intended to balance out the *natural* production of fixed nitrogen. These natural processes are now burdened or overburdened by the addition of industrially fixed nitrogen in the form of ammonia, ammonium nitrate, and urea fertilizers. ☛ It has been estimated that the conversion of fixed nitrogen from the biosphere to atmospheric N_2 falls about 10 million metric tons *behind* the total production of fixed nitrogen by natural means plus the Haber process. The excess fixed nitrogen in the biosphere can be expected to speed up eutrophication of lakes and rivers. The problem of excess CO_2 from the burning of fossil fuels described earlier is less serious because increased photosynthesis can absorb large amounts of CO_2, but not nearly so much fixed nitrogen is needed for plant growth, because plants are mostly cellulose.

Urea, $H_2N{-}\overset{O}{\overset{\|}{C}}{-}NH_2$, is produced by reaction of NH_3 with CO_2:

$$2NH_3 + CO_2 \rightarrow H_2N{-}\overset{O}{\overset{\|}{C}}{-}NH_2 + H_2O.$$

Urea is a white crystalline solid that can be applied directly on the ground.

There are two other problems connected with the heavy use of nitrogen fertilizers to increase crop yields. Most of the fixed nitrogen ends up as nitrites (NO_2^-) and nitrates (NO_3^-) in the soil and ground water through the action on ammonia of what are called *nitrifying bacteria* in the soil. Nitrites and nitrates are harmful to animals and humans, particularly children, and these chemicals are suspected of being carcinogens (cancer-causing agents). The nitrite and nitrate concentrations are already approaching the danger level in the ground water in some heavily fertilized agricultural areas. Both nitrite and nitrate ions can be removed from drinking water by ion exchange (Section 11.3), but the process is expensive.

The second problem that arises when crop yields are increased by using fixed-nitrogen fertilizers results from rapid removal of other essential elements and compounds from the soil by the increase in the per acre yield of crops. Phosphates and the transition metals molybdenum and cobalt are needed for plant growth. Phosphates go into the DNA and RNA of plant cells, and molybdenum and cobalt are part of the *nitrogenase* enzymes used by bacteria for their fixation of nitrogen. These elements and others must sooner or later be replaced in the soil if agricultural crops are to grow properly under heavy fertilization. The full list of necessary trace elements includes B, Si, V, Cr, Fe, Cu, Zn, Se, Sn, and I, in addition to the more obvious ones, such as N, P, S, Ca, and Mg.

Metabolism of fixed-nitrogen compounds There are *denitrifying* bacteria (*Pseudomonas denitrifican*, for example) in the soil that use the oxygen in soil nitrates instead of molecular oxygen to oxidize glucose or other carbohydrates for their life energy. Their overall effect is to reduce the amount of nitrate in the soil. Equations such as the following ones summarize the metabolism of nitrates, using potassium nitrate (KNO_3) as reactant:

$$5C_6H_{12}O_6 + 24KNO_3 \rightarrow 30CO_2 + 18H_2O + 24KOH + 12N_2 \qquad \Delta G^0 = -570 \text{ kcal/mole of glucose}$$

$$C_6H_{12}O_6 + 6KNO_3 \rightarrow 6CO_2 + 3H_2O + 6KOH + 3N_2O \qquad \Delta G^0 = -545 \text{ kcal/mole of glucose}$$

The catabolism of glucose to water and CO_2 yields 686 kcal/mole, so the denitrifying bacteria at 570 and 545 kcal per mole of glucose are doing pretty well without molecular oxygen. Since these bacteria *reduce* the amount of fixed nitrogen in the biosphere, their increased growth will help offset the fixed nitrogen poured in through use of fertilizer.

Biochemists have found that building fixed nitrogen into amino acids by soil bacteria and certain algae starts by reducing any nitrites or nitrates to ammonium ion:

$$NO_2^- + 8H^+ + 6e^- \rightarrow 2H_2O + NH_4^+$$
$$NO_3^- + 10H^+ + 8e^- \rightarrow 3H_2O + NH_4^+$$

One reaction that puts the nitrogen fixed in the ammonium ion into the $—NH_2$ group of an amino acid uses **NADPH** as a reducing agent:

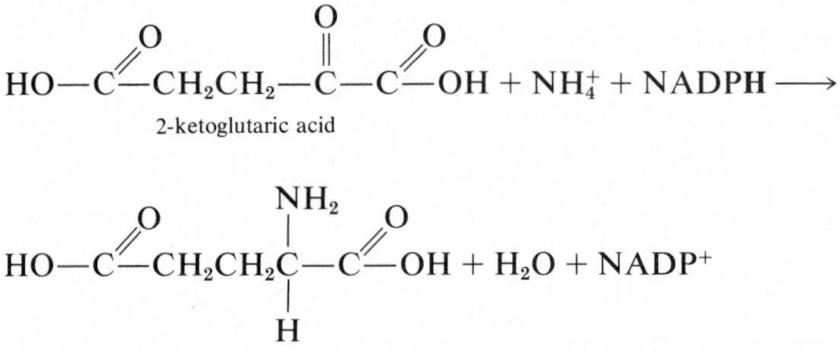

2-ketoglutaric acid

Glutamic acid

Nitrous oxide, N_2O, although a form of fixed nitrogen, is not useful to plants. The atmosphere contains 0.5 ppm of N_2O, partly from the denitrification reaction, partly from the reaction of oxygen atoms with nitrogen molecules in the upper atmosphere.

The 2-ketoglutaric acid is in Kreb's cycle (Figure 14.4) of glucose metabolism and can act as an amino group (NH_2) carrier in the *trans-amination* of other 2-keto acids to produce their corresponding 2-amino acids (Section 14.2) for protein synthesis.

You have seen several times that the N_2 molecule is very stable. To cause N_2 to react directly with hydrogen to form ammonia requires a catalyst and high temperatures and pressures. It is remarkable that the enzymes (nitrogenases) in some bacterial and algal cells can produce the ammonia from N_2 and H_2O at ordinary temperatures and pressures.

Exercises

4. The skeleton equations for the reactions in the preparation of ammonium nitrate, starting with N_2 and H_2, are as follows:

$$N_2 + H_2 \not\rightarrow NH_3 \qquad \text{(Haber process)}$$
$$NH_3 + O_2 \overset{Pt}{\not\rightarrow} NO + H_2O \qquad \text{(Ostwald process)}$$
$$NO + O_2 \not\rightarrow NO_2$$
$$NO_2 + H_2O \not\rightarrow HNO_3 + NO$$
$$HNO_3 + NH_3 \not\rightarrow NH_4^+NO_3^-$$

Balance these equations where they are not balanced, and calculate the weight of ammonium nitrate that would be produced from 1 metric ton (1000 kg) of N_2 assuming 100% efficiency.

5. To get some idea of the scale upon which nitrogen fixation is carried out using the Haber process, calculate the volume in liters of N_2, measured at STP, in 30 million metric tons of the element. What is this volume in cubic meters? (Use one liter = 1000 cm^3.)

6. Taking the human population of earth as 3.5 billion, calculate the weight in pounds of nitrogen fixed per person in one year. (Use data from Exercise 5. One metric ton = 1000 kg = 2205 lb.)

NO_X—photochemical smog Depending on where you live you may have seen on the weather page of your newspaper a table labeled "Air Pollution Index" with an entry something like "NO_X = 25 pphm" (pphm means parts per hundred million). The entry means that for every hundred million molecules in the air in your city there was a total of 25 molecules of oxides of nitrogen, NO and NO_2, when the measurement was made. Most of this fixing of nitrogen in the oxides results from explosions of the gasoline–air mixture in the cylinders of automobile, bus, and truck engines. Like flashes of lightning, the explosions provide the sudden high temperature and quick cooling required to make NO by the reaction $N_2 + O_2 \rightarrow 2NO$. Further reaction of NO with air gives NO_2. Nitrogen fixed by car, truck, and bus engines amounted in 1973 to something of the order of 20 *thousand* metric tons per million vehicles per year. This quantity seems small compared to the 30 *million* metric tons per year fixed by the Haber process and the 7–8 million tons from lightning and other processes in the atmosphere. However, with more than 100 million cars, buses, and trucks in the United States in 1972, the amount of nitrogen fixed per year comes to 2 million metric tons, a figure comparable to the others. Clearly, the operation of motor vehicles is adding to the amount of fixed nitrogen in the biosphere. And, as standards of living rise, more fertilizer will be used to provide food, and, in the absence of reasonable alternatives, more vehicles will be operated. It seems certain that the natural processes for converting fixed nitrogen to N_2 will be unable to prevent a continuous increase in fixed nitrogen in the biosphere unless positive preventive action is taken.

EXAMPLE

The maximum emission of nitrogen oxides permitted under law for new cars sold in 1974 in California was 3 g per car per mile. Taking 10,000 mi per year as the average distance a car is driven, we see that the maximum emission of nitrogen oxides comes to $3 \times 10{,}000 = 30{,}000$ g, or 30 kg per car per year. If the California law on emissions were applied to all of the approximately 100 million motor vehicles in the whole U.S., NO_X emission would not exceed $100{,}000{,}000 \times 30 = 3 \times 10^9$ kg or $(3 \times 10^9/1000) = 3 \times 10^6$ metric tons per year. Taking NO_X as NO, which is $(14/30) \times 100 = 47\%$ N, the maximum amount of fixed nitrogen added to the biosphere under law would be $3 \times 10^6 \times 0.47 = 1{,}410{,}000$ metric tons per year. This represents a reduction to $(2 \times 10^6/1.41 \times 10^6) \times 100 = 71\%$ of the nitrogen fixed by cars in the absence of legal control. However, 1,410,000 tons represents significant pollution by fixed nitrogen, relative to the 8 million tons from lightning and the 30 million tons from fertilizer.

Catalytic exhaust converters required on all 1975 automobiles promise a further reduction in NO_X.

Automobile exhaust is a mixture of nitrogen, carbon dioxide, water vapor, NO and NO_2, unburned gasoline hydrocarbons, and carbon monoxide (CO). Under certain conditions the hydrocarbons and nitrogen oxides interact to produce the unhealthy and unpleasant mixture of chemicals called *photochemical smog*. There are requirements for the generation of this kind of smog:

1. Still air, usually in a *temperature inversion layer* (Figure 15.3) causing pollutants to remain concentrated in the air instead of being thinned out by wind or rising air currents (Photo 39, p. 620)

Other types of smog are formed from soot and SO_2 that result from burning coal that has a high sulfur content. London smog is of this type and is rapidly being eliminated by requiring the use of natural gas instead of high-sulfur coal for home heating.

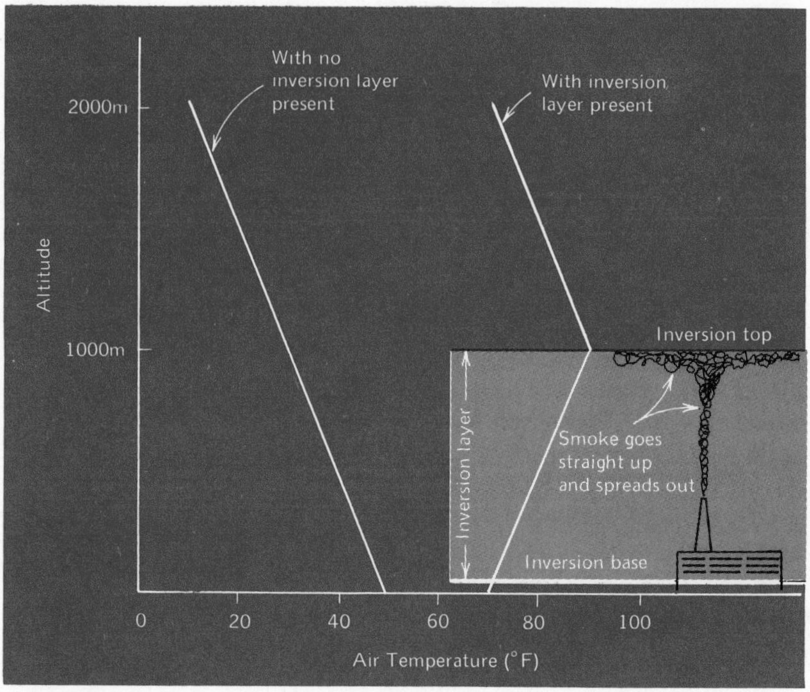

FIG. 15.3 Atmospheric temperature changes with altitude with and without an inversion layer present.

2. Oxides of nitrogen and unburned hydrocarbons from auto engines or other sources
3. Sunlight to provide the energy needed by the reactions producing smog

When a temperature inversion layer is present in the atmosphere, air temperature rises with increasing altitude up to the level of the inversion (300–600 m, 1000–2000 ft) and then suddenly begins to fall (Figure 15.3). The term "inversion layer" is used because under normal atmospheric conditions, the air temperature constantly *decreases* at the rate of about 1°C for every 100 m of additional altitude. ☛ Pollutants discharged into the air collect under the top of the inversion layer, which acts like a "lid," keeping them concentrated close to the ground. You can tell whether there is an inversion layer on any particular day if you watch how smoke from smokestacks rises in the air. If it goes straight up and then spreads out in a flat cloud, an inversion layer is present (Figure 15.3).

At an altitude of 11 km (6.8 mi) the temperature levels out, only to start rising at 25 km (15 mi) as the *ozone layer* in the atmosphere is reached.

Oxides of nitrogen and unburned hydrocarbons alone do not produce the complex mixture of polluting chemicals that make up photochemical smog. Sunlight is needed, and this requirement is seen in the heavy smog attacks in the coastal cities of California—particularly

PHOTO 39 Smog trapped on the ground by an inversion layer (*Los Angeles County Air Pollution Control District.*)

in Los Angeles, where photochemical smog first appeared as the Los Angeles area filled up with people and cars, starting about 1942.

The formation of photochemical smog starts with nitrogen dioxide, NO_2, a poisonous, yellow-brown gas. Because it is colored, it must absorb light in the visible range of wavelengths. The absorption of wavelengths less than 4000 Å causes the NO_2 to dissociate into an NO molecule and *atomic* oxygen:

Photochemical activation	$NO_2 + h\nu \rightarrow NO_2^*$
Unimolecular decomposition	$NO_2^* \longrightarrow NO + O$

The free oxygen atoms produced by this photochemical reaction react with oxygen molecules to form ozone (O_3) molecules, $O_2 + O \rightarrow O_3$. The air on a smoggy day in Los Angeles or San Francisco may contain as much as 20 pphm of ozone (20 O_3 molecules per 100 million of $N_2 + O_2$ molecules). The ozone concentration would be much higher than this if ozone were not constantly being used up by reactions with the unburned gasoline hydrocarbons from auto exhausts. Ozone attacks and cleaves carbon–carbon double bonds, producing ketones

Home incinerators add NO_X to the atmosphere—their use, and backyard burning in general, were made illegal several years ago in Los Angeles and more recently in the San Francisco Bay Area and other population centers.

and aldehydes in the process. Letting R stand for a hydrocarbon residue (CH_3, C_2H_5, and so on), we can show the reaction as follows:

$$\underset{R}{\overset{R}{>}}C{=}CHR \xrightarrow{O_3} \underset{R}{\overset{R}{>}}C{=}O + O{=}CHR$$

Ketone Aldehyde

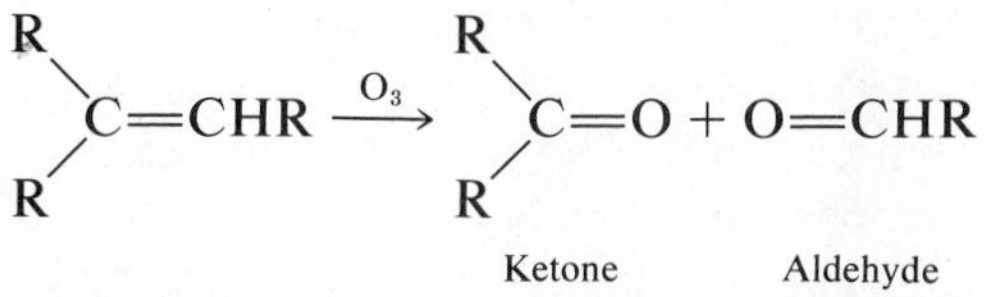

In a way this is a helpful reaction because ozone itself is quite poisonous (toxic at 100 pphm). However, the ketones and the aldehydes get further oxidized by oxygen atoms to *peroxyacyl* radicals that capture NO_2 molecules to give compounds called peroxyacyl nitrates or "PAN."

$$R{-}\overset{\overset{\displaystyle \ddot{O}.}{\|}}{C}{-}H + :\ddot{O}: \rightarrow R{-}\overset{\overset{\displaystyle \ddot{O}.}{/\!\!/}}{C}. \quad + H{-}\ddot{O}\cdot$$

Acyl Hydroxyl
radical radical

$$R{-}\overset{\overset{\displaystyle \ddot{O}.}{/\!\!/}}{C}. \quad + O_2 \rightarrow R{-}\overset{\overset{\displaystyle \dot{O}}{/\!\!/}}{C}{-}\ddot{O}{-}\ddot{O}\cdot$$

Peroxyacyl radical

$$R{-}\overset{\overset{\displaystyle \ddot{O}.}{/\!\!/}}{C}{-}\ddot{O}{-}\ddot{O}\cdot + .\overset{\overset{\displaystyle \ddot{O}.}{}}{\underset{\ddot{O}:}{N}} \rightarrow R{-}\overset{\overset{\displaystyle O}{/\!\!/}}{C}{-}\ddot{O}{-}\ddot{O}{-}\overset{\overset{\displaystyle \ddot{O}.}{}}{\underset{\ddot{O}:}{N}}$$

A peroxyacyl nitrate (PAN)

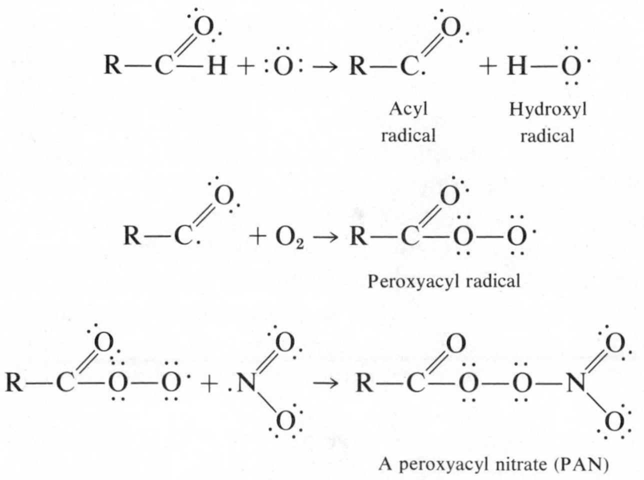

This reaction removes NO_2, another poison in smog, and might also seem helpful by decreasing both the NO_2 and the O_3 concentrations. Unfortunately, the peroxyacyl nitrates are very unpleasant and dangerous compounds. They are mainly responsible for the eye, nose, and lung irritation in heavy smog. At times smog has been so bad in the Los Angeles area that school children were not permitted out of doors at recess. Lung irritation from smog is particularly hard on elderly persons who have heart or respiratory problems such as asthma and emphysema.

The chemicals in the Los Angeles photochemical smog are also harmful to plants. An extreme example may be seen in the San Bernardino National Forest, located in the San Bernardino Mountains 40 mi east of Los Angeles. Prevailing winds carry smog from Los Angeles over hills and mountains and deposit it in the forest. Many species of trees are dying, and the U.S. Forest Service is now planting other types of trees not so susceptible to the poisons carried through the air.

EXAMPLE Complex chemical problems are often not susceptible to simple solutions, as shown by the results of the first attack on smog—reduction of hydrocarbon emissions from cars. One big source of emissions is "blowby" of unburned gasoline past the pistons into the crankcase of the engine. From there, the gasoline hydrocarbons escape into the air through the oil-filler cap. This source of hydrocarbons was closed off by using "positive crankcase ventilation" (PCV), in which air is slowly drawn through the crankcase, using the vacuum at the carburetor, to carry the crankcase hydrocarbon fumes with the gas–air mixture to be burned in the cylinders. A second and important source of hydrocarbon emissions is the gasoline vapor displaced from gasoline tanks whenever they are filled.

But it soon became clear that cutting down on hydrocarbon emissions alone was not going to solve the smog problem. It was *increasing* the NO_X and O_3 concentrations by removing the hydrocarbons that were trapping them. The next step was to try to reduce the oxides of nitrogen in auto exhaust. There are two ways to do this—redesigning the engine so that combustion takes place at a lower temperature, or attaching to the exhaust system a converter containing a catalyst that speeds up the conversion of NO into its elements. We saw in Section 7.4 that the equilibrium constant for the reaction $2NO \rightleftarrows N_2 + O_2$ has the huge value of 2.36×10^{30} at 25°C. A good catalyst for the reaction working at ordinary temperatures would reduce the NO_X concentrations to the vanishing point. At this writing both approaches to solving the smog problem were being pursued, and the Wankel rotary gasoline engine, as opposed to piston engines, gave promise of solving the NO_X problem, as did the Honda CVCC, a stratified-charge engine.

The ozone layer, nitric oxide, and the SST The ozone layer is the part of the atmosphere at the top of the stratosphere and between about 25 and 75 km (from 15 to 45 mi) above the earth's surface. In this layer there is an unusually high concentration of ozone that absorbs and screens out the intense, short-wavelength radiation (~2900 Å and less) pouring in from the sun. Living things on earth owe their lives to the screening effect of the O_3 layer because radiation of such short wavelength and intensity is deadly to both plants and animals. The ozone molecules in the layer absorb the short-wavelength rays in the wavelength region 2000–3500 Å, and decompose into O_2 and free oxygen atoms. The free atoms react with O_2 to re-form O_3 molecules with emission of heat. Using $h\nu$ to stand for the radiation absorbed, we can show the mechanism of ozone screening as follows:

Photochemical activation	$O_3 + h\nu \rightarrow O_3^*$
Unimolecular decomposition	$O_3^* \rightarrow O_2 + O$
Bimolecular combination	$O_2 + O \rightarrow O_3 +$ heat energy

In the first step the light energy absorbed is converted to electronic energy to make a photoexcited or "hot" ozone molecule (starred). The extra energy then breaks bonds, giving O_2 molecules and oxygen atoms, as shown in the equation. When O_2 molecules and O atoms recombine in the bimolecular step to form O_3, the energy of the new O—O bond formed is released as heat. The heat energy keeps the ozone layer about 80°C warmer than the air layers just above and below it. By this mechanism the O_3 layer converts the lethal radiation from the sun into heat energy.

From the cyclic nature of the mechanism for energy absorption by ozone, it would appear that once some ozone was in the layer it could go through the protective cycle indefinitely. However, there is a slow reaction between ozone molecules and oxygen atoms that destroys ozone:

$$O_3 + O \rightarrow 2O_2$$

With this reaction going on, however slowly, the O_3 layer would finally disappear, leaving earth at the mercy of the deadly, short-wavelength radiation. This does not happen because O_3 is constantly being generated by a photochemical reaction of the O_2, also present in the ozone layer:

Photochemical activation	$O_2 + h\nu \rightarrow O_2^*$
Unimolecular decomposition	$O_2^* \rightarrow 2O$
Bimolecular combination	$O_2 + O \rightarrow O_3 +$ energy

The energy levels for electrons in the molecular orbitals of the O_2 molecule are spaced farther apart than in the O_3 molecule, so radiation of very short wavelength ($\approx$ 1800–2000 Å) is needed to decompose O_2 molecules into O atoms. This radiation from the sun is present in the upper part of the ozone layer, so any loss of O_3 in the layer is constantly being made up, and the "O_3 radiation screen" doesn't disappear.

The SST (supersonic transport) is an airliner designed to fly at speeds of twice to three times the sea-level speed of sound. To fly that fast, an SST must escape the frictional drag of the "thick air" at ordinary altitudes and fly at altitudes of about 20 km (about 12 mi). This part of the atmosphere is close enough to the ozone layer that there is concern about what the exhaust gases from the jet engines of an SST (or a fleet of them) might do to the vital ozone layer. The exhaust pollution problem is even more serious there than it is at sea level, because there is very little mixing of air where SSTs would fly, and exhaust pollutants could remain in concentrated form, possibly for as long as 5 years, according to one estimate.

To judge the impact of SSTs on the ozone layer we have to look at what kinds of chemicals there are in the exhaust of a jet engine, then at any effects they might have on the O_3 layer. Jet exhaust contains the same types of compounds as auto exhaust – CO_2, H_2O, NO, CO, and hydrocarbons. It turns out that *all* of these compounds (except CO_2) could reduce the concentration of ozone in the layer. Hydrocarbons, especially ones with carbon–carbon double bonds, react with ozone and destroy it in the way described earlier in the discussion of photochemical smog. Carbon monoxide destroys ozone because it is oxidized by O_3 to CO_2: $CO + O_3 \rightarrow CO_2 + O_2$, and H_2O molecules can convert ozone to oxygen by way of a complex series of chain reactions. But the most serious effects on O_3 concentration are predicted for the nitric oxide (NO) in jet exhaust – it sets off a fast chainlike reaction that destroys ozone:

$$NO + O_3 \rightarrow NO_2 + O_2$$

$$NO_2 + O \rightarrow NO + O_2$$

The oxygen atoms needed for the second step are constantly being produced by the photodecompositions of O_3 and O_2 described earlier. An SST is estimated to produce as much as a ton of NO per hour, and by taking the rates of all the ozone-forming and ozone-destroying reactions into account, it was estimated that a regularly operating fleet of 500 SSTs could cut the concentration of O_3 in the ozone layer to about half of its natural value. The result would be a substantial increase in the amount of short-wavelength (2900 Å and less) radiation reaching the surface of earth, with disastrous effects on life in the biosphere.

At this writing a British-French group and the Soviet Union had built a few experimental SSTs and were flying them. Earlier the U.S. government decided *not* to build SSTs, and a large part of the argument against building them was concern for their effects on the life-protecting ozone layer, brought to light by chemists.

Exercises

7. Given the following oxidation-potential data:

$$H_2SO_3 + H_2O \rightleftarrows SO_4^{2-} + 4H^+ + 2e^- \qquad \mathscr{E}^0 = -0.17 \text{ V}$$

$$O_2 + H_2O \rightleftarrows O_3 + 2H^+ + 2e^- \qquad \mathscr{E}^0 = -2.07 \text{ V}$$

Photochemical smog is termed an "oxidizing smog," while smog in which SO_2 (the anhydride of H_2SO_3) is the main pollutant is called "reducing smog." Explain why the terms "oxidizing" and "reducing" are used.

8. Give the reaction rate equations for each of the following simple (elementary) reactions (Section 9.3):

$$NO_2 \rightarrow NO + O$$

$$R{-}C({=}O)\cdot + O_2 \rightarrow R{-}C({=}O){-}O{-}O\cdot$$

$$NO + O_3 \rightarrow NO_2 + O_2$$

9. There is only one O—N bond length in NO_2. Explain why this should be so, using the theory of resonance. What value would you predict for the O—N—O bond angle in NO_2, using the electrostatic repulsion rules? (The skeleton is O—N—O.)

15.4 Nonpolluting Energy Sources High-temperature combustion of fossil fuels for energy is bound to produce oxides of nitrogen and, if the fuel contains sulfur, sulfur dioxide. Pollution of the atmosphere by these gases is most easily prevented when the power plant is stationary and the stack gases from the steam boilers can be "scrubbed" by contact with alkaline or other solutions to absorb NO_X and SO_2. Progress in this cleanup of stack gases has already been made, though at considerable cost—as much as 13% of the plant cost. One large modern steel plant in Texas adds nothing but CO_2 and water vapor (and steel) to its environment, and new steam generating plants are being designed to minimize pollution by their stack gases.

Moving vehicles (cars, trucks, aircraft) present a more difficult problem because the devices for removing polluting emissions from their exhaust are bulky and have to travel with the vehicle. Storage batteries or fuel cells using hydrogen or methanol would solve the problems for cars and trucks. Cutting down emissions from aircraft engines remains the most difficult problem. Use of methanol as fuel in jet engines would help reduce hydrocarbon emissions, but nitrogen oxides would still remain a problem.

What possible energy sources are there that would be totally nonpolluting? Nuclear power is nonpolluting provided that no accidents release radioactive materials, and that a safe way can be found to dispose of their radioactive wastes. There is no reason to think that nuclear power plants present any greater danger to the environment than the pollution accompanying the use of fossil fuels. No person has been injured by radioactivity in the existing nuclear power plants, and nuclear-powered submarines have traveled millions of miles under demanding conditions without any incidents involving radioactive materials. It is estimated that by the year 2000, most (90%) of the electrical energy in the U.S. will be made by nuclear-powered steam generating plants (Chapter 16) or by hydroelectric plants using water power.

Solar energy There are three nonpolluting energy sources that depend finally on energy from the sun – sunlight itself, temperature differences in the ocean, and wind. All three add nothing to the environment that wasn't going to be added naturally anyway, and all three sources have been used in a practical way to generate power. Applications of the Claude process for generating power from ocean waters at higher and lower temperatures have been shown to work. The process uses huge turbines driven by the difference in the vapor pressures of water over warm surface water and cold water from the depths of the ocean. Solar energy drives the wind (Section 15.5), and windmills have been known for centuries. Windmills generating power at the rate of 100 kilowatts (kw) are operating in Vermont.

Georges Claude (1870–1960) was a French chemist and physicist who thought of the sea as an inexhaustible source of energy. His process has since been shown to work at an installation on the west coast of Africa.

EXAMPLE

The watt (W) is the unit of electrical *power*, the time rate of delivery or consumption of electrical energy. For an electric current of A amperes flowing under the influence of an emf of V volts, the power in watts is given by the equation watts = volts × amperes = VA. A light bulb passing a current of 0.625 amp at 120 V is using electrical power at the rate of $0.625 \times 120 = 75$ watts. Electrical *energy* is measured in watt hours, equal to (power in watts) × (time in hours). Operation of a 75-watt light bulb for 100 hr consumes $75 \times 100 = 7500$ watt hours, or 7.5 kilowatt hours (kWh). Bills for electricity are based on the number of kWh of electrical energy used.

A household using gas for cooking and heating consumes about 600 kWh of electrical energy per month. A 100-kW windmill generator operating at peak output would have to run only 6 hr to produce the 600 kWh of electrical energy for the house for a month. With storage batteries to store excess electrical energy generated during low load periods and to provide electrical energy during peak load periods or when there is no wind, a 100-kW windmill could supply lighting and appliance energy for 10 houses by operating at peak output for only 2 hr a day.

Another way to use the energy in sunlight to provide pollution-free energy is to convert the sunlight directly to energy, using solar cells. Solar cells utilizing silicon or other semiconductors generate an emf when light falls on them. They can turn the energy of sunlight ($h\nu$) directly into electrical energy with 65% efficiency, and have been used for years to provide the power for telephone message repeater stations in remote areas. They were spectacularly successful as the main source of electrical energy for the orbiting capsule of the Skylab missions in 1973–1974.

EXAMPLE

⊸ Let's calculate what area of solar cell surface operating at 65% efficiency would be required to provide the 20 kWh per day (600 kWh per month) of electricity required for lighting and appliances in a typical home. First you need to know the value of the *solar constant.* This is the energy falling on one cm^2 of a surface that is at right angles to the sun and outside the earth's atmosphere, when the earth is at its average distance from the sun. The value of the solar constant is 2.00 cal min^{-1} cm^{-2}, which translates to 1.40 kWh per m^2 per hour. This is the value outside the atmosphere. Because of absorption by the atmosphere (in the O_3 layer, particularly), only about 1.00 kWh of solar energy is received per hour by 1 m^2 at the surface of the earth. Assuming 6 hr of effective daylight, each square meter at the earth's surface receives 6 kWh of solar energy on an average day. Solar cells convert about 65% of the light to electrical energy, so 1 m^2 of cell surface would produce $0.65 \times 6 = 4$ kWh per day. With a daily demand of 20 kWh for lighting and powering household appliances, about $\frac{20}{4} = 5$ m^2 (6 sq yd) of cell surface would be needed. A solar panel 2 yd by 3 yd would be sufficient under optimum circumstances. ⊸

The use of solar cells to convert the energy in sunlight directly into electricity has been so successful that imaginative scientists have suggested the use of huge solar-cell panels, moving in synchronous orbits outside earth's atmosphere, to collect solar energy and transmit it by way of microwaves to receivers on earth. A synchronous orbit is one in which the orbiting object moves so that it remains always directly above the same point on earth. When you see "via satellite" on a TV screen, the microwave TV signal has been relayed from a distant station by a synchronously orbiting Telstar satellite. Photo 38 is an artist's conception of a solar energy "power plant."

EXAMPLE

⊸ Expressed in terms of potential by available electrical power, the solar constant in space just outside the earth's atmosphere is 1.40 kW per square meter (m^{-2}) of cell surface perpendicular to the light rays from the sun. Assuming only 25% efficiency for a solar-panel, microwave broadcasting-receiving system using a synchronous orbit, a 100,000-kW panel would have an area given by the following:

$$\frac{100{,}000 \text{ kW}}{0.25 \times 1.40 \text{ kW m}^{-2}} = 2.86 \times 10^5 \text{ m}^2 = 0.286 \text{ km}^2$$

Using this result, a solar panel 1 km square (0.62 mi on a side) would deliver power in the amount of

$$\frac{1}{0.286} \times 100{,}000 = 350{,}000 \text{ kW}$$

This amount of electrical power is about 3% of the power generated from fossil fuels, hydroelectric plants, and nuclear reactors by Pacific Gas and Electric Company in California, and enough to serve about 400,000 homes using natural gas rather than electricity for heating and cooking.

At the present time, costs for the highly purified silicon needed for solar energy cells are prohibitive, and the cost of putting a large panel in orbit is astronomical, so synchronously orbiting solar-cell arrangements are well in the future, except for their use in powering Telstar radio and TV signal repeaters.

At the time of this writing, a new process for growing pure silicon "ribbons" promised drastic reduction in cost of solar cells.

You might think that large solar-panel power plants in the deserts of the southwest U.S. would function at less cost than panels in orbit. Well they might, but only at noon *and* on the equator is the sun ever directly overhead in a position to shine straight down on a solar panel. In other regions of earth the sun's rays not only come in at a slant but also lose energy in coming through the atmosphere. A solar panel in space can readily be positioned so that the rays of sunlight it receives are perpendicular to it.

Another efficient way to use the energy in sunlight is for heating water. In an efficient rooftop solar-energy absorber, water can be heated to temperatures as high as 90°C (194°F). Solar hot-water heaters were popular in California before natural gas was introduced. They are still used extensively in Japan, where the government has allocated 20 billion dollars for solar-energy research during the next 20 years.

Solar energy is being used successfully to heat several houses in the U.S., some even in areas where the winters are cold. In this application the thermal energy in the hot water coming from the rooftop absorber is stored by using it to melt sodium sulfate decahydrate [$Na_2SO_4 \cdot 10H_2O$, mp 32.3°C (88°F)] stored in large, well-insulated tanks. When heat is needed for warming the house, cold water from the hot-water radiators in the house is pumped through coils in the tanks of melted sodium sulfate. Sodium sulfate decahydrate begins to crystallize from the melt when cooled to 32.3°C and, by giving up its heat of fusion, warms the water in the coils to 32.3°C. The warmed water is circulated back to the radiators to heat the house. The only continuing cost is for electricity to run the circulating pumps, and even that energy could be obtained from solar cells.

Summing up, the future for utilization of solar energy seems optimistic. It is an absolutely clean source—using it puts nothing into the environment except heat energy which would have been absorbed by the ground anyway. Using solar energy falling on the roof of a house to heat the house instead of the outside air has no polluting impact on the environment.

Geothermal energy The interior of the earth is a molten material called magma that at places has worked up through cracks to within a few miles below the crust of the earth. Tapping this source of thermal energy provides *geothermal energy* (Greek, *geos*: earth). At present, geothermal energy is being put to use only in places where water seeps down to a confining dome of solid rock heated by the magma. There the water is converted into steam that ordinarily works its way to the surface through geysers and vents (Figure 15.4). By drilling wells (some as deep as 9000 ft), the steam is brought to the surface to drive steam turboelectric generators. Geothermal generation of electricity has been carried out at Larderello in Italy since 1904 (400,000 kW currently), since 1958 in Wairekei, New Zealand (193,000 kW, 10% of the total power supply there in 1972), and since 1960 at "the Geysers," about 100 mi north of San Francisco (502,000 kW in 1974). At the Geysers, the steam comes out of the wells at a temperature of 177°C (350°F) and a pressure of about 100 psi (above atmospheric pressure). There are plans to develop the Geysers geothermal energy site to a production of 900,000 kW by 1977, at which time it will be producing about 10% of the electric power generated by the Pacific Gas and Electric Company in California. The Geysers site (Photo 40) will then be the largest geothermal power installation in the world.

Geothermal steam itself is free, of course, but it has some disadvantages. At the Geysers the steam contains traces of rock dust

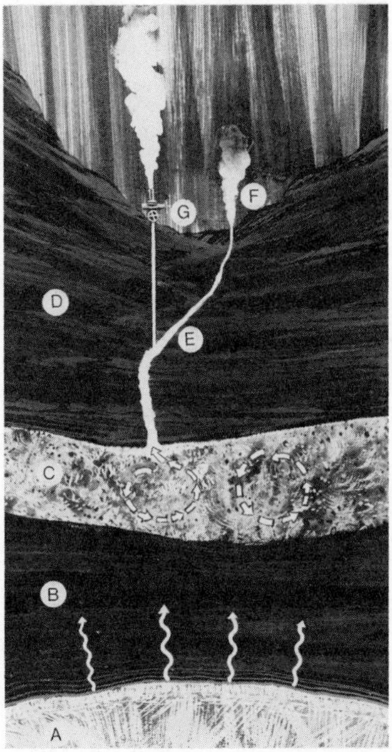

FIG. 15.4 Geothermal field: (a) magma—molten mass, still in the process of cooling; (b) solid rock—conducts heat upward; (c) porous rock—contains water that is boiled by heat from below; (d) solid rock—prevents steam from escaping; (e) fissure—allows steam to escape; (f) geyser, fumarole, hot spring; (g) well—taps steam in fissure. (*Pacific Gas and Electric Company.*)

PHOTO 40 Units 1 and 2 of Pacific Gas and Electric Company's Power Plant in Sonoma County, California. These units went into operation in 1960 and 1963 respectively. The units have a combined capacity of 24 million watts. (*Pacific Gas and Electric Company.*)

carried up from the wells, corrosive hydrogen sulfide gas (0.05%), and boric acid, H_3BO_3, so expensive abrasion- and corrosion-resistant piping and turbine blades have to be used. The operation at the Geysers presents no threat of thermal or chemical pollution of the local streams or area because the steam coming from the turbines is condensed to water, using water that has been cooled in an evaporation tower, and then pumped back into the ground. In an evaporation tower, water sprays down a framework through a blast of air coming from huge fans at the bottom of the tower. The water cooled by evaporation is then circulated through the turbine condensers. In fact, more water is produced in the condensers than is evaporated in the tower, so the process actually produces water. Poisonous impurities (H_2S, boric acid) in the water make it useless for drinking or irrigation and polluting if discharged into a stream, so the excess water is pumped back down through separate wells to the hot rock to be made back into steam.

There are many places on earth where hot rocks are accessible from the surface through wells, but not many where there is a natural source of water flowing down to the rocks to be made into steam. Dry-rock geothermal energy may one day be developed by pumping water down one set of wells to be made into steam that comes up another set.

15.5 Effects of the Unusual Properties of Water on the Biosphere

The biosphere contains an estimated 1.5×10^9 km^3 (3.6×10^8 cu mi) of water. ☛ More than 97% is salt water in the oceans, seas, and inland salt lakes. Less than 3% of all the water on earth is fresh, and approximately 77% of *all* fresh water is locked in polar ice caps and glaciers. The remaining fresh water (less than 1% of the total volume of water on earth) is in lakes and rivers, in the ground, and in the atmosphere in the form of $H_2O(g)$. Water has unusual properties that affect the atmosphere, hydrosphere, and lithosphere profoundly. This section connects the properties and structure of water with its effects on the biosphere and examines water pollution, sanitation, and the desalting of seawater.

This volume expressed in *milliliters* is 1.5×10^{24} ml, only about twice the number of particles in one mole.

Water as a solvent Water dissolves many substances, both inorganic and organic—it is the *carrier* for the molecules and ions needed for the reactions of life. The ability of water to dissolve so many different kinds of compounds can be traced to its molecular structure. The water molecule is "bent" or nonlinear, and because of this geometry and the difference in electronegativity between hydrogen and oxygen, water molecules have a permanent electric dipole moment (μ) directed as shown by the arrow:

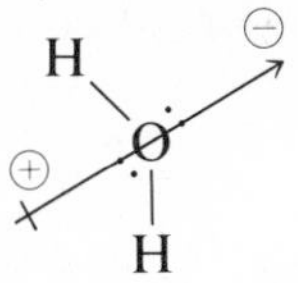

Angle H—O—H = 104.5°
Dipole moment (μ) = 1.85 debye units

The permanent dipole moment makes water what is called a "polar" liquid, a liquid capable of dissolving most ionic compounds to a considerable extent. To dissolve an ionic compound, the solvent must overcome the attraction between the oppositely charged ions in the solid. Water does this by forming clusters of water molecules around the ions, as shown for sodium and hydroxide ions in Figure 15.5. The (+) ends of the water dipoles point toward the negatively charged OH^- ions, the (−) ends of the dipoles toward the positively charged Na^+ ions. The clustering of water molecules around the ions is called "solvation by water" or *hydration*.

The layers of water molecules around ions "cover up" (spread out or dissipate) the charges on the ions, making it easier to separate the ions and bring the ionic compound into solution. Related to the dipole moment of water molecules is the high value (80 at 20°C) for the *dielectric constant* (*D*) of liquid water. The dielectric constant of a vacuum is 1.00, and the dielectric constant of any matter between charged ions appears in the *denominator* of the equation for Coulomb's

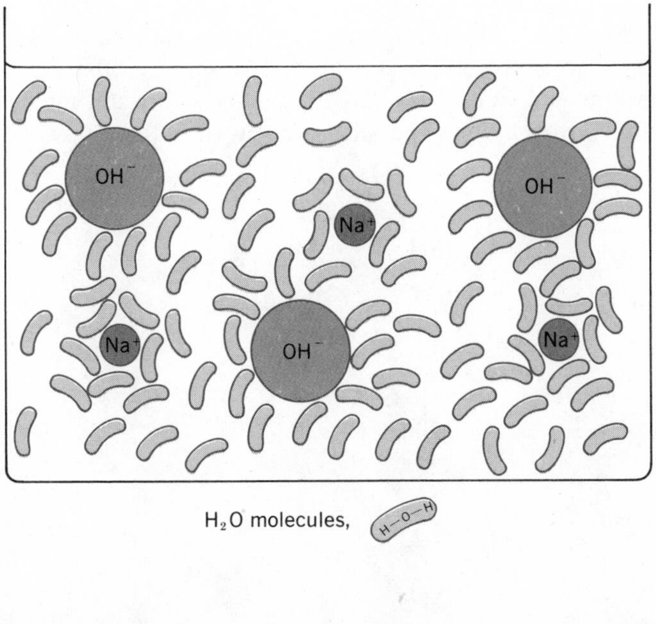

FIG. 15.5 Schematic representation of hydrated sodium [$Na^+(aq)$] and hydroxide [$OH^-(aq)$] ions in an aqueous solution of sodium hydroxide.

law (Introduction), which gives the attractive force between unlike charges of q_1 and q_2 on the ions:

$$f = \frac{q_1 q_2}{D r^2}$$

With a dielectric constant of 80, water greatly decreases the force of attraction between oppositely charged ions, further helping to separate ions and dissolve ionic compounds. The dielectric constants for hydrocarbons are about 4; consequently they do not dissolve ionic compounds.

Water also dissolves many compounds that are not ionic because of its ability to form hydrogen bonds (Section 4.6) to oxygen or nitrogen atoms in the compound. Acetone, ethanol, acetic acid, and pyridine dissolve in water in any proportions because of hydrogen bonding that "drags" their molecules into solution. Using dashed lines for hydrogen bonds, we can show the following examples:

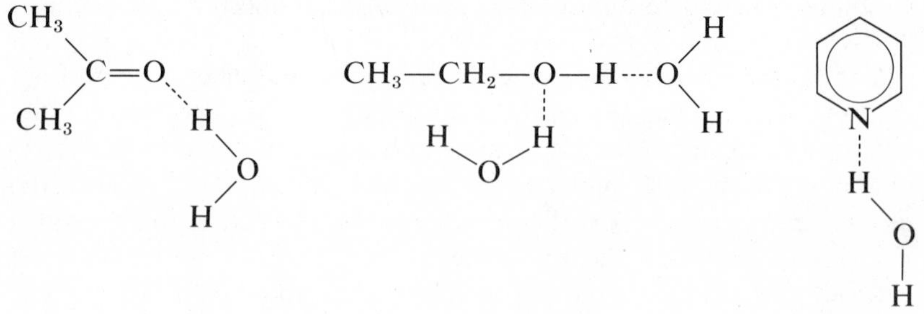

Acetone Ethanol Pyridine

Molecules of sugars such as glucose, sucrose, fructose, and ribose are also very soluble in water because, like ethanol, they have OH groups and can form hydrogen bonds with water molecules. Some of the hydrogen bonds possible are shown in the following structure:

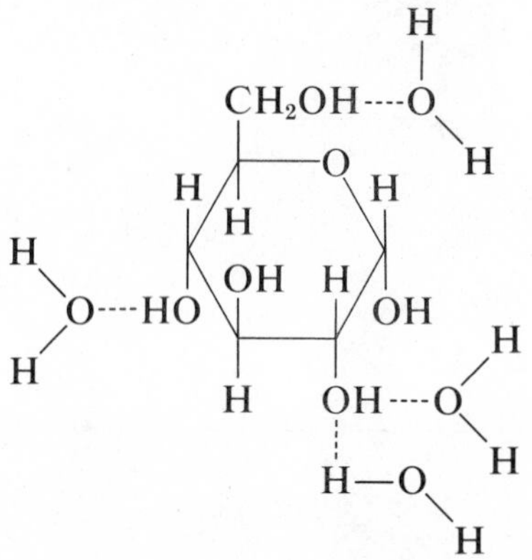

Protein molecules are kept in solution by hydrogen bonding of —NH— and $>C{=}O$ groups to water molecules. Among the biopolymers of glucose, however, cellulose does not dissolve in water, so you have to be somewhat careful about predicting solubilities where hydrogen bonding is a possibility. In cellulose, hydrogen bonding between adjacent molecules is stronger than hydrogen bonding between cellulose molecules and water molecules, so cellulose does not dissolve. Hydrocarbons are only very slightly soluble in water because C—H bonds do not form hydrogen bonds, and hydrocarbons are covalently rather than ionically bonded.

Exercises

10. What weight of water is needed to provide the hydrogen atoms needed for the photosynthesis of one mole of glucose?

11. The solubility of O_2 in water at 25°C and 1 atm pressure is 3.16 ml per 100 ml of water. Taking the partial pressure of O_2 (p_{O_2}) in the atmosphere at sea level as 159 torr, calculate the concentration in moles per liter of O_2 in water in equilibrium with the atmosphere at 25°C (Henry's law, Section 5.4). Assuming that the gills of a fish are 100% effective in removing this dissolved oxygen from water, what volume of water must pass through the gills of a fish to give it the oxygen needed for the catabolism of one mole of glucose?

12. How many cubic miles of water are there per person in the biosphere? (Use 3.50 billion for the population of the earth.)

13. Show all of the different kinds of hydrogen bonds possible for acetic acid in water solution.

14. Would you predict glycerol (1,2,3-propanetriol) to be soluble in water? What about the solubility of decanol? Of stearic acid?

15. Which would you predict to have the greater solubility, a DNA nucleotide or an RNA nucleotide? (*Hint*: consider hydrogen bonding.)

Hydrogen bonding in ice and water Hydrogen bonding between H_2O molecules gives water and ice some unusual properties that have powerful effects on the biosphere. Among these properties are relatively high bp and mp for a molecule of its size, high heat of vaporization, large heat capacity, and a temperature of maximum density. The properties and their effects are discussed in the numbered paragraphs that follow.

1. The boiling and freezing points of water are high compared to these temperatures for other molecules of about the same molecu-

lar weight (Figure 4.30). Water is liquid between 0°C and 100°C because of hydrogen bonding, so it can provide a liquid solvent for the reactions of life, most of which go on between 0°C and 50°C.

2. The unusually high heat of vaporization of water, also due to hydrogen bonding (Section 5.7), makes H_2O a good "heat-transfer" substance. Water evaporates from oceans, lakes, and rivers at ordinary temperatures, absorbing heat provided by the sun in the process, and the heat absorbed remains in the water vapor in the air until cooling causes precipitation in the form of rain, snow, or hail. When precipitation occurs, the heat absorbed during evaporation is released as the heat of condensation. The result is transfer of thermal energy from place to place around the globe. The transfers change temperatures and produce air-pressure differences that make winds. The winds are driven by the sun and ride "on the back of H_2O."

EXAMPLE An awesome example of water acting as a heat-transfer agent is the hurricane. A hurricane is a kind of "heat engine" and to understand how a hurricane works you need to know about *absolute* and *relative humidity*. The absolute humidity is the mass (weight) of water vapor present in a given volume of air. It can be given in grams per cubic meter or simply as the partial pressure of H_2O in the air sample. Relative humidity is the ratio of the quantity of water vapor in the air to the amount required to saturate the air with water vapor at the temperature of the sample. Letting p_{H_2O} stand for the partial pressure of water vapor in the air and P_T^0 stand for the vapor pressure (Figure 5.4) of water at the temperature (T) of the air, we have

$$\text{Relative humidity} = \frac{p_{H_2O}}{P_T^0} \times 100$$

When it's raining hard, the relative humidity is 100%, because with all the liquid water present, the air is saturated with water vapor. Relative humidity depends on the temperature and is more important than absolute humidity in the part played by H_2O in weather.

A hurricane builds up over the warm ocean close to the equator, with the formation of a large body of air with high (~90%) relative humidity and full of potential energy from the heat of vaporization the air mass contains. The body of warm air drifts northward and begins to cool. Cooling causes the relative humidity to rise and when it reaches 100% rain begins to fall. As the water vapor condenses to rain, the heat of condensation released raises the temperature of the air mass. The warmed air rises, and in doing so it expands because atmo-

Because P_T^0 falls more rapidly with decreasing temperature than p_{H_2O} does.

spheric pressure drops with increasing altitude. The expansion cools the air, increasing the relative humidity even further and causing more rain to fall. Soon the whole body of the humid air mass is rushing toward the central "hole in the sky" through which the air, warmed by condensation of water vapor to rain, is rising. The winds blowing toward the hole take on a circular pattern (counterclockwise in the northern hemisphere, clockwise in the southern) and reach speeds as high as 200 mph. The high winds and torrential rains are the "trademark" of a hurricane. As a heat engine the hurricane generates winds, using thermal energy stored in water vapor evaporated at a higher temperature and carried to a lower temperature where the energy is released by condensation of vapor to liquid. A hurricane also produces a large mass-transfer of water originally in the tropical ocean where the storm was born.

The "hole in the sky" actually appears at the center of the hurricane and is called the "eye" of the hurricane. In the eye there is little wind, and the sun may even be shining. But if you've ever experienced a hurricane you know that the eye will pass and the terrible wind and rain that are at the storm's edges will return. Hurricanes reaching the southeastern coast of the U.S. originate in the Gulf of Mexico or the Caribbean and usually follow a northeasterly course across the U.S. Once the center of a hurricane "comes ashore" and the storm is inland, it begins to lose its energy because there is no more warm, water-saturated tropical air coming in to fuel it. However, high winds and heavy rainfall accompany the storm until it goes out to sea in the Atlantic.

Hurricanes in the Pacific are called typhoons. Tornadoes are hurricanes in miniature form. They result when a mass of cold air at high altitudes flows over warm, humid air over land, mainly in the midwestern states of the U.S. The eye of a tornado is the "twister" itself (Photo 41). ⊸

3. Because of hydrogen bonding, water has a large capacity for absorbing thermal energy without rapid temperature rise. Much of the energy absorbed goes into breaking hydrogen bonds, rather than raising temperature. Water has a high *heat capacity* or *specific heat*, defined as the number of calories of heat energy required to raise the temperature of 1 g of a substance by 1°C. The specific heat of water at 15°C is 1.0 cal g^{-1} deg^{-1} and stays close to this value over the whole liquid range 0–100°C. For comparison the specific heats of ethanol and mercury at 25°C are 0.58 and 0.03 cal g^{-1} deg^{-1}, respectively.

   The effect of the large specific heat of water on the environment is to soften or "temper" weather changes. A body of water cools the surrounding land during the hot summer because the water

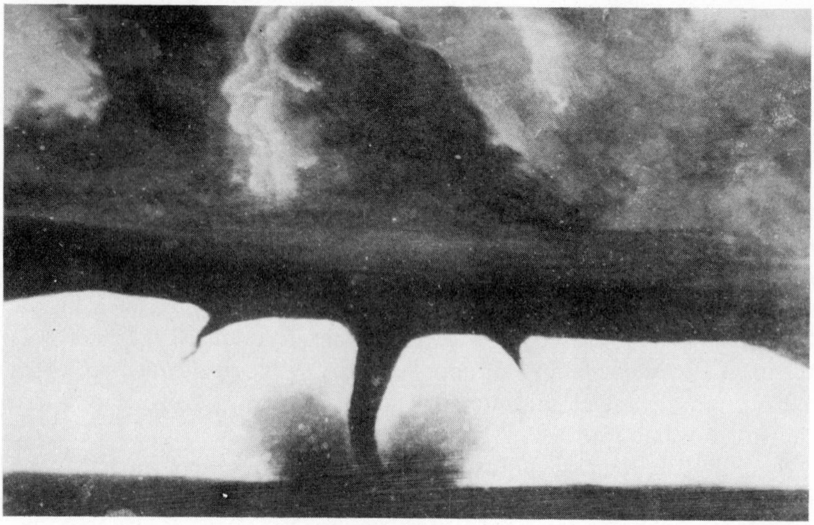

PHOTO 41 The tornado—a natural "heat engine." This is the first known photo of a tornado (it was taken by F. N. Robinson in the late 1880s). (*National Oceanic and Atmospheric Administration.*)

doesn't warm up as fast as the land. In the winter a lake helps to keep the surrounding land warm because it doesn't cool off as quickly as the land. Whole regions of the earth are made more livable by the control of temperature extremes by bodies of water or warm ocean currents such as the Gulf Stream in the Atlantic.

4. Because of hydrogen bonding, there is a temperature (3.98°C) at which water has its maximum density (1.00000 g/ml). Hydrogen bonding is also responsible for the fact that ice floats. Both of these properties of water control the part of the biosphere that is in fresh-water lakes and streams in the cooler zones of earth.

As the water in a lake begins to cool when winter approaches, the water at the surface (where the cooling takes place) finally drops to the temperature of maximum density, 3.98°C (39.16°F). Because the surface layer is now more dense than the warmer underlying water, the surface layer sinks to the bottom. This sinking of the surface to the depths or "autumn turnover" continues as the lake cools down until the water in the whole lake reaches 3.98°C. Now, with further cooling, the surface layer becomes less and less dense, so it remains where it is, and when 0°C (32°F) is reached ice begins to form. Ice is less dense than water at 0°C ($d^{0°C}_{water} = 0.9998$ g/ml, $d^{0°C}_{ice} = 0.9170$ g/ml), so when ice forms it *floats*, and the *lake freezes from the top downward.* For most substances, the solid form is *more dense* than the liquid at the freezing point. If H_2O behaved like most substances, ice would sink as

Like water, antimony (Sb) and bismuth (Bi) are exceptions that have solid forms that are less dense than their liquids at their freezing points. This property is used to good advantage when typesetters put some antimony with the lead in type metal. When the type is cast, the expansion of the type metal as it solidifies ensures a sharp image.

it formed and lakes would *freeze from the bottom up*. You can see that this direction of freezing would result in complete freezing of the lake, killing the fish and plant life in it. It would also mean that, unless the summer was very, very hot, the ice at the bottom of the lake would never melt completely, so there would be no soil available for aquatic plants to grow in, and the lake water would remain icy cold.

When spring comes, the ice at the surface melts and the water starts warming from 0°C toward 3.98°C, the temperature of maximum density—and again the more dense surface layers begin to sink and displace the colder but *less* dense bottom water toward the top. This "spring turnover" goes on until the whole lake reaches 3.98°C, when further warming makes the surface layer less dense than the depths, and exchange of surface and deep water stops. The two turnovers in fresh-water lakes provide a vertical mixing of surface water rich in dissolved oxygen with corresponding benefit to the fish life in the lake. ☛

Experienced fresh-water fishermen know that the fishing is likely to be good at the time of the turnover.

Salt water, because of the effect of dissolved electrolytes (such as NaCl and $MgCl_2$), does not have a temperature of maximum density like fresh water. This means that there is no "turnover" of the water in the oceans. Instead, the colder water in the polar regions sinks because of its greater density, carrying oxygen and nutrients into the depths and forcing warmer surface water in the tropical regions toward the poles. The result is a very different circulation from that in lakes.

When ice is frozen from seawater, the salt does not solidify with the ice, so the polar ice caps are fresh-water ice. They contain the largest quantity of fresh water on earth.

Water quality The quality of water in the biosphere depends on what is in it other than H_2O molecules and dissolved atmospheric gases. ☛ You have seen in earlier chapters the effects of several of the substances put into the biosphere by human activities. Among those that harm water quality are excesses of phosphates and fixed-nitrogen compounds, detergents from domestic wastes, mercury compounds, and other industrial wastes. Insecticides of the DDT or chlorinated hydrocarbon type are particularly dangerous to the biosphere, because they remain and accumulate for decades.

Freshly distilled water tastes "flat" because it contains little dissolved oxygen.

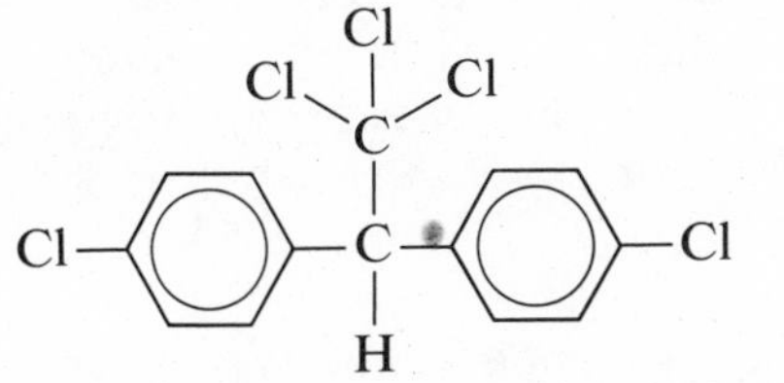

4,4′-Dichlorodiphenyl-2,2,2-trichloroethane (DDT)

Furthermore, they get into the food chains that connect lower organisms (phytoplankton, algae, sea plants), intermediate organisms that feed on them (oysters, fish), and higher organisms (birds, animals, humans. Insecticides of the DDT (chlorinated hydrocarbon) type are not water soluble to any great extent but are soluble in oils and fat. Because of this property they are not rapidly excreted by humans and other animals and so accumulate in body fat. Each link in a food chain *concentrates* DDT because the higher organism is larger, eats more of the lower organism, and can't effectively excrete the traces of DDT present in its food. We all carry in our bodies measurable amounts of DDT that has accumulated this way. DDT is not toxic to humans in the short term as far as is known. But it is toxic to birds, and their eggs, and has been blamed for the decreasing population of certain kinds of fish-eating birds (eagles, ospreys, falcons). The dangers of using DDT have been recognized, and it is being replaced with other insecticides that decompose more quickly to harmless compounds and so do not accumulate in the biosphere. The naphthalene derivative Sevin is one:

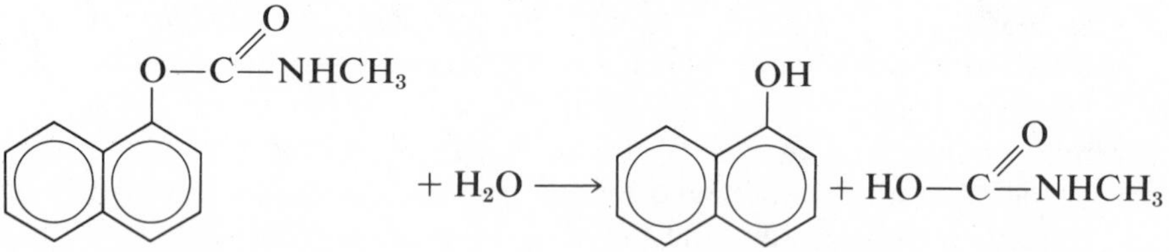

$$\text{Sevin} + H_2O \longrightarrow \alpha\text{-naphthol} + HO-\overset{O}{\underset{}{C}}-NHCH_3$$

Sevin is more expensive to use than DDT because naphthalene is scarce compared to benzene, and Sevin must also be applied more frequently because it hydrolyzes rather rapidly in the way shown and loses its potency. The products of the hydrolysis are water soluble and not toxic in small amounts.

Insect pheremones and hormones Insect control without harming water quality will soon be possible through the use of insect hormones and *pheremones*. Pheremones are volatile compounds secreted by insects and used as a means of communication between them, through the atmosphere. Among pheremones are sex-attractants excreted by the female to stimulate the males. Most sex-attractants are simple compounds that can be made in the laboratory. Baiting a poisoned trap with a sex-attractant is a sure and harmless way to cut down a particular insect population by killing off the males before they mate. As far as is known, pheremones are species-selective, and harmless to other organisms.

EXAMPLE ⊷ The larvae (caterpillars) of the gypsy moth (Photo 42) are among the worst plant predators. There have been times when the attacking line of gypsy moth larvae could be detected by aerial photograph because of their devastating effect on vegetation. In 1971 alone, 582,433 hectares (2250 sq mi) were devastated (defoliated) by gypsy moth larvae in the eastern U.S. The female moth desiring to mate excretes a slightly volatile sex-attractant (**1**) having the structure (+)-10-acetoxy-*cis*-7-hexadecen-1-ol. The attractant is carried through the air to the males and is almost unbelievably effective in attracting males and exciting them to mate.

(Photo 14, Chapter 4)

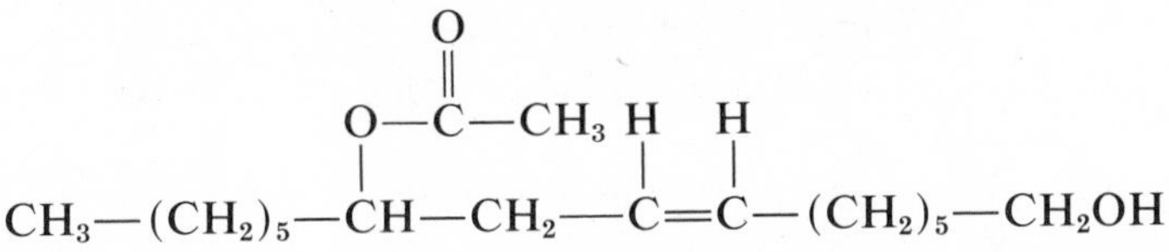

1

Female gypsy moth sex-attractant

It has been found that 0.01 microgram (0.00000001 g), the amount of the attractant in one female moth, if distributed properly, would excite more than a billion male moths.

A synthetic compound called Disparlure, *cis*-7,8-epoxy-2-methyloctadecane (**2**), is easier and cheaper to synthesize than the gypsy moth sex-attractant and has approximately the same activity as the natural pheremone.

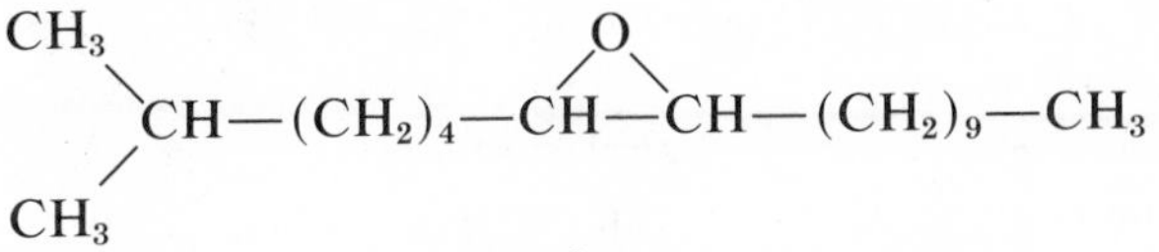

2

Disparlure

Distribution of Disparlure at the rate of 15 g per hectare (2.47 acres) is successful in disrupting mating between gypsy moths to an extent that prevents any increase or spread in the gypsy moth population.

PHOTO 42 Female (left) and male gypsy moths, which are responsible for widespread destruction of trees (see Photo 14). The male is very sensitive to the sex pheremone of the female and can thus detect her at great distances. Note the male's highly developed antennae—the organs of smell—and the female's very simple antennae. (*U. S. Department of Agriculture.*)

Exercises

16. What are the differences in structure between the natural gypsy moth sex-attractant and Disparlure?

17. Calculate the number of molecules of the gypsy moth sex-attractant needed to excite one male to mate.

Pheremones also serve as a means of communication between insects that are either harmless to vegetation or even necessary. Honey bees are needed to perform the essential step of pollination in the production of many useful plants. When the queen bee takes off on her nuptial flight, she excretes a volatile sex-attractant (**3**) that is carried on the wind to let the males know of her intentions and location.

$$CH_3-\overset{\overset{\displaystyle O}{\|}}{C}-CH_2CH_2CH_2CH_2CH_2-\overset{\overset{\displaystyle H}{|}}{C}=\overset{\overset{\displaystyle H}{|}}{C}-\overset{\overset{\displaystyle O}{\|}}{C}-OH$$

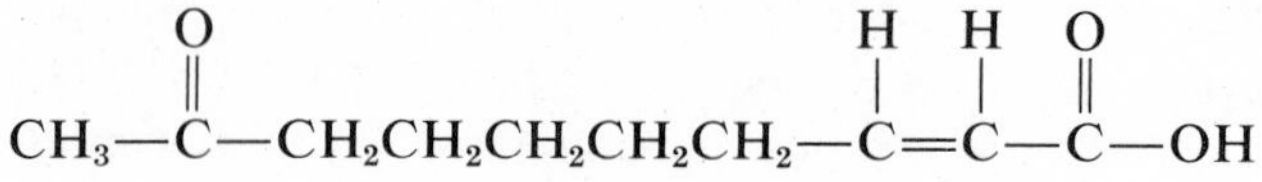

3

Honey bee queen substance

The queen substance also regulates the reproductive cycle of the colony of bees. When worker bees absorb the substance provided by the queen, their ovaries go into a nonproductive state so that production of "royal cells" from which new queen bees develop stops. In this case the queen substance plays the part that the contraceptive "pill" plays for women.

Sex-attractant pheromones have been found for other species, including ants, silkworm moths, cockroaches, fruit flies, and even animals as large as musk deer. All these pheromones are relatively low-molecular-weight, fairly simple compounds that are volatile enough to communicate chemically through the air. Such pheromones are highly species-selective and thus can be used to attract and destroy the males of one kind of insect without harming others. 🖎

The structure of Grandlure, a mating pheromone of the boll weevil (secreted by the male) has been determined and synthesized. Its use will control the devastation of cotton by the weevil.

Pheromones also have the advantage that the insect they affect does not develop an immunity to the agent. With DDT and other chlorinated hydrocarbon insecticides, insects become resistant after a few generations, even to the point at which the chlorinated hydrocarbon insecticide becomes totally inactive.

The development of a silkworm moth from the fertilized egg to the complete adult is controlled by two insect *hormones*, juvenile hormone and ecdysone. Hormones are regulatory substances secreted by glands and carried through the circulatory system of the organism.

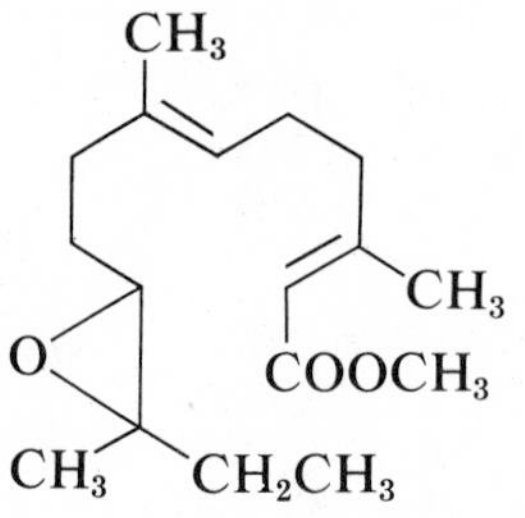

Juvenile hormone

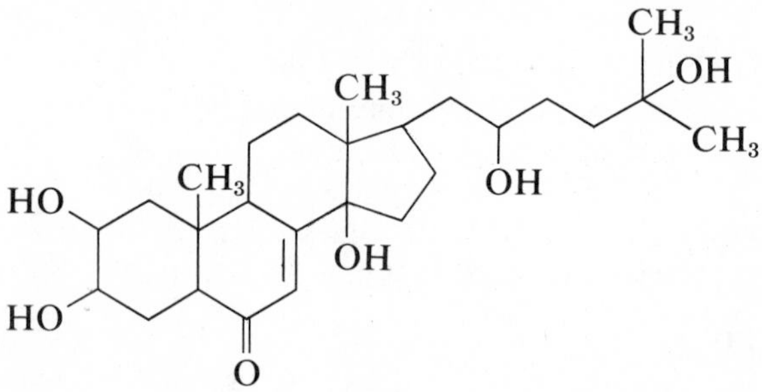

Ecdysone

Ecdysone initiates the molts through which the moth larva passes, and juvenile hormone must be present if the larva is to develop completely. On the other hand, juvenile hormone must be absent (or at very low concentration) when the larva starts to develop into a mature, fertile adult. Furthermore, juvenile hormone must be absent from the insect eggs if they are to hatch normally. Spraying eggs or larva with juvenile hormone leads to the death of the insect from "natural causes." Juvenile hormone is so potent that a gram of the substance can prevent the development of about a billion adult insects. Juvenile hormone and ecdysone are broad-spectrum insecticides, so spraying crops with them could wipe out helpful as well as harmful insects. Nonetheless, they represent a way of attacking insect predators alone—as far as is known, they are harmless to other organisms.

Insect pheremones and juvenile hormones belong to what are called "third-generation insecticides." The first-generation insecticides were poisons such as lead arsenate [$Pb_3(AsO_4)_2$] and nicotine (from tobacco), which are toxic not only to insects but to fish, birds, and mammals, including humans. The chlorinated hydrocarbons, toxic to almost all insects (and some animals), were the second-generation insecticides.

There are three other ways of killing harmful insects without polluting the environment. In one, the males of the species are trapped in moderate numbers (using pheremones) and irradiated with X rays to make them sterile. They are then released to satisfy the mating urge of the females, producing eggs that will not hatch. An insect population is in a delicately balanced steady state—any significant reduction in their numbers means that natural events that destroy them will gain the upper hand and destroy the whole population. This absolutely species-selective and nonpolluting method was successfully used to destroy the screw worm population threatening the raising of beef cattle in Florida a few years ago.

Another way to kill a predatory insect is to feed the plant that he chews on a substance lethal to the insect but harmless to plants and the animals who eat them. The insecticidal substance is only in the juice of the plant and will kill only those insects that prey on the plant.

The third way to kill a predatory insect is to introduce another predatory insect that feeds on it. For example, ladybugs and praying mantises are used to control aphids and other harmful insects.

Pollution by sewage Sewage is the great spoiler of water quality around the world. Even though sewage-treatment plants to improve water quality are required by law in the U.S. and other developed countries, their discharges are not water of high quality except in a very few cases. Also, many cities do not have separate storm sewers to handle heavy rainfall, and during rainstorms the extra volume of water added to the regular sewage overcomes the capacity of the treatment plant. When this happens, untreated raw sewage is discharged into the nearest body of water with immediate heavy pollution. San Francisco, when rains are heavy, dumps hundreds of thousands of gallons per hour of untreated sewage into the bay waters, endangering many species of fish and polluting the water for swimming and water sports.

In addition to phosphates, nitrogen-fixed compounds, detergents, and industrial wastes, sewage contains fats, proteins, carbohydrates, and nitrogen compounds found in agricultural wastes and in human feces and urine. These compounds serve as food for *aerobic* (air- or oxygen-needing) bacteria in water, and the seriousness of pollution by human wastes can be measured by using what is termed the "biological oxygen demand" (BOD) of the water containing the pollutants. The BOD of a water sample is the number of milligrams of O_2 consumed during 5 days at 20°C by the action of the aerobic bacteria on wastes present in a 1-liter sample of the water. The BOD of domestic sewage is about 200 mg per liter, but values may run as high as 12,000 mg (12 g) per liter for cellulose wastes from canneries and paper mills. Water with a BOD in excess of 5 mg per liter is considered polluted. (The concentration of O_2 dissolved in water in equilibrium with air at 25°C is only 8 mg per liter.)

In water with a BOD of 5 mg per liter or more, the aerobic bacteria living on the organic wastes in sewage use up the dissolved oxygen to the point at which fish die. In dying they add their fats, carbohydrates, and proteins to the burden of organic waste, further increasing the BOD of the water. Finally, dissolved oxygen gives out completely and the anaerobic bacteria that don't need oxygen to live take over. Anaerobic bacteria catabolize glucose in a different, more reductive, way than aerobic bacteria, and produce methane (CH_4) instead of water:

$$C_6H_{12}O_6 \xrightarrow[\text{bacteria}]{\text{anaerobic}} 3CO_2 + 3CH_4 + \text{energy}$$

$$C_6H_{12}O_6 + 6O_2 \xrightarrow[\text{bacteria}]{\text{aerobic}} 6CO_2 + 6H_2O + \text{energy}$$

Methane is called "marsh gas" because it bubbles up in swamps and marshes where the BOD of the bottom water is naturally high from the mass of dead vegetable matter there and the anaerobic bacteria have taken over. Anaerobic bacteria also convert sulfate ion to hydrogen sulfide (H_2S). The foul, rotten-egg smell released by digging in a bog or swamp is due to H_2S from this source. Experimental aeration of bottom water in small, polluted, eutrophic lakes, using air or pure oxygen, has raised the O_2 concentration to 4 mg/liter. The effect is to return the lake to a "healthy" condition quite rapidly by supplying the oxygen needed to meet the large BOD. Aerobic bacterial decomposition of wastes is encouraged, and the lake clears itself.

Proper treatment of sewage should have as its objective the production of water of as high (or higher) quality than the domestic water supply of the city producing the sewage. In other words, you should be able to drink with complete safety and pleasure the water discharged by the treatment plant into the biosphere. This kind of treatment has already been achieved in the small California towns of Santee, South Lake Tahoe, and Los Gatos, among other forward-looking communities. In fact, Santee and Los Gatos use the water discharged from their treatment plants to maintain lakes for boating and water sports.

Organic pollutants that bacteria can live on are naturally removed from water by aerobic bacteria in a rushing, plunging stream, where dissolved oxygen is constantly renewed. It is said ☛ that a stream of this sort purifies itself in about 7 mi of its length, but, of course, the distance needed depends on how heavy the pollution is. One method of treating sewage uses this natural method—basically the sewage is agitated or stirred with compressed air or oxygen (aeration) to encourage rapid digestion of the organic solids by the aerobic bacteria already present. The digested sewage then stands in large, open basins to complete bacterial action and allow the solids to settle. Ion exchange or other chemical methods are next applied to remove phosphates, nitrates, and other impurities. ☛ Finally, the water is chlorinated or treated with ozone to kill bacteria and is allowed to stand in large ponds exposed to air and sunshine to complete purification. Sewage that ends up this way is said to have had *tertiary treatment*. Primary treatment involves simply letting the solids settle out of the sewage before discharging the water into the biosphere. In secondary treatment, bacterial action is used to decompose the organic matter as just described, using a natural bacterial process under forcing conditions.

(Photo 33, Chapter 12)

The "backwash" water from regenerating the ion exchangers (Section 11.3) is in demand as fertilizer because of its phosphate and fixed-nitrogen content.

An alternative secondary treatment uses anaerobic bacterial action, rather than aerobic, to digest the organic matter. Here the products of bacterial activity are methane and H_2S. After separating the H_2S, the methane is used to provide heat and electricity for the sewage-treatment plant or put into the city natural gas supply.

Tertiary treatment is a chemical one that must often be designed to take out specific as well as general types of pollutants, depending on the type of industry in the area. Most of the treatment plants in the U.S. carry the sewage only through the secondary stage. Tertiary treatment of sewage is becoming more and more necessary with increasing industrial activity, and in many areas tertiary treatment is now required by law.

Exercises

18. A relative humidity of about 55% is comfortable. Calculate the partial pressure of H_2O (p_{H_2O}) in the air with a relative humidity of 55% at 21°C, where the vapor pressure of water is 18.7 torr. What is the absolute humidity in g/liter?

19. Humans and other mammals maintain their own characteristic body temperature no matter what the temperature of the surroundings. Fish take on a body temperature equal to the temperature of their surroundings. Taking the heat capacity of water and the solubility of oxygen in water (see Exercise 11 for data) into consideration, explain why fish *have to* take on the temperature of the surrounding water.

20. Given the following data:

	ΔG_f^0, kcal/mole
$C_6H_{12}O_6(s)$, glucose	−220.0
$CH_4(g)$, methane	−12.1
$CO_2(g)$	−94.3

Calculate ΔG^0 for the anaerobic bacterial catabolism of glucose according to the following equation:

$$C_6H_{12}O_6(s) \rightarrow 3CO_2(g) + 3CH_4(g)$$

How does this value of ΔG^0 compare to the value for the aerobic catabolism of glucose to CO_2 and water? (*Hint*: see Standard Free Energy of Formation of Compounds in Section 10.5.)

Desalting sea water To a farmer of arid land next to the ocean, salt (NaCl) is a "pollutant" that prevents him from using the unlimited quantities of sea water to irrigate his crops. There are huge land areas that could be used to raise food for the growing population of earth if only fresh water for irrigation were available. Several methods for making fresh water out of sea water have been developed in recent years, but all involve technical problems or require energy input that makes the fresh water produced too expensive for agricultural purposes with food prices at their present levels.

Distillation of sea water—evaporation followed by condensation—gives water free from salts (such as NaCl and $MgCl_2$) because salts are ionic compounds that have no vapor pressure at the temperature of boiling water. They are therefore left behind in a distillation (Sec-

Potatoes are being grown successfully and economically in desert sand in eastern Washington simply by putting the necessary plant nutrients in the water used for irrigation. With fresh water available, even the Sahara Desert could be used for this kind of agriculture, called *hydroponics*.

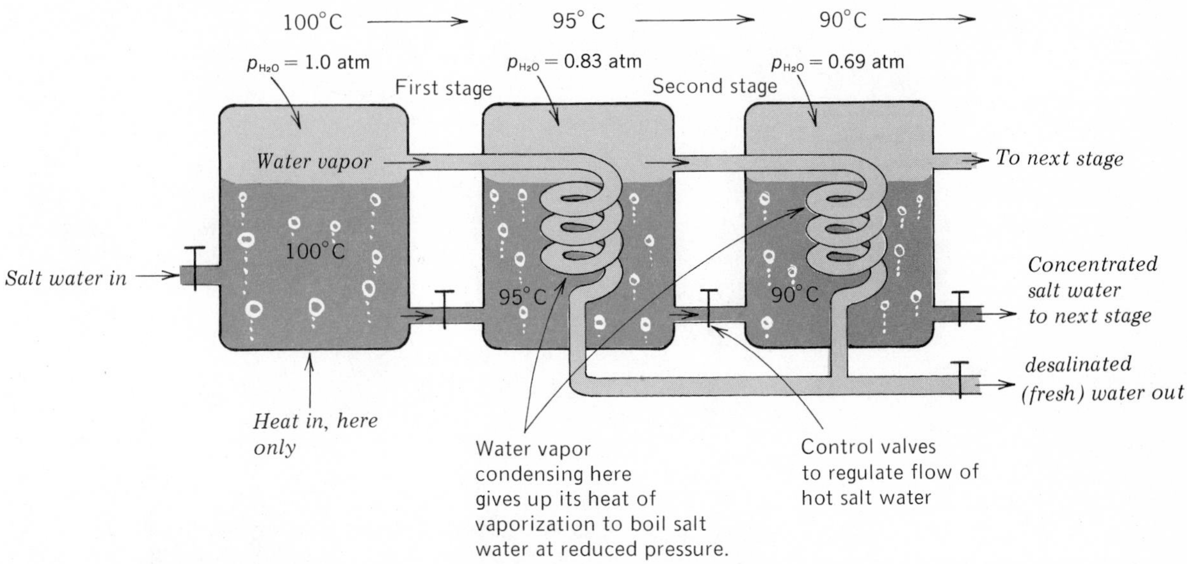

FIG. 15.6 Multiple-stage (effect) distillation to desalinate water. The last stage terminates in a vacuum pump to maintain reduced pressure as indicated by values of p_{H_2O}.

tion 2.2). Simple distillation of sea water on ships produces fresh water for drinking and to supply boilers, but the process is too wasteful of energy for making the volumes required for irrigation. The trouble with simple or "one-stage" distillation (Figure 2.2) is that the heat energy put in to evaporate the water in the distilling flask (or boiler) is thrown away through warming the cooling water in the condenser. Distillation of sea water is made much more economical in terms of energy use when sea water is used as cooling agent in the condenser, and the sea water warmed in this way is then distilled at a lower temperature and pressure in a second apparatus ☛ (Figure 15.6). When this process is repeated again and again, the process is called "multiple-effect distillation." By multiple effects, most of the heat energy put into the original batch of sea water can be used over and over again by distillation at successively lower temperatures and pressures, so that the original energy is not completely thrown away as it is in simple distillation. In spite of its greater efficiency in using energy, multiple-effect distillation still cannot economically produce fresh water from salt water for extensive agricultural use. ☛

A liquid will boil if its vapor pressure is greater than the applied pressure. The *normal* bp of a liquid is the temperature at which the vapor pressure is 1 atm.

(Photo 10, Chapter 6)

The thermal energy in sunshine can be used to drive a single-effect "solar still" (Figure 15.7). Although the energy in this application is without cost, ☛ the process is so slow that huge areas of sea water have to be warmed by the sun to get much distilled water.

Such a "still" operates without harm to the environment, in contrast to the thermal energy and CO_2 additions that result from burning fossil fuels. Thermal energy in sunshine is absorbed by the ground or plants anyway.

Reverse osmosis (Section 6.3) is used to "filter" the salts out of sea water, and experimental installations capable of producing thousands of gallons of fresh water per hour by this method have been operated. Pumping costs, and plugging of the semipermeable mem-

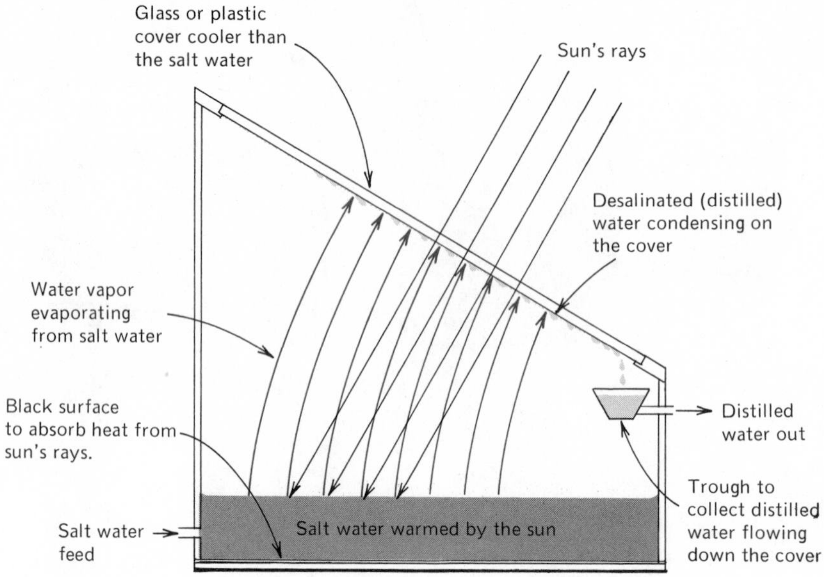

FIG. 15.7 A simple solar still.

brane by suspended solids in the sea water so that even H_2O under excess pressure can't get through, make desalting by this method too expensive for making irrigation water. A modification of reverse osmosis, called electroosmosis or *electrodialysis*, is used to remove the low concentrations of ions (PO_4^{3-}, NO_3^-) present in water coming from secondary sewage treatment. Here membranes constructed so that they are permeable to anions on the one hand and cations on the other are arranged with electrodes and an applied emf (voltage) to take the ions out of the water (Figure 15.8). But this method also becomes too expensive in terms of electrical energy when applied to removing the high concentrations of ions in sea water.

Ion exchange used in the "demineralized water" way (Section 11.3) will make fresh water out of sea water, but the cost of the acid and base for regenerating the ion exchange resins is too high for making irrigation water. Survival kits for those downed at sea from aircraft often contain "one-use" ion exchange kits that make fresh water out of sea water for drinking.

As you can see, the problems in turning sea water into agricultural irrigation water are largely matters of economics rather than chemistry. If food becomes precious enough, sea water will be desalted by some means or another to open up more land for raising crops.

15.6 Summary Guide The biosphere, where life goes on, is a layer about two miles thick around the earth consisting of the hydrosphere (water), parts of the atmosphere (air), and the lithosphere (crust).

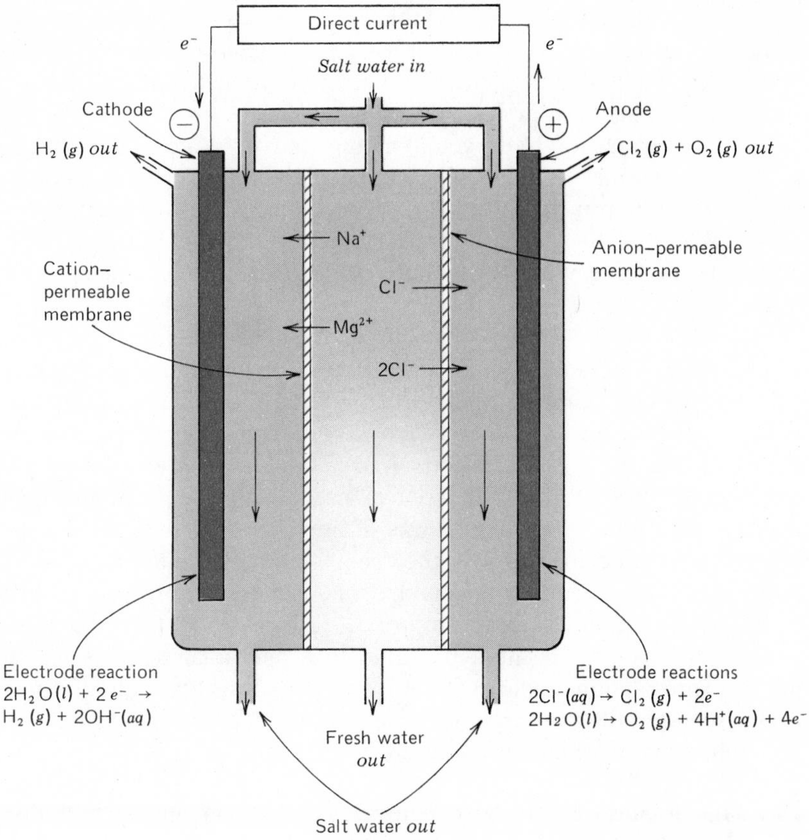

FIG. 15.8 Principles of electrodialysis for desalting sea water, using ion-selective membranes.

The biosphere has a limited capacity to handle the chemicals dumped into it by an industrial society. The effects of industry on the biosphere can be evaluated, using the cycles key elements go through.

Before the emergence of technological civilization, carbon was balanced between photosynthesis and animal metabolism. By releasing, in combustion, energy stored millions of years ago in fossil fuels by photosynthesis, human activities are adding to the biosphere excessive amounts of carbon dioxide, hydrocarbons, and nitrogen and sulfur oxides. These additions are unbalancing to the biosphere – smog in cities is one result of the unbalance, water pollution from the dumping of phosphates and fixed-nitrogen fertilizers is another. The "greenhouse" effect of excess CO_2 in the atmosphere is a possible third.

Energy stored by photosynthesis in the form of starch in cereal grains is largely degraded to heat in the food chain plant → animal meat → human.

Nuclear reactors, fuel cells, and geothermal and solar energy are nonpolluting energy sources that can relieve our dependence on fossil fuels for energy, thus sparing their dwindling reserves for making the petrochemicals necessary for the high standards of living in modern society. The limited natural reserves of metals and chemicals in the lithosphere must also be spared for use by future generations by using recycling and sensible priority programs.

Water, because of hydrogen bonding and its effects, has unique properties that make the biosphere congenial to life. Water in the form of water vapor transfers sunlight energy from place to place around earth. Water is the solvent for the reactions of life because of its ability to dissolve ionic compounds and covalent sugar and protein molecules.

Water is also the most important carrier of pollutants put into the biosphere by human activities. Insecticides of the chlorinated hydrocarbon type have already been spread around the globe by water, and their residues, which are in all of us, have shown toxic effects in some forms of life other than insects. Insect pheremones and hormones that are species-specific will control predatory insects in the near future, without harming other forms of life.

Sewage carried by water is the most important pollutant of fresh and salt water. Adequate sewage-treatment methods promise an end to this kind of pollution in the future.

Seawater can be desalted by multistage and solar distillation, reverse osmosis, ion exchange, and electrodialysis, but all of these processes are still too expensive in energy or in chemicals to provide large amounts of water for agriculture.

Exercises

21. One important source of hydrocarbons producing air pollution is the hydrocarbon vapor that is displaced from vehicle gasoline tanks every time they are filled. Taking octane (C_8H_{18}) to represent gasoline, calculate the weight in grams of octane vapor displaced into the atmosphere when 10 gal (37.9 liters) of gasoline are put into the tank of a car at 19°C, where $P^0_{\text{octane}} = 10$ torr. How many metric tons (1 metric ton = 1000 kg) of hydrocarbons are introduced when the tanks of 2 million cars are similarly filled? Can you suggest a way to stop this pollution?

22. Power engineers distinguish between "wet" and "dry" or superheated steam. In wet steam at a given temperature, the steam pressure (p_{H_2O}) is equal to the vapor pressure of water ($P^0_{H_2O}$) at that temperature. For dry steam, the p_{H_2O} is less than the $P^0_{H_2O}$ of water at the temperature of the steam. Dry steam is preferred for running turboelectric generators. The geothermal steam at the Geysers in California is at a pressure of 115 psi (absolute) and 177°C, where $P^0_{H_2O} = 7016$ torr. Is the Geysers' steam of the preferred (dry) type? (1 atm = 14.7 psi.)

23. A wind-driven electric generator with a 125-ft-diameter blade is being built near Sandusky, Ohio. The power output is expected to be 100 kW of electricity with winds of 18 mph or more. How many joules per second of energy will the machine produce? (One joule = 1 volt coulomb = 1 volt amp sec^{-1}.)

24. What metabolic intermediate is almost certainly used by the cells of the female gypsy moth to synthesize her sex-attractant? (Chapter 14.)

25. Which of the naturally occurring compounds, β-carotene or cholesterol, would be more suitable as a starting point for making ecdysone?

26. There are 2.80 g of combined phosphorus in the amount of a dishwashing powder recommended for one load of dishes in the washer. If the phosphorus is present as sodium trimetaphosphate, $(NaPO_3)_3$, what amount of the phosphate is present in the recommended amount of washing powder? Assuming that the dishwasher is used once a day and that the monthly use of water by the household totals 10,000 gal, approximately how many parts per million (ppm) of $(NaPO_3)_3$ are there due to dishwashing alone in the sewage from the house? How does this concentration compare to the 0.04×10^{-3} g/kg of H_2O needed by growing algae?

Key Words

Aerobic bacteria bacteria that require oxygen to live.

Anaerobic bacteria bacteria that can live without oxygen.

Atmosphere air surrounding the earth.

Biosphere the life region of earth.

BOD biological oxygen demand of polluted water.

Catabolism series of reactions by which carbohydrates and fats are converted to CO_2 and H_2O with release of energy in living organisms.

Chlorinated hydrocarbon insecticides an insecticide of the DDT type, soluble in fat but not in water.

Denitrifying bacteria bacteria using the oxygen in soil nitrates to oxidize glucose for their life energy and return fixed nitrogen to the atmosphere as N_2.

Eutrophication a natural process changing a lake into a swamp that is accelerated by an oversupply of nutrients.

Geothermal energy thermal energy obtained from hot rocks beneath the earth's surface.

Heat capacity (specific heat) of a substance number of calories of heat energy required to raise the temperature of one gram of the substance by 1°C.

Hydrosphere water in oceans, seas, lakes, rivers, glaciers, and polar ice caps.

Insect hormones regulatory substances secreted by glands in insects and carried through their circulatory system.

Inversion layer layer in the atmosphere close to the ground, where temperature increases with altitude instead of decreasing.

Kilowatt (kW) a measure of electrical power, volts × amps/1000.

Kilowatt hour (kWh) measure of the quantity of electrical energy equal to kW × hours.

Lithosphere the solid outer crust of earth (from Greek, *lithos*: stone).

Magma the molten rock beneath the crust of earth.

NADPH-NADP$^+$ an enzyme cofactor that transfers electrons and protons from H_2O to CO_2 in photosynthesis.

Nitrogen fixation conversion of molecular nitrogen into compounds usable by plants.

PAN peroxyacyl nitrates present in photochemical smog—responsible for eye irritation and plant damage.

Pheremones volatile organic compounds secreted by insects and some animals for communication through the air.

Photochemical smog mixture of ozone, nitrogen oxides, hydrocarbons, and peroxyacyl nitrates trapped in an inversion layer and activated by sunlight.

Photosynthesis conversion of CO_2 and H_2O to glucose in plants using energy in sunlight.

Pollutants harmful substances put into the biosphere by human activities.

Power the time rate of production or consumption of energy.

Solar cell a device using semiconductors for converting the energy of sunlight directly into electricity.

Solar constant the rate at which radiant energy from the sun strikes a surface perpendicular to the sun's rays—1.40 kW m^{-2} outside the atmosphere, 1.00 kW m^{-2} at the surface of the earth.

Solar still a device for distilling fresh water from seawater using energy from sunlight.

Synchronous orbit orbital path keeping a satellite directly over a fixed point on earth.

Tertiary treatment rendering waste water (sewage) fit to drink.

Suggested Readings

Beroza, Morton, "Insect Sex Attractants," *Am. Sci.*, **59**:320–325 (May-June 1971).

Carson, Rachel, *Silent Spring*, Boston, Houghton Mifflin Co., 1972.

Env. Sci. and Tech., Vol. 8, No. 5 (May 1974). The issue is devoted to water-treatment methods.

Gillett, J. D., "The Mosquito: Still Man's Worst Enemy," *Am. Sci.,* **61**: 430–436 (July–Aug. 1973).

Hutchinson, G. Evelyn, "Eutrophication," *Am. Sci.,* **61**:269–279 (May–June 1973).

Jackson, W., *Man and the Environment,* 2nd Ed., Dubuque, Iowa, William C. Brown Co., 1971.

Lappé, Frances Moore, *Diet for a Small Planet,* New York, Ballantine, 1971.

"Life's Dependence on Earth's Ultraviolet Screen," *Science News,* **103**: 101 (Feb. 17, 1973).

"New Oxygen Technology Revives Dead Lakes," *Chem. and Eng. News,* Dec. 10, 1973, p. 20.

Othmar, Donald F., and Roels, Oswald A., "Power, Fresh Water, and Food from Cold, Deep Sea Water," *Science,* **182**:121–125 (Oct. 12, 1973).

"Ozone and the SST," *Chemistry,* **46**:21–22 (April 1973).

Simpson, R. H., "Hurricane Prediction: Progress and Problem Areas," *Science,* **181**:899–901 (Sept. 7, 1973).

Smith, F. G. Walton, "What the Oceans Means to Man," *Am. Sci.,* **60**(1):16–19 (Jan.–Feb. 1972).

Stevens, Leonard A., "Breakthrough in Water Pollution," *Reader's Digest,* June 1971, pp. 167–175.

Weeks, W. F., and Campbell, W. J., "Towing Icebergs to Irrigate Arid Lands," *Bulletin of Atom. Sci.,* **29**:35–39 (May 1973).

Wiener, Aaron, "The Development of Israel's Water Resources," *Am. Sci.,* **60**:466–473 (July–Aug. 1972).

Witmer, Fred E., "Reusing Water by Desalination," *Env. Sci. and Tech.,* **7**:314–318 (April 1973).

Young, Patrick, "Pulling the Plug on Lake Erie," *Harpers,* **245**:48–49 (Aug. 1972).

chapter 16

Old World cave drawings of mammoths, probably drawn 15,000 years ago. (*Photo 43, P. S. Martin, Science, vol. 179, cover, 9 March 1973. Copyright 1973 by the Americal Association for the Advancement of Science.*)

Radioactivity and Nuclear Reactions

16.1 Introduction Samples of naturally radioactive materials send out fast-moving streams or rays of particles, which may be α rays ($^4_2He^{2+}$ ions), β rays (electrons, $^{0}_{-1}e$), γ rays (photons, $h\nu$), or a combination of them. The behavior of the three kinds of rays in the electric field between two electrically charged plates is shown in Chapter 1 in Figure 1.2. The α rays are bent toward the negatively charged plate because it attracts their positive charge, the β rays bend toward the positively charged plate, showing their negative charge, and the γ rays go straight through because they carry no charge. The particles making up the rays come from the nuclei of radioactive atoms. The particles forming the rays ($^4_2He^{2+}$, $_{-1}^{0}e$, and photons) are not present as particles in the nucleus. The situation is like the emission of a photon (light) by a photoexcited atom (Section 3.2). The photon comes out of the atom, but it wasn't "in" the atom; excess energy was "in" the atom. As explained in Section 1.4, emission of an α particle decreases both mass number (M) and atomic number (Z) by 2, emission of a β particle increases the atomic number by 1, and emission of a γ-ray photon causes no change in either M or Z. You can *think* of the β-ray electron as being produced by the change $^1_0n \rightarrow {}^1_1H + {}_{-1}^{0}e$ going on *inside* the nucleus, even though the particles as such are not present.

Radioactivity is strictly a *nuclear* property, so the energy and kind of rays for a particular type of radioactive atom do not depend at all on the state of chemical combination of the atom. Because most elements have isotopes, different kinds of nuclei may belong to the same *chemical element,* so when talking about radioactive atomic nuclei, the mass number as well as the atomic number or symbol must be given. For example, potassium, as it occurs naturally in compounds on earth, is made up of three isotopes: $^{39}_{19}K$ (93.10%), $^{40}_{19}K$ (0.018%), and $^{41}_{19}K$ (6.88%). Only the rarest isotope, $^{40}_{19}K$, is radioactive—its nucleus shoots out a β particle ($_{-1}^{0}e$) and changes to a $^{40}_{20}Ca$ nucleus, which is not radioactive. This is a *nuclear* (or radioactive) *decay process,* or a nuclear reaction, for which the equation and energy released are as follows:

Rules for balancing equations for nuclear reactions are given in Section 1.4. Balancing the equations is easier if both mass number and atomic number are attached to the chemical symbols for the atom.

$$^{40}_{19}K \longrightarrow {}^{40}_{20}Ca + {}_{-1}^{0}e + 1.32 \times 10^6 \text{ electron volts} \qquad (1.32 \text{ MeV})$$

Radioactive decay processes as in this example are *spontaneous* and always release energy. Energies are usually given in millions of electron volts (MeV), and the equation means that each time a $^{40}_{19}K$ nucleus "erupts," the electron ejected flies off with a kinetic energy of 1.32 MeV. This is a fairly typical energy value for radioactive decay processes. If you recall that ordinary chemical reactions have energies in the 1–10 electron-volt (eV, not MeV) range, you can see that nuclear reaction energies are millions of times larger than chemical reaction energies. The decay of one *mole* of $^{40}_{19}K$ yields 3.04×10^7 kcal. For comparison, the formation of one *mole* of H_2 from its atoms yields only 1.04×10^2 kcal.

The rates of radioactive decay processes have a huge range, and it is not possible to say when any one nucleus will erupt in radioactive decay. The decay of the individual nuclei of a radioactive sample has to be handled statistically, and the most useful way of describing the rate of a decay process is to give what is called its *half-life*. The half-life ($t_{1/2}$) of a radioactive decay process is the time required for one-half of the radioactive atoms present in a sample to decay. For the decay of $^{40}_{19}K$, $t_{1/2} = 1.3 \times 10^9$ years. This means that however many $^{40}_{19}K$ atoms are on earth at this moment, only half will be left 1.3×10^9 years in the future. Put another way, if you have 1.00 g of pure $^{40}_{19}K$ today, there will be only 0.50 g left after 1.3×10^9 years; the other 0.50 g will be $^{40}_{20}Ca$. Many decay processes have shorter or much shorter half-lives. Examples of half-lives for some isotopes that occur naturally are shown in the *radioactive decay series* in Figure 16.1. The range of half-lives is from 10^9 years to 0.0001 second. There are four main radioactive decay series like the one in Figure 16.1 that have been worked out in detail. Three of the series (including the one in Figure 16.1) end, after many intermediate steps, at different, stable (nonradioactive) isotopes of lead (Pb), and the fourth ends at the only stable isotope of bismuth ($^{209}_{83}Bi$). Using arrows and dots to show many intermediate radioactive isotopes, we can list the four series as follows:

Uranium–radium series: $^{238}_{92}U \rightarrow \rightarrow \rightarrow \cdots \longrightarrow {}^{206}_{82}Pb$

Uranium–actinium series: $^{235}_{92}U \rightarrow \rightarrow \rightarrow \cdots \longrightarrow {}^{207}_{82}Pb$

Thorium series: $^{232}_{90}Th \rightarrow \rightarrow \rightarrow \cdots \longrightarrow {}^{208}_{82}Pb$

Neptunium series: $^{237}_{93}Np \rightarrow \rightarrow \rightarrow \cdots \longrightarrow {}^{209}_{83}Bi$

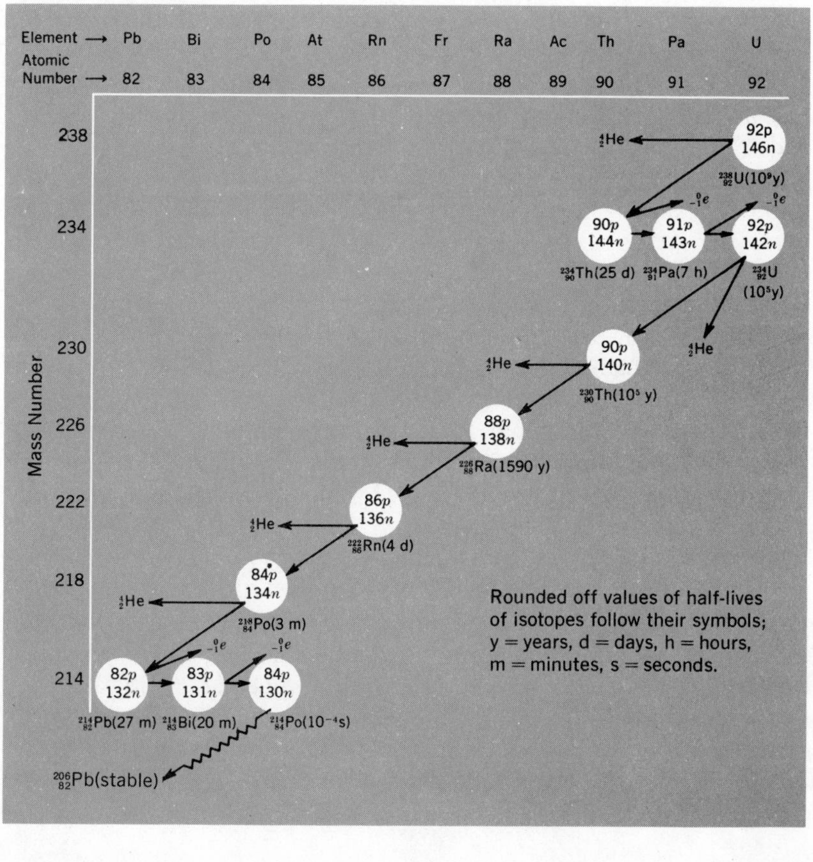

FIG. 16.1 Abbreviated portion of the uranium–radium radioactive decay series.

Exercises

1. If a radioactive decay series starts with an isotope of *even* mass number, all of the atoms in the series must also have even mass numbers. Why is this so? Can an isotope of even mass number be in a decay series that starts with an isotope of *odd* mass number?

2. Complete and balance the nuclear equation $^{238}_{92}U \rightarrow {}^{234}_{90}Th + \_\_\_\_$. The decay of one $^{238}_{92}U$ atom in this way yields 4.2 MeV of energy. Calculate the energy yield in kcal/mole. (See inside back cover for conversion factors.)

Half-lives for radioisotopes are found by counting (with a Geiger or scintillation counter) the number of particles per minute coming out of a weighed sample of the isotope. See Exercise 3 for a sample calculation.

If lead had somehow "leaked into" the sample, the other isotope of geonormal lead, $^{208}_{82}Pb$, would also be present.

16.2 Dating Using Radioisotopes The age of the earth can be calculated by using the half-lives of certain radioisotopes ☚ combined with mass spectrometric analysis of rocks that contain the isotopes. The reasoning goes like this: suppose you analyze a sample taken from the inner part of a rocky piece of uranium ore for the amounts of uranium and lead isotopes present. You find that the isotopes of lead present are $^{206}_{82}Pb$ and $^{207}_{92}Pb$. Furthermore, you find that the numbers of $^{206}_{82}Pb$ atoms and $^{238}_{92}U$ atoms in the sample *are equal*. Since the $^{206}_{82}Pb$ atoms must have come from the radioactive decay of $^{238}_{92}U$ atoms (Figure 16.1) during the time since the rock was formed, ☚ the age of the rock sample is the same as one half-life for the overall decay of $^{238}_{92}U$ to $^{206}_{82}Pb$. The age is one half-life because it is the time required for one-half the radioactive atoms to decay to give the *equal* numbers of $^{238}_{92}U$ and $^{206}_{82}Pb$ atoms found by analysis. The half-life for the decay series $^{238}_{92}U \rightarrow \rightarrow \rightarrow \cdot \cdot \cdot \longrightarrow {}^{206}_{82}Pb$ is 4.5×10^9 (4.5 billion) years, so the rock sample containing equal numbers of $^{238}_{92}U$ and $^{206}_{82}Pb$ atoms is 4.5×10^9 years old. Actually, the numbers of the isotopic atoms found are not quite equal for the oldest rocks found on earth—the $^{206}_{82}Pb/^{238}_{92}U$ ratio is slightly less than 1.00 and gives an age of 3.0×10^9 years for the oldest earth rocks. However, the isotope ratio in samples of meteorites and rocks brought back from the moon is very close to 1.00, so the conclusion is that the solar system is 4.5×10^9 years old and that the earth took 1.5×10^9 years to cool down from the molten condition in which it was originally formed, so minerals could crystallize. The same age for earth results if we use $^{208}_{82}Pb/^{232}_{90}Th$ ratios from the thorium decay series for thorium-bearing minerals.

Almost the same kind of reasoning is applied in *radiocarbon dating* of wooden objects used by people in the prehistoric times up to 50,000 years ago. Radiocarbon dating depends on there being a trace of the radioactive isotope $^{14}_{6}C$ in geonormal carbon. The isotope is produced by the nuclear reaction of neutrons with nitrogen atoms in the upper atmosphere. Using $^{1}_{0}n$ to show the zero charge and unit mass number of the neutron, we can write the equation for the formation of $^{14}_{6}C$ as follows:

$$^{14}_{7}N + ^{1}_{0}n \rightarrow ^{14}_{6}C + ^{1}_{1}H$$

The $^{14}_{6}C$ atoms formed react with oxygen to form $^{14}_{6}CO_2$, which mixes with the nonradioactive $^{12}_{6}CO_2$ and $^{13}_{6}CO_2$ in the atmosphere. The neutrons are constantly being formed by the bombardment of atoms in the upper atmosphere by cosmic rays ☛ coming in from outer space. Radiocarbon decays to a stable $^{14}_{7}N$ atom by loss of a β particle ($_{-1}^{0}e$) in a reaction that has a half-life of 5760 years:

Cosmic rays are very-high-energy streams of heavy nuclei, such as those of iron (Fe). When they collide with atoms, they produce showers of nuclear debris, including neutrons.

$$^{14}_{6}C \xrightarrow{t_{1/2} = 5760\text{ yr}} ^{14}_{7}N + _{-1}^{0}e + 0.16\text{ MeV}$$

Because $^{14}_{6}C$ is being formed at a constant rate and is decaying to $^{14}_{7}N$ at a constant rate, the concentration of $^{14}_{6}CO_2$ in the CO_2 of the atmosphere remains the same. In radiocarbon dating, it is assumed that the $^{14}_{6}C/^{12}_{6}C$ atomic ratio in atmospheric CO_2 hasn't changed from its present value of one $^{14}_{6}C$ atom in 10^{12} $^{12}_{6}C$ atoms for at least 100,000 years. Next comes the realization that when a tree dies it stops taking in CO_2 from the air for photosynthesis to make wood. At death, the $^{14}_{6}C/^{12}_{6}C$ ratio in the most recently formed wood (cellulose) in the tree corresponds to the steady-state concentration in the atmosphere, one $^{14}_{6}C$ atom to 10^{12} $^{12}_{6}C$ carbon atoms. Once the tree is dead, it stops taking up any more CO_2, so the $^{14}_{6}C$ present at death is not replenished from the atmospheric pool and continues to decay to $^{14}_{7}N$. The age of the wood can be told from how much $^{14}_{6}C$ remains.

EXAMPLE

⊸ Suppose you have a sample of wood from the coffin of an Egyptian mummy and the ratio of $^{14}_{6}C$ carbon atoms to $^{12}_{6}C$ carbon atoms in the sample is measured ☛ and found to be 1 in 2×10^{12} or one-half the ratio in the atmosphere. This ratio means the wood sample must have been formed one half-life for $^{14}_{6}C$ (5760 years) ago, placing the life of the mummy at about 4000 B.C. Radiocarbon dating has been checked independently by using it to determine the life of the heartwood (first-formed wood) of a California redwood tree, which was also dated by counting the yearly growth rings in a cross section of the trunk. Counting rings gave the age as 2928 years; radiocarbon analysis gave it as 2900 years. ⊸

Special particle counters are used to find out how much $^{14}_{6}C$ is present in the sample.

Radiocarbon dating covers the most significant part of the prehistoric period because it can be applied to once-living materials that are up to about 50,000 years old. Using this method, it has been shown that the last glaciation of the Western hemisphere occurred about 11,000 years ago, and that the volcanic explosion that blew the hole now filled by Crater Lake in Oregon occurred about 6500 years ago. Charcoal samples from the Lascaux caves in France, where the famous "cave drawings" are located, were found by radiocarbon dating to be 15,000 ±900 years old, indicating human habitation at that early date (Photo 43).

Exercise **3.** The half-life for the decay of $^{232}_{90}Th$ to $^{228}_{88}Ra$ is 1.39×10^{10} years. Give the nuclear equation for the decay process. Approximately what fraction of the $^{232}_{90}Th$ present when the earth was formed is present today? (Use 4.5×10^9 years for the age of the earth.)

EXAMPLE In describing dating with radioisotopes, periods of time equal to isotopic half-lives were used because they make the calculations and logic of the method clearer. It is not necessary to deal only with half-lives.

Suppose you are a radiologist treating a cancer internally with radiation from injected radioisotopes. Cancer of the thyroid gland is treated in this way, using β rays from $^{131}_{53}I$, which has a half-life of 8 days:

$$^{131}_{53}I \xrightarrow{t_{1/2} = 8 \text{ days}} {}^{131}_{54}Xe + {}^{0}_{-1}e$$

You would want to be able to calculate just how much of the isotope remained at any time after injection. To develop a method for this kind of problem we start with the fact that radioactive decay is a first-order process or unimolecular reaction (Section 9.3) in which the number of nuclei decaying at any time is directly proportional to the number of nuclei that haven't decayed yet. Using $\Delta N/\Delta t$ to stand for the number of nuclei that decay in the short time interval Δt, and N for the number of nuclei left to decay, the rate law can be written as follows:

$$\text{Rate} = \frac{\Delta N}{\Delta t} = kN$$

In this equation the proportionality constant k is called the *decay constant* and gives the probability that a nucleus will decay in a fixed time interval. The decay constant has units of time^{-1}: decays per second, per minute, and so on. Applying calculus to the rate equation gives the value of the decay constant k in terms of the half-life of the isotope as

This equation is derived at the end of Appendix 1.

$$k = \frac{0.693}{t_{1/2}}$$

The rate law for any radioactive decay process then takes the following form:

$$\frac{\Delta N}{\Delta t} = \frac{0.693N}{t_{1/2}}$$

Suppose you have a mole of geonormal carbon and want to calculate the number of $^{14}_{6}C$ atoms in it that decay each second. Only one atom in 10^{12} of geonormal carbon is $^{14}_{6}C$, so the sample contains

$$\frac{1}{10^{12}} \times 6.023 \times 10^{23} = 6.023 \times 10^{11}\ ^{14}_{6}\text{C atoms}$$

The half-life of $^{14}_{6}C$ is 5760 years or 1.817×10^{11} sec. ☛ Substituting the values of N and $t_{1/2}$ into the rate expression gives the number of $^{14}_{6}C$ nuclei decaying each second:

One year = 3.154×10^{7} sec.

$$\frac{\Delta N}{\Delta t} = \frac{0.693 \times 6.023 \times 10^{11}}{1.817 \times 10^{11}} = 2.30 \text{ decays per second}$$

In radiocarbon dating special particle counters that are shielded from all outside α, β, and γ rays must be used because the amount of $^{14}_{6}C$ in geonormal carbon is so small. By contrast, in a mole of *pure* $^{14}_{6}C$, there would be 3.31×10^{12} nuclei decaying each second.

Applying calculus to the rate law for radioactive decay gives the following equation:

$$\Delta t = 3.32\ (t_{1/2}) \log \frac{N_0}{N_t}$$

Here, N_0 is the number of radioactive atoms present at the beginning of the time interval Δt, N_t is the number left at the end of the time interval, and $t_{1/2}$ is the half-life of the radioisotope. Applying the equation to the decay of $^{131}_{53}I$ ($t_{1/2} = 8$ days) with which this example started, the radiologist would calculate the time required for 90% of the radioisotope to decay in the following way:

$$t = (3.32)(8) \log \frac{100}{10} = 27 \text{ days}$$ ⊸

16.3 Stability of Nuclei and Radioactivity Atoms with stable nuclei are not radioactive—radioactive decay occurs when an unstable nucleus emits particles whose loss leaves the new nucleus more stable than the original one was. Sometimes emission of a single particle leaves a completely stable nucleus—this is the case with the natural radioisotopes $^{40}_{19}K$ and $^{14}_{6}C$, which have already been discussed, and with the hydrogen isotope tritium, $^{3}_{1}H$, which decays to a stable helium isotope, ${}^3_1H \rightarrow {}^3_2He + {}_{-1}^{0}e$. On the other hand, emission of a particle may produce, instead of a stable nucleus, one that itself is radioactive. This is the situation for the radioactive decay series discussed in Section 16.1.

What makes a nucleus stable is not completely understood, mainly because the forces of attraction between nucleons (protons, neutrons, and mesons ☛) come into play only at the very short (10^{-13} cm or less) distances between nucleons—the particles making up atomic nuclei—and therefore cannot be measured directly with laboratory-size samples. Coulomb's law of electrostatic attraction, which works so well for understanding the electronic structures of atoms and molecules, can be tested in the laboratory, using charged metal balls. This kind of experiment cannot be done with nucleons. There are, however, some nuclear properties that can be used to predict not only which nuclei are stable but—if they are not stable—what is the most likely type of radioactivity to lead them toward stability.

Mesons are particles that appear when atomic nuclei are blasted apart by cosmic rays. They have masses between those of the electron and proton and may carry a charge.

First come the "magic numbers" of nucleons that are found in stable nuclei. These numbers come out of applying wave mechanics to the nucleons in much the same way that the stable duet (2) and octet (8) of electrons come out of wave mechanics applied to nuclei and electrons

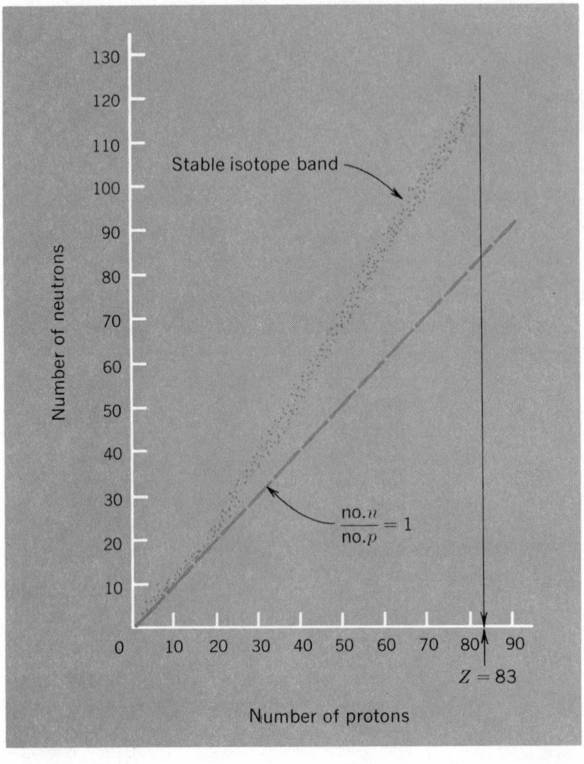

FIG. 16.2 Schematic representation of the stability band for nuclides.

in the electron cloud of the atom. For nucleons the "magic numbers" are 2, 8, 20, 50, 82, and 126—nuclei with these numbers of protons or neutrons are nonradioactive and are not easily blasted apart by cosmic rays or other high-energy particles. The nuclei of $^{4}_{2}He$ (2 protons + 2 neutrons), $^{16}_{8}O$ (8 protons + 8 neutrons), $^{40}_{20}Ca$ (20 of each), and $^{208}_{82}Pb$ (82 protons + 126 neutrons) are unusually stable to disintegration when struck by high-energy particles.

Stability is favored when a nucleus has even numbers of protons and neutrons. These nuclei tend to be nonradioactive, and there are more stable nuclei of the even–even type than of the odd–odd or even–odd types, as the following tabulation shows:

Neutron number	*Proton number*	*Number of stable isotopes*
even	even	162
odd	even	56
even	odd	52
odd	odd	4

The odd–odd combination of neutrons and protons is particularly unstable—the four odd–odd but stable nuclei are $^{1}_{1}H$, $^{6}_{3}Li$, $^{10}_{5}B$, and $^{14}_{7}N$. The radioactive isotope $^{40}_{19}K$ described earlier belongs to the odd–odd type (19 protons, 21 neutrons).

Finally, the ratio of the numbers of neutrons to protons—the n/p ratio—is an important factor in stability, particularly with the heavier elements beyond lead (Pb). Neutrons are clearly necessary to bind protons together in nuclei because $^{2}_{2}He$ (two protons only), $^{3}_{3}Li$ (three protons only), and the like do not exist. The particularly stable nuclei $^{4}_{2}He$ and $^{40}_{20}Ca$ have n/p ratios of 1, but when the n/p ratio gets to be 1.54 or greater, radioactivity usually appears. For $^{157}_{88}Ra$ the n/p value is 1.57, and for $^{238}_{92}U$ it is 1.59—both of these isotopes are radioactive, as you have seen earlier. With the exception of $^{209}_{83}Bi$, all atoms with Z greater than 82 (Pb) are radioactive. Figure 16.2 shows how the neutron/proton ratio for the stable nuclei is very nearly 1 (approximately equal numbers of neutrons and protons) for atoms up to $Z = 20$ (Ca) but then begins to grow larger and larger, giving the "stability band" of isotopes or nuclides.

The isotopes with Z greater than 82 (Pb; 83 for $^{209}_{83}Bi$) have too many neutrons for the number of protons to be stable. These nuclei move toward stability by decreasing the n/p ratio through emitting a β particle. The following are examples near the beginning of the decay series in Figure 16.1:

$$ {}^{234}_{90}\text{Th} \xrightarrow[\text{25 days}]{\nearrow {}^{0}_{-1}e} {}^{234}_{91}\text{Pa} \xrightarrow[\text{7 hr}]{\nearrow {}^{0}_{-1}e} {}^{234}_{92}\text{U} $$

$$ \frac{N}{P} = \frac{234 - 90}{90} = 1.60 \qquad \frac{234 - 91}{91} = 1.57 \qquad \frac{234 - 92}{92} = 1.54 $$

Many natural radioisotopes of mass number greater than 209 decay by loss of an α particle. This process is necessary to reduce the mass number to 208–209, where the first stable isotopes appear (remember, β emission does not change M). Emission of α particles drives the n/p ratio up, but as it increases, the half-lives of the isotopes get shorter and shorter, showing that they are more and more unstable. You can also see this trend in Figure 16.1, from which the following series is taken:

The stable isotope ${}^{208}_{82}\text{Pb}$ also has $n/p = 1.54$, but its nucleus contains magic numbers of both protons (82) and neutrons (126), making the nucleus particularly stable.

$$ {}^{234}_{92}\text{U} \xrightarrow[10^6\text{ yr}]{\nearrow {}^{4}_{2}\text{He}} {}^{230}_{90}\text{Th} \xrightarrow[10^5\text{ yr}]{\nearrow {}^{4}_{2}\text{He}} {}^{226}_{88}\text{Ra} \xrightarrow[\text{1590 yr}]{\nearrow {}^{4}_{2}\text{He}} {}^{222}_{86}\text{Rn} \xrightarrow[\text{4 days}]{\nearrow {}^{4}_{2}\text{He}} {}^{218}_{84}\text{Po} \xrightarrow[\text{3 min}]{\nearrow {}^{4}_{2}\text{He}} {}^{214}_{82}\text{Pb} $$

$$ \frac{N}{P} = 1.54 \qquad 1.56 \qquad 1.57 \qquad 1.58 \qquad 1.60 \qquad 1.61 $$

When the n/p ratio gets too big—as it is at ${}^{214}_{82}\text{Pb}$—β-ray emission starts again, giving ${}^{214}_{83}\text{Bi}$ (Figure 16.1).

Exercise

4. The radioisotope ${}^{214}_{83}\text{Bi}$ actually decays by both β emission (99.96%) and α emission (0.04%). What are the two isotopes formed? They further decay to the same isotope, ${}^{210}_{82}\text{Pb}$. What kinds of emissions are involved for each?

16.4 Synthetic Isotopes—Induced Radioactivity Only about 50 radioactive isotopes occur naturally on earth. They were formed when the planet was formed. Beginning about 50 years ago, methods were discovered for converting one kind of atom into another—for the *transmutation* of atoms. Some atoms produced in this way by bombarding nuclei with rapidly moving particles from radioactive atoms are nonradioactive, familiar ones. Examples are the first transmutation observed, conversion of nitrogen atoms to oxygen atoms, using α particles from radium, and the reaction that led to the discovery of the neutron:

Some isotopes, such as ${}^{90}_{38}\text{Sr}$, were not originally present but are now as a result of testing nuclear (atomic) bombs (Section 16.5).

In writing nuclear equations ionic charges are ordinarily omitted from symbols, so the α particle is ${}^{4}_{2}\text{He}$ rather than ${}^{4}_{2}\text{He}^{2+}$, which is, strictly speaking, the correct symbol.

$$ {}^{14}_{7}\text{N} + {}^{4}_{2}\text{He} \rightarrow {}^{17}_{8}\text{O} + {}^{1}_{1}\text{H} $$

$$ {}^{9}_{4}\text{Be} + {}^{4}_{2}\text{He} \rightarrow {}^{12}_{6}\text{C} + {}^{1}_{0}n $$

Neutrons outside of the nucleus are unstable (radioactive) and break down with a half-life of 12 min into an electron, a proton, and a particle called a neutrino ($\bar{\nu}$): ${}^{1}_{0}n \rightarrow {}^{1}_{1}\text{H} + {}^{0}_{-1}e + \bar{\nu}$.

With the development of particle accelerators such as the cyclotron (Figure 16.3) and, more importantly, the nuclear "pile" or reactor (Figure 16.5) that provides large quantities of neutrons, it became possible to carry out hundreds of transmutations. At this writing, more than 1200 different kinds of atoms had been produced or "synthesized," using the methods of nuclear chemistry. Most of the atoms synthesized are not found naturally on earth and most are fiercely radioactive, with fairly short half-lives. ☛ Cyclotrons and other particle accelerators (such as bevatrons and linear accelerators) are being used mainly to probe the secrets of nuclear structure. The production of synthetic radioisotopes in quantity for use in medicine, industry, and "tracer" experiments uses neutrons from nuclear reactors and the fission products (Section 16.5) that are the "ashes" of nuclear fuel in the reactor. The neutron is more suitable as a bombarding particle because, having no charge, it is not repelled by the positive charge on the nucleus being bombarded and can "drift" in close enough to be captured by the nucleus. Among the hundreds of isotopes made available using neutron bombardment are these: ${}^{14}_{6}C$, used in tracer work in following organic reactions such as the fate of ${}^{14}_{6}CO_2$ in photosynthesis; ${}^{59}_{26}Fe$, used in tracing blood circulation; ${}^{131}_{53}I$, used in treatment of thyroid cancer; ☛ and ${}^{60}_{27}Co$, used in radiation therapy for other cancers, in vitamin B_{12} absorption studies, ☛ in food sterilization, and to make cross-linked synthetic polymers by irradiation. The equations for the formations and decays of a few radioisotopes are as follows:

The synthetic isotopes were probably present when the earth was formed but have all undergone radioactive decay since that time because of their short half-lives.

The hormone thyroxine (Figure 4.4) contains iodine and is synthesized in the thyroid gland, so the ${}^{131}_{53}I$ concentrates there, where its radiation kills off cancer cells.

Vitamin B_{12} (cobalamine) contains a Co atom.

Formation	*Decay ($t_{1/2}$ on arrow)*
${}^{14}_{7}N + {}^{1}_{0}n \longrightarrow {}^{14}_{6}C + {}^{1}_{1}H$	${}^{14}_{6}C \xrightarrow{5760\text{ yr}} {}^{14}_{7}N + {}^{0}_{-1}e$
${}^{58}_{26}Fe + {}^{1}_{0}n \longrightarrow {}^{59}_{26}Fe + \gamma\text{ ray}$	${}^{59}_{26}Fe \xrightarrow{45\text{ days}} {}^{59}_{27}Co + {}^{0}_{-1}e$
${}^{130}_{52}Te + {}^{1}_{0}n \longrightarrow {}^{131}_{53}I + {}^{0}_{-1}e$	${}^{131}_{53}I \xrightarrow{8\text{ days}} {}^{131}_{54}Xe + {}^{0}_{-1}e$
${}^{59}_{27}Co + {}^{1}_{0}n \longrightarrow {}^{60}_{27}Co + \gamma\text{ ray}$	${}^{60}_{27}Co \xrightarrow{5.3\text{ yr}} {}^{60}_{28}Ni + {}^{0}_{-1}e + \gamma\text{ ray}$

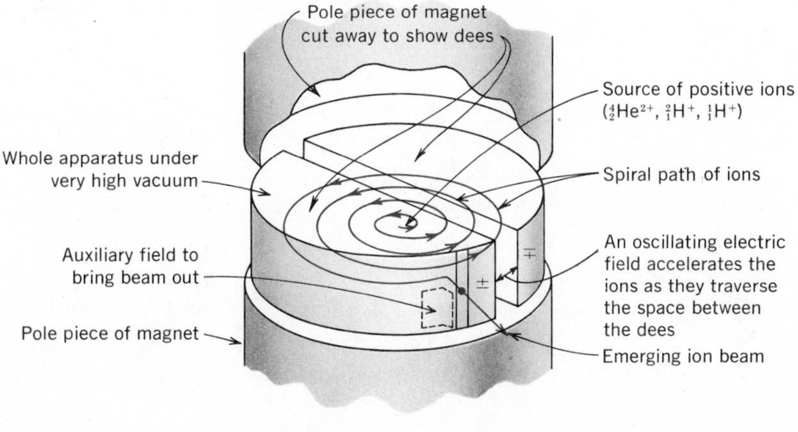

FIG. 16.3 Acceleration of positive ions in a cyclotron.

In all of these cases the nucleus formed by the radioactive decay process is stable. Some synthetic isotopes decay by positron ($_{+1}^{0}e$) emission from the nucleus:

$$^{38}_{19}K \longrightarrow ^{38}_{18}Ar + _{+1}^{0}e$$

The positron, as the symbol $_{+1}^{0}e$ shows, is like an electron in mass but carries a positive charge. Its ejection from the nucleus causes a *decrease* of one unit in atomic number without changing the mass number. Positrons quickly react with electrons to produce γ rays. Another path for radioactive decay found with synthetic isotopes is called *electron capture*. Here an electron in the 1*s* orbital of the cloud is captured and goes into the nucleus, with a corresponding decrease in atomic number by one unit.

Because the electron comes from the *K* ($n = 1$) shell, electron capture is sometimes called *K* capture.

$$^{37}_{18}Ar \xrightarrow{35\ days} ^{37}_{17}Cl + energy\ (h\nu)$$
$$^{7}_{4}Be \xrightarrow{54\ days} ^{7}_{3}Li + energy\ (h\nu)$$

The energy released in electron capture comes out in two kinds of photons—a γ-ray photon from the nucleus and an X-ray photon generated when an electron from an outer orbital drops down to fill the vacancy left in the 1*s* orbital by the captured electron. The natural radioisotope of potassium $^{40}_{19}K$ decays by electron capture to form $^{40}_{18}Ar$, as well as by β emission to produce $^{40}_{20}Ca$, as described earlier.

$$^{40}_{19}K \xrightarrow{1.3 \times 10^{9}\ yr} ^{40}_{18}Ar + energy\ (h\nu)$$

This alternative decay pathway for $^{40}_{19}K$ is the basis for yet another geological dating method, which uses the $^{40}_{19}K/^{40}_{18}Ar$ ratio in rocks. Even though argon (Ar) is a gas, it remains trapped inside of rocks until they are broken up and heated. Mass spectrometry gives the $^{40}_{19}K/^{40}_{18}Ar$ ratio, and calculation of the age of the rock follows the pattern for dating by using the $^{206}_{82}Pb/^{238}_{92}U$ ratio. The two methods give the same age for the earth.

EXAMPLE *Isotope dilution analysis* is a way of getting information, particularly on living things, that cannot be obtained any other way. Suppose you want to determine the volume of blood in a man swollen with edema from diseased kidneys, without harming him. One way is to inject a known amount of a colored substance (a dye) into his bloodstream and wait a few minutes for the dye to mix with the blood. Then you withdraw a measured volume of blood and analyze it for the quantity of dye it contains. Suppose a 10-ml sample is taken and is found to contain $\frac{1}{400}$ the amount of dye injected. If 10 ml of blood contains $\frac{1}{400}$

the dye injected, the total blood volume must be 400 times 10 ml or 4000 ml. We say the *dilution factor* is 400 in this case. But using a dye has serious drawbacks—it doesn't stay in the bloodstream and the body starts to excrete it once it is injected. Isotope dilution analysis using the radioisotope $^{59}_{26}Fe$ avoids both of these difficulties. If it is injected in the form of $^{59}_{26}Fe^{2+}SO_4^{2-}$, the radioactive $^{59}_{26}Fe^{2+}$ ions are very rapidly built into hemoglobin, ☛ the iron-containing compound in red blood cells that transports oxygen from lungs to tissue and is present *only* in the bloodstream. It is not necessary to inject a solution of *isotopically pure* $^{59}_{26}FeSO_4$—ordinary (geonormal) iron(II) sulfate containing a tiny amount of $^{59}_{26}Fe^{2+}SO_4^{2-}$ is all that is needed. Suppose you have a solution of $Fe^{2+}SO_4^{2-}$ that contains enough $^{59}_{26}Fe^{2+}SO_4^{2-}$ so that 1.00 ml of the solution gives off 5000 β particles per minute. You inject 90.00 ml of this solution into the bloodstream of the patient, wait a few minutes for the $^{59}_{26}Fe^{2+}$ to mix with the blood and be incorporated in hemoglobin, then draw a 1.00-ml sample of blood and count the number of β particles it gives off. You get a count of 100 per minute. The dilution factor is 50 in this case:

Sometimes $^{51}_{24}Cr$ is also used—it rapidly replaces Fe^{2+} in hemoglobin.

$$\text{Dilution factor} = \frac{5000 \text{ counts min}^{-1} \text{ ml}^{-1}}{100 \text{ counts min}^{-1} \text{ ml}^{-1}} = 50$$

Since 90.00 ml of the radioiron sulfate solution was injected, the blood diluted it 50 times and the blood volume of the patient is $90 \times 50 = 4500$ ml. ⊸

Neutron activation analysis takes advantage of the production of radioactive isotopes when almost any stable atom is bombarded with neutrons. A sample to be analyzed is exposed to a neutron source. ☛ The energies of the rays, mainly γ rays, that come out of the radioactive nuclei produced are characteristic of the atoms activated. Determining the γ-ray energies, using special counters, identifies the radioactive nuclei and the stable nuclei from which they were formed in the neutron bombardment. The intensity of the γ rays also allows calculation of the amount of the particular element in the sample. Neutron activation analysis is the most sensitive way of finding how much of many elements are present in a sample of matter, even when there are only tiny traces. [Remember that the success of $^{14}_{6}C$ (radiocarbon) dating depends on the presence of one (or fewer) $^{14}_{6}C$ atoms per 10^{12} of nonradioactive carbon atoms.] Children with the disease called cystic fibrosis, in which the lungs fill up with thick mucus, can be diagnosed to have the disease by measuring the amount of copper in their fingernails, using neutron activation analysis. The amount of copper is unusually high for those unfortunate enough to have the disease. Early treatment is often successful.

Such as $^{9}_{4}Be$ + a radioactive source of α particles: $^{9}_{4}Be + ^{4}_{2}He \rightarrow ^{12}_{6}C + ^{1}_{0}n$.

Exercises

5. Assuming that the atoms in a $^{14}_{6}CO_2$ molecule hold together through the radioactive decay of the $^{14}_{6}C$ nucleus, what would be the formula of the molecule left after decay occurred? Where have you seen the molecule before? Is it an atmospheric pollutant?

6. Even though the radioactive decay of $^{40}_{19}K$ produces $^{40}_{20}Ca$, the $^{40}_{19}K/^{40}_{20}Ca$ ratio in rocks is not useful for geological dating. Why is this so? (*Hint*: consider where Ca comes from.)

16.5 Nuclear Fission and Fusion—Power from the Stars The stars shine on and on for billions of years—our sun is a star. Where does the energy "to shine the stars" come from? Here on earth, science is close to discovering the secret of their limitless source of energy. When the nuclear reactions that make it possible for the stars to go on and on can be used on earth it will no longer be necessary to use up fossil fuels (oil, coal, natural gas), and we will have unlimited energy reserves for the future. Nuclear *fusion* and *fission* provide energy for the stars to shine. To appreciate the huge amounts of energy available from these two kinds of nuclear reactions, it is best to start with the amounts of energy released by the natural radioactive decay of the heavier isotopes (Z greater than 83) found on earth.

The nuclear reactions of radioactivity have energies in the MeV (million-electron-volt) range. One MeV is only 3.83×10^{-17} kcal but a *mole* of MeV is $6.023 \times 10^{23} \times 3.83 \times 10^{-17} = 2.307 \times 10^{7}$ kcal. Let's try to put this huge energy quantity in scale, using an example.

EXAMPLE

⊸ The radium isotope with the longest half-life (1622 years) is $^{226}_{88}Ra$. It decays by emitting an α particle with energy of 4.78 MeV:

$$^{226}_{88}Ra \rightarrow {}^{222}_{86}Rn + {}^{4}_{2}He + 4.78\ \text{MeV}$$

Decay of a mole of $^{226}_{88}Ra$ releases energy amounting to $2.307 \times 10^{7} \times 4.78 = 1.10 \times 10^{8}$ kcal. Half this amount of energy, or 5.5×10^{7} kcal, would be released during the first half-life period of 1622 years. The average rate of energy release during this time is $(5.5 \times 10^{7}/1622) = 3.39 \times 10^{4}$ kcal per year. There are 8760 hours in a year, making the average rate of energy emission per mole $(3.39 \times 10^{4}/8760) = 3.9$ kcal/hour. Some of the energy appears as heat and some as light, produced when molecules in air that have been made energy-rich by collision with the fast-moving α particles return to their ground state. As a result of the energy released, a sample of radium is warm and glows in the dark. The polonium isotope $^{210}_{84}Po$ ($t_{1/2}$ = 138 days) is so fiercely radioactive that a 0.50-g sample reaches a temperature of 500°C from the energy release. ⊸

The average rate must be used because the instantaneous rate of energy emission is constantly decreasing as the number of $^{226}_{88}Ra$ left to decay becomes smaller and smaller.

Radioactive decay processes clearly yield much larger energies than ordinary exothermic chemical reactions. But even larger energy quantities are released in the nuclear reactions of *fission* and *fusion*.

Some nuclei are on the edge of self-destruction—they split approximately in half immediately after absorbing a single neutron. The process is called *nuclear fission* and occurs with the uranium isotope ${}^{235}_{92}\text{U}$ and the plutonium isotope ${}^{239}_{94}\text{Pu}$. The nuclear fission of these isotopes produces a variety of atomic nuclei with atomic numbers (Z) in the range 30–60. An example of two fission processes is the following:

$$
{}^{235}_{92}\text{U} + {}^{1}_{0}n \rightarrow \underset{\text{(unstable)}}{\left[{}^{236}_{92}\text{U}\right]} \xrightarrow[t_{1/2} \approx 10^{-10}\ \text{sec}]{\text{Fission}} \begin{cases} {}^{90}_{38}\text{Sr} + {}^{143}_{54}\text{Xe} = 3\,{}^{1}_{0}n + \text{Energy} \\ {}^{140}_{56}\text{Ba} + {}^{93}_{36}\text{Kr} + 3\,{}^{1}_{0}n + \text{Energy} \end{cases}
$$

The sum of the nuclear binding energies (Section 2.5) of the products of fission is *greater* than the nuclear binding energy of the reactants, so fission occurs with energy release representing the difference. The difference in binding energies shows up as the difference between the matter mass (Section 2.5) of a $[{}^{236}_{92}\text{U}]$ atom and the sum of the matter mass of the fission products. The breakup of one mole or 236.05 g of $[{}^{236}_{92}\text{U}]$ gives only 235.85 g of material products—atoms and neutrons. The missing $(236.05 - 235.85) = 0.20$ g of matter appears as 0.20 g of energy in the form of heat and radiation. Using the Einstein equation, $E = mc^2$, we can express this quantity of energy in more familiar units. The equation gives the energy in units of *ergs* when m is in grams and c, the velocity of light, is in cm/sec:

$$
\left.\begin{array}{l}\text{Energy from fission} \\ \text{of one mole of } {}^{235}_{92}\text{U}\end{array}\right\} = mc^2 = (0.20)(2.998 \times 10^{10})^2 = 1.80 \times 10^{20}\ \text{ergs}
$$

$$
= 1.80 \times 10^{20} \times 2.390 \times 10^{-11}\ \text{kcal}
$$

$$
= 4{,}300{,}000{,}000\ \text{kcal}
$$

One erg $= 2.390 \times 10^{-11}$ kcal.

EXAMPLE

To put the huge amount of energy released in the fission of one mole of ${}^{235}_{92}\text{U}$ in scale, it is about the quantity of heat produced by burning 1.3 million lb (650 tons) of coal. To make the comparison another way,

(Photo 7, Chapter 2)

Yield of 1 lb of coal = 3,275 kcal of energy
Yield of 1 lb of ${}^{235}_{92}\text{U}$ = 8,300,000,000 kcal of energy

Put yet another way, the energy in 1 lb of coal but in the form of food would take care of one person working vigorously for one day, but the energy from fission of 1 lb of uranium would be equivalent to the same diet for 2,534,000 people for one day.

The fission of ${}^{235}_{92}U$ requires one neutron at the start, but between two and three neutrons are released every time a [${}^{236}_{92}U$] nucleus "bursts in half." We have here the ingredients for a *branching* chain reaction, which could cause a whole chunk of ${}^{235}_{92}U$ to undergo fission. Using ○ for ${}^{235}_{92}U$ atoms, ⊖ for [${}^{236}_{92}U$], and ◠ and ◡ for the lighter nuclei produced by fission, we can show the branching chain reaction something like this, assuming three ${}^{0}_{1}n$ are produced in each fission:

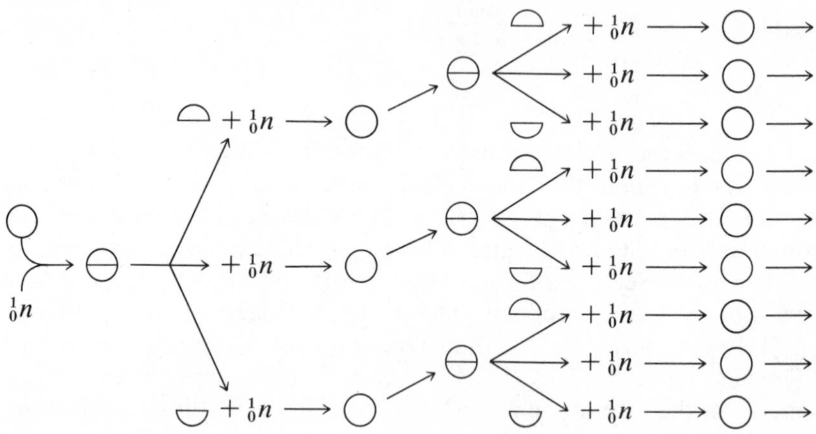

There is a constant flow of a few neutrons produced by cosmic rays to provide the neutron needed to start the chain.

After the first fission, the number of nuclei that split increases very rapidly, as you can see. All that is necessary is to make sure that none of the neutrons escape, and then the whole sample of ${}^{235}_{92}U$ can be made to undergo fission. This situation results when the piece of ${}^{235}_{92}U$ reaches a certain size, called the *critical mass,* which is about the size of a grapefruit. The earliest fission type of nuclear bomb was simply a device containing two smaller, separated pieces of ${}^{235}_{92}U$ that were driven together by a chemical explosive to form a critical mass. In later bombs a much smaller, subcritical mass of fissionable material was made critical by compressing it instantaneously from all directions. The mass is held together by a heavy outer steel casing for the instant required for the whole piece of uranium to undergo fission (Figure 16.4). The result is the "atomic bomb" explosion with its "mushroom cloud." More than 80% of the energy released appears as heat (kinetic energy of the atoms and neutrons produced). The "daughter" nuclei, as the fission products are called, are highly radioactive and contaminate the atmosphere. The part played by ${}^{90}_{38}Sr$ in bone cancer has already been told, but *any* general increase in the radioactivity to which

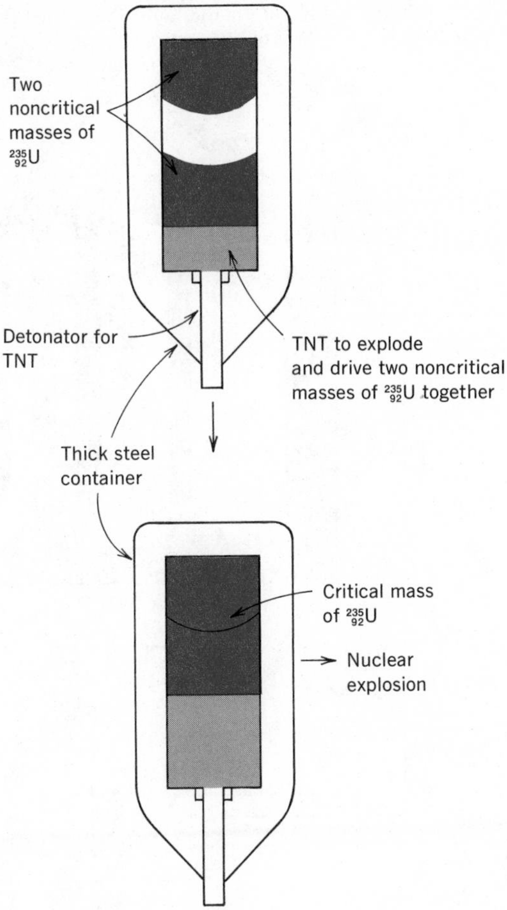

FIG. 16.4 Principle of a nuclear fission bomb.

Linus Carl Pauling (b. 1901) was born in Portland, Oregon. He has won two Nobel prizes—the first in 1954 for his work on electrochemical theories of valence and resonance, the second in 1963 for his work against nuclear testing.

(Photo 26, Chapter 7)

we are exposed increases the number of genetic mutations that may affect future generations of people forever. Atmospheric testing of nuclear bombs by the U.S. and the U.S.S.R. was stopped many years ago to avoid these dangers. Linus Pauling, then professor of chemistry at the California Institute of Technology, was a prime mover in getting the ban. Bombs are now tested deep underground so that none of the radioactive products escape. ☛ Hopefully, one day the tests will be stopped forever, and the thousands of bombs on hand dismantled, and their fissionable material put to constructive work in nuclear reactors.

Nuclear fission can be and is controlled in *nuclear reactors* for the generation of electricity. At the heart or "core" of a nuclear reactor (Figure 16.5) is a slightly greater than critical mass of $^{235}_{92}U$. Uncontrolled fission is prevented by neutron-absorbing "control rods" made of boron carbide (B_4C) or tantalum metal (Ta). Boron and tantalum nuclei absorb neutrons very effectively and with the control rods fully

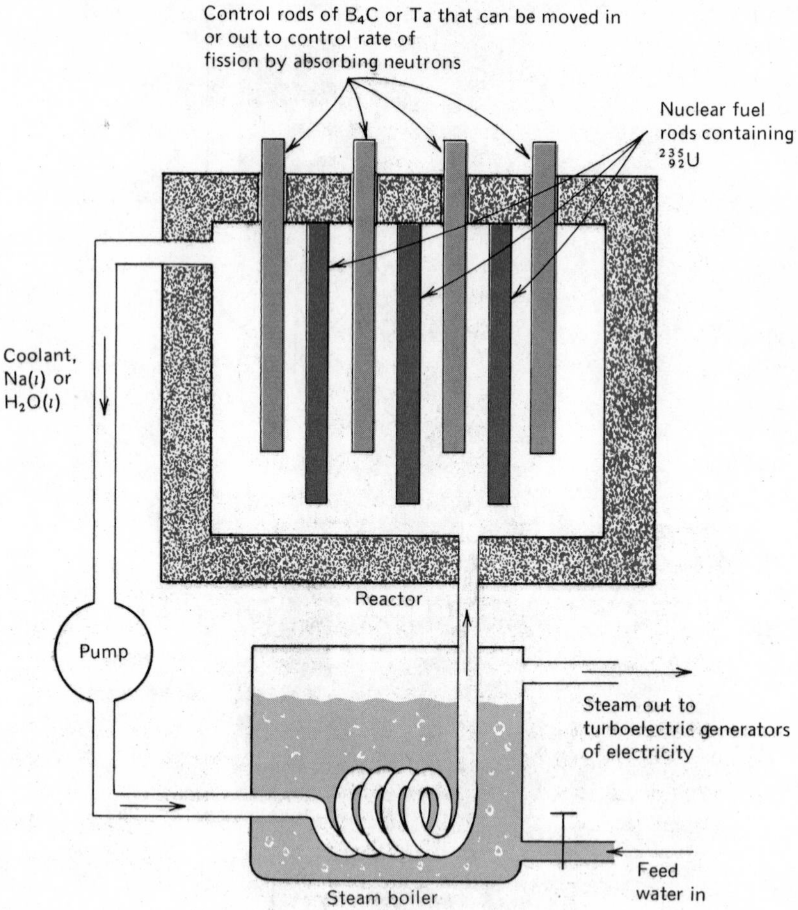

FIG. 16.5 Principles of a fission nuclear reactor.

inserted into the core, fission stops because there are no neutrons left to carry on the chain. By drawing the rods partly out, fission is controlled so that it doesn't become a branched chain reaction. This means that the control rod materials absorb two of the three neutrons produced by fission, leaving one neutron to carry on fission in a *non-branched* chain reaction. This control can be shown, using the symbols we used earlier for the branched chain reaction and vertical lines ‖ for the control rods that absorb neutrons.

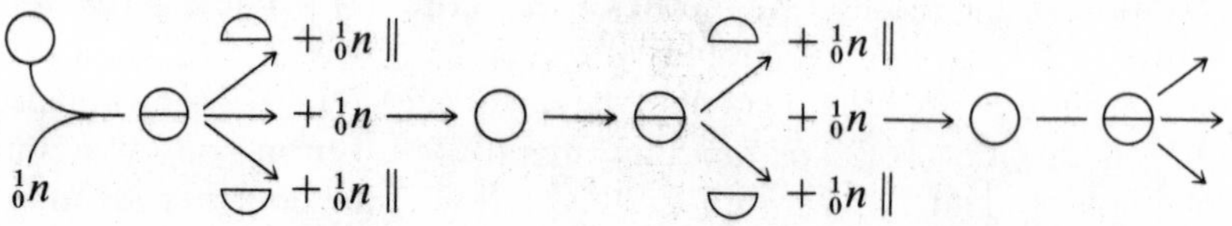

Controlled fission produces enormous quantities of heat in the core of the reactor. The heat is transferred by a circulating liquid or gas to boilers for making steam to run turboelectric generators. In some reactors, water under pressure is circulated through the core as the heat-transfer liquid. Liquid sodium metal (mp 96°C) is also used as the heat-transfer liquid. You can get some idea of the amount of heat released in the core from the fact that for one reactor, 400,000 watts (400 kw) or 5732 *kcal* are released per second in *each liter* of volume in the core. To get such enormous quantities of heat out of the core, the rates of circulation of the heat-transfer liquid are almost unbelievable. In a sodium-cooled reactor, pumping rates are as large as *5 million gal per hour.* The volume of the core is a few cubic meters in such a reactor. Any interruption in the removal of heat results in a "meltdown" of the core from the heat released by fission unless the control rods are inserted instantly. ☞ Melting of the core does *not* result in an atomic bomb type of explosion because the fissionable material is not held compactly together in a critical mass. However, reactor failure could result in radioactive poisoning of a large surrounding area if the nuclear fuel and fission products accumulated in the core escaped. To guard against this, the whole reactor is encased in a thick-walled "pressure vessel."

The control rods are inserted automatically if the reactor gets "out of hand."

The fissionable uranium isotope $^{235}_{92}U$ is only 0.72% geonormal uranium; 99.72% is $^{238}_{92}U$, which does not undergo fission. This means that geonormal uranium must be enriched in $^{235}_{92}U$ by isotope separation methods before it can serve as a nuclear reactor fuel. However, non-fissionable $^{238}_{92}U$ does have a very valuable property—it absorbs a neutron to form $^{239}_{92}U$, which is converted in fairly rapid steps to an isotope of plutonium, $^{239}_{94}Pu$ (half-lives under arrows).

$$^{238}_{92}U \xrightarrow[\ ^{1}_{0}n\]{} \ ^{239}_{92}U \xrightarrow[24 \text{ min}]{^{0}_{-1}e} \ ^{239}_{93}Np \xrightarrow[2.3 \text{ days}]{^{0}_{-1}e} \ ^{239}_{94}Pu \xrightarrow[\ ^{1}_{0}n\]{} [^{240}_{94}Pu] \xrightarrow{} \begin{array}{l} \text{Fission products} \\[4pt] \text{More than } 2^{1}_{0}n \end{array}$$

more than $1^{1}_{0}n$

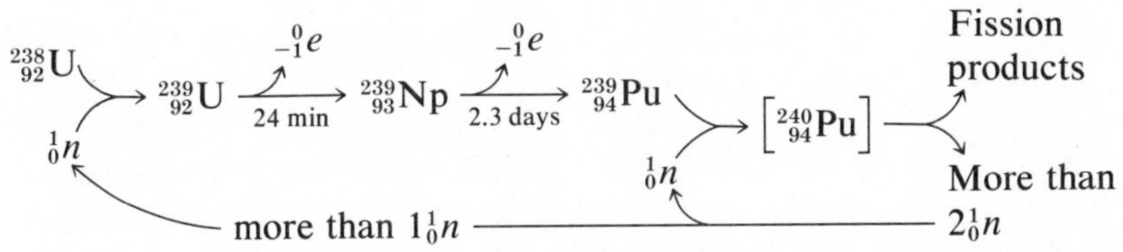

As shown, $^{239}_{94}Pu$ does undergo fission. The important point is that the fission not only produces a neutron to carry itself on, but it also produces between one and two *extra* neutrons that are absorbed by $^{238}_{92}U$ to make more $^{239}_{92}Pu$—not only more but *more fissionable material than is used up in the fission of* $^{235}_{92}U$. The result is the production of more fissionable fuel in the form of plutonium than is used in the form of $^{235}_{92}U$. The amount of fissionable fuel doubles in about 10 years. Reactors of this type are called "breeder reactors" for good reason. They have a central core of $^{235}_{92}U$ surrounded by a "blanket" of $^{238}_{92}U$, where the extra neutrons from the fission of ${}^{235}_{92}U$ carry on the formation of plutonium. Use of breeders allows all of the uranium isotopes in the geonormal element to become fuel for nuclear fission reactors. The most abundant thorium isotope, ${}^{232}_{90}Th$ ($t_{1/2} = 1.4 \times 10^{10}$ years), is also a "fertile" isotope like ${}^{238}_{92}U$ and goes through the same kind of cycle, producing the fissionable isotope ${}^{233}_{92}U$, which plays the part of ${}^{239}_{92}Pu$ in the ${}^{238}_{92}U$ breeder cycle. There is enough uranium and thorium on earth to provide fuel for breeder reactors to satisfy our needs for energy for thousands of years.

Nuclear reactors of the fission type, whether "breeders" or not, are not without certain disadvantages. There is always the possibility of accidental leakage of radioactive material, however unlikely and well guarded against. An accident, whether leakage or a meltdown, could make it necessary to evacuate the surrounding area for at least 10,000 years. Further, disposal of the fiercely radioactive fission products without harming life on earth presents very serious problems. Some of the fission products have half-lives in the tens of thousands of years. Radioactive wastes would not be a problem with fusion reactors, to be described shortly.

Exercises

7. Absorption of a neutron by the thorium isotope ${}^{232}_{90}Th$ gives ${}^{233}_{90}Th$, which undergoes radioactive decay to ${}^{233}_{92}U$. Give the chemical symbols for the intermediate nuclei and the type of particle emissions in their radioactivity.

8. In the fission of ${}^{233}_{92}U$ ($M = 233.0395$ amu) after capturing a neutron ($M = 1.0087$ amu) the total matter mass of the products is 233.8182 amu. Calculate the energy released from the fission of one mole of ${}^{233}_{92}U$ in ergs, MeV, and kcal.

Nuclear fusion Nuclear fusion takes place when two atomic nuclei collide with so much energy that they "melt" or fuse together to form a new nucleus. The new nucleus has an atomic number equal to the sum of the atomic numbers of the reactant nuclei, but may eject a neutron, or undergo fission. Some examples of possible fusion reactions, one of which used both heavy isotopes of hydrogen (deuterium ${}^{2}_{1}H$, and tritium ${}^{3}_{1}H$), are the following:

$${}^{2}_{1}H + {}^{3}_{1}H \longrightarrow {}^{4}_{2}He + {}^{1}_{0}n + 17.60 \text{ MeV}$$

$${}^{2}_{1}H + {}^{2}_{1}H \longrightarrow [{}^{4}_{2}He] \text{ ☚ } \longrightarrow {}^{3}_{1}H + {}^{1}_{1}H + 4.00 \text{ MeV}$$

$${}^{7}_{3}Li + {}^{1}_{1}H \longrightarrow 2{}^{4}_{2}He + 17.50 \text{ MeV}$$

As with the [${}^{236}_{92}U$] in the ${}^{235}_{92}U$ reactor, the ${}^{4}_{2}He$ nucleus formed by fusion is in such a high energy state that it breaks up into fission products instantly.

In all cases the reactions are exothermic because the high nuclear binding energy of the products makes their matter mass less than the matter mass of the reactants. As in fission, the mass difference appears as energy. Nearly all of the energy released is in the form of the kinetic energy (heat energy) of the products.

So far it has been possible to get fusion reactions to go only by using a fission type of nuclear bomb to provide the high temperatures ($\sim 10^8$ degrees C) needed to overcome the huge activation energy (Section 9.5) for fusion. Atomic nuclei, because of their like charges, repel each other strongly at short distances and for them to fuse together, the repulsion must be overcome by giving the nuclei very high kinetic energy. Temperatures in the hundreds of millions of degrees centigrade are necessary, and the arrangement of fission and fusion materials to promote fusion is shown in Figure 16.6. The combination is called a "hydrogen bomb" or "H bomb" and has explosive force equal to as much as 60 million *tons* (60 megatons) of ordinary chemical explosives such as TNT (trinitrotoluene). *One* H bomb can have explosive force about ten times that of *all* the chemical explosive bombs dropped in the seven years of World War II and in the Vietnam War. A few H bombs could destroy most of the human population of any state in the U.S.

Physicists and nuclear chemists hope to be able to carry on fusion reactions in a controlled way before too long. Success of these efforts would mean that an unlimited amount of energy for generating electricity would be available to us virtually forever, using the deuterium in seawater (Photo 44).

PHOTO 44 The superconducting magnet of the "Baseball II" experimental fusion reactor, suspended above the reactor chamber. The magnetic field acts as a "magnetic bottle" that holds a plasma of electrons and nuclei at 360 million degrees F long enough to ignite a fusion reaction. (*University of California Lawrence Livermore Laboratory.*)

EXAMPLE

Let's see just how much energy would become available if deuterium (^2_1H) separated from seawater were used in fusion reactors. First we need to know how much deuterium is available as "fuel" from seawater, and then how much energy it would yield on fusion with itself or other isotopes.

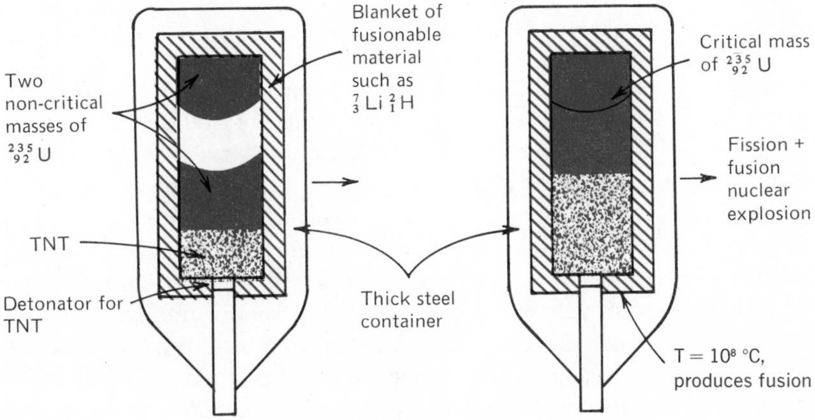

FIG. 16.6 Principle of a nuclear fusion bomb. Explosion of the TNT drives two noncritical masses of $^{235}_{92}U$ together to produce fission that raises the temperature of fusionable material to 10^8 degrees C where nuclear fusion occurs.

The oceans of the earth occupy about 1.50×10^9 km^3 (Section 15.5) and weigh about 1.50×10^{24} g. Correcting for the salt present leaves 1.40×10^{24} g of water. Hydrogen is 11.10% of water by weight and ${}^2_1H$ is 2 parts per 5000 by weight. Putting these figures together gives the weight of ${}^2_1H$ in the oceans as

One ${}^2_1H$ atom ($M = 2.014$) in 5000 atoms of H ($M = 1.008$).

$$\text{Wt } {}^2_1H \text{ in oceans} = \frac{1.4 \times 10^{24} \times 0.111 \times 2}{5000}$$
$$= 6.2 \times 10^{19} \text{ g or } 6.8 \times 10^{13} \text{ tons}$$

Now we want to find how much energy can be obtained by fusion of 6.8×10^{13} tons of deuterium. Suppose it became possible to carry out both of the first two examples of fusion, using the ${}^3_1H$ (tritium) produced in the first reaction to fuel the second. The reactions and the overall reaction are as follows:

$${}^2_1H + {}^2_1H \longrightarrow [{}^4_2He] \longrightarrow {}^3_1H + {}^1_1H + 4.00 \text{ MeV}$$
$${}^3_1H + {}^2_1H \longrightarrow {}^4_2He + {}^1_0n + 17.60 \text{ MeV}$$
$$\text{Sum } 3{}^2_1H \longrightarrow {}^4_2He + {}^1_1H + {}^1_0n + 21.6 \text{ MeV}$$

Three moles of ${}^2_1H$ is 6.042 g and would release $6.023 \times 10^{23} \times 21.6$ MeV or 1.30×10^{25} MeV, according to the overall equation. In kcal, the energy released is 5.00×10^8 kcal per 6.042 g of ${}^2_1H$. This energy quantity is only about $\frac{1}{10}$ that produced by fission of one mole of ${}^{235}_{92}U$ (4.3×10^9 kcal/mole), but on an equal weight basis the deuterium fusion comes out well ahead because the molecular weight of deuterium is so small.

1 MeV = 3.83×10^{-17} kcal.

$$\text{Energy from 1 lb } {}^2_1H = (453.6/6.042) \times 5.00 \times 10^8 = 3.75 \times 10^{10} \text{ kcal}$$
$$\text{Energy from 1 lb } {}^{235}_{92}U = 8.30 \times 10^9 \text{ kcal}$$
$$\text{Energy from 1 lb coal} = 3.28 \times 10^3 \text{ kcal}$$
$$\text{Energy from } {}^2_1H \text{ in the oceans} = 2000 \times 6.8 \times 10^{13} \times 3.75 \times 10^{10} = 5.1 \times 10^{27} \text{ kcal}$$

One way of looking at these figures is to note that 1 lb of ${}^2_1H$ gives about as much energy on fusion as burning 11,400,000 lb of coal [$(3.75 \times 10^{10}/3.28 \times 10^3) = 11.4 \times 10^6$]. Coal reserves are in the hundreds of trillions of tons and are estimated to last about 400 years at present rates of consumption. Since, weight for weight, deuterium fusion releases 11 million times as much energy as coal, deuterium reserves in the sea would provide energy for human activities for 11,400,000 × 400 years or *4.6 billion years*—about the age of earth. Controlled nuclear fusion would forever free us from dependence on fossil fuels (coal, oil, natural gas) for energy, and save the fossil fuels for higher use as starting materials for drugs, plastics, fibers, and all the other products of organic chemical technology. Nuclear fusion reactions of the type described do not produce large quantities of radioactive products—they are "cleaner" in this sense than fission reactions.

Our sun and all stars get their energy from nuclear fusion reactions. In the sun, the main energy source is believed to be the fusion of hydrogen to helium. The reaction goes in steps but the equation for the overall reaction is

$$4\,{}^{1}_{1}\mathrm{H} \rightarrow {}^{4}_{2}\mathrm{He} + 27\ \mathrm{MeV}$$

This reaction goes on so fast that about 600 million tons of hydrogen are converted into helium every second. Even at such an enormous rate, the sun is estimated to have consumed only about $\frac{1}{2}$ of its hydrogen in this way in the last 5 billion years.

All of the fusion reactions described earlier in this section are strongly exoenergetic, but it is not required that nuclear fusion reactions be so. The synthesis of the *transuranium* elements ${}_{93}\mathrm{Np}$ through ${}_{105}\mathrm{X}$ was achieved by nuclear fusion reactions. You have already seen how ${}^{239}_{93}\mathrm{Np}$ (neptunium) and ${}^{239}_{94}\mathrm{Pu}$ (plutonium) are formed from ${}^{238}_{92}\mathrm{U}$ on bombardment with neutrons. Bombardment with deuterium nuclei or α particles of high energy produced in a cyclotron also brings about fusion. Californium (Cf) was made in this way from curium (Cm):

The transuranium elements of $Z = 104$ and 105 have been prepared but not yet named.

$${}^{242}_{96}\mathrm{Cm} + {}^{4}_{2}\mathrm{He} \rightarrow {}^{244}_{98}\mathrm{Cf} + 2\,{}^{1}_{0}n$$

The as-yet-unnamed element 104 was prepared by bombarding another californium isotope with carbon nuclei that had been accelerated to very high velocities (and kinetic energies) in a heavy-ion accelerator:

$${}^{249}_{98}\mathrm{Cf} + {}^{12}_{6}\mathrm{C} \rightarrow {}^{257}_{104}\mathrm{X} + 4\,{}^{1}_{0}n$$

Element 104 is called kurchatovium by Russians, honoring their physicist Igor Kurchatov, and rutherfordium by Americans, honoring Sir Ernest Rutherford, father of the nuclear atom. IUPAC will decide on the name.

Among all of the syntheses of transuranium elements, those of neptunium, plutonium, and americium are the only ones that produced weighable amounts of the elements. In the other cases tiny amounts of the isotopes were obtained—just enough to determine their chemical properties by special microchemical methods, including ion exchange chromatography. In some cases only a few thousand atoms were produced, but they could be detected by their characteristic radioactivity.

The transuranium elements are highly radioactive with very short half-lives (seconds to months) in most cases. A few, such as ${}^{237}_{93}\mathrm{Np}$ (neptunium), ${}^{242}_{94}\mathrm{Pu}$, and ${}^{247}_{96}\mathrm{Cm}$, have half-lives of 10^{5}–10^{7} years. Presumably all of the transuranium elements were present when the earth was formed but have almost completely decayed in the 4.5×10^9 years since that time. Some have been found in the debris of H-bomb explosions, where they are produced by the very high density of neutrons.

Exercises

9. Taking the half-life of $^{247}_{96}Cm$ (curium) as 4.5×10^7 years, how many half-lives for the isotope are there in the age of the earth (4.5×10^9 years)? For each gram of curium present at the formation, what amount remains? [First express your answer in the form $X = (1/n)^m$, then take the logarithms of both sides to get the equation $\log X = m \log 1/n$, and remember $\log 1/n = -\log n$.]

10. Given the data on atomic weights for $^{242}_{96}Cm$, 242.0588 amu, and $^{244}_{98}Cf$, 244.0659 amu, and using any necessary data from Table 2.2, find out whether the following reaction is endo- or exothermic and by how much in MeV (one amu = 932 MeV):

$$^{242}_{96}\text{Cm} + ^{4}_{2}\text{He} \rightarrow ^{244}_{98}\text{Cf} + 2^{1}_{0}n$$

What minimum kinetic energy for the $^{4}_{2}He$ particle would be necessary for the reaction to go? (Express in MeV.)

11. The radium isotope with the longest half-life is $^{226}_{88}Ra$, $t_{1/2} = 1590$ years, yet it was isolated from the mineral *pitchblende,* where it is present to about $1.6 \times 10^{-5}\%$. How can it be there at all when its half-life is so short? (*Hint*: pitchblende is an ore of uranium; see Figure 16.1.)

16.6 Summary Guide Radioactivity is a property of individual, isotopic atoms, not of elements. It is a nuclear property and goes on unaffected by changes in the electron cloud of the isotope. The nuclei of radioactive isotopes shoot out electrons (β particles, $_{-1}^{0}e$), positrons (positive electrons, $_{+1}^{0}e$), the nuclei of helium atoms (α particles, $^{4}_{2}He$), or γ-ray photons (γ rays, $h\nu$); γ rays often accompany or follow the other particles shot out.

The particles emitted are moving very rapidly and have kinetic energies in the millions of electron volts (MeV). Each radioisotope has its own rate of decay and half-life ($t_{1/2}$). The half-life of an isotope is the time required for one-half of the atoms in a sample to decay. Half-lives range from microseconds (10^{-6} sec) to billions of years. Isotopic analysis of uranium–lead ores shows that the earth is about 4.5 billion years old. Radiocarbon dating based on the $^{14}_{6}C$ content of once-living things permits dating them back to prehistoric times, as long ago as 50,000 years.

Most of the naturally occurring radioactive isotopes are heavy atoms with atomic numbers (Z) greater than 83; $^{40}_{19}K$, $^{3}_{1}H$, and $^{14}_{6}C$ in their respective geonormal elements are exceptions. The heavier radioisotopes have neutron/proton ratios (n/p) that are too large for stability. They decay in a series of steps to stable isotopes of lead, emitting α, β, and γ particles on their way.

By bombardment with neutrons, energetic α particles, or other ions, hundreds of radioisotopes not found on earth have been prepared. Some, the transuranium elements, extend the Periodic Table to element 106. Others find use in the treatment of cancer by radiation therapy, in medical diagnosis, in sterilization, and in other industrial applications.

Huge amounts of energy are liberated in the splitting, or fission to lighter atoms, of uranium, plutonium, and thorium isotopes. The fusion of very light atoms into heavier ones also releases energy. Fission gives an ordinary nuclear bomb its explosive power; fission-promoted fusion produces the H-bomb explosion. Controlled fission in nuclear reactors is becoming important as a means of making steam to generate electric power. Breeder reactors manufacture more fissionable isotopes than they consume. By this means, the reserves of thorium and uranium isotopes can provide energy for human activities for thousands of years. Controlled fusion, using deuterium from water, has the promise, when achieved, of supplying our energy needs for billions of years.

Exercises

12. Assuming that your body is 18% geonormal carbon, how many ${}^{14}_{6}C$ atoms decay in your body in 24 hr? Will this number change as you grow older? (Use data from Section 16.2.)

13. The half-life ($t_{1/2}$) for ${}^{40}_{19}K$ is 1.3×10^9 years, and the isotope makes up 0.0118 atom percent ☛ of geonormal potassium. Taking $[K^+]$ in the fluid inside a cell as 4.5×10^{-3} mole/liter, calculate the number of ${}^{40}_{19}K$ decays per hour in 1.00 ml of cell fluid (8760 hour = 1 year).

Atoms per hundred atoms.

14. A red blood cell (erythrocyte) is a disk ☛ typically 7.2 μ (7.2 microns; 1 micron = 10^{-6} meter) in diameter and 2.00 μ thick. Calculate the volume of one red cell. ☛ Using data from Exercise 13, calculate the chance that one ${}^{40}_{19}K$ decay will occur in a given red cell during its 100-day lifetime in the blood.

(Photo 24, p. 308)

The volume of a cylinder of diameter d and height h is $(d/2)^2 \times \pi \times h$.

15. Taking the fission of one mole of ${}^{235}_{92}U$ as yielding 4.3×10^9 kcal, at what rate, in grams of the isotope per second, must the fission be going on inside a 100,000-kW nuclear reactor (1 watt = 1 joule/sec)? In grams per hour?

Key Words

Branching chain reaction a nuclear fission reaction (initiated by neutron capture) that produces two or more neutrons to carry on the fission of other nuclei.

Breeder reactor a nuclear reactor using fission that produces more fissionable isotopes than it consumes.

Critical mass the mass of a fissionable isotope large enough to sustain a branching chain nuclear fission reaction—principle of the nuclear bomb and the source of energy in nuclear reactors.

Electron capture (*K* capture) radioactive decay process in which an inner shell (1*s*) electron is captured by the nucleus, accompanied by emission of X and γ rays.

Half-life period of time required for one-half of the atoms in a sample of a radioactive isotope to decay.

Isotope dilution analysis determination of the amount of an element in a sample by adding a known amount of a radioactive isotope of the element to the sample, and then using a particle counter to measure the extent to which the radioisotopic atoms are diluted by the nonradioactive atoms of the element.

MeV energy unit equal to 10^6 electron volts.

Neutron activation analysis a very sensitive method of analysis for elements, using irradiation with neutrons to convert the atoms into radioactive isotopes whose characteristic γ radiation can be measured.

Nuclear decay process in which the nucleus of a radioactive isotope moves to a state of greater stability by ejecting electrons ($_{-1}^{0}e$, β particles), positrons ($_{1}^{0}e$), helium nuclei ($_{2}^{4}He$, α particles), or energy as γ rays of electromagnetic radiation.

Nuclear fission spontaneous separation of a heavy atom nucleus into two lighter atoms after neutron capture.

Nuclear fusion the production of a heavier nucleus by the combination of two lighter nuclei on collision; principle of the H bomb.

Nuclear reactor a device for generating thermal energy that uses a nuclear fission reaction as source.

Positron ($_{1}^{0}e$) the very short-lived, positively charged counterpart of the electron.

Radiocarbon dating a method for dating once-living materials based on the amount of the radioisotope $_{6}^{14}C$ they contain.

Radioisotope dating method of finding the age of rocks by measuring the relative amounts of radioisotopes and their stable decay products, using mass spectrometry.

Transmutation conversion of one kind of atom to another.

Transuranium elements elements produced by nuclear bombardment and having atomic number greater than 92, that for uranium.

Suggested Readings

Bair, W. J., and Thompson, R. C., "Plutonium: Biomedical Research," *Science*, **183**:715–721 (Feb. 22, 1974).

Farmer, John G., and Murdoch, Baxter S., "Atmospheric Carbon Dioxide Levels as Indicated by the Stable Isotope Record in Wood," *Nature,* **247**: 273–274 (Feb. 1, 1974).

Feld, Bernard T., "The Menace of a Fission Power Economy," *Sci. and Pub. Aff.,* **30**(4):32–34 (April 1974).

Gofman, John W., M.D., "The Case Against Nuclear Power," *Catalyst,* **2**(3):18–20 (1972).

Hammond, Allen L., "Stable Isotopes: Expanded Supplies May Lead to New Uses," *Science,* **176**:1315–1317 (June 23, 1972).

Hammond, R. Philip, "Nuclear Power Risks," *Am. Sci.,* **62**:155–160 (March–April 1974).

Kubo, Arthur S., and Rose, David J., "Disposal of Nuclear Wastes," *Science,* **182**:1205–1211 (Dec. 21, 1973).

Luce, Charles F., "Nuclear Energy Least Damaging," *Catalyst,* **2**(3):15–17 (1972).

Maisel, Albert Q., "Big Push to Atomic Breeder Reactors," *Reader's Digest,* April 1972, pp. 164–168.

Matwyoff, N. A., and Ott, D. G., "Stable Isotope Tracers in the Life Sciences and Medicine," *Science,* **181**:1125–1132 (Sept. 21, 1973).

Metz, W. D., "Laser Fusion: A New Approach to Thermonuclear Power," *Science,* **177**:1180–1182 (Sept. 29, 1972).

Miller, James Nathan, "Just How Safe Is a Nuclear Power Plant?" *Reader's Digest,* June 1972, pp. 95–100.

Scheinman, Lawrence, "Safeguarding Nuclear Materials," *Sci. and Pub. Aff.,* **30**(4):34–36 (April 1974).

Sullivan, Walter, "Experimental Findings Challenge Accepted Theories on Atomic Physics and Cause Confusion in Science," The New York *Times,* April 29, 1974.

APPENDIX 1
Mathematical Review

Only a few simple mathematical operations are required in this text. Among them are addition, subtraction, multiplication, division, raising to a power, finding the logarithm of a number, and finding a number, given its logarithm. These mathematical operations are covered in high school algebra.

When these operations are applied to solving chemical problems, the "unknown" quantities usually represented in algebra by x, y, and z are replaced by quantities describing systems of matter. The greatest difficulty for the student frequently arises in translating from the x, y, z of "pure mathematics" to the symbols for quantities of chemical interest, rather than in performing the arithmetical operations prescribed by the algebraic signs $+$, $-$, $\sqrt{\ }$, and the like.

In making the translations, it is helpful to regard the algebraic signs as *operators*. Perhaps the simplest operator is the plus ($+$) sign. Like all operators it is a direction to perform a certain arithmetical operation, in this case that of adding together the two quantities it connects ($2 + 2 = 4$). The directions given by the operators $+$, $-$, $/$, $\sqrt{\ }$ are too familiar to require much comment; but some of the others used in this text, although inherently simple in their directions for arithmetical operations, may seem strange if you have not recently used your algebra in a scientific context. These operations are described in the following sections.

Exponents and Exponential Notation An expression like r^2 is readily understood. The exponential operator 2 directs us to multiply the quantity by itself. In an expression such as $f = k/r^6$ the exponential operator 6 tells us to multiply r together six times, that is, $r^6 = r \times r \times r \times r \times r \times r$; and the slant bar operator (equivalent to $\div$) instructs us to divide the value thus found into the value of k to obtain the value of f.

Not so familiar, perhaps, are the operations called for by expressions such as 6.023×10^{23} and 4.8023×10^{-10}, which are read, respectively, as "six point oh two three times ten to the twenty-third power" and "four point eight oh two three times ten to the negative tenth power." The combined operator ($\times 10^{23}$) is an instruction to multiply by 100,000,000,000,000,000,000,000, or by 1 followed by 23 zeros. The operation amounts to adding zeros to the number preceding the times sign until there are 23 decimal places to the right of the original decimal point, then moving the decimal point to the end. Thus

$$6.023 \times 10^{23} = 6\underbrace{02{,}300{,}000{,}000{,}000{,}000{,}000{,}000}_{\text{23 decimal places}}$$

The operator ($\times 10^{-10}$) in the second example of exponential expressions is an instruction to multiply the value preceding it by ten raised to the negative tenth power, that is, to *divide* by 10^{10}. Thus the expression 4.8023×10^{-10} means $4.8023/10^{10}$. It is especially simple to handle such expressions if they are arranged so that *there is only one digit, zero excluded, to the left of the decimal point.* If you do this, the exponent is the number of zeros *including* one placed in front of the decimal point before the first digit of the number preceding the times sign appears. Thus

$$4.8023 \times 10^{-10} = \underbrace{0.0000000000}_{\text{10 zeros}}48023 \qquad \underbrace{0.0}_{\text{2 zeros}}825 = 8.205 \times 10^{-2}$$

You should form the habit of placing a zero in front of the decimal point in all decimal expressions less than unity, not only because of its convenience but also because it is standard scientific notation. Thus, .23 is incorrect notation, 0.23 is correct.

Changing the number of digits preceding the decimal point can make the location of the decimal point easy in a calculation, especially where a slide rule is used. Suppose you need to divide 1.750×10^{17} by 4.802×10^{-10}:

$$\frac{1.750 \times 10^{17}}{4.802 \times 10^{-10}} = ?$$

Since the value of 1.750/4.802 is less than 1, it is convenient to take one power of 10 out of the factor 10^{17} in the numerator and multiply it into 1.750, obtaining for the numerator 17.50×10^{16}. This makes locating the decimal point easier:

$$\frac{1.750 \times 10^{17}}{4.802 \times 10^{-10}} = \frac{17.50 \times 10^{16}}{4.802 \times 10^{-10}} = 3.644 \times 10^{26}$$

This type of manipulation can readily be used to find the value of V in the following:

$$V = \frac{0.2537}{760} \times 82.05 \times 300$$

Rewriting with only one digit before the decimal point in each factor, you get

$$V = \frac{(2.537 \times 10^{-1}) \times (8.205 \times 10) \times (3.00 \times 10^{2})}{7.60 \times 10^{2}}$$

Taking care of the exponents, remembering that $10 = 10^1$, and $10^0 = 1$, gives $(10^{-1} \times 10 \times 10^2)/10^2 = 10^{(-1+1+2-2)} = 10^0 = 1$ and

$$V = \frac{2.537 \times 8.205 \times 3}{7.60} = \frac{62.45}{7.60} = 8.217$$

That $10^0 = 1$ is not unique. Any number raised to the zero power is unity according to the laws for exponents. Thus if $n^a = n^a$, then $n^a/n^a = 1$, but

$$\frac{n^a}{n^a} = (n^a)(n^{-a}) = n^{(a-a)} = n^0 = 1$$

Logarithms A system of logarithms is defined by the equation

$$N = a^{\log_a N}$$

where N is the number obtained by raising the base a to the power $\log_a N$. For ordinary logarithms, $a = 10$, so we may write

$$N = 10^{\log_{10} N}$$

Thus if $N = 100$, $\log_{10} N = 2$, since $10^2 = 100$. From the fact that $10^0 = 1$ it follows that $\log_{10} 1 = 0$. We will not attempt to review here the use of logarithm tables covered early in the student's mathematical training. A review may be found preceding the logarithm tables in the *Handbook of Chemistry and Physics* published by the Chemical Rubber Company. The logarithms in Appendix 6 are adequate for most calculations.

In chemistry, calculations with logarithms are made with ordinary or "Briggsian" logarithms whose base is 10. Following common practice, "log" stands for $\log_{10}$ in this text. Logarithms to the base

$e = 2.71828 \ldots$, called "Napierian" (or natural) logarithms, arise in the mathematical treatment of many chemical systems. To distinguish them from ordinary logarithms the abbreviation used is ln, that is,

$$N = e^{\ln N} = (2.71828 \ldots)^{\ln N}$$

and the natural logarithm of a number is the power to which 2.71828 . . . must be raised to produce the number. Conversion between the two types of logarithmic systems is very simple. *Wherever the operator* ln *appears it is replaced by* 2.303 *log*. Thus the expression $RT \ln K$ becomes $2.303RT \log K$. A logarithmic relationship particularly useful in chemistry is

$$\log \frac{1}{N} = -\log N$$

For example,

$$\log \frac{1}{3.00 \times 10^{-4}} = -\log(3.00 \times 10^{-4}) = -[\log 3 + \log 10^{-4}]$$
$$= -\log 3 - \log 10^{-4}$$
$$= -0.4771 + 4.0000 \ldots = 3.5229$$

The value for $(-\log 10^{-4}) = -(-4) = 4.0000 \ldots$; the value of 0.4771 for log 3 is obtained from the log table in Appendix 6.

The reverse calculation, that of finding a number N when its logarithm is given, is simple as long as the logarithm is a positive number. Thus if $\log N = 2.5600$, $N = 10^{2.5600} = 10^{(0.5600+2.000)} = 10^{0.5600} \times 10^2$. Looking up the antilogarithm of 0.5600 in the log table, we find that $10^{0.5600} = 3.63$, so $N = 3.63 \times 10^2$. However, the logarithms of numbers between 0 and 1 are *negative* in sign (remember, $\log 1 = 0$, so log 0.50, for example, must be negative). Negative logarithms are common in chemistry because numbers in the range from 0 to 1 frequently arise. Suppose we wish to find the value of the number N for which $\log N$ has the value -7.6396. The *number* will be *positive* in sign and somewhere between 0 and 1. Direct use of the log table is impossible because the smallest entry is 1.000, and $\log 1 = 0.000$; negative logs for the range 0 to 1 are not tabulated. Problems of this sort are most readily solved by using a combination of exponential notation and logarithms. For the example under discussion, you first write, using the definition of a logarithm,

$$N = 10^{-7.6396}$$

Using the law of exponents that $y^a y^b = y^{(a+b)}$, you can separate the expression for N into a more manageable form. Since $-8.0000 + 0.3604 = -7.6396$,

$$N = 10^{-7.6396} = (10^{-8.000})(10^{+0.3604})$$

The log table shows that the antilog of 0.3604 is 2.293; that is, $10^{0.3604} = 2.293$. Using exponential notation, we obtain the value $N = 2.293 \times 10^{-8}$.

Indexes (i), Differences (Δ), and Multiple Summation (Σ) It is often convenient to use *indexes* in generalizing from individual experimental results to a law in general form. An index in chemistry is usually a formula written as a subscript to a quantity of chemical interest. A weight of sodium can thus be written in the general form W_{Na}, of chlorine W_{Cl}. The quantitative results of all electrolytic decompositions of sodium chloride into sodium and chlorine are summarized, using indexes, in the general equation $W_{\mathrm{Na}}/W_{\mathrm{Cl}} = 0.648$.

Indexes are very convenient in "shorthand" descriptions of systems. Thus, instead of having to describe a system as containing "ten moles of water, three moles of sugar ($C_{12}H_{22}O_{11}$), and one-half mole of sodium chloride," we can describe the system using the symbol n_i for the number of moles of each substance, i being the index showing the substance to which the n refers. Using index notation, we can write

$$n_{\mathrm{H_2O}} = 10.00 \qquad n_{\mathrm{C_{12}H_{22}O_{11}}} = 3.00 \qquad n_{\mathrm{NaCl}} = 0.50$$

The summation sign Σ (Greek, *sigma*) is often used in generalized statements. Thus, the total number of moles of substances in a system can be written as

$$\sum_i n_i = \text{total number of moles}$$

For the case of water, salt, and sugar,

$$\sum_i n_i = n_{\mathrm{H_2O}} + n_{\mathrm{C_{12}H_{22}O_{11}}} + n_{\mathrm{NaCl}} = 10.00 + 3.00 + 0.50 = 13.50$$

Such notation permits concise mathematical statements of many laws. Examples will be found in the text.

The Greek letter Δ (*delta*) is used to indicate changes in quantities. Thus $\Delta n_{\mathrm{H_2O}}$ means the change in the number of moles of water. *The value of a change symbolized by Δ is always determined by subtracting the value of the quantity* ***before*** *the process occurs from its value*

***after** the process has occurred.* Thus, if ten moles of molten sodium chloride are electrolyzed long enough to decompose half of it into its elements, five moles of NaCl will remain, and Δn_{NaCl} = number of moles of NaCl *remaining* minus number of moles of NaCl *originally* present. Or

$$\Delta n_{NaCl} = 5.00 - 10.00 = -5.00$$

On the other hand, since there was no sodium or chlorine originally present as such, $\Delta n_{Na} = 5.00 - 0.00 = 5.00$ and $\Delta n_{Cl_2} = 2.50 - 0.00 = 2.50$ (two moles of NaCl yield only one of Cl_2: $2NaCl \rightarrow 2Na + Cl_2$). Generalizing, we obtain

$$\Delta \text{ (of a quantity)} = \text{final value} - \text{initial value}$$

Other mathematical operations and symbols are explained in the text where they are useful to the discussion. You should pay particular attention to understanding them when they are first introduced, for you will encounter them later. Most of the difficulty in making chemical calculations arises from lack of full understanding of the meaning of the symbols, not from the arithmetic called for by the operators.

Radioactive Decay Half-life (Section 16.2) Using calculus, we write the rate law for the decay of a radioisotope as

$$\frac{-dN}{dt} = kN$$

where N = number of atoms of radioisotope present at time t. Rearranging gives

$$\frac{dN}{N} = -k\,dt$$

Integrating this equation between the limits $t = 0$ and $t_{1/2}$ (the half-life) and between N_0, the number of atoms of the radioisotope present at t_0, and $N_0/2$, the number left at $t_{1/2}$, gives

$$\int_{N_0}^{N_0/2} \frac{dN}{N} = -k \int_{t=0}^{t=t_{1/2}} dt$$

$$\left(\ln \frac{N_0}{2} - \ln N_0\right) = -k(t_{1/2} - 0)$$

$$-\ln 2 + \ln N_0 - \ln N_0 = -kt_{1/2}$$
$$\ln 2 = kt_{1/2}$$

Replacing ln 2 by 2.303 log 2 to convert to ordinary logarithms gives

$$2.303 \log 2 = kt_{1/2}$$
$$(2.303)(0.30103) = kt_{1/2}$$
$$k = \frac{(2.303)(0.30103)}{t_{1/2}} = \frac{0.693}{t_{1/2}}$$

This equation applies to any first-order process—any unimolecular or first-order chemical equation—as well as to radioactive decay.

APPENDIX 2
Force, Energy, and Charge

Newton's Laws of Motion Newton's first law of motion is that *a moving mass-particle will continue to move at constant speed in a straight line unless it is acted upon by a force.* The speed *and* direction of motion of a particle are both given by specifying the *velocity* of the particle. Thus the law may also be stated: *The velocity of a particle remains unchanged unless a force acts on the particle.* It is important to distinguish velocity from speed. Velocity is called a *vector* quantity because it has both magnitude and direction; speed has magnitude only and is called a *scalar* quantity. Two automobiles moving at the same *speed* may collide; they can never collide if they have the same *velocity*.

Newton's second law is a quantitative statement about the effect of a force in changing velocity. *For a particle of given mass, the change in velocity is directly proportional to the force acting on the particle.* In an equation, the law is

$$f = ma$$

where f is the value of the force, m is the mass of the particle, and a is the rate of change of the velocity with time, or the *acceleration* of the mass produced by the force. One measuring unit of force is the *dyne*, which is the force acting for 1 sec required to increase the velocity of a particle of mass 1 g by 1 cm per sec in the line of motion of the particle.

The SI ☛ unit of force is the *newton*, the force that, acting for 1 sec on a mass of 1 kg, gives the mass a velocity of 1 m per sec.

The SI (for Systėme Internationale) units are agreed upon by scientists the world over. They are described in Appendix 8.

According to Newton's first law, a particle does not depart from straight-line motion unless acted on by a force. For a particle to move in a curved path a force must be acting; remove the force and the particle resumes its straight-line motion. A particle moving at constant speed in a circle suffers a continuous change in velocity. According to Newton's second law, the force causing the circular motion must be constant in magnitude, though its direction of action must be constantly changing (force is a vector quantity). When an object is swung in a circle at the end of a rope, the pull on the rope maintains the circular path. The pull may be constant, but the *direction* of pulling is constantly changing. The force required for circular motion is called the *centripetal* force and its direction of application is at all times toward the center of the circle of motion (Figure A.2.1). It is therefore at all times exerted perpendicular to a line tangent to the circle. Remove the centripetal force and the particle "flies off on a tangent"; that is, it resumes straight-line motion. The velocity with which it resumes straight-line motion is called the *tangential velocity*.

Closely related to the quantification of force by Newton's second law is the definition of *work*. Work is done when a force acts, as when a weight is lifted or a spring compressed. Work is defined as *the product of the force and the distance through which the force operates*. One *erg* is the amount of work done when a force of 1 dyne acts through a distance of 1 cm (0.394 in.); the erg is sometimes called a *dyne centimeter* for this reason. It is a very small unit of work, and a more convenient one is the *joule*, which equals 10^7 ergs. The joule is the SI unit of work and energy. It is the work done by a force of 1 newton acting through a distance of 1 m. The work done in raising a 1-lb weight 1 ft (1 "foot pound") is 1.356 joules.

Energy is the capacity to do work, and its units are therefore those of work, that is, ergs or joules. Both work and energy can be expressed

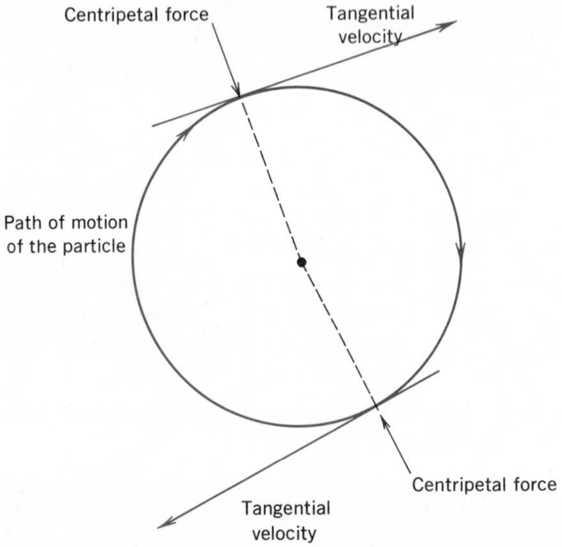

FIG. A.2.1 Newton's laws and circular motion.

in other units. These are defined as they are introduced in the text, and conversion factors are provided on the inside of the back cover.

Coulomb's Law of Electrostatics Bodies carrying electric charges of the same sign repel each other; if the charges are of opposite sign they attract each other. Experiment shows that the *force of attraction or repulsion is directly proportional to the product of the charges and inversely proportional to the square of the distance between them.* This is Coulomb's law. The force is given for charged particles in a vacuum by the equation

$$f = \frac{q_1 q_2}{r^2}$$

where q_1 and q_2 are the electrical charges with sign and r is the distance between them. It follows that f is positive for repulsions, negative for attractions. If the force is to be in dynes, then the distance must be in centimeters and the charges must be measured in *electrostatic units,* abbreviated esu. One esu is defined as the quantity of charge that repels the same charge, or attracts an equal but opposite charge, with a force of 1 dyne when the charges are separated in a vacuum by a distance of 1 cm (Figure A.2.2).

Point charge
q_1

Vacuum

Point charge
q_2

$\longleftarrow r$ cm \longrightarrow

$$f = \frac{q_1 q_2}{D r^2}$$

$f = 1$ dyne if $q_1 = q_2 = 1$ esu
and $r = 1$ cm

(a) Coulomb's Law in a vacuum, $D = 1$

Charges in a
vacuum

$q_1 \bullet$ $\longleftarrow 1 \times 10^{-8}$ cm \longrightarrow $\bullet q_2$

Charges in
water at 20°C
$D = 80.36$

$q_1 \bullet$ $\longleftarrow 1 \times 10^{-8}$ cm \longrightarrow $\bullet q_2$

$$f = \frac{(-4.803 \times 10^{-10})(-4.803 \times 10^{-10})}{(1 \times 10^{-8})^2}$$

$$= 2.307 \times 10^{-3} \text{ dyne}$$

$$f = \frac{(-4.803 \times 10^{-10})(-4.803 \times 10^{-10})}{80.36 \times (1 \times 10^{-8})^2}$$

$$= 2.871 \times 10^{-5} \text{ dyne}$$

(b) Effect of dielectric constant on repulsion of two
electronic charges ($q_1 = q_2 = -4.803 \times 10^{-10}$ esu)

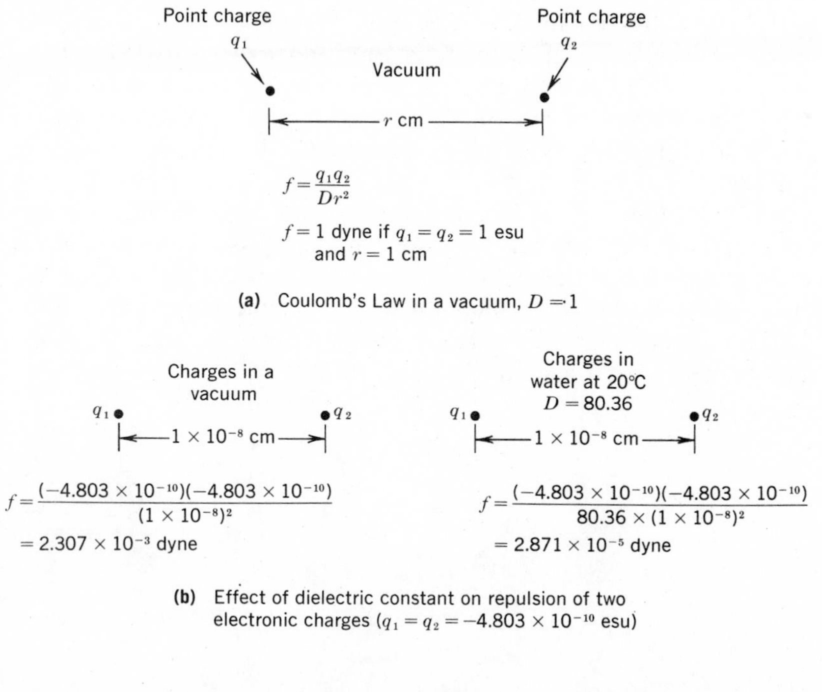

FIG. A.2.2 Coulomb's law.

The charge on the electron is -4.803×10^{-10} esu, and represents the smallest amount of charge known. It follows that 1 esu of charge is the total charge carried by $1/(4.803 \times 10^{-10}) = 2.082 \times 10^{9}$ electrons, a number equal to roughly half the human population of the earth.

If the charges are not in a vacuum—that is, if there is matter between them—the weakening effect of the matter on the force between the charges must be taken into account in the equation for Coulomb's law, and the equation becomes

$$f = \frac{q_1 q_2}{Dr^2}$$

Here, D is the *dielectric constant* of the matter in which the charges are immersed. The value of D for a vacuum is unity. It is very nearly unity for gases, being 1.00059 for air at atmospheric pressure, but has larger values for liquids and solids. For carbon tetrachloride (CCl_4), $D = 2.238$ at 25°C; for water, $D = 80.36$ at 20°C; for solid sodium chloride, $D = 6.12$ at 25°C. The effect of the dielectric constant in weakening the force of attraction between two electronic units of charge of opposite sign treated as point charges is shown in Figure A.2.2(b).

Electrostatic units are used in discussing forces between charged particles; but with electric currents (flows of electrons) the *coulomb*, a larger unit of charge, is more convenient. One coulomb, the SI unit of electrical charge, equals 2.998×10^{9} esu, or alternatively, 1 esu $= 1/(2.998 \times 10^{9})$ coulomb $= 3.336 \times 10^{-10}$ coulomb. The value of the charge of the electron in coulombs is $(-4.803 \times 10^{-10}) \times (3.336 \times 10^{-10}) = -1.602 \times 10^{-19}$ coulomb. The charge represented by one mole (6.023×10^{23}) of electrons is thus $(6.023 \times 10^{23}) \times (-1.602 \times 10^{-19}) = -96{,}487$ coulomb. This quantity of charge, regardless of sign, is called *one faraday*.

Newton's third law is that for every *action* there is an equal but opposite *reaction*. The law explains why the pressure of a gas on the walls of the vessel containing the gas results from collisions of the gas molecules with the walls. The colliding gas molecule exerts a force on the wall and the wall exerts an equal but opposite force on the gas molecule, sending it back to the body of the gas.

APPENDIX 3
The Atomic Weights and Numbers of the Elements

Element	*Symbol*	*Atomic Number*	*Atomic Weight*[a]
Actinium	Ac	89	(227)
Aluminum	Al	13	26.9815
Americium	Am	95	(243)
Antimony	Sb	51	121.75
Argon	Ar	18	39.948
Arsenic	As	33	74.9216
Astatine	At	85	(210)
Barium	Ba	56	137.34
Berkelium	Bk	97	(247)
Beryllium	Be	4	9.0122
Bismuth	Bi	83	208.980
Boron	B	5	10.811
Bromine	Br	35	79.904
Cadmium	Cd	48	112.40
Calcium	Ca	20	40.08
Californium	Cf	98	(249)
Carbon	C	6	12.01115
Cerium	Ce	58	140.12
Cesium	Cs	55	132.905
Chlorine	Cl	17	35.453
Chromium	Cr	24	51.996
Cobalt	Co	27	58.9332
Copper	Cu	29	63.546
Curium	Cm	96	(247)
Dysprosium	Dy	66	162.50
Einsteinium	Es	99	(254)
Erbium	Er	68	167.26

[a]Parentheses indicate mass numbers of the most stable or the best-known isotopes.

Element	Symbol	Atomic Number	Atomic Weight
Europium	Eu	63	151.96
Fermium	Fm	100	(257)
Fluorine	F	9	18.9984
Francium	Fr	87	(223)
Gadolinium	Gd	64	157.25
Gallium	Ga	31	69.72
Germanium	Ge	32	72.59
Gold	Au	79	196.967
Hafnium	Hf	72	178.49
Helium	He	2	4.0026
Holmium	Ho	67	164.930
Hydrogen	H	1	1.00797
Indium	In	49	114.82
Iodine	I	53	126.9044
Iridium	Ir	77	192.2
Iron	Fe	26	55.847
Krypton	Kr	36	83.80
Kurchatovium	Ku	104	(260)
Lanthanum	La	57	138.91
Lawrencium	Lw	103	(257)
Lead	Pb	82	207.19
Lithium	Li	3	6.939
Lutetium	Lu	71	174.97
Magnesium	Mg	12	24.312
Manganese	Mn	25	54.9380
Mendelevium	Md	101	(256)
Mercury	Hg	80	200.59
Molybdenum	Mo	42	95.94
Neodymium	Nd	60	144.24
Neon	Ne	10	20.183
Neptunium	Np	93	(237)
Nickel	Ni	28	58.71
Niobium	Nb	41	92.906
Nitrogen	N	7	14.0067
Nobelium	No	102	(254)
Osmium	Os	76	190.2
Oxygen	O	8	15.9994
Palladium	Pd	46	106.4
Phosphorus	P	15	30.9738
Platinum	Pt	78	195.09
Plutonium	Pu	94	(244)
Polonium	Po	84	(210)
Potassium	K	19	39.102
Praseodymium	Pr	59	140.907
Promethium	Pm	61	(145)
Protactinium	Pa	91	(231)

Element	*Symbol*	*Atomic Number*	*Atomic Weight*
Radium	Ra	88	(226)
Radon	Rn	86	(222)
Rhenium	Re	75	186.2
Rhodium	Rh	45	102.905
Rubidium	Rb	37	85.47
Ruthenium	Ru	44	101.07
Samarium	Sm	62	150.35
Scandium	Sc	21	44.956
Selenium	Se	34	78.96
Silicon	Si	14	28.086
Silver	Ag	47	107.868
Sodium	Na	11	22.9898
Strontium	Sr	38	87.62
Sulfur	S	16	32.064
Tantalum	Ta	73	180.948
Technetium	Tc	43	(97)
Tellurium	Te	52	127.60
Terbium	Tb	65	158.924
Thallium	Tl	81	204.37
Thorium	Th	90	232.038
Thulium	Tm	69	168.934
Tin	Sn	50	118.69
Titanium	Ti	22	47.90
Tungsten	W	74	183.85
Uranium	U	92	238.03
Vanadium	V	23	50.942
Xenon	Xe	54	131.30
Ytterbium	Yb	70	173.04
Yttrium	Y	39	88.905
Zinc	Zn	30	65.37
Zirconium	Zr	40	91.22

APPENDIX 4
The Electron Configurations and First Ionization Potentials of Gaseous Atoms

Atomic Number (*Z*)	*Atom*	*Electron Configuration*	*First Ionization Potential* (eV)
1	H	$1s^1$	13.60
2	He	$1s^2$	24.58
3	Li	$[He]2s^1$	5.39
4	Be	$[He]2s^2$	9.32
5	B	$[He]2s^22p^1$	8.30
6	C	$[He]2s^22p^2$	11.26
7	N	$[He]2s^22p^3$	14.54
8	O	$[He]2s^22p^4$	13.61
9	F	$[He]2s^22p^5$	17.42
10	Ne	$[He]2s^22p^6$	21.56
11	Na	$[Ne]3s^1$	5.14
12	Mg	$[Ne]3s^2$	7.64
13	Al	$[Ne]3s^23p^1$	5.98
14	Si	$[Ne]3s^23p^2$	8.15
15	P	$[Ne]3s^23p^3$	11.00
16	S	$[Ne]3s^23p^4$	10.36
17	Cl	$[Ne]3s^23p^5$	13.01
18	Ar	$[Ne]3s^23p^6$	15.76
19	K	$[Ar]4s^1$	4.34
20	Ca	$[Ar]4s^2$	6.11

Atomic Number (*Z*)	*Atom*	*Electron Configuration*	*First Ionization Potential* (eV)
21	Sc	$[Ar]4s^2 3d^1$	6.56
22	Ti	$[Ar]4s^2 3d^2$	6.83
23	V	$[Ar]4s^2 3d^3$	6.74
24	Cr	$[Ar]4s^1 3d^5$	6.76
25	Mn	$[Ar]4s^2 3d^5$	7.43
26	Fe	$[Ar]4s^2 3d^6$	7.90
27	Co	$[Ar]4s^2 3d^7$	7.86
28	Ni	$[Ar]4s^2 3d^8$	7.63
29	Cu	$[Ar]4s^1 3d^{10}$	7.72
30	Zn	$[Ar]4s^2 3d^{10}$	9.39
31	Ga	$[Ar]4s^2 3d^{10} 4p^1$	6.00
32	Ge	$[Ar]4s^2 3d^{10} 4p^2$	7.88
33	As	$[Ar]4s^2 3d^{10} 4p^3$	9.81
34	Se	$[Ar]4s^2 3d^{10} 4p^4$	9.75
35	Br	$[Ar]4s^2 3d^{10} 4p^5$	11.84
36	Kr	$[Ar]4s^2 3d^{10} 4p^6$	14.00
37	Rb	$[Kr]5s^1$	4.18
38	Sr	$[Kr]5s^2$	5.69
39	Y	$[Kr]5s^2 4d^1$	6.50
40	Zr	$[Kr]5s^2 4d^2$	6.95
41	Nb	$[Kr]5s^1 4d^4$	6.77
42	Mo	$[Kr]5s^1 4d^5$	7.10
43	Tc	$[Kr]5s^2 4d^5$	7.28
44	Ru	$[Kr]5s^1 4d^7$	7.36
45	Rh	$[Kr]5s^1 4d^8$	7.46
46	Pd	$[Kr]4d^{10}$	8.33
47	Ag	$[Kr]5s^1 4d^{10}$	7.57
48	Cd	$[Kr]5s^2 4d^{10}$	8.99
49	In	$[Kr]5s^2 4d^{10} 5p^1$	5.79
50	Sn	$[Kr]5s^2 4d^{10} 5p^2$	7.34
51	Sb	$[Kr]5s^2 4d^{10} 5p^3$	8.64
52	Te	$[Kr]5s^2 4d^{10} 5p^4$	9.01
53	I	$[Kr]5s^2 4d^{10} 5p^5$	10.45
54	Xe	$[Kr]5s^2 4d^{10} 5p^6$	12.13
55	Cs	$[Xe]6s^1$	3.89
56	Ba	$[Xe]6s^2$	5.21
57	La	$[Xe]6s^2 5d^1$	5.61
58	Ce	$[Xe]6s^2 4f^1 5d^1$	6.91
59	Pr	$[Xe]6s^2 4f^3$	5.76
60	Nd	$[Xe]6s^2 4f^4$	6.31
61	Pm	$[Xe]6s^2 4f^5$	
62	Sm	$[Xe]6s^2 4f^6$	5.60
63	Eu	$[Xe]6s^2 4f^7$	5.67
64	Gd	$[Xe]6s^2 4f^7 5d^1$	6.16
65	Tb	$[Xe]6s^2 4f^9$	6.74

Atomic Number (Z)	*Atom*	*Electron Configuration*	*First Ionization Potential* (eV)
66	Dy	$[Xe]6s^2 4f^{10}$	6.82
67	Ho	$[Xe]6s^2 4f^{11}$	
68	Er	$[Xe]6s^2 4f^{12}$	6.08
69	Tm	$[Xe]6s^2 4f^{13}$	5.81
70	Yb	$[Xe]6s^2 4f^{14}$	6.20
71	Lu	$[Xe]6s^2 4f^{14} 5d^1$	5.00
72	Hf	$[Xe]6s^2 4f^{14} 5d^2$	7.00
73	Ta	$[Xe]6s^2 4f^{14} 5d^3$	7.88
74	W	$[Xe]6s^2 4f^{14} 5d^4$	7.98
75	Re	$[Xe]6s^2 4f^{14} 5d^5$	7.87
76	Os	$[Xe]6s^2 4f^{14} 5d^6$	8.70
77	Ir	$[Xe]6s^2 4f^{14} 5d^7$	9.00
78	Pt	$[Xe]6s^1 4f^{14} 5d^9$	9.00
79	Au	$[Xe]6s^1 4f^{14} 5d^{10}$	9.22
80	Hg	$[Xe]6s^2 4f^{14} 5d^{10}$	10.43
81	Tl	$[Xe]6s^2 4f^{14} 5d^{10} 6p^1$	6.11
82	Pb	$[Xe]6s^2 4f^{14} 5d^{10} 6p^2$	7.42
83	Bi	$[Xe]6s^2 4f^{14} 5d^{10} 6p^3$	7.29
84	Po	$[Xe]6s^2 4f^{14} 5d^{10} 6p^4$	8.43
85	At	$[Xe]6s^2 4f^{14} 5d^{10} 6p^5$	9.50
86	Rn	$[Xe]6s^2 4f^{14} 5d^{10} 6p^6$	10.75
87	Fr	$[Rn]7s^1$	4.00
88	Ra	$[Rn]7s^2$	5.28
89	Ac	$[Rn]7s^2 6d^1$	6.90
90	Th	$[Rn]7s^2 6d^2$	6.95
91	Pa	$[Rn]7s^2 5f^2 6d^1$	
92	U	$[Rn]7s^2 5f^3 6d^1$	6.10
93	Np	$[Rn]7s^2 5f^4 6d^1$	
94	Pu	$[Rn]7s^2 5f^6$	5.10
95	Am	$[Rn]7s^2 5f^7$	6.00
96	Cm	$[Rn]7s^2 5f^7 6d^1$	
97	Bk	$[Rn]7s^2 5f^9$	
98	Cf	$[Rn]7s^2 5f^{10}$	
99	Es	$[Rn]7s^2 5f^{11}$	
100	Fm	$[Rn]7s^2 5f^{12}$	
101	Md	$[Rn]7s^2 5f^{13}$	
102	No	$[Rn]7s^2 5f^{14}$	
103	Lw	$[Rn]7s^2 5f^{14} 6d^1$	
104	Ku	$[Rn]7s^2 5f^{14} 6d^2$	

APPENDIX 5
Derivation of Molecular Orbitals for Homonuclear Diatomic Molecules

The energy sequence of molecular orbitals for homonuclear diatomic molecules (H_2, N_2, O_2, and F_2 are examples) given in Figure 4.8 is obtained by using linear combinations of atomic orbitals (LCAO method). An orbital, the probability pattern for the location of an electron in a given energy state, is calculated by what is called the wave equation for the electron. Wave equations for electrons are complex mathematical equations that come out of applying the principles of wave mechanics to the behavior of electrons in atoms and molecules. These equations are called ψ (psi) functions, and a subscript is attached to show which orbital the ψ function describes, as in ψ_{1s} for the wave function of an electron in the $1s$ atomic orbital.

For the $1s$ electron in the hydrogen atom the wave function is

$$\psi_{1s} = \sqrt{\frac{1}{\pi a^3}}\, e^{-r/a}$$

where $a = 0.53 \times 10^{-8}$ cm and r = distance in cm between nucleus (proton) and electron. The wave functions for the higher orbitals become more and more complicated as the orbitals depart from simple spherical shapes.

The ψ functions are equations for probability waves, and when ψ functions for atomic orbitals are combined to make molecular orbitals, the wave nature of the functions must be taken into account. When

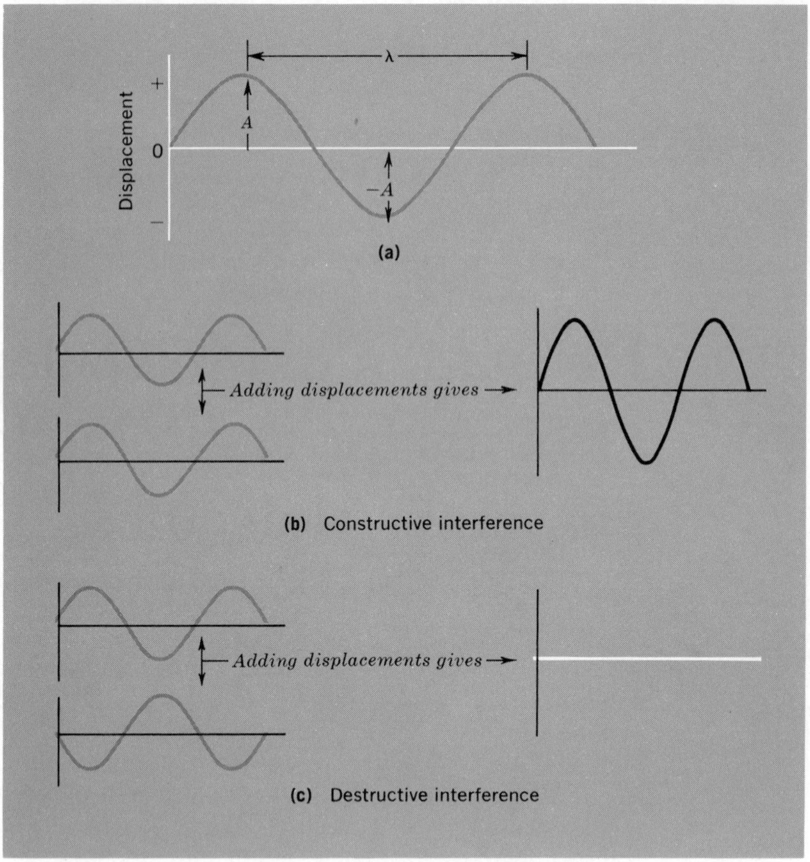

FIG. A.5.1 Transverse waves: (a) amplitude (A) and wavelength (λ); (b) constructive and (c) destructive interference of two superimposed transverse wave trains.

two wave trains meet they may add or cancel. *Constructive interference* occurs when the displacements of the waves add, destructive interference when they cancel each other (Figure A.5.1).

When the ψ functions for atomic orbitals are combined ("overlapped") to make molecular orbitals, account is taken of the wave nature of the ψ functions by permitting them to interfere both constructively and destructively. Constructive interference of atomic orbitals gives bonding molecular orbitals; destructive interference gives antibonding molecular orbitals. Using diagonal lines to show the sign of ψ in two ψ_{1s} orbitals, we show the results of combining them to give bonding σ_1 and antibonding σ_1^* (sigma) molecular orbitals in Figure A.5.2. The antibonding orbital of any pair of molecular orbitals made in this way by subtracting the ψ atomic orbital functions is always higher in energy (less stable). Higher energy results from the nodal plane in the antibonding orbitals that keeps the electron(s) away from the region between the atomic nuclei, where attraction of positive charge on the nuclei for electrons is greatest. In general, the larger the

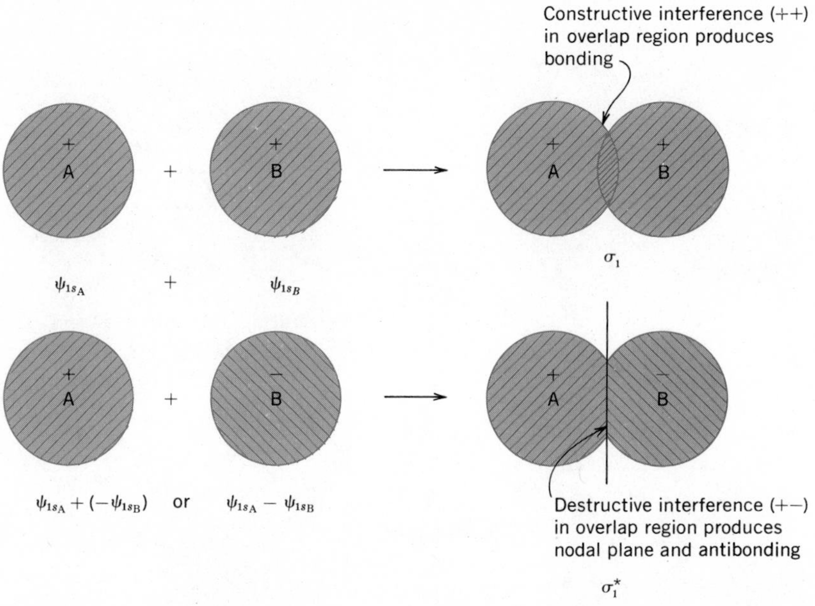

FIG. A.5.2 Constructive and destructive interference of two 1s atomic wave functions in overlap region, producing the σ_1 and σ_1^* molecular orbitals.

number of nodes in a wave function, the higher the energy of an electron described by the function.

The linear combinations $\psi_{2sA} + \psi_{2sB}$ and $\psi_{2sA} - \psi_{2sB}$ gives the bonding σ_2 and antibonding σ_2^* orbitals (Figure A.5.3). Atomic orbitals of the 2*p* type can be combined in three different geometrical ways. Figure

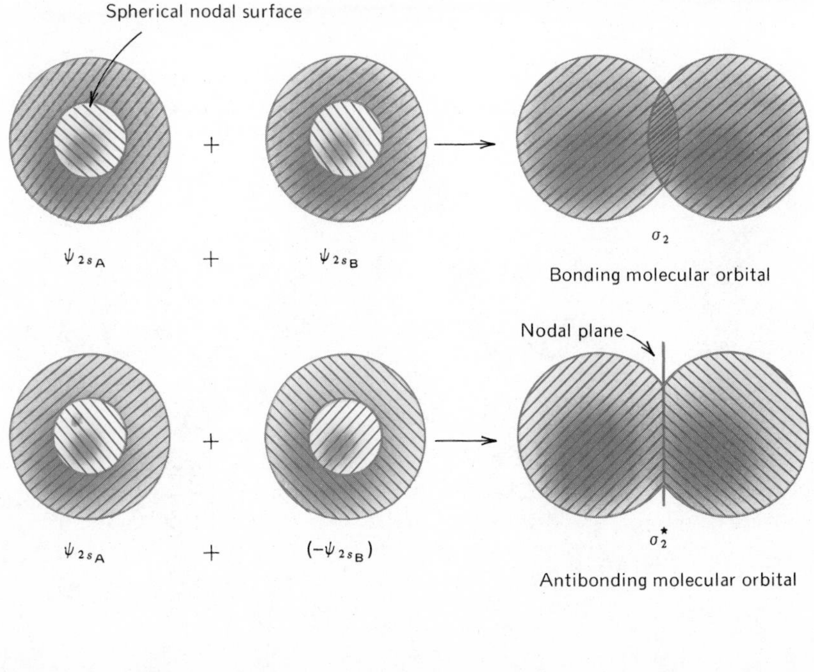

FIG. A.5.3 Overlap of two 2*s* atomic wave functions to produce the σ_2 and σ_2^* molecular orbitals of the l-shell (spherical nodal surfaces in 2*s* atomic orbitals not carried into resulting molecular orbitals).

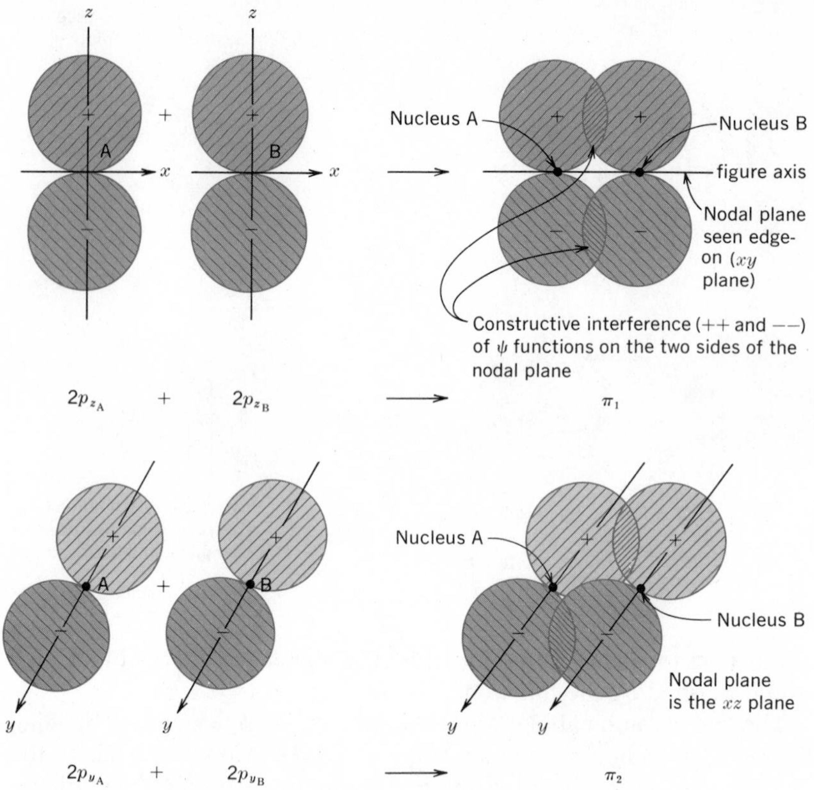

FIG. A.5.4 Combination of $2p$ wave functions to yield the bonding π_1 and π_2 molecular orbitals.

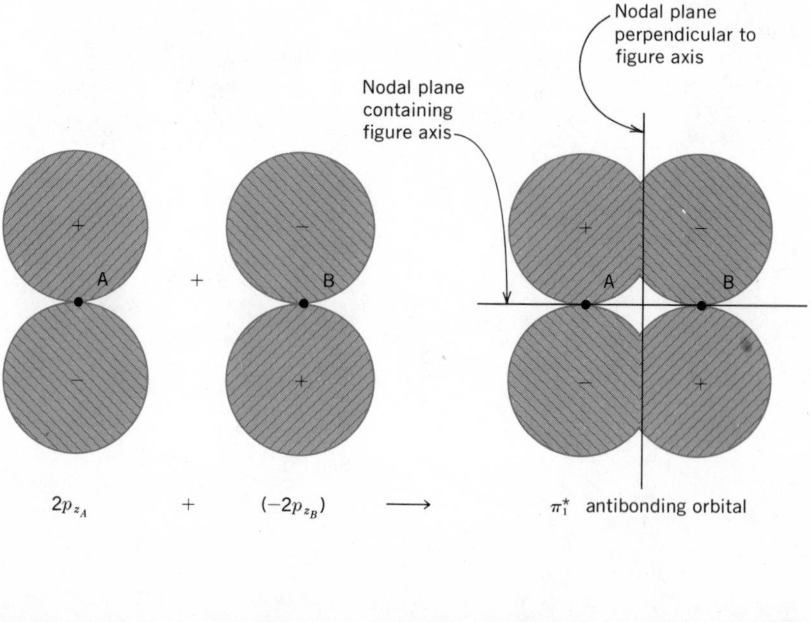

FIG. A.5.5 Antibonding combination of two $2p_z$ orbitals (the π_2^* orbital is similarly formed in the combination of two $2p_y$ orbitals).

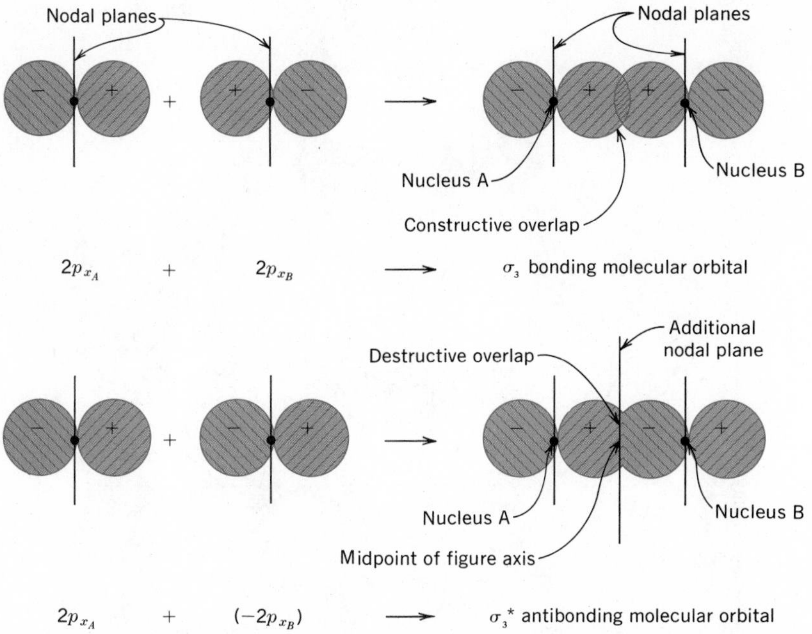

FIG. A.5.6 Overlap of two $2p_x$ orbitals on adjacent nuclei to produce the σ_3 and σ_3^* molecular orbitals.

A.5.4 shows the combinations that give the bonding π_1 and π_2 (pi) molecular orbitals. The antibonding π_1^* and π_2^* orbitals come from the combinations $2p_{zA} - 2p_{zB}$ and $2p_{yA} - 2p_{yB}$. Figure A.5.5 shows the antibonding combination of the $2p_z$ orbitals that gives the π_1^* molecular orbital.

Atomic $2p_x$ orbitals combined in the ways shown in Figure A.5.6 produce the σ_3 and σ_3^* orbitals. These orbitals are of the σ type (rather than π) even though they are made by linear combinations of $2p$ atomic orbitals. Although they have nodal planes, none of the planes contains both atomic nuclei as required for an orbital of the π type.

Figure A.5.7 summarizes the ways in which linear combinations of atomic orbitals (LCAO) give the types and relative energies of the molecular orbitals for homonuclear diatomic molecules.

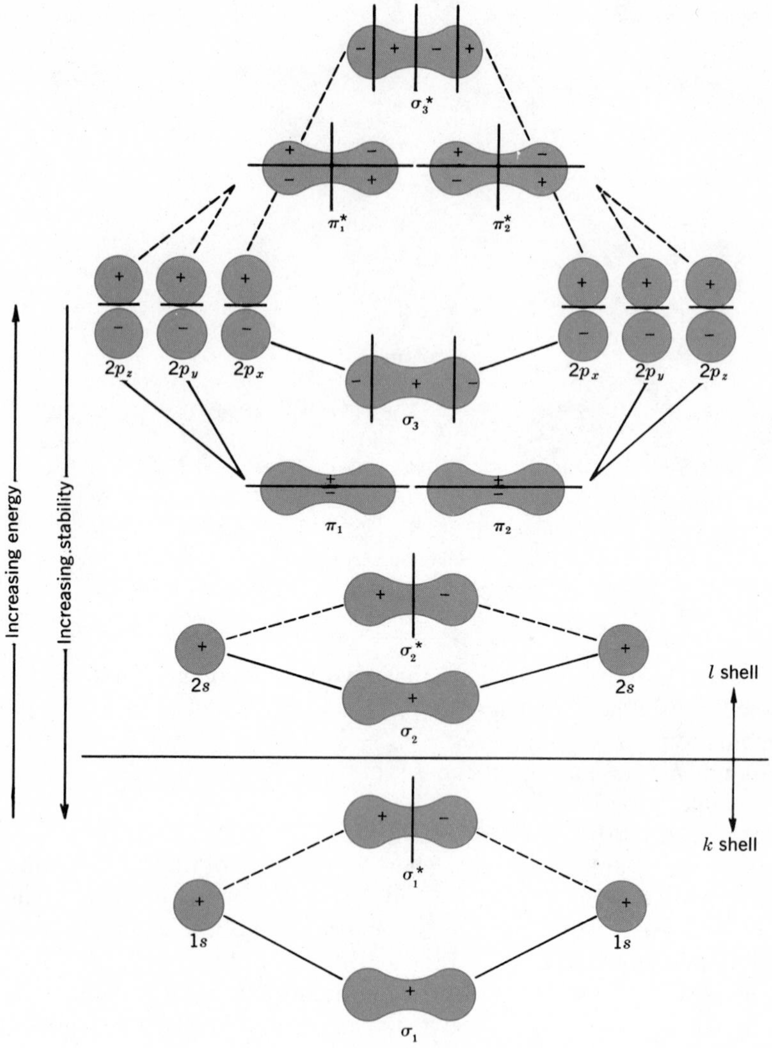

FIG. A.5.7 Molecular orbitals from linear combinations of atomic orbitals. (———, ψ functions added; – – – – – –, ψ functions subtracted. The energies of the π_1, π_2, and σ_3 orbitals are very close; their order is reversed for some molecules.)

APPENDIX 6
Common Logarithms

In using this table of logarithms you will find it convenient to mentally insert a decimal point between the two digits of the numbers (characteristics) in the N column as in 1.0, 1.1, 1.2, . . . , and a decimal point and a zero in front of the numbers (mantissas) in the body of the table as in 0.0000, 0.0043, 0.0086, ... The numbers across the top of the table (0–9) add one digit to the right of the numbers in the N column. Thus, log 1.30 = 0.1139, log 1.31 = 0.1173, log 1.39 = 0.1430.

To find a number, given its log, find the decimal part of the log in the body of the table, read off the corresponding value of N, including the number (0–9) from the top row, and put the digits that are to the left of the decimal point in the value of the logarithm as an exponent of 10. For example, if $\log N = 0.2380$, $N = 1.73$; if $\log N = 3.2380$, $N = 1.73 \times 10^3$. If the value of the decimal part of the log falls between two values in the body of the table, use the number at the top that has the closest decimal value. Thus, suppose $\log N = 0.0575$; 0.0575 is closer to the value for four at the top (0.0569) than the value for five at the top (0.0607) so $N = 1.14$. Similarly, if $\log N = 5.7295$, $N = 5.36 \times 10^5$.

A useful equation is $N = 10^{\log N}$. For information on more accurate use of a log table see explanations at the beginning of the tables in *The Handbook of Chemistry and Physics* published by the Chemical Rubber Company.

N	0	1	2	3	4	5	6	7	8	9
10	0000	0043	0086	0128	0170	0212	0253	0294	0334	0374
11	0414	0453	0492	0531	0569	0607	0645	0682	0719	0755
12	0792	0828	0864	0899	0934	0969	1004	1038	1072	1106
13	1139	1173	1206	1239	1271	1303	1335	1367	1399	1430
14	1461	1492	1523	1553	1584	1614	1644	1673	1703	1732
15	1761	1790	1818	1847	1875	1903	1931	1959	1987	2014
16	2041	2068	2095	2122	2148	2175	2201	2227	2253	2279
17	2304	2330	2355	2380	2405	2430	2455	2480	2504	2529
18	2553	2577	2601	2625	2648	2672	2695	2718	2742	2765
19	2788	2810	2833	2856	2878	2900	2923	2945	2967	2989

N	0	1	2	3	4	5	6	7	8	9
20	3010	3032	3054	3075	3096	3118	3139	3160	3181	3201
21	3222	3243	3263	3284	3304	3324	3345	3365	3385	3404
22	3424	3444	3464	3483	3502	3522	3541	3560	3579	3598
23	3617	3636	3655	3674	3692	3711	3729	3747	3766	3784
24	3802	3820	3838	3856	3874	3892	3909	3927	3945	3962
25	3979	3997	4014	4031	4048	4065	4082	4099	4116	4133
26	4150	4166	4183	4200	4216	4232	4249	4265	4281	4298
27	4314	4330	4346	4362	4378	4393	4409	4425	4440	4456
28	4472	4487	4502	4518	4533	4548	4564	4579	4594	4609
29	4624	4639	4654	4669	4683	4698	4713	4728	4742	4757
30	4771	4786	4800	4814	4829	4843	4857	4871	4886	4900
31	4914	4928	4942	4955	4969	4983	4997	5011	5024	5038
32	5051	5065	5079	5092	5105	5119	5132	5145	5159	5172
33	5185	5198	5211	5224	5237	5250	5263	5276	5289	5302
34	5315	5328	5340	5353	5366	5378	5391	5403	5416	5428
35	5441	5453	5465	5478	5490	5502	5514	5527	5539	5551
36	5563	5575	5587	5599	5611	5623	5635	5647	5658	5670
37	5682	5694	5705	5717	5729	5740	5752	5763	5775	5786
38	5798	5809	5821	5832	5843	5855	5866	5877	5888	5899
39	5911	5922	5933	5944	5955	5966	5977	5988	5999	6010
40	6021	6031	6042	6053	6064	6075	6085	6096	6107	6117
41	6128	6138	6149	6160	6170	6180	6191	6201	6212	6222
42	6232	6243	6253	6263	6274	6284	6294	6304	6314	6325
43	6335	6345	6355	6365	6375	6385	6395	6405	6415	6425
44	6435	6444	6454	6464	6474	6484	6493	6503	6513	6522
45	6532	6542	6551	6561	6571	6580	6590	6599	6609	6618
46	6628	6637	6646	6656	6665	6675	6684	6693	6702	6712
47	6721	6730	6739	6749	6758	6767	6776	6785	6794	6803
48	6812	6821	6830	6839	6848	6857	6866	6875	6884	6893
49	6902	6911	6920	6928	6937	6946	6955	6964	6972	6981
50	6990	6998	7007	7016	7024	7033	7042	7050	7059	7067
51	7076	7084	7093	7101	7110	7118	7126	7135	7143	7152
52	7160	7168	7177	7185	7193	7202	7210	7218	7226	7235
53	7243	7251	7259	7267	7275	7284	7292	7300	7308	7316
54	7324	7332	7340	7348	7356	7364	7372	7380	7388	7396
55	7404	7412	7419	7427	7435	7443	7451	7459	7466	7474
56	7482	7490	7497	7505	7513	7520	7528	7536	7543	7551
57	7559	7566	7574	7582	7589	7597	7604	7612	7619	7627
58	7634	7642	7649	7657	7664	7672	7679	7686	7694	7701
59	7709	7716	7723	7731	7738	7745	7752	7760	7767	7774

N	0	1	2	3	4	5	6	7	8	9
60	7782	7789	7796	7803	7810	7818	7825	7832	7839	7846
61	7853	7860	7868	7875	7882	7889	7896	7903	7910	7917
62	7924	7931	7938	7945	7952	7959	7966	7973	7980	7987
63	7993	8000	8007	8014	8021	8028	8035	8041	8048	8055
64	8062	8069	8075	8082	8089	8096	8102	8109	8116	8122
65	8129	8136	8142	8149	8156	8162	8169	8176	8182	8189
66	8195	8202	8209	8215	8222	8228	8235	8241	8248	8254
67	8261	8267	8274	8280	8287	8293	8299	8306	8312	8319
68	8325	8331	8338	8344	8351	8357	8363	8370	8376	8382
69	8388	8395	8401	8407	8414	8420	8426	8432	8439	8445
70	8451	8457	8463	8470	8476	8482	8488	8494	8500	8506
71	8513	8519	8525	8531	8537	8543	8549	8555	8561	8567
72	8573	8579	8585	8591	8597	8603	8609	8615	8621	8627
73	8633	8639	8645	8651	8657	8663	8669	8675	8681	8686
74	8692	8698	8704	8710	8716	8722	8727	8733	8739	8745
75	8751	8756	8762	8768	8774	8779	8785	8791	8797	8802
76	8808	8814	8820	8825	8831	8837	8842	8848	8854	8859
77	8865	8871	8876	8882	8887	8893	8899	8904	8910	8915
78	8921	8927	8932	8938	8943	8949	8954	8960	8965	8971
79	8976	8982	8987	8993	8998	9004	9009	9015	9020	9025
80	9031	9036	9042	9047	9053	9058	9063	9069	9074	9079
81	9085	9090	9096	9101	9106	9112	9117	9122	9128	9133
82	9138	9143	9149	9154	9159	9165	9170	9175	9180	9186
83	9191	9196	9201	9206	9212	9217	9222	9227	9232	9238
84	9243	9248	9253	9258	9263	9269	9274	9279	9284	9289
85	9294	9299	9304	9309	9315	9320	9325	9330	9335	9340
86	9345	9350	9355	9360	9365	9370	9375	9380	9385	9390
87	9395	9400	9405	9410	9415	9420	9425	9430	9435	9440
88	9445	9450	9455	9460	9465	9469	9474	9479	9484	9489
89	9494	9499	9504	9509	9513	9518	9523	9528	9533	9538
90	9542	9547	9552	9557	9562	9566	9571	9576	9581	9586
91	9590	9595	9600	9605	9609	9614	9619	9624	9628	9633
92	9638	9643	9647	9652	9657	9661	9666	9671	9675	9680
93	9685	9689	9694	9699	9703	9708	9713	9717	9722	9727
94	9731	9736	9741	9745	9750	9754	9759	9763	9768	9773
95	9777	9782	9786	9791	9795	9800	9805	9809	9814	9818
96	9823	9827	9832	9836	9841	9845	9850	9854	9859	9863
97	9868	9872	9877	9881	9886	9890	9894	9899	9903	9908
98	9912	9917	9921	9926	9930	9934	9939	9943	9948	9952
99	9956	9961	9965	9969	9974	9978	9983	9987	9991	9996

APPENDIX 7 Answers to Selected Exercises

Introduction *1.* heavier cyclist in both cases. *3.* at the bottom of the valley C. *6.* 37°C. *8.* −40°. *10.* increases force 3^2 or 9 times. *14.* 2.9×10^{-9} in. *15.* 2.9979×10^8 m sec^{-1}, 2.9979×10^{10} cm sec^{-1}. *17.* 997 g/liter. *19.* (a) 5, (d) 2, (f) 3, (h) 5, (j) 4, (l) 3. *21.* 401.18 ml. *23.* 3.6×10^8 $mile^3$, 1.5×10^{24} ml, 2.5 times the Avogadro number. *25.* 40%.

Chapter 1 *2.* 453.6 g. *4.* 3.817×10^{-26} kg, 23.00 amu. *5.* 5.45×10^{-4} calculated either way. *7.* ${}^{40}_{20}Ca^{2+}$, ${}^{32}_{16}S^{2-}$, ${}^{35}_{17}Cl^{-}$; ${}^{40}_{20}Ca^{2+}$ is a cation, the other two are anions. *10.* 3.829×10^{-23} kcal eV^{-1}. *11.* 1.727×10^{-22} kcal per bond. *12.* 1.83×10^{-16} kcal. *13.* 10^8 Å cm^{-1}, 10^{10} Å m^{-1}. *16.* 3.75×10^{10} cm. *18.* 35.45. *19.* 1.992×10^{-23} g per atom. *20.* 6.02×10^{23} atoms. *21.* 6.02×10^{23} atoms.

Chapter 2 *2.* ${}^{35}_{17}Cl{-}{}^{35}_{17}Cl$, ${}^{37}_{17}Cl{-}{}^{37}_{17}Cl$, ${}^{35}_{17}Cl{-}{}^{37}_{17}Cl$. *3.* ${}^{1}_{1}H^{+}$, ${}^{1}_{1}H_2^{+}$. *6.* 15.999 amu. *9.* 15.999 amu. *11.* F has only one isotope (${}^{19}_{9}F$) but O has three (${}^{16}_{8}O$, ${}^{17}_{8}O$, ${}^{18}_{8}O$), so which isotope must be specified. *12.* O_3, $3O_2 \rightarrow 2O_3$. *14.* 16.34 g of Zn. *16.* 1.2046×10^{24} H, O. *18.* 8.48 g of O_2, 0.265 mole of O_2, 0.530 mole of O. *19.* CH_2O, glucose $(CH_2O)_6$ or $C_6H_{12}O_6$, an unlimited number, $C_2H_4O_2$. *22.* 9.65×10^4 coulombs. *23.* 110×10^6 kcal. *24.* 394 kcal.

Chapter 3 *1.* 235 kcal/mole. *3.* 912 Å. *6.* two electrons in the same orbital repel each other more strongly than two electrons in different orbitals. *8.* Equal. *9.* n^2. *10.* $2n^2$. *13.* Ca has two valence electrons, O needs two to complete its octet. *15.* $Li\cdot + {}^{2}_{1}H\cdot \rightarrow Li^{+}[{}^{2}_{1}H:]^{-}$. *17.* the two 5*s* electrons. *19.* (a) ionic, (b) covalent, (c) ionic, (d) ionic, (e) covalent. *21.* $1s^2 2s^2 2p^6 3s^2 3p^4$, covalent.

Chapter 4 *1.* 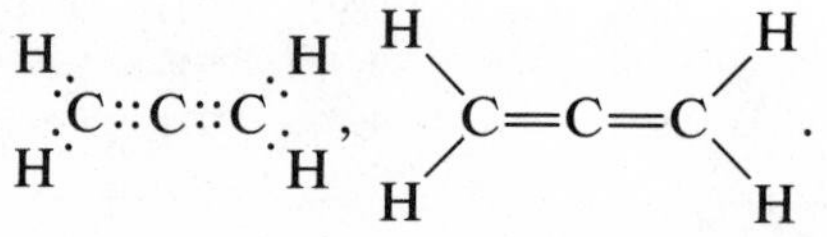

3. (a) 32, (b)

$$H:\overset{\displaystyle :\ddot{O}:}{\underset{\displaystyle :\ddot{O}:H}{\ddot{O}:P:\ddot{O}:H}}$$

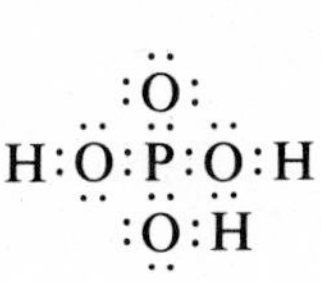

5. No,

⟷ .

7. $\sigma_1^2[\sigma_1^\star]^2\sigma_2^2[\sigma_2^\star]^2$,

unstable. 9. $H:\ddot{\ddot{F}}:$. 13. O—N—O angle $> 120°$, paramagnetic (odd number of electrons). 16. $109.5°$ 18. Methyl alcohol molecules

hydrogen-bond to each other. 19.

$$CH_3—C\overset{\displaystyle O\cdots H—O}{\underset{\displaystyle O—H\cdots O}{\Big\langle\Big\rangle}}C—CH_3.$$

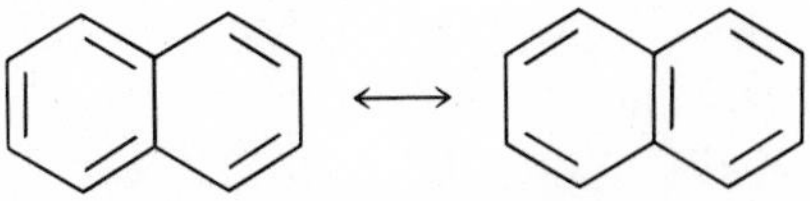

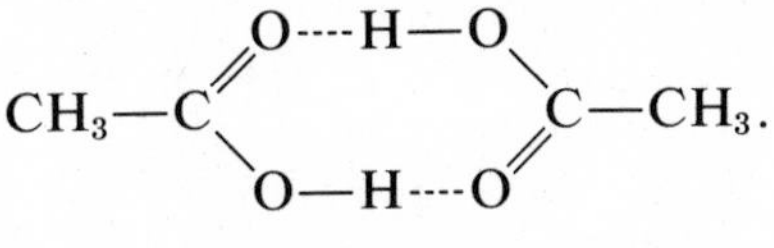

22. The electron configuration of F_2^- is $\sigma[\sigma_1^\star]^2\sigma_2^2[\sigma_2^\star]^2\pi_1^2\pi_2^2\sigma_3^2[\pi_1^\star]^2$ $[\pi_2^\star]^2[\sigma_3^\star]^1 - 10$ bonding electrons, 9 antibonding electrons, so F_2^- should be held together by a one-electron bond.

23.

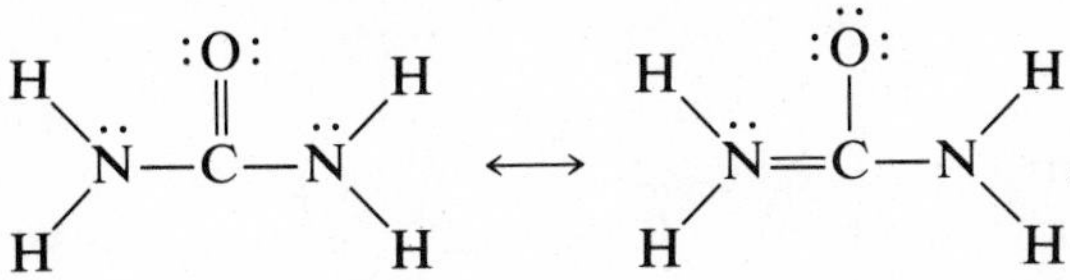

$$\underset{H}{\overset{H}{\diagdown}}\ddot{N}—\overset{\displaystyle :\overset{\|}{\ddot{O}}:}{C}—\ddot{N}\underset{H}{\overset{H}{\diagup}} \quad\longleftrightarrow\quad \underset{H}{\overset{H}{\diagdown}}\ddot{N}=\overset{\displaystyle :\overset{|}{\ddot{O}}:}{C}—\ddot{N}\underset{H}{\overset{H}{\diagup}}\;,\ \text{yes}.$$

25. $:N:::N:\ddot{O}:,\ :N\equiv N—\ddot{O}:,\ :N:::N:\ddot{O}: + Ca: \rightarrow :N:::N: +$

$Ca^{2+}:\ddot{O}:^{2-}$

Chapter 5 2. volume must be halved. 3. 27.3 K, $-245.7°C$. 6. 20.18, neon. 8. 1.224 atm, $p_{He} = 0.306$ atm, $p_{N_2} = 0.918$ atm. 10. 0.208 mole, partial pressures doubled. 12. He less likely to produce "bends." 14. $2NO(g) + O_2(g) \rightarrow 2NO_2(g)$, $3NO_2(g) + H_2O(l) \rightarrow 2HNO_3(l)$ $+ NO(g)$, 6.3 g HNO_3. 16. 278 moles of O_2, 151 atm, 2220 lb in.$^{-2}$. 17. Takes off like a rocket. 18. 5170 cal/mole. 20. yes, in the Sea Lab, $p_{O_2} = 1906$ torr. 21. $X_{O_2} = 0.0174$, $X_{He} = 0.9826$. 24. H_2 by a factor of 3.1. 25. yes, $\Delta S_v = 27.5$ cal mole^{-1} deg^{-1}.

Chapter 6 1. 1.293 g/liter. 3. 490 g. 5. $X_{\text{acetic acid}} = 0.0155$, X_{H_2O} $= 0.9845$. 7. $Al_2Cl_6(s) + 12H_2O(l) \rightarrow 2Al(H_2O)_6^{3+}(aq) + 6Cl^-(aq)$, 53.34 g, $[Al_2Cl_6] = 0$, $[Al(H_2O)_6^{3+}] = 0.40$ mole/liter, $[Cl^-] = 1.2$ mole /liter. 8. 100.00 ml. 11. 23.5 torr. 14. $H_3O^+(aq) + CH_3COO^-(aq)$ $\rightarrow CH_3COOH(aq) + H_2O(l)$, ($Na^+$ and Cl^- "bystander" ions). 15. 0.050 mole/liter. 17. 0.0115 mole, 6.95×10^{21} molecules. 19. $X_{\text{glycol}} = 0.244$, $X_{H_2O} = 0.756$, $p_{H_2O} = 574$ torr. 21. 7.2 atm, 5.5×10^3 torr, cell membranes are permeable to glucose. 25. 1.05 liter.

Chapter 7 1. $2SO_2(g) + O_2(g) \rightleftarrows 2SO_3(g)$. 2. $N_2(g) + 3H_2(g)$ $\rightleftarrows 2NH_3(g)$. 3. $CaCO_3(s) \rightleftarrows CaO(s) + CO_2(g)$. 5. No, open system. Yes, gas $\rightleftarrows$ liquid. *6.* No, open system—exchanging matter through breathing air. *7.* (b) $N_2(g) + O_2(g) \rightleftarrows 2NO(g)$ $K = [NO]^2/[N_2][O_2]$, (d) $H_2SO_4(aq) + 2H_2O(l) \rightleftarrows 2H_3O^+(aq) + SO_4^{2-}(aq)$, $K = [H_3O^+]^2[SO_4^{2-}]/[H_2SO_4][H_2O]^2$. *8.* (a) product, (b) 64,700 liters/mole, (c) yes, yes, (d) liters/mole. *12.* Would increase as H_2O boils away. *14.* Q halved in value, to the right. *16.* Endothermic, no. *18.* $1/\sqrt{4.8 \times 10^5} = 1.4 \times 10^{-3}$. *20.* $(1.5 \times 10^{-4})(2 \times 1.5 \times 10^{-4})^2 = 1.35 \times 10^{-11}$ mole^3 l^{-3}. *22.* $[Ag^+] = 4 \times 10^{-3}$ mole/liter, much less than 3.2×10^{-2} mole/liter, so principle applies. *24.* $K_p = (p_{CO} \times p_{H_2})/p_{H_2O}$, high temperature and low pressure. No. *25.* 0.00116 g in H_2O layer, 0.09884 in CCl_4. *26.* $\frac{1}{2}$, $\frac{1}{4}$, $(\frac{1}{2})^n$, an infinite number, but 10 in this case leaves only 0.0098 percent of the original amount in liquid *A*.

Chapter 8 *1.* 3.26×10^{-18}, $[H_2O] = 55.4$ moles/liter. *4.* 1.58×10^{-2} mole/liter for both, better, increase because of additional number of ions. *6.* 3, 5.5, 9, 0, −1. *7.* 3.16×10^{-2}, 1.26×10^{-6}, 6.31×10^{-8}, $1 = (10^0)$, 10. *8.* $[OH^-] = 2 \times 10^{-2}$, 12.30. *11.* 1.0×10^{-4}, 4. *13.* 1 *M* $Na^+HSO_4^-$. *15.* $CN^-(aq) + CH_3COOH(aq) \rightarrow HCN(aq) + CH_3COO^-(aq)$. *16.* $CH_3CHOHCOO^-(aq) + H_2O(l) \rightleftarrows CH_3CHOHCOOH(aq) + OH^-(aq)$, greater, 9.17×10^{-9}, 8.04. *18.* 4.78, methyl orange. *20.* 4.74. *21.* (Vol of acetic acid solution)/(vol of sodium acetate solution) = 0.556. *22.* Fe(s) will dissolve. *23.* Hg(s) will not dissolve appreciably. Over geological time FeS will have dissolved in acidic groundwater. *24.* Urine more acidic, pH = 6.3. *25.* ΔpH = −0.02.

Chapter 9 *4.* 3.72×10^{-3} mole/liter. *5.* $p_{H_2O} = 105$ torr, $p_{O_2} = 53$ torr. *7.* Twice as fast. *9.* rate $= kC_AC_B$, $k = 2 \times 10^{-1}$ liter mole^{-1} sec^{-1}. *13.* rate $= kC_{NO} \times C_{OONO}$ bimolecular.

14. $\cdot\ddot{\text{N}}::\underset{\cdot\cdot}{\ddot{\text{O}}}$, $\underset{\cdot\cdot}{\ddot{\text{O}}}::\dot{\text{N}}:\underset{\cdot\cdot}{\ddot{\text{O}}}:$, $:\underset{\cdot\cdot}{\ddot{\text{O}}}:\underset{\cdot\cdot}{\ddot{\text{O}}}:\dot{\text{N}}::\underset{\cdot\cdot}{\ddot{\text{O}}}$, all.

19. During combustion in air, the N_2 acts as a coolant decreasing temperature and as a diluent of O_2 slowing combustion. *20.* rate $= kC_{\text{pinene}}$ mole liter^{-1} sec^{-1}, unimolecular. *21.* 1.2×10^{15}. *22.* (a) 44 kcal exothermic, (b) steps 3 and 4, (c) H—H bond broken in step 3 is stronger than Cl—Cl bond broken in step 4, (d) 1 and 2, (e) 5, 6 and 7, (f) no, it's not a simple reaction. *23.* rate $= kC_{RCl} \times C_{OH^-}$. Rates respectively: $1 \times 10^{-3} \times C_{RCl}$, $1 \times 10^{-9} \times C_{RCl}$, $1 \times 10^{-12} \times C_{RCl}$ all in mole liter^{-1} sec^{-1}. RCl will last a long time. *24.* $\frac{1}{2}$, $\frac{1}{4}$, $(\frac{1}{2})^{24} = \frac{1}{6} \times 10^{-8}$.

Chapter 10 *1.* 1.602×10^{-19} coulomb. *3.* 96,484 sec. *4.* 2. *7.* 1320 joule sec^{-1}, 315.5 cal sec^{-1}, 0.3155 kcal sec^{-1}. *8.* (a) and (b) no reaction, (c) $Cl_2(g) + 2Br^- \rightarrow 2Cl^- + Br_2(l)$, (d) $2Al(s) + 6H^+ \rightarrow 2Al^{3+} + 3H_2(g)$, (e) $2Fe^{3+} + 2I^- \rightarrow 2Fe^{2+} + I_2(s)$. *9.* 1.23 V, 89. *11.* Sn^{2+}, Sn^{4+}. *13.* Right to left, 0.29 V. *15.* $Zn(s) + I_2(s) \rightarrow Zn^{2+} + 2I^-$, yes, 1.30 V, formation of $ZnI_2(s)$. *17.* Decrease it to a value less than zero. Yes, up. *18.* 1.05 V. *19.* $1 \times 10^{-9.18} = 6.6 \times 10^{-10}$, none. *20.* $K = [Fe^{2+}]^2/[I^-]^2[Fe^{3+}]^2$, $K = [Fe^{3+}]^2[ClO_3^-]/[Fe^{2+}]^2[ClO_4^-][H^+]^2$, the second because H^+ is a reactant. *23.* $\Delta\mathscr{E}^0 = 0.35$ V, $k = 7.94 \times 10^{11}$. Any Fe^{2+} formed is reduced back to $Fe(s)$, $Zn(s)$ sacrificing itself to oxidation to protect the iron. *25.* 90,280 kcal with H_2, 245,000 kcal with C_8H_{18}. *26.* 10.5 kg (23 lb) of LiH, 0.45 ft³. *27.* 0.41 V, less.

Chapter 11 *2.* $3Ag(s) + NO_3^- + 4H^+ \rightleftarrows 3Ag^+ + 2NO(g) + 2H_2O(l)$. $K = 10^{5.44} = 2.75 \times 10^5$. *5.* $[4Li^+ + 4e^- \rightleftarrows 4Li(s)\mathscr{E}^0 = -3.05] + [4OH^- \rightleftarrows 2H_2O(l) + O_2(g) + 4e^- \mathscr{E}^0 = -0.40]$, $\Delta\mathscr{E}^0 = -3.45$ V, $K = 10^{-235}$. Negative sign, reverse reaction is spontaneous. *6.* 174 liters. *7.* For the water. *9.* Not a good idea, would make fire even worse. *11.* 0.30 *M* HCl. *12.* Found mol. wt of 323 shows Fe_2Cl_6 is correct.

15. Li—⁺O: with H above and H below (H / O \ H). *16.* BF_3 because F more electron-withdrawing

(electronegative) than Cl. *17.* $H—He^+Cl^- \rightarrow H—Cl + He$. *20.* $FeSO_4$ and NaCN or any solutions of an iron(II) (ferrous) compound and of a soluble cyanide. *21.* $Cu(H_2O)_4^{2+} + Zn(NH_3)_4^{2+} \rightarrow Cu(NH_3)_4^{2+} + Zn(H_2O)_4^{2+}$. *24.* S +6, N +5, P +5, S +4, Cr +6, Mn +7. *27.* No. *28.* 2.22×10^5 liters. *30.* 0.0100 *M*. *33.* $Cr_2(SO_4)_3$, the same. *37.* Discontinue use, adds Na^+ to replace Ca^{2+} and Mg^{2+} giving dilute NaCl solution. *39.* One having the same glucose concentration, the same electrolyte concentrations, and the same osmotic pressure as blood to avoid unwanted changes in blood composition.

Chapter 12 *2.* +4, +4, +4, +5, +7 respectively. *4.* $CO_2 + 2NaOH \rightarrow Na_2CO_3 + H_2O$, Solvay process. *5.* third- and higher-period elements do not form strong π bonds so GeO_2 and SiO_2 are σ-bonded crystals similar to diamond. *6.* −23 kcal, yes since negative, $10^{16.9} = 7.2 \times 10^{16}$. *9.* 120°, 120°, 109.5° respectively. *10.* $HClO < HClO_2$

$< HClO_3 < HClO_4$. *13.* +5, $HO—P(=O)(OH)—O—P(=O)(OH)—OH$, in ADP and

ATP. *16.* SnH_4, PbH_4, tetrahedral like CH_4, higher since their mol. wt are larger. *18.* The reaction for SiH_4. ΔG_f^0 (SiO_2) is negative and about twice ΔG_f^0 (CO_2). *20.* P_2O_3, P_4O_6, +3. *22.* 6.1 atm, 90 lb in.⁻². *24.* Exothermic, LeChatelier's. *27.* Covalent, van der Waals (secondary), octahedral. *31.* Yes, S—O bond length is shorter than Xe—O bond length because of resonance. *33.* pH = 5.6 in 0.01 *M*

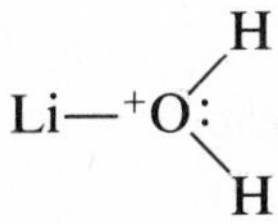

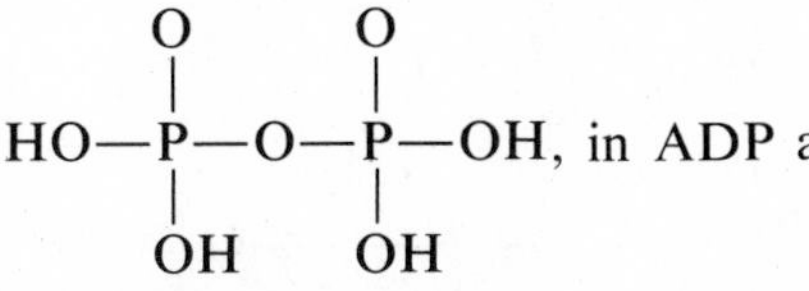

$NH_4^+NO_3^-$, use increases soil acidity. *34*. $K = 10^{17 \times 2 \times 0.65} = 1.25 \times 10^{22}$. *36*. $1s^22s^22p^63s^23p^64s^23d^{10}4p^2$, metalloid (Ge). *39*. They decrease.

Chapter 13 *2*. By the "other way" no ^{18}O would appear in the ester, contrary to the fact. *5*. Going across and down 109.5°, 120°; 109.5°, 109.5°; 109.5°, 120°; 109.5°, 109.5°; 120°, 120°; 180°, 180°; 120°. *6*. Repulsion between large electron clouds of I atoms.

7. 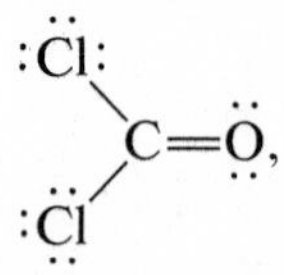, 120° (measured 124°). *9*. (a) 2-butanol, (b) 2-pentene, (c) 4-ethyl-3-octanone, (d) butanoic acid (butyric acid), (e) 2-methylbutanoic acid, (f) 2-hexyne, (g) triiodomethane (iodoform). *16*. All are correct structures. *17*. Functional $H—C(=O)—OC_2H_5$, ethyl formate. *19*. (b), (e). *25*. Mostly propane because it has a higher vapor pressure (lower bp). *26*. 280 liters of CH_4, 560 liters of O_2, 2680 liters of air. *27*. $CH_3CH_2CH_3(g) + 5O_2(g) \rightarrow 3CO_2(g) + 4H_2O(l)$, $2CH_3CH_2CH_2CH_3(g) + 13O_2(g) \rightarrow 8CO_2(g) + 10H_2O(g)$. *29*. Three, 1,2-, 1,3-, and 1,4-dimethylbenzenes. *32*. 320 ml, 2.84×10^5 kcal. *34*. $CH_3CH_2CH(OH)CH_3$, 2-butanol. *37*. $CH_3C(=O)—O—CH_3$, methyl acetate. *40*. Octadecanoic acid. *41*. $HO—CH_2—CH_2—OH$, $HO—CH_2—CH(OH)—CH_2—OH$, ethene (ethylene), oxidation. *45*. Going across and down: 1,2-dibromobenzene (*o*-dibromobenzene), 3-nitroaniline (*m*-nitroaniline), 2-aminophenol (*o*-aminophenol), 2,4,6-trinitrophenol ("picric acid"), diphenylacetic acid, 3-phenyl-2-butanoic acid. *48*. To cross-link chains. *49*. Methane (mol. wt 16) is safer because it's lighter than air and doesn't accumulate in the bilges to present danger of explosion. *52*. Contains no C=C to be cleaved by O_3 in smog, so doesn't crack. *53*. Acetic acid is hydrogen-bonded. *54*. pH = 11.5, $pK_a = 10.9$.

Chapter 14 *1*. 2.09×10^{15}. *2*. 6, 64.

4. 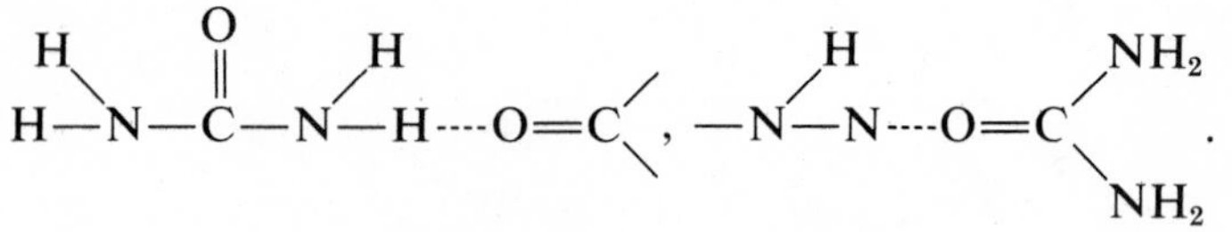

5. —S—S— link between two cysteine residues. *7*. No. *9*. Yes, no, no, no. *11*. Tristearin. *14*. Hydrogen bonding between water molecules and the ether oxygens in carbowax make is soluble.

20.

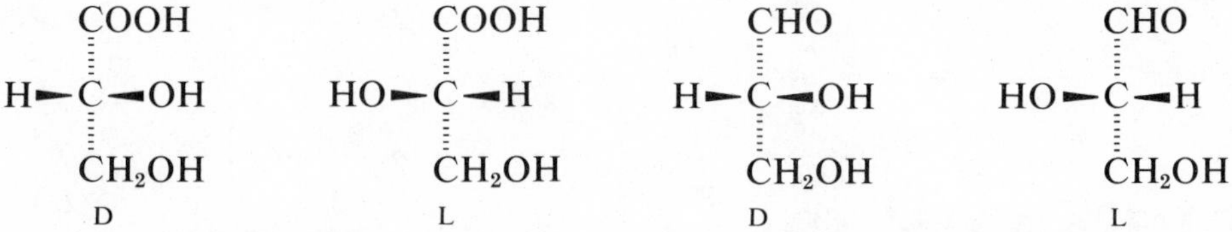

$$
\begin{array}{cccc}
\text{COOH} & \text{COOH} & \text{CHO} & \text{CHO} \\
| & | & | & | \\
\text{H---C---OH} & \text{HO---C---H} & \text{H---C---OH} & \text{HO---C---H} \\
| & | & | & | \\
\text{CH}_2\text{OH} & \text{CH}_2\text{OH} & \text{CH}_2\text{OH} & \text{CH}_2\text{OH} \\
\text{D} & \text{L} & \text{D} & \text{L}
\end{array}
$$

21. Yes, no. *23.* $\searrow\!\!\text{C}=\text{O} + \text{H}^+ + \text{NADH} \rightarrow \searrow\!\!\text{CHOH} + \text{NAD}^+$

25. All +5, no. *29.* 6, 54. *31.* -546 kcal, 0.80 (80%). *33.* Yes.
36. 1.55. *37.* Acetylation breaks up strong intermolecular hydrogen bonds that make cellulose insoluble. *39.* $2\text{CoASH} + \text{Pb}^{2+} \rightleftarrows (\text{CoAS})_2\text{Pb} + 2\text{H}^+$. *40.* Diastereoisomers. No, it has the mirror-image configuration at C-5. $2[\text{C}_6\text{H}_8\text{O}_6(s) \rightleftarrows \text{C}_6\text{H}_6\text{O}_6(s) + 2\text{H}^+ + 2e^-]$ $+ [\text{O}_2(g) + 4\text{H}^+ + 4e^- \rightleftarrows 2\text{H}_2\text{O}(l)]$ gives $2\text{C}_6\text{H}_8\text{O}_6(s) + \text{O}_2(g) \rightleftarrows 2\text{C}_6\text{H}_6\text{O}_6(s) + 2\text{H}_2\text{O}(l)$.

Chapter 15 *2.* H_2O oxidized: $2\text{H}_2\text{O}(l) \rightleftarrows \text{O}_2(g) + 4\text{H}^+ + 4e^-$. NADP^+ reduced: $\text{NADP}^+ + \text{H}^+ + 2e^- \rightleftarrows \text{NADPH}$. NADPH oxidized: $\text{NADPH} \rightleftarrows \text{NADP}^+ + \text{H}^+ + 2e^-$. CO_2 reduced: $6\text{CO}_2(g)$ $+ 24\text{H}^+ + 24e^- \rightleftarrows \text{C}_6\text{H}_{12}\text{O}_6 + 6\text{H}_2\text{O}(l)$. *3.* $\text{SO}_3^{2-} + \text{H}_2\text{O}(l) \rightleftarrows \text{SO}_4^{2-}$ $+ 2\text{H}^+ + 2e^-$, $\text{O}_2(g) + 4\text{H}^+ + 4e^- \rightleftarrows 2\text{H}_2\text{O}(l)$. *4.* 2858 kg. *8.* rate $= kC_{\text{NO}_2}$, rate $= kC_{\text{RCO}} \times C_{\text{O}_2}$, rate $= kC_{\text{NO}} \times C_{\text{O}_3}$.

9.

$$
\cdot\ddot{\text{N}} \quad \longleftrightarrow \quad \ddot{\text{N}} \quad ,
$$

about 120°. *11.* 2.70×10^{-4} mole/liter, 22,200 liters. *14.* Soluble, slightly soluble, very slightly soluble. *15.* RNA, has additional —OH for hydrogen bonding. *17.* 20,000. *18.* 10.3 torr, 0.010 gram/liter.
20. -99.2 kcal, about $\frac{1}{7}$. *21.* 2.4 g, 4.8 metric tons. *22.* 5946 torr < 7016 torr, yes. *23.* 100,000 joule sec^{-1}. *24.* Acetyl-CoA. *25.* Cholesterol. *26.* 9.2 g, 183 times as much.

Chapter 16 *1.* Because only σ emission (^4_2He) changes mass number. No. *3.* $^{232}_{90}\text{Th} \rightarrow ^{228}_{88}\text{Ra} + ^4_2\text{He}$, about $\frac{4}{5}$. *5.* $^{14}_6\text{C} \rightarrow ^{14}_7\text{N} + _{-1}^{0}e$, NO_2, in smog and HNO_3 production. *8.* 2.07×10^{20} erg mole^{-1}, 4.95×10^9 kcal/mole, 1.29×10^{26} MeV mole^{-1}. *9.* 100, 8×10^{-31} g.
13. 19 per hour. *14.* 1 chance in 4 million. *15.* 0.00131 g sec^{-1}, 4.70 g hr^{-1}.

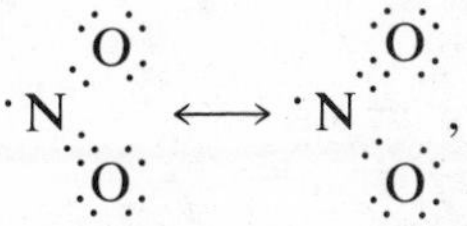

APPENDIX 8
International System (SI) of Units

Basic Units

Length The *meter* (m), the length equal to 1,650,763.37 wavelengths of the red-orange line in the spectrum of the $^{86}_{36}Kr$ atom. 1 m = 39.37 in.

Mass The *kilogram* (kg), equal to the mass of the standard kilogram, a block of platinum-iridium alloy kept in the French Bureau of Standards near Paris. 1 g = 10^{-3} kg.

Time The *second* (sec is used in this book; s is the official abbreviation), the duration of 9,192,631,770 periods of the electromagnetic radiation corresponding to the transition between the two hyperfine energy levels of the ground state of the $^{133}_{55}Cs$ atom.

Electric current The *ampere* (amp), the constant current that, flowing in two parallel wires of negligible cross section and 1 meter apart in a vacuum, produces between the wires a force of 2×10^{-7} newton (N) per meter of wire length. A current of 0.833 amps flows through the filament of a 100-watt light bulb in a 120-V house circuit.

Absolute (thermodynamic) temperature The *kelvin* (K), 1/273.16 of the thermodynamic temperature of the triple-point of water (the temperature at which the solid, liquid, and gas forms of H_2O are in equilibrium). The Celsius ("centigrade") degree is equal to the kelvin in size.

Amount of substance The *mole* (M), the amount of substance that contains as many elementary entities of specified nature as there are atoms in exactly 0.012 kg (12 g) of $^{12}_{6}C$.

Derived Units

hertz (Hz) The unit of frequency; 1 hertz = 1 sec^{-1} or one cycle per second.

newton (N) The unit of force; 1 newton = 1 kg m sec^{-2}, the force that, acting on a mass of 1 kg for 1 sec, gives the mass a velocity of 1 m sec^{-1}, or acting on a mass of 1 kg, gives it an acceleration of 1 m sec^{-2}.

pascal The unit of pressure; 1 pascal = 1 newton m^{-2}, 1 pascal = 9.869×10^{-6} atm = 7.500×10^{-3} torr.

joule (J) The unit of work, energy, or quantity of heat; 1 joule = 1 newton m. The joule is the work done by a force of 1 newton acting through a distance of 1 meter.

calorie (cal) Unit of energy; 1 cal = 4.1840 joules.

watt (W) Unit of power (time rate of flow of energy); 1 watt = 1 joule sec^{-1}.

coulomb (c) The unit of quantity of electricity; 1 coulomb = 1 amp sec^{-1}. One coulomb of electricity plates out 0.0011180 g of silver from a silver nitrate solution.

volt (V) The unit of electromotive force (emf); 1 V = 1 watt amp^{-1} (or watts = volts × amperes).

electron volt (eV) The kinetic energy acquired by an electron in passing through a potential difference of 1 V in a vacuum;

1 eV = 1.6021×10^{-19} joule.

ohm (Ω) The unit of electrical resistance; 1 Ω = 1 V amp^{-1}.

liter The unit of volume; 1 liter = 10^{-3} m^3 = 1000 milliliters (ml), 1000 cm^3, 0.03532 cu ft, 0.2642 U.S. gal (liquid).

centimeter (cm) A unit of length; 1 cm = 10^{-2} m, 0.3937 inches (in.).

Index

Figures and Photos are in *italics*.
Key words or "definition of" are in **boldface** type.
Tables are followed by (t).